Elementary Statistics for Geographers

Elementary Statistics for Geographers

Third Edition

JAMES E. BURT
GERALD M. BARBER
DAVID L. RIGBY

THE GUILFORD PRESS
New York London

A Division of Guilford Publications, Inc.
72 Spring Street, New York, NY 10012
www.guilford.com

Printed in the United States of America

This book is printed on acid-free paper.

Last digit is print number: 9 8 7 6 5 4 3 2 1

Library of Congress Cataloging-in-Publication Data

Burt, James E.
Elementary statistics for geographers / James E. Burt, Gerald M. Barber, David L. Rigby. — 3rd ed.
p. cm.
Includes bibliographical references and index.
ISBN 978-1-57230-484-0 (hardcover)
1. Geography—Statistical methods. I. Barber, Gerald M. II. Rigby, David L. III. Title.
G70.3.B37 2009
519.5024'91—dc22 2008043409

Preface

Readers who know our book will quickly see that this edition represents a significant revision, containing both a substantial amount of new material and extensive reorganization of topics carried over from the second edition. However, our purpose remains unchanged: to provide an accessible algebra-based text with explanations that rely on fundamentals and theoretical underpinnings. Such an emphasis is essential if we expect students to utilize statistical methods in their own research or if we expect them to evaluate critically the work of others who employ statistical methods. In addition, when students understand the foundation of the methods that are covered in a first course, they are far better equipped to handle new concepts, whether they encounter those concepts in a more advanced class or through reading on their own. We acknowledge that undergraduates often have a limited mathematical background, but we do not believe this justifies a simplified approach to the subject, nor do we think that students are well served by learning what is an inherently quantitative subject area without reference to proofs and quantitative arguments. It is often said that today's entering students are less numerate than previous generations. That may be. However, in our 20-plus years of teaching undergraduates we have seen no decrease in their ability or in their propensity to rise to an intellectual challenge. Like earlier versions, this edition of *Elementary Statistics for Geographers* is meant for instructors who share this outlook, and for their students, who—we trust—will benefit from that point of view.

The Descriptive Statistics section of this edition greatly expands the coverage of graphical methods, which now comprise a full chapter (Chapter 2). This reflects new developments in computer-generated displays in statistics and their growing use; also, students increasingly seem oriented toward visual learning. It is likely, for example, that a student who obtains a good mental image of skewness from Chapter 2 can use that visual understanding to grasp more readily the quantitative measures presented in Chapter 3. A second new chapter appearing in the descriptive section is Chapter 4, Statistical Relationships. It introduces both concepts of and measures for correlation and regression. This is somewhat nonstandard, in that most books postpone these topics until after the discussion of univariate methods. We have found that earlier introduction of this material has several advantages. First, correlation and regression are large topics, and some students do better learning them in two parts. Second, the concept of association is useful when explaining certain aspects of probability theory such as independence, conditional probability, and joint probability. Finally, it is easier to discuss nonparametric tests such as chi-square when the idea of statistical association has already been presented. Of course, instructors who prefer to cover correlation and regression in one section of their course can postpone Chapter 4 and cover it as part of a package with Chapters 12 and 13.

The Inferential Statistics section has also been heavily revised. We merged basic probability theory with the treatment of random variables to create more streamlined coverage in a single chapter (Chapter 5, Random Variables and Probability Distributions). Gone is the Computer-Intensive Methods chapter, with much of that material incorporated into the Nonparametric Methods chapter. As bootstrapping and related techniques have become mainstream, it is appropriate to locate them in their natural home with other nonparametric methods. Chapter 11, Analysis of Variance, is a new chapter, which covers both single- and two-factor designs. Also new is Chapter 13, Extending Regression Analysis, which treats diagnostics as well as transformations and more advanced regression models (including multiple regression).

The last section, Patterns in Space and Time, contains a revised version of the Time Series Analysis chapter from the second edition, and the entirely new Chapter 14, Spatial Patterns and Relationships. The latter is an overview of spatial analysis, and covers point patterns (especially nearest neighbor analysis), spatial autocorrelation (variograms, join counts, Moran's *I*, LISA, and *G*-statistics), and spatial regression (including an introduction to geographically weighted regression).

Additionally, there are lesser changes too numerous to itemize. We've placed greater emphasis on worked examples, often with accompanying graphics, and the datasets that we refer to throughout the book are available on the website that accompanies this book. On the website, readers can also find answers to most of the end-of-chapter exercises. See *www.guilford.com/pr/burt* for the online resources.

We have said already that this new edition adheres to the previous editions' emphasis on explanation, rather than mere description, in its presentation of quantitative methods. Several other aspects are also unchanged. We have retained the coverage of time series, which of course is seldom covered in this type of book. Time series data are extremely common in all branches of geography; thus, geographers need to be equipped with at least a few tools of analysis for temporal data. Also, once students get to linear regression, they are well positioned to understand the basics of time series modeling. In other words, ability to handle time series can be acquired at little additional cost. Because time series are so common, geographers will likely have occasion to deal with temporal data regardless of their formal training in the subject. We believe that even simple operations like running means should not be undertaken by individuals who do not appreciate the implications of the procedure. Because most students will not take a full course in time series, minimal coverage, at least, is essential in an introductory text. Also, we've received strong positive feedback on this material from instructors.

We have continued our practice from the second edition, of not tying the book to any particular software package. We believe that most instructors use software for teaching this material, but no package has emerged as an overwhelming favorite. We might gain a friend by gearing the book to a particular package, but we would alienate half a dozen more. Also, since statistical software is becoming increasingly easy to use, students require less in the way of instruction. And we want the book to stay current. We have found that even minor changes in the formatting of output can confound students who have been directed to look for particular tables of values or particular terminology in software packages.

Finally, in keeping with the trend from edition to edition, what was a long book is even longer. Unless it is used in a year-long course, instructors will have to be very selective with regard to what they assign. With this in mind, we have attempted to make the chapters as self-contained as possible. Except for the chapter on probability and sampling theory, a "pick-and-choose" approach will work reasonably well. For example, we know from experience that some instructors leave out the Nonparametric Methods chapter altogether, with no downstream effects, whereas others skip various chapters and subsections within chapters. If some students complain about having to skip around so much, most appreciate a book that covers more than what is taught in the course. Later, when confronted with an unfamiliar method in readings or on a research project, they can return to a book whose notational quirks have already been mastered, and can understand the new technique in context with what was presented in the course. As we reflect on our own bookshelves, it is precisely that kind of book that has proved most useful to us over the years. We wouldn't presume to claim that our work will have similar lasting utility, but we offer it in the belief that it is better to cover too much than too little.

Many people deserve our thanks for their help in preparing this book. We are particularly grateful to students and teaching assistants at UCLA, Queen's University, and the University of Wisconsin–Madison for telling us what worked and what didn't. Thanks also to the panel of anonymous reviewers for their comments on previous versions of the manuscript. You improved it greatly. We also very much appreciate the hard work by everyone at The Guilford Press involved with the project, especially our editor, the ever-patient and encouraging Kristal Hawkins. Our production editor William Meyer also deserves particular mention for his careful attention to both the print and digital components of the project. Most of all, we thank our families for so willingly accepting the cost of our preoccupation. To them we dedicate the book.

Contents

Elementary Statistics for Geographers

I

INTRODUCTION

1

Statistics and Geography

Most of us encounter probability and statistics for the first time through radio, television, newspapers, or magazines. We may see or hear reports of studies or surveys concerning political polls or perhaps the latest advance in the treatment of cancer or heart disease. If we were to reflect on it for a moment, we would probably notice that statistics is used in almost all fields of human endeavor. For example, many sports organizations keep masses of statistics, and so too do many large corporations. Many companies find that the current production and distribution systems within which they operate require them to monitor their systems leading to the collection of large amounts of data. Perhaps the largest data-gathering exercises are undertaken by governments around the world when they periodically complete a national census.

The word "statistics" has another more specialized meaning. It is the methodology for collecting, presenting, and analyzing data. This methodology can be used as a basis for investigation in such diverse academic fields as education, physics and engineering, medicine, the biological sciences, and the social sciences including geography. Even traditionally nonquantitative disciplines in the humanities are finding increasing uses for statistical methodology.

DEFINITION: STATISTICS

Statistics is the methodology used in studies that collect, organize, and summarize data through graphical and numerical methods, analyze the data, and ultimately draw conclusions.

Many students are introduced to statistics so that they can interpret and understand research carried out in their field of interest. To gain such an understanding, they must have basic knowledge of the procedures, symbols, and *vocabulary* used in these studies.

No matter which discipline utilizes statistical methodology, analysis begins with the collection of data. Analysis of the data is then usually undertaken for one of the following purposes:

1. To help *summarize* the findings of some inquiry, for example, a study of the travel behavior of elderly or handicapped citizens or the estimation of timber reforestation requirements.
2. To obtain a better understanding of the phenomenon under study, primarily as an aid in generalization or *theory validation,* for example, to validate a theory of urban land rent.
3. To make a *forecast* of some variable, for example, short-term interest rates, voter behavior, or house prices.
4. To *evaluate* the performance of some program, for example, a particular form of diet, or an innovative medical or educational program or reform.
5. To help *select* a course of action among a set of possible alternatives, or to plan some system, for example, school locations.

That elements of statistical methodology can be used in such a variety of situations attests to its impressive versatility.

It is convenient to divide statistical methodology into two parts: *descriptive statistics* and *inferential statistics.* Descriptive statistics deals with the organization and summary of data. The purpose of descriptive statistics is to replace what may be an extremely large set of numbers in some dataset with a smaller number of summary measures. Whenever this replacement is made, there is inevitably some loss of information. It is impossible to retain *all* of the information in a dataset using a smaller set of numbers. One of the principal goals of descriptive statistics is to minimize the effect of this information loss. Understanding *which* statistical measure should be used as a summary index in a particular case is another important goal of descriptive statistics. If we understand the derivation and use of descriptive statistics and are aware of its limitations, we can help to avoid the propagation of misleading results. Much of the distrust of statistical methodology derives from its misuse in studies where it has been inappropriately applied or interpreted. Just as the photographer can use a lens to distort a scene, so can a statistician distort the information in a dataset through his or her choice of summary statistics. Understanding what descriptive statistics can tell us, *as well as what it cannot,* is a key concern of statistical analysis.

In the second major part of statistical methodology, *inferential statistics,* descriptive statistics is linked with probability theory so that an investigator can generalize the results of a study of a few individuals to some larger group. To clarify this process, it is necessary to introduce a few simple definitions. The set of persons, regions, areas, or objects in which a researcher has an interest is known as the *population* for the study.

DEFINITION: STATISTICAL POPULATION

A statistical population is the total set of elements (objects, persons, regions, neighborhoods, rivers, etc.) under examination in a particular study.

For instance, if a geographer is studying farm practices in a particular region, the relevant population consists of all farms in the region on a certain date or within a certain time period. As a second example, the population for a study of voter behavior in a city would include all potential voters; these people are usually contained in an eligible voters list.

In many instances, the statistical population under consideration is *finite;* that is, each element in the population can be listed. The eligible voters lists and the assessment rolls of a city or county are examples of finite populations. At other times, the population may be *hypothetical.* For example, a steel manufacturer wishing to test the quality of output may select a batch of 100 castings over a few weeks of production. The population under study is actually the *future* set of castings to be produced by the manufacturer using this equipment. Of course, this population does not exist and may have an infinitely large number of elements. Statistical analysis is relevant to both finite and hypothetical populations.

Usually, we are interested in one or more characteristics of the population.

DEFINITION: POPULATION CHARACTERISTIC

A population characteristic is any measurable attribute of an element in the population.

A fluvial geomorphologist studying stream flow in a watershed may be interested in a number of different measurable properties of these streams. Stream velocity, discharge, sediment load, and many other characteristic channel data may be collected during a field study. Since a population characteristic usually takes on different values for different elements of the population, it is usually called a *variable.* The fact that the population characteristic does take on different values is what makes the process of statistical inference necessary. If a population characteristic does not vary within the population, it is of little interest to the investigator from an inferential point of view.

DEFINITION: VARIABLE

A variable is a population characteristic that takes on different values for the elements comprising the population.

Information about a population can be collected in two ways. The first is to determine the value of the variable(s) of interest for each and every element of the population. This is known as a *population census* or *population enumeration.* Clearly, it is a feasible alternative only for finite populations. It is extremely difficult, some would argue even impossible, for large populations. It is unlikely that a national decennial Census of Population in a large country actually captures all of the individuals in that population, but the errors can be kept to a minimum if the enumeration process is well designed.

DEFINITION: POPULATION CENSUS

A population census is a complete tabulation of the relevant population characteristic for all elements in the population.

The second way information can be obtained about a population is through a *sample.* A sample is simply a subset of a population, thus in sampling we obtain values for only selected members of a population.

DEFINITION: SAMPLE

A sample is a subset of the elements in the population and is used to make inferences about certain characteristics of the population as a whole.

For practical considerations, usually time and/or cost, it is far more convenient to sample rather than enumerate the entire population. Of course, sampling has one distinct disadvantage. Restricting our attention to a small proportion of the population makes it impossible to be as accurate about population characteristics as is possible with a complete census. The risk of making errors is increased.

DEFINITION: SAMPLING ERROR

Sampling error is the difference between the value of a population characteristic and the value of that characteristic inferred from a sample.

To illustrate sampling error, consider the population characteristic of the average selling price of homes in a given metropolitan area in a certain year. If each and every house is examined, it is found that the average selling price is $150,000. However, if only 25 homes per month are sampled and the average selling price of the 300 homes in the sample (12 months × 25 homes), the average selling price in the sample may be $120,000. All other things being equal, we could say that the difference of $150,000 – $120,000 = $30,000 is due to sampling error.

What do we mean by *all other things being equal*? Our error of $30,000 may be partly due to factors other than sampling. Perhaps the selling price for one home in the sample was incorrectly identified as $252,000 instead of $152,000. Many errors of this type occur in large datasets. Information obtained from personal interviews or questionnaires can contain factual errors from respondents owing to lack of recall, ignorance, or simply the respondent's desire to be less than candid.

DEFINITION: NONSAMPLING OR DATA ACQUISITION ERRORS

Errors that arise in the acquisition, recording, and editing of statistical data are termed nonsampling or data acquisition errors.

In order that error, or the difference between the sample and the population can be ascribed solely to sampling error, it is important to *minimize* nonsampling errors. Validation checks, careful editing, and instrument calibration are all methods used to reduce the possibility that nonsampling error will significantly increase the total error, thereby distorting subsequent statistical inference.

The link between the sample and the population is probability theory. Inferences about the population are based on the information in the sample. The quality of these inferences depends on how well the sample reflects, or represents, the population. Unfortunately, short of a complete census of the population, there is no way of knowing how well a sample reflects the population. So, instead of selecting a *representative* sample, we select a *random* sample.

DEFINITION: REPRESENTATIVE SAMPLE

A representative sample is one in which the characteristics of the sample closely match the characteristics of the population as a whole.

DEFINITION: RANDOM SAMPLE

A random sample is one in which every individual in the population has the same chance, or probability, of being included in the sample.

Basing our statistical inferences on random samples ensures unbiased findings. It is possible to obtain a very unrepresentative random sample, but the chance of doing so is usually very remote if the sample is large enough. In fact, because the sample has been randomly chosen, we can *always determine the probability* that the inferences made from the sample are misleading. This is why statisticians always make probabilistic judgments, never deterministic ones. The inferences are always qualified to the extent that random sampling error may lead to incorrect judgments.

The process of statistical inference is illustrated in Figure 1-1. Members, or units, of the population are selected in the process of sampling. Together these units comprise the sample. From this sample, whereas inferences about the population are made. In short, sampling takes us from the population to a sample, statistical inference takes us from the sample back to the population. The aim of statistical inference is to make statements about a population characteristic based on the information in a sample. There are two ways of making inferences: *estimation* and *hypothesis testing.*

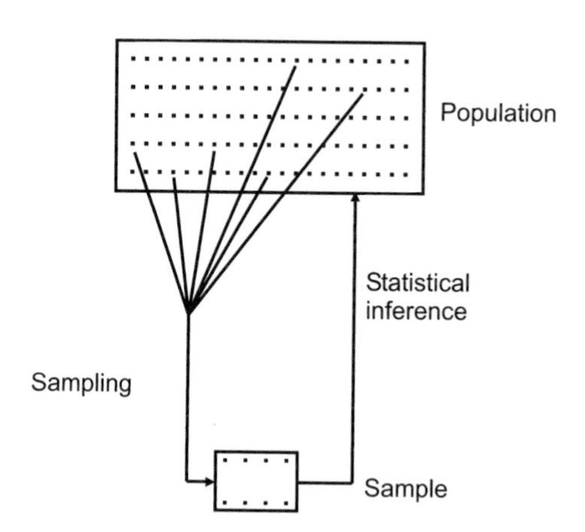

FIGURE 1-1. The process of statistical inference.

DEFINITION: STATISTICAL ESTIMATION
Statistical estimation is the use of the information in a sample to estimate the value of an unknown population characteristic.

The use of political polls to estimate the proportion of voters in favor of a certain party or candidate is a well-known example of statistical estimation. *Estimates* are simply the statistician's best guess of the value of a population characteristic. From a *random sample* of voters, we try and guess what proportion of *all* voters will support a certain candidate.

Through the second way of making inferences about a population characteristic, *hypothesis testing,* we hypothesize a value for some population characteristic and then determine the degree of support for this hypothesized value from the data in our random sample.

DEFINITION: HYPOTHESIS TESTING
Hypothesis testing is a procedure of statistical inference in which we decide whether the data in a sample support a hypothesis that defines the value (or a range of values) of a certain population characteristic.

As an example, we may wish to use a political poll to find out whether some candidate holds an absolute majority of decided voters. Expressed in a statistical way, we wish to know whether the proportion of voters who intend to vote for the candidate exceeds a value of 0.50. We are not interested in the actual value of the population characteristic (the candidate's exact level of support), but in whether the candidate is likely to get a majority of votes. As you might guess, these two ways of making inferences are intimately related and differ more at the conceptual level. The relation between them is so intimate that, for most purposes, both can be used to answer any problem. No matter which method is used, there are two fundamental elements of any statistical inference: the inference itself and a measure of our faith, or confidence in it. A useful synopsis of statistical analysis, including both descriptive and inferential techniques, is illustrated in Figure 1-2.

1.1. Statistical Analysis and Geography

The application of statistical methods to problems in geography is relatively new. Only for about the last half-century has statistics been an accepted part of the academic training of geographers. There are, however, earlier references to uses of descriptive statistics in literature cited by geographers. For example, several 19th-century researchers, including H. C. Carey (1858) and E. G. Ravenstein (1885), used statistical techniques in their studies of migration and other interactions. Elementary methods of descriptive techniques are commonly seen in the geographical literature of the early 20th century. But for the most part, the three paradigms that dominated academic

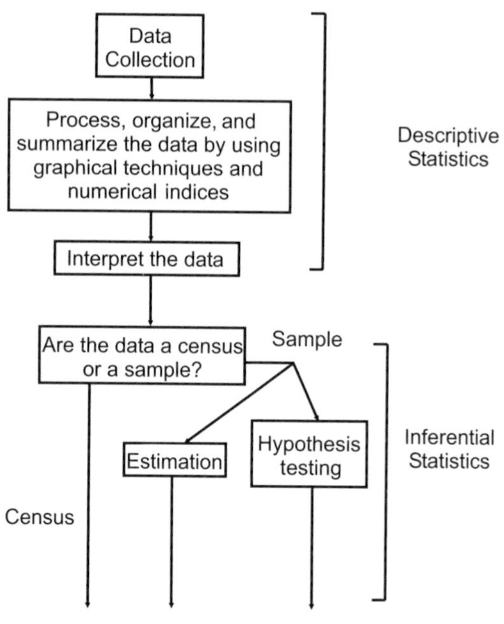

FIGURE 1-2. Statistical analysis.

geography in the first half of the 20th century—exploration, environmental determinism and possibilism, and regional geography—found few uses for statistical methods. Techniques for statistical inference were emerging at this time but were not applied in the geographical literature.

Exploration

This paradigm is one of the earliest used in geography. Unexplored areas of the earth continued to hold the interest of geographers well into the current century. Explorations, funded by geographical societies such as the Royal Geographical Society (RGS) and the American Geographical Society (AGS), continued the tradition of geographers collecting, collating, and disseminating information about relatively obscure and unknown parts of the world. The research sponsored by these organizations helped lead to the establishment of academic departments of geography at several universities. But, given only a passing interest in generalization and an extreme concern for the unique, little of the data generated by this research were ever analyzed by conventional statistical techniques.

Environmental Determinism and Possibilism

Environmental *determinists* and *possibilists* focused on the role of the physical environment as a controlling variable in explaining the diversity of the human impact on the landscape. Geographers began to concentrate on the physical environment as a control of human behavior, and some determinists went so far as to contend that environmental factors drive virtually all aspects of human behavior. Possibilists held a less extreme view, asserting that people are not totally passive agents of the environment, and had a long, and at times bitter debate with determinists. Few geographers studied human–environment relations outside this paradigm; and very little attention was paid to statistical methodology.

Regional Geography

Reacting against the naive lawmaking attempts of the determinists and possibilists were proponents of regional geography. Generalization of a different character was the goal. According to this paradigm, an *integration* or *synthesis* of the characteristics of areas or regions was to be undertaken by geographers. Ultimately, this would lead to a more or less complete knowledge of the areal differentiation of the world. Statistical methodology was limited to the systematic studies of population distribution, resources, industrial activity, and agricultural patterns. Emphasis was placed on the data collection and summary components of descriptive statistics. In fact, these systematic studies were seen as preliminary and subsidiary elements to the essential tasks of regional synthesis. The definitive work establishing this paradigm at the forefront of geographical research was Richard Hartshorne's *The Nature of Geography,* published in 1939.

Many of the contributions in this field discussed the problems of delimiting homogeneous regions. Each of the systematic specializations produced its own regionalizations. Together, these regions could be synthesized to produce a regional geography. A widely held view was that regional delimitation was a personal interpretation of the findings of many systematic studies. Despite the fact that the map was considered one of the cornerstones of this approach, the analysis of maps using quantitative techniques was rarely undertaken. A notable exception was Weaver's (1954) multiattribute agricultural regionalization; however, his work was not regarded as mainstream regional geography at the time.

Beginning in about 1950, the dominant approach to geographical research shifted away from regional geography and regionalism. To be sure, the transition took place over the succeeding two decades and did not proceed without substantial opposition. It was fueled by the increasing dissatisfaction with the regional approach and the gradual emergence of an acceptable alternative. Probably the first indication of what was to come was the rapid development of the systematic specialties of geography. The traditional systematic branches of physical, economic, historical, and political soon were augmented with urban, marketing, resource management, recreation, transportation, population, and social geography. These systematic specialties developed very close links with related academic disciplines—historical geography with history, social geography with sociology, and so forth. Economic geographers in particular looked to the discipline of economics for modern research methodology. Increased training in these so-called parent disciplines was suggested as an appropriate means of improving the quality of geographical scholarship. Throughout the 1950s and 1960s, the teaching of systematic specialties and research in these fields became much more important in university curricula. The historical subservience of the systematic fields to regional geography was reversed during this period.

The Scientific Method and Logical Positivism

The new paradigm that took root at this time focused on the use of the *scientific method.* This paradigm sought to exploit the power of the scientific method as a vehicle to establish truly geographical laws and theories to explain spatial patterns. To some, geography was reduced to pure spatial science, though few held this rather extreme view. As it was applied in geography, the scientific method utilized the deductive approach to explanation favored by *positivist* philosophers.

The deductive approach is summarized in Figure 1-3. The researcher begins with a perception of some real-world structure. A pattern, for example, the distance decay of some form of spatial interaction, leads the investigator to develop a model of the phenomenon from which a generalization or hypothesis can be formulated. An experiment or some other kind of test is used to see whether or not the model can be verified. Data are collected from the real world, and verification of the hypothesis or speculative law is undertaken. If the test proves successful, laws and then theories can be developed, heightening our understanding of the real world. If these tests prove successful in many different empirical applications, then the *hypothesis* gradually comes to be accepted as a *law.* Ultimately, these laws are combined to form a theory.

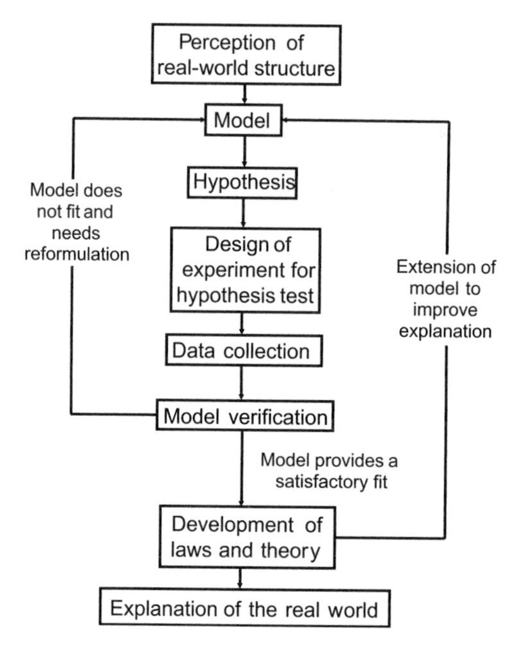

FIGURE 1-3. The deductive approach to scientific explanation.

This approach obviously has many parallels to the methodology for statistics outlined in the introduction to this chapter.

The deduction-based scientific method began to be applied in virtually all fields of geography during the 1950s and 1960s. It remains particularly important in most branches of physical geography, as well as in urban, economic, and transportation geography. Part of the reason for this strength is the widespread use of the scientific method in the physical sciences and in the discipline of economics.

Quantification is essential to the application of the scientific method. Mathematics and statistics play central roles in the advancement of geographic knowledge using this approach. Because geographers have not viewed training in mathematics as essential, the statistical approach has been dominant and is now accepted as an important research tool by geographers. That is not to say that the methodology has been accepted throughout the discipline. Historical and cultural geographers shunned the new wave of quantitative, theoretical geography. Part of the reason for their skepticism was that early research using this paradigm tended to be long on quantification and short on theory. True positivists view quantification as only a means to an end—the development of theory through hypothesis testing. It cannot be said that this viewpoint was clear to all those who practiced this approach to geographic generalization. Too often, research seemed to focus on what techniques were available, not on the problem or issue at hand. The methods *themselves* are clearly insufficient to define the field of geography.

Many researchers advocating the use of the scientific method also defined the discipline of geography as *spatial science.* Human geography began to be defined in terms of *spatial* structures, *spatial* interaction, *spatial* processes, or *spatial* organization. Distance was seen as the key variable for geographers. Unfortunately, such a narrow view of the discipline seems to preclude much of the work undertaken by cultural and historical geographers. Physical geography, which had been brought back into geography with the onset of the quantitative revolution, was once again set apart from human geography. Reaction against geography as a spatial science occurred for several reasons. Chief among these reasons was the disparity between the type of model promised by advocates of spatial science and what they delivered. Most of these theoretical models gave adequate descriptions of reality only at a very general level. The axioms on which they were based seemed to provide a rather poor foundation for furthering the development of geographical theory.

By the mid-1960s, a field now known as *behavioral geography* was beginning to emerge. It was closely linked with psychology and drew many ideas from the rich body of existing psychological research. Proponents of this approach did not often disagree with the basic goals of logical positivism—the development of theory-based generalizations—only with how this task could be best accomplished. Behavioral geographers began to focus on individual spatial cognition and behavior, primarily from an inductive point of view. Rather than accept the unrealistic axioms of perfect knowledge and perfect rationality inherent in many models developed by this time, behavioral geographers felt that the use of more realistic assumptions about behavior might provide deeper insights into spatial structures and spatial behavior. Their inductive approach was seen as a way of providing the necessary input into a set of richer models based on the deductive mode. Statistical methodology has a clear role in this approach.

Postpositivist Approaches to Geography

Although statistics and quantitative methods seemed to dominate the techniques used during the two decades in the period 1950–1970, a number of new approaches to geographical research began to emerge following this period. First, there were approaches based on *humanistic philosophies*. Humanistic geographers take the view that people create *subjective* worlds in their own minds and that their behavior can be understood only by using a methodology that can penetrate this subjectivity. By definition then, there is no *single, objective* world as is implicit in studies based on positivist, scientific, approaches. The world can only be understood through people's intentions and their attitudes toward it. Phenomenological methods might be used to view the diversity and intensity of experiences of place as well as to explore the growing "placelessness" in modern urban design, for example. Such approaches found great favor in cultural and historical geography.

Structuralists reject both positivist and humanistic methodologies, arguing that explanations of observed spatial patterns cannot be made by a study of the pattern itself, but only by the establishment of theories to explain the development of the societal condition within which people must act. The structuralist alternative, exemplified by Marxism, emphasizes how human behavior is constrained by more general societal processes and can be understood only in those terms. For example, patterns of income segregation in contemporary cities can be understood only within the context of a class conflict between the bourgeoisie on one hand and the proletariat, or workers, on the other. Understanding how power and therefore resources are allocated in a society is a prerequisite to comprehending its spatial organization.

Beginning as *radical geography* in the late 1960s, much of the early effort in this subfield was also directed at the shortcomings inherent in positivist-inspired research. To some, Marxist theory provided the key to understanding capitalist production and laid the groundwork for the analysis of contemporary geographical phenomena. For example, the emergence of ghettos, suburbanization, and other urban residential patterns was analyzed within this framework. More recently, many have explored the possibilities of geographical analysis using variants of the philosophy of structuralism. Structuralism proceeds through an examination of dynamics and rules of systems of meaning and power.

Interwoven within these views were critiques of contemporary geographical studies from feminist geographers. The earliest work, which involved demonstrating that women are subordinated in society, examined gender differences in many different geographical areas, including cultural, development, and urban geography. The lives, experiences, and behavior of women became topics of legitimate geographical inquiry. This foundation played a major role in widening the geographical focus to the intersection of race, class, and sexual orientation, and to how they interact in particular spaces and lives under study.

Human geography has also been invigorated by the impact of *postmodern* methodologies. Postmodernism represents a critique of the approaches that dominated geography from the 1950s to the 1980s and that are therefore labeled as *modernist.* Postmodern researchers stress textuality and texts, deconstruction, reading and interpretation as elements of a research methodology. Part of the attraction of this approach is the view that postmodernism promotes differences and eschews conformity to the modern style. As such its emphasis on heterogeneity, particularity, or uniqueness represents a break with the search for order characteristic of modernism. A key concern in postmodern work is *representation*—the complex of cultural, linguistic, and symbolic processes that are central to the construction of meaning. Interpreting landscapes, for example, may involve the analysis of a painting, a textual description, maps, or pictures. *Hermeneutics* is the task of interpreting meaning in such texts, extracting their embedded meanings, making a "reading" of the landscape. One set of approaches focuses on deconstruction of these texts and analysis of *discourses.* The importance of language in such representations is, of course, paramount. The world can only be understood through language that is seen as a method for transmitting meaning.

The Rise of Qualitative Research Methods in Geography

One consequence of the emergence of this extreme diversity to the approach of human geography is a renewed focus on developing suitable tools for this type of research. These so-called qualitative methods serve not as a competitor but more of a complement to the toolbox, which statistical methods offer to the researcher. The three most commonly used qualitative methods are interviews, techniques for analyzing textual materials (taken in the broadest sense), and observational techniques.

The use of data from interviews is familiar to most statisticians since the development of survey research was closely linked to developments in probability theory and sampling. However, most of the work in this field has focused on one form of interview—the personal interview, which uses a relatively structured format of questions. This method can be thought of as a relatively limiting one, and qualitative geographers tend to prefer more *semistructured* or *unstructured* interview techniques. When used properly, these methods can extract more detailed and personal information from interviewees. Like statisticians, those who employ qualitative methods encounter many methodological problems. How many people should be interviewed? How should the interview be organized? How can the transcripts from an interview be coded to elicit understanding? How can we search for commonalities in the transcripts? Would a different analyst come up with the same interpretations? These are not trivial questions.

In *focus groups,* 6 to 10 people are simultaneously interviewed by a moderator to explore a topic. Here, it is argued that the group situation promotes interaction among the respondents and sometimes leads to broader insights than might be obtained by individual interviews. Statisticians have employed focus groups to help design questionnaires. Marketing experts commonly use them to anticipate consumer reaction to new products. Today focus groups are being used in the context of many different types of research projects in human geography.

Textual materials, whether in the format of written text, paintings or drawings, pictures, or artifacts, can also be subjected to both simple and complex methods of analysis. At one end, simple *content analysis* can be used to extract important information from transcripts, often assisted by PC-based software. Simple word counts or coding techniques are used to analyze textual materials, compare and contrast different texts, or examine trends in a series of texts. Increasingly, researchers are interested in "deconstructing" texts to reveal multiple meanings, ideologies, and interpretations that may be hidden from simple content analysis.

Finally, qualitative methods of observing interaction in a geographical environment are increasingly common. Attempting to understand the structure and dynamics of certain geographic spaces at both the micro level (a room in a building) or in a larger context (a neighborhood or shopping mall) by observing how participants behave and interact can provide useful insights. Observers with weak or strong participation in the environment are possible. Compare, for example, the data likely to be available from a hidden camera recording pedestrian activity in a store, to the data obtained by a researcher living and observing activity in a small remote village. Clearly, one's *positioning* to the observed is important.

All of these techniques have their role in the study of geography. Some serve as useful complements to statistically based studies. For example, when statisticians make interpretations based on the results of surveys, it is often useful to use in-depth unstructured interviews to assess whether such interpretations are indeed valid. A focus group might be used to assess whether the interpretations being made are in agreement with what people actually think. It is easy to think of circumstances where one might wish to use quantitative statistical methods, purely qualitative techniques, or a mixture of the two.

The Role of Statistics in Contemporary Geography

What then is the role of statistics in contemporary geography? Why should we have a good understanding of the principles of statistical analysis? Certainly, statistics is an important component of the research methodology of virtually all systematic branches of geography. A substantial portion of the research in physical, urban, and economic geography employs increasingly sophisticated statistical analysis. Being able to properly evaluate the contributions of this research requires us to have a reasonable understanding of statistical methodologies.

For many geographers, the map is a fundamental building block of all research. Cartography is undergoing a period of rapid change in which computer-based methods are continuing to replace much conventional map compilation and production. Microcomputers linked to a set of powerful peripheral data storage and graphical devices are now essential tools for contemporary cartography. Maps are inherently mathematical and statistical objects, and as such they represent one area of geography where dramatic change will continue to take place for some time to come. This trend has forced many geographers to acquire better technical backgrounds in mathematics and computer science, and has opened the door to the increased use of statistical and quantitative methods in cartography. Geographic information systems (GIS) are one manifestation of this phenomenon. Large sets of data are now stored, accessed, compiled, and subjected to various cartographic display techniques using video display terminals and hard-copy devices.

The analysis of the spatial pattern of a single map and the comparison of sets of interrelated maps are two cartographic problems for which statistical methodology has been an important source of ideas. Many of the fundamental problems of displaying data on maps have clear and unquestionable parallels to the problems of summarizing data through conventional descriptive statistics. These parallels are discussed briefly in Chapter 3, which focuses on descriptive statistics.

Finally, statistical methods find numerous applications in *applied geography*. Retail location problems, transportation forecasting, and environmental impact assessment are three examples of applied fields where statistical and quantitative techniques play a prominent role. Both private consulting firms and government planning agencies encounter problems in these areas on a day-to-day basis. It is impossible to underestimate the impact of the wide availability of microcomputers on the manner in which geographers can now collect, store and retrieve, analyze, and display the data fundamental to their research. The methodologies employed by mathematical statisticians themselves have been fundamentally changed with the arrival and diffusion of this technology. No course in statistics for geographers can afford to omit applied work with microcomputers in its curriculum.

In sum, statistical analysis is commonplace in contemporary geographical research and education, as it is in the other social, physical, and biological sciences. It is now being more thoughtfully and carefully applied than in the past and includes an ever widening array of specific techniques. Moreover, research using *both* quantitative and qualitative methods is increasingly common. Such an approach exploits the advantages of each class of tools, and minimizes their disadvantages when relying on either alone.

1.2. Data

Although Figure 1-2 seems to suggest that statistical analysis begins with a dataset, this is not strictly true. It is not unusual for a statistician to be consulted at the earliest stages of a research investigation. As the problem becomes clearly defined and questions of appropriate data emerge, the statistician can often give invaluable advice on sources of data, methods used to collect them, and characteristics of the data themselves. A properly executed research design will yield data that can be used to answer the questions of concern in the study. The nature of the data used should never be overlooked. As a preliminary step, let us consider a few issues relating to the sources of data, the kinds of variables amenable to statistical analysis, and several characteristics of the data such as measurement scales, precision, and accuracy.

Sources of Data

A useful typology of data sources is illustrated in Figure 1-4. At the most basic level, we distinguish between data that already exist in some form, which can be termed

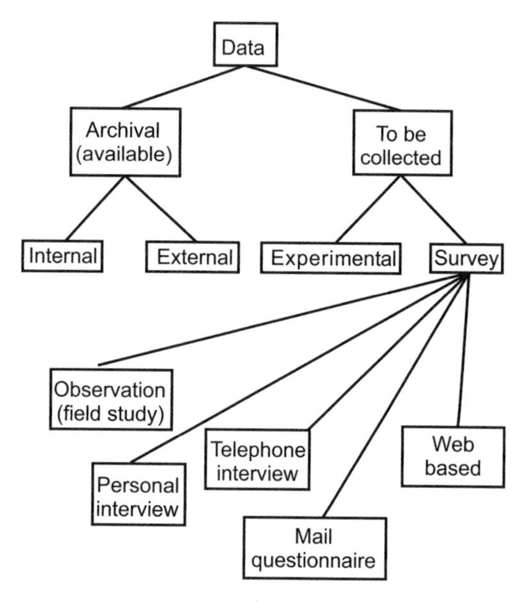

FIGURE 1-4. A typology of data sources.

archival, from data that we propose to collect ourselves in the course of our research. When these data are available in some form in various records kept by the institution or agency undertaking the study, the data are said to be from an *internal source.*

DEFINITION: INTERNAL DATA

Data available from existing records or files of an institution undertaking a study are data from an internal source.

For example, a meteorologist employed by a weather forecasting service normally has many key variables such as air pressure, temperature, and wind velocity from a large array of computer files that are augmented hourly, daily, or other predetermined frequency. Besides the ready availability of this data, the meteorologist has the added advantage of knowing a great deal about the instruments used to collect the data, the accuracy of the data, and possible errors. In-depth practical knowledge of many factors related to the methods of data collection, is often invaluable in statistical analysis. For example, we may find that certain readings are always higher than we might expect. When we examine the source of the data, we might find that all the data were collected from a single instrument that was incorrectly calibrated.

When an *external* data source must be used, many important characteristics of the data may not be known.

DEFINITION: EXTERNAL DATA

Data obtained from an organization external to the institution undertaking the study are data from an external source.

Caution should always be exercised in the use of external data. Consider a set of urban populations extracted from a statistical digest summarizing population growth over a 50-year period. Such a source may not record the exact areal definitions of the urban areas used as a basis for the figures. Moreover, these areal definitions may have changed considerably over the 50-year study period, owing to annexations, amalgamations, and the like. Only a primary source such as the national census would record all the relevant information. Unless such characteristics are carefully recorded in the external source, users of the data may have the false impression that no anomalies exist in the data. It is not unusual to derive results in a statistical analysis that cannot be explained without detailed knowledge of the data source. At times, statisticians are called upon to make comparisons among countries for which data collection procedures are different, data accuracy differs markedly, and even the variables themselves are defined differently. Imagine the difficulty of creating a snapshot of world urbanization collecting variables taken from national census results from countries in every continent. Organizations such as the United Nations spend considerable effort integrating data of this type so that trends and patterns across the globe can be discerned.

Another useful distinction is between *primary* and *secondary* external data.

DEFINITION: PRIMARY DATA

Primary data are obtained from the organization or institution that originally collected the information.

DEFINITION: SECONDARY DATA

Secondary data are data obtained from a source other than the primary data source.

If you must use external data, *always* use the primary source. The difficulty with secondary sources is that they may contain data altered by recording or editing errors, selective data omission, rounding, aggregation, questionable merging of datasets from different sources, or various *ad hoc* corrections. For example, never use an encyclopedia to get a list of the 10 largest cities in the United States; use the U.S. national census. It is surprising just how often research results are reported with erroneous conclusions—only because the authors were too lazy to utilize the primary data source or were unaware that it even existed.

Metadata

It is now increasingly common to augment any data source with a document or database that provides essential information about the data themselves. These so-called metadata or simply "data about data" provide information about the content, quality, type, dates of creation, and usage of the data. Metadata are useful for *any* information source including pictures or videos, web pages, artifacts in a museum, and of course statistical data. For a picture, for example, we might wish to know details about the exact location where it was taken, the date it was taken, who took the picture, detailed physical characteristics of the recording camera such as the lens used, and any post-production modifications such as brightness and toning applied to it.

DEFINITION: METADATA

Metadata provide information about data, including the processes used to create, clean, check, and produce it. They can be presented in the form of a single or multiple set of documents and/or databases, and they can be made available in printed form or through the web.

The items that should be included in the metadata for any object cannot be precisely and unambiguously defined. While there is considerable agreement about what is to be included in the metadata, many providers augment these basic elements with other items specific to their own interests. In general, the goal of providing metadata *is to facilitate the use and understanding of data.* For statistical data, a metadata document includes several common components:

1. *Data definitions.* This component includes the content, scope, and purpose of the data. For a census, for example, this should include details of the questions asked to recipients, coding used to indicate invalid or missing responses, and so forth. These pieces of information can be obtained by examining the questionnaire and the instructions given to those who apply the questionnaire or the set of instructions given to the interviewers on how to code responses from the recipients.

Several different documents can be included in the metadata for potential users. If the data has been collected from a survey, the original questionnaire is particularly useful since it will contain the exact wording used by interviewers. Since responses are highly sensitive to the wording choices of researchers, this is an essential component of any metadata. After it has been collected, many items are *coded* and assigned numerical or alphabetical codes representing the actual responses by the subject. For example, responses where the subject was unwilling to answer can be coded differently than those where the subject did not know the answer. In addition, responses might be simply *missing* or *invalid.*

Another useful set of information is sometimes available when the data are stored as a database. A *data dictionary* describes and defines each field or variable in the database, provides information on how it is stored (as text, integer, date, or floating point number with a given number of decimal places), and the name of the table in which it is placed. The file types, database schema, processing checks, transformations, and calculations undertaken on the raw data should be included.

If you wish to compare a number of different statistical studies on the same topic, you may find it essential to compare the background information on each data element used in the study. For example, suppose you want to compare vacancy rates in the apartment rental market in several different places. You may find that this task is particularly difficult when different studies have employed different definitions for both rental units and vacancies. While it is generally agreed that a vacancy rate measures the proportion of rental units unoccupied, there will undoubtedly be variations on how this statistic was actually calculated. Were all rental units visited? Were postal records used to verify occupancy? Were landlords contacted to verify occupancy? As you can see, knowing how the data were collected is almost as valuable as the number itself!

2. *Method of sampling.* Many sources of data are based on samples from populations. How was the sample undertaken? Exactly what sampling procedures were used? How large was the sample? Were some items sampled more intensively than others? For example, when we estimate a vacancy rate, we inevitably combine data from different types of rental units, varying from large residential complexes of over 100 or even 500 units to small apartments rented (perhaps even illegally) by individual homeowners. Sometimes the results of the study may reflect the differential sampling used to uncover these units.

The size of the sample and the size of the population themselves are extremely important characteristics of the data source. A sample of 500 units from a population of a potential 100,000 units in some city is less useful than a sample of 500 taken from a city where the estimated number of rental units is only 10,000. The exact dates of the survey are also important, as vacancy rates vary considerably over the year. It is important to know the currency of the data as well. Situations change rapidly over time. Public opinion data are particularly problematic because they are sometimes subject to radical change in an extremely short period of time.

Sometimes the objects under study are stratified by type, and sampling within each stratum is undertaken independently and at differential sampling fractions. In order to combine the objects into a single result, they must be properly weighted to reflect this differential sampling. For example, in a vacancy rate study we might differentially sample types of units, spending more resources on units with lower rents than on those with higher rents. To combine the results in order to come up with a single measure for the vacancy rate, we apply weights to each type to reflect their relative abundance in the overall housing stock.

3. *Data quality.* When measures of data quality are available, they are also an important indicator of the usefulness of a data source. As we shall see in Section 1.3, we should examine our data for *accuracy, precision,* and *validity.* Suppose a study collected some data in the field and used a GPS to determine the location of the phenomenon. Depending on what type of GPS was used, its potential internal error, and the time period over which the coordinates of the location were determined, we may have data of different quality. The quality of the data collected may reflect the precision of the recording instruments, the training and experience of the interviewers, the ability of the survey instrument to yield the answers to questions of interest, and the care taken to verify and clean the data collected.

4. *Data dissemination and legal issues.* Information on how the data can be obtained and how they are distributed is also an important component of metadata. In an era when data are increasingly being distributed electronically, it is now common to specify the procedure for obtaining the data and the particular file formats used. Sometimes data analysis may be undertaken using a statistical package that imports the data provided in one format and alters it to make it compatible with the data commonly analyzed by the program. At times the import process can truncate the data or

change the number of decimal places. Errors can be introduced by file manipulations that truncate rather than round numbers if the number of decimal places is reduced. If the data are disseminated by the original organization that collected the data, this will often ensure that the data used in a study are the best available. This should be apparent in the metadata.

Not all data can be made publicly available, and a considerable number of data sources must deal with legal issues related to privacy and ownership. This is particularly true for data collected on individuals or households where it is possible to suppress the distribution of data that can lead to the identification of an individual household or small group of individuals. For example, figures on incomes earned by households are sensitive and are not normally made available except for large groups of households, for example, census tracts.

5. *Lists of studies based on the data.* It is no longer unusual for data collection agencies or providers to also include in their metadata a bibliography of studies and reports that have utilized the data. These may be internal reports, academic journal articles, research monographs, or other published documents. Being able to see how others have used the data and their conclusions can tell us a lot about the potential issues that may arise in our own study. Suppose an analyst using housing data to estimate vacancy rates feels that the study underestimated the vacancy rate since it placed to much emphasis on high-income properties and ignored low-rent properties that were often advertised only locally in particular neighborhood markets. It would be foolish of us to ignore this result if it might possibly affect the interpretations we developed in our study, which used the data to determine the length of time typical units were vacant.

6. *Geographic data.* Data are at the core of GIS, and metadata are now commonly provided for spatial data so that users can know the spatial extent, locational accuracy and precision, assumed shape of the earth, and projection used to develop a map integral to some dataset. It is obvious that when we are describing *areal* data, we need to know the exact boundaries of places and any changes to these areal definitions over time. For example, several GIS software packages contain a *metadata editor* so that the characteristics of any layer of spatial information can be completely detailed. Developing suitable official standards for geographic metadata is becoming increasingly imprtant.

7. *Training.* Some data collection agencies provide courses that introduce users to data sources, particularly large complex data collection exercises such as a national census. Training and help files are now provided online so that users can know a great deal about the data before beginning their analysis.

More and more, the need for the exchange of statistical data is creating a demand for the effective design, development, and implementation of parallel metainformation systems. As data become increasingly distributed using web-based dissemination tools, software tools that document metadata for statistical data will become increasingly important. As this trend continues, users will be able to undertake statistical analysis of data with a better understanding of the strengths and weakness of the data itself.

Data Collection

When the data required for a study cannot be obtained from an existing source, they are usually collected during the course of the study. It should be clear that any data collection procedure should be undertaken in parallel with an exercise in metadata creation. As our data collection takes place we continually augment our metadata file or document to reflect all characteristics that may be important to users. When, where, what, and how were the data collected? by whom? where? It is almost as difficult to provide accurate metadata as it is to provide the data themselves!

Data acquisition methods can be classified as either *experimental* or *nonexperimental.*

DEFINITION: EXPERIMENTAL METHOD OF DATA COLLECTION

An experimental method of data acquisition is one in which some of the factors under consideration are controlled in order to isolate their effects on the variable or variables of interest.

Only in physical geography is this method of data collection prominent. Fluvial geomorphologists, for example, may use a flume to control such variables as stream velocity, discharge, bed characteristics, and gradient. Among the social sciences, the largest proportion of experimental data is collected in psychology.

DEFINITION: NONEXPERIMENTAL METHOD OF DATA COLLECTION

A nonexperimental method of data collection or *statistical survey* is one in which no control is exercised over the factors that may affect the population characteristic of interest.

There are five common survey methods. *Observation* (or field study) requires the monitoring of an ongoing activity and the direct recording of data. This form of data collection avoids several of the more serious problems associated with other survey techniques, including incomplete data. While techniques based on observation are well developed in anthropology and psychology, their use within geographical research is more recent and limited.

In addition to observation, three other methods of data collection are often used to extract information from households, individuals, or other entities such as corporations or organizations: *personal interviews, telephone interviews,* and *web-based interviews.* In a personal interview, a trained interviewer asks a series of questions and records responses on a specially designed form. This procedure has obvious advantages and disadvantages. An alternative, and often cheaper, method of securing the data from a set of households is to send a *mail questionnaire.* This method is often termed *self-enumeration* since the individual completes the questionnaire without assistance from the researcher. The disadvantages of this method include nonresponse, partial response, and low return rates for completed questionnaires. Factors affecting the quality of data from mail surveys include appropriate wording, proper question order, question types, layout, and design. For telephone and personal interviews there is the added impact of the rapport developed between the interviewer and the subject.

Over time, technological change has had an immense impact on these techniques. Computer-assisted telephone interviewing (CATI) is now the norm with random-digit dialing. Some interviews are now conducted using e-mail or web browser-based collection pages. Important issues related to these techniques include variations in coverage, privacy concerns, and accuracy. Groves et al. (2004) is an especially useful overview of the issues related to all types of survey techniques.

Characteristics of Datasets

Statistical analysis cannot proceed until the available data have been assembled into a usable form.

DEFINITION: DATASET

A dataset is a collection of statistical information or values of the variables of interest in a study.

Geographers collect or analyze two typical forms of *datasets*. An example of the first type, sometimes known as a *structural matrix,* is shown in Table 1-1. In this example, the observational units are climatic stations. Five variables are contained within the dataset. The information on every variable from one observational unit is often termed an observation; it is also common to speak of the data value for a single variable as an *observation* since it is the observed value. In this case, the rows of the dataset represent different locations, and the columns represent the different variables available for analysis. These places might represent areas or simply fixed locations such as cities and towns. Such a matrix allows us to examine the *spatial variation* or *spatial structure* of these individual variables.

Table 1-2 illustrates the second typical form of datasets, an *interaction matrix,* in which the variable of interest is expressed as the flow or interaction between various locations (A through G), which are both the row and column headings of the

TABLE 1-1
A Geographical Dataset

Climatic station	Days per year with precipitation	Annual rainfall, cm	Mean January temperature, °C	Mean July temperature, °C	Coastal or inland
A	114	71	6	16	C
B	42	48	12	16	C
C	54	32	–4	21	I
D	32	28	–8	20	I
E	41	129	16	18	C
F	26	18	1	22	I
G	3	8	24	29	I

TABLE 1-2
An Interaction Matrix

		To						
		A	B	C	D	E	F	G
From	A	58	49	60	91	92	34	14
	B	42	48	12	16	68	25	72
	C	54	32	72	73	63	82	81
	D	45	60	20	28	57	20	46
	E	41	12	17	48	33	99	14
	F	26	18	30	22	10	66	29
	G	13	18	24	19	29	77	29

matrix. Each entry in this matrix represents one observation. Looking across any single row allows us to see the outflows from a single location. Similarly, a single column contains all the inflow into one location. It is easy to see that this matrix can be analyzed in any number of ways in order to search for patterns of spatial interaction.

The variables in a dataset can be classified as either *quantitative* or *qualitative.* Quantitative values can be obtained either by counting or by measurement and can be ordered or ranked.

DEFINITION: QUANTITATIVE VARIABLE
A quantitative variable is one in which the values are expressed numerically.

Discrete variables are those variables that can be obtained by counting. For example, the number of children in a family, the number of cars owned, the number of trips made in a day are all counting variables. The possible values of counting variables are the ordinary integers and zero: 0, 1, 2, . . . , *n*. Quantities such as rainfall, air pressure, or temperature are measured and can take on *any* continuous value depending upon the accuracy of the measurement and recording instrument. *Continuous* variables are thus inherently different from discrete variables. Since continuous data must be measured, they are normally *rounded* to the limits of the measuring device. Heights, for example, are rounded to the nearest inch or centimeter, and temperatures to the nearest degree Celsius or Fahrenheit.

Qualitative variables are neither measured nor counted.

DEFINITION: QUALITATIVE VARIABLE
Qualitative variables are variables that can be placed into distinct nonoverlapping categories. The values are thus non-numerical.

Qualitative variables are sometimes termed *categorical* variables since the observational units can be placed into categories. Male/female, land-use type, occupation, and plant species are all examples of qualitative variables. These variables are defined by the set of classes into which an observation can be placed. In Table 1-1, for example, climatic stations are classified as either coastal (C) or inland (I).

Numerical values are sometimes assigned to qualitative variables. For example, the yes responses to a particular question in a survey may be assigned the value 1 and the no responses a value of 2. The variable gender may be identified by males = 1 and females = 0 (or vice versa!). In both of these examples, each category has been assigned an *arbitrary* numerical value. As we shall see, it is improper to perform most mathematical operations on qualitative variables expressed in this manner. Consider, for example, the operation of addition. Although this operation is appropriate for a quantitative variable, it makes no sense to add the numerical values assigned to the variable gender.

Besides being described as qualitative or quantitative, variables can also be classified according to the *scale of measurement* on which they are defined. This scale defines the amount of information the variable contains and what numerical operations can be meaningfully undertaken and interpreted. The lowest scale of measurement is the nominal scale. Nominal scale variables are those qualitative variables that have no implicit ordering to their categories. Even though we sometimes assign numerical values to nominal variables, they have no meaning. Consider the variable in Table 1-1 that distinguishes coastal climatic stations (C) form inland ones (I). All that we can really do is distinguish between the two types of stations. In other words, we know that stations A, B, and E are coastal and therefore different from stations C, D, F, and G. Also, stations A,B, and E are all alike according to this variable. One way of summarizing a nominal variable is to count the number of observations in each category. This information can be summarized in a bar graph or a simple table of the following form:

Category	Count or frequency	Proportion	Percentage
C	3	0.429	43
I	4	0.571	57
	7	1.000	100

The use of percentages or relative proportions to summarize such data is quite common. For example, we would say that 0.429, or 43%, of the climatic stations are coastal and 0.571, or 57%, are inland.

Proportional summaries are often extremely useful in comparing responses to similar questions from two or more different surveys, or from different classes of respondents to the same question in a survey. For example, studies of migrants to cities of the Third World often include questions concerning the sources of information used by individual migrants in selecting their destination. A question of this type might be phrased in the following way: In reaching your decision to come here, you must have had some information about job possibilities, living conditions, income, and the like. Which of the following gave you the *most* information? The question would then list a variety of potential sources of information: relatives, friends, newspapers, radio, or

TABLE 1-3
Percentage Distribution of Responses Concerning Primary Information Source Concerning Migration Destinations Used by Migrants

Sources of information	Males	Females
Newspapers	13	7
Radio	3	2
Government labor office	2	3
Family members	40	28
Friends	27	41
School teacher	4	1
Career counselor	1	1
Other sources	10	17
Total	100	100

other. We could compare the proportional use of these information sources from different studies, or by gender or educational attainment of the respondent.

Table 1-3 summarizes the responses to this question and differentiates by gender of the respondent. Clearly, about two-thirds of the respondents cite family or friends as the dominant information source. However, men seem to rely more on family and less on friends in comparison to women. Also, men tend to use newspapers more frequently, and female migrants seem to use other sources more often. These results indicate the significant role played by kin and friends in the rural-urban migration process, but also point to potentially important differences in primary sources of information by gender. An interesting research question is whether or not these observed differences are truly important. If the respondents to the questionnaire were randomly selected, we might be able to use an inferential technique to determine whether this observed difference is due to random sampling or is indicative of an important difference between males and females. Of course, we would also wish to compare the results of a number of similar studies from different cities in different countries before confirming the importance of this hypothesis as a *general* rule. The hypothesis may be limited to the current study.

If the categories of the *qualitative* variable can be put into order, then the scale of measurement of the variable is said to be *ordinal.* An example of an ordinal variable is the strength of opinion measured in responses to a question with the following categories:

Strongly agree	Agree	Neutral	Disagree	Strongly disagree
2	1	0	−1	−2
1	2	3	4	5
5	4	3	2	1

Three different numerical assignments are given, and each is consistent with an ordinal scale for this variable. In all cases, the stronger the agreement with the question, the higher the value of the numerical assignment. In fact, any assignment can be used as long as the values assigned to the categories maintain the ordering implicit in the wordings attached to the categories. The assignment of values –200, –10, 270, 271, 9382 meets this criterion. Note that it doesn't matter which end of the scale is assigned the lowest values, nor does it matter if they are given negative values. All of the categories could be given negative values, all positive values, or a mixture of the two. Of course, the scales defined in the table above have the added advantage of simplicity.

The numerical difference between the values assigned to different categories has no meaning for an ordinal variable. We can neither subtract nor add the values of ordinal variables. Notice, however, that the numerical assignments used to define ordinal variables often use a constant difference or unit between successive categories. In the first scale defined above, each category differs from the next by a value of 1. This does not mean that a respondent who checks the box for strongly agree is 2 units higher than a respondent who checks the box for neutral. We can only say that the first respondent agrees more with the question than does the second respondent. Only statements about *order* can be made using the values assigned to the categories of an ordinal variable.

Quantitative, or numerical variables, whether discrete or continuous, can be classified into two scales of measurement. *Interval* variables differ from ordinal variables in that the interval-scale of measurement uses the concept of *unit distance.* The difference between any two numbers on this scale can always be expressed as some number of units. Both Fahrenheit and Celsius temperature scales are examples of interval-scale variables. Although it makes sense to compare differences of interval scale variables, it is not permissible to take ratios of the values. For example, we can say that 90°F is 45°F hotter than 45°F, but we cannot say that it is twice as hot. To see why, let us simply convert these temperatures to the Celsius scale: 90°F = 32°C and 45°F = 7°C. Note that 32°C is *not* twice as hot as 7°C.

At the highest level of measurement are ratio-scale variables. Any variable having the properties of an interval scale variable as well as a natural origin of zero is measured at the ratio scale of measurement. Distance measured in kilometers or miles, rainfall measured in centimeters or inches, and many other variables commonly studied by geographers are measured on the ratio scale. It is possible to compute ratios of such variables as well as to perform many other mathematical operations such as logarithms, powers, or roots. Because we can take ratios of distances, we can therefore say that a place 200 miles from us is twice as far as one 100 miles from us. While there is a logical and theoretical distinction between ratio and interval variables, this distinction rarely comes into play in practice.

Of far greater interest is the specification of the level of measurement of a variable measured indirectly. For example, the scale of measurement of a variable constructed from the responses to a question or a set of questions in an interview may not be easily identified. This variable may be ordinal, interval, or even ratio. That is, respondents may treat the categories of scale with the labels *strongly agree, agree,*

TABLE 1-4
Levels of Measurement: A Summary

Level	Permissible operations[a]	Examples
Nominal	A = B or A ≠ B, counting	Presence or absence of a road linking two cities or towns Land-use types Gender
Ordinal	A < B or A > B or A = B	Preferences for different neighborhoods in rank order Ratings of shopping center attractiveness
Interval	Subtraction (A – B)	Temperatures °F
Ratio	Addition (A + B) Multiplication (A × B) Take ratios A/B and compare Square roots Powers Logarithms Exponentiation	Distances (imperial or metric) Density (persons per unit area) Stream discharge Shopping center square footage Wheat or corn yield

[a]Permissible mathematical operations at each level of measurement include all operations valid at lower levels of measurement.

neutral, disagree, strongly disagree as if they were part of an interval, not ordinal, scale. A significant amount of research in psychology has examined methods for deriving scales with interval properties from test questions that are, strictly speaking, only ordinal in character. Attitude scales and some measures of intelligence are two examples where interval properties are desirable. An investigator must sometimes decide what sorts of operations on the variables collected are meaningful, or whether the operations exceed the information contained in the variable. A summary of the scales of measurement, along with a list of permissible mathematical operations and examples, is given in Table 1-4.

1.3. Measurement Evaluation

The utility of any statistical analysis ultimately rests on the quality of the data used, regardless of how sophisticated the analysis itself is. We may be less likely to overstate the significance of our results if we first closely examine the nature of our data. For example, suppose our data suggest that men travel 10 minutes longer than females on the journey to work. On the face of it, this appears to be important. However, when we examine the source of our data, we find that it comes from a self-administered questionnaire in which respondents *reported* their travel times in a simple question expressed as "How long do you spend on your journey to work?" How did respondents answer this question? Did they report their journey time on the day of the survey, or was it based on some sort of average time? Suppose we independently found that over one-half of respondents report travel times that differ from the true travel time by 20 minutes or more. What does this say about our difference of 10 minutes by gender?

It is therefore *essential* to rigorously evaluate the quality of our data with respect to several key principles—before we undertake any statistical analysis. These principles include measurement *validity, accuracy, and precision.*

Validity

Measurement validity is perhaps the most abstract of these principles. Loosely speaking, it is the degree to which a variable measures what it is supposed to measure. In many physical science applications, validity is usually not an issue. For example, the variable temperature is commonly used as a measure of thermal energy, or heat. The relation between heat and temperature is well known; thus there is not likely to be much debate about whether or not temperature measures an object's energy content. Furthermore, the principles of liquid thermometers are understood, thus it is clear that measurements taken by liquid thermometers do in fact reflect temperature.

DEFINITION: MEASUREMENT VALIDITY

Measurement validity refers to the degree of correspondence between the concept being addressed and the variable being used to measure that concept.

There are, however, instances in both physical and social sciences where the situation is less clear, and measurement validity is questionable. In the first case, it may be that the concept being studied is imperfectly defined. To pick just a few examples, we could mention intelligence, social status, drought, and ecosystem diversity. Everybody has a rough idea of what these concepts are about, but precise definitions are elusive. To take another example, perhaps we want to know if air pollution control policies have improved air quality. Many air quality measures are available, including average pollutant level, maximum level, and number of days above some threshold. If the definition of "air quality" is vague, there will likely be questions about the validity of the variable chosen to measure it. It may be that several different *dimensions* of the term *air quality* need to be addressed, including its maximum, persistence, and regularity. Issues of measurement validity are rarely given the emphasis that they deserve. This often leads to the publication of research results that are misleading or erroneous, or simply wrong.

Accuracy

Another important consideration is *accuracy.*

DEFINITION: MEASUREMENT ACCURACY

Measurement accuracy refers to the absence of error, or the degree of agreement between a measured and true value.

We would say, for example, that a thermometer is accurate if the measured value is close to the true temperature. Note that this does not imply that temperature is a good measure of heat; thus *accuracy* need not imply *validity.* Note also that the definition implies the existence of a "true" value that is at least potentially observable when the measurement is made. Without this assumption, the concept of accuracy has little meaning.

It is convenient to describe the concept of accuracy within an analysis of the errors possible in the measurement process. Let us divide the total error of measurement into two components: *systematic* and *random* error.

$$\text{Total error} = \text{Systematic error} + \text{Random error}$$

Ideally then, we would like our total error to be small, which is accomplished by minimizing each of the two components. Let us examine each of the components separately. The first component, *systematic* error, arises if the instrument consistently gives high or low values. Of course, we would like our systematic error to be zero. If it is, we say that our measurements are *unbiased.*

The second component is error that is not attributable to poor calibration of the measurement instrument leading to systematic error, but appears to be random or unpredictable in nature. The actual physical process used to estimate temperature is subject to a range of effects, each of which is individually small but together appear random in nature. Better instruments may be able to reduce this error by controlling for these effects. In this way, our total error is also reduced.

Precision

In general, the *precision* of a measurement refers to the level of exactness or to the range of values possible in the measurement process. A thermometer that can provide estimates within a tenth of a degree is more precise than one that can only provide estimates within a whole degree. In terms of our error equation, precision then is related to the random component of error. It is easy to see that our total error is also reduced by making this as small as possible.

The difference between these two measurement errors is best illustrated within the example of repeated temperature measurements. Let us suppose that four different thermometers A, B, C, and D are used to measure an air temperature known to be 20°C. Five readings are taken with each thermometer over the course of an hour. The recordings from each of the four instruments are illustrated in the temperature scales of Figure 1-5. We can see that thermometer A has no systematic error and leads to readings that may be above or below the true temperature within a range of random error of about one-half a degree. Thermometer B has a systematic error of about +1° since the average reading seems to be around 21°, but the measurements also vary within about one-half a degree. We would say that both of these thermometers are equally *precise,* but thermometer A is more accurate.

Consider now thermometers C and D. Like thermometer A, thermometer C has no bias, but it is much less accurate than A. The problem with C is its low precision,

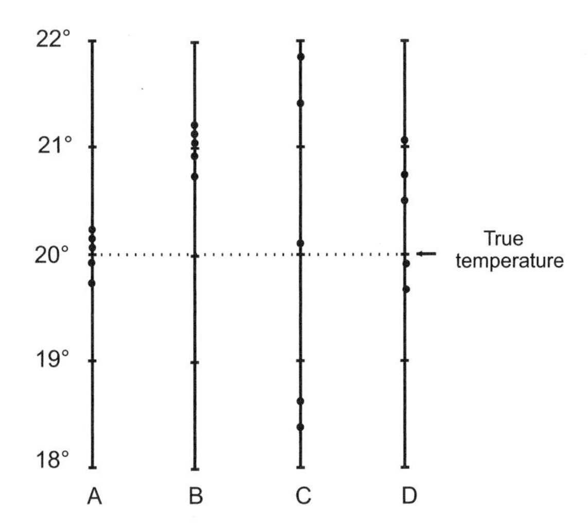

FIGURE 1-5. Differentiating between random and systematic error. • indicates recorded temperature.

since readings from the thermometer vary between roughly 18° and 20°. This is almost four times less precise than A. Thermometer D is biased but has a small random error, leading to an accuracy higher than the unbiased thermometer C. In the search for high overall accuracy, one must consider both sources of error. In particular, there are times when one will prefer a biased instrument over an unbiased alternative simply because the overall errors are smaller. In still other cases the concern is almost exclusively with bias, so that random error components are very much less important.

When the results of an empirical study are analyzed, a good first step is always to closely examine the operational definitions chosen for the variables. Are they suitable? Are they unambiguously defined? Are they the best dataset that could have been used? Are they sufficiently precise? Are all variables unbiased? Could they be responsible for any misleading inferences? The need for caution in the use of data applies to all data sources, including those that are based on experiments as well as those derived from surveys or observation.

1.4. Data and Information

Some historians and other commentators describe the age we live in as the *information age.* This age is characterized by the widespread use and adoption of information communications and technologies. Some argue that one of the fundamental keys to economic success at the firm or institution level—and indeed going all the way to the national level—is the ability to create information and successfully analyze it. *Information technology* is a general term for the technologies involved in collecting, processing, organizing, interpreting, and presenting data. Where does data and statistical analysis fit into this paradigm?

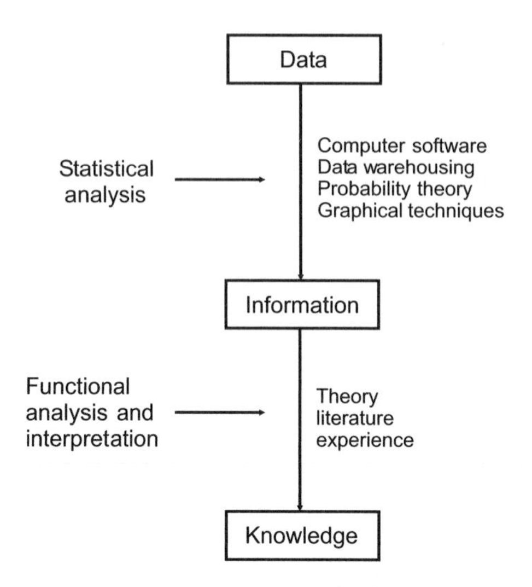

FIGURE 1-6. Statistics and information.

Figure 1-6 displays the concept of the *hierarchy of information* as described by many information system analysts. At the lowest level of the hierarchy is *data.* The meaning of data in this view is quite similar to what we have described in Section 1.3: specific observations of measured numbers. At the next level of the hierarchy, we have *information.* Information is developed from the data by processing and summarizing the data to yield facts and ideas.

DEFINITION: INFORMATION

Data that have been processed into a meaningful form, one that has value to the user, have been transformed into information.

In turn, these ideas are turned into *knowledge.* Knowledge is developed by converting this information using specific theories or understandings based on the techniques and analysis drawn from the specific subject areas. It is selectively organized information that provides understanding, recommends courses of action, and provides the basis for many decisions.

Statistical analysis can thus be viewed as the link *between* data and information. Using concepts of probability theory and also computer software, we examine the available data and search for patterns or facts that will aid us in expanding knowledge within the experience, literature, or theory available to us. Viewed in this way, we can see parallels between, for example, production workers who monitor the quality and productivity of their machines in order to operate and control them, and researchers interested in development studies who seek to understand the role of capital accumulation in economic growth. In each case, data are collected, processed, and summarized in order to create information, and then they are drawn together and interpreted to increase the knowledge of the subject area.

It is now not even uncommon to view information, and therefore data as *power.* Those who can best collect, store, manipulate, analyze, and understand their data will be more powerful. In a global economy, it is argued, such power will be translated to economic success, and firms, or even national states, that embrace information technology and analysis will be at a distinct advantage over those who do not. Data is now being accumulated at an unprecedented rate. Attempting to automate the statistical analysis of this data is leading to the development of new techniques of statistical analysis such as *data mining, knowledge discovery,* and *online analytical processing.*

The explosion of data in response to the growth of *automatic* data collection tools means that we may drown in data unless we can develop related automatic tools to turn these data into useful information. The use of conventional statistical analysis software is becoming too time consuming and labor intensive to keep up with the growth of data. Soon, the discovery of patterns, regularities, and simple rules from these databases will only be feasible using a more robust set of tools that are able to uncover previously unknown, nontrivial, and useful information from these extremely large data repositories. To this point, most of the advances in this area have focused on business applications, but these advances will soon spread to other research areas including geography.

1.5. Summary

Statistical analysis includes methods used to collect, organize, present, and analyze data. Descriptive statistics refers to techniques used to describe data, either numerically or graphically. Inferential statistics includes methods used to make statements about a population characteristic on the basis of information from a sample. Statistical inference includes both hypothesis testing and estimation methods.

Within geography, most applications of statistical methodology are rather recent, having become a significant part of the research literature only after the "quantitative revolution" of the 1950s and 1960s. Statistical methodology is most commonly utilized by geographers advocating a scientific approach to the discipline, an approach that is now common in many of the systematic branches of the field. Historical and cultural geographers find fewer uses for the methodology. Recent advocates of humanistic and structuralist approaches tend to be particularly critical of the basis of the scientific method and often reject statistical methodology. Even in these fields, however, there are several instances where statistical methodology can be fruitfully employed in applied research. Recently, the toolbox of human geographers has been enriched by the addition of qualitative techniques. These techniques can often be used hand-in-hand with statistical techniques in many research problems.

One of the first tasks of the statistician is to evaluate the data being used or being proposed in a research inquiry. If suitable data are not already available, or the limitations of existing data sources preclude their confident use, the researcher must collect new data. In some instances, an experimental approach is possible. It is mostly in various systematic specialties within physical geography that this is a feasible alternative, and statistical surveys are more commonly used to collect suitable data.

Whenever statistical surveys are undertaken, it is necessary to proceed with caution, recognizing the limitations of the data collected in this manner. The greater the control exercised in the design of data collection procedures, the better the data ultimately available to the researcher. It follows that the ability of geographers to make sound judgments in their research often rests on the very first steps taken in their research design. Generating precise, accurate, and valid sets of variables can only assist the development of theory and explanation in geography.

REFERENCES

H. Carey, *Principles of Social Science* (Philadelphia: Lippincott, 1858).

R. Hartshorne, *The Nature of Geography* (Lancaster, PA: Association of American Geographers, 1939).

E. Ravenstein, "The Laws of Migration," *Journal of the Royal Statistical Society* 48 (1885), 167–235.

J. C. Weaver, "Crop Combinations in the Middle West," *Geographical Review* 44 (1954), 175–200.

FURTHER READING

Nearly every introductory statistics textbook for social scientists includes a presentation of much of the material discussed in this chapter. See, for example, McGrew and Monroe (2000) for a slightly different approach from that taken here. Bluman (2004) is a slightly more mathematical presentation than this text but contains many simple examples drawn from many fields. For additional information on information science and statistics, see Hand et al. (2000) and Han and Kamber (2001). The book on survey methodology by Groves et al. (2004) provides an excellent overview of the current status of survey methodology. Finally, students wishing to compare statistical methodology to the rapidly growing area of qualitative methods should consult either Hay (2005) or Limb and Dwyer (2001).

Allan G. Bluman, *Elementary Statistics: A Step by Step Approach, 6th ed.* (Boston: McGraw-Hill, 2007).

Robert M. Groves, Floyd J. Fowler, J., Mick P. Couper, James M. Lepkowski, Eleanor Singer, and Roger Tourangeau, *Survey Methodology* (Hoboken, NJ: Wiley-Interscience, 2004).

Jiawei Han and Micheline Kamber, *Data Mining: Concepts and Techniques* (San Francisco: Morgan Kauffman, 2001).

David J. Hand, Heikki Mannila, and Padhraic Smyth, *Principles of Data Mining* (Boston: MIT Press, 2000)

Iain Hay, *Qualitative Research Methods in Human Geography, 2nd ed.* (London: Oxford University Press, 2005).

Melanie Limb and Claire Dwyer, *Qualitative Methodologies for Geographers: Issues and Debates* (London: Arnold, 2001).

J. Chapman McGrew and Charles B. Monroe, *An Introduction to Statistical Problem Solving in Geography, 2nd ed.* (New York: McGraw-Hill, 2000).

PROBLEMS

1. Explain the meaning of the following terms:
   - Descriptive statistics
   - Inferential statistics
   - Statistical population
   - Population characteristic
   - Variable
   - Population census
   - Sample
   - Sampling error
   - Nonsampling error
   - Representative sample
   - Random sample
   - Statistical estimation
   - Hypothesis test
   - Measurement precision
   - Metadata
   - Inductive approach
   - Deductive approach
   - Internal data
   - External data
   - Primary data
   - Secondary data
   - Experiments
   - Surveys
   - Quantitative and qualitative variables
   - Scale of measurement
   - Ordinal, interval, and ratio scales
   - Measurement validity
   - Measurement accuracy
   - Qualitative methods
   - Data as information

2. Under what conditions might it be advantageous to undertake a population census rather than a sample?

3. What impacts do you think the Internet might have on the distribution and analysis of statistical data?

4. What is the level of measurement of the following variables?
   a. SAT score
   b. Number of tests or quizzes in your statistics course
   c. Acres of land devoted to corn
   d. Number of break-ins in 2004 by neighborhood
   e. Social insurance or Social Security number
   f. Impression of a certain place selected by recipients from a scale of 1 to 5
   g. Name of birthplace
   h. Year of birth

5. Explain why human geographers have undertaken so few experimental studies in their research.

6. Use the Internet to search the government data collection agency in a specific country (e.g., *www.statcan.ca* [Canada], *www.census.gov* [United States], or *www.abs.gov.au* [Australia]). Examine the depth and availability of data from this source. Find another website where the some of the same data can be obtained. Can you locate the metadata for these data, or at least some elements of them?

7. Use the Internet to locate five different sources of data that can be used in statistical analyses. For each source:
   a. Identify the source.

   b. Evaluate the data according to the typology of data sources discussed in section 1.3 of this chapter.
   c. Edentify any questions you might have before you think the data should be used in any inferential study.

8. The United Nations typically undertakes studies that require the evaluation of data collected in many different countries in many different formats. Locate a study of this type and describe the statistical issues that need to be addressed in this form of analysis.

9. Some research projects are developed using the case study approach. Using this approach, we choose a single observation for detailed study, but we still want to generalize the results of our survey to wider populations. For example, we might choose to study one inner city neighborhood, rather than the inner city as a whole, but we are interested in making conclusions relevant across the city or even to other cities. When do you think case studies are potentially more valuable than surveys? What can be done to generalize the results of the study?

II

DESCRIPTIVE STATISTICS

2

Displaying and Interpreting Data

In Chapter 1 you were introduced to several types of data geographers encounter in their research. We discussed some of the characteristics of those data and how they influence the form of statistical investigation. We continue this theme in Chapter 2 by examining how to display and how to read or interpret different kinds of data.

A set of statistics, or numeric information, can tell many stories. One story involves what the data represent. For example, what variable is being measured, and what is the unit of measurement? These questions are not incidental to subsequent analysis. For example, suppose an economic geographer wished to compare unemployment rates over time across a number of countries. He or she would have to be careful that the unemployment rate was defined in the same way within each country and that the definition was consistent over time. A second story might focus on the source of the data and on how the data were collected. A climatologist might regard precipitation or temperature data reported by a government agency of higher quality (more accurate and more consistent) than that reported by a casual observer. Many of us might be guilty of assuming that data gathered by mechanical recording devices are more reliable than data gathered by human recorders. Social scientists undoubtedly face different data considerations than natural scientists. Gathering information from human subjects may generate various forms of bias—cultural, gender, economic, political—introduced unwittingly by the questions asked or by the characteristics of the interviewer, or introduced on purpose by the interviewee. It is important to be aware of the potential limitations of data prior to analysis.

Once we have obtained some data and have carefully documented how they were collected and what they measure, we confront the question of how to present them. This is yet a third story, for one set of data can be presented in many different ways, some good and some bad. Good presentation typically involves displaying the numeric information in a summary form such that the main characteristics of the data are easy to identify. For very small datasets with a limited number of observations, a summary may hardly be necessary. However, for even moderate-sized problems, it is usually quicker and easier to distill information from a well-structured table, or more specifically, from a plot or a graph, than from a disorganized table of numbers.

A variety of statistical projects may be completed with the data presentation stage. For problems in which more detailed statistical analysis is required, invariably it is good practice to "look" at your data first. Data visualization often reveals patterns or trends that guide subsequent investigation or that constrain the types of analysis that may be performed. We cannot emphasize too strongly the importance of looking at data prior to further statistical interrogation.

Our first encounter with a new dataset is often as a list of different values of a variable. That first impression is often negative, especially when the dataset contains a large number of observations. No matter how numerate we are and no matter how long we stare at a table of numbers, making sense of what we see is difficult. When we look at the values of a variable of interest, we hope to get an overall impression of the distribution of those values.

DEFINITION: DISTRIBUTION

The distribution of a variable indicates the different values that the variable might assume and how often each value occurs within a set of observations.

Generally, we characterize a distribution by asking the following questions:

- Where are the end points of the distribution, the minimum and maximum values?
- How are the observations distributed between those end points: do they cluster around certain values, do they tend to avoid others, or are they evenly spread across the range of values?
- Where is the center of the distribution or the average value of the variable of interest?
- Are there any observations with values that are very different from others, and if so, where are they located?

The way that we describe the distribution of values of a variable also depends on the type of data that we encounter. In Chapter 1 we argued that qualitative or categorical variables will be described differently from quantitative variables. We will see why this is the case later in this chapter.

The focus of this chapter, then, is how to summarize data. That summary is usually the first step in statistical exploration. To keep things simple, for the moment, we consider a single variable at a time or what we call univariate analysis. Though tables and graphs provide useful summaries and insights about the distribution of a variable, they do not yield precise measures of a distribution or allow us to compare distributions except in the simplest of ways. In Chapter 3, therefore, we explore quantitative measures that more accurately describe the distributions of different kinds of variables. In Chapter 4 we push further to explore relationships between variables.

In Section 2.1 we begin discussion of the display and summary of data and describe how to organize observations on categorical variables and how to display those observations in simple graph form. In Section 2.2 attention turns to the development of summary tables and the range of graphs that can be used to display numeric information. Section 2.3 looks at the special case of a variable for which observations are gathered over time. Section 2.4 extends the discussion further to examine different types of spatial data and some of the ways that geographical information can be organized and displayed.

2.1. Display and Interpretation of the Distributions of Qualitative Variables

Organizing qualitative data in a summary table and displaying that information in a graph or chart are relatively straightforward tasks because of the character of qualitative data. The values of qualitative variables are discrete labels or categories that provide "natural" breakpoints or divisions that can be exploited by the researcher. Consider the following example.

EXAMPLE 2-1. Vegetation Type. A team of biogeographers sampled 57 lakes in the eastern Sierra Nevada. They were interested in the distribution of chironomids (nonbiting midges) in lake surface sediment. The chironomid assemblies were statistically linked to lake surface temperature. This linkage allows researchers to estimate past climatic (temperature) conditions for the late-Glacial and early Holocene periods, about 14,000–8,000 years before present, from subfossil chironomid remains. Along with the chironomid and lake temperature data, measurements were gathered on a series of other variables, including the dominant vegetation type around each lake. The lake vegetation data are presented in Table 2-1. These data are reproduced on the book's website as *lakeveg.html.* It is important to note that these data have a spatial component, the location of the individual lakes from which measurements on a series of variables were recorded. Mapping these lakes and the values of the variables of interest would be one way of representing the data. More explicit discussion of spatial data is found at the end of Chapters 2, 3, and 4 and in Chapter 14.

The abbreviations in Table 2-1 refer to one of five vegetation types: US = upper subalpine; S = subalpine; UM = upper montane; JP = Jeffery pine woodland; PP = pinyon pine–juniper woodland. The number of observations in this sample is $n = 57$. Note that the observed values of the variable of interest are categories of vegetation

TABLE 2-1
Vegetation Types around Sierra Nevada Lakes

PP	PP	JP	JP	JP	JP	US	S	S	S	US	S
S	US	US	US	US	US	S	US	S	US	US	US
UM	JP	PP	S	PP	PP	UM	UM	UM	UM	UM	UM
S	PP	UM	S	S	S	S	S	S	US	S	S
S	S	US	US	S	S	S	S	US			

Source: Porinchu et al. (2002).

TABLE 2-2
Frequency Table for Lake Vegetation Type

Vegetation type	Count or frequency	Percentage
S	23	40.4
US	15	26.3
UM	8	14.0
PP	6	10.5
JP	5	8.8
Total	57	

type: thus, the data are categorical or qualitative. It is impossible to rank the different vegetation types, and so these dataare nominal. Note that even with the relatively small data set of Table 2-1 it is difficult to gain a quick appreciation of the distribution of lake vegetation type. With more observations or categories of the variable of interest this task would be considerably harder still. Hence there is a need for techniques to organize and display the data values to aid their effective summary.

Describing the distribution of a nominal variable is not a simple task because it is impossible to identify the end points of the distribution. Similarly, it is impossible to define an average value. All we can do in this case is to count the number of observations, or lakes, that are associated with different categories of vegetation and report whether or not the observations are evenly allocated between categories. Counting the number of lakes that have one of the vegetation types in our sample above yields a frequency table. A frequency table for a categorical variable lists the different values that the variable may assume and notes the number of observations that fall into each category. The order of the categories in a frequency table for nominal data is arbitrary, though sometimes categories are ranked by the number of observations they contain as in Table 2-2. It is common in such tables to convert the counts to percentages in order to show the relative frequency, or the weight, of observations in the different categories. The percentage for a particular category is found by dividing the absolute frequency of observations in that category by the total number of observations.

Table 2-2 shows that the distribution of lakes in terms of vegetation type is unbalanced or uneven: approximately 40% of the lakes sampled have subalpine vegetation characteristics, and two-thirds of the lakes have vegetation consistent with subalpine or upper subalpine categories.

Describing the distribution of a nominal variable is more difficult when the variable has a large number of values or categories. In this situation the researcher might attempt to group sets of the original values. This is possible only if such grouping makes sense. For example, another part of the chironomid dataset identifies the number of chironomid species found within each of the lakes: there are almost 40 species in total. To examine the distribution of chironomid species across the lakes, we could replace the vegetation types in a frequency table with the different kinds of chironomids. The counts, or absolute frequencies, would then list the number of lakes in which each of the different species of chironomid was present. However, with 40 different species such a table would not efficiently summarize the data. In this case, it is possible to proceed by grouping the chironomid species into warm-water taxa, cold-water taxa, and others. The resulting frequency table with only three categories might provide a useful summary of the chironomid distribution.

The tasks involved in describing the distribution of an ordinal variable are a little different. The following example focuses on educational attainment in the U.S. adult population.

EXAMPLE 2-2. Educational Attainment. Educational attainment measures the highest level of education that a person has reached. It is possible to measure a person's education in different ways. One way might be to count the number of years that an individual has been in school or college. This would provide a quantitative measure of the amount of education a person has received. The disadvantage of this measure is that one year of elementary school education is valued in the same way as a year of university education. It is unclear which of these years should be more highly valued and thus it is, perhaps, inappropriate to measure education in this numeric form. The more common way of measuring educational attainment is by asking whether a person has reached a particular category or threshold in their education. The U.S. Bureau of the Census measures educational attainment in this way using the categories: not a high school graduate; high school graduate; some college but no degree; associate degree; bachelor's degree; advanced degree. Because these categories are non-numeric, this measure of educational attainment is qualitative. Unlike the last example, however, we can rank the different categories of educational attainment from lowest to highest. Clearly, having a bachelor's degree is more education than having a high school diploma. Thus, in this example, educational attainment is an ordinal variable. The U.S. Bureau of the Census has gathered information on educational attainment for all persons aged 25 or over in the United States. The data are presented in Table 2-3.

TABLE 2-3
Frequency Table of Educational Attainment in the United States, 2000

Educational attainment	Count or frequency (000)	Percentage	Cumulative percentage
Not a high school graduate	27,716	15.8	15.8
High school graduate	58,058	33.1	48.9
Some college but no degree	30,871	17.6	66.5
Associate degree	13,682	7.8	74.3
Bachelor's degree	29,818	17.0	91.3
Advanced degree	15,085	8.6	99.9[a]
Total	175,230		

Note: The count or frequency is expressed in thousands (000), so the total count of 175,230 represents 175,230,000, or a little over 175 million.

[a]Does not sum to 100 because of rounding.

Source: U.S. Bureau of the Census (2000).

The raw data in this example are individual observations, n = 175,230,000, across a number of categories as in Table 2-1. In Table 2-3 these observations have been distributed across six categories, forming a frequency table of educational attainment for the U.S. adult population. Note that the educational categories in Table 2-3 are arranged from the lowest level of educational attainment to the highest. How can we describe this distribution? Even though the variable of interest is categorical, because the categories can be ordered from "not a high school graduate" to "advanced degree" we do have some indication of the spread of values of educational attainment in the United States. We know that almost 28 million U.S. adults (15.8% of the U.S. total) aged 25 or over have not completed their high school education. Similarly, we know that about 15 million U.S. adults (8.6%) have an advanced degree. The rest of the adult population has a level of education somewhere between these extremes. With ordinal data we can also identify a midpoint, or an average value, using the information in the cumulative percentage column. The cumulative percentage, or the cumulative relative frequency, records the proportion of the population that has a level of educational attainment equal to or less than the category specified. Thus, the cumulative percentage for the value high school graduate is 48.9%. This means that 48.9% of the population studied have attained at most a high school diploma. The category that contains the 50th cumulative percentile denotes the median or average level of the variable of interest. Thus, for this example, the average level of educational attainment for U.S. adults in 2000 was some college but no degree. A more precise definition of the median is provided in Chapter 3.

Bar Charts and Pie Charts

The distribution of qualitative or categorical variables can be readily understood from a frequency table. To aid interpretation or presentation of such data, frequency tables are often displayed in the form of bar charts or pie charts. Figures 2-1 and 2-2 show bar charts and pie charts of the lake vegetation and educational attainment data, respectively. Bar charts show either the number or the relative frequency (percentage) of observations across the different categories of the variable of interest. Bar charts may be organized with variable categories arranged along the horizontal or x-axis or along the vertical or y-axis. In Figures 2-1a and 2-2a the categories of the variable are found on the horizontal axis. In Figure 2-1a the heights of the bars represent the number of lakes associated with each vegetation type. Figure 2-2a shows the relative frequency of different levels of educational attainment in the U.S. population. For nominal variables, it is good practice to arrange variable categories in a bar chart according to the frequency of the observations they contain. This is especially important when the number of categories is relatively large. For ordinal data, the categories should follow the values of the variable. Be careful to make the width of the bars in a bar chart the same so that wider/narrower bars do not appear to have more/less weight than they should. Some additional concerns about the structure of bar charts and histograms are left until the next section.

Pie charts for the categorical variables are displayed in Figures 2-1b and 2-2b. Pie charts resemble a pie that has been divided into a number of slices. The number

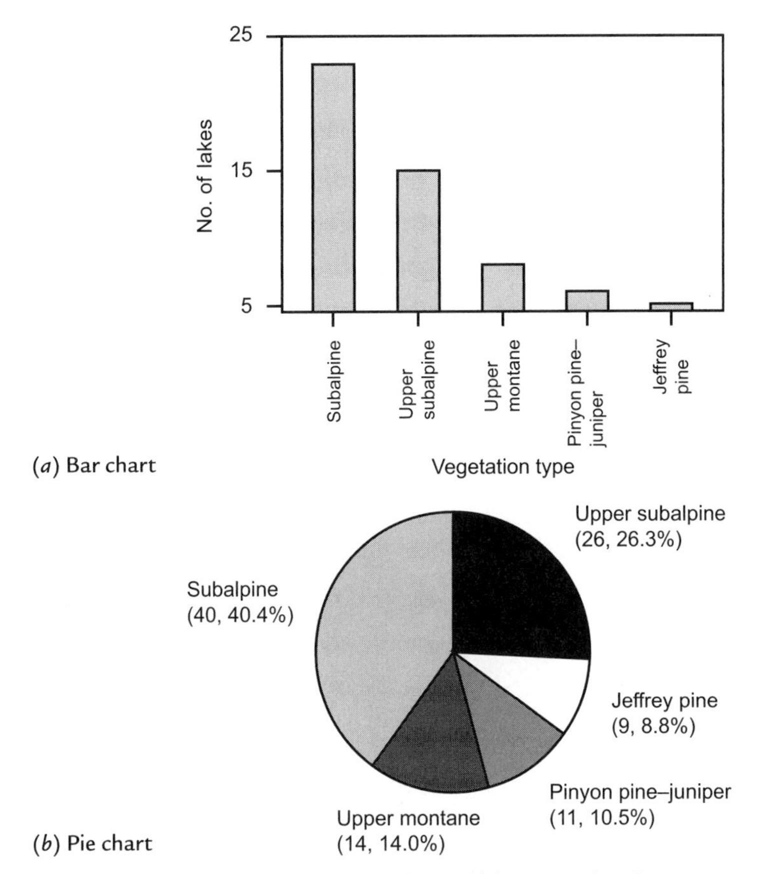

FIGURE 2-1. Bar chart and pie chart of lake vegetation data.

of slices is given by the number of categories into which observations on a variable fall. The size of each slice is determined by the relative frequency, or the percentage, of the total number of observations in that category. To aid interpretation, the pie chart often includes relative frequencies for its different values. Like the bar chart, the pie chart provides a quick visual impression of the distribution of a variable. However, the principal advantage of the bar chart over the pie chart is the ease with which the relative weight of different categories can be compared. For example, in Figure 2-2b it is unclear from the graphic alone whether the three categories of some high school but no degree, some college but no degree and bachelor's degree have relative frequencies that are significantly different from one another. The differences in relative frequencies of these categories are easier to see in the bar chart of Figure 2-2a.

Adding colors or a shading scheme in order to more clearly differentiate bars or slices in these graphs is a matter of preference. For nominal data, choice of colors or a shading scheme is not so critical, though care should be taken to avoid garish

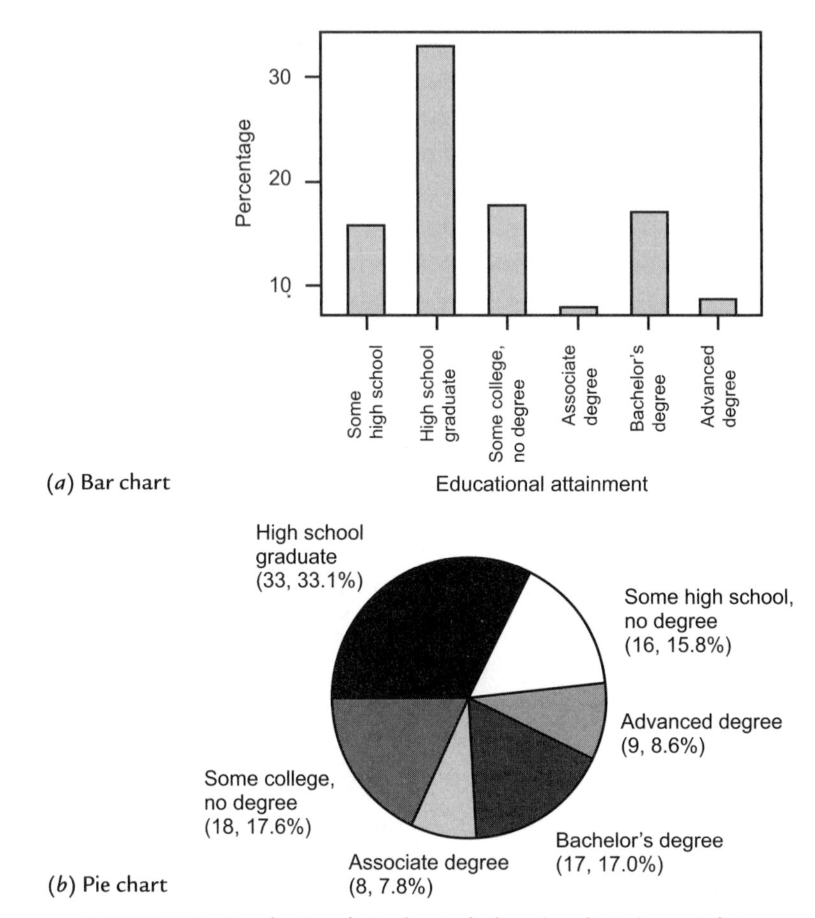

FIGURE 2-2. Bar chart and pie chart of educational attainment data.

colors and patterns. For ordinal data, it is good practice to ensure that progressions of color intensity or of shading scheme follow the ranking of the data values.

2.2. Display and Interpretation of the Distributions of Quantitative Variables

Describing the distribution, or summarizing the character, of a quantitative variable follows the same general pattern as for a qualitative variable, though the task is a little more complex. What makes the job more difficult with numeric data is that the variable of interest is rarely distributed across a series of discrete values. Thus, we do not have "ready-made" categories that we can employ to construct frequency tables or bar charts. Furthermore, all the values of a quantitative variable may be unique, or there may be only a small number of repetitions. How do we proceed in this situation? Consider the following example where measures of dissolved oxygen have been taken from 50 lakes.

Frequency Tables

EXAMPLE 2-3. Dissolved Oxygen. Dissolved oxygen (DO) is one measure of the quality of water in a lake or stream. As wastes are discharged into lakes and streams, the amount of DO in the water declines. This oxygen is used in the biochemical reactions by which the nutrients in the wastes are broken down. Public water supplies should have a DO content of at least 5 milligrams per liter, or 5 mg/L at any time. Table 2-4 contains a list of DO values estimated from a sample of $n = 50$ lakes. DO is a ratio level variable.

Table 2-4 shows that the water samples from some lakes have DO values below the public water supply standard of 5 mg/L, and others have levels above it. Closer examination reveals that there are 25 different DO values in the 50 lakes. It is extremely difficult to scan this list of numbers and develop a general understanding of water quality. Our task is to try and summarize these values: to develop a general picture of the distribution of DO values across the 50 observations. Just as we proceeded for qualitative variables, the first step is to construct a frequency table that displays the raw data in a more organized form. Immediately, however, we confront the problem, noted earlier, that while categorical variables have classes into which observations readily may be placed, quantitative variables do not. If we were to let each value recorded in Table 2-4 of the DO data represent a separate category or class we would end up with Table 2-5. In this table there are 25 classes into which the 50 different DO values fall. Imagine how large a frequency table might be for datasets with thousands of observations across hundreds of different values of a variable. While Table 2-5 does make clearer the end points of the distribution, it does not summarize the data in a way that gives us a much greater understanding of the overall character of DO values than Table 2-4.

Figure 2-3 displays the data of Table 2-5 in a bar chart. The figure helps us visualize the distribution of DO values, but with the data arranged in so many classes it does not provide much of a summary. Furthermore, the horizontal axis of Figure 2-3 omits DO values for which there are no observations. Consequently, interpretation of the spread of observations along the axis is complicated. We do not want these sorts of complications in a chart that is supposed to reveal the general characteristics of the distribution of a variable simply and accurately. The histogram in Figure 2-3 is

TABLE 2-4
Dissolved Oxygen Values in Lakes

5.1	6.4	5.6	6.4	4.4	4.9	5.6	5.8	5.6	6.1
5.6	4.3	6.8	5.9	5.1	6.6	6.7	5.3	6.2	5.1
5.3	5.9	6.9	6.0	5.6	5.7	5.4	5.7	4.2	5.9
5.7	5.4	4.8	5.5	5.8	5.4	4.8	6.3	5.2	5.5
5.8	4.7	5.6	5.4	5.7	5.9	6.4	4.5	5.8	4.7

TABLE 2-5
An Ungrouped Frequency Table

DO value, mg/L	Frequency	DO value, mg/L	Frequency	DO value, mg/L	Frequency
4.2	1	5.3	2	6.2	1
4.3	1	5.4	4	6.3	1
4.4	1	5.5	2	6.4	3
4.5	1	5.6	6	6.6	1
4.7	2	5.7	4	6.7	1
4.8	2	5.8	4	6.8	1
4.9	1	5.9	4	6.9	1
5.1	3	6.0	1		
5.2	1	6.1	1		

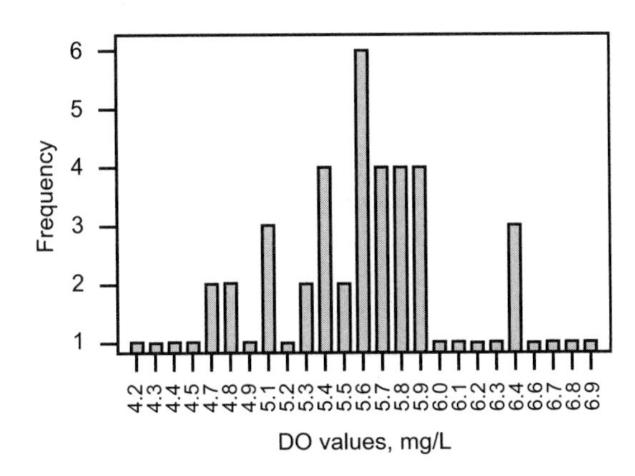

FIGURE 2-3. Bar chart of DO data from Table 2-5.

analogous to a dotplot, where values assumed by a variable of interest are listed on the horizontal axis and a series of dots, arranged vertically, record the number of observations at each of those values.

How can we improve the summary and representation of the DO data? The answer lies in grouping together observations that have similar values so that we end up with fewer categories in our frequency tables and charts. How many categories or classes should we use, and how do we determine what values of a variable to place in each class? There is no single correct answer to these questions, but a few simple rules should be kept in mind when designing frequency tables for quantitative data.

RULE 1: Select an appropriate number of classes.

Usually, the most difficult decision to make is the number of classes to be used in a frequency table. Too few classes will lump observations together so that we don't get an accurate impression of the variation in a set of data, making it difficult to determine the approximate location of a central value or average. Too many classes and we end up with a bar chart like Figure 2-3, which provides little summary information. A general rule of thumb is to use somewhere between 5 and 15 classes depending on the number of observations. As the number of observations increases, more classes can usefully be employed. Some statisticians offer firmer guidelines. Sturges (1926), for example, proposes the rule

$$k = 1 + 3.3 \log_{10} n$$

where k represents the number of classes or intervals and n is the number of observations in the dataset. For the DO data this suggests $(1 + 3.3(\log_{10} 50)) = 6.6$, thus six or seven classes. While Sturges's rule usually works quite well, we stress that it is only a guideline. Indeed, for datasets with more than 200 observations, Sturges's rule tends to *oversmooth* the distribution. In this case, Scott's (1979) rule provides a more robust alternative

$$k = 3.5sn^{-1/3}$$

where s represents the standard deviation of the data (see Chapter 3).

There are other considerations in choosing the number of classes. First, when observations are grouped together, we are saying that they are alike in some way. As a general rule, then, never select a number of groupings or intervals that force unlike observations into the same class. We break this rule if we place natural breakpoints within a class and not at a class boundary. For example, with temperature data, setting class limits of –5° to +5°Celsius might be unwise, tantamount to claiming that temperatures above freezing can logically be grouped with those below freezing.

The second consideration to bear in mind is that grouping is a form of data *aggregation,* and there is a trade-off between the gains from simplicity derived from aggregation and the loss of information that it necessitates. A frequency table with six classes means we have replaced the total number of observations, 50 in the case of our water quality example, with 12 new numbers—the six class midpoints and the six frequencies of the classes. (We only require 12 numbers because we can generate the class limits that are halfway between midpoints.) There is a sizeable gain in simplicity, but some information has been lost. Consider the first interval in Table 2-6 where the DO data are grouped into six classes. If we know only the class midpoint of 4.25 and the frequency of 3, then a good estimate of the sum of the observations in the interval is $3 \times 4.25 = 12.75$. Part of the information loss is the exact values of the three observations in the interval. Yet we are still able to approximate characteristics of the three observations. The actual values of those observations sum to $4.2 + 4.3 + 4.4 = 12.9$, very close to the estimate made using the frequency table. In this case, the gains in simplicity seem to outweigh the slight loss in information that occurs. In general, the loss of information becomes more serious if we group our observations into too few intervals. We will see this shortly in the analysis of histograms, the graphical counterparts to frequency tables for quantitative variables.

TABLE 2-6
A Frequency Table for the DO Data

Class	DO values, mg/L		Frequency or count	Relative frequency
	Interval	Midpoint		
1	4.0–4.49	4.25	3	0.06
2	4.5–4.99	4.75	6	0.12
3	5.0–5.49	5.25	10	0.20
4	5.5–5.99	5.75	20	0.40
5	6.0–6.49	6.25	7	0.14
6	6.5–6.99	6.75	4	0.08
			$\sum = 50$	$\sum = 1.0$

RULE 2: Class intervals should not overlap and must include all observations.

An observation can appear in one and only one class. Classes are mutually exclusive: placement in one class therefore precludes placement in any other interval. Do not specify overlapping interval widths such as 3–5 and 4–6. An observation of 4.5 would appear in both intervals and thus be double-counted, having twice as much weight as an observation found in only one class. Although it is permissible to have classes such as <3 (less than 3) or 5+ (greater than 5, or >5), these should ordinarily be avoided. If they are selected because there is an extreme value, an observation whose value is much smaller or much larger than any others, then they are surely misused. The difficulty with such open intervals is that it is impossible to tell what the highest or lowest values might be. As we will soon see, extreme values, or outliers, are important characteristics of some data sets that should not be hidden.

To avoid ambiguity and to define mutually exclusive intervals, set class limits carefully. Note that the upper bound of one interval should not equal the lower bound of the next interval. In Table 2-6, for example, the upper bound of the first interval for the DO data is 4.49, while the lower bound of the second interval is 4.5. An observed value of 4.5 in the data would therefore be placed in the second interval of the frequency table. If the upper bound of the first category and the lower bound of the second category were both set as 4.5, we would not know where to place an observation with the value 4.5. Defining limits with one more decimal place than the most precise observation in the data set is a useful general rule.

RULE 3: Use intervals with simple bounds.

It is easier to use intervals with a common width of 0.5, 1.0, 10, 100, or 1000 rather than peculiarly selected widths such as 7.28, 97.63, or 729.58. Simplicity is the key in terms of retrieving information from a frequency table or a graph. Similarly, the frequency table proves much easier to interpret if the lower bound or class limit of the first interval is a suitably chosen integer, or round number. If the interval width is conveniently chosen, this leads to regularly spaced, easy to interpret class limits and midpoints. The values selected in Table 2-6 serve this purpose. A good first step in constructing a frequency table is to order the data from lowest to highest value. Then, select a lower limit for the first class at a round number just below the lowest valued observation and an upper limit for the last class at a round number just above the highest valued observation. In Table 2.6, for example, 4.0 is slightly below 4.2, the smallest value in the data set, and 7.0 is slightly above the maximum value of 6.9. Given these limits, we know that our frequency table must span the range 7.0 – 4.0 = 3 units. Sturges's rule suggests six or seven class intervals. If we choose six intervals, then class widths of 0.5 work well for they are easy to interpret.

RULE 4: Respect natural breakpoints.

For many variables, certain values within the observed range represent natural values for a class limit. For a series of temperature readings expressed in degrees Celsius, it is appropriate to use 0° or 100° as class limits. For some variables expressed as percentages, such as the proportion of the vote in an election, 50% represents a natural class limit because it is the breakpoint for a majority. For pH values, 7.0 marks the border between acidic and alkaline substances and is a natural breakpoint. In the example of water quality, 5.0 makes a convenient class limit since it is the breakpoint between lakes meeting the public water supply standard of 5.0 mg/L of dissolved oxygen and those lakes not meeting this standard.

RULE 5: All intervals should be the same width.

It may seem obvious why this rule is important. It certainly simplifies matters. When the information from a frequency table is displayed in a bar chart or histogram, equal class widths are strongly recommended. When we examine such graphs, our brain responds to a hierarchy of visual clues, and area takes precedence over height. Thus, for two intervals with the same height and same relative frequency, we would interpret the wider bar as being more significant. In some cases such interpretation may be warranted, but we must take care if that interpretation is undesirable.

Table 2-6 shows a frequency table with well-designed intervals for the DO data, following the rules above. Compared to the ungrouped frequency data in Table 2-5, this table more effectively summarizes the DO data. The end points of the DO values span the range from 4.0 to 7.0 mg/L, the center of the distribution is around 5.75 mg/L, most observations are found in the interval 5.5 to 5.99, and the remaining values appear to be relatively evenly spread around this interval.

Frequency Histograms

Often, it is easier to understand the information in a frequency table when it is presented in graphical form. There appears to be some truth in the adage "A picture is worth a thousand words," or in this case, perhaps, a thousand observations. Bar charts and pie charts are conventionally used to display the values in a frequency table for qualitative data. For quantitative data the histogram is widely used. Figure 2-4 shows a histogram corresponding to the data in Table 2-6. Rules for constructing histograms

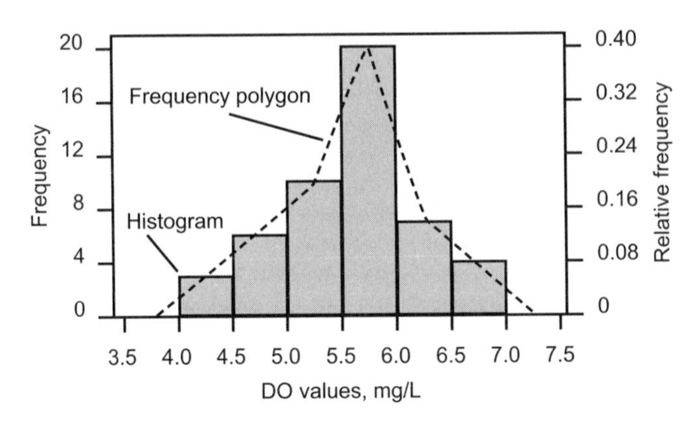

FIGURE 2-4. Histogram of the DO data.

follow those outlined above for the frequency table. The horizontal axis of the histogram lists the class intervals into which the observations of the variable of interest are grouped. Bars are then drawn for each group on the *X*-axis. The height of the bars is determined by the number, or frequency, of observations that fall into the respective intervals. The vertical axis on the histogram can be labeled with absolute frequencies, relative frequencies, or both. In this respect the histogram shows the relative concentration, or density, of observations of a variable across its range of values. Histograms are useful for comparing the shape of distributions of different variables: they illustrate the range of the variable under study, they locate a central value, and they show the distribution of observations across the chosen class intervals. The shape of the histogram in Figure 2-4 confirms the interpretation of the distribution of dissolved oxygen values in Table 2-6.

It is important that intervals be of equal width so that the histogram does not give a misleading impression of the data. Suppose, for example, that the last two classes shown in Table 2-6 are grouped together, or aggregated, to create a single class with bounds 6.0 to 6.99. The resulting histogram, shown in Figure 2-5, gives the impression that there are as many lakes with oxygen levels greater than 6.0 than there are with oxygen levels less than 5.5. In fact, there are almost twice as many lakes with oxygen levels below 5.5 as there are lakes with oxygen levels greater than 6.0, as Table 2-6 and Figure 2-4 reveal. There is another advantage in selecting intervals of equal widths. Assume that the width of classes is standardized to one unit and the relative frequencies are expressed as proportions. The total area under the histogram (area = height × width) is

$$0.06(1) + 0.12(1) + 0.20(1) + 0.40(1) + 0.14(1) + 0.08(1) = 1.0$$

The histogram of Figure 2-5 does not have this property.

In many instances, bar graphs or histograms give the impression that the variable is not really continuous. For categorical variables this is not a problem. However, for quantitative data we can often give a better representation of the distribution

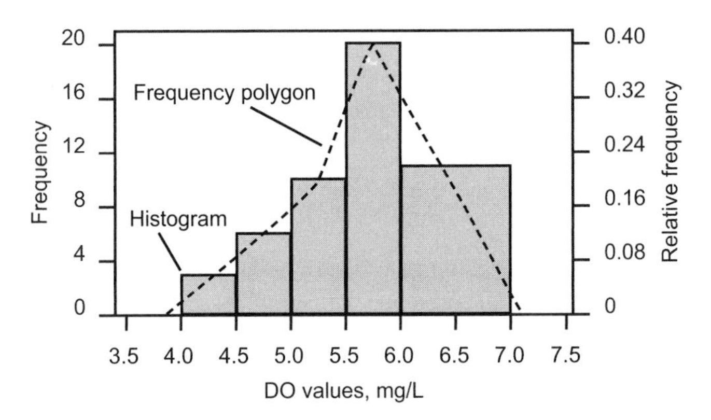

FIGURE 2-5. Histogram with unequal intervals.

of a continuous variable if we construct a frequency polygon. A frequency polygon is obtained by connecting the midpoints at the top of each interval of a frequency histogram. By convention, we usually extend the end points of the polygon beyond the upper and lower bounds to what normally would have been the midpoint of the next interval. In Figure 2-4 the frequency polygon begins at 3.75 and ends at 7.25. By using this convention, it is easy to show that the area under the histogram and the area of the frequency polygon are equal. For every triangular area not in the histogram but under the frequency polygon, there is a corresponding triangle above the frequency polygon but beneath the histogram. Thus, both representations have equal areas.

We can think of these graphical representations as crude approximations to smooth curves. If we are graphing the distribution of a sample, we might think of this sample as an approximation to a more general distribution—the distribution of the variable for the population. As a consequence, it is usual to smooth the histogram and frequency distribution of a simpler representation as in Figure 2-6. This curve is only a graphical approximation to what we believe the actual population to be: it is an ideal shape. If there are a large number of observations in a sample and we construct

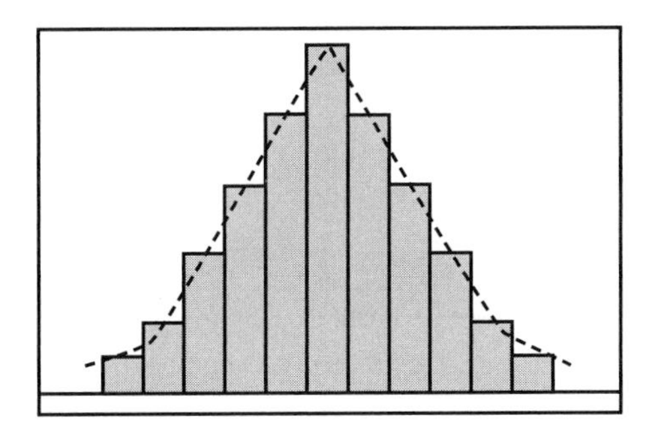

(*a*) Histogram

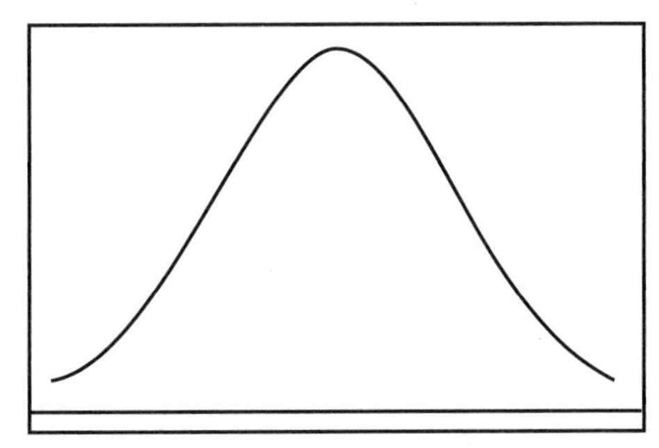

(*b*) Continuous distribution

FIGURE 2-6. Histogram as a simplification of a continuous distribution.

a histogram with many intervals, we can probably generate a close approximation to this smooth curve. The distributions in Figure 2-6 are symmetric, unimodal, and bell-shaped. These are the classic features of the *normal distribution.* The meaning of these characteristics is discussed in the next section.

Characteristics of Histograms

To aid the summary and interpretation of the distribution of quantitative variables, it is useful to refer to the location and shape of histograms or smoothed frequency polygons. The location of the histogram specifies the range of values of the variable of interest across which observations fall. A central value, or an average, is used to locate the center of a histogram. The shape of the histogram specifies whether the histogram is symmetric, whether it has a pronounced tail to the left or right, whether it is peaky or flat, and how many peaks it contains. The following figures highlight some of these characteristics.

Figure 2-7a and Figure 2-7b show two histograms, with the same shape, that are located at different points along the horizontal axis or at different values of the variable of interest, *X*. The histograms overlap across parts of their range, but in general the values assumed by variable *X* in Figure 2-7b are greater than those assumed by variable *X* in Figure 2-7a. Figure 2-8 illustrates a histogram that is *symmetric* around a central value of variable *X*. A symmetric distribution is one in which each half of the distribution is a mirror image of the other half. For many inferential statistical tests, a symmetric distribution is required. When symmetry is absent, the distribution is said to be *asymmetric* or *skewed.* If the distribution has a tail that extends to the left, as in Figure 2-9a, it is said to be *negatively skewed.* The histogram of the DO data in Figure 2-6 shows slight negative skew. If the tail of the histogram extends to the right, as in Figure 2-9b, the distribution is *positively skewed.* The distribution of household income tends to be positively skewed: there are a relatively small number of households with extremely high incomes.

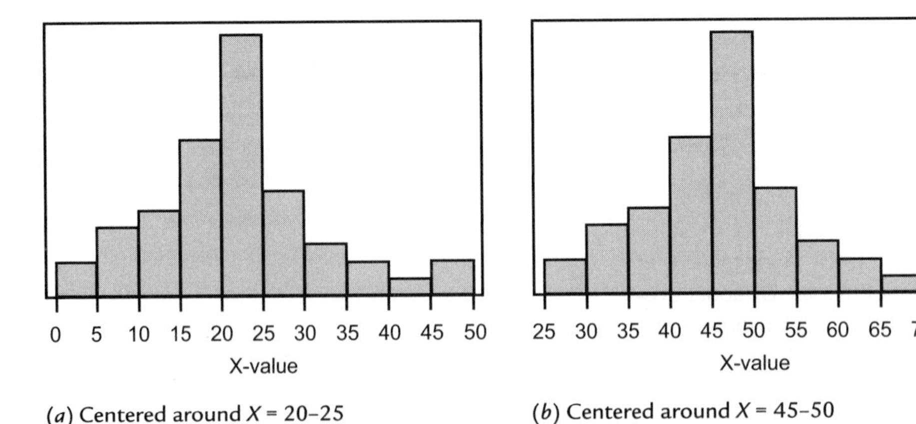

(*a*) Centered around X = 20–25

(*b*) Centered around X = 45–50

FIGURE 2-7. Histograms in different locations.

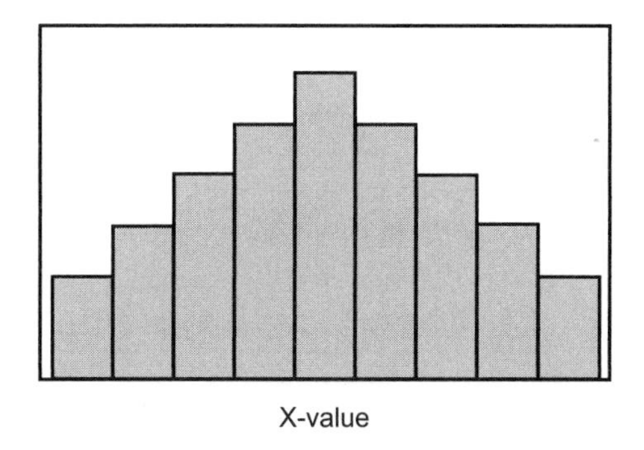

FIGURE 2-8. Symmetric histogram.

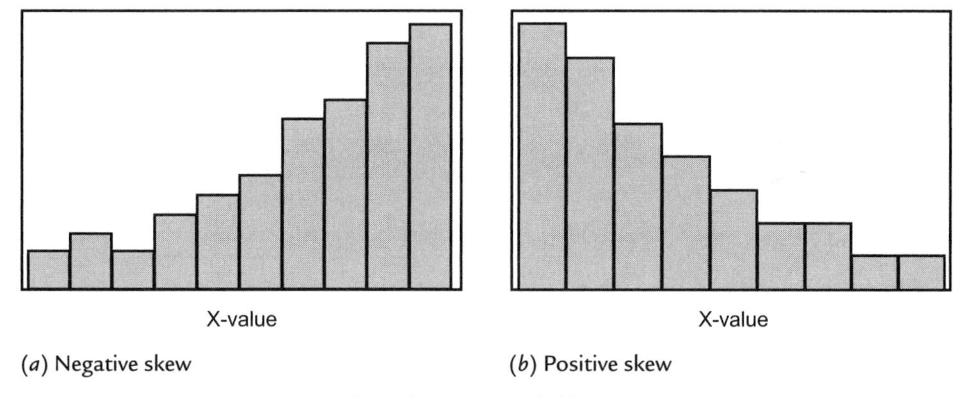

(*a*) Negative skew (*b*) Positive skew

FIGURE 2-9. Asymmetric histograms.

Kurtosis measures how peaky or flat a distribution is. It is common to identify the number of distinct peaks in a histogram. Peaks indicate values of the variable X around which observations tend to concentrate or cluster. Histograms with very sharp peaks are said to be peaky or *leptokurtic*. Histograms with no distinct peaks, where the tops of the different intervals are more or less in line with one another, are *platykurtic* or tending to rectangular. This implies that there is no typical value: any value in the range is almost equally likely. Figure 2-10a shows a histogram with one distinct peak or *mode*. This histogram is said to be *unimodal*. A mode is the value, or interval, of a variable that occurs most frequently in a set of observations. The DO data of Figure 2-6 has a single mode. Figure 2-10b illustrates a *bimodal* histogram, one with two distinct modes. Bimodal distributions tend to occur when two separate factors contribute to the distribution of a variable. For example, the heights of all 18-year-olds might contain two modes, one for females and one for males. The distribution of heights for each sex taken individually would probably be unimodal. Figure 2-10c shows a distribution characterized by multiple modes. Many inferential statistics are invalid when they are applied to variables with bimodal or *multimodal* distributions.

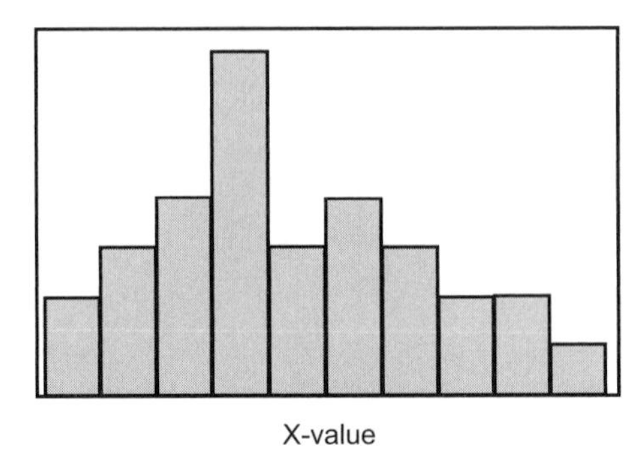

(*a*) Unimodal histogram

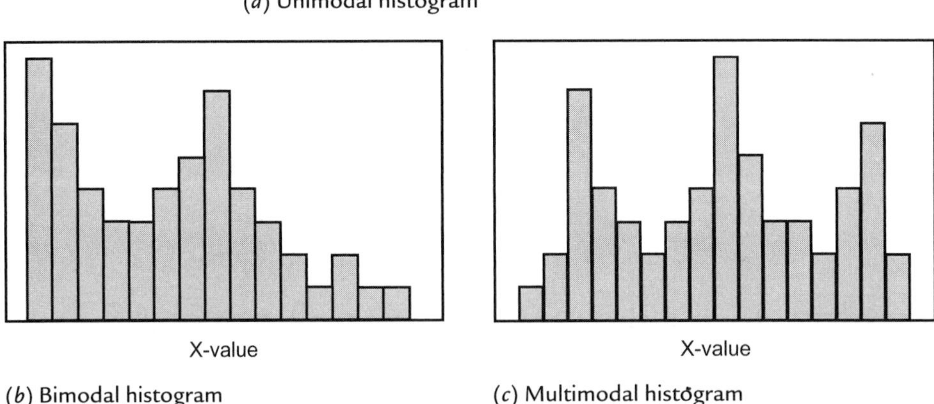

(*b*) Bimodal histogram (*c*) Multimodal histogram

FIGURE 2-10. Modality of a distribution.

We use these terms to characterize the shape of histograms and thus the distribution of the variables they portray. Histograms are also useful for identifying values that are peculiar in their location, or values that are located a long way from others. These stray values are called *outliers.* Figure 2-11 shows an outlier left of the main part of the distribution of the variable *X*. This outlier signals the presence of a relatively small value of the variable. When we detect such unusual values in a dataset, it is useful to check the data to see that the outlier represents a real observation rather than a typographical error such as a decimal point or a zero in the wrong place. Real outliers, rather than data errors, often provide important information about the processes that influence data values.

Although histograms are widely used to display the distribution of a single variable, they can be misleading. Two features of histograms have a significant impact on their shape: first, interval width, or what sometimes is referred to as bin width; second, the origin of class intervals, or bins. Table 2-7 lists values of median household income for U.S. states in 1999. In Figure 2-12, these data are used to generate three histograms quite different in their characteristics resulting from different choices of interval width and origin.

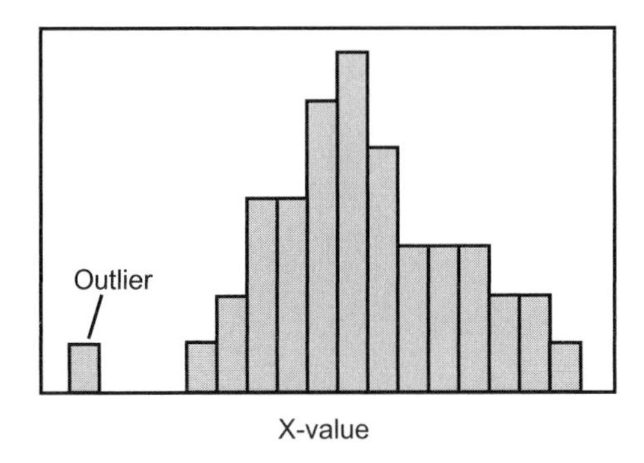

FIGURE 2-11. Histogram with an outlier.

TABLE 2-7
Average Household Income ($) in U.S. States, 1999

AL	34,315	HI	49,820	MA	50,502	NM	34,133	SD	35,282
AK	51,571	ID	37,572	MI	44,667	NY	43,393	TN	36,360
AZ	40,558	IL	46,590	MN	47,111	NC	39,184	TX	39,927
AR	32,182	IN	41,587	MS	31,330	ND	34,604	UT	45,726
CA	47,493	IA	39,469	MO	37,934	OH	40,956	VT	40,856
CO	47,203	KS	40,624	MT	33,024	OK	33,400	VA	46,677
CT	53,935	KY	33,672	NE	39,250	OR	40,916	WA	45,776
DE	47,381	LA	32,566	NV	44,581	PA	40,106	WV	29,696
DC	40,127	ME	37,240	NH	49,467	RI	42,090	WI	43,791
FL	38,819	MD	52,868	NJ	55,146	SC	37,082	WY	37,892
GA	42,433								

Source: U.S. Bureau of the Census (2000).

EXAMPLE 2-4. Average Household Income by State. Income is a general measure of economic well-being. The income values presented here are medians for U.S. households by state. The median is a measure of average such that 50% of the households in a state have incomes below this value and 50% have incomes above this value. (The median as a measure of central tendency, or average, is discussed in greater length in Chapter 3.) A household is generally defined by the U.S. Bureau of the Census as all people who occupy a particular housing unit: a house, an apartment, or a group of rooms used as separate living quarters. Household income is the sum of pretax money income earned in the last year by occupants of a housing unit who are 16 years of age or older. Already you can see that while the variable average household income appears relatively straightforward, precise interpretation of what it represents is quite complex. Prior to any examination of data, the researcher should ensure that she has a good understanding of what the variable of interest is measuring.

In Figure 2-12a, seven intervals are used, following Sturges's rule, and histogram classes originate at $25,000 for median household income. This histogram broadly appears symmetric about an average income value in the range $40,000–$45,000. The

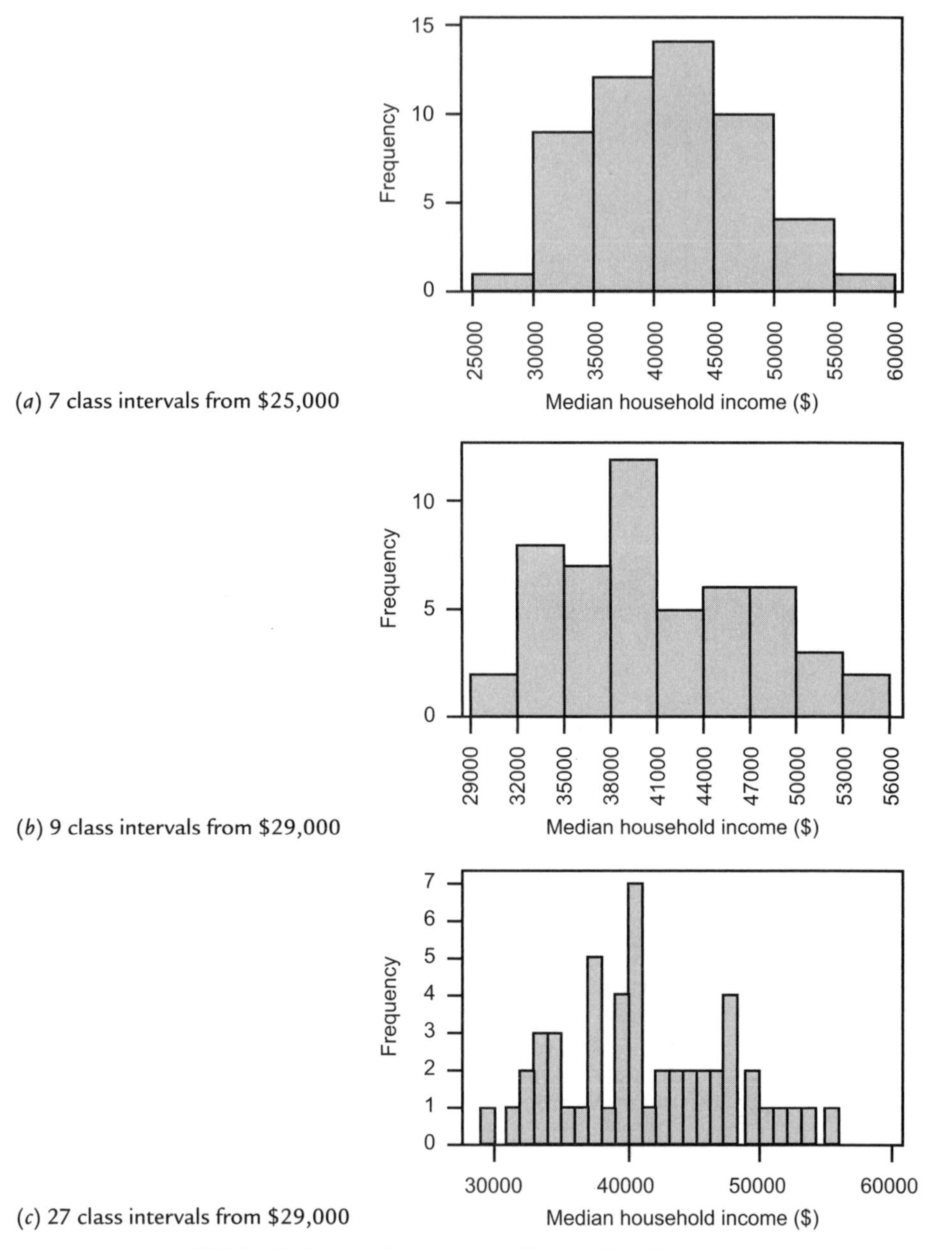

FIGURE 2-12. Median household income in U.S. states.

overwhelming majority of states appear to have median household income values between $30,000 and $50,000. In Figure 2-12b, nine intervals are used, originating at $29,000. This histogram is asymmetric with positive skew. It is also leptokurtic: it has a pronounced peak in the range $38,000 to $41,000. Note that to avoid congestion on the horizontal axis in Figure 2-12b, the household income values have been reoriented. Apart from the modal class and extreme values, below $32,000 and above $5,000, the distribution of median household income values is relatively flat or even. The histogram in Figure 2.12c is different again. This histogram uses 27 intervals originating at an income value of $29,000. Based on the rules outlined earlier, this is far too many classes for only 50 observations: we are using a large number of categories in this example to show how the shape of the histogram is influenced. Figure 2.12c is also peaky or leptokurtic, and it is multimodal as opposed to unimodal like Figures 2.12a and 2.12b.

The Cumulative Frequency Table and Histogram

For some purposes, it is convenient to summarize and illustrate data in a slightly different way. Suppose, for example, that we are interested in determining the number or proportion of lakes with DO values less than or equal to some given level. How many lakes have DO values less than 5.0? What proportion of the lakes have DO levels less than 5.5? Quick answers to these questions can be derived by portraying the data in cumulative form, as in Table 2-8. The cumulative frequency column is calculated by summing the frequencies of all intervals below and including the interval under consideration. At the end of the first interval, for a value of $X = 4.49$, there are three observations. There are nine observations below a value of 4.99—three in the first interval and six in the second interval. The table is completed in this way. The last entry in the cumulative frequency column is always equal to the total number of observations, or the sample size, n. In Table 2-8, the last entry in the forth column is 50.

Sometimes it is useful to express this cumulative frequency in relative terms, as in the last column. We calculate the cumulative percentage as the cumulative frequency divided by the total number of observations. For the first row, this is 3/50 = 0.06 (or

TABLE 2-8
A Cumulative Frequency Table for the DO Data

Class	Interval	Midpoint	Frequency	Cumulative frequency	Cumulative relative frequency
1	4.0–4.49	4.25	3	3	0.06
2	4.5–4.99	4.75	6	9	0.18
3	5.0–5.49	5.25	10	19	0.38
4	5.5–5.99	5.75	20	39	0.78
5	6.0–6.49	6.25	7	46	0.92
6	6.5–6.99	6.75	4	50	1.00
			$\sum = 50$		

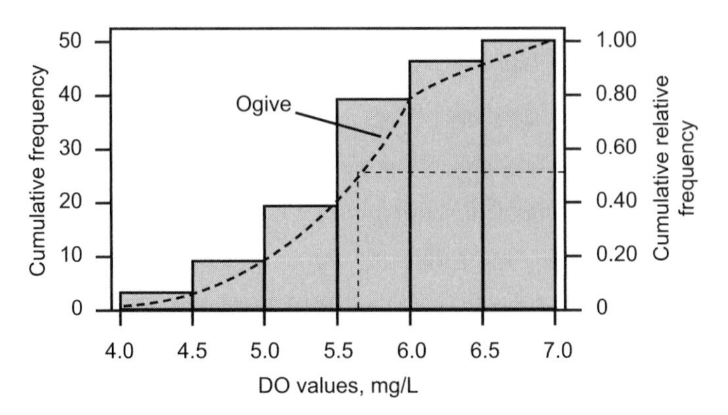

FIGURE 2-13. Cumulative frequency histogram.

6%): for the second row 9/50 = 0.18 (18%); and so on. Expression of this information as relative frequencies in either proportions or percentages is extremely helpful when we have a large number of observations and the simple cumulative frequencies are not as easily interpretable. Suppose, for example, there are 2200 lakes and the cumulative frequency at less than a DO level of 5.0 is 550. This is much easier to interpret if we say that 25% of the lakes have DO levels below 5.0 mg/L. The information from the cumulative frequency table is shown in Figure 2-13 in the form of a cumulative frequency histogram.

Figure 2.13 also plots a *cumulative frequency polygon*, an *ogive*. The ogive is drawn by connecting the lower bound of the first interval to the cumulative frequency at the end of the first interval, at the end of the second interval, and so on. The final point is always at the end of the last interval with the cumulative frequency equal to the total number of observations. The ogive provides estimates of the number of observations, or the proportion of observations, less than a specified value of the variable of interest. Thus, suppose we are interested in the DO value that is exceeded by 50% of the observations (lakes). Draw a horizontal line from the cumulative relative frequency axis at 0.50 to the ogive and then extend the line vertically down to the horizontal axis. This yields a DO value of approximately 5.6 to 5.7. This procedure is illustrated in Figure 2-13.

Dot Plots and Quantile Plots

One of the problems with histograms, which we discussed earlier, is that their shape changes quite dramatically as we alter the width and size of class intervals. This is a result of trying to fit what is often a continuous distribution into a series of discrete classes. To some extent, the class intervals misrepresent the variability in the data: they do not show variation between observations within a class interval; and they suggest discontinuities in the data, represented by the sharp edges of the histogram blocks, or bins, that may not exist. These are the penalties of summarizing a dataset in graphical form. There are other types of graphs that represent the distribution of a

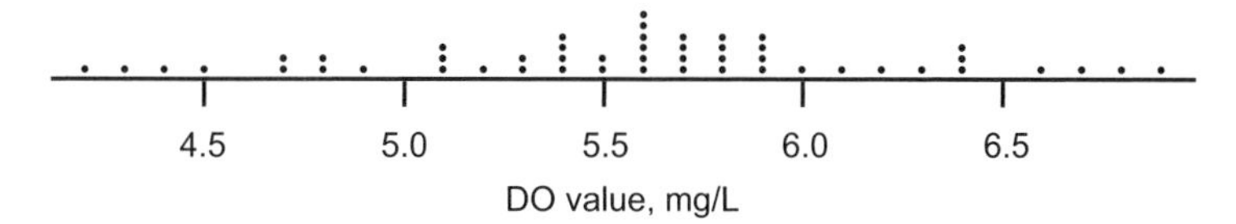

FIGURE 2-14. Dot plot of the DO data.

variable. These graphs present data visually, even if they do not provide summaries. Two such examples are the *dot plot* and the *quantile plot.*

The dot plot is similar to the histogram in that the vertical bars are replaced by dots, each representing an observation from a dataset. The horizontal axis of the dot plot records the range of values of the variable being mapped. The dot plot is simple to use and requires little explanation. It is therefore a useful technique for presenting information to a nonspecialized audience. The dot plot has the advantage over the histogram of displaying all observations on a variable and its portrayal of a distribution is not distorted by choice of the number of classes and their widths. However, for large datasets with continuous variables the dot plot can get a little messy because of overplotting. Figure 2-14 presents a dot plot of the DO data. Note that dot plots can be used to display qualitative as well as quantitative data.

In a quantile plot, each observation on a variable is represented by a point in a two-dimensional scatterplot. By convention, the vertical axis records the values of the variable and the horizontal axis denotes the probability that the variable, X, is equal to or less than some specified value. The quantile plot is thus a cumulative distribution function, just like the ogive. Unlike the ogive, however, the quantile plot is based on the distribution of individual observations and is thus independent of any class intervals that might be constructed from the data. The p quantile is the value in a dataset such that $100(p)\%$ of the observations are smaller than this value and $100(1-p)\%$ are larger. For example, the $p = 0.25$ quantile is the value in a data that is greater than $100(0.25) = 25\%$ of all observations and less than $100(0.75) = 75\%$ of all observations. This value corresponds to the first, or lower, quartile in the data. The $p = 0.5$ quantile corresponds to the second quartile, or median. We will examine the quartiles in more detail in the next chapter. A quantile plot for the DO data of Figure 2-4 is shown in Figure 2-15.

The quantile plot can be interpreted much like the ogive in Figure 2-13. Of most importance is the slope of the quantile plot for different values of the variable of interest. Notice in Figure 2-15 that the slope of the plot for low and high values of dissolved oxygen is steep. This indicates that there are relatively few values in these ranges of the distribution. DO values tend to cluster in the 5.3–5.8 range, and across these values the gradient of the quantile plot is gentle. Flat portions of quantile plots indicate clustering of observations. The $p = 0.5$ quantile for the DO data suggests a median value of approximately 5.6 mg/L. This is found by tracing a vertical line from the p-value of 0.5 up to the quantile plot and then reading off the value of that observation from the vertical axis.

A better feel for the information provided by quantiles is provided in Figure 2-16, which illustrates the relationship between shapes of common distributions and

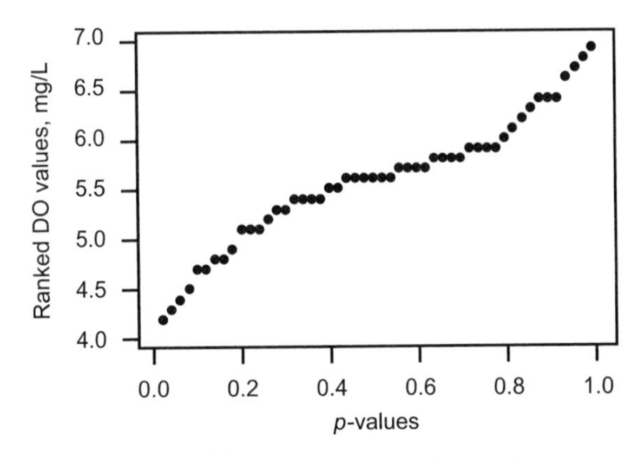

FIGURE 2-15. Quantile plot of the DO data.

their associated quantile plots. Figure 2-16a shows a symmetric histogram with few observations in the tails of the distribution, and an increasing frequency of observations close to the center. The associated quantile plot shows a characteristic S-shape with a flat spot toward the center, indicating a relatively high density of observations, and increasingly steep end points, characteristic of a low density of observations. The uniform or rectangular distribution in Figure 2-16b produces an almost linear quantile plot as the density of observations across the histogram is constant. In both Figures 2-16a and 2-16b, the $p = 0.5$ quantile, associated with the median, is found in the middle of the values of the variable of interest arranged on the vertical axis. In Figure 2-16c the positively skewed distribution results in a quantile plot that resembles the upper half of the S-shaped quantile in Figure 2-16a. The gradient of this plot increases throughout its range, as the density of observations decreases. The $p = 0.5$ quantile in Figure 2-16c is associated with a relatively small value of the variable of interest, indicating that the median is located toward the left side of the positively skewed histogram. Note that a negatively skewed distribution would produce a quantile plot resembling the lower half of the S in Figure 2-16a as the density of observations in the left-hand tail of the distribution results in a steep quantile map, gradually flattening as values of the variable increase in size and number. In this case, the $p = 0.5$ quantile would be associated with a relatively high value of the variable X, indicating that the median is located toward the right side of a histogram displaying negative skew.

Tools of Exploratory Data Analysis: Stem-and-Leaf Displays and Box Plots

The tools of data display discussed so far in this chapter represent what might be called the traditional core of descriptive graphical statistics. Over the past 20 or so years, a wealth of novel methods of analyzing data have been developed to supplement these techniques. These approaches are now usually termed *exploratory data analysis* (EDA), a name taken from the influential book with the same title written by John Tukey

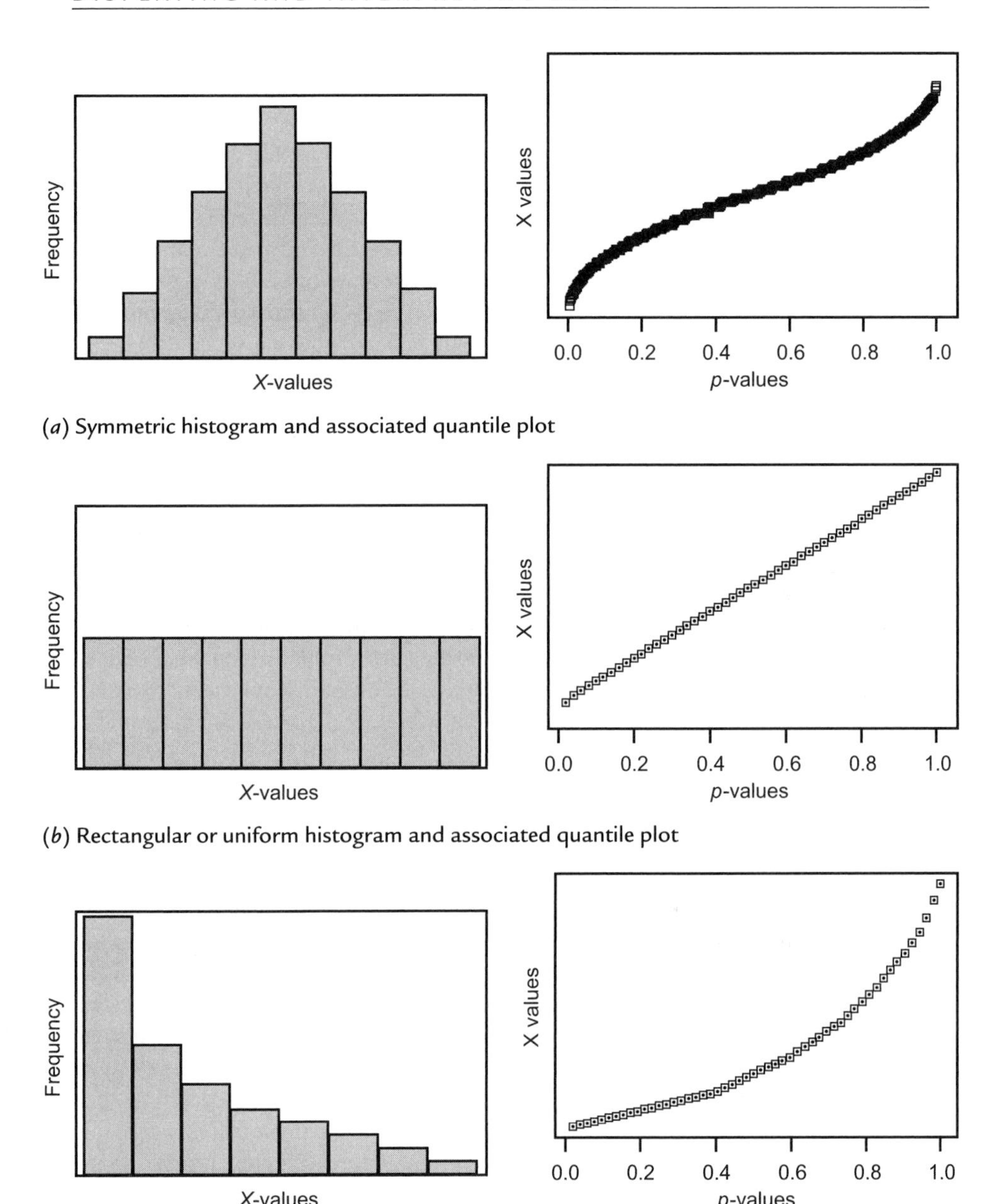

(*a*) Symmetric histogram and associated quantile plot

(*b*) Rectangular or uniform histogram and associated quantile plot

(*c*) Histogram with positive skew and associated quantile plots

FIGURE 2-16. Common histogram shapes and associated quantile plots.

(1977). The motivation underlying the development of EDA was the establishment of a new suite of methods that could be used to uncover important features of datasets that may go unnoticed if traditional techniques are employed.

Many introductory textbooks give the false impression that there is a single, correct way to analyze data. In EDA, by contrast, there is an explicit recognition that there may be several different ways to confront numerical data, each having particular strengths and weaknesses. EDA is more of a practical approach to data investigation. Rather than being limited to a set of conventional techniques derived from a particular philosophy of statistics, the data themselves are supposed to guide students in the choice of techniques for the problem at hand, not vice versa.

Because their origin is still relatively recent, EDA techniques are not as widely available in standard computer packages for statistical analysis. In fact, some would argue that many of the advantages to be gained from confronting a set of data with these techniques would be lost if they were forced into the mechanical mold of computer programming algorithms. Since these techniques remain exploratory in nature, there are fewer rules of thumb to guide students in their application. Analysis of data can thus be thought of as part art and part science. As experience with these techniques is gained, more "rules" may emerge. In a sense, however, the development of "rules" of analysis is almost against the spirit of EDA: they are like handcuffs limiting our reach and inhibiting our freedom to understand the important qualities of data. In the following two subsections we briefly outline the essential elements of two EDA techniques, the stem-and-leaf display and the box plot.

STEM-AND-LEAF DISPLAY

Consider a collection of values on some variable, or what an exploratory data analyst prefers to call a *batch.* It is not usual to call our data a *sample,* since we are seldom interested in inferential techniques when we apply EDA. The first step in conventional descriptive statistics is to reduce a large set of data values to a convenient framework for their summary, usually a frequency table and histogram. The first step in EDA has the same purpose, but uses different techniques. These graphical techniques are called displays. The stem-and-leaf display, sometimes called a stem plot, is used to illustrate the following:

1. The range of values within the batch
2. Where concentrations of values occur
3. How many concentrations of values are there within the batch
4. Whether there are gaps in the range of values
5. Whether there is symmetry in the batch
6. Whether there are extreme values, or outliers, that differ markedly from the majority of values in the batch
7. Any other data peculiarities

This is very similar for the rationale we discussed earlier for characterizing any set of data. Like dot plots and quantile plots, stem-and-leaf displays show each observation in a set of data or a batch. The unique feature of the stem-and-leaf display is that the actual digits that make up the data values in the batch are used to construct the display. This means that it is possible to reconstruct the original data from the stem-and-leaf display, a construction not possible from most other graphical tools. Also, the actual digits of the data batch are used to guide the sorting and display process.

Stems and Leaves

Four steps are required to construct a stem-and-leaf display:

1. Sort the data batch from lowest to highest value
2. Separate each observation into stem, leaf and ignored portion (if any)
3. List the stems vertically in increasing order from top to bottom
4. For each observation, or data value, write the display digit(s) on the same row as its stem. When all data values in the batch are displayed, there should be one leaf for each observation.

A series of examples should clarify the meaning of these steps.

EXAMPLE 2-5. Stem-and-Leaf Display Using Average Household Income Data for U.S. States. Let us use the average household income data for US states to illustrate the steps involved in making a stem-and-leaf plot. Table 2-9 shows these data sorted from lowest to highest value. Next we must divide the observed values into three classes: stems or sorting digits; leaves or display digits; and digits that can be ignored. Using the first observation, we can divide the income data as follows:

Sorting digits:	Display digits:	Digits to be ignored:
2	9	696

In this example the income values in the tens of thousands represent the sorting digits, the income value in the thousands represents the display digit, and the final three digits are ignored. Just as we make a choice for the number and size of intervals in constructing a histogram to best illustrate the properties of a variable, so, too, must the correct choice of sorting and display digits be made in a stem-and-leaf display. When the digits are divided 2-9-696, we are saying that income in thousands of dollars is

TABLE 2-9
Sorted State Average Household Income Data

29,696	33,672	37,082	39,184	40,558	42,090	45,726	47,381	52,868
31,330	34,133	37,240	39,250	40,624	42,433	45,776	47,493	53,935
32,182	34,135	37,572	39,469	40,856	43,393	46,590	49,467	55,146
32,566	34,604	37,892	39,927	40,916	43,791	46,677	49,820	
33,024	35,282	37,934	40,106	40,956	44,581	47,111	50,502	
33,400	36,360	38,819	40,127	41,587	44,667	47,203	51,571	

TABLE 2-10
Stem-and-Leaf Display of State Median Household Income Data

Stems	Leaves
2	9
3	1
3	22333
3	4445
3	677777
3	89999
4	00000001
4	2233
4	4455
4	667777
4	99
5	01
5	23
5	5

Note: $n = 51$; display digit = $1,000

important, but income values below a thousand dollars are unimportant. If we divide the digits 29-6-96 we are indicating that income values greater than one hundred dollars are important and smaller values are unimportant. This might be desirable if all the state income values were close to $30,000. The effects of using different sorting and display digits on the appearance of stem-and-leaf displays are illustrated throughout this section.

Continuing our example, using the state income data, we arrange the stem values vertically and list the display digits along the appropriate row of the display. Table 2-10 shows the resulting stem-and-leaf display, when the digits are divided 2-9-696, or when the display digit is $1,000. The stem-and-leaf display of the household income data resembles a histogram turned on its side. This display clearly shows the end points of the income data, it indicates a central value around $40,000–41,000 and reveals that the distribution exhibits a slight positive skew. It also shows that the minimum and maximum income values are slightly more distant from their nearest neighbors than any other observations in the dataset. This information is not so readily gleaned from any of the histograms in Figure 2-12.

EXAMPLE 2-6. Stem-and-Leaf Display of the Lake DO Data. To form a stem-and-leaf display of the DO data, we first sort the observations and then must decide how to partition the actual batch values. The data are divided at the decimal point, yielding only three stems 4, 5, and 6. The tenths' digit is used as the display digit, forming the leaves of the display, and no digits are ignored. When all the values in the batch have only two digits, this is the only logical construction. The resulting stem plot for the DO data is shown in Table 2-11. There are 9 leaves for stem 4, cor-

TABLE 2.11
Stem-and-Leaf Display of DO Lake Batch with Three Stems

Stems	Leaves
4	234577889
5	111233444455666666777788889999
6	01234446789

Note: n = 50; display digit = tenths mg/L.

responding to the first nine values of the DO data (see Table 2.5): 4.2, 4.3, 4.4, 4.5, 4.7, 4.7, 4.8, 4.8, and 4.9. Note that we can generate these nine values from the first row of the stem-and-leaf display itself. There are 30 leaves for stem 5 and 11 leaves for stem 6. This is not a good example of a stem-and-leaf display, as there are too few stems for the 50 leaves. Consequently, it is difficult to identify an average DO value. Nevertheless, we can still see that this batch of values appears symmetric, that it ranges from 4.2 to 6.9, and that most values are concentrated at values between 5.0 and 5.9. A number of ways of improving the appearance of this display are illustrated next.

Multiple Lines per Stem

Sometimes, the stem-and-leaf display for a batch can be ineffective if it is elongated or compressed. The information in Table 2-11 is too compressed to be of much use. There are two ways of constructing a display to overcome this problem. First, we can use two lines per stem, doubling the number of lines in the display. Normally, this also helps to alleviate the appearance of overcrowding in the stem plot. Each stem is converted to two lines by associating leaf values of 0, 1, 2, 3, and 4 with the first line of the stem and the values 5, 6, 7, 8, and 9 with the second line of the stem. By convention, the first line is labeled with an asterisk and the second line with a dot. Table 2-12 illustrates a revised version of the stem-and-leaf display of the DO batch. Expressed in this way, the symmetry of the batch of dissolved oxygen values is much more apparent than in the display of Table 2-11. Of course, histograms with too few intervals have the same difficulty in adequately supplying a useful graphical summary of the distribution of a variable.

An alternative to the two lines per stem format is the five lines per stem display illustrated in Table 2-13. In this stemplot, the five lines for each stem are denoted by asterisks for leaf values 0 and 1, by a T for leaves 2 and 3, (i.e. Two and Three), an F for values 4 and 5 (Four and Five, etc.), an S for 6 and 7, and a dot for leaves 8 and 9. The symmetry of the DO batch is again prominent, as is the concentration of values between 5.4 and 5.9. In this example, there appears to have been a definite advantage to moving from a display of one line per stem to two lines per stem. However, there appears little to be gained from extending the display yet further to five lines per

TABLE 2-12
A Two-Line Stem-and-Leaf Display of the Lake DO Batch

Stems	Leaves
4*	234
4·	577889
5*	1112334444
5·	5566666677778888 9999
6*	0123444
6·	6789

Note: n = 50; display digit = tenths mg/L; the first line is labeled with an asterisk, and the second line is labeled with a dot.

TABLE 2-13
A Five-Line Stem-and-Leaf Display of the Lake DO Batch

Stems	Leaves
4T	23
4F	45
4S	77
4·	889
5*	111
5T	233
5F	444455
5S	6666667777
5·	88889999
6*	01
6T	23
6F	444
6S	67
6·	89

Note: n = 50; display digit = tenths mg/L.

stem. Another basic principle of EDA is to use the simplest version of any display that adequately illustrates the salient characteristics of the distribution of batch values.

Listing Stray Values

Observed values that are far removed from the remainder of the batch are known as strays or outliers. That a batch contains such values seems a sufficiently important feature to warrant a special designation in a stem-and-leaf display. Imagine, for the moment, that the DO data contain two more observations: 2.8 and 7.6. To include these in the stem-and-leaf display of Table 2-12 data would require five additional two-line stems, and three of these would have no leaves. Once it is determined that

TABLE 2.14
A Re-expression of Table 2-12 Using the Convention for Strays

Stems	Leaves
LO	2.8
4*	234
4·	577889
5*	1112334444
5·	5566666677778888999
6*	0123444
6·	6789
HI	7.6

Note: $n = 50$; display digit = tenths mg/L.

there are strays in a batch, it is useful to identify them with labels HI and LO, depending on which end of the display they belong. To emphasize the gap between these values and the rest of the batch, one blank line is left between the strays and the remaining observations. Table 2-14 shows how Table 2-12 would be amended to include the two hypothetical stray values.

Positive and Negative Values

If the batch contains both negative and positive values, then the stems near zero require special treatment. Numbers slightly less than zero are attached to stem –0, and numbers slightly greater than zero are given stem +0. For example, the display

Stems	Leaves
–1	2
–0	78
+0	456
1	3

Note: Display digit = 0.1.

contains the values –1.2, –0.7, –0.8, 0.4, 0.5, 0.6, and 1.3. A value of exactly 0.0 could be placed on either zero stem. If there are several different 0.0 values, then it is best to divide them as equally as possible between the two zero stems. This preserves the appearance of the display.

BOX-STYLE PLOTS

Here we shall examine two related displays, the box-and-whiskers plot and the box plot. Unlike the quartile plot and the stem-and-leaf display, both box-style plots present summary characteristics of a distribution. Those characteristics are sometimes referred to as the *five-number summary* and comprise the end points of a distribution (the minimum and maximum values), the median and the quartiles. Like the median, the quartiles will be discussed in more detail in Chapter 3. For the moment, they can be understood as the observations corresponding to the $p = 0.25$ and $p = 0.75$ quantiles. These are the observations in the middle of the bottom half and in the middle of the top half of the distribution. The median, of course, is the observation that splits the entire distributioninto two halves. Actually, when Tukey (1977) introduced the

DEFINITION: FIVE-NUMBER SUMMARY

The five-number summary of a distribution comprises the minimum and maximum values, the median and the quartiles.

box-and-whiskers plot, the five-number summary included *hinges* rather than quartiles. Hinges and quartiles are typically similar in value and often the same. However, it is now conventional to use quartiles.

The Skeletal Box Plot or the Box-and-Whiskers Box Plot

To produce the box-and-whiskers box plot requires calculation of the five-number summary values. The following example shows how these values are used to form the plot.

EXAMPLE 2-7. Average Public School Teacher Wages by State, 1989. The data in this example are taken from the National Education Association. The observed values represent the average (mean) wages of public school teachers in the 50 states and the District of Columbia. The numbers are in U.S. dollars and represent gross wages—wages prior to taxation and various other deductions. The data are listed, along with state abbreviations, in Table 2-15.

The five-number summary values for the teachers' salaries are shown in the top part of Figure 2-17 where the observations in the batch are presented as a simple dot plot. In the next chapter we discuss how to find these summary values. The bottom part of Figure 2-17 shows the box-and-whiskers diagram of the salary batch. The limits of the "box" in the lower part of the figure correspond to the lower (first) quartile and the upper (third) quartile.

The line inside the box denotes the position of the median (the second quartile). The ends of the "whiskers" represent the minimum and maximum observations. This box-and-whiskers diagram suggests that the wage batch is asymmetric and positively skewed. This can be confirmed by looking at the values of observations in the dot plot. Skew is indicated by the unequal lengths of the two whiskers and by the position of the median, which is pulled to the left of the box center. The box itself clearly shows the position of most of the batch values.

TABLE 2-15
Average Public School Teacher Wages ($), 1989

AL	22,934	HI	25,845	MA	26,900	NM	22,644	SD	18,095
AK	41,480	ID	20,969	MI	30,168	NY	30,678	TN	21,800
AZ	24,640	IL	27,170	MN	27,360	NC	22,795	TX	25,160
AR	19,538	IN	24,274	MS	18,443	ND	20,816	UT	22,341
CA	29,132	IA	21,690	MO	21,974	OH	24,500	VT	20,325
CO	25,892	KS	22,644	MT	22,482	OK	21,419	VA	23,382
CT	26,610	KY	20,940	NE	20,939	OR	25,788	WA	26,015
DE	24,624	LA	20,460	NV	25,610	PA	25,853	WV	20,267
DC	33,990	ME	19,583	NH	20,263	RI	29,470	WI	26,525
FL	22,250	MD	27,186	NJ	27,170	SC	21,570	WY	27,224
GA	22,080								

Source: National Education Association.

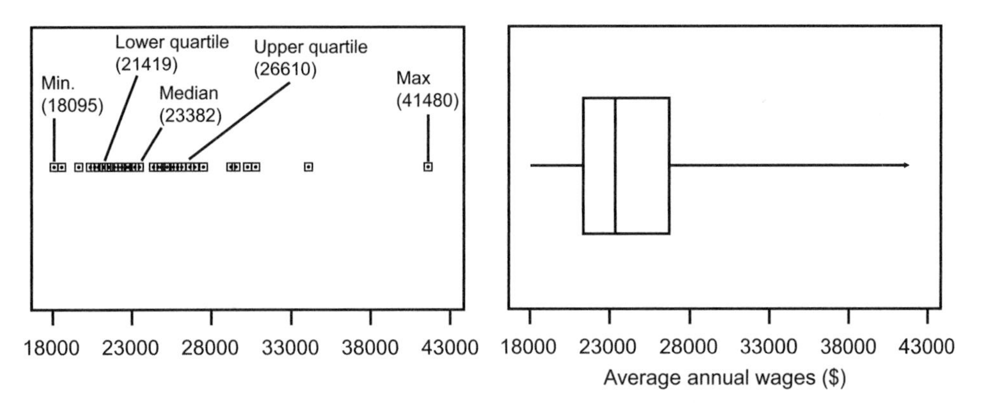

FIGURE 2-17. Dot Plot and Box-and-Whiskers Diagram of Teachers' Wages.

Outliers

One of the defects of the box-and-whiskers diagram is that it fails to treat outliers adequately. Remember that outliers are values that are either so low or so high that they appear to be isolated from the rest of the batch. A simple modification of the box-and-whiskers plot is needed to direct our attention to these outlying values. The question we now face is how to identify outliers? One simple rule that has been adopted to identify outliers is an expression based on the location of the first and third quartiles.

DEFINITION: INNER FENCES AND OUTER FENCES

Inner fences are defined as

$$\text{First quartile} - (1.5)\ \text{IQR}$$
$$\text{Third quartile} + (1.5)\ \text{IQR}$$

Outer fences are defined as

First quartile – (3) IQR
Third quartile + (3) IQR

where IQR is the interquartile range or third quantile – first quantile

---

Note that the fences are not real observations in the dataset or batch; rather, they are values constructed to assign observations to categories within the overall distribution of the variable of interest. Any data value that lies beyond the inner fences is defined as *outside,* and any data value that lies beyond the outer fences is defined as *far outside.* The outermost data values, one to the left of the median and one to the right, that lie just inside the inner fences are known as *adjacent values.* Any observations outside the adjacent values are defined as outliers.

Box Plot

A box plot is a modified version of the box-and-whiskers diagram that provides a better representation of outliers or stray values in a distribution. The box is constructed in an identical fashion: solid lines mark the boundaries of the box with a length from the lower quartile to the upper quartile, and a solid line across the box denotes the median. The whiskers are drawn as dashed or solid lines, and, unlike the box-and-whiskers plot, are drawn from each quartile out to the corresponding *adjacent value.* We do not use the fences because they are not actual data points, whereas we know that adjacent values are always members of a batch. Outliers, if present, are shown individually and labeled appropriately. Outside values might be designated with dots, and far outside values might be shown as asterisks, for example.

Figure 2-18 shows a modified box plot of the teachers' wages data. Note that the end of the lower whisker is in the same position as in the box-and-whisker diagram in Figure 2-17. This is because the lower fence is well below the minimum value

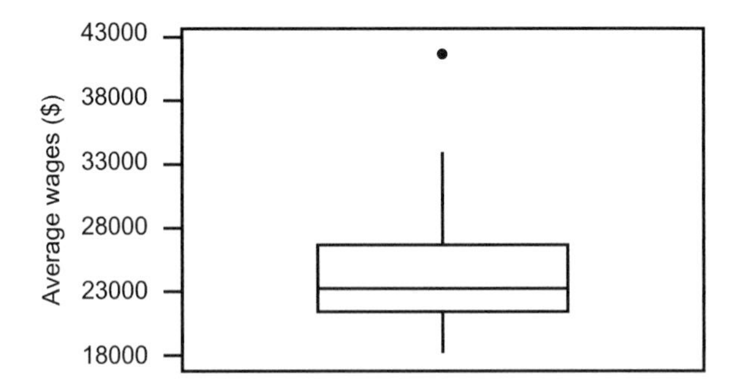

FIGURE 2-18. Box plot of teachers' wages. • denotes an outside value.

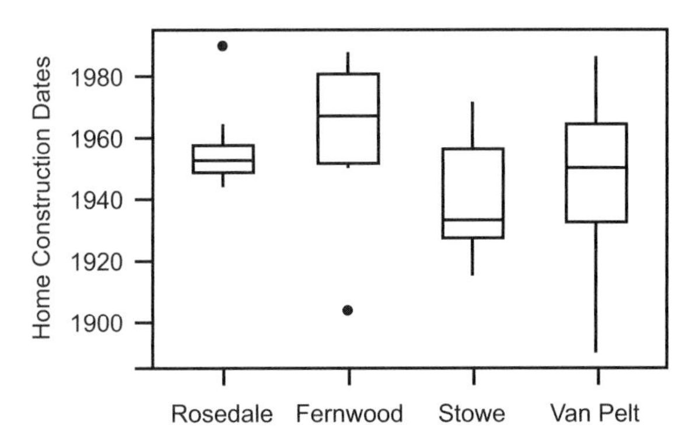

FIGURE 2-19. Box plots of construction dates for different neighborhoods.

in the wages batch, and thus the adjacent value is the smallest observation. The upper whisker in Figure 2-18 stops far short of its position in Figure 2-17. The modified box plot conveys the same information as the box-and-whiskers plot, but it also shows that the batch contains an outlier. This outlier is an outside value. On checking the data value, the outlier represents average public school teacher wages in the state of Alaska. This value does not represent an error, but rather reflects the relatively high cost of living in Alaska.

Comparing Box Plots

In order to compare several different batches, a number of box plots may be placed side by side. For example, Figure 2-19 compares construction histories for houses in different neighborhoods by developing individual box plots of building dates in each neighborhood and placing them on an identical scale. Several comparative statements can be drawn from this figure. First, we note that Fernwood is the most recently developed of the four neighborhoods, except for a single house built well before a more or less continuous periods of house construction. The Rosedale neighborhood is noteworthy because it was developed over a relatively short period and has the appearance of being a post–World War II suburban bedroom community. The recent outlier in this neighborhood represents a new home, rebuilt after a fire. By contrast, Stowe and Van Pelt have longer development histories, with no apparent outliers. The development history of Van Pelt is so long one might conjecture that it is a neighborhood with quite different vintages of housing. Some older houses may have been demolished and replaced by newer housing, leading to a neighborhood with a mixed housing stock. This could be checked by more detailed EDA, perhaps using stem-and-leaf displays. In sum, comparison of box plots by placing several batches on an identical scale appears to be a useful technique for illustrating differences and similarities among sets of data.

2.3. Displaying and Interpreting Time-Series Data

The graphical techniques that we have examined to this point show the frequencies with which observations in a dataset take on particular values. Geographers and other scientists are often interested in exploring data that have been gathered over time. Observations on a variable that are arranged temporally, or in a time sequence, are called a time series. Examples of time series are daily temperatures at 2:00 p.m. at a particular weather station for the past 20 years; the daily closing value of the Dow Jones stock market index; the annual economic output (gross domestic product) of a particular country; tree-ring widths; and monthly measurements of atmospheric CO_2.

Time series are analyzed to describe, understand or explain, predict, and sometimes even control the underlying process generating the observations in the time series. Those observations may consist of a number of different components such as a long-term trend, a cycle, a seasonal influence, and irregular or random fluctuations. The particular component in which we are interested often depends on the nature of the problem being studied. For example, a climatologist interested in long-term climatic change may focus on the trends in a time series and ignore seasonal and other cycles in the data. Economic geographers, in contrast, might be interested in unemployment rates and how they vary over the business cycle (approximately 7- to 10-year periods of upswings and downswings in the economy). Still others may be concerned with energy price fluctuations between summer and winter seasons, or with the seasonal behaviors of particular animal species.

In this section we discuss the conventional way of displaying time-series data and we think about how to interpret such data using a number of examples drawn from different subfields of geography. In Chapter 3 we extend this brief discussion to incorporate different ways of summarizing times-series data. Chapter 15 focuses on more sophisticated procedures for modeling time series.

What Is a Time Series?

Let us begin by providing a formal definition for a time series.

DEFINITION: TIME SERIES

A time series is a sequence of observations $Y_1, Y_2, Y_3, \ldots, Y_t, \ldots, Y_n$ that are collected or recorded over successive increments in time.

Notice that for a time series we have used Y rather than X as the symbol for the observed variable. We do so to emphasize that the order of the observations is crucial in time-series analysis. Throughout most of statistics, the observations can be scrambled with no effect. For example, X_7 and X_{22} could be switched with no effect on the results. By contrast, in time series the temporal structure is paramount, and it is essential that we preserve the time order of the observations.

Notice also that the definition of a time series specifies measurements or readings taken at equally or almost equally spaced time intervals. Very little theoretical work has been done on time series observed at unequal intervals, though it is not uncommon to see a time-series analysis with a few missing observations or with an extended gap somewhere in the series. These interruptions in time series occur for various reasons. For example, physical geographers often rely on devices such as data loggers that automatically collect and record data, and these devices occasionally break down. In other cases researchers might be prevented from accessing study sites at the beginning of a field season because of late snowfall or delayed ice melt. Other scientists confront different problems. The U.S. Bureau of the Census interrupted its regional data collection activities between 1979 and 1981 because of budgetary problems. Recoding or redefining the way a variable is measured also causes time series to be truncated.

Relatively small gaps within a time series can often be treated as missing values. In other cases it is quite common to *interpolate* the values of a few missing observations so that the data can be treated as a conventional time series. When the data series is long and contains hundreds or even thousands of observations, this should have an insignificant impact on the results of the analysis and any interpretations we make. However, caution is in order. Extreme observations can often have a substantial impact on time-series analysis, just as they can lead to significant changes in the values of any simple descriptive statistics. While we may have some justification in eliminating extreme observations, or outliers, that we suspect to be invalid, it is important to remember that every observation in a dataset tells us something about the variable being analyzed. Replacing observations just because they look strange is unwise.

Some methods of interpolating values in a time series are discussed in the next chapter. These methods are based on the concept of smoothing. In short, we replace an observation by the average values of similar observations. In a time series, this normally means we generate a missing value from neighboring observations. However, in some instances this procedure itself may be biased. For example, we could not wish to replace an average monthly rainfall figure for July with, say, the average of June and August; it may be better to use average July values from previous and subsequent years. A visual examination of the time series is usually an important guide in deciding what kind of substitution to perform. Considerable care also must be taken with the first and last few observations in a time series. Choice of the time frame, the start and end points of the investigation, can have a major impact on identification of long-term trends and other components within temporal data.

Time-series data normally arise in one of two ways:

1. First, we can generate such a series by taking measurements of a specific continuous variable at equally spaced time intervals. Such series are often defined as *instantaneously recorded.* For example, if we were to take the ambient air temperature at 3:00 p.m. every day for a year, we would generate a set of 365 observations of a continuous time series.

2. Second, we can generate a discrete time series through *accumulation* or *aggregation* of data. Specific examples of accumulated series include monthly rainfall at Santiago, Chile; quarterly industrial production in Canada; the number of hurricanes making landfall each year in the United States; the number of armed robberies each day in Los Angeles; and the reported cases of rubella by week at a health clinic in New York City.

Examining Time-Series Data

A time-series graph plots each observation on a variable of interest together with the moment in time to which the value corresponds. By convention, time is always recorded along the horizontal axis, while the variable of interest is plotted against the vertical axis.

EXAMPLE 2-8. Dow Jones Industrial Average Stock Market Index. The Dow Jones Industrial Average is an index of stock market prices based on the share values of 30 large companies that are listed on the New York Stock Exchange. The index provides one of many barometers of the state of the economy. Table 2-16 shows a portion of the Dow Jones time series. The stock market values are shown for four quarters (the end of each three-month period) in each year. A single observation in the time series records the index value and the quarter and year to which that observation corresponds. Thus, the first four values in the time series, 211.2, 229.2, 228.2, and 236.5, show the Dow Jones Industrial Average over the four quarters of 1950.

The index values that comprise the Dow Jones data could be displayed and summarized with the techniques outlined in the last section. However, we typically gather time-series data to examine questions about the movement, or change in value, of a variable over time. The most straightforward way of presenting such information is in a time-series graph. Figure 2-20 plots the quarterly closing value of the Dow Jones index between 1950 and 2003. Figure 2-20 shows the Dow Jones index value on the vertical axis, and the time of each observation is recorded on the horizontal axis.

In this figure we have linked the quarterly index values together with a solid line. Alternatively, a time-series graph might comprise a series of individual observations represented by a symbol. In other cases the symbols denoting each observation in the series are joined together by a line. Symbols are often added to a line that displays a time series to emphasize the values of individual observations. However, in time series with a very large number of observations, showing symbols can make the graphs look rather busy.

When we examine a time series, or a plot of the values of a variable against time, we are looking for a general pattern. Figure 2-20 shows that the Dow Jones index increased relatively slowly between 1950 and 1970; it then remained more or

TABLE 2-16
Sample of the Dow Jones Industrial Average Stock Market Index

Year	Quarter	Index	Year	Quarter	Index	Year	Quarter	Index
1950	1	211.2	1951	1	254.4	1952	1	270.4
	2	229.2		2	255.2		2	275.0
	3	288.2		3	277.5		3	278.2
	4	236.5		4	270.0		4	293.5

Notes: The four quarters correspond to consecutive three-month periods during the year. For example, the first quarter comprises January, February, and March. The stock market index value recorded is the peak, or maximum, closing day value during the last month of each quarter.

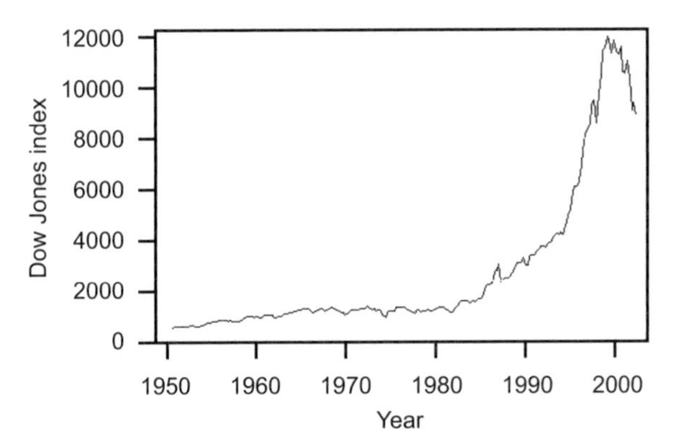

FIGURE 2-20. Time-series plot of the Dow Jones Industrial Average, 1950–2003.

less constant until 1980, after which it increased rapidly, reaching a peak in 1999. Since then it has decreased sharply. Overall, the Dow Jones time series exhibits a strong upward *trend,* a general increase in value over time. It is worth noting here that when we graph a variable whose value changes markedly, small fluctuations in the value of the variable tend to get dampened. For example, Figure 2-21 illustrates the Dow Jones index values between 1950 and 1980, in the first frame using the same vertical scale as Figure 2-20 and in the second frame using an expanded vertical scale.

Unlike Figure 2-20, Figure 2-21b reveals that between 1950 and 1980 the Dow Jones index was anything but constant, or flat. Indeed, when plotted against an expanded vertical scale, the Dow Jones index also appears to have a cyclical component, an alternating series of relatively rapid increases followed by decreases, superimposed on the general upward trend. These cycles, though somewhat irregular, appear to have a *duration,* or a periodicity, of around four years. We typically measure the length of a cycle by the distance between successive peaks or troughs. Cycles in stock market indices are commonly identified, though less commonly explained. Some link the four-year cycle illustrated in Figure 2-21b to presidential elections.

Within the fields of medicine and public health, a number of familiar examples of variables measured over time can be found. Electrocardiagrams and electroencaphalograms are well-known medical procedures that generate time series of elements of bodily activity. Time-series data that trace the paths of various epidemics also are extremely common. Analyzing the incidence of HIV infection or death due to AIDS may yield insights into the requirements for providing public health facilities and allocating resources to limit the spread of this disease.

Figure 2-22 presents a time series drawn from the medical field, showing the weekly number of deaths from heart disease in New Zealand between 1994 and 1997. The average number of heart-related deaths per week in New Zealand is about 150 over the four-year period studied. However, as Figure 2-22 shows, relatively few weeks conform to that average. These data exhibit no general trend or cycle, but rather, they display marked *seasonality.* The figure reveals that the number of weekly heart-related

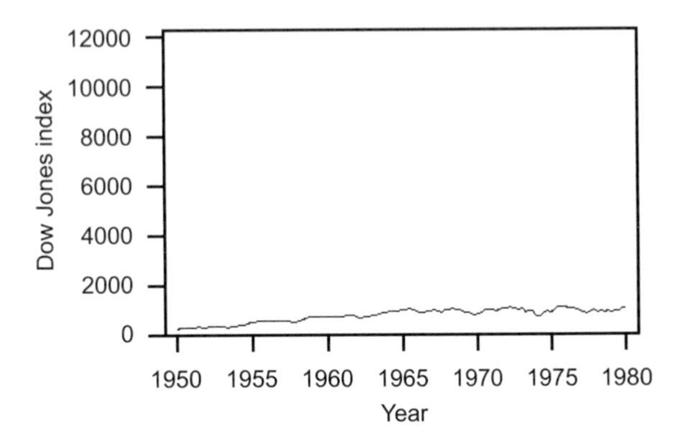

(*a*) Original vertical scale

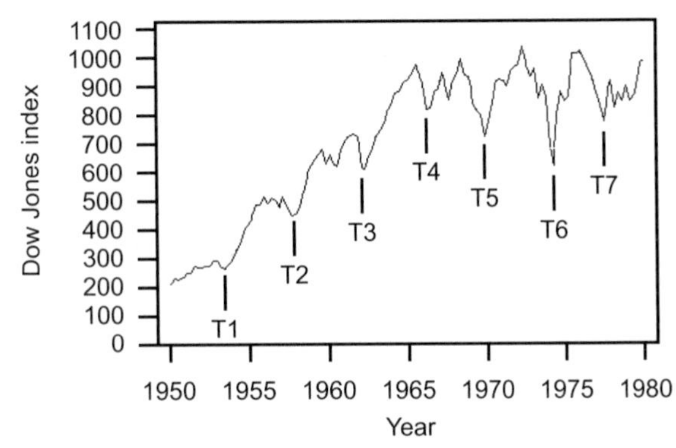

(*b*) Expanded vertical scale

FIGURE 2-21. Dow Jones Index with different vertical scales. Cycle troughs are indicated by *T1, T2, . . . , T6* in Figure 2-21b.

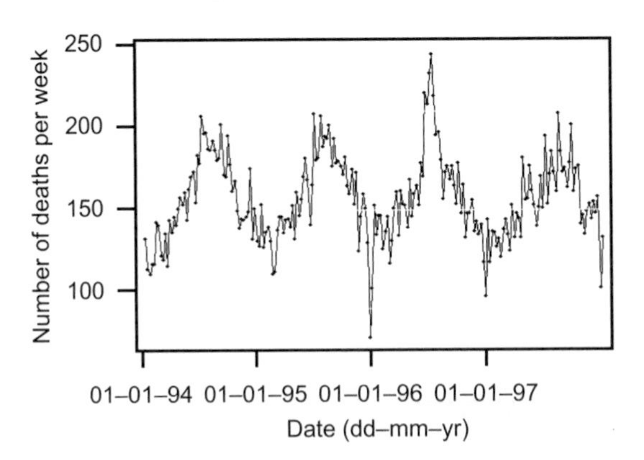

FIGURE 2-22. Weekly deaths from heart disease in New Zealand. *Source:* New Zealand Health Information Service, 2001.

deaths is almost twice as high during the winter (June–August) than it is during the summer (January–February) and that this pattern repeats itself year after year. This seasonality can be explained by the fact that heart disease is a big killer of the elderly, and during winter the elderly body is under significantly more strain. Poorer weather in winter also contributes by increasing inactivity.

In the next chapter we will examine a series of techniques that can be used to summarize time-series data and to decompose the various components of time series illustrated above. Our attention now turns to the display and interpretation of spatial data.

2.4. Displaying and Interpreting Spatial Data

One of the primary reasons for gathering data is to help us understand the process or processes that may have generated them. Thus, the epidemiologist collects data in hopes that their analysis will reveal something about the nature of a disease. Similarly, a political scientist examines election returns in order to gain some understanding of the base(s) of support for a political party. In any research project, the kind of data collected will usually depend on the question(s) being asked. For example, scientists interested in global warming will seek temperature data that are organized in the form of a time series. For geographers, the types of questions frequently asked have to do with the spatial characteristics of various phenomena: the locations of cities; geographical differences in temperature and precipitation; population density; the orientation of drumlins; the number of plant species at a particular location; regional differences in economic growth or unemployment. To examine these sorts of questions, geographers seek data that combine observations on a variable, or variables, of interest with references to the geographical locations that those observations represent.

In this section, we highlight the special character of spatial data, examine different forms of spatial data, and show how those forms conventionally are displayed and interpreted. In the following chapter, we discuss various techniques for summarizing spatial data numerically.

What Are Spatial Data?

Just as a time series combines temporal information with observations on a variable of interest, spatial data combine geographical information with observed values of other variables.

DEFINITION: SPATIAL DATA

Spatial data combine observations on a variable of interest with reference to their geographical location.

Spatial data, or geographical data, do not differ from a time series simply by the addition of another dimension. In fact, trying to model or capture the main characteristics of a time series is a lot more straightforward than trying to model a spatial process, for at least two reasons. First, most time series are gathered over regular intervals of time, such as days, weeks, or years, whereas spatial data typically represent phenomenona that are located at irregular distances from one another. Second, time can be considered as moving only in one direction, such that the future values of a variable only depend on values that have been observed in the past. *Dependence* in spatial data is far more complex, reflecting Waldo Tobler's (1970) first law of geography that everything is related to everything else, but nearby things are more related than distant things.

Spatial data can assume any of the different types or levels of measurement discussed in Chapter 1. The presence or absence of a university in a particular city represents an observation of a nominal spatial variable. Similarly, the lake vegetation data of Example 2-1, if linked to the geographical location of each of the 57 lakes, would represent a spatial dataset comprising observations on a nominal variable of interest. The average level of educational attainment measured across cities, counties, or states in the United States would represent a *spatial distribution* of an ordinal variable. Elevation data or precipitation data gathered at specific points on the earth's surface are examples of quantitative spatial data. The direction of the wind at a particular site, and the distance from one location to another, also represent examples of spatial data.

Types of Spatial Data and Interpretation

The spatial component of spatial data refers to a geographical unit or object. There are three main types of geographical objects: points, lines, and areas. As a component of spatial data, each of these objects is located using some spatial reference or coordinate system. The great majority of techniques used in geography assume the observations lie on a plane. Thus there is an implicit assumption that if the data were originally measured in a geodetic (latitude, longitude) system, they have been projected to the plane. *Spatial data analysis* focuses both on the spatial arrangement of geographical objects in the mapping plane and on the values of the phenomena they represent.

POINT DATA

On the one hand, *points* represent discrete locations in space that have zero area. For example, a pair of geographical coordinates such as longitude and latitude refer to a point location. The more precise those coordinates, the smaller the area of the location identified. On the other hand, for us to be able to identify them, points must be represented by a symbol that has a nonzero area. In practice, points are often used to refer to objects that occupy a significant area such as cities or lakes. This is common when such objects are portrayed on a map showing a very large area. When the spatial scale of geographical investigation gets finer, it may not be possible to sustain the point representation of some objects.

The dot map is the most commonly used cartographic display for point data. Analysis of spatial point distributions, including the search for both pattern and underlying process, has been a subject of continuing interest to geographers. Some of the techniques used for point pattern analysis are discussed in Chapter 14. Here we focus on simple interpretation of dot maps. Dot maps may represent individual people, members of a plant species, the location of rain gauges, towns and cities, and many other things.

EXAMPLE 2-9. Earthquake Data. An earthquake is a sudden movement of the earth's surface typically caused by the shifting of tectonic plates or volcanic eruption. The focus of an earthquake is the point inside the earth's crust where the earthquake occurs. The earthquake's epicenter is the location on the earth's surface directly above the focus. Earthquake data usually include timing, the location of the epicenter, and size, typically expressed in units of the Richter scale. Table 2-17 shows a sample of earthquake data taken from California over the last 100 years.

Mapping earthquake epicenters in Figure 2-23 provides an example of spatial point data. These data report the largest earthquakes of the last 100 years located in California or just off the California coast. These earthquakes all have magnitudes greater than 5.2 on the Richter scale. In Figure 2-23a we ignore the timing and size of individual earthquakes simply to report their location. The figure shows a marked cluster, or spatial concentration, of earthquake epicenters just off the northern California coast and a more widespread scatter across much of the southern portion of the state. Smaller clusters also are evident just south of San Francisco Bay and around Mammoth Mountain to the east.

To interpret a spatial point pattern such as Figure 2-23a, we pay attention to the location of data points relative to one another. We focus on whether the data points are clustered together or widely dispersed over the study area, and on whether they are distributed in a uniform (regular) or a random pattern. We might also look for obvious departures from a general pattern that may be indicative of spatial outliers that

TABLE 2-17
Sample of Historical Earthquake Data from California

Date	Time	Latitude (N)	Longitude (W)	Magnitude
4–18–1906	4:12 A.M.	37.70	122.50	8.3
1–22–1923	1:04 A.M.	34.00	117.30	7.2
11–4–1927	4:50 A.M.	34.70	120.80	7.3
5–18–1940	8:36 P.M.	32.73	115.50	7.1
7–21–1952	3:52 P.M.	35.0	119.17	7.7
11–8–1980	2:27 A.M.	41.12	124.67	7.2
10–17–1989	4:04 P.M.	37.37	121.88	7.1
8–17–1991	2:12 P.M.	41.68	126.50	7.1

Note: Date is month–day–year; time is Pacific Standard Time; geographical coordinates are shown in decimal degrees; magnitude is expressed as units on the Richter scale.
Source: U.S. Geological Survey: *earthquake.usgs.gov.*

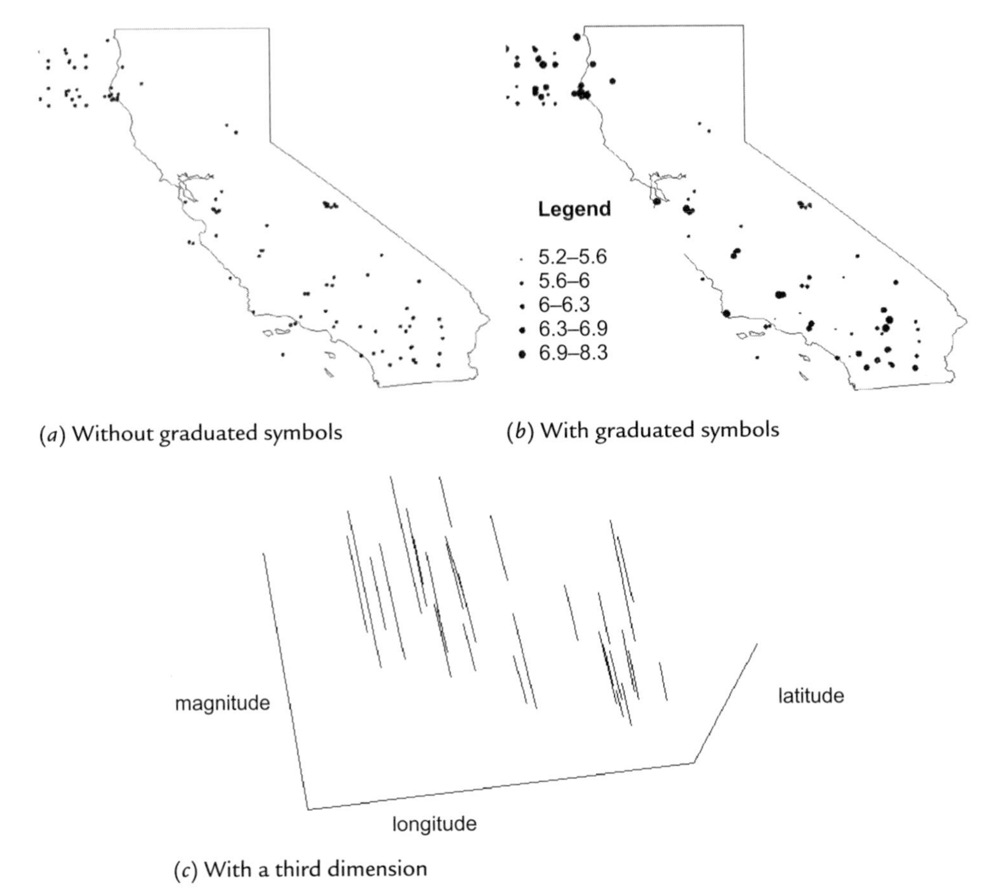

(*a*) Without graduated symbols

(*b*) With graduated symbols

(*c*) With a third dimension

FIGURE 2-23. Earthquake epicenters in California.

perhaps mark unusual events. In Chapter 4 we discuss some more formal ways of examining these sorts of questions.

In Figure 2-23b, we take advantage of additional information about the variable of interest. Here we use dots that are graduated in size to reflect the magnitude of each earthquake. The legend, or key, illustrates a series of intervals on the Richter scale into which the earthquakes may be placed. In this map we can see the overall spatial distribution of earthquakes as well as the geography of earthquakes of different magnitudes. Choice of the number of data intervals in a dot map follows the same general guidelines as for constructing frequency tables and histograms. Care must be taken to use enough intervals to show the range of values of a variable, but not so many that the figure becomes difficult to interpret.

The graduated symbols of Figure 2-23b provide one way of displaying additional information in a map-based diagram. Another possibility is to use a third dimension as in Figure 2-23c. In this case, vertical lines illustrate the magnitudes of earthquake events for a subsample of the California earthquake data.

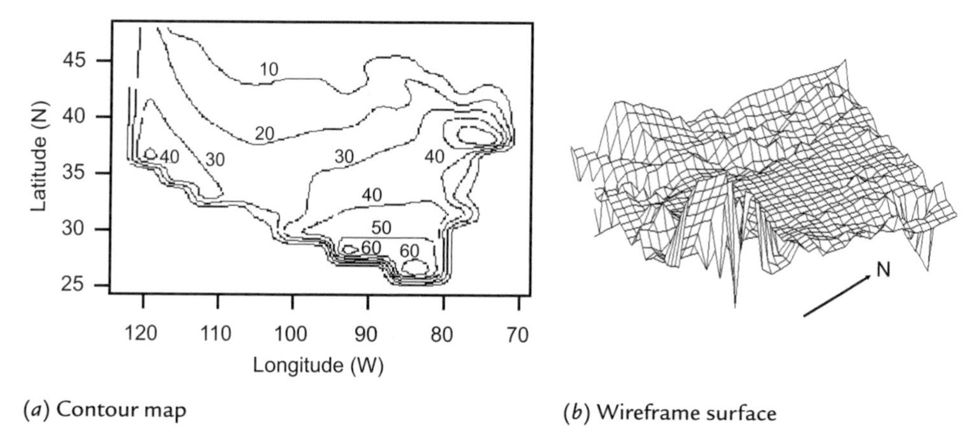

(*a*) Contour map (*b*) Wireframe surface

FIGURE 2-24. Average minimum January temperature surfaces for contiguous states in the United States.

Point data may represent individual spot measurements of a variable that is continuously distributed over space, such as precipitation or temperature. The individual data points thus represent a sample of observations that could be drawn from an infinitely large population. It is common to represent spatially continuous variables using an *isarithmic* map, often called a *contour* map (Figure 2-24a), or using a *surface plot* such as the wireframe (Figure 2-24b). In Figure 2-24a, isolines are drawn to connect points that share the same average minimum January temperature across the United States. Figure 2-24b displays the same variable at a much greater resolution: the contour map is fitted using only 50 data points, whereas the wireframe map is based on more than 1000 data points. These surface maps are useful for detecting spatial trends in data values. Clearly, with fewer observations, as in Figure 2-24a, only the most general patterns are evident. With a larger number of observations, as in Figure 2-24b, smaller scale variations also can be detected.

AREAL DATA

Areal data comprise observations on a variable across two-dimensional subdivisions of a study region. Geographical subdivisions can be natural spatial units such as drainage basins, but more commonly they represent political boundaries like census tracts, counties, provinces or states, and countries. Data for these spatial zones are gathered from individuals, or other types of point observations, and aggregated to the level required for analysis. For instance, the household income data of Example 2-4 were gathered at the level of individual households. The sampled households were then grouped together, or aggregated, at the state level in order to show state average household income. Because areal data are aggregates of individual observations, caution must be exercised when interpreting and analyzing such information because patterns will often change as the spatial scale of the geographical subdivisions is altered. This

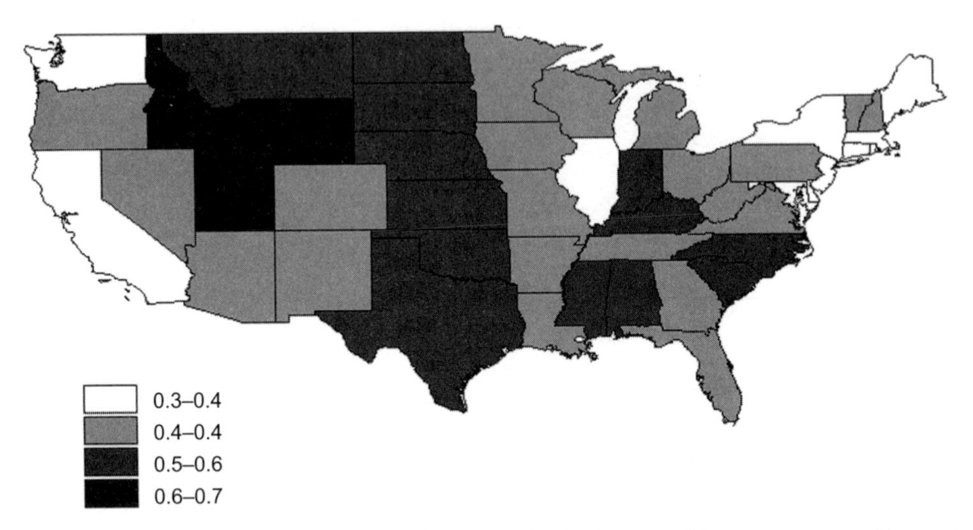

FIGURE 2-25. State-level percentage vote for George Bush. *Source: spatialnews.geocomm.com/features/election2000.*

is part of what is called the Modifiable Areal Unit Problem, or MAUP, which we will examine in more detail in Chapter 3.

We usually display areal data using choropleth maps. Figures 2-25 and 2-26 show choropleth maps of the percentage of the vote for George Bush in the 2000 presidential election at state and county levels. For clarity, the vote is reported for the 48 contiguous states only. The highly aggregated state map suggests a relatively clear picture of support for George Bush from Texas up through the Plains to the northern Mountain states, and in isolated pockets elsewhere. The disaggregated county map contains much more detailed information than the state map. For example, while the state map suggests that all Texans voted for George Bush, the county map shows that this was not the case, as a number of counties bordering Mexico tended to support the democratic presidential candidate, Al Gore. The main disadvantage of the county map is the difficulty of interpretation: some broad trends are evident, but others, perhaps, are lost in the detail.

While the patterns observed in Figure 22-25 and 2-26 are artifacts of the spatial units chosen to display the election data, these patterns also depend on the number of intervals employed to divide the range of data values. In the presidential voting maps four intervals were used with a cutoff at 0.5 that seems to make sense when examining electoral returns. However, just as with frequency tables and frequency histograms, choropleth maps look quite different when we change the number of intervals or the classification scheme across which data values are grouped. This is demonstrated in Figure 2-27, which maps the state average income data of Example 2-4. In Figure 2-27a, four equal-width income classes are used. The choice of four intervals is arbitrary, and it should be noted that because the intervals have not been chosen with care the boundaries are located at peculiar data values. Figure 2-27b uses seven equal-

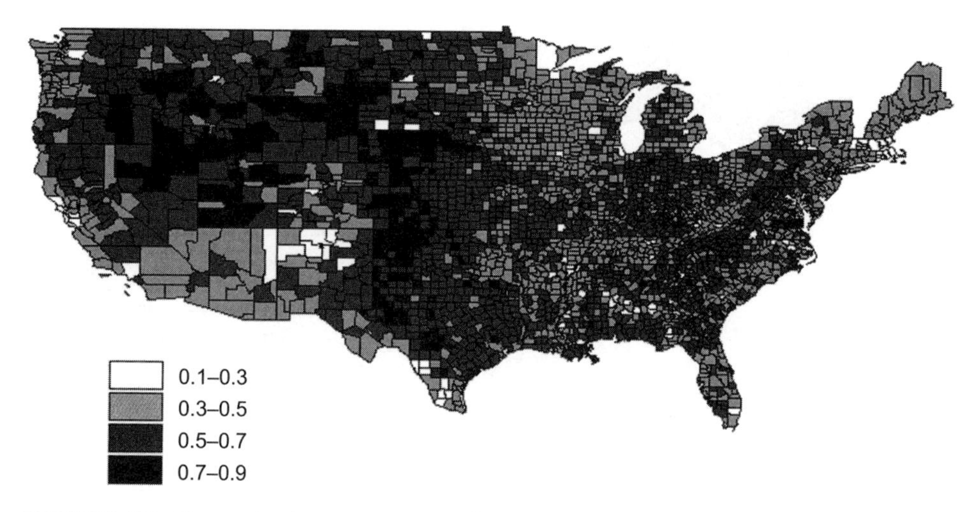

FIGURE 2-26. County-level percentage vote for George Bush. *Source: spatialnews.geocomm.com/features/election2000.*

spaced data intervals, following Sturges's rule, and these are located at readily interpretable data values. Figure 2-27b reveals much more of the state-to-state variation in average household income values than does Figure 2-27a. An altogether different classification scheme is employed in Figure 2-27c. Here five unequal-width categories separate the state income observations. Boundaries between the categories are located at *natural breaks* in the dataset. These breaks are chosen so that the *variance* of state income values is minimized within each of the classes. In Figure 2-27d, states are allocated to one of five classes depending on how many *standard deviations,* or how far, the state's median household income is from the average of all state income values. (Chapter 3 gives precise definitions for the variance and standard deviation.)

A large number of mapping schemes can be used with choropleth maps; Figure 2-27 only shows a sample of these. The different techniques for categorizing data have various strengths and weaknesses and should be employed according to the task at hand. Extended discussion of these issues is beyond the scope of the current text, but the interested reader might consult Slocum et al. (2003), Unwin (1981), Tufte (1983), and Monmoneier (1991). In general, the map is a summary device much like the plots discussed earlier in this chapter and thus should be drawn to reveal the most information in a parsimonious fashion. However, while the map may provide an indication of the range of possible values of a variable of interest, it goes much further to tell us something about the spatial organization of those values. We will return to this issue in subsequent chapters.

LINE DATA

The last major type of spatial data is line data. Lines are often used to show a variety of geographical features such as roads and rivers. They are also usefully employed in

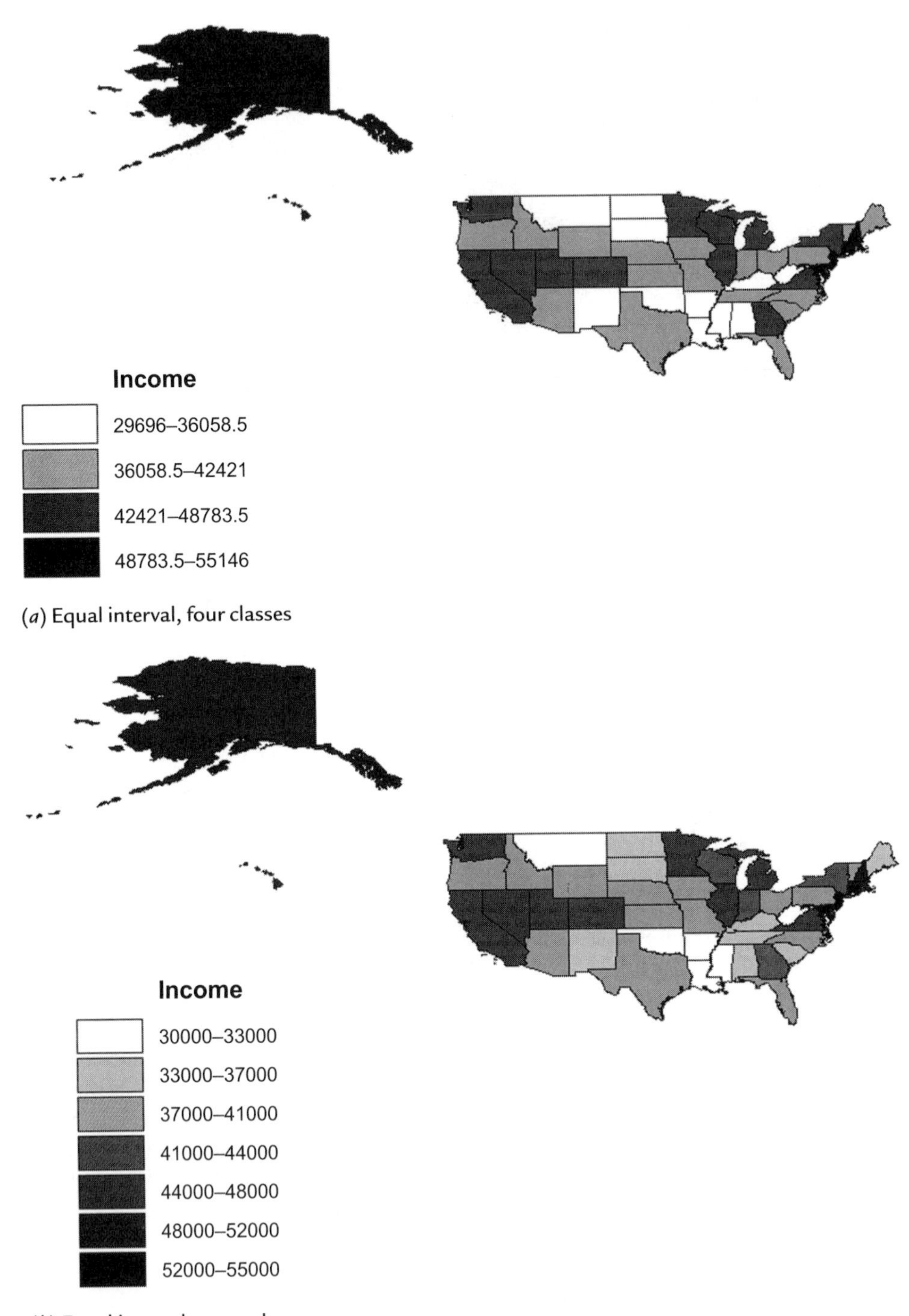

FIGURE 2-27. State differences in median household income using different choropleth map schemes.

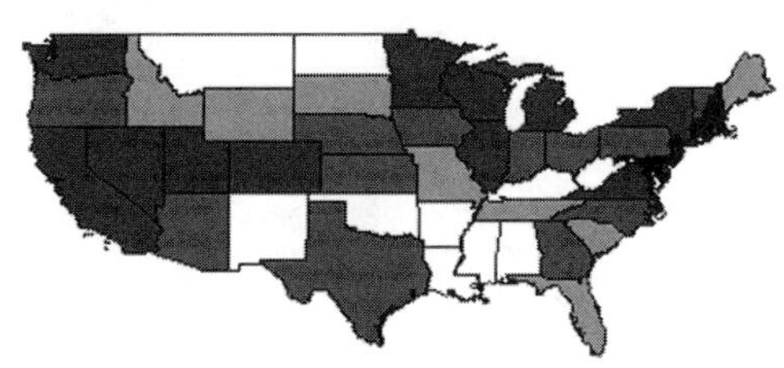

Income

29696–34604

34604–38819

38819–42433

42433–47493

47493–55146

(*c*) Natural breaks, five classes

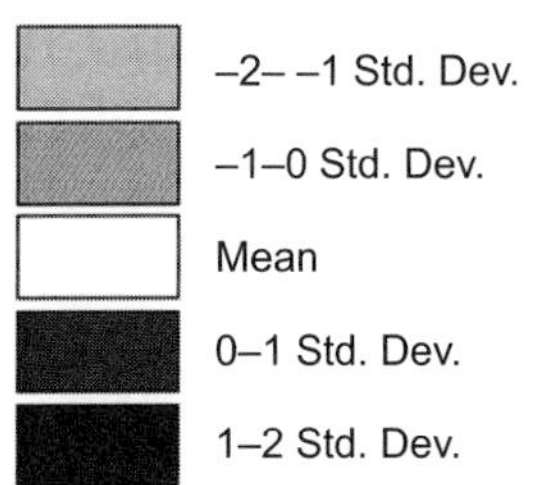

(*d*) Standard deviations, five classes

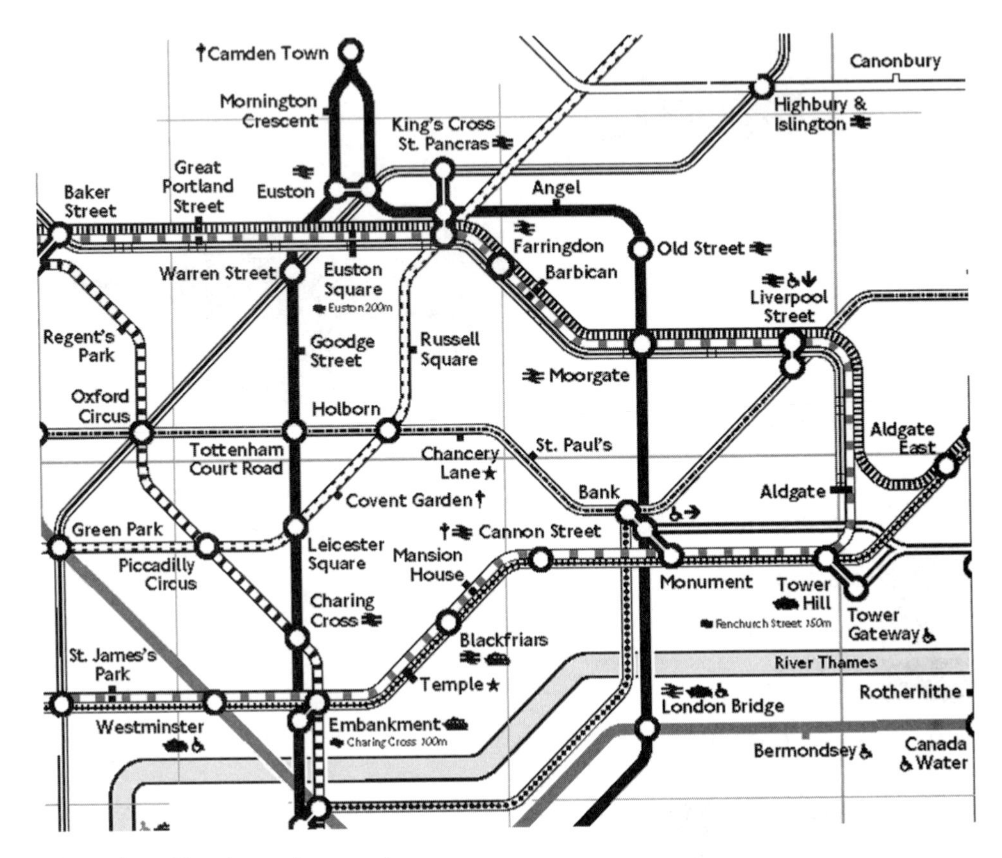

(*a*) Portion of London underground

(*b*) Major U.S. rivers

FIGURE 2-28. Examples of line data. *Source for c:* Erickson and Hayward (1991).

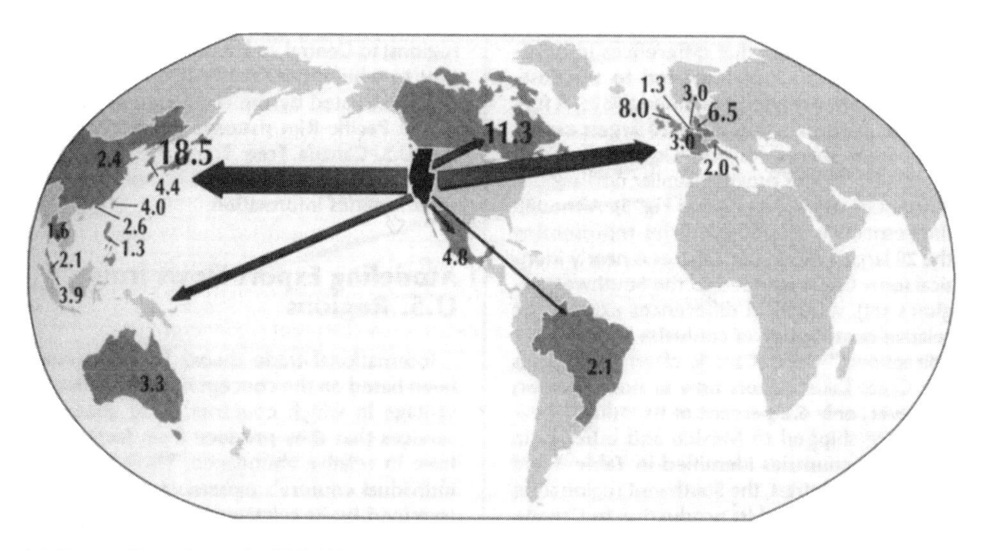

(*c*) Export flows from the U.S. West

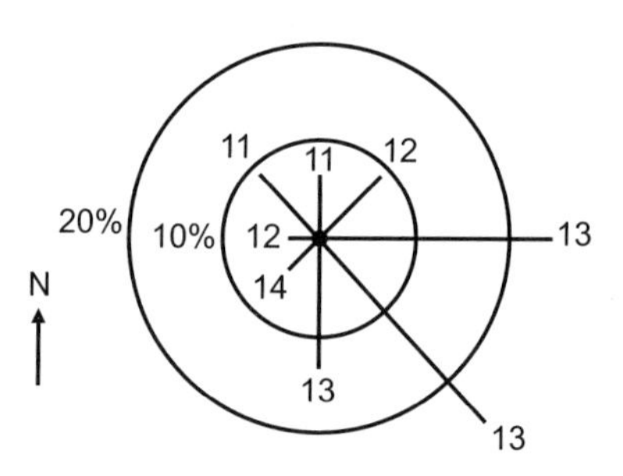

(*d*) Wind rose for Rapa Nui

the analysis of distance, direction, connectivity, and flow volume. A variety of specialized techniques for analysis of the individual paths or branches of line data and for analysis of the characteristics of line networks have been developed. Figure 2-28 shows some examples of line data.

Figure 2-28a illustrates part of the London underground transport network. The network comprises a series of points or nodes connected by regular or straight-line routes. This type of transportation map shows the geometric positions of network features, their *topology,* but does not provide a faithful representation of their real geography. The different line symbols represent different railway lines, showing one way that we can represent nominal level information with a line map. The US river system in Figure 2-28b is an irregular network, comprising lines that cannot be described by a simple mathematical function. In Figure 2-28c, lines of different weight or thickness are used to demonstrate the volume of exports from the United States to different parts of the world. Different line weights, such as these, can show interval or ratio data. In this map the direction of the lines is illustrative only: they show only the general direction of export flows. Figure 2-28d does a better job of showing direction. It is a diagram called a *wind rose.* A conventional wind rose identifies wind direction on a 16-point compass: N, NNE NE, ENE, E, and so on. The length of the straight lines on the wind rose shows the relative frequency with which winds blow from a particular direction over a specified period of time. Concentric circles, in this case showing 10% and 20% bands, are used as visual clues. The numbers at the end of each of the straight-line wind vectors represent the average wind speed for that vector. Figure 2-28d shows that the prevailing winds through the month of September at Rapa Nui, in the south Pacific, generally fall into the east to south quadrant of the wind rose. Note that average wind speed could be represented by different line weights.

Spatial Data Conversion

The distinction between continuous and discrete areal data and point data is sometimes blurred. Cartographers may argue that a continuous areal representation is appropriate if the phenomenon exists everywhere on the maps, both at and between observation points. This argument appears to be valid for many physical phenomena such as rainfall and temperature, but is less compelling for a variable such as population density. To map this variable in a continuous form on an isarithmic map, the assumption must be made that people are found everywhere and not just at discrete points. Remember that it is possible to convert data expressed in one form to most other forms, if we are willing to make a few assumptions.

First, let us consider one way in which point data can be converted to areal data. The area around each point is associated with it in such a way that mutually exclusive and collectively exhaustive subdivisions are created. That is, every map location is associated with a single data point. The locations associated with a point define a polygon surrounding that point. A common method is the construction of *Thiessen polygons.* These areas or polygons are drawn in order to satisfy the condition that *any location within the polygon is closer to the polygon's data point than to any other data point.* These polygons were defined by climatologist A. H. Thiessen to create regions around rainfall stations in such a way that station totals could be weighted by their surrounding areas to compute an "average" rainfall for the entire area. His solution is based on the assumption that observed rainfall at any location is likely to be most similar to the nearest point for which a rainfall total is known.

In abbreviated form, a process for constructing Thiessen polygons is

1. Join each point to neighboring points to form a network of triangles. (Many such triangulations are possible; Guibas and Stelfi (1985) describe one way to find the network appropriate for Thiessen polygons.)
2. Bisect the lines comprising the network.
3. Draw the regions.

It is characteristic of the construction that the bisectors created in step 2 will either meet in 3s at a single point or else end at the border of the map. The construction is illustrated for a five-point problem in Figure 2-29. The bisectors meet at three differ-

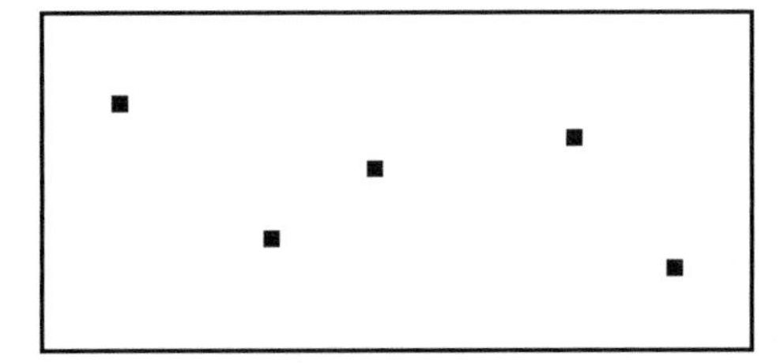

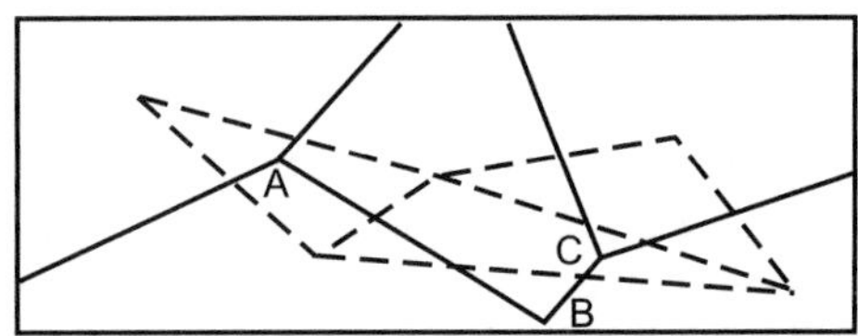

FIGURE 2-29. Conversion of point to areal data using Thiessen polygons.

ent intersections surrounding the central point, A, B, and C. The other ends of the bisectors terminate at the map boundary. Besides the original application in climatology, Thiessen polygons also find application in human geography. Suppose the points represent competing retail outlets in a city. Thiessen polygons define market areas that minimize distance traveled by consumers. Each goes to the closest retail outlet.

Other data conversions are also possible. For instance, areal data can be converted to point data by selecting a representative point in each area. The *center of gravity* or the *centroid,* the geometric center of a polygon, are obvious choices. The implicit assumption is that all the activity or phenomenona are concentrated at this point and are not spread throughout the area. Point data also can be converted to a set of isolines by using the process of *interpolation.* The process is illustrated in Figure 2-30. In this figure, the rainfall totals for four points, *A, B, C,* and *P,* are given as well as their locations on a map. The position of the isoline with a value of 50 between point *P* and the three points *A, B,* and *C* can be determined by simple geometry. Since point *A* has a value of 55, the difference between the rainfall at points *A* and *P* is 10. Therefore, on the line *AP* the isoline of 50 should lie midway between *A* and *P.* Between *B* and *P* the split is in the ratio 5:7, and between *C* and *P* the split is in the ratio 5:1. The isoline of 50 is then approximated as a smooth curve running through these points. This is a linear interpolation since it assumes the value of the variable of interest changes linearly between the observed points. Other *nonlinear* interpolations are possible.

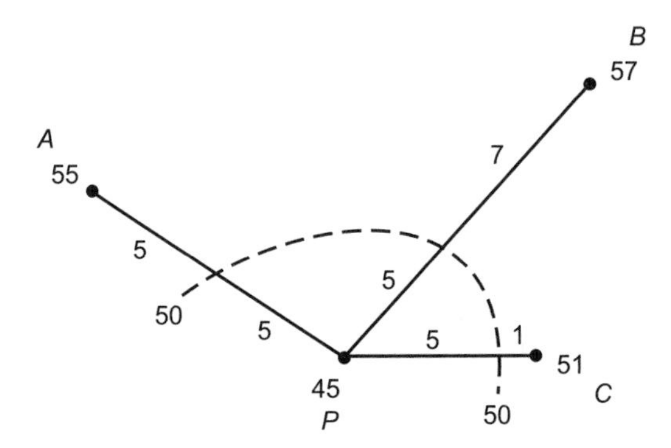

FIGURE 2-30. Linear interpolation of an isoline from point data.

2.5. Summary

The summary and graphical display of data represent a critical stage in all statistical analysis. Visualizing "what the data look like" provides useful clues that guide subsequent investigation. Indeed, the distribution of a variable often limits the range of statistical techniques that can be applied. The techniques discussed in this chapter illustrate many of the ways that univariate data may be represented graphically. Typically, there is no single best solution to data display. The different graphing methods have various strengths and weaknesses and, in different applications, each may provide greater insight than others. In most cases it is worth experimenting with these methods to find the one most suited to the task at hand. Time series and spatial data provide additional information that is used to organize and display the observed values of a variable of interest. The temporal and geographical attributes of such data pose significant challenges to the researcher. We confront some of these challenges in Chapter 4.

REFERENCES

R. A. Erickson and D. J. Hayward, "The International Flows of Industrial Exports from U.S. Regions," *Annals of the Association of American Geographers* 81 (1991), 381–390.

L. Guibas and J. Stolfi, "Primitives for the Manipulation of General Subdivisions and the Computation of Voronoi Diagrams," *ACM Transactions on Graphics,* 4 (1985), 74–123.

M. Monmoneier, *How to Lie with Maps* (Chicago: Chicago University Press, 1991).

D. F. Porinchu, G. M. MacDonald, A. M. Bloom, and K. A. Moser, "The Modern Distribution of Chironimid Sub-Fossils (Insecta: Diptera) in the Sierra Nevada, California: Potential for Paleoclimatic Reconstructions," *Journal of Paleolimnology* 28 (2002), 355–375.

D. W. Scott, "On Optimal and Data-Based Histograms," *Biometrika* 66 (1979), 605–610.

T. A. Slocum, R. B. McMaster, F. C. Kessler, and H. H. Howard, *Thematic Cartography and Geographic Visualization* (New York: Prentice Hall, 2003).

H. Sturges, "The Choice of a Class-Interval," *Journal of the American Statistical Association* 21 (1926), 65–66.

W. R. Tobler, "A Computer Movie Simulating Urban Growth in the Detroit Region," *Economic Geography* 46 (1970), 234–240.

E. R. Tufte, *The Visual Display of Quantitative Information* (Cheshire, CT: Graphics, 1983).

J. W. Tukey, *Exploratory Data Analysis* (Reading, MA: Addison-Wesley, 1977).

D. Unwin, *Introductory Spatial Analysis* (London: Methuen, 1981).

FURTHER READING

Tukey provides an excellent introduction to many of the techniques discussed in this chapter. The bible, in terms of graph design, remains Tufte. A good introductory time series text is that by Bowerman and O'Connell. Bailey and Gatrell and Fotheringham et al provide excellent introductions to spatial data and its analysis.

T. C. Bailey and A. C. Gatrell. *Interactive Spatial Data Analysis* (Harlow, UK: Longman, 1995).

B. L. Bowerman and R. T. O'Connel, *Times Series Forecasting, 2nd ed.* (Boston: Duxbury, 1987).

A. S. Fotheringham, C. Brunsdon, and M. Charlton, *Quantitative Geography.* (Thousand Oaks, CA: Sage, 2000).

DATASETS USED IN THIS CHAPTER

Most of the data sets used in this chapter can be found at the web site for the book. The data are stored in html format.

lakeveg.html—Lake vegetation data
dodata.html—Dissolved oxygen data
income.html—Median household income by state
teachrwage.html—Average wages of teachers by state
dow.html—Dow Jones Industrial Average stock market index
earthqk.html—California earthquake data
vote2000.html—Presidential election data for 2000 by state
nzdeaths.html—New Zealand heart disease deaths

PROBLEMS

1. Explain the meaning of the following terms:
   - Frequency table
   - Distribution of a variable
   - Unimodal, bimodal, and multimodal distributions
   - Outlier
   - Spatial data

2. Explain the difference between a frequency polygon and an ogive.

3. Outline the five main rules used in the construction of a frequency table.

4. What are the main components of a time series?

5. The following table below shows part of a dataset containing information about students:

Name	Major	Status	GPA
Flores, Vera	Physics	Junior	3.3
Jones, Sean	English	Freshman	2.9
Trinh, Nga	Geography	Sophomore	3.8
:	:	:	:

   a. What constitutes an observation or a record in this dataset?
   b. How many variables are identified?
   c. List the type (quantitative or qualitative) and the level of measurement of each variable.

6. A large city is divided into 60 police precincts. The number of burglaries in the last 12 months in each precinct is as follows:

   200 251 182 191 219 195 224 171 204 205 186 221 193 171 206 225 170 242 200 231
   196 188 219 180 224 208 205 184 236 182 207 209 193 225 209 194 219 176 236 234
   160 186 203 201 190 201 173 213 258 200 221 180 209 279 172 161 211 241 211 181

   a. Construct a frequency table of these data. Justify your choice of categories.
   b. Draw a frequency histogram and polygon.
   c. Describe the distribution.
   d. Does the distribution have any outliers? How did you identify them?

7. With the California earthquake data (*earthqk.html*) from the website for this book:
   a. Produce a stem-and-leaf plot of earthquake magnitudes.
   b. Describe the distribution of earthquake magnitudes.
   c. Using a sheet of graph paper, design a suitable grid and then plot the locations of earthquake epicenters with the latitude and longitude coordinates.

8. Take any 2 years of the deaths from heart disease data for New Zealand (*nzdeaths.html*):
   a. Draw separate frequency histograms for the number of deaths per week in each of the 2 years. Explain your choice of the number of intervals.
   b. How do the frequency distributions differ?

   (Is the comparison more straightforward if you make sure that you use the same intervals for the 2 years?)

9. Retrieve the state teachers' wage data (*teachrwage.html*) and construct the following graphical displays:
   a. Frequency histogram, stem-and-leaf plot, box plot, and dot plot.
   b. Assess the character of the wage distribution from each of these plots.
   c. Which of the plots seems to convey most information?

10. Draw a cumulative frequency ogive and a quantile plot with the wage data. Answer the following questions:
    a. Across U.S. states, 80% of teachers earn less then what wage?
    b. Across U.S. states, 30% of teachers earn more than what wage?
    c. Across U.S. states, 20% of teachers earn more than what wage?

    (*Hint:* think about your answer to question a.)

    d. Across U.S. states, what is the value of the wage such that 50% of teachers earn more than this value and 50% of teachers earn less than this value?

3

Describing Data with Statistics

In Chapter 2 we reviewed a number of tabular and graphical techniques used to summarize and display the distribution of a variable. While these methods can tell us a great deal about a variable and even allow us to make simple comparisons between two or more variables, they are never precise. To make more precise statements about the distribution of a variable, quantitative measures or descriptive statistics are used. Descriptive statistics provide a compact way of summarizing the characteristics of a variable, and they enable exact comparisons between pairs of variables to be made. In addition, descriptive statistics are an essential component of statistical inference with which much of the latter half of this book is concerned.

In this chapter, our review of descriptive statistics begins in Section 3.1 where we examine how to calculate different measures of central tendency or location. We infuse that discussion with an examination of the properties of the different measures and an assessment of when they should be used. In Section 3.2 we turn our attention to a series of measures of the spread or the variability of a distribution. Again we discuss the characteristics of the different measures, and how and when to use them. Section 3.3 examines other numerical measures of the characteristics of frequency distributions. Descriptive statistics that may be used to summarize and simplify time series data are discussed in Section 3.4. Descriptive statistics for spatial data are reviewed in Section 3.5. Section 3.6 summarizes the main arguments of the chapter.

3.1. Measures of Central Tendency

One important characteristic of a distribution is the location of the typical value, or center, of the variable. In fact there are several different measures of *central tendency,* each having certain advantages and disadvantages as well as specific mathematical properties. Each measure usually identifies a single number that can be considered the "middle" of a set of observations. The fact that we can compute several different measures of central tendency implies we have to fully understand each one in order to know when each is an appropriate summary statistic. Although each measure is in some sense the "average" or center of a set of numbers, none is considered to be *universally* superior to all others. As we shall see, however, there is one that is more commonly used.

Midrange

The *range* of a variable is simply the difference between the highest and lowest valued observations. Let the lowest valued observation be denoted X_{min} and the highest valued observation X_{max}. The *midrange* lies exactly halfway between these two observations. Although it is easy to calculate, it has the disadvantage of being inordinately affected by the two extreme observations. Neither of these can be considered typical, so there is no reason to think that any calculation based on them would be either! For this reason, this measure is seldom used. For the dissolved oxygen (DO) data introduced in Chapter 2, the maximum is 6.9, and the minimum is 4.2; thus, the range is 2.7, and the midrange is $(6.9 + 4.2)/2 = 5.55$.

DEFINITION: MIDRANGE

The midrange is the arithmetic mean of the two extreme observations:

$$\text{Midrange} = \frac{X_{max} + X_{min}}{2} \tag{3-1}$$

Mode

Strictly speaking, the mode of a distribution is the value of the variable that appears most frequently. For many continuous-valued, interval-scale variables, there may be no exact repeated values in the data. Therefore, we could say there is no mode, or we might even say that there are *n* different modes. This measure of central tendency is not very useful in this case. If the data are grouped, the *modal category* or *modal class* is defined as the interval with the highest frequency; the midpoint of the modal class is termed the *crude mode*. Unfortunately, this approach does not define a *unique* mode or interval because the value depends on the number and location of the intervals selected in the grouping procedure. Loosely speaking, the value associated with a peak of a histogram or frequency polygon is termed a *mode*. In this sense, there may be one or more modes of a distribution.

DEFINITION: MODE

The mode is the value of a variable that occurs most frequently within a sample or population of data values.

When is the mode useful? Under certain conditions, this measure is useful for nominal and ordinal variables and for discrete-valued interval variables. For nominal variables, the mode is the category with the greatest frequency. Using our example of lake vegetation type from Chapter 2, we can say that the modal vegetation category

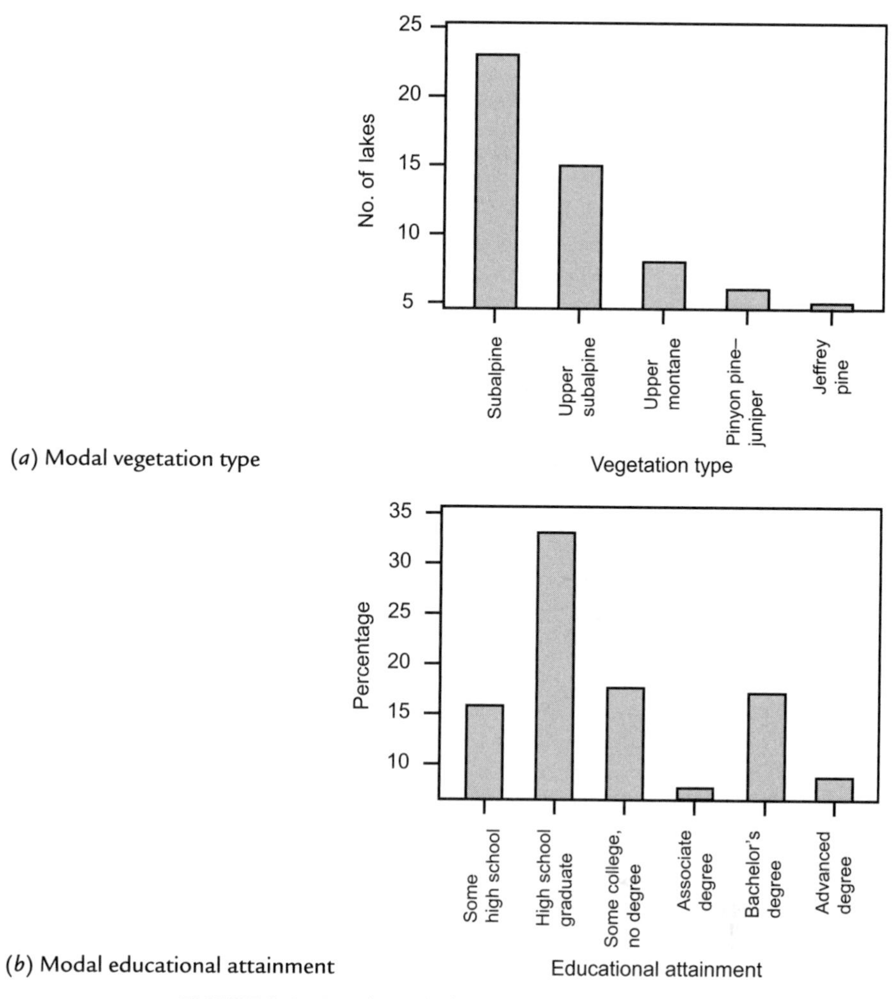

(*a*) Modal vegetation type

(*b*) Modal educational attainment

FIGURE 3-1. Bar charts indicating modal categories.

is subalpine. Subalpine vegetation is found around 23 of 57 lakes sampled in the Sierra Nevada, more than any other type of vegetation (see Figure 3-1a). Note that the mode is not really a measure of central tendency in this case since a nominal variable is not ordered and hence can have no center. For ordinal variables and discrete-valued quantitative variables, which can be ordered, a mode does provide a useful measure of central tendency when one value or category has a frequency greater than others. Figure 3-1b shows that "high school graduate" was the modal level of educational attainment for U.S. adults in 1999. Table 3-1 shows the distribution of a discrete-valued variable, household size, in the United States. The mode is 2 persons per household. This seems preferable to saying that there are 2.512 persons in the typical U.S. household, which is the arithmetic average.

TABLE 3-1
Distribution of Household Size in the U.S., 2001

Number of household members	Frequency
1	28,505,040
2	35,397,324
3	17,303,550
4	14,803,838
5	6,912,785
6	2,145,945
≥7	1,360,375
Total	106,428,857

Notes: Based on 12 monthly estimates during 2001.
Source: U.S. Census Bureau: *www.census.gov.*

Median

The median of a set of observations is the value of the variable that divides the observations so that one-half are less than or equal to the median and one-half are greater than or equal to it. To find the median value of a distribution containing n observations:

1. Arrange the observations in ascending order or rank.
2. Calculate the median rank = $(n + 1)/2$.
3. Find the value of the median that is given by the observation with rank $(n + 1)/2$.

If the distribution contains an odd number of observations, the median rank will be an integer and the value of the corresponding observation can be read off easily. However, if the distribution contains an even number of observations, the median rank will be a real-valued number. For example, suppose a dataset contained $n = 100$ observations. The median rank of this dataset would be $(100 + 1)/2 = 50.5$. The value of the median would then be found as the arithmetic mean (see below) of the observations ranked 50th and 51st. Note that the value of the median is not $(n + 1)/2$; this value only locates the rank or the position of the median in the ordered list of observations.

DEFINITION: MEDIAN
The median of a distribution is the value that is larger than half of the observations in the distribution and smaller than the other half.

To find the median level of dissolved oxygen in our sample of 50 lakes, we first arrange the DO values in rank order. This is done in Table 3-2. The median rank is $(50 + 1)/2 = 25.5$, and so the median is found as the arithmetic mean of the observations ranked 25th and 26th. Therefore, the median DO value is $(5.6 + 5.6)/2 = 5.6$.

TABLE 3-2
DO Values Listed in Ascending Order Together with Their Rank

Rank	DO value	
1	4.2	
2	4.3	
3	4.4	
4	4.5	
5	4.7	
6	4.7	
7	4.8	
8	4.8	
9	4.9	
10	5.1	
11	5.1	
12	5.1	
13	5.2	→ 1st (lower) quartile = 25th percentile = P_{25}
14	5.3	
15	5.3	
16	5.4	
17	5.4	
18	5.4	
19	5.4	
20	5.5	
21	5.5	
22	5.6	
23	5.6	
24	5.6	
25	5.6	
} 25.5	(5.6 + 5.6)/2 = 5.6	→ Median = 2nd quartile=5th decile = 50th percentile = P_{50}
26	5.6	
27	5.6	
28	5.7	
29	5.7	
30	5.7	
31	5.7	
32	5.8	
33	5.8	
34	5.8	
35	5.8	
36	5.9	
37	5.9	
38	5.9	→ 3rd (upper) quartile = 75th percentile = P_{75}
39	5.9	
40	6.0	
41	6.1	
42	6.2	
43	6.3	
44	6.4	
45	6.4	
46	6.4	
47	6.6	
48	6.7	
49	6.8	
50	6.9	

The median also is called a positional measure because it locates the position of an observation relative to the set of other observations. Just as it is possible to define the median, or middle term, it is also possible to define *quartiles* as the values that divide the set of observations into quarters. The *first quartile* divides the lower half of a distribution into two equal sets. The *median* is the *second quartile,* and the *third quartile* divides the upper half of a distribution into two equal sets. Thus, 25% of the observations are below the first quartile, 50% are below the median, and 75% are below the third quartile. The values of the quartiles may be found in similar fashion to the median. First, we find the ranks of the quartiles and then the corresponding observations. Formulae for the ranks of the quartiles are:

1. First quartile rank = $(n + 2)/4$
2. Second quartile (median) rank = $(2n + 2)/4$
3. Third quartile rank = $(3n + 2)/4$

Using the DO data example, once again, we find that the values of the quartiles are

Quartile	Rank	Value
First	$(n + 2)/4 = 13$	5.2
Second	$(2n + 2)/4 = 25.5$	5.6
Third	$(3n + 2)/4 = 38$	5.9

We can generalize this procedure to define *deciles,* which divide the observations into tenths, and *percentiles,* which divide the data into hundredths. The median is the second quartile, the fifth decile, and the 50th percentile. The nth percentile is usually specified as P_n, and the three quartiles as Q_1, Q_2, and Q_3. Therefore, $P_{25} = Q_1 = 5.2$, $P_{50} = Q_2 = 5.6$, and $P_{75} = Q_3 = 5.9$. Positional terms are commonly associated with educational testing procedures and are used to locate a student among a large set of student scores on an examination. Percentiles have been infrequently applied in geographical research.

A general formula to find the nth percentile, P_n, is to find the rank

$$k = \frac{P_n(n + 1)}{100}$$

and then obtain the corresponding value. Thus, P_{80}, or the 80th percentile of the DO data may be found by first obtaining the rank of the observation denoting the 80th percentile

$$k = \frac{80(n + 1)}{100} = 40$$

and then finding the value of the 40th ranked observation in the DO data. From Table 3-2, this value is 6.0.

TABLE 3-3
Cavendish's Density of the Earth Measurements

4.88	5.26	5.29	5.34	5.42	5.47	5.55	5.61	5.65	5.79
5.07	5.27	5.30	5.36	5.44	5.50	5.57	5.62	5.68	5.85
5.10	5.29	5.34	5.39	5.46	5.53	5.58	5.63	5.75	

Notes: Density is measured here as a multiple of the density of water.
Source: Stigler, S.M. (1977).

EXAMPLE 3-1. Density of the Earth. In 1798 the scientist Henry Cavendish measured the density of the earth 29 times using a torsion balance. The results of his measurements are presented in Table 3-3. With $n = 29$ observations, the median rank is $(n + 1)/2 = 15$. The median estimate of the earth's density, the value of the 15th observation, is thus 5.46.

Mean

The mean of a set of observations is the most commonly used measure of central tendency. In most instances the term *mean* is the shortened version of *arithmetic mean,* and it is the sum of the values of all observations, divided by the number of observations. In common use this is simply the average of a set of observations. The mean DO value of the 50 lakes is $(4.2 + 4.3 + \ldots + 6.8 + 6.9)/50 = 279/50 = 5.58$. The three dots, or ellipses, are shorthand for noting that the summation omits some terms in the set. The middle 46 observations of DO are omitted. To write them out is unnecessary since the meaning of the summation is clear.

To simplify many of the formulas in descriptive and inferential statistics, it is convenient to utilize a system of notation. If a variable is denoted by the uppercase X, then we can use the subscript i to refer to individual observations of X. Thus when $i = 1$, refers to the first observation of X. When $i = 2$, $X_{i=2}$ or X_2 refers to the second observation of X. When $i = n$, $X_{i=n}$ or X_n refers to the last or the nth observation of X. Using this notation, we can express the formula for the mean of a sample as $(X_1 + X_2 + \ldots + X_n)/n$. An even more efficient way of writing this expression makes use of the *summation operator* or what is sometimes called *sigma notation.* The uppercase Greek symbol sigma, or Σ, is a mathematical shorthand that has considerable use in statistics. It is known as the summation operator since it means that the items following the symbol are to be summed. Thus, ΣX_i means sum all observations of the variable X, where the different observations are each represented by the subscript i.

Sometimes, we need to be a little more formal. So you might also encounter the term $\Sigma_{i=1}^{n} X_i$, which is read "sigma X_i for i equals 1 to n". This means sum values of the variable X through all successive values of the *index of summation i.* Below the sigma, the index of summation specifies the start of the summation, in this case $i = 1$. This is the usual case, but it is possible to start the summation at any point within a series of observations, say $i = 3$, $i = 21$, or a general point such as $i = k$. Above the sigma, the index defines the last observation to be included in the summation, here $i = n$.

Using this notation, we can write

$$\sum_{i=1}^{n} X_i = X_1 + X_2 + \ldots + X_n$$

The sigma operator greatly simplifies the specification of many different operations in statistics. Students unfamiliar with summation notation should consult the Appendix to this chapter, which contains a brief review of summation notation. Included in this review are several simple rules for summations that are used extensively to simplify many of the formulas encountered in statistical analysis.

DEFINITION: SAMPLE MEAN $\bar{X}$

The mean of a set of observations X_i for $i = 1, 2, \ldots, n$ is defined as

$$\bar{X} = \frac{\sum_{i=1}^{n} X_i}{n} \tag{3-2}$$

The symbol $\bar{X}$, read "*X* bar," denotes the sample mean of a variable *X*. At times, we find it necessary to distinguish the mean of a *sample* from the mean of a *population.* For a sample, the observations are numbered consecutively from 1 to *n,* and the mean is denoted by $\bar{X}$. For a finite population consisting of *N* elements, the notation and the formula for the mean are modified.

DEFINITION: POPULATION MEAN μ

The mean of a population with *N* members is defined as

$$\mu = \frac{\sum_{i=1}^{n} X_i}{N} \tag{3-3}$$

The symbol μ is the lowercase Greek letter mu. The formulas for the sample mean and the population mean are *not* interchangeable. There are many different sample means. Depending on which *n* observations are selected from a population or universe with *N* members, we get a different value for $\bar{X}$. However, if we compute the mean from all *N* observations in the population, there is only one possible value for the mean, μ. Generally, a sample size *n* is only a small portion of the members of the population, so that $n < N$.

Two important conventions have been introduced in these formulas that are followed throughout this text. First, lowercase *n* always refers to a sample size, and upper case *N* always refers to the size of a finite population. Second, characteristics of populations are defined by Greek letters such as μ, but samples are not. These distinctions become particularly important when we try to make inferences about a population from a sample. For example, we might wish to use $\bar{X}$ to estimate a value for μ. When $n = N$ and all members of the population are included in the sample, $\bar{X} = \mu$.

TABLE 3-4
University Professor Salaries, 2001–2002 ($1000)

118.8	109.0	108.8	106.2	105.7	100.9	98.2	98.1
96.7	94.5	93.4	92.9	92.8	90.5	88.7	87.4

Notes: Each of the salary figures represents the average salary of full professors at a particular university, in thousands of U.S. dollars.
Source: American Association of University Professors (2002).

We might expect that, on average, the value of the sample mean will get progressively closer to the population mean as the sample size increases. It turns out that this is exactly the case. We return to this issue of the exact relationship between the sample and population means in Chapter 6.

EXAMPLE 3-2. Professor Salaries. The American Association of University Professors publishes annual data on faculty salaries. A sample of salary data for full professors drawn from, $n = 16$, leading U.S. universities is shown in Table 3-4. The mean of this sample is

$$\bar{X} = \frac{\sum_{i=1}^{n} X_i}{n} = \frac{X_1 + X_2 + X_3 + \ldots + X_{16}}{n}$$

$$= \frac{118.8 + 109.0 + 108.8 + \ldots + 87.4}{16}$$

$$= \frac{1582.6}{16} = 98.913$$

In other words, the mean annual salary of full professors for the sample of 16 universities is $98, 913. Note that this is the average of the 16 university values in the sample, not the average salary of individual full professors at those universities. Only if all universities were of the same size could we be sure the two are the same.

Choosing an Appropriate Measure of Central Tendency

For the dissolved oxygen (DO) data listed in Table 3-2, a number of different measures of central tendency can be obtained. These are shown in Figure 3-2 in a series of dot plots of the DO data. All of these measures have very similar values, but this is not necessarily a usual outcome. Let us review the derivation of each of these measures.

The mode is associated with a peak of the distribution; it is the value that occurs most frequently within a set of data, as Figure 3-2a shows. The midrange, Figure 3-2b, divides the length of the distribution into two equal parts. The median is the value that splits the observations of a distribution into two equal-sized groups. In Figure 3-2c, there are 22 observations with values less then the median and 22

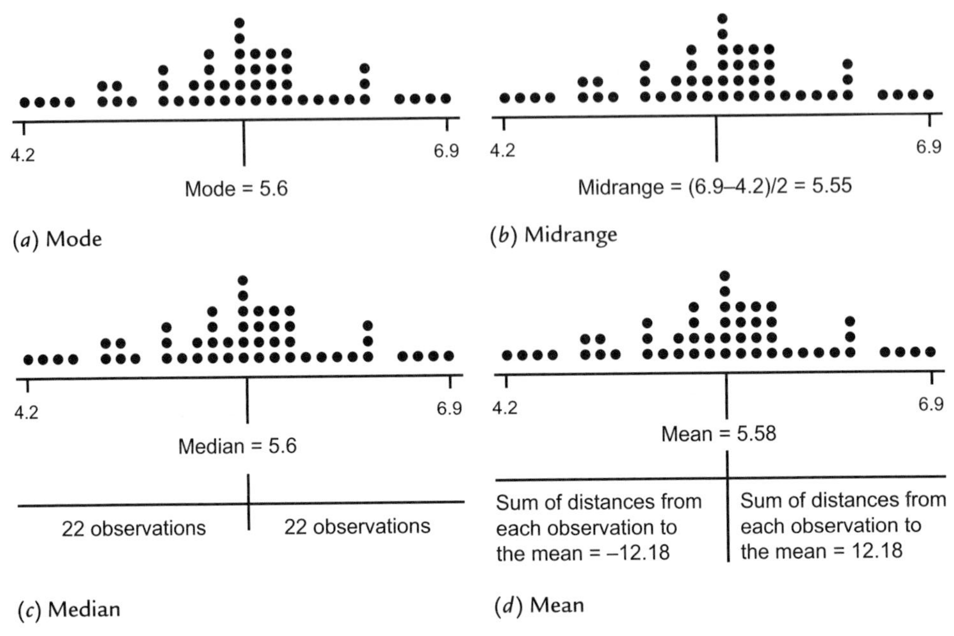

FIGURE 3-2. Measures of central tendency using the DO data.

observations with values greater than the median. Note that the DO data sample comprises 50 observations: the remaining 6 observations all have values identical to the median. You can consider the median as a balance point, such that the number of observations on each side of the median is the same.

What about the mean? The mean can also be considered a balance point. Think of the observations in the dot plot as a set of weights placed along a line that is centered on the mean. The weight of each observation depends on its distance from the mean. Observations near the endpoints of the distribution exert the most weight, and those closest to the mean the least. The sum of the weights to the left and right side of the mean must be equal for the line to be in balance. In Figure 3-2d, for example, the observations to the left of the mean have a collective weight of –12.18. In other words, if you sum the distances between each of these observations and the mean, the total equals –12.18. If you sum the distance to the mean for all observations with values greater than the mean, the total equals 12.18. Note that in the case of the median as a balance, all observations have the same weight regardless of their location along the line.

When should the different measures of central tendency be used? With nominal data, the mode is the only measure of central tendency available. For ordinal data, the median can be calculated as well as the mode. Typically, the median would be the preferred measure, as the mode can be an unreliable measure of the average. For quantitative data, all four measures of central tendency are available. Because the midrange is based only on two extreme observations it is rarely used. This leaves the median and the mean as the two most widely used measures of central tendency.

The mean and the median both have advantages as a measure of average, depending on the nature of a distribution. The principal advantage of the mean is that it is sensitive to a change in any of the observations. Consider once more the DO values in Figure 3-2 and imagine that the highest DO value of 6.9 was changed to 10.0. In this case, the median would remain 5.6, because there would still be 22 observations with values less than the median, and 22 observations with values greater than the median. However, the mean would change from 5.58 to 5.642, reflecting the change in weight of one of the observations in the DO distribution. In most instances, any change in the observed values that comprise a distribution is reflected in the mean, though not necessarily in the other measures of central tendency.

However, there are several instances for which the mean is not a good measure of central tendency:

1. For *bimodal distributions.* Consider Figure 3-3a. Note that the mean and median are coincident, but neither picks up the "typical value" of this variable. Neither

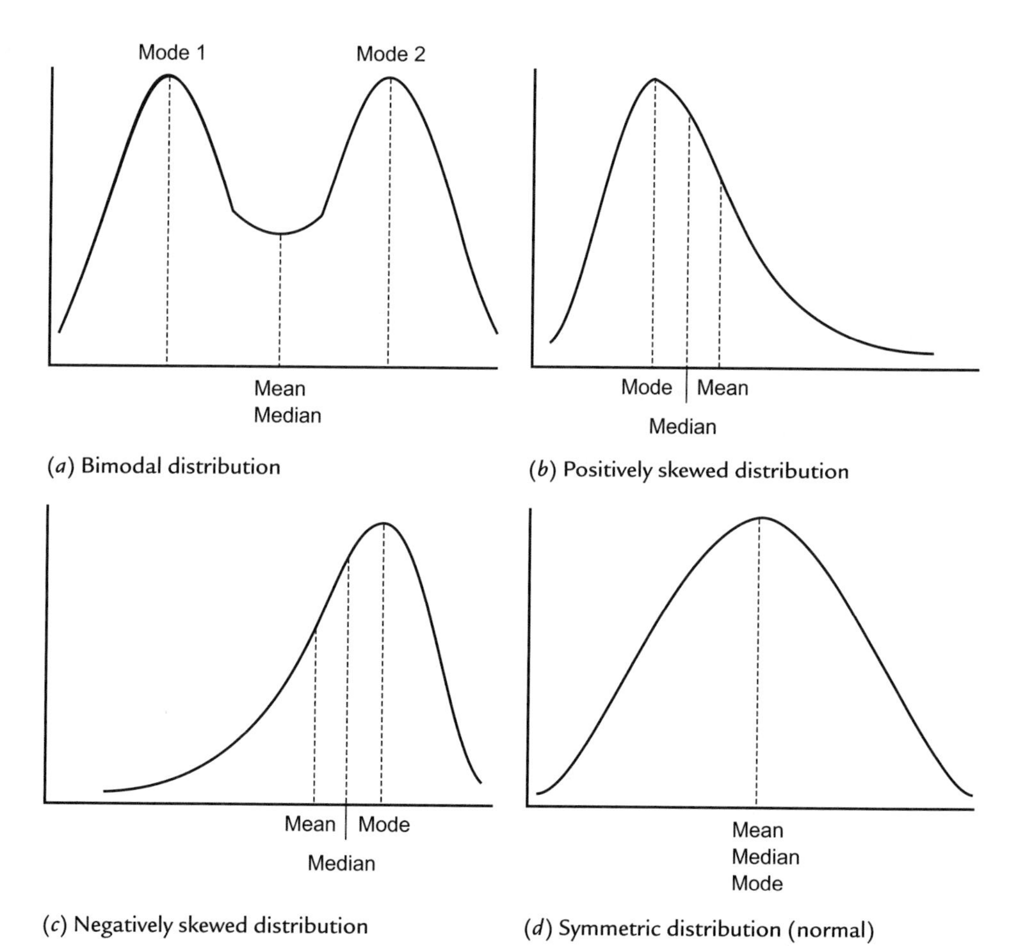

(*a*) Bimodal distribution

(*b*) Positively skewed distribution

(*c*) Negatively skewed distribution

(*d*) Symmetric distribution (normal)

FIGURE 3-3. Measures of central tendency in different distributions.

would the midrange, which is at the same location. Because all these measures provide one value, neither can be useful in this case. The best summary statistic in this case is the *mode,* since two "typical" values are clearly necessary to characterize the distribution.

2. For *skewed distributions.* The mean is not very useful for variables that are either positively or negatively skewed. Figure 3-3b and Figure 3-3c illustrate the relative location of the mean, median, and mode for skewed distributions. For a positively skewed distribution, the mean is greater than the median, which is greater than the mode. For a negatively skewed distribution the reverse is true: the mean is less than the median, which is less than the mode. Although the median is pulled toward the tail of the distribution, it is not nearly as sensitive as the mean. Note that for symmetric distributions the three measures are equal (see Figure 3-3d). As a distribution deviates from this shape, the three become increasingly different. The median is usually the best measure of central tendency for skewed distributions.

3. For *a distribution with an extreme value (or a small number of extreme values).* We can think of a distribution with an extreme value as highly skewed. The median is also the best measure in this case. Consider a simple set of five observations: 2, 2, 3, 5, and 8. The mean of these data is 4, the median is 3, the mode is 2, and the midrange is 3. If the value of 8 is changed to 200, then the mode remains at 2, the median stays at 3, the mean increases to 44, and the midrange climbs to 96! The mean is clearly no longer even close to any of the observations and cannot be described as a typical value. However, the median remains central to four of the five observations. House prices and household incomes are two examples of highly skewed (positive) distributions. For these variables, the median is the preferred measure of central tendency. The median is sometimes said to be a *robust* measure of central tendency because it is insensitive to extreme values and to changes in the value of individual observations.

In sum, the mean is the most commonly used measure of central tendency. It is sensitive to all changes in the distribution and makes use of all the information in a set of observations. Unless one of the conditions above applies, it is the best measure of central tendency for a distribution.

Properties of the Mean and Median

There are three important properties of the mean and median:

PROPERTY 1

The sum of the deviations of each observation from the mean is zero.

$$\sum_{i=1}^{n}(X_i - \bar{X}) = 0 \qquad (3\text{-}4)$$

This is another sense in which the mean is the middle of a set of observations, as Figure 3-2d shows. The property is easily proved using the simple rules for summation reviewed in Appendix 3b. Using Rule 1, we can express Equation 3-4 as

$$\sum_{i=1}^{n} X_i - \sum_{i=1}^{n} \bar{X} = 0$$

And since $\bar{X}$ is a constant (for any given set of observations), we can write $\Sigma_{i=1}^{n} \bar{X} = n\bar{X}$ from Rule 4. Finally, we note from our definition of the sample mean in Equation 3-2 that

$$n\bar{X} = n\left(\frac{\sum_{i=1}^{n} X_i}{n}\right) = \sum_{i=1}^{n} X_i \tag{3-5}$$

and Equation 3-5 can be rewritten as

$$\sum_{i=1}^{n} X_i - \sum_{i=1}^{n} X_i = 0$$

This property is easily illustrated using any set of observations. To keep matters simple, consider once again the five numbers 2, 2, 3, 5, and 8 for which $\bar{X} = 4$. Summing the deviation yields $(2 - 4) + (2 - 4) + (3 - 4) + (5 - 4) + (8 - 4) = -2 + (-2) + (-1) + 1 + 4 = 0$.

PROPERTY 2

The sum of squared deviations of each observation from the mean is a minimum, that is, less than the sum of squared deviations from any other number. Formally, we say

$$\sum_{i=1}^{n} (X_i - M)^2$$

is minimized when $M = \bar{X}$. This is often called the least squares property of the mean.

The quantity

$$\sum_{i=1}^{n} (X_i - \bar{X})^2$$

is the total variation of variable X. Although it is slightly more difficult to prove, this property is easy to illustrate. Using the same set of five observations, we calculate the sum of squared deviations as follows:

Observation	Mean = 4	Median = 3
2	$(2 - 4)^2 = 4$	$(2 - 3)^2 = 1$
2	$(2 - 4)^2 = 4$	$(2 - 3)^2 = 1$
3	$(3 - 4)^2 = 1$	$(3 - 3)^2 = 0$
5	$(5 - 4)^2 = 1$	$(5 - 3)^2 = 4$
8	$(8 - 4)^2 = 16$	$(8 - 3)^2 = 25$
	$\sum = 26$	$\sum = 31$

We see that the sum of squared deviations is 26 for the mean and 31 for the median. Replacing the mean with any other value would lead to a sum greater than 26. This property suggests yet another sense in which the mean is the center of a distribution.

A closer examination of Equation 3-6 reveals why the mean is a poor measure of central tendency when the set of observations is highly skewed or contains an extreme value. To achieve the minimization property 2, a value of M must be selected that minimizes the squared deviations. A deviation of $X - M = 1$ is counted as $1^2 = 1$, but a deviation of $X - M = 5$ is counted as $5^2 = 25$ and is given 25 times as much weight. Squaring the deviations penalizes larger deviations more than smaller ones. As a result, the mean is "drawn" toward extreme values, or values in the tail of the distribution. In this way, the high penalties associated with large deviations are minimized. Figure 3-3 illustrates this result graphically. The location of the mean is the most extreme of the measures of central tendency in a highly positively or negatively skewed distribution.

PROPERTY 3

The sum of absolute deviations is minimized by the median. That is, the expression

$$\sum_{i=1}^{n} |X_i - M| \tag{3-6}$$

is minimized when M is equal to the median of the X-values.

This can also be illustrated by using the small sample of $n = 5$ observations 2, 2, 3, 5, and 8. For comparative purposes let us once again compute the sum of absolute deviations for two values: M equal to the mean 4 and M equal to the median 3.

Observation	Mean = 4	Median = 3
2	$|2 - 4| = 2$	$|2 - 3| = 1$
2	$|2 - 4| = 2$	$|2 - 3| = 1$
3	$|3 - 4| = 1$	$|3 - 3| = 0$
5	$|5 - 4| = 1$	$|5 - 3| = 2$
8	$|8 - 4| = 4$	$|8 - 3| = 5$
	$\sum = 10$	$\sum = 9$

The sum of absolute deviations from the median is 9, and from the mean it is 10. Because it has been drawn toward the extreme value of 8 in order to minimize the squared deviations, the mean is in an inferior position to the median for absolute deviations. Although the median is 5 units away from this extreme observation, it is penalized far less by using this criterion than by using the sum of squared deviations.

Besides explaining why the mean and the medium are superior measures of central tendency in particular cases, these properties have interesting interpretations when we consider spatial data in Section 3.4 of this chapter. In particular, they have important implications for locating the center of a two-dimensional point pattern.

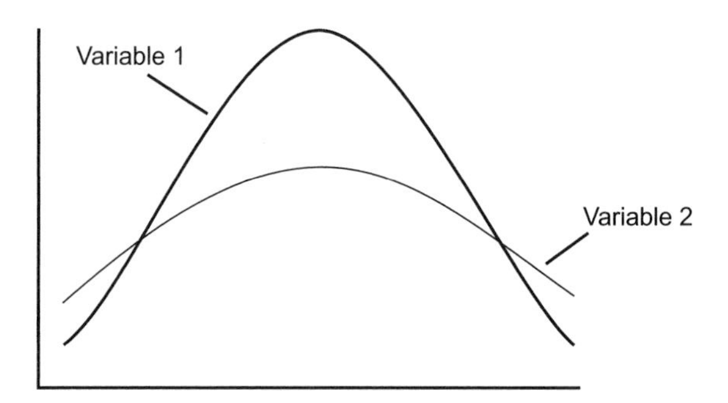

FIGURE 3-4. Variables with identical means but different dispersions.

3.2. Measures of Dispersion

The measures of central tendency discussed in Section 3.1 are very useful for calculating a summary statistic that represents an average, or typical, value of variable X. However, these statistics measure only one characteristic of a frequency distribution. Compare the two distributions shown in Figure 3-4. Both variables have identical means but very different *dispersions,* or *spread,* about the mean. Variable 1 is much more compact than variable 2. When describing distributions with numbers, it is common to provide a measure of dispersion as well as central tendency. Just as for the concept of central tendency, several different measures can be used to evaluate the spread or dispersion of a distribution.

Range

In defining the *midrange* of a variable as a measure of central tendency, it was necessary to define the first measure of dispersion of a distribution, the *range.*

DEFINITION: RANGE

The range of a set of observations is the difference between the highest and lowest valued observations:

$$\text{Range} = X_{\text{max}} - X_{\text{min}} \qquad (3\text{-}7)$$

For the dissolved oxygen variable, for example, we find the range to be 6.9 – 4.2 = 2.7. The minimum DO value gives a fairly good estimate of the lower bound on current lake oxygen levels. Over time, analysis of the lower bounds, upper bounds, and range would provide us with a useful snapshot of variation in lake DO levels. There is one important feature about the range that must not be forgotten when it is used as a measure of dispersion: it only takes into account the two most extreme observations in the dataset. When one, or both, of these extreme values is unusual and removed from other observations, then the range may be misleading. We should also keep in mind that larger samples almost invariably have greater ranges than smaller samples, since they are more likely to contain the rare or unusual members of a population.

Interquartile Range

Whereas the range only depends on the extremes of a distribution, the interquartile range uses the middle half of a distribution to measure dispersion. Consequently, the interquartile range is generally a more stable measure of dispersion. Different samples from the same population may have widely different ranges but are likely to have fairly similar interquartile ranges. The interquartile range of the DO data, presented in Table 3-2 is 5.9 – 5.2 = 0.7.

DEFINITION: INTERQUARTILE RANGE

The interquartile range is the difference between the first and third quartiles of a distribution:

$$\text{Interquartile range} = P_{75} - P_{25} \tag{3-8}$$

where P_{75} is the value of the 75th percentile or the third quartile, and P_{25} is the value of the 25th percentile or the first quartile.

Unfortunately, neither the range nor the interquartile range takes full advantage of the entire set of observations and their values. Three other measures of dispersion are available, each of which uses the values of all observations in a distribution. These three measures are related in that they are all based on the deviations or the distances of each observation from the mean $\bar{X}$. The mean is used as the center of the distribution because it is the single most satisfactory measure of central tendency.

Mean Deviation

As formalized in Property 1, the sum of the deviations about the mean is zero. This is true because the positive deviations offset the negative deviations as Figure 3.2d illustrates. This causes a problem in calculations of dispersion that are based on deviations from the mean. No matter how dispersed a variable, the sum of its deviations about its mean will always be zero. To get around this problem the mean deviation focuses on the absolute value of the deviations.

The calculation of the mean deviation for the dissolved oxygen data is illustrated in Table 3-5. The second column contains the signed deviations and sums to 0.0. The absolute deviations sum to 24.36, leading to a mean (or average) absolute deviation of 24.36/50 = 0.4872. There is one disadvantage to the mean deviation. Many statistical results depend on easy algebraic manipulation of formulas such as Equation 3-9. Note how we were able to prove that the deviations summed to zero by manipulating Equation 3-4. The algebraic simplification of absolute value terms is

TABLE 3-5
Work Table for the Mean Deviation, Standard Deviation, and Variance

Observation	X_i	$X_i - \bar{X}$	$\lvert X_i - \bar{X} \rvert$	$(X_i - \bar{X})^2$	X_i^2
1	5.1	–0.48	0.48	0.2304	26.01
2	5.6	0.02	0.02	0.0004	31.36
3	5.3	–0.28	0.28	0.0784	28.09
4	5.7	0.12	0.12	0.0144	32.49
5	5.8	0.22	0.22	0.0484	33.64
6	6.4	0.82	0.82	0.6724	40.96
7	4.3	–1.28	1.28	1.6384	18.49
8	5.9	0.32	0.32	0.1024	34.81
9	5.4	–0.18	0.18	0.0324	29.16
10	4.7	–0.88	0.88	0.7744	22.09
11	5.6	0.02	0.02	0.0004	31.36
12	6.8	1.22	1.22	1.4844	46.24
13	6.9	1.32	1.32	1.7424	47.61
14	4.8	–0.78	0.78	0.6084	23.04
15	5.6	0.02	0.02	0.0004	31.36
16	6.4	0.82	0.82	0.6724	40.96
17	5.9	0.32	0.32	0.1024	34.81
18	6.0	0.42	0.42	0.1764	36.00
19	5.5	–0.08	0.08	0.0064	30.25
20	5.4	–0.18	0.18	0.0324	29.16
21	4.4	–1.18	1.18	1.3924	19.36
22	5.1	–0.48	0.48	0.2304	26.01
23	5.6	0.02	0.02	0.0004	31.36
24	5.8	0.22	0.22	0.0484	33.64
25	5.7	0.12	0.12	0.0144	32.49
26	4.9	–0.68	0.68	0.4624	24.01
27	6.6	1.02	1.02	1.0404	43.56
28	5.7	0.12	0.12	0.0144	32.49
29	5.4	–0.18	0.18	0.0324	29.16
30	5.9	0.32	0.32	0.1024	34.81
31	5.6	0.02	0.02	0.0004	31.36
32	6.7	1.12	1.12	1.2544	44.89
33	5.4	–0.18	0.18	0.0324	29.16
34	4.8	–0.78	0.78	0.6084	23.04
35	6.4	0.82	0.82	0.6724	40.96
36	5.8	0.22	0.22	0.0484	33.64
37	5.3	–0.28	0.28	0.0784	28.09
38	5.7	0.12	0.12	0.0144	32.49
39	6.3	0.72	0.72	0.5184	36.69
40	4.5	–1.08	1.08	1.1664	20.25
41	5.6	0.02	0.02	0.0004	31.36
42	6.2	0.62	0.62	0.3844	38.44
43	4.2	–1.38	1.38	1.9004	17.64
44	5.2	–0.38	0.38	0.1444	27.04
45	5.8	0.22	0.22	0.0484	33.64
46	6.1	0.52	0.52	0.2704	37.21
47	5.1	–0.48	0.48	0.2304	26.01
48	5.9	0.32	0.32	0.1024	34.81
49	5.5	–0.08	0.08	0.0064	30.25
50	4.7	–0.88	0.88	0.7744	22.09

(continues)

TABLE 3-5 (continued)

$n = 50$

$n - 1 = 49 \quad \sum_{i=1}^{n} X_i = 279 \quad \sum_{i=1}^{n}(X_i - \bar{X}) = 0 \quad \sum_{i=1}^{n}|X_i - \bar{X}| = 24.36 \quad \sum_{i=1}^{n}(X_i - \bar{X})^2 = 20.02 \quad \sum_{i=1}^{n} X_i^2 = 1576.84$

$$\text{Mean} = \frac{\sum_{i=1}^{n} X_i}{n} = \frac{279}{50} = 5.58$$

$$\text{Mean deviation} = \frac{\sum_{i=1}^{n}|X_i - \bar{X}|}{n} = \frac{24.36}{50} = 0.4872$$

$$\text{Variance} = s^2 = \frac{\sum_{i=1}^{n}(X_i - \bar{X})^2}{n-1} = \frac{20.02}{49} = 0.4086$$

$$\text{Standard deviation} = \sqrt{\frac{\sum_{i=1}^{n}(X_i - \bar{X})^2}{n-1}} = \sqrt{\frac{20.02}{49}} = 0.6392$$

Or from computational formulas:

$$s^2 = \frac{1}{n-1}\left(\sum_{i=1}^{n} X_i^2 - n\bar{X}^2\right) = \frac{1}{49}(1576.84 - 50(5.58)^2) = 0.4086$$

$$s = \sqrt{\frac{1}{n-1}\left(\sum_{i=1}^{n} X_i^2 - n\bar{X}^2\right)} = \sqrt{\frac{1}{49}(1576.84 - 50(5.58)^2)} = 0.6392$$

quite cumbersome and generally does not lead to simple results. For this and other reasons, an alternative measure of dispersion, the standard deviation, is more commonly used.

DEFINITION: MEAN DEVIATION

The mean deviation of a variable X is defined as

$$\text{Mean deviation} = \frac{\sum_{i=1}^{n}|X_i - \bar{X}|}{n} \qquad (3\text{-}9)$$

Standard Deviation

DEFINITION: SAMPLE STANDARD DEVIATION

The standard deviation of a sample of observations is the square root of the average of the squared deviations about their mean:

$$s = \sqrt{\frac{\sum_{i=1}^{n}(X_i - \bar{X})^2}{n-1}} \qquad (3\text{-}10)$$

DEFINITION: POPULATION STANDARD DEVIATION

The standard deviation of a population is the square root of the average of the squared deviations about the population mean μ:

$$\sigma = \sqrt{\frac{\sum_{i=1}^{n}(X_i - \mu)^2}{N}} \tag{3-11}$$

The notation is thus $\bar{X}$ and s for samples, and μ and σ for populations.

There are two significant differences between Equations 3-10 and 3-11. First, the squared deviations for the sample standard deviation s are summed over the n different observations, whereas the squared deviations for the population standard deviation are summed over all N members of the population. Second, the denominator for s is $n - 1$, but it is N for σ. When n is very large, (say, 100 or more), the difference between s and σ is quite small; and when $n = N$, the values of s and σ are equal. That is, when the sample contains all the members of the population, there can be no difference between s and σ.

So why does the equation for the sample standard deviation have $n - 1$ in the denominator? It turns out that if we divide by n we tend to underestimate the true value of the population standard deviation σ. By dividing by $n - 1$, the standard deviation becomes larger, and the underestimate is corrected. A more detailed explanation is given in Chapter 7 where we deal with the statistical estimation of population parameters. However, this anomaly can also be explained by using two much simpler arguments.

First, let us recall the second property of the mean, which states that $\Sigma_{i=1}^{n}(X_i - M)^2$ is minimized for any set of observations when $M = \bar{X}$. That is, substituting any other number for M other than $\bar{X}$ will necessarily result in a larger sum of this expression. Even if we substitute the true population mean μ, at best we can do as well as with $M = \bar{X}$, and generally we will do worse. However, if we know μ, it makes sense to estimate s as

$$\sqrt{\frac{\sum_{i=1}^{n}(X_i - \mu)^2}{n}} \tag{3-12}$$

When we do not know μ, we replace it by $\bar{X}$ to obtain

$$\sqrt{\frac{\sum_{i=1}^{n}(X_i - \bar{X})^2}{n}} \tag{3-13}$$

However, since the sum of the squared deviations around $\bar{X}$ must be less than that around μ, Equation 3-13 must be smaller than Equation 3-12. It turns out that by replacing n by $n - 1$, we divide by a smaller denominator and get a larger value for s. So, the correction of n to $n - 1$ turns out to be just right.

There is also a second explanation of this correction. One other important property of the mean is that the sum of deviations about the mean is zero: $\Sigma_{i=1}^{n}(X_i - \bar{X}) = 0$. Because of this identity, only $n - 1$ of the n deviations are independent quantities. If we know $n - 1$ of the deviations, we can always use Equation 3-4 to determine the other. Consider once again the five observations 2, 2, 3, 5 and 8. The first four deviations sum to $(2 - 4) + (2 - 4) + (3 - 4) + (5 - 4) = (-2) + (-2) + (-1) + 1 = -4$. For Equation 3-4 to hold, the last deviation must equal 4. This is verified by calculating $8 - 4 = 4$. In general, then, among all the quantities $X_1 - \bar{X}, X_2 - \bar{X}, \ldots, X_n - \bar{X}$, there are only $n - 1$ independent quantities. This is another reason we divide by $n - 1$ and not n. The remaining observation offers no new information. In fact, for estimation purposes, we should also divide the sample mean absolute deviation by $n - 1$.

Equation 3-11 does not represent the most computationally efficient formula for determining the population standard deviation. Use of Equation 3-11 requires N different subtractions (one for each observation), N different squares, one division, and one square root operation. By using simple rules for summations, we can show that Equation 3-11 is equivalent to

$$\sigma = \sqrt{\frac{1}{N}\left(\sum_{i=1}^{n} X_i^2 - N\mu^2\right)} \tag{3-14}$$

Equation 3-14 requires $N + 1$ squares (one for each observation and one for the mean), one subtraction, one multiplication, one division, and one square root. This compares favorably with the computational requirements of Equation 3-11. The analogous formula for s is

$$s = \sqrt{\frac{1}{n-1}\left(\sum_{i=1}^{n} X_i^2 - n\bar{X}^2\right)} \tag{3-15}$$

The calculation of the standard deviation for the oxygen data is illustrated in Table 3-5. At the bottom of the table, the numerical equivalence of Equations 3-10 and 3-15 is shown for the DO variable.

Variance

A final measure of dispersion is the variance, which is simply the average of the squared deviations about their mean. The sample variance has a divisor of $n - 1$, and the

DEFINITION: SAMPLE VARIANCE

$$s^2 = \frac{\sum_{i=1}^{n}(X_i - \bar{X})^2}{n-1} \tag{3-16}$$

DEFINITION: POPULATION VARIANCE

$$\sigma^2 = \frac{\sum_{i=1}^{n} (X_i - \mu)^2}{N} \qquad (3\text{-}17)$$

population variance has a divisor of N. Variants of these formulas for computational purposes are easily derived as the squares of Equations 3-14 and 3-15. The variance is seldom used as a descriptive summary statistic because the squaring of the deviations means that s^2 has units that are the square of X and is therefore difficult to interpret. For example, if X is income in dollars, s^2 has units of "squared dollars." If X is temperature, in degrees, s^2 has units of "squared degrees." With no social or physical interpretation, such quantities don't carry much meaning. By taking the square root of the variance to obtain the standard deviation, we compensate for the initial squaring of the observations and produce a measure that is usually much easier to interpret. For inferential purposes we will see that the variance is an extremely important measure of a variable.

Interpretation and Use of the Standard Deviation

Both the standard deviation and the variance give an indication of how typical of a whole distribution the mean actually is. The larger these measures, the greater the spread of the observations and the less typical the mean. When all observations are equal to the mean, all deviations are zero and both s and s^2 are zero. Just like the arithmetic mean, the standard deviation is very sensitive to extremes. Because it involves the squares of deviations, the standard deviation gives more relative weight to large deviations. A highly skewed distribution, or one with a few extreme observations, would have a relatively large standard deviation.

Perhaps more importantly, a skewed distribution has two tails of different shape, and the spread of these two tails cannot be described by a single summary statistic. Thus, skewed distributions are not effectively summarized by either the mean or the standard deviation. The median and interquartile range are more representative summary statistics in this case; the five-number summary discussed in Chapter 2 would be even more useful. The mean and standard deviation are most suited to describing the characteristics of symmetric distributions.

Together, the mean and standard deviation can be used to locate individual observations within the distribution of a variable as a whole. This is an alternative to the percentile method, which relies on ordering our observations. For example, we might speak of an observation or value of X that is between one and two standard deviations above the mean, or between the mean and one standard deviation below the mean. It is very common to identify atypical or extreme observations as those greater than a certain number of standard deviations above or below the mean. Observations located more than two standard deviations from the mean are usually said to be in the "tails" of the distribution. These observations can thus be considered atypical.

Coefficient of Variation

To compare the dispersion of two frequency distributions having different means, we usually cannot simply compare the two standard deviations. Of course, if the means of the two variables are equal, then the variable with the smaller standard deviation is less dispersed. If, however, the means of the two variables are unequal, then it is misleading to rely on the standard deviation alone. A variable with a mean value in the thousands is likely to have a larger standard deviation than a variable with a mean value less than 100. The relative variability of a frequency distribution is measured by the ratio of the standard division to the mean.

DEFINITION: COEFFICIENT OF VARIATION

The coefficient of variation (CV) of a distribution with sample mean $\bar{X}$ and standard deviation s is defined as

$$CV = \frac{s}{\bar{X}} \tag{3-18}$$

For populations, an equivalent measure is obtained by substituting μ for $\bar{X}$ and σ for s. Because both $\bar{X}$ and s are measured in the same units (that is, units of variable X), their quotient must be a dimensionless measure. So, we can use the CV to compare variables measured in different units or scales.

To illustrate the advantages of the coefficient of variation, examine the following rainfall data collected from three climatological stations over a 60-year period.

	Annual rainfall		
Parameter	Station A	Station B	Station C
$\bar{X}$	92.6	97.3	38.8
s	16.6	12.8	9.1
CV	0.179	0.132	0.235

Stations A and B can be directly compared. Their means are almost the same, but station B has a much lower standard deviation and therefore is less dispersed. The coefficients of variation verify this. However, even though station C has a lower standard deviation than either station A or B, an examination of the CV for these three stations indicates that station C has the most variable rainfall of the three. Relatively less dispersed variables have lower coefficients of variation. The lower limit to this measure is zero. Because $\bar{X}$ appears in the denominator, the CV is not defined if the sample mean is zero. Perhaps more importantly, it is an unstable measure for samples whose mean is near zero: small changes in s can give rise to large changes in CV.

3.3. Higher Order Moments or Other Numerical Measures of the Characteristics of Distributions

The parameters of a statistical population are sometimes called the *moments* of the population. The rth moment about the mean is defined as

$$m^r = E(X - \mu)^r$$

The second moment about the mean is the variance of the distribution

$$m^2 = E(X - \mu)^2 = \sum_{i=1}^{n} (X_i - \mu)^2/N$$

The mean and the variance are the most frequently used measures of the characteristics of a univariate frequency distribution. Occasionally, however, we resort to *higher order moments* of a distribution to describe other characteristics of a distribution's shape. In particular, measures of *skewness* are based on the third moment about the mean, and measures of *kurtosis* are based on the fourth moment about the mean.

In Chapter 2 we introduced skewness as an indicator of the asymmetry of a distribution about its mean. A measure of the skewness of a distribution is

$$\text{Skewness} = \frac{\sum_{i=1}^{n} (X_i - \mu)^3}{N\sigma^3} \tag{3-19}$$

where μ is the population mean and σ is the population standard deviation. A normal distribution (see Chapter 5) has a skewness value of zero, as does every other symmetric distribution. Values of skewness greater than zero indicate positive skew, where most of the observations of a variable of interest are found to the left of the mean and where relatively few observations have values that are greater than the mean. Values of skewness less than zero indicate negative skew, where most of the observations of a variable of interest are larger than the mean and where relatively few observations have values less than the mean.

Kurtosis, an indicator of the peakiness of a distribution was also introduced in Chapter 2. The kurtosis of a distribution is measured as

$$\text{Kurtosis} = \frac{\sum_{i=1}^{n} (X_i - \mu)^4}{N\sigma^4} \tag{3-20}$$

Kurtosis measures the extent to which the observations in a frequency distribution cluster around particular values of the variable of interest. When observations are tightly clustered, a histogram exhibits a distinct peak, and measures of kurtosis tend to be greater than three. When observations are evenly spread across the different values of the variable of interest, a histogram appears relatively flat and measures of kurtosis tend to be smaller than three. Measures of kurtosis for a normal distribution are approximately equal to three. Such a distribution is said to be mesokurtic, or not markedly peaky or flat. Measures of skewness and kurtosis for the dissolved oxygen data in Table 3-5 are –0.115 and 2.665, respectively. Note that these calculations use the sample mean and standard deviation and replace N in the denominator of Equations 3-19 and 3-20 with $n - 1$. Thus, the dissolved oxygen data exhibit slight negative skew and is relatively mesokurtic, though tending to platykurtic. (In some texts, the value three is subtracted from the measure of kurtosis given in Equation 3-20, such that a normal distribution has a kurtosis value of zero.)

A number of classical tests of statistical inference assume that the data under study are normally distributed (see Chapter 5). When measures of skewness depart significantly from zero, or when measures of kurtosis are significantly different from three, a distribution is probably not normal. Though such statistics are beyond the scope of this book, tests of the normality of a distribution such as the Jarque-Bera test, focus on measures of skewness and kurtosis.

3.4. Using Descriptive Statistics with Time-Series Data

The descriptive statistics outlined above can be used to calculate average values, dispersion, or higher order moments of a variable measured over time. While such statistics can be useful, they do not necessarily lead to a greater understanding of the time series. Quite often, however, we can use descriptive statistics to gain additional insight into the characteristics of a time series. In particular, measures of central tendency, calculated for subsets of the observations on a variable, may be employed to *smooth* the time series.

Smoothing of a time series is a statistical procedure used to dampen the fluctuations in the value of a variable recorded over time. Smoothing is often performed in order to identify an underlying trend or cyclical component in a time series. The difference between the smoothed series and the original time series is often identified as the random component. More often than not, it is the smoothed or systematic component of a time series in which we are most interested, not the apparently random fluctuations around it. Different measures of central tendency are used to perform data smoothing. Let us consider the topic of smoothing within the context of a simple example.

EXAMPLE 3-3. A Time Series of Traffic Counts. Traffic counts were recorded at an intersection in a residential neighborhood where complaints of excessive traffic noise had been registered over the previous several months. City officials decided to take daily vehicle counts between 5:00 and 5:30 P.M. for an initial baseline period of four weeks. At this time, a traffic amelioration scheme was introduced, and the survey continued for an additional four weeks. Data on the daily vehicle counts over the entire eight-week period are shown in Figure 3-5. A transportation geographer was asked to analyze the data with a view to evaluating the effectiveness of the traffic diversion. Two characteristics of the vehicle count series are apparent in Figure 3-5:

1. There are marked fluctuations within each seven-day period. There is significantly less traffic on weekends. To some extent, Mondays and Fridays

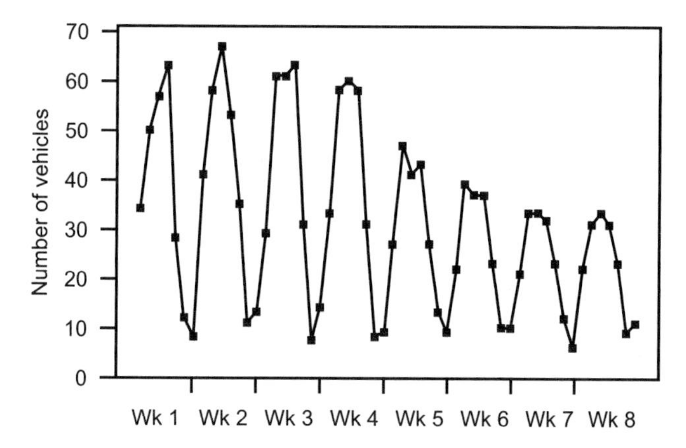

FIGURE 3-5. Intersection vehicle counts of Example 3-3.

seem to have lower traffic volumes than midweek. Upon investigation, the analyst found that this was due to the flex-time scheduling at a nearby major manufacturing enterprise where employees typically program their off day on either side of the weekend.

2. A downward shift in the underlying direction of the series *seems* to coincide with the introduction of the traffic reduction program.

To present the evidence effectively, the analyst wished to average the observations in the time series in order to obtain the clearest view of the overall pattern of traffic during the study period.

Moving Average

One of the simplest ways of smoothing the fluctuations in a time series is to calculate the average of successive, overlapping groups of observations.

DEFINITION: MOVING AVERAGE

A moving average of a time series is developed by replacing each successive sequence of T observations by the mean of the sequence. The first sequence contains the observations $Y_1, Y_2, \ldots, Y_T$, the second sequence contains the observations $Y_2, Y_3, \ldots, Y_{T+1}$, and so on. The value of T is the *term* of the moving average.

The original time series is thus replaced by a new series of means that run or move as we go along the series from the first to the last observation.

For a portion of the traffic count data, the calculations leading to the seven-term moving average are shown in Table 3-6. The full dataset is available from the website for the book. The actual vehicle counts over the study period are given in the third

TABLE 3-6
Calculation of Seven-Term Moving Average for the Vehicle Count Data

Week and day	t	Number of vehicles	Seven-term total	Seven-term moving average	Fluctuating component
1M	1	34			
T	2	50			
W	3	57			
T	4	63	252	36.00	27.00
F	5	28	259	37.00	–9.00
S	6	12	267	38.14	–26.14
S	7	8	277	39.57	–31.57
2M	8	41	267	38.14	2.86
.	.	.	.	.	.
8M	50	22	158	22.57	–0.57
T	51	31	158	22.57	8.43
W	52	33	155	22.14	10.86
T	53	31	160	22.86	8.14
F	54	23			
S	55	9			
S	56	11			

column. We begin calculation of the moving average at $t = 4$, since this is the first value of t over which seven surrounding values are observed. The sum of these values is shown in the fourth column of the table. For $t = 4$, the value 252 is equal to 34 + 50 + 57 + 63 + 28 + 12 + 8. The moving average is found by dividing this sum by 7. Thus, for $t = 4$, the seven-term moving average is calculated as 252/7 = 36.0. Note that the seven-term moving average series ends at $t = 53$, because there are no longer seven surrounding observations for $t = 54$, 55, and 56.

What does the smooth sequence look like? Figure 3-6 superimposes the smoothed seven-term moving average sequence on the original data series. As we might expect, the smoothed series has markedly less variation that the original time series, and it is much easier to see the trend of traffic over the eight-week survey period. First, it is clear that traffic flows decrease exactly when the traffic amelioration scheme was introduced at the end of week 4. Second, there appears to be another slight decline in the second week after the introduction of the amelioration scheme, possibly as more and more drivers find the diversion to be slow and seek alternative routes. Finally, there is some evidence that traffic patterns have stabilized. The path of the seven-term moving average becomes horizontal in weeks 7 and 8. This suggests that there may be no further declines in traffic volumes and that further measures will have to be introduced if this traffic volume is considered unacceptable.

Once we have determined the smoothed part of the time series using the moving average, we can define the fluctuating component of the series as the *difference* between the values of the moving average and the values of the original time series given in column 3 of Table 3-6. This represents the day-to-day fluctuation in traffic levels and illustrates the significance of both midweek peaks, weekend troughs, and the reduced volumes on Mondays and Fridays. The fluctuating component is graphed

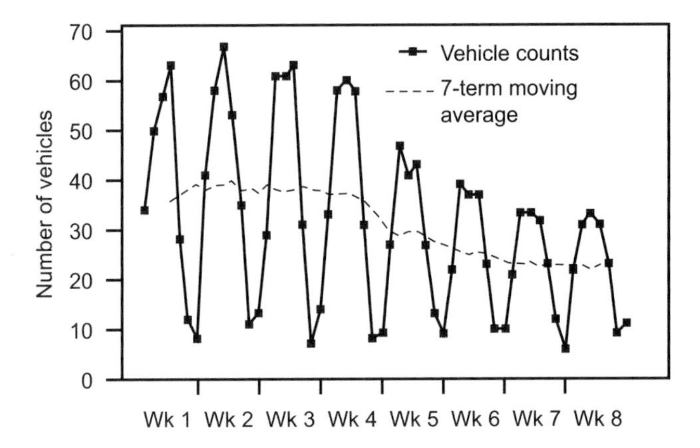

FIGURE 3-6. Seven-term moving average of vehicle counts in Example 3-3.

in Figure 3-7. The series is horizontal over the entire eight-week study period, though the overall level of the fluctuation is reduced in the "after" period.

Effect of Term on Moving Average

The path of the moving average is particularly sensitive to the length of the term chosen for the calculation. It was natural to use a seven-term moving average for the vehicle count data; three- or five-term moving averages would not be as effective in portraying the systematic component of the series. Figure 3-8 illustrates the three-, five-, and seven-term moving averages for the vehicle count data. Note that the three- and five-term series do not smooth the data as well as the seven-term moving average. In general, the best choice for the term of the moving average series is the length of the fluctuating component. The normal weekly pattern of traffic in a North American city leads us to the selection of the $T = 7$-term moving average. Unfortunately, many time

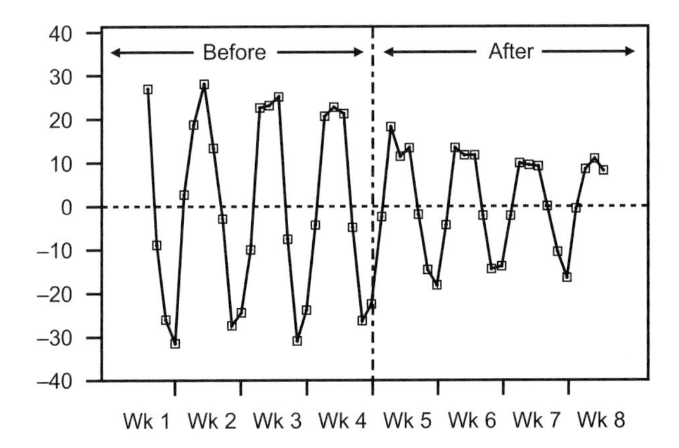

FIGURE 3-7. The fluctuating component of the vehicle count series.

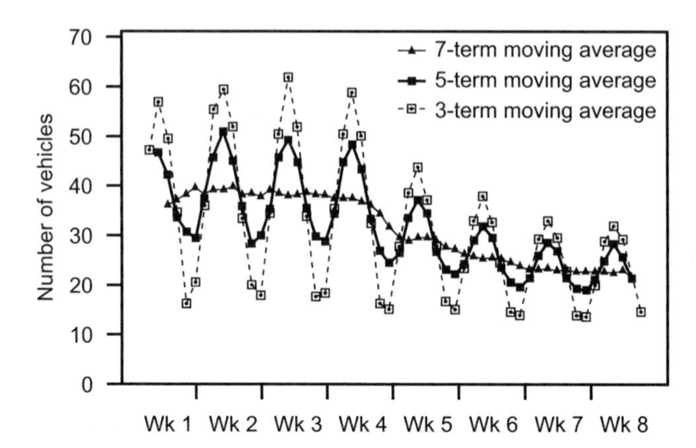

FIGURE 3-8. Three-term, five-term, and seven-term moving averages of the traffic data.

series in the social sciences oscillate with considerable frequency, and it is difficult to look at the series and pick out the *single* term length that best smoothes the series. Typically, one must use a computer program to generate moving average series for increasing values of T and examine the paths of these series to isolate the best value of T. The smallest value of T that dampens the fluctuations in a time series is then selected.

A word of caution is in order, however. An unusually large or small observation can introduce oscillations into a time series where none exists in the original series. Why? Remember that any time series observation Y_T enters T successive terms of the moving average series. As an example, let us change the value of a single original observation and compare the resulting moving average. Suppose the traffic count for Wednesday in the second week of the original series is changed from 67 to vehicles. This value clearly is unusual for a midweek observation and resembles typical weekend traffic levels in the neighborhood for this time of day. Figure 3-9 illustrates the seven-term moving average for both series. Note the dramatic change in the moving average series during the second week. Because we are using a seven-term moving average, seven consecutive data points are sharply shifted downward. The movement of the series is now markedly different, and the overall impact of the traffic reduction scheme is not nearly so obvious. If there were a second unusual observation in the third week, there would be two U-shaped segments in the series, and it would be even more difficult to interpret the impact of the traffic amelioration plan. The point is clear: where unusual observations exist in a time series, the paths of many smoothed moving average series may have erratic sequences.

Centered Moving Average

When an even number of terms is required in a moving average, for example, a four-term or a 12-term calculation, a problem arises. Consider the following case in which a four-term moving average is taken:

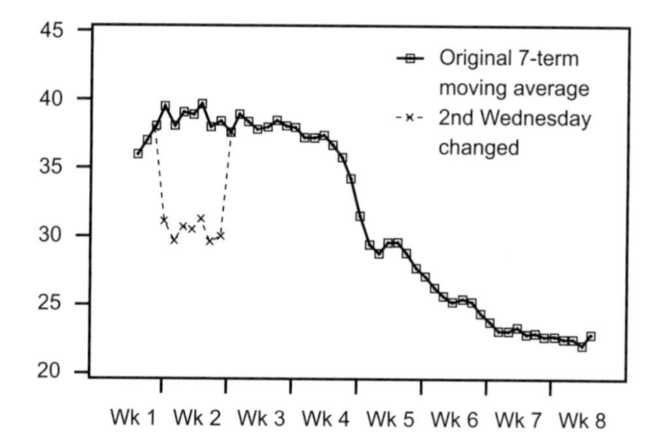

FIGURE 3-9. The effect of an extreme observation on a moving average series.

t	1	2		3		4		5		6	7
Y_t	2	7		6		8		4		6	2
Moving total $T = 4$			23		25		24		20		
Moving average $T = 4$			5.75		6.25		6.00		5.00		

The four-term totals and averages, placed at the centers of their sequences, are located at $t = 2.5$, 3.5, and so on. The difficulty with these locations is that we can't easily and directly compare these moving averages to the original observations. We remedy this situation by centering the moving averages as follows.

t	1	2		3		4		5		6	7
Y_t	2	7		6		8		4		6	2
Moving total $T = 4$			23		25		24		20		
Moving average $T = 4$			5.75		6.25		6.00		5.00		
Centered moving average				6.0		6.125		5.5			

For example, the first centered moving average of 6.0 at $t = 3$ is the average of the first two moving averages of $T = 4$; that is, $6.0 = (5.75 + 6.25)/2$. For $t = 4$, we find the second centered average to be $(6.25 + 6.00)/2 = 6.125$. Note that we shorten the number of values in the series when we use centered moving averages. In general, we lose moving average values for the first and the last $T/2$ time periods. In this series $T = 4$, and we lose the first two and the last two elements in the time series.

It is possible to view a centered moving average as a *weighted moving average* when the observations in the series are given unequal weights. The centered moving average of 6.0 for $t = 3$ can be calculated by weighting the first five observations by the sequence 1, 2, 2, 2, 1:

$$[1(2) + 2(7) + 2(6) + 2(8) + 1(4)]/8 = 48/8 = 6.0$$

Similarly, two-term centered moving averages can be calculated as weighted three-term moving averages, and six-term moving averages can be determined as weighted seven-term moving averages.

3.5. Descriptive Statistics for Spatial Data

Most of the descriptive statistics discussed in Sections 3.1–3.3 of this chapter can be applied directly to spatial data. In this section we highlight some more specialized variants of descriptive statistics that may be used when the researcher confronts geographical data that refer to areas or points.

Areal Data

Unfortunately, unless we are willing to convert areal data to point data, by using area centroids or some other representative point, descriptive statistics for areas or regions usually fail to capture the spatial component of the data. Nonetheless, there are graphical and statistical summaries for data published across discrete areal units such as census tracts, counties, states or provinces, and many other administrative units. These summary techniques are particularly applicable to choropleth and stepped statistical surface maps.

LOCATION QUOTIENTS

The location quotient is frequently employed in economic geography and locational analysis, but it has much wider applicability. The location quotient (LQ) is an index for comparing a region's share of a particular activity with the share of that same activity found at a more aggregate spatial level. Suppose, for example, that the following information is taken from a larger dataset that details the employment structure of a country divided into four areas or regions, denoted D, E, F and G:

	Regional employment				
Sectors	D	E	F	G	Nation
Manufacturing	5	70	15	10	100
Services	40	220	60	80	400
Other	105	210	125	60	500
Total	150	500	200	150	1,000

Let us assume that we are interested in determining whether employment in manufacturing or services is concentrated in some area(s), or whether it is evenly distributed across the country? Location quotients compare the distribution of an activity to some *base,* or *standard*—in this case to employment in that activity in the nation as a whole. The question can be therefore be rephrased as follows: Is manufacturing employment more or less spatially concentrated than employment in services?

From the preceding sample data, the location quotient for manufacturing employment in region D is (5/150)/(100/1000) = 0.0333/0.1 = 0.333. That is, manufacturing constitutes 3.33% of total employment in region D, but 10% of total employment in the base region, or the nation. In total, region D contains 15% of the nation's employment. On the basis of this national employment share, we might expect region D to have 15% of the nation's total manufacturing workforce, or (0.15)(100) = 15 manufacturing employees. However, region D has only five employees in the manufacturing sector, only one-third of the expected number. As the ratio of two percentages, the location quotient is dimensionless.

DEFINITION: LOCATION QUOTIENT

The location quotient for activity i in region j is equal to the percentage of total activity in region j devoted to activity i divided by the percentage of the base region's total activity devoted to activity i, or

$$LQ_i^j = \frac{A_i^j / \sum_{i=1}^{n} A_i^j}{B_i / \sum_{i=1}^{n} B_i} \qquad (3\text{-}21)$$

where A_i^j represents the level of activity i in region j, B_i represents the level of activity i in the base region, and n denotes the number of activities.

Location quotients can be interpreted by using the following conventions:

1. If LQ > 1, this indicates a relative concentration of activity i in the region of interest compared to the base region.
2. If LQ = 1, the region's share of activity i mirrors the base region's share of that activity.
3. If LQ < 1, the region's share of activity i is lower than that generally observed, or lower than that found in the base region.

The table of location quotients for employment in the manufacturing and services sectors of our four regions is as follows:

	Location quotients	
Region	Manufacturing	Services
D	0.333	0.667
E	1.400	1.100
F	0.750	0.750
G	0.677	1.333

For manufacturing, the location quotients reveal a concentration in region E and less than expected shares in the other three regions. For services, the spatial distribution of employment is less concentrated, with regions E and G containing more workers than expected and regions D and F containing fewer workers than expected. Mapping these location quotients might reveal the existence of spatial patterns in the employment data.

The selection of the base or standard distribution used in the denominator of the location quotient is subject to choice. Usually, if the regions or activities are part of a meaningful aggregate, then the aggregate is used as the base. With employment distributed over different sectors and areas, it makes sense to use total employment in the region as a whole as the base distribution in our example. Changing the base or standard distribution can have a significant impact on location quotients.

COEFFICIENT OF LOCALIZATION

One of the drawbacks to use of the location quotient is that one value is calculated for each area of the region being analyzed. For a city with perhaps 300 or more census tracts, this would be a very inefficient form of summary, even if the location quotients themselves were mapped. A more efficient alternative, the coefficient of localization (CL) describes the relative concentration of an activity by a single number.

DEFINITION: COEFFICIENT OF LOCALIZATION

The coefficient of localization (CL) is a measure of the relative concentration of an activity in relation to some base. To calculate CL:

1. Calculate each region's share of activity i within the base or the study area as a whole

$$A_i^j / B_i$$

for all j where, where $j = 1, \ldots, m$ denotes the region.

2. Calculate each region's share of total activity within the study area or base

$$\sum_{i=1}^{n} A_i^j / \sum_{i=1}^{n} B_i$$

for all j.

3. For each region j subtract the value in step 2 from that in step 1, and add *either* all the positive differences or all the negative differences.

Using our example of manufacturing employment across four regions, we can calculate the coefficient of localization as follows.

Region	Share of manufacturing employment	Share of total employment	Difference –	Difference +
D	0.05	0.15	0.1	
E	0.70	0.50		0.2
F	0.15	0.20	0.05	
G	0.10	0.15	0.05	
Total	1.00	1.00	0.2	0.2

Therefore the CL for manufacturing is 0.2. Similarly, the CL for service employment across the four regions is 0.1. Note that calculation of the CL requires that the shares of employment for the activity of interest and for the sum of all activities should add up to 1.0.

The CL ranges from 0 to 1. This differs from the location quotient that has a lower limit of zero but an upper limit approaching positive infinity. If CL = 0, the distribution of the given activity is evenly spread across the regions in accordance with the distribution of total activity. As CL approaches 1, the activity of interest becomes increasingly concentrated in one region. The results above indicate that both services and manufacturing employment are relatively evenly spread over the four regions, though services employment is more evenly distributed.

LORENZ CURVE

The Lorenz curve is another way to index the distribution of a variable among spatial units, although it is commonly used to measure the extent of inequality of a variable distributed over aspatial categories. For example, if all residents of a country have the same income, then income is distributed equally. However, if a country contains some rich and poor individuals, then there is some inequality in the income distribution. There is an obvious analogy to areal data. An activity may be concentrated in one or a few areas, or it may be spread evenly throughout a region. The Lorenz curve is a graphical display of the degree of inequality.

As with location quotients and the coefficient of localization, the Lorenz curve compares the spatial distribution of some activity to a base distribution. The Lorenz curve is constructed by using the following rules:

1. Calculate the location quotients for the various regions that comprise the study area. Reorder the regions in decreasing order of their location quotients.
2. Cumulate the percentage distributions of both the activity of interest and the base activity in the order determined in step 1.
3. Graph the cumulated percentages for the activity of interest and the base activity, and join the points to produce a Lorenz curve.

The necessary calculations for the manufacturing employment data are shown in Table 3-7. The Lorenz curves for both manufacturing and service employment are illustrated in Figure 3-10.

A Lorenz curve has several important properties. If the activity of interest is evenly distributed across the regions of the study area, that is, in proportion to a base activity, then the Lorenz curve is a straight line following the 45° diagonal shown in Figure 3-10. The more concentrated the activity, the further the Lorenz curve is from the diagonal. In the limiting case, the curve follows the X-axis to the point (100, 0) and then proceeds vertically to the point (100, 100). Except for the case where the activity is perfectly evenly distributed, the slope of the Lorenz curve is always increasing. This follows directly from the ordering of the areas by LQ. The Lorenz curves in Figure 3-10 suggest that manufacturing employment is more concentrated than service

TABLE 3-7
Work Table for Lorenz Curve Calculations for Manufacturing Data

Step 1. The location quotient for the four regions are D = 0.333, E = 1.400. F = 0.750, G = 0.667. So the order of the regions in the cumulative table should be E, F, G and D.

Step 2.

Region	LQ	Percentage of manufacturing employment	Percentage of total employment	Cumulative percentage	
				Manufacturing	Total
E	1.400	70	50	70	50
F	0.750	15	20	85	70
G	0.667	10	15	95	85
D	0.333	5	15	100	100
		100	100		

Step 3. See Figure 3-10.

employment. Researchers who prefer graphical summaries to statistical measures generally favor the Lorenz curve.

The most common summary measure of inequality used in conjunction with the Lorenz curve is the *Gini coefficient,* or, as it is sometimes called, the *index of dissimilarity.* The range of the Gini coefficient is from 0 to 100%. There are two other ways of calculating the value of this index. First, it can be determined by identifying the largest difference between the cumulated percentages of the activity of interest and the base activity. From Table 3-7, the maximum difference between the two cumulative columns for manufacturing and total (base) employment is 70 – 50 = 20. This is exactly equal to the vertical deviation between the Lorenz curve and the diagonal at this point. The point (70, 50) differs from the diagonal (70, 70) by 20%.

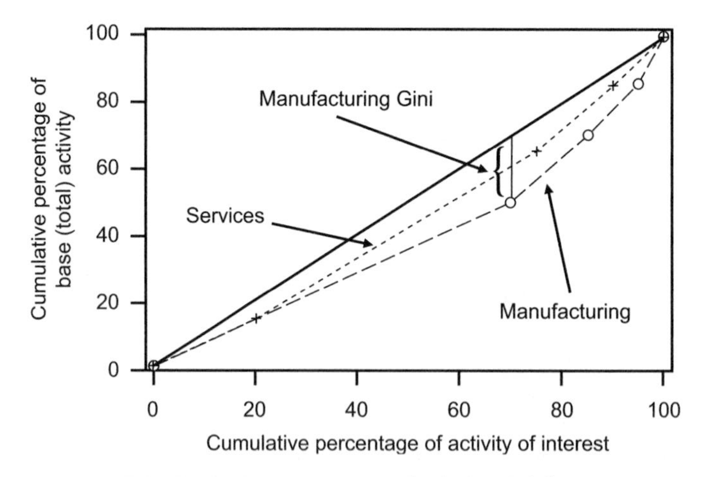

FIGURE 3-10. Lorenz curves for industrial data.

DEFINITION: GINI COEFFICIENT, OR INDEX OF DISSIMILARITY
The Gini coefficient, or index of dissimilarity, is defined graphically as the maximum vertical deviation between the Lorenz curve and the diagonal.

Second, we might note that the CL for manufacturing was previously calculated as 0.20. The Gini coefficient can be obtained easily by multiplying the coefficient of localization by 100. For manufacturing, we see that the Gini coefficient is (0.20)(100) = 20, and for service employment (0.10)(100) = 10. Another equivalent method for calculating the Gini coefficient is to take one-half of the sum of the *absolute value* of the differences between the uncumulated percentage distributions of the activity of interest and the base activity. Using the data for the manufacturing sector given in Table 3-7, we calculate the Gini coefficient as $\frac{1}{2}(|70-50| + |15 - 20| + |10 - 15| + |5 - 15|) = \frac{1}{2}(20 + 5 + 5 + 10) = 20$. Smaller values of both the Gini coefficient and the CL indicate similarity between the spatial distribution of the activity of interest and the base activity. This is why the Gini coefficient is sometimes called the index of dissimilarity.

The Lorenz curve and Gini coefficient also can be used to measure the degree of similarity of the percentage distributions of any two activities, neither of which is necessarily the base. For example, it is possible to compute the Gini coefficient between manufacturing and service employment distributions. Using the original data introduced for calculating location quotients, we see the Gini coefficient is $\frac{1}{2}(|5 - 10| + |70 - 55| + |15 - 15| + |10 - 20|) = \frac{1}{2}(30) = 15$. This application of the Gini coefficient is utilized in urban social geography to compare the spatial distribution of ethnic groups in cities. The similarity of the spatial distribution of ethnic groups is a useful indicator of their integration or assimilation into the host society. Over time, as assimilation occurs, the residential segregation of many ethnic groups becomes less pronounced. Gini coefficients can be used to test this hypothesis.

All the procedures described for areal data have one common problem that limits their utility in assessing the similarity of two maps. They are inextricably tied to the exact areal subdivisions used in their calculations and can be interpreted only in this light. As we see later in this chapter, this leads to several different specific problems. The values of the coefficients are very sensitive to the size and areal definitions used as the basis for their calculation. These issues reappear in several later chapters of the book.

Point Data

All the techniques used to describe areal distributions presented in the last subsection are applicable to spatial as well as aspatial data. Unfortunately, each of the measures fails to explicitly incorporate the spatial dimensions through a variable related to one of the fundamental spatial concepts—distance, direction, or relative location. This is not true of the statistical methods designed for the analysis of point data. Distance is either explicitly or implicitly included within these measures. This branch of statistics is termed, appropriately enough, *geostatistics*.

The first step in analyzing a set of point data is to overlay the map with a Cartesian grid and determine the coordinates of each point on the map. Figure 3-11 illustrates this procedure for a point distribution with nine observations. Each of the nine points is given an *X* and *Y* coordinate on the basis of a 100 × 100 grid placed over the map. The origin of the grid is usually placed at the southwest corner of the map, so that all coordinates have positive values. Observation 1, for example, has coordinates (20, 40). The set of nine observations can thus be represented by the following point coordinate data:

Observation No.	*X* coordinate	*Y* coordinate
1	20	40
2	30	60
3	34	52
4	40	40
5	44	42
6	48	62
7	50	10
8	60	50
9	90	90

Depending on the type of point distribution involved, we might add a third variable to these data. For example, if the dots represent towns, villages, and cities, we might associate a "weight" such as the population, to each of the points. In any case, at least two, and possibly three, variables characterize the dot map. The first step in geostatistics is to summarize this map in an efficient way. The standard statistical concepts of central tendency and variability are commonly used for this purpose. Each concept has a distinctly spatial interpretation.

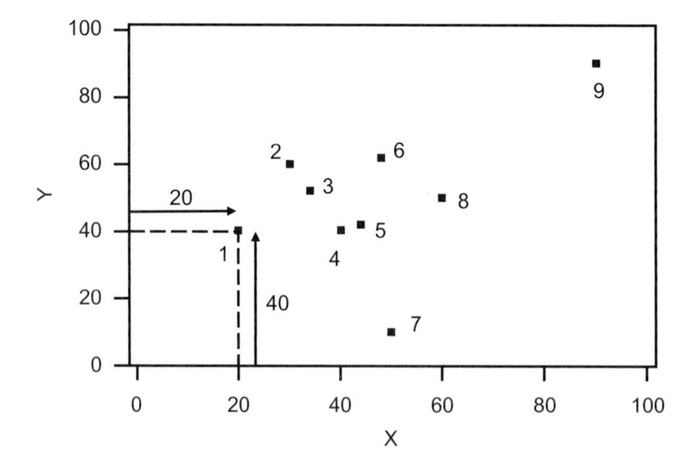

FIGURE 3-11. Obtaining grid coordinates from a map.

MEASURES OF CENTRAL TEDENCY

The first question to answer is, what is the center, or middle, or average, of this point distribution? Translated into geostatistics, the question is more appropriately phrased, where on the map is the center of this point distribution? Five different measures can be used to identify the center of the map:

1. Mean center
2. Weighted mean center
3. Manhattan median
4. Euclidean median
5. Weighted medians

Since the properties of these five measures are so often incorrectly stated and insufficiently identified in the geographic literature, they are discussed in detail in this section.

Mean Center

The mean center can be thought of as the "center of gravity" of a point distribution and is a simple generalization of the familiar arithmetic mean. It is easy to calculate.

DEFINITION: MEAN CENTER

Let (X_i, Y_i), $i = 1, 2, \ldots, n$, be the coordinates of a given set of n points on a map. The mean center of this point distribution is defined as $(\bar{X}, \bar{Y})$ and is given by

$$\bar{X} = \sum_{i=1}^{n} X_i / n \quad \text{and} \quad \bar{Y} = \sum_{i=1}^{n} Y_i / n \tag{3-22}$$

Note that the mean center defines a point or location on the map with coordinates $(\bar{X}, \bar{Y})$. Using the coordinate data for the set of nine points of Figure 3-11, we find the mean center has coordinates

$$\bar{X} = \frac{20 + 30 + 34 + 40 + 44 + 48 + 50 + 60 + 90}{9} = 46.22$$

$$\bar{Y} = \frac{40 + 60 + 52 + 40 + 42 + 62 + 10 + 50 + 90}{9} = 49.56$$

This location is identified by the circular symbol in Figure 3-12.

Weighted Mean Center

The mean center can also be generalized to include the case for which each of the points on the map has an associated frequency, magnitude, or "weight." For example,

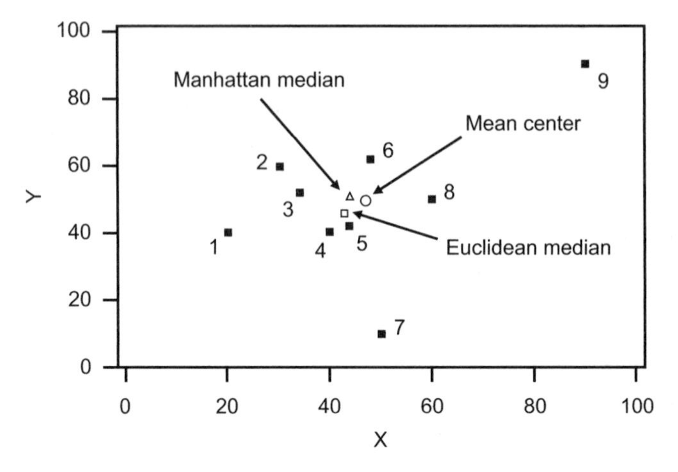

FIGURE 3-12. Measures of central tendency for the point pattern of Figure 3-11.

suppose that each of the points in Figure 3-11 represents a city with a given population. The location of the weighted mean center should be drawn toward those points with the largest populations and away from those points with the smallest populations. If the populations of the cities represented by the points in Figure 3-11 are 10, 20, 10, 20, 10, 80, 10, 90, and 100, respectively, we expect the weighted mean center to be drawn toward the locations of observations 6, 8, and 9. Since these three observations all lie in the northeast corner of the map, the weighted mean center will be drawn in this direction. The values of both the X and Y coordinates of the weighted mean center should be greater than the coordinates of the mean center, (46.22, 49.56).

DEFINITION: WEIGHTED MEAN CENTER

Let (X_i, Y_i), $i = 1, 2, \ldots, n$, be the coordinates of a set of n points, and let w_i be the weight attached to the ith point. The weighted mean center has coordinates $(\bar{X}_w, \bar{Y}_w$ given by

$$\bar{X}_w = \frac{\sum_{i=1}^{n} w_i X_i}{\sum_{i=1}^{n} w_i} \quad \text{and} \quad \bar{Y}_w = \frac{\sum_{i=1}^{n} w_i Y_i}{\sum_{i=1}^{n} w_i} \tag{3-23}$$

One simple application of the weighted mean center is to trace the center of gravity of a population distribution over time. To do this, the weighted mean center of a population dot map (where observations are cities of different size) is calculated for a series of maps of the same region at regular intervals of time. The movement of the mean center over the study period summarizes the change in the distribution of population in the region. Figure 3-13 shows the westward trend of the mean population center of the US between 1790 and 1990. The weighted mean center for the nine points of Figure 3-11 is determined in the following way:

$$\bar{X}_w = \frac{10 \cdot 20 + 20 \cdot 30 + 10 \cdot 34 + 20 \cdot 40 + 10 \cdot 44 + 80 \cdot 48 + 10 \cdot 50 + 90 \cdot 60 + 100 \cdot 90}{10 + 20 + 10 + 20 + 10 + 80 + 10 + 90 + 100}$$

$$= \frac{21{,}120}{350} = 60.34$$

$$\bar{Y}_w = \frac{10 \cdot 40 + 20 \cdot 60 + 10 \cdot 52 + 20 \cdot 40 + 10 \cdot 42 + 80 \cdot 62 + 10 \cdot 10 + 90 \cdot 50 + 100 \cdot 90}{10 + 20 + 10 + 20 + 10 + 80 + 10 + 90 + 10}$$

$$= \frac{21{,}900}{350} = 62.57$$

As expected, the weighted mean center with coordinates (60.34, 62.57) is located to the northeast of the mean center (46.22, 49.56).

The formula for calculating the mean center, Equation 3-22, is equivalent to the formulae used to calculate the sample arithmetic mean, Equation 3-2. There is also one important property of the arithmetic mean that is shared with the mean center and has interesting implications for a spatial distribution. Recall that the arithmetic mean of a set of numbers minimizes $\Sigma_{i=1}^{n}(X_i - \bar{X})^2$. This property, the so-called least squares property of the mean, was discussed in Section 3.1. The equivalent property for the mean center is that $(\bar{X}, \bar{Y})$ minimizes

$$\sum_{i=1}^{n}(X_i - \bar{X})^2 + (Y_i - \bar{Y})^2$$

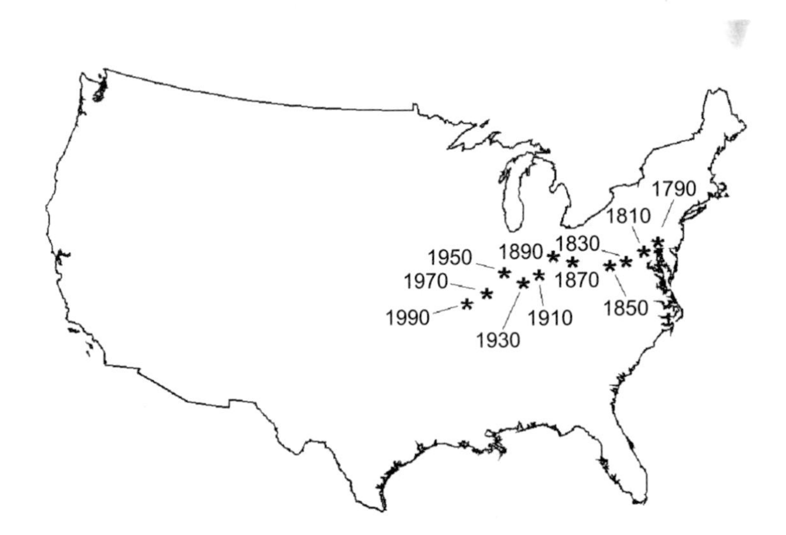

FIGURE 3-13. Tracing the mean population center of the United States.

Note that, by the Pythagorean theorem, the distance between points (X_i, Y_i) and $(\bar{X}, \bar{Y})$ is

$$d = \sqrt{(X_i - \bar{X})^2 + (Y_i - \bar{Y})^2}$$

Therefore, the mean center has the property that it minimizes the sum of squared distances from itself to the n other points. Also, like the mean, the mean center is very sensitive to the existence of extreme observations. Extreme observations in a point distribution are distant points set apart from the others. To see this sensitivity, let us augment the nine points of Figure 3-11 with a tenth point having coordinates (1,000, 1,000). The new mean center is (141.6, 144.6). This point lies between the existing observations and the new extreme point.

Manhattan Median

The concept of the median can also be applied to point distributions. For a set of n observations, the median is defined as the "middle," or $[(n + 1)/2]$th observation in an ordered array of values of X. How can we find the middle, or median, of a point distribution? The spatial median in this sense is the point of intersection of two perpendicular lines—one that divides the distribution of points in a north–south direction into two equal parts and the other that divides it into two equal parts in an east–west direction. Consider again the nine points of Figure 3-11. In the north-south direction, the line must pass through observation 8 since it has four points below it (observations 1, 4, 5, and 7) and four observations above it (observations 2, 3, 6, and 9). A horizontal line through observation 8 can thus be drawn. In an east-west direction, observation 5 is the middle point. A vertical line is drawn through this observation. The intersection of these two perpendicular lines is shown as the triangular symbol in Figure 3-12. It is close to, but not coincident with, the mean center.

The Manhattan median presents two distinct problems. First, let us see what happens when we try to determine the median for a set with only eight observations, such as in Figure 3-14. Using this definition, we would have to find a line in the north–south direction having four observations on each side. As in Figure 3-14, *any vertical line* between points A and C will have four points on either side. Similarly, *any horizontal line* between points A and B divides the set into two equal parts. The shaded area in Figure 3-14 encloses all points having the property of the Manhattan median. It is obviously not a unique point. What went wrong? It turns out that the Manhattan median is *never* unique when there is an even number of points and is *always* unique when there is an odd number of points.

The second problem with this measure of central tendency can also be illustrated with Figure 3-14. Suppose the coordinate system of the map is shifted to the position shown by the dashed lines. What happens to the shaded area defining the median? It must be rotated in the same way. This will define another rectangular area—for the same points! The location of the Manhattan median is therefore not unique under axis rotation. This is somewhat undesirable from a statistical point of view.

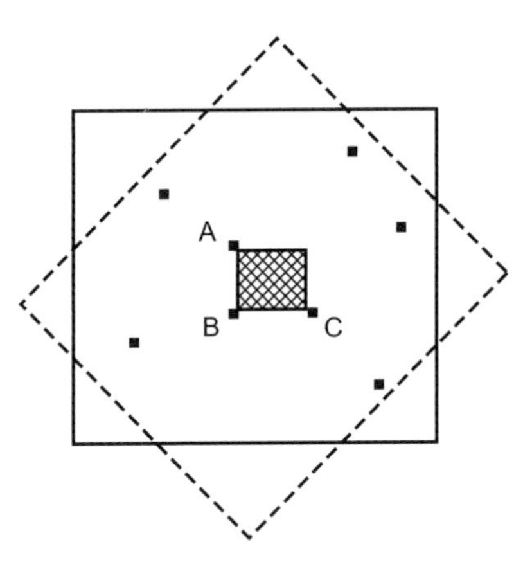

FIGURE 3-14. Manhattan median for an even number of points.

The Manhattan median shares one important property with the median of a frequency distribution. From Equation 3-6, we know that the median minimizes the sum of the absolute deviations between itself and the other n points in the frequency distribution. The equivalent property for the Manhattan medium is formalized as follows:

DEFINITION: MANHATTAN MEDIAN

Let $(X_i, Y_i, i = 1, 2, \ldots, n$, be the coordinates of a set of n points. The Manhattan median (X_m, Y_m) minimizes

$$\sum_{i=1}^{n} |X_i - X_m| + |Y_i - Y_m| \tag{3-24}$$

That is, (X_m, Y_m) minimizes the sum of the absolute Manhattan deviations from itself to the other n points of the distribution.

Why is it called the Manhattan median? Imagine we are trying to locate some facility in a city where travel is limited to the north–south and east–west directions. It is impossible to travel "as the crow flies." As shown in Figure 3-15, the Manhattan "distance" between the two points (X_i, Y_i) and (X_m, Y_m) is $|X_i - X_m| + |Y_i - Y_m|$. The absolute value signs are used to ensure that the calculated distance is non-negative. This measure, or *metric,* of spatial separation bears a close resemblance to the movement possibilities in a dense rectangular grid street network of a large city, in particular Manhattan. Hence the name *Manhattan metric* and the term *Manhattan median.* Note that the Manhattan distance between two points is always greater than the distance calculated by using the Pythagorean theorem, except when the two points lie on a north–south or east–west line. We term the distance metric calculated by using Pythagoras's theorem the *Euclidean metric.*

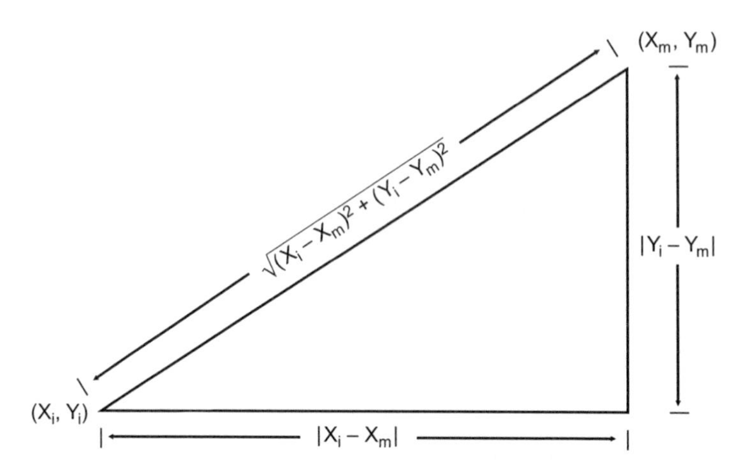

FIGURE 3-15. Distances between points in the Manhattan and Euclidean metrics.

Euclidean Median

Knowing that the Manhattan median minimizes the sum of Manhattan distances from itself to the n points of a spatial distribution, we might ask, which point minimizes the Euclidean distances from itself to these same n points?

DEFINITION: EUCLIDEAN MEDIAN

Let (X_i, Y_i), $i = 1, 2, \ldots, n$, be the coordinates of a set of n points. The Euclidean median has coordinates (X_e, Y_e) and is defined as the location that minimizes

$$\sum_{i=1}^{n} \sqrt{(X_i - \bar{X})^2 + (Y_i - \bar{Y})^2} \tag{3-25}$$

Unfortunately, there is no direct method of determining the coordinates (X_e, Y_e). Appendix 3A describes one iterative numerical algorithm developed by Kuhn and Kuenne (1962) which can be used to solve this problem. Only small problems can be solved without the aid of a computer or programmable calculator. For the nine points of Figure 3-11, the location of the Euclidean median is found to be (43.98, 42.05). As illustrated in Figure 3-12, the location of the Euclidean median (the square symbol) is very close to, but distinct from, the mean center and Manhattan median.

Weighted Medians

Just as the mean center can be generalized to the weighted mean center, so, too, can the Euclidean and Manhattan medians. The weighted Manhattan median is defined to have equal weights above and below, to the left and to the right. Unlike its unweighted counterpart, problems of nonuniqueness are not nearly so common. The weighted Euclidean median has drawn much more attention from geographers than the weighted Manhattan median.

DEFINITION: WEIGHTED EUCLIDEAN MEDIAN

Let (X_i, Y_i), $i = 1, 2, \ldots, n$, be the coordinates of a set of n points distributed in the plane. To each of these points there is an attached weight w_i. The weighted Euclidean median (X_{we}, Y_{we}) minimizes

$$\sum_{i=1}^{n} w_i \sqrt{(X_i - X_{we})^2 + (Y_i - Y_{we})^2} \tag{3-26}$$

In other words, the distances between the median and each point are weighted by the value w_i. These weights are defined in the context of the problem at hand. This problem also can be solved using the iterative algorithm proposed by Kuhn and Kuenne. The locations of the weighted mean center, weighted Manhattan median, and weighted Euclidean median for the points in Figure 3-11 are illustrated in Figure 3-16. The weights used in each case are the populations set for the weighted mean center problem. Again, all three locations are reasonably close, although we would expect divergences when there are extreme points (or weights in the spatial distribution).

Interpreted in one way, the weighted Euclidean median can be shown to be the solution to the classical location problem of Alfred Weber. *Weber's problem* is to find the best, or optimal, location for a factory—one that minimizes the sum of transport costs between the factory and two sources of raw materials and between the factory and the market. The situation is illustrated in Figure 3-17. If we make the additional assumptions that (1) transportation costs are a linear function of distance traveled, (2) the weights attached to the points represent the weights of raw material required to produce 1 ton of the finished product, and (3) the weight at the market is 1, then

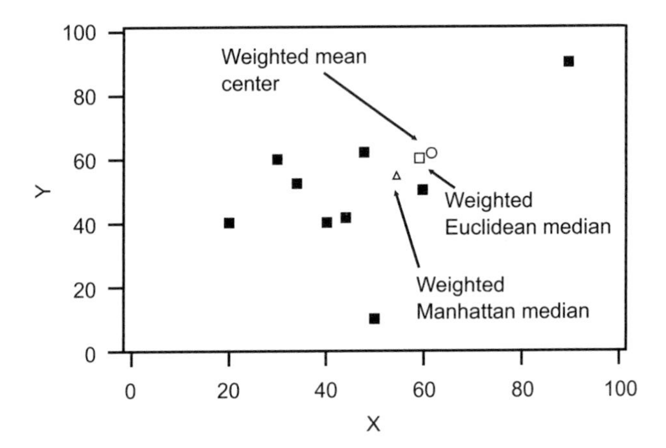

FIGURE 3-16. Weighted measures of central tendency for point distribution in Figure 3-11.

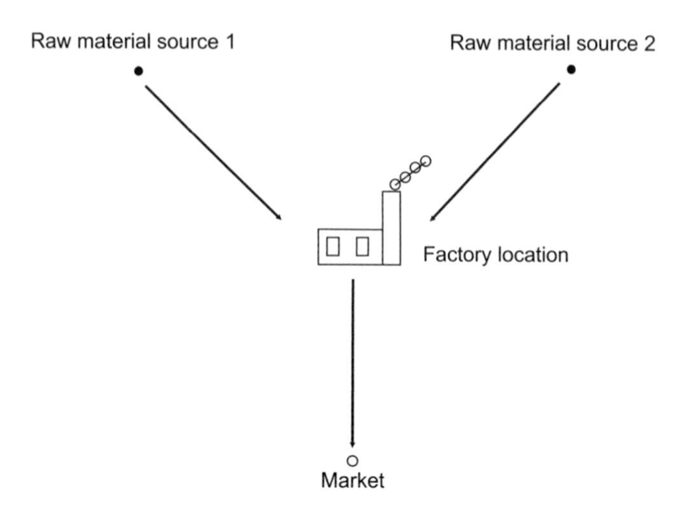

FIGURE 3-17. Weber's problem as the weighted Euclidean median.

the optimal location for the factory coincides with the weighted Euclidean median. The location of the factory represents the resolution of the forces "pulling" the factory toward each of the three locations.

This problem also can be viewed as a public facility location decision. Given the locations of groups of facility users, for example, schoolchildren, where is the best location for a public facility such as a school? The weights might be defined as the number of school-age children living on a block. The objective of minimizing (3–26) can be interpreted as minimizing the total distance traveled by schoolchildren. Note that this also minimizes the average distance traveled by the children. The weighted mean center minimizes the square of the distances traveled by the schoolchildren. Since we are interested in the length of their walk to school, not the *square* of the length of their trip, the weighted Euclidean median is the best solution.

The problem of locating a spatial median is a cornerstone of many private and public facility location problems. One generalization that has drawn a great deal of attention from geographers is the complex problem of locating an entire system of facilities, say five schools, within a spatial distribution of potential users. This is one variant of what are now termed *location-allocation* problems. Once a set of facilities is *located,* consumers or patrons are *allocated* to the appropriate, usually the closest, facility. The optimal locations for a system of five facilities are illustrated in Figure 3-18. This simple extension of the Weber problem is known as the *multiple-source Weber problem.* There is now a rich body of literature on both the theoretical aspects of location-allocation problems and their applications to school, hospital, and other facility systems. Note that across much of this literature, variants of network distances are often used because Euclidean distances are impractical in many real-world situations.

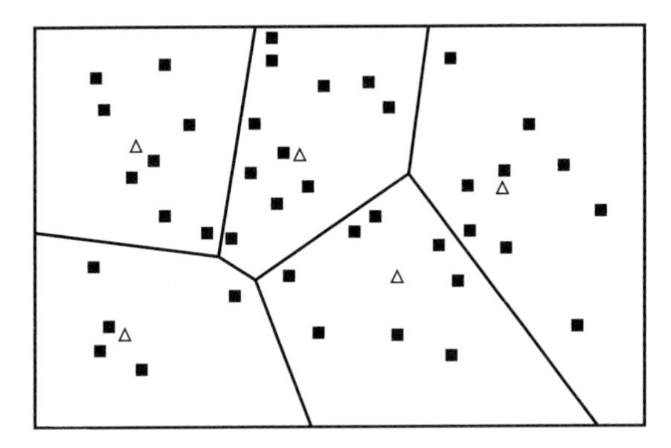

FIGURE 3-18. Optimal locations for a system of five facilities.

MEASURES OF DISPERSION

After central tendency, the second important characteristic of a spatial distribution is its dispersion. Measures of dispersion in descriptive statistics are usually based on the notion of a deviation, the difference in value of an observation from a central value such as the mean or median. For spatial distributions, the notion of a deviation is the actual distance between an observation and the central point.

Standard Distance

The *standard distance* (SD) is the spatial equivalent to the standard deviation.

DEFINITION: STANDARD DISTANCE

Let (X_i, Y_i), $i = 1, 2, \ldots, n$, be the coordinates of a set of n points. Then, the standard distance is defined as

$$SD = \sqrt{\frac{\sum_{i=1}^{n}(X_i - \bar{X})^2}{n} + \frac{\sum_{i=1}^{n}(Y_i - \bar{Y})^2}{n}} \tag{3-27}$$

Alternatively, this formula can be approximated as $SD \approx \sqrt{S_x^2 + S_y^2}$

More dispersed point patterns will have larger standard distances. For example, consider the two point distributions in Figure 3-19. The distribution in Figure 3-19b is clearly more dispersed than the point distribution in Figure 3-19a. The standard distance can be represented graphically, as in Figure 3-19, by drawing circles with a radius equal to the standard distance around the mean center of a distribution. It is also possible to use weighted observations by computing a standard distance around the weighted mean center. Together with the mean center, the standard distance can be

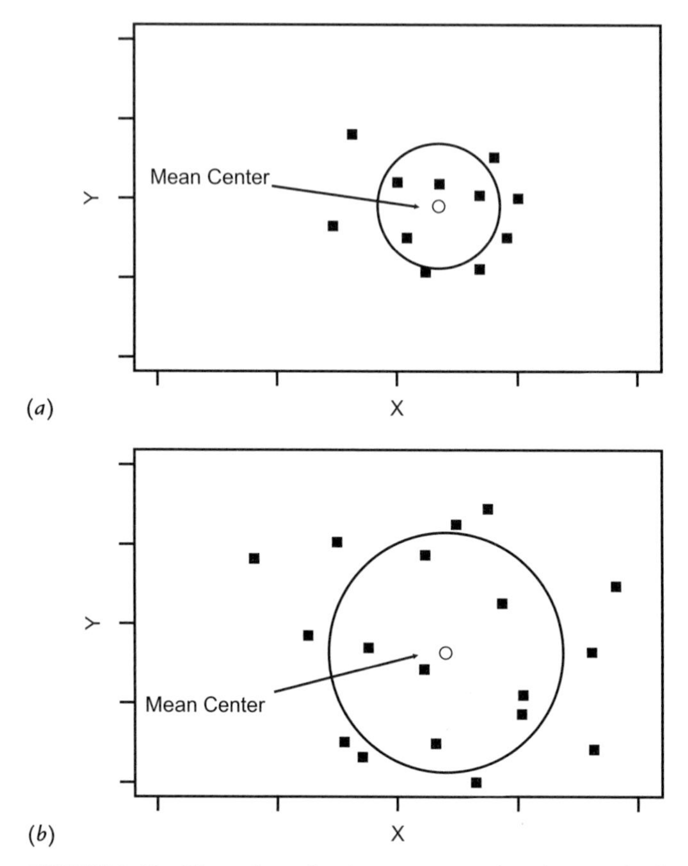

FIGURE 3-19. Dispersion of point patterns using the standard distance.

used to compare and contrast point distributions. Both measures are very sensitive to extreme observations.

Quartilides

Dispersion about the Manhattan median center can be graphically displayed by using the spatial equivalent of the interquartile range. A *quartilide* divides a point distribution into quarters. Surrounding the median are eastern, western, northern, and southern quartilides. Together, these define a rectangle, whose size clearly depends on the dispersion of the set of points. An example is shown in Figure 3-20 for a distribution with $n = 16$ points. In this case, problems of nonuniqueness arise because the median of a set of points with an even number of observations is an area. The median should have eight points to the east and eight to the west. Many lines satisfy this criterion; the one selected here lies *midway* between the two central observations. Similarly, the line locating the median in a north-south direction lies midway between

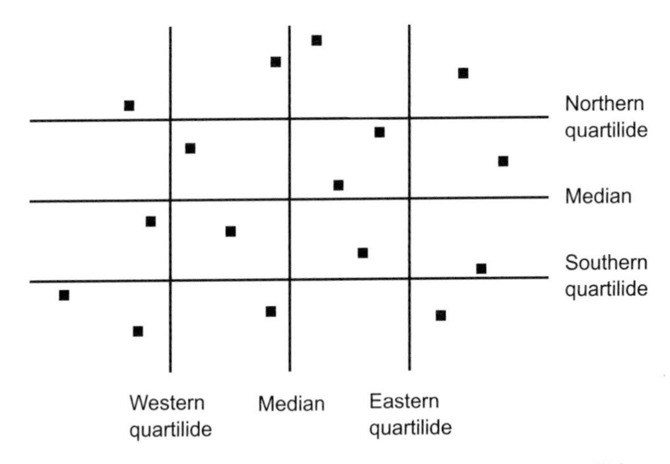

FIGURE 3-20. Dispersion of a point distribution using quartilides.

the central points in this direction. The same convention is used to define the quartilides. The eastern quartilide lies midway between the fourth and fifth easternmost points, the western quartilide lies midway between the fourth and fifth westernmost points, and so on. The larger the rectangle defined by these quartilides, the more dispersed the set of points. Quartilides seem to have been used rarely in the geographical literature.

An Empirical Approach

A useful empirical approach to describe the dispersion of a point pattern is to graph the cumulative frequency of points around some central location in bands of distance. For example, the point data of Figure 3-21a are summarized in this manner in Figure 3-21b. It is easy to see that 80% of the points lie within 6 miles of the mean center, 20 percent within 2 miles, and so on.

One useful benchmark is the uniform distribution—the percentages that would be expected if the points were evenly distributed around the mean center. The number of points within a circle of a given radius from the mean center should be proportional to the area within that circle. Since the area of a circle is proportional to the radius squared, the uniform distribution is not represented as a straight line on the graph in Figure 3-21b. What would a uniform distribution having 80% of its observations within 6 miles of the mean center be like? Since the area within a distance of 4 miles is $\pi(4^4) = 16\pi$ and the area within 6 miles is $\pi(6^2) = 36\pi$, we would expect $16\pi/36\pi = 4/9$ as many points within this radius. Similarly, a uniform distribution would have one-ninth as many points within 2 miles as it does within 6 miles. By comparison, then, the observed point distribution is more concentrated than a uniform distribution within the first 3 miles and less concentrated within the next 3 miles.

Although the measures of central tendency and dispersion can help us to make simple comparisons between different point distributions, they yield very limited

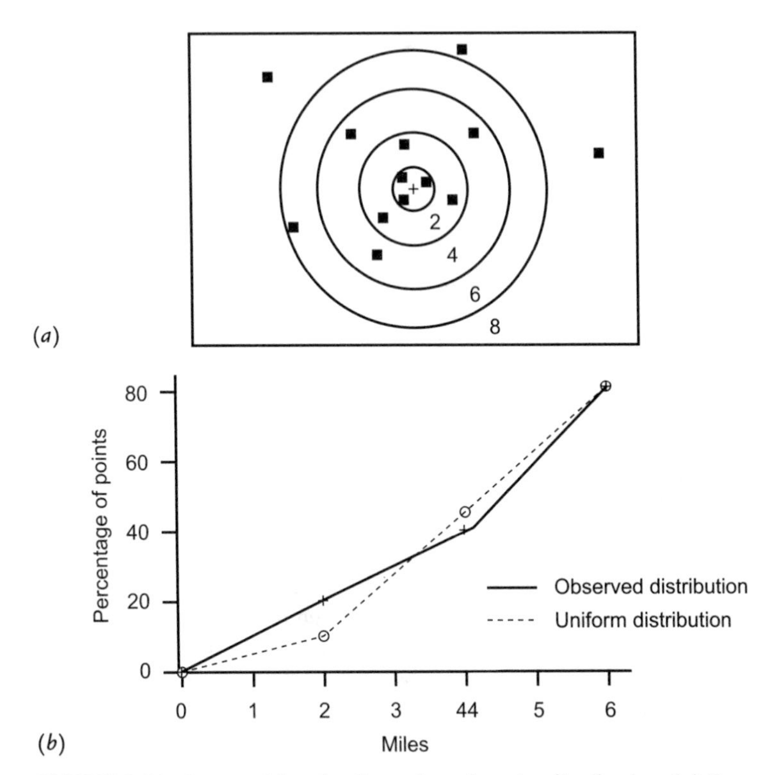

FIGURE 3-21. Summarizing the dispersion of a point distribution. (a) Concentric circles with distances. (b) Graphical comparison to a uniform distribution.

information about the nature of the point patterns. Some point patterns are uniform, consisting of a more or less even distribution of points on the map. Other patterns are very clustered, with many points located in one area of the map and large areas without any points. A random pattern contains elements of both clustered and uniform patterns. In Chapter 14, more sophisticated methods are used to assess the characteristics of a point pattern.

Descriptive Statistics and Spatial Data: Three Problems

Three recurring problems arise when virtually any descriptive statistic is calculated using spatial data. Therefore, it is necessary to be extremely careful in interpreting the results of a statistical analysis based on location observations. In fact, many of the recent advances in spatial statistics have been developed in response to the need to overcome the limitations imposed by these problems. The three problems are often referred to as the boundary problem, the problem of modifiable units, and the problem of pattern.

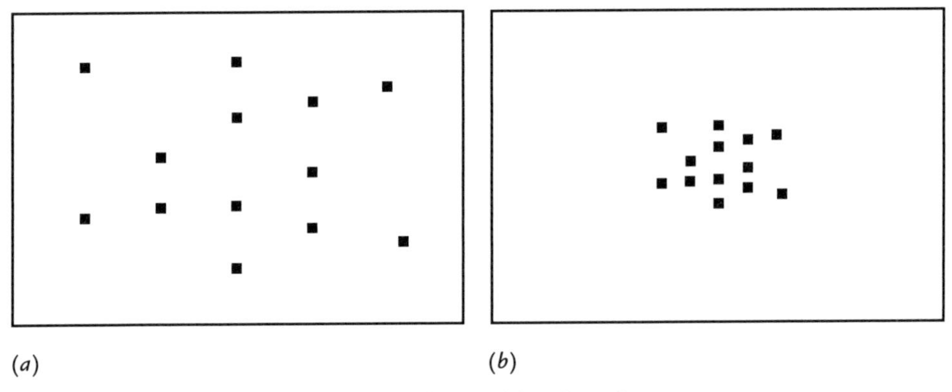

FIGURE 3.22. Boundary problem for point patterns.

BOUNDARY PROBLEM

The location of the boundary of the study area, as well as the placement of the internal boundaries in an areal design, is often a crucial question in geographical research. Poorly chosen designs can lead to several problems. Let us illustrate the potential difficulties with two examples. First, consider the two point patterns of Figure 3-22. Both are identical and would yield identical values for any measure of central tendency or dispersion. However, in relation to their boundaries these patterns are clearly different. The distribution in (a) could be described as a reasonably regular, dispersed pattern. In (b), the pattern could be described as clustered. This means that the standard distance, or any other measure of dispersion, cannot be interpreted independent of the study area.

As a second example, consider the problem of locating areal boundaries within some study area. Suppose the location of some phenomenon, for example, a particular ethnic group in a city, is depicted as the shaded area in the two maps of Figure 3-23.

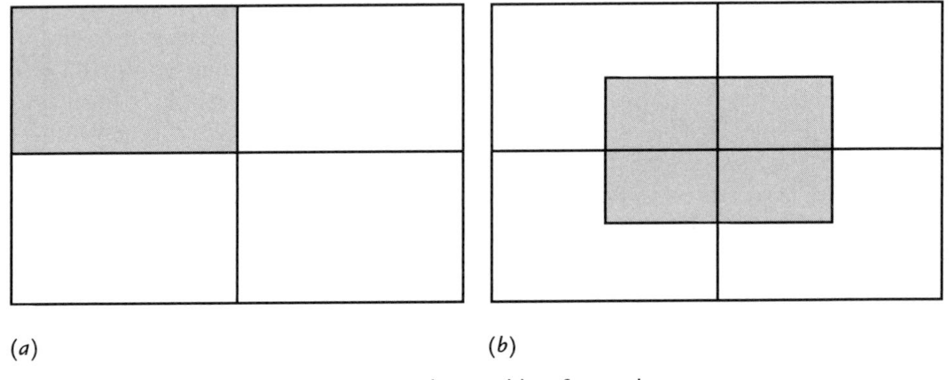

FIGURE 3-23. Boundary problem for areal patterns.

Despite the fact that the "map" is the same in each case, the values of location quotients, coefficients of localization, and Gini coefficients would be markedly different for these two cases. All measures for Figure 3-23a would indicate significant areal concentration of the ethnic group, since the boundary of one zone completely enclosed the neighborhood occupied by the ethnic group. For the boundaries in (b), these same measures would tend to indicate a rather even distribution of the ethnic group in this study area. In short, the map in (a) suggests ethnic *segregation,* and the map in (b) suggests *integration.*

Although the results may not always be as dramatic as in this case, boundary locations can mask certain map patterns. The moral is clear—summary statistics can be interpreted only for the particular areal divisions on which their calculation is based. If this framework is poorly chosen, the results may be at the minimum misleading and possibly even false. Where there is some control over the placement of areal boundaries, these problems can sometimes be minimized by the use of a fine areal breakdown within an appropriately delimited study area.

MODIFIABLE UNITS PROBLEM

A second problem, termed the *modifiable units problem,* arises because data are often collected over arbitrary areal divisions or spatial units. For example, income might be collected by census tract, by metropolitan area, by township, by county, and so on. We might be interested in household income, but our data are most likely available on some other unit of analysis that has little or no theoretical justification. Unfortunately, the conclusions reached are often affected by the way the areal units are defined. Because the units are more or less arbitrary, the problem is termed the *modifiable units problem.*

This problem manifests itself in two ways, which we refer to as "scale" and "shape." The scale issue is one in which the values for many descriptive statistics vary systematically when increasingly aggregated areal data are used. An example of the effects of areal aggregation is shown in Figure 3-24. In (a), the areal distribution of some variable X is shown in an area with 16 cells. The mean value of the map is $\mu =$ 75, and the variance is $\sigma^2 = 9.75$. Aggregate these data by joining neighboring cells to create eight zones, as in Figure 3-24b. In each two-cell aggregation, the value of X for each zone is the average value of the two smaller cells from which it has been created. For example, the zone in the northwest corner of (b) has a value of $(4 + 8)/2 =$ 6. What does change with this aggregation? The mean remains constant at 7.5, but the variance declines significantly, to $\sigma^2 = 1.75$. Much of the variation in X is now lost. Since geographers are often interested in spatial variation, this is particularly unfortunate. All observations on the eight-zone map are within a range $6 < X < 9$, but the values on the original map have a wider range of $2 < X < 12$. At an even higher level of aggregation, note that the four-district representation in Figure 3.24c has a variance of zero! This is an obvious distortion of the real situation.

In many instances, then, spatial aggregation tends to reduce the variation depicted on a map. Comparisons of maps of a variable at different levels of aggregation must take into account the ramifications of this variance reduction. The problem be-

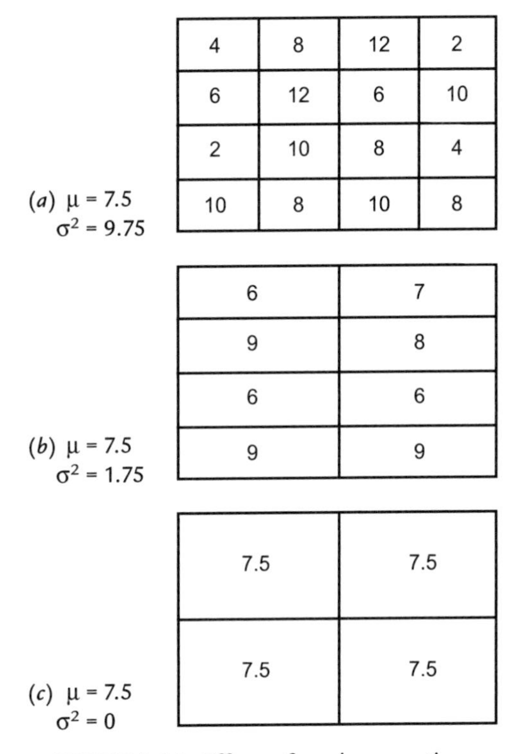

FIGURE 3-24. Effects of areal aggregation.

comes even more acute when we try to examine the *relationship* between two maps of different variables.

The shape issue arises because even at the same scale of analysis, different areal definitions can have a substantial impact on the values of most descriptive statistics. To see this, suppose the aggregation of Figure 3-24a into eight zones from 16 cells is accomplished by joining contiguous north-south rather than east-west neighbors, as in Figure 3-25a, or by using a mixed pattern of north-south and east-west aggregations, as in Figure 3-25b. Again, both systems lead to the same map mean of 7.5. However, although each of these aggregations has the same variance, it is higher than that of the eight-zone map of Figure 3-24b. The effects of using differently shaped areal units are not nearly so predictable as the effects of aggregation. Although all aggregations decrease the variance (it is possible to maintain the variance if the zones joined are exactly alike), some systems will result in significantly lower variances, others not. In this sense, the aggregations in Figure 3-25 are superior to the one in Figure 3-24b. They both are more representative of the more detailed map in Figure 3-24a. But what if the map in Figure 3-24a were not known? Can we have faith that the eight-zone map used in our analysis is truly representative of the actual variation of the variable across the map? We cannot be sure.

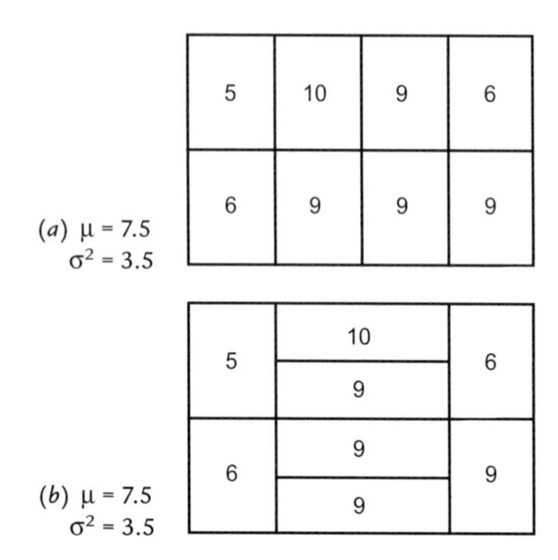

FIGURE 3-25. Problem of modifiable units.

One conclusion we can draw from this analysis is that it is always better to join similar zones when we must aggregate the data. This will preserve the variation in the original map as much as possible. Since one of the "laws" of geography is that closer places are more alike than distant places, contiguous areal aggregations are likely to be less disruptive than aggregations of areas that are not close together.

PATTERN PROBLEM

All the methods described in this section share one shortcoming: they are generally incapable of assessing the type of pattern that exists on a map. Consider the two contrasting patterns of Figure 3-26a and b. Again, suppose the shaded areas represent residential concentrations of some particular ethnic group in this city. The coefficient of localization, Gini coefficient, or Lorenz curve would indicate a significant level of

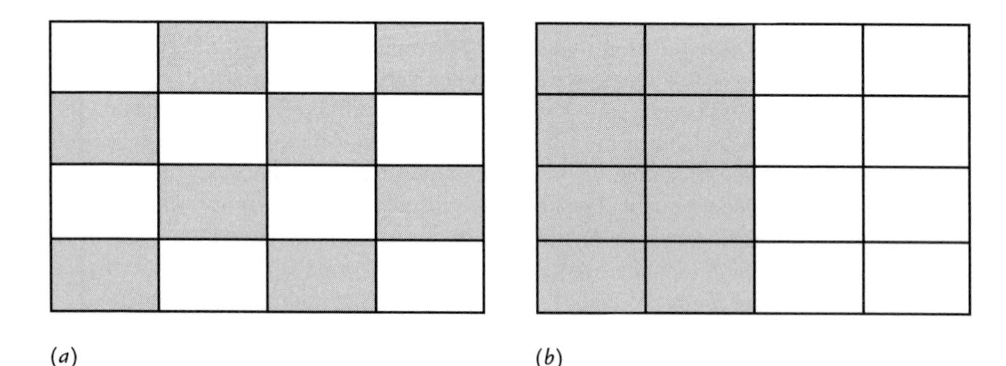

FIGURE 3-26. Illustrating the problem of map pattern.

areal concentration of this ethnic group in both maps. There are eight zones completely populated by the group and eight zones where they are completely absent. Moreover, and even more important, because these measures are insensitive to pattern, both maps would have identically valued coefficients of areal concentration. But are these maps similar? Do both maps have about the same degree of concentration of this ethnic group? One could argue that there is much more segregation in the city mapped in Figure 3-26b. The host population excludes them not only from their neighborhoods, but also from nearby neighborhoods. Techniques capable of distinguishing between areal patterns and point distributions are discussed in Chapter 14.

3.6. Summary

In the first part of this chapter we were concerned with describing a set of observations using statistics. These statistics provide numerical summaries of the distributions of variables that were explored graphically in Chapter 2. These methods are useful for describing datasets whether they represent statistical populations or samples from these populations.

The average or typical value of a variable usually can be best summarized by the arithmetic mean, although the mode and the median are preferred in particular situations. For making inferences, the mean is preferred. The standard deviation is the most commonly used measure of the variability of a dataset, though its square, the variance, is most useful in inferential statistics. Under certain conditions, the interquartile range proves a better descriptive measure than the standard deviation.

The standard deviation and the mean are the best summary measures of a variable. They also can be used to locate an observation within a frequency distribution. For example, we might speak of an extreme observation as being more than two standard deviations from the mean. The mean and standard deviation are sometimes referred to as the first and second moments of a distribution. The skewness and kurtosis of a variable can be calculated as higher moments of a distribution.

We then explored how time-series data can be smoothed using the mean as a moving average. Smoothing a time series is used to uncover patterns in temporal data and to separate a systematic trend in the series from random fluctuations.

Finally, we discussed more specialized descriptive statistics that are frequently used to characterize spatial data. Geographers share a common interest in the analysis of spatial distributions. Whether the data they collect are for areas, points, directions, or networks, specialized techniques can be used to summarize the maps of spatially varying phenomena. In fact, some of the techniques geographers often use to analyze spatial patterns are used to analyze aspatial variables. For areal data, the relative concentration of a variable across a map can be measured by the location quotient, the Gini coefficient, the coefficient of localization, or the Lorenz curve. For point data, measures of central tendency and dispersion such as the mean center and standard distance can be used to summarize these distributions. These measures have properties similar to conventional statistical measures such as the mean and standard deviation.

Virtually all the descriptive statistics that are used with spatial data have several important shortcomings. First, they are extremely sensitive to the choice of boundaries. Second, they are not independent of the way areal units are defined. In particular, both the scale or degree of areal aggregation and the shape of the areal units significantly affect the values of many statistics. In fact, one could almost take some maps and determine the areal subdivision that result in any desired value for some statistics! Finally, most methods are not capable of assessing the nature of the pattern exhibited on the map. More specialized methods presented in later chapters address several of these issues.

Appendix 3a. Review of Sigma Notation

Just as we use plus (+) to denote addition or minus (–) to indicate subtraction, we can use the symbol Σ (pronounced sigma) to denote the repeated operation of addition. The symbol Σ is known as the *summation operator,* since it literally means "take the sum of". In conjunction with the summation operator, we normally utilize an *index of summation* to specify the elements to be summed. For example,

$$\sum_{i=1}^{4} X_i$$

is interpreted as the sum of X_1 to X_4, or $X_1 + X_2 + X_3 + X_4$. If $X_1 = 1$, $X_2 = 7$, $X_3 = 9$, and $X_4 = 16$, then $\Sigma_{i=1}^{4} X_i = 1 + 7 + 9 + 16 = 36$. The index of summation can be placed below the sigma or just to the right of the summation. Also, any lowercase letter can be used as the index of summation. Usually, the variable i is used in a single summation, although j and k are also commonly used.

The sigma operator can also be used when an undetermined number of elements are to be summed. Thus, $X_1 + X_2 + \ldots + X_n$ can be expressed as $\Sigma_{i=1}^{n} X_i$. The length of the summation depends on the value of n. When there is no possibility of confusion, it is convenient to suppress the index and limits of the summation. We may thus write ΣX when we mean the sum is to be taken over "all" values of X. The specific interpretation of *all* should be evident from the context of the summation. In this text we find it convenient to suppress these indices when the meaning is $\Sigma_{i=1}^{n}$.

Several rules for simplifying summations periodically are used in the text. All the following rules can be verified using elementary algebra.

RULE 1: If X and Y are two variables, then

$$\sum_{i=1}^{n} (X_i + Y_i) = \sum_{i=1}^{n} X_i + \sum_{i=1}^{n} Y_i$$

To see this, expand the left-hand side:

$$\begin{aligned}\sum_{i=1}^{n} (X_i + Y_i) &= (X_1 + Y_1) + (X_2 + Y_2) + \ldots + (X_n + Y_n) \\ &= (X_1 + X_2 + \ldots + X_n) + (Y_1 + Y_2 + \ldots + Y_n) \\ &= \sum_{i=1}^{n} X_i + \sum_{i=1}^{n} Y_i\end{aligned}$$

RULE 2: If k is a constant and X is a variable, then

$$\sum_{i=1}^{n} kX_i = k\sum_{i=1}^{n} X_i$$

Again, expand the left-hand side:

$$\begin{aligned}\sum_{i=1}^{n} kX_i &= kX_1 + kX_2 + \ldots + kX_n \\ &= k(X_1 + X_2 + \ldots + X_n) \\ &= k\sum_{i=1}^{n} X_i\end{aligned}$$

RULE 3: For any two variables X and Y,

$$\sum_{i=1}^{n} X_i Y_i = X_1Y_1 + X_2Y_2 + \ldots + X_nY_n$$

Note that this does not equal $(\Sigma_{i=1}^{n} X_i)(\Sigma_{i=1}^{n} Y_i)$. Consider the following simple example:

i	X	Y
1	2	4
2	6	8
3	10	12

First, we calculate $\Sigma_{i=1}^{3} X_i Y_i = 2 \cdot 4 + 6 \cdot 8 + 10 \cdot 12 = 176$. Then we calculate $(\Sigma_{i=1}^{3} X_i)(\Sigma_{i=1}^{3} Y_i)$ as $(2 + 6 + 10) \cdot (4 + 8 + 12) = 18 \cdot 24 = 432$.

RULE 4: If k is a constant, then

$$\sum_{i=1}^{n} k = nk$$

Note that $\sum_{i=1}^{n} k = \underbrace{(k + k + k + \ldots + k)}_{n \text{ times}} = nk$

RULE 5: If m is an exponent and X is a variable, then

$$\sum_{i=1}^{n} X_i^m = X_1^m + X_2^m + \ldots + X_n^m$$

RULE 6: If m is an exponent and X is a variable, then

$$\left(\sum_{i=1}^{n} X_i\right)^m = (X_1 + X_2 \ldots X_n)^m$$

Suppose that $m = 2$, $n = 3$, $X_1 = 2$, $X_2 = 6$, $X_3 = 10$. Then

$$\sum_{i=1}^{3} X_i^2 = 2^2 + 6^2 + 10^2 = 140$$

$$\left(\sum_{i=1}^{3} X_i\right)^2 = (2 + 6 + 10)^2 = 324$$

In general, we see from Rules 5 and 6 that

$$\sum_{i=1}^{n} X_i^m \neq \left(\sum_{i=1}^{n} X_i\right)^m$$

We can also work with double summations, that is, those with two separate subscripts. When we encounter the expression $\Sigma_{i=1}^{n}\Sigma_{j=1}^{n}X_{ij}$ we work from inside out. First we use the j subscript and then the i subscript so that

$$\begin{aligned}\sum_{i=1}^{n}\sum_{j=1}^{n} X_{ij} &= \sum_{i=1}^{n}(X_{i1} + X_{i2} + \ldots X_{in}) \\ &= (X_{11} + X_{12} + \ldots X_{1n}) + (X_{21} + X_{22} + \ldots X_{2n}) \\ &\quad + \ldots + (X_{n1} + X_{n2} + \ldots X_{nn})\end{aligned}$$

If X is an array or matrix of the form

	$j = 1$	$j = 2$	...	$j = n$
$i = 1$	X_{11}	X_{12}	...	X_{1n}
$i = 2$	X_{21}	X_{22}	...	X_{2n}
:	:	:	:	:
$i = n$	X_{n1}	X_{n2}	...	X_{nn}

then the expression $\Sigma_{i=1}^{n}\Sigma_{j=1}^{n}X_{ij}$ leads to the sum of all the entries in the matrix. Note that we can also use summation notation to compute the sum of certain rows or columns of this matrix. For example,

$$\sum_{j=1}^{n} X_{2j} = X_{21} + X_{22} + \ldots + X_{2n}$$

specifies the sum of the second row. Similarly, the sum of column 1 is

$$\sum_{i=1}^{n} X_{i1} = X_{11} + X_{21} + \ldots + X_{n1}$$

Appendix 3b. An Iterative Algorithm for Determining the Weighted or Unweighted Euclidean Median

One of the most efficient algorithms for solving the weighted or unweighted Euclidean median problem was developed by Kuhn and Kuenne (1962). The following algorithm is a simple variant of their method.

Let (X_i, Y_i), $i = 1, 2, \ldots, n$, be the set of n given points with weights w_i, $i = 1, 2, \ldots, n$. The location of the current estimate of the median in the tth iteration is (X^t, Y^t). As the algorithm proceeds, (X^t, Y^t) gradually converges to (X_e, Y_e). The algorithm stops whenever the estimates of the coordinates in successive iterations are less than some predetermined tolerance level TOL. For example, if we wish the coordinates of

(X_e, Y_e) found by the algorithm to be within 0.01, we set TOL = 0.01. Since it is an approximating algorithm, the method is more efficient if a "good" starting point is selected. A good point is one that is close to the answer. The bivariate mean $(\bar{X}, \bar{Y})$ and weighted mean $(\bar{X}_w, \bar{Y}_w)$ are good and obvious choices for a starting point. So $(X^1, Y^1) = (\bar{X}, \bar{Y})$. The algorithm can be described as the following sequence of steps:

1. Calculate the distance from each point (X_i, Y_i) to the current estimate of the median location

$$\text{where } d_i^t = \sqrt{(X_i - X^t)^2 + (Y_i - Y^t)^2}$$

   where d_i^t is the distance from point i to the median during the tth iteration.
2. Determine the values $K_i^t = w_i / d_i^t$. Note that in the unweighted case all values of w_i are set equal to 1.
3. Calculate a new estimate of the median from

$$X^{t+1} = \frac{\sum_{i=1}^{n} K_i^t X_i}{\sum_{i=1}^{n} K_i^t} \qquad Y^{t+1} = \frac{\sum_{i=1}^{n} K_i^t Y_i}{\sum_{i=1}^{n} K_i^t}$$

4. Check to see whether the location has changed between iterations. If $|X^{t+1} - X^t|$ and $|Y^{t+1} - Y^t| \leq$ TOL, stop. Otherwise, set $X^t = X^{t+1}$ and $Y^t = Y^{t1}$ and go to step 1.

The algorithm usually converges in a small number of iterations. For the nine points of Figure 3-11, for the unweighted case, the following steps summarize the results of the algorithm initialized with TOL = 0.10.

Iteration	X^t	Y^t
1	44.867	45.517
2	43.865	44.060
3	43.672	43.204
4	43.710	42.742

The location of the unweighted median is thus estimated as (43.71, 42.74).

REFERENCES

American Association of University Professors, 2002.

H. W. Kuhn and R. E. Kuenne, "An Efficient Algorithm for the Numerical Solution of the Generalized Weber Problem in Spatial Economics," *Journal of Regional Science* 4 (1962), 21–33.

S. M. Stigler, "Do Robust Estimators Work with Real Data?," *Annals of Statistics* 5 (1977), 1055–1078.

FURTHER READING

Virtually all elementary statistics textbooks for geographers include a discussion of at least part of the material presented in this chapter. Students often find it useful to consult other textbooks to reinforce their understanding of key concepts. Four textbooks intended for geography students are those by Clark and Hosking (1986), Ebdon (1985), Griffith and Amrhein (1991), and O'Brien (1992). A useful first reference on geostatistics is Neft (1966), while Isaaks and Srivastava (1989) is more comprehensive and more complex. Extensions of statistical analysis to directional data are provided by Mardia (1972). Geographical applications of graph theory and network analysis are reviewed in Tinkler (1979) and Haggett and Chorley (1969), respectively.

W. A. V. Clark and P. L Hosking, *Statistical Methods for Geographers* (New York: Wiley, 1986).
D. Ebdon, *Statistics in Geography, 2nd ed.* (Oxford: Basil Blackwell, 1985).
D. A. Griffith and C. G. Amrhein *Statistical Analysis for Geographers* (Englewood Cliffs, NJ: Prentice-Hall, 1991).
P. Haggett and R. J. Chorley, *Network Analysis in Geography* (London: Edward Arnold, 1969).
E. H. Isaaks and R. Mohan Srivastava, *Applied Geostatistics* (Oxford: Oxford University Press, 1989).
D. Neft, *Statistical Analysis for Areal Distributions* (Philadelphia: Regional Science Research Unit, 1966).
L. O'Brien, *Introducing Quantitative Geography* (London: Routledge, 1992).
K. J. Tinkler, "Graph Theory," *Progress in Human Geography* 3 (1979), 85–116.

DATASETS USED IN THIS CHAPTER

The following data sets employed in this chapter can be found on the website for this book.

cavendish.html
dodata.html
traffic.html

PROBLEMS

1. Explain the following terms:
   - Mean
   - Median
   - Percentile
   - Interquartile range
   - Standard deviation
   - Coefficient of variation
   - Skewness
   - Kurtosis
   - Location quotient
   - Lorenz curve
   - Gini coefficient
   - Mean center
   - Standard distance
   - Moving average
   - Smoothing

2. When is the mean not a good measure of central tendency? Give an example of a variable that might best be summarized by some other measure of central tendency.

3. Consider the following five observations: –8, 14, –2, 3, 5.
   a. Calculate the mean.
   b. Find the median.
   c. Show that the Properties 1, 2, and 3 hold for this set of numbers.

4. Life expectancy at birth (in years) is shown for a random sample of 60 countries in 1979. The data are drawn from the United Nations Human Development Report:

   76.7 64 71.1 40.1 52.3 64.5 74.4 78.6 69.5 62.6 57.9 42.4 69.5 73 50.8 78 71 66.6 48.6 76.9 70 68.9 70 72.4 65.1 39.3 69.6 47.9 73.1 72.9 58 69.9 44.9 51.7 74 69.4 70.5 57.3 45 65.4 78.5 73.9 73 64.4 69.5 45.2 78.1 78.5 71.4 64 48.5 64 66.3 72.9 78.1 47.2 72.4 68.2 72 76 78.1

   a. Calculate the following descriptive statistics for these data: mean, variance, skewness and kurtosis.
   b. Write a brief summary of your findings about the distribution of life expectancy in these countries.

5. Using the data of Problem 4, determine
   a. Range
   b. First and third quartiles
   c. The interquartile range
   d. The 60th percentile
   e. The 10th percentile
   f. The 90th percentile

6. The following data describe the distribution of annual household income in three regions of a country:

Region	$\bar{X}$	s
A	$42,000	$16,000
B	33,000	13,000
C	27,000	11,000

   a. In which region is income the most evenly spread?
   b. In which region is income the least evenly spread?

7. The following data represent a time series developed over 45 consecutive time periods (read the data by rows):

   569 416 422 565 484 520 573 518 501 505 468 382 310 334 359 372 439 446 349 395 461 511 583 590 620 578 534 631 600 438 534 467 457 392 467 500 493 410 412 416 403 433 459 467

a. Smooth these data using a three-term moving average.
b. Smooth these data using a five-term moving average.
c. Graph the original data and the two smoothed sequences from (a) and (b).
d. Graph the fluctuating component from each of the two sequences.

8. Consider a geographic area divided into four regions, north, south, east, and west. The following table lists the population of these regions in three racial categories: Asian, Black, and White:

	Population			
Region	Asian	Black	White	Total
North	600	200	700	1500
South	200	300	400	900
East	150	150	250	550
West	100	300	200	600
Total	1050	950	1550	3550

a. Calculate location quotients for each region and each racial group in the population.
b. Calculate the coefficient of localization for each group.
c. Draw the Lorenz curves for each of the racial groups on a single graph.
d. Write a brief paragraph describing the spatial distribution of the three groups using your answers to (a), (b), and (c).

9. What is the smallest possible value for a location quotient? When can it occur? What is the largest possible value for a location quotient?

10. Consider the following 12 coordinate pairs and weights:

X coordinate	Y coordinate	Weight
60	80	4
45	45	5
70	60	6
55	60	7
65	75	4
70	45	3
80	60	2
45	75	2
30	70	2
55	50	1
70	65	1
0	40	1

a. Locate each point, using the given coordinates, on a piece of graph paper.
b. Calculate the mean center (assume all weights are equal to 1). Calculate the weighted mean center. Calculate the Manhattan median. Locate each measure on the graph paper.

c. Calculate the standard distance. Draw a circle with radius equal to the standard distance centered on the mean center.
d. (Optional) Use the Kuehn and Kuenne algorithm (see Appendix 3b) to determine the Euclidean median.

11. For the census of the United States (*www.census.gov*) or Canada (*www.statisticscanada.ca*), or an equivalent source, generate the value of a variable for a region containing at least 30 subareas. For example, you could look at some characteristic of the population for the counties within a state or province.
    a. Calculate the mean and variance for the variable of interest across the subareas.
    b. Group contiguous subareas in pairs so that there are 15 or more zones; calculate the mean and variance for the variable of interest across the newly constructed zones.
    c. Group contiguous zones in part (b) to get eight or more areas; find the mean and variance of the variable of interest across these areas.
    d. Continue the process of aggregating regions into larger geographical units until there are only two subregions. At each step calculate the mean and variance of the variable of interest across the areas.
    e. The mean of the variable at each step should be equal. Why?
    f. Construct a graph of the variance of the variable (Y-axis) versus the number of subareas (X-axis). What does it reveal?

12. For the same data used in Problem 11b, generate five different groupings in which subareas are joined in pairs. Calculate the variance of the variable of interest for each grouping. Do your results confirm the existence of the problem of modifiable units?

4

Statistical Relationships

Chapters 2 and 3 examined how to display, interpret, and describe the distribution of a single variable. In this chapter, we extend the discussion by exploring statistical relationships between variables. A few moments of reflection should confirm the importance of *bivariate analysis,* or thinking about how the values assumed by one variable are related to those assumed by another variable. Examples abound:

1. Is human activity, measured perhaps by global economic output, related to the atmospheric temperature of the earth?
2. Is the rate of taxation related to economic growth?
3. Is drug X an effective treatment for disease Y?
4. Is the value of the Dow Jones Industrial Average stock market index on one day linked to its value on another day?
5. Is average household income in a state related to average household income in neighboring states? Alternatively, do states with high (low) average incomes tend to cluster in space?

Just as we used graphical and numerical techniques to explore the distribution of values of a single variable, we can use related techniques to investigate and describe the relationship between variables. While the nature of a relationship, or the statistical dependence, between two quantitative variables will be our focus, we will also explore how to examine the relationship between a quantitative variable and a qualitative variable. Investigation of the interaction between more than two variables, or what we call multivariate analysis, occupies later chapters of the book.

Bivariate data comprise a set of observations each one of which is associated with a pair of values for two variables, X and Y. For example, college applicants often are characterized by their SAT scores and their GPA. Atmospheric scientists are interested in tracking levels of carbon dioxide and the temperature of the earth's atmosphere over a series of observational units, typically years. Development economists might compare per capita income with life expectancy across countries. In this case, the observations would be individual countries.

The analysis of bivariate data in this chapter is organized as follows. Section 4.1 reviews the question of statistical dependence and highlights some of the main issues that will concern us throughout much of the rest of the chapter. In Section 4.2 we show how the graphical techniques introduced in Chapter 2 can be used to compare two distributions. These distributions may represent observations from different variables, or they may represent different samples on a single variable. We end the section looking at scatterplots that display the joint distributions of two variables. The *correlation coefficient* is introduced in Section 4.3 as a measure of the strength of the relationship between two quantitative variables. Section 4.4 discusses the closely related concept of *regression.* In Section 4.5 we extend the concept of correlation to investigate the relationship between successive values of a single time-series variable, what we call temporal or serial *autocorrelation.* Exploration of spatial autocorrelation, the relationship between values of a single variable distributed over space, is left to Chapter 14.

It is important to look at your data before engaging in bivariate analysis. Looking at your data can be done efficiently using the graphical methods discussed in Chapter 2. Examining the distributions of each of your variables independently often provides insights into how they might be employed in bivariate studies. Such examination can also alert you to potential issues that limit which bivariate techniques can be used, for example, if a variable has a non-normal distribution.

4.1. Relationships and Dependence

In statistics, when we say that two variables are related, we mean that the value(s) assumed by one of the variables provides at least some information about the value(s) assumed by the other variable. It might be that one variable has a causal influence on the other variable, or it might be that both variables are influenced by yet another variable or by a combination of other variables. Statistics can't speak to how relationships arise, it merely provides methods for detecting and analyzing relationships. In what follows, therefore, we will use the word "influence" simply to mean that one variable is connected to another, either directly or indirectly through the action of other variables.

Consider Table 4-1, which shows the values of two *discrete,* integer-valued, quantitative variables, X and Y. Variable X can assume the values 1, 2, and 3, while variable Y can assume the values 1, 2, 3, and 4. For each value of X there are 100

TABLE 4-1
Dependent Variables: A Bivariate Sample of 300 Observations

	$Y = 1$	$Y = 2$	$Y = 3$	$Y = 4$	All Y
$X = 1$	5	15	25	55	100
$X = 2$	25	30	30	15	100
$X = 3$	35	30	30	5	100
All X	65	75	85	75	300

TABLE 4-2
Independent Variables: A Bivariate Sample of 300 Observations

	$Y = 1$	$Y = 2$	$Y = 3$	$Y = 4$	All Y
$X = 1$	25	30	30	15	100
$X = 2$	25	30	30	15	100
$X = 3$	25	30	30	15	100
All X	75	90	90	45	300

observations as shown in the row totals. Notice that for $X = 1$, the most common value of Y is 4, occurring 55 out of 100 times. If we knew that $X = 1$, we could guess $Y = 4$ and be right more often than not. By contrast, for $X = 3$, that value of Y occurs only 5% of the time. Guessing that $Y = 4$ when $X = 3$ would be a mistake; we would be better off guessing that $Y = 1$. If we had no information about X, we would use the bottom row of the table. In that case, the best guess we can make is $Y = 3$, but that value occurs only slightly more often than $Y = 2$ and $Y = 4$. This is an example where the value of one variable (X) provides information about another (Y). These variables are dependent, in the sense that the distribution of one variable depends on the value of the other variable.

Suppose that the two variables X and Y are independent. In that case we might see the distribution shown in Table 4-2. Notice that the distribution of Y is the same for all values of X. Thus, the value of X provides no information about the value of Y: the variables are statistically independent. Note that this does not mean that all values of Y are equally likely. Indeed, in Table 4-2 a value of $Y = 3$ occurs twice as often as $Y = 4$. Independence means that there is no change in the relative occurrence of the different values of one variable as values of the other variable change.

DEFINITION: STATISTICAL DEPENDENCE
When the probability of a variable taking a particular value is influenced by the value assumed by another variable, then the two variables are statistically dependent.

Bivariate analysis typically is undertaken to determine whether or not a statistical relationship exists between two variables, and if it does, to explore the direction and the strength of that relationship. The statistical dependence between two variables might be positive or negative or too complex to describe with a single word, and it might be strong or weak. In the next few sections we examine how to look for relationships between variables and how to characterize them.

4.2. Looking for Relationships in Graphs and Tables

When we explore relationships, or statistical dependence between variables, often we begin by using some of the principles of describing and summarizing data outlined in Chapters 2 and 3:

1. Display the data in a graph or table that allows comparison between the variables.
2. Summarize the general patterns in the data and look for obvious departures, or outliers, from that pattern.
3. Describe the direction and strength of statistical dependence numerically.

We start here with the first two of these tasks, examining how to use some of the graphical techniques presented in Chapter 2 to explore the relationship between two variables. In the following examples, we examine how to use qualitative information to get a better understanding of the distribution of a quantitative variable. We then move on to investigate scatterplots, the most common way of exploring the relationship between two quantitative variables.

Qualitative versus Quantitative Variables

We commonly encounter sets of data that include both qualitative and quantitative variables. Areal spatial data provide an example, where we might have quantitative information on a variable *X*, distributed across a number of areas or regions. Geographers use the qualitative, areal data to map the quantitative variable in the search for spatial patterns. In the biomedical field, the health-care workers often have quantitative data on patient health, together with categorical or qualitative data on whether a particular drug has been administered. When examined across a series of patients, researchers can determine the effectiveness of a drug in combating an illness. Qualitative variables are frequently used to *stratify* or to group observations on a quantitative variable, in order to gain understanding of the nature of the distribution of that variable. The graphical techniques of Chapter 2 can be used for this task, as we demonstrate below.

EXAMPLE 4-1. Fisher's Irises. Ronald Fisher was a geneticist and statistician who developed many of the foundations of modern mathematical statistics. In the early 1930s, he published a dataset on the characteristics of three different species of iris to illustrate the principles of discriminant analysis. Table 4-3 contains a portion of Fisher's iris data set. (The complete dataset can be found on the book's website.)

TABLE 4-3
A Portion of Fisher's Iris Data

Species name	Petal width	Petal length	Sepal width	Sepal length
I. setosa	2	14	33	50
I. virginica	24	56	31	67
I. virginica	23	51	31	69
I. versicolor	20	52	30	65
I. versicolor	19	51	27	58

Note: $n = 150$; only five observations are shown here; all measurements are in millimeters.
Source: Fisher (1936).

TABLE 4-4
Stem-and-Leaf Plot of Sepal Length

a. All species combined

Stems	Leaves
4*	3444
4·	566667788888899999
5*	000000000111111111222223444444
5·	5555555666666777777778888888999
6*	000000111111222233333333334444444
6·	5555566777777778889999
7*	0122234
7·	677779

b. Species separated

I. setosa		*I. virginica*		*I. versicolor*	
Stems	Leaves	Stems	Leaves	Stems	Leaves
4T	3	4T		4T	
4F	4445	4F		4F	
4S	666677	4S		4S	
4·	888889999	4·	9	4·	9
5*	000000011111111	5*		5*	001
5T	22223	5T		5T	2
5F	4444455	5F		5F	455555
5S	77	5S	67	5S	6666677777
5·	8	5·	8889	5·	88899
6*		6*	0011	6*	00001111
6T		6T	22333333	6T	22333
6F		6F	444445555	6F	445
6S		6S	77777	6S	66777
6·		6·	88999	6·	89
7*		7*	1	7*	0
7		7T	2233	7T	
7		7F	4	7F	
7		7S	67777	7S	
7·		7·	9	7·	

Note: Display digits = millimeters.

Sepal length, taken from Fisher's iris dataset, is displayed in stem-and-leaf plots in Table 4.4. The sepal is the small leaf-like structure at the base of flower petals. The stem-and-leaf plot of Table 4.4a, for all iris species combined, shows that sepal length ranges from a minimum value of 43 mm to a maximum value of 79 mm. Most of the observations in the table are found between 50 mm and 64 mm. In Table 4.4b the sepal length observations are separated by species. It should be clear from the table that iris characteristics vary from one species to the next. The species *I. setosa* has a considerably shorter and less dispersed sepal length than the other two species. In turn, the species *I. virginica* has a sepal length that is on average a little longer and more dispersed than that of the species *I. versicolor.* Using a qualitative variable to separate the observations in a data set can reveal useful information, as this example illustrates.

Calculating the mean and standard deviation of sepal length for the three samples of iris species confirms the differences alluded to above:

Species	Mean ($\bar{X}$)	Standard deviation (s)
I. setosa	50.10	3.54
I. virginica	65.88	6.36
I. versicolor	59.36	5.16

In Chapter 9, we show how to test whether differences between samples and differences, between different groups of observations in a dataset are statistically significant.

Box plots provide another way of separating observations on a quantitative variable by the categories of a qualitative variable. We look for differences between groups of observations in order to gain a better understanding of the distribution of a quantitative variable, or because theory suggests a particular pattern of dependence between two variables. For example, theoretical arguments within economic geography support the claim that productivity (output per unit of input) should increase with city size. In Figure 4-1, box plots illustrate the distribution of U.S. manufacturing productivity levels in 1963 across four, somewhat arbitrary, categories of city size.

In order to construct Figure 4-1, the population sizes and productivity values for 230 U.S. cities were recorded. The cities were then divided into four size classes. For each of these four size classes, a boxplot was produced. Figure 4-1 shows that the distribution of labor productivity does indeed vary for cities of different size. The box

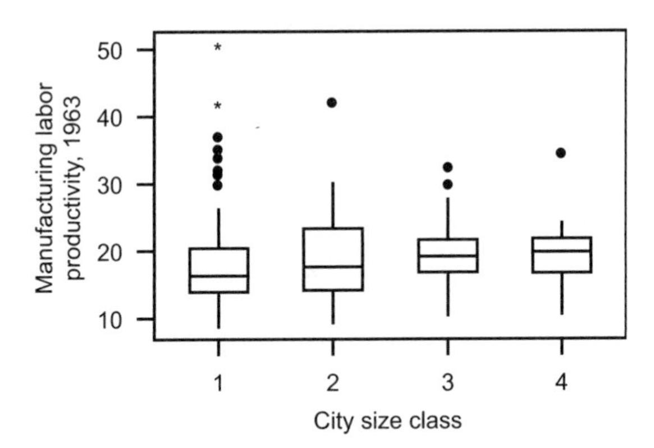

FIGURE 4-1. Labor productivity and city size. n = 230. City size classes are based on population: 1 = 0–333,333; 2 = 333,334–666,666; 3 = 666,667–999,999; 4 = 1,000,000 or larger. denotes outside values; * denotes far outside values or outliers. *Source:* U.S. Census Bureau.

plots show that median productivity levels increase consistently across the four city size classes, as theory predicts. The range of productivity values gets smaller with city size. This probably reflects greater specialization in smaller cities and the considerable variation in productivity between different manufacturing sectors. Separating the observations on a variable of interest, in this case, metropolitan productivity, across the categories of a second variable, yields greater insight into the distribution of the original variable.

Quantitative versus Quantitative Variables

Public health professionals long have argued that smoking poses serious health risks. One way to demonstrate these risks is to show data that link cigarette smoking to certain health outcomes. Figure 4-2 does this, plotting the average number of cigarettes smoked per person each day against the death rate from lung cancer across $n = 44$ U.S. states in 1960. Death rates are measured as the annual number of deaths per 100,000 people. Figure 4-2 is a *scatterplot;* it is the standard way of displaying the relationship between two quantitative variables. Each point in the scatterplot represents a U.S. state for which information is recorded on two variables, cigarette consumption, plotted on the horizontal, or X-axis, and the lung cancer death rate, plotted on the vertical, or Y-axis. The arrangement of the points, the scatter, tells us if and how the two variables are related.

As Figure 4-2 shows, the number of cigarettes smoked increases, so the death rate from lung cancer increases. The two variables appear to be statistically dependent on one another. How do we know this? Well, we could imagine splitting the data into two groups, one group of states with relatively low levels of cigarette consumption and the other group comprising states with relatively high levels of cigarette use. Then we might ask the question of whether or not the probability of observing relatively low or relatively high death rates from lung cancer varies between the two groups.

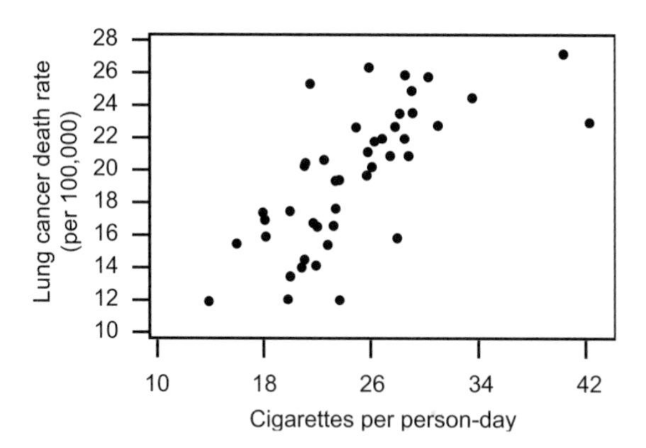

FIGURE 4-2. Scatterplot of cigarettes smoked per person per day against annual lung cancer death rates in U.S. States, 1960. *Source:* Fraumeni (1968)

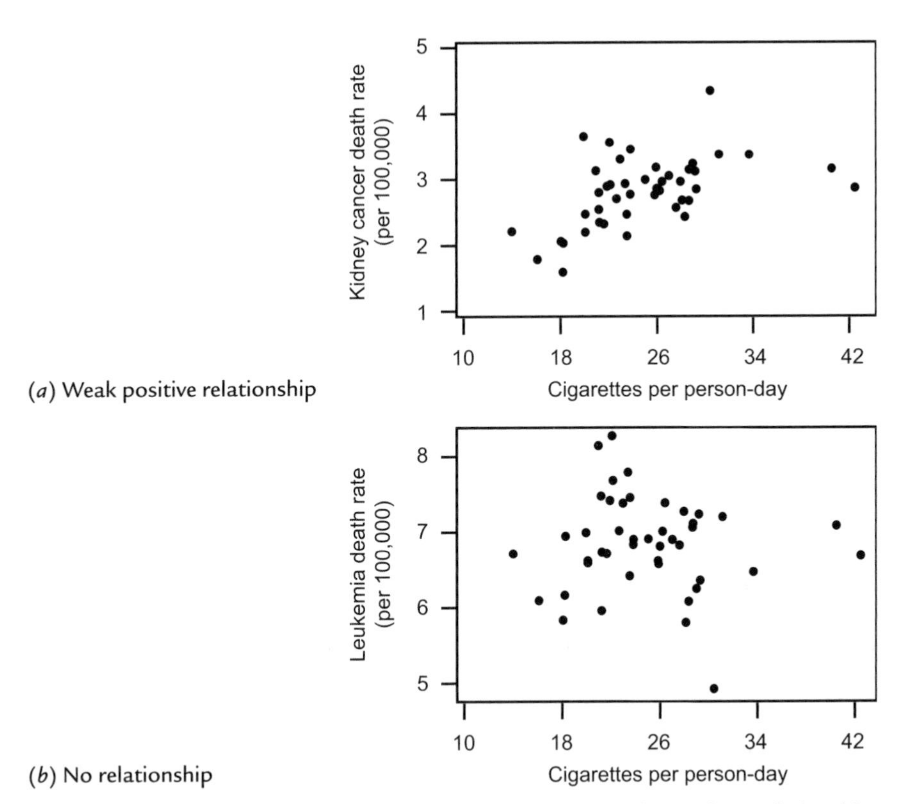

FIGURE 4-3. Scatterplots showing a weak positive relationship and no relationship. *Source:* Fraumeni (1968).

The answer is yes. States with higher levels of cigarette use are clearly those states with a higher rate of lung cancer deaths. From Section 4.1, if the values of one variable are influenced by the values taken by another variable, then the two variables are dependent: they *covary* with one another. Furthermore, because the two variables in Figure 4-2 covary in the same direction, as cigarette use increases lung cancer death rates increase, we say that they are *positively* related to one another. Two variables are *negatively* related when above average values of one variable are associated with below average values of the second variable.

Figure 4-3 shows two more scatterplots from the same smoking and cancer dataset. Figure 4-3a illustrates the relationship between the death rates from kidney cancer and smoking. Once more, the general scatter of points is oriented from the bottom-left of the scatterplot toward the top-right. This is again indicative of a positive relationship between the two variables, though the *strength* of that association is lower than that in Figure 4-2. Visually, the strength of statistical dependence between two variables can be gauged by how closely the scatterplot resembles a line. Points that cluster along a line are evidence of a strong (negative or positive) relationship, and points that are widely scattered across the plot are evidence of a weak relationship.

The widely dispersed scatterplot of Figure 4-3b shows that there is virtually no statistical dependence between leukemia and smoking. In the next section, we show how to calculate the strength of a linear relationship between two variables.

EXAMPLE 4-2. Income Convergence. Over the last 10 to 15 years, the question of national income convergence has figured prominently in academic debates within economics, regional science, and economic geography. A number of authors have claimed that there is evidence to support the argument that the poorer countries of the world are catching up to the richer nations in terms of average income. Others have claimed that the convergence of national incomes is limited to the advanced industrialized countries of the world. A simple test of the income convergence hypothesis involves plotting the rate of growth of average income within a group of countries over a specified time period against the levels of average income in those countries at the start of the time period. If the poorest countries at the start of the study period record the fastest growth rates, this is evidence that they are catching up, or converging, toward the richer nations.

Figure 4-4a plots average income and income growth rates for a sample of countries over the period 1960–1990. Each point in the scatterplot represents an individual country and a pair of values on the level of income per worker and income growth. Note that we have added marginal box plots to this figure. This provides a way of examining the univariate distributions of the two variables that make up the scatterplot. While the growth rate data look roughly symmetric, the average income data (GDP/worker) is positively skewed, as is usually the case with the distribution of incomes. The box plots also show the presence of some outside values, but not outliers. The scatterplot of Figure 4-4a shows little evidence of a relationship between average income and growth rates and thus lends little support to the convergence hypothesis.

In Figure 4-4b we show how to use an additional categorical variable to separate observations in the scatterplot. Here two groups of countries are identified: those that belong to the OECD and those that do not. Membership in the OECD might be considered a crude surrogate of a richer country. It is quite clear from Figure 4-4b that for OECD countries there is a negative relationship between growth rates and initial levels of income. That is, OECD members with the lowest levels of average income in 1960 have generally experienced the fastest income growth rates between 1960 and 1990. This is clear support for the convergence hypothesis. Overall, the scatterplot suggests that while the rich countries have become more alike one another, in terms of average income, the poorer countries of the world are not catching up.

4.3. Introduction to Correlation

The scatterplot is a useful tool for visualizing the pattern and strength of association between two quantitative variables. Like the graphical techniques of Chapter 2, however, the scatterplot is not very precise. Just as we employed descriptive statistics to yield more exact measures of the characteristics of a univariate distribution, so we

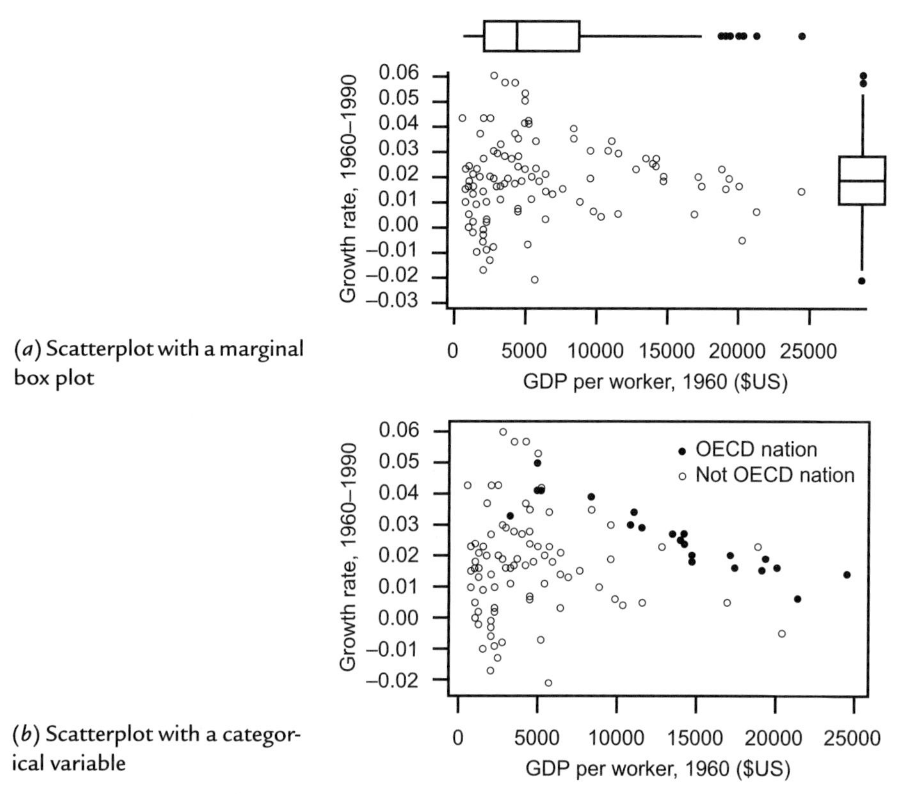

FIGURE 4-4. Scatterplot of income convergence. • denotes an outside value in Figure 4-6a; growth rates are average annual compound; GDP measures gross domestic product; OECD stands for the Organization of Economic Cooperation and Development. *Source:* Summers and Heston (1991): Penn World Tables.

can deploy numerical measures of the direction and the strength of the *covariation* of two variables. The most commonly used measure of the relationship between two interval or ratio variables is *Pearson's product moment correlation coefficient.* In this section we discuss how to calculate and interpret Pearson's correlation coefficient. Indices of the covariation of nominal and ordinal variables are discussed in Chapter 10.

In Chapter 3, we defined the variance of a single variable X as a function of the distribution of the values of X about its mean. With bivariate data we have observations on two variables, X and Y. Following Chapter 3, we can measure the sample variance of each variable separately as

$$s_X^2 = \frac{\sum_{i=1}^{n}(X_i - \bar{X})^2}{n-1} \quad \text{and} \quad s_Y^2 = \frac{\sum_{i=1}^{n}(Y_i - \bar{Y})^2}{n-1}$$

where s_X^2 and s_Y^2 represent the variance of X and Y, respectively.

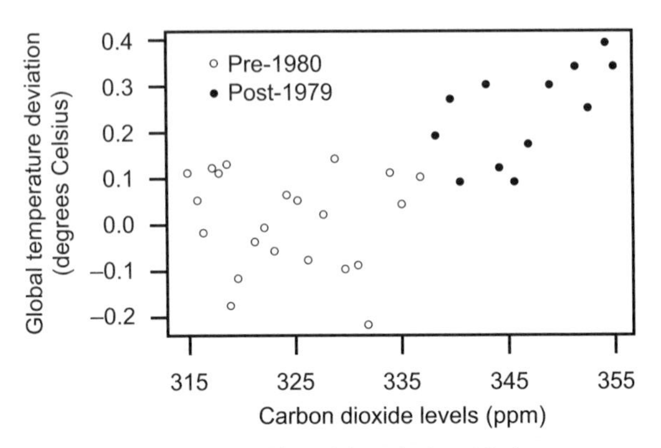

FIGURE 4-5. Scatterplot of the relationship between atmospheric CO_2 and global temperature variations, 1958–1991. ppm is parts per million; the observations are split into years up to 1979 and after 1979; the temperature anomalies are global temperature deviations from the mean value of the period 1950–1979. *Sources:* Jones et al. (1986); Keeling and Whorf (2005).

These measures of the variance of X and Y are independent of one another. That is, the measure of the variance of X is unrelated to the measure of the variance of Y and vice versa. However, what we need here is a measure of how the two variables covary together about their respective means. To clarify the problem, consider Figure 4-5, which shows the relationship between levels of carbon dioxide (CO_2) in the atmosphere and annual global temperature variations (deviations from an average (mean) value) between 1958 and 1991. Each point in the scatterplot represents a year for which we have a pair of values of the CO_2 level and the global temperature deviation. Imagine that we identify a number of years in Figure 4-5 where atmospheric CO_2 levels are greater than average for the period under study. Some of these observations are indicated by the solid dots on the right-hand side of the figure. The question of interest here is whether the global temperature variations for these same years are greater than average, lower than average, or distributed above and below the average over the study period.

In order to answer this question, we must take each observation in turn and compare the CO_2 level of that year to its mean, and then compare the temperature deviation of that same year to its mean. Examining the signs of the differences between each pair of values and their respective means tells us whether the two variables have a positive relationship and tend to move together (both variables have values greater or lower than their means in the same years), or whether they have a negative relationship and tend to move inversely to one another (one variable has a value greater [lower] than its mean while the other variable has a value lower [greater] than its mean in the same year), or whether they show no relationship (the variables move below and above their means in no apparent order). This is precisely what the covariance does for us.

DEFINITION: SAMPLE COVARIANCE

$$s_{XY} = \frac{1}{n-1} \sum_{i=1}^{n} (X_i - \bar{X})(Y_i - \bar{Y})$$

The calculation of the covariance is explained geometrically in Figure 4-6, which again displays the scatterplot of the CO_2 (X) and temperature data (Y). The scatterplot reveals a positive relationship, and so we would expect s_{XY} to be positive. The plot area of Figure 4-6 is divided into four quadrants, labeled I, II, III, and IV, by drawing two lines representing the means $\bar{X}$ and $\bar{Y}$. Let us examine the sign and magnitude of $(X_i - \bar{X})(Y_i - \bar{Y})$ in each of these four quadrants. If both X_i and Y_i are *greater* than their respective means, then the observation is in quadrant I. The product of the deviation from the means is *positive,* since both deviations are positive. This holds true for all observations in quadrant I. In quadrant III, both deviations are negative since the values of X_i and Y_i are below their means. However, the product of the two deviations $(X_i - \bar{X})(Y_i - \bar{Y})$ is positive since it is the product of two negative values. In quadrants II and IV the deviations are of opposite signs and therefore the product of the deviations must be negative.

The value of s_{XY} is based on the sum of these products for all observations. When there is a positive relation between the variables, as in Figure 4-6, $s_{XY} > 0$. This is because there are many points in quadrants I and III with large positive values for $X_i - \bar{X})(Y_i - \bar{Y})$, but only a few points in quadrants II and IV, where the product is negative. If, however, the scatter suggested a negative relation between X and Y, there would be many points in quadrants II and IV and few in quadrants I and III. In this case the covariance would be negative. If the scatterplot of X and Y values reveals a near equal distribution of points in all the quadrants, then the covariance s_{XY} would be near zero.

Using the covariance as a measure of the association between two variables has a significant drawback. The value of the covariance is dependent on the units in which the variables X and Y are measured. To remedy this problem, we divide the covariance

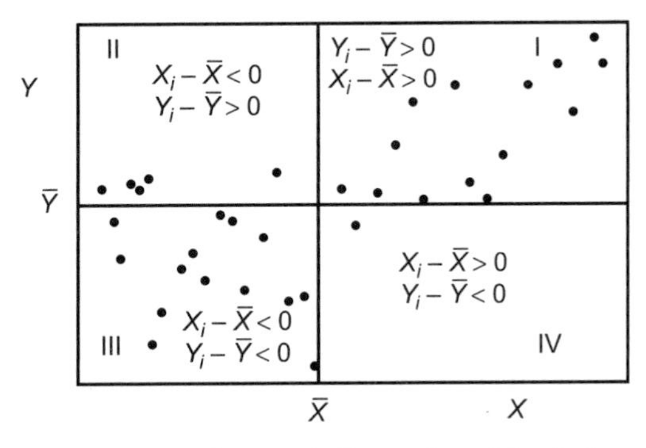

FIGURE 4-6. Geometric illustration of the derivation of the covariance of X and Y.

by the standard deviations of the variables. The standard deviation is measured in the same units as the original variables, and so the resultant ratio is *dimensionless*—it is simply a number. This ratio is called Pearson's product–moment correlation coefficient, or the correlation coefficient for short.

DEFINITION: SAMPLE CORRELATION COEFFICIENT

$$r = \frac{s_{XY}}{s_X s_Y} = \frac{[1/(n-1)]\sum_{i=1}^{n}(X_i - \bar{X})(Y_i - \bar{Y})}{\sqrt{[1/(n-1)]\sum_{i=1}^{n}(X_i - \bar{X})^2}\sqrt{[1/n-1)]\sum_{i=1}^{n}(Y_i - \bar{Y})^2}} \tag{4-1}$$

Equation 4-1 can be simplified by canceling the term $1/(n-1)$ that appears in both the numerator and denominator. This yields

$$r = \frac{\sum_{i=1}^{n}(X_i - \bar{X})(Y_i - \bar{Y})}{\sqrt{\sum_{i=1}^{n}(X_i - \bar{X})^2}\sqrt{\sum_{i=1}^{n}(Y_i - \bar{Y})^2}} \tag{4-2}$$

For computing purposes, especially when the calculations are performed by hand, it is helpful to rearrange Equation 4-2 as

$$r = \frac{\sum_{i=1}^{n}X_iY_i - \left(\sum_{i=1}^{n}X_i\right)\left(\sum_{i=1}^{n}Y_i\right)/n}{\sqrt{X_i^2 - \left(\sum_{i=1}^{n}X_i\right)^2/n}\sqrt{\sum_{i=1}^{n}Y_i^2 - \left(\sum_{i=1}^{n}Y_i\right)^2/n}} \tag{4-3}$$

This is the most convenient formula for calculating Pearson's r.

Characteristics of Pearson's Correlation Coefficient

The correlation coefficient measures the direction and strength of a linear relationship between two variables. The correlation coefficient, r, may take any value between –1 and 1. A positive value of r indicates that the variables X and Y are positively related: an increase (decrease) in one of the variables is associated with an increase (decrease) in the other. When r is negative, there is a negative relationship between X and Y such that when the value of one of the variables increases (decreases), the value of the other variable decreases (increases). Values of r close to zero indicate a very weak linear relationship between variables. The correlation coefficient gets progressively closer to the value –1 or 1 as the strength of the linear relationship gets stronger. A value $r = 1$ represents a situation in which all observations fall along a straight line with positive slope (see Figure 4-7a) This is known as perfect positive correlation. A value or $r = -1$ represents perfect negative or inverse correlation, where all observations fall along a line with negative slope (see Figure 4-7b). Note that the Pearson's product–moment correlation coefficient only measures the *linear* association between

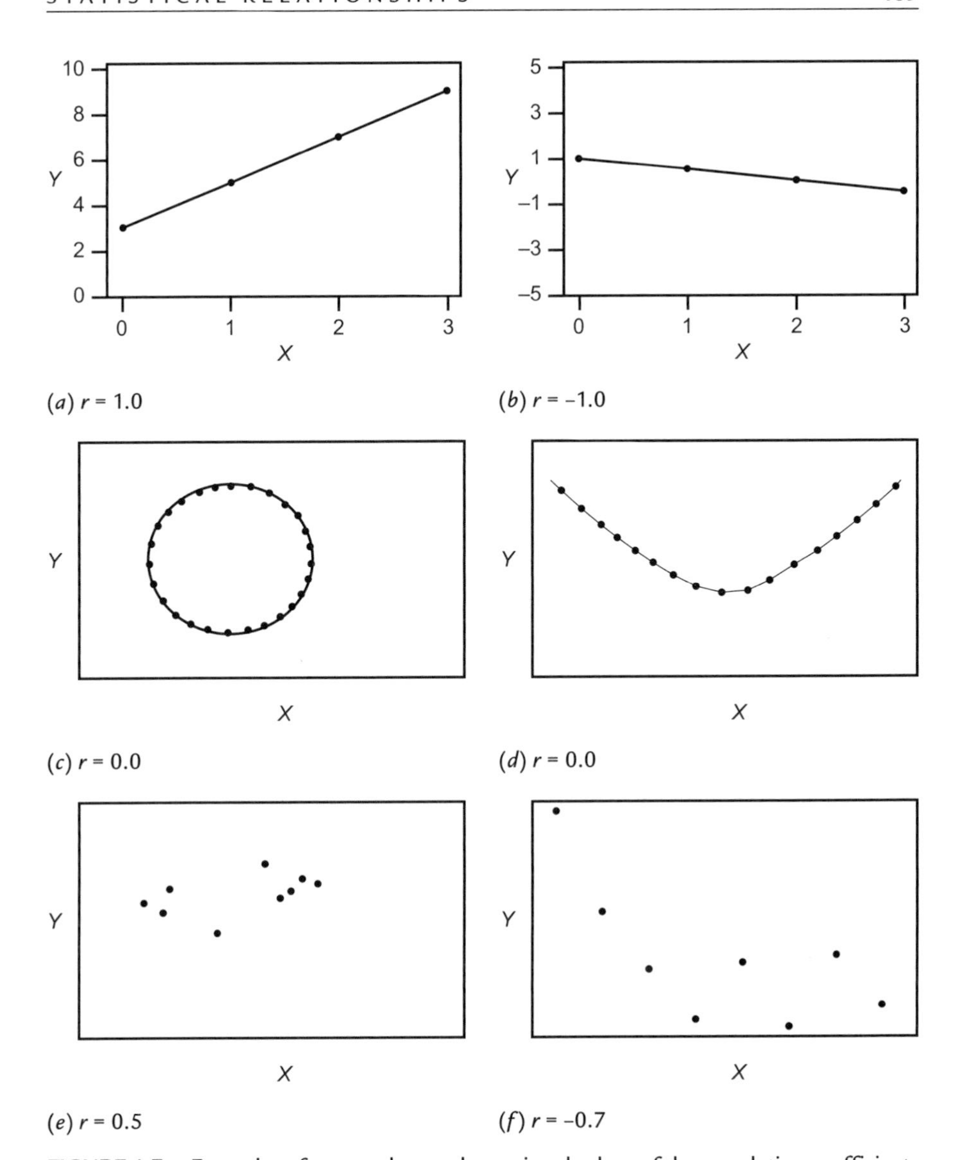

FIGURE 4-7. Examples of scatterplots and associated values of the correlation coefficient.

two variables. Two variables that have a *nonlinear* relationship, even if perfect (*i.e.* the relationship could be described by an equation), would not produce a correlation coefficient of −1 or 1 and could have a correlation coefficient of 0 (see Figure 4-7c and Figure 4-7d). It should be clear that zero correlation and statistical independence mean different things. Figure 4-7e and Figure 4-7f show two more scatterplots with correlation coefficients between 0 and ±1.0.

TABLE 4-5
Calculation of Pearson's r for Data of Figure 4-7a

Point (X_i, Y_i)	$X_i - \bar{X}$	$(X_i - \bar{X})^2$	$Y_i - \bar{Y}$	$(Y_i - \bar{Y})^2$	$(X_i - \bar{X})(Y_i - \bar{Y})$
(0, 3)	−1.5	2.25	−3	9	4.5
(1, 5)	−0.5	0.25	−1	1	0.5
(2, 7)	0.5	0.25	1	1	0.5
(3, 9)	1.5	2.25	3	9	4.5
	$\Sigma = 0$	$\Sigma = 5.0$	$\Sigma = 0$	$\Sigma = 20$	$\Sigma = 10.0$

$$r = \frac{s_{XY}}{s_X s_Y} = \frac{10/3}{\sqrt{20/3}\sqrt{5/3}} = \frac{10}{\sqrt{100}} = 1$$

Examples of Calculating the Correlation Coefficient

To illustrate the calculation of the correlation coefficient using the alternative formulas in Equations 4-2 and 4-3, consider the following examples.

EXAMPLE 4-3. The four points (0, 3), (1, 5), (2, 7), (3, 9) fall on the line $Y = 2X + 3$. As Figure 4-7a shows, there is a perfect linear relationship between X and Y for these points. The sample correlation coefficient should be $r = 1$. To confirm this, first we calculate $\bar{X} = (0 + 1 + 2 + 3)/4 = 1.5$ and $\bar{Y} = 3 + 5 + 7 + 9)/4 = 6$. The calculations for Pearsons r; obtained using the covariance formulation in Equation 4-2, are summarized in Table 4-5.

EXAMPLE 4-4. In this example we use Equation 4-3 to calculate the correlation coefficient between the CO_2 data and the global temperature data of Figure 4-5. (Not all 34 observations are listed in the table: the omitted data can be found on the website for the book.) The calculations are displayed in Table 4-6. The resulting r value of 0.68 indicates that CO_2 levels and global temperature variations are positively correlated. This means that higher levels of carbon dioxide in the earth's atmosphere are positively related to warmer global temperatures. Alternatively, we might say that higher global temperatures are related to higher levels of CO_2 in the atmosphere.

Table 4-7 presents a *correlation matrix* for cigarette smoking and death rates from various forms of cancer. These data were gathered for a number of US states. The individual cells in the matrix report the correlation coefficient between the row and column variables. There are several important features of correlation matrices. First, the correlation between a variable and itself is always 1.0, so that the elements along the *principal diagonal* of the matrix always contain the value $r = 1.0$. To see why this is so, examine Equation 4-3. If we substitute X for Y in this equation, the numerator and the denominator are identical; thus the ratio must be 1.0. Second, correlation matrices are symmetric. The correlation between two variables X and Y

TABLE 4-6
Calculation of Pearson's *r* for the Data of Figure 4-5

X_i	Y_i	X_i^2	Y_i^2	X_iY_i
314.67	0.11	99017.2	0.0121	34.61
315.59	0.05	99597.0	0.0025	15.78
316.19	–0.02	99976.1	0.0004	–6.32
317.01	0.12	100495.3	0.0144	38.04
317.69	0.11	100926.9	0.0121	34.95
318.36	0.13	101353.1	0.0169	41.39
.	.	.	.	.
354.99	0.34	126017.9	0.1156	120.70
$\sum_{i=1}^{n} X_i = 11292.5$	$\sum_{i=1}^{n} Y_i = 2.97$	$\sum_{i=1}^{n} X_i^2 = 3755872$	$\sum_{i=1}^{n} Y_i^2 = 1.0275$	$\sum_{i=1}^{n} X_iY_i = 1029.6$

$n = 34.$

$$r = \frac{1029.6 - (11292.5)(2.97)/34}{\sqrt{3755872 - (11292.5)^2/34}\ \sqrt{1.0275 - (2.97)^2/34}} = \frac{43.17}{\sqrt{5283.8}\ \sqrt{0.7681}} = 0.68$$

TABLE 4-7
Correlation Matrix for Smoking and Death Rates from Various Forms of Cancer

	Cigarettes	Bladder	Lung	Kidney	Leukemia
Cigarettes	1.0	0.704	0.697	0.487	–0.068
Bladder	0.704	1.0	0.659	0.359	0.162
Lung	0.697	0.659	1.0	.283	–0.152
Kidney	0.487	0.359	0.283	1.0	0.189
Leukemia	–0.068	0.162	–0.152	0.189	1.0

Note: $n = 43$.
Source: Fraumeni (1968).

necessarily equals the correlation between variables *Y* and *X*. Note in Equation 4-3 that the identical numerical result is obtained if variable *X* is substituted for variable *Y* and vice versa.

Table 4-7 reveals a relatively strong positive relationship between cigarette smoking and death rates from the three different forms of urinary cancer. There is almost no correlation between smoking and leukemia. Death rates from urinary cancers are generally uncorrelated with those from leukemia. Bladder and lung cancer are relatively strongly positively related, though death rates from both these forms of cancer are only moderately related to death rates from kidney cancer. Comparing these data to Figure 4-2 and Figure 4-3 should help to confirm the linkages between the patterns of a scatterplot and the value of the correlation coefficient.

4.4. Introduction to Regression

Pearson's product–moment correlation coefficient measures the direction and strength of the linear relationship between two quantitative variables. As such, the correlation coefficient provides an indication of how closely the paired observations in a scatterplot resemble a straight line. This suggests that another way to summarize a scatterplot is to draw a straight line that represents, on average, the relationship between the values of one variable and those of a second. This is precisely what a regression line is.

DEFINITION: REGRESSION LINE
A regression line is a straight line that describes how the values of a response variable (Y) depend on the values of an independent, or an explanatory variable (X).

More generally, regression analysis is the study of the statistical relationship between a dependent, or response, variable and a series of independent, or explanatory, variables. In this section we restrict attention to simple linear regression, or analysis of the statistical relationship between a dependent variable and a single independent variable. After some more preliminary remarks, the discussion shifts to the process of fitting a line through a scatterplot and to the benefits of the *ordinary least squares* fitting procedure. We then examine how to interpret the resulting regression equation and how to evaluate the *goodness of fit* of the regression line. This brief discussion of regression is extended and deepened in Chapters 12 and 13.

Although correlation and regression are closely related, there is an important difference between the two concepts. In correlation analysis, the two variables whose association is being measured are treated interchangeably. The association between variable one and variable two is identical to the association between variable two and variable one. However, in regression analysis, there is a clear asymmetry to the relationship between variables: the value of the dependent variable, always denoted by Y, is said to depend on the value of the independent variable, always denoted by X. This asymmetry exists because when we estimate the regression line, we typically assess its goodness of fit using the distance between the regression line and the observed values of the dependent variable (Y) in the scatterplot. If we reversed the order of the variables in the regression, the shape of the scatterplot would change, and thus the regression line summarizing the scatterplot would also change. That line would now be fit according to the distance between the line and the observed values of the original independent variable (X).

Figure 4-8 illustrates this, first showing a regression line fit to global temperature (Y) data using carbon dioxide as an explanatory variable (X), and second showing a regression line fit to carbon dioxide data (Y) using global temperature as an explanatory variable (X). It is clear that the gradients of the regression lines and their intercepts vary between the two plots. Indeed they must, because they are measured in different units. For example, in Figure 4-8a the slope of the line is 0.0082°C per ppm of carbon dioxide. In Figure 4-8b, the slope is 56.175 ppm of carbon dioxide per degree C. What might not be so clear is that the inferred relationships differ between

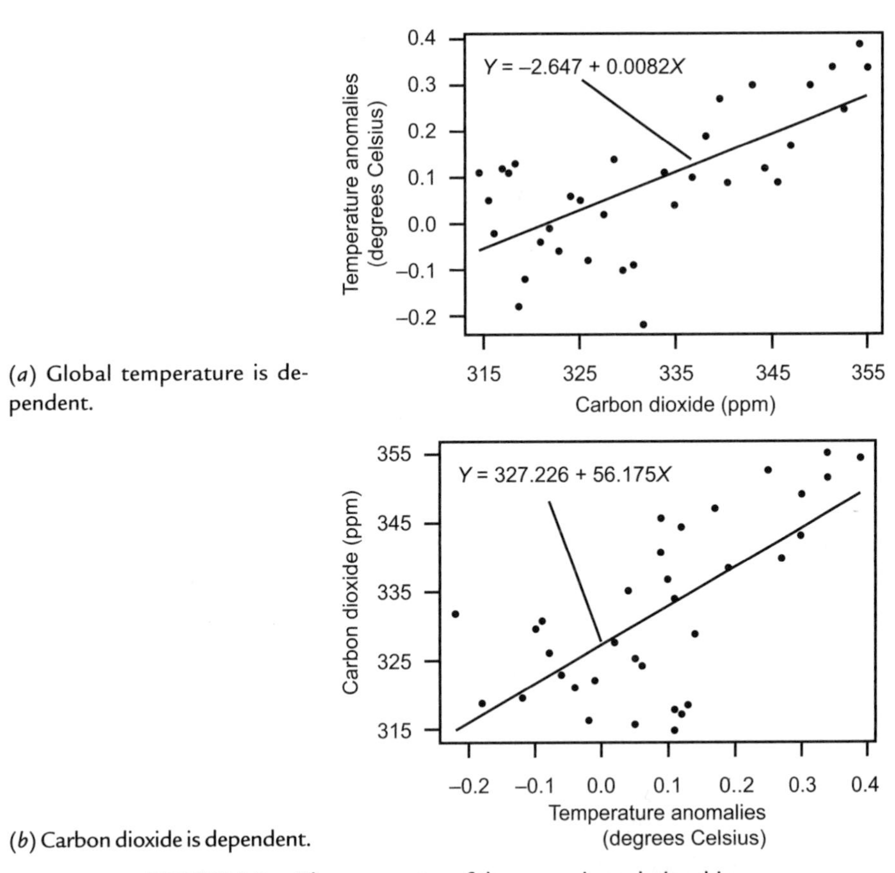

(*a*) Global temperature is dependent.

(*b*) Carbon dioxide is dependent.

FIGURE 4-8. The asymmetry of the regression relationship.

the two plots. That is, the slopes are not simply reciprocals of one another. (See Appendix 4a for a review of the elementary geometry of a straight line.)

It is this asymmetry between variables that leads us to designate one of the variables in the simple linear regression model dependent and the other independent. A crucial question in regression analysis is identifying the dependent variable. Statistical analysis cannot help us with this question. While regression examines the dependence of one variable on others, it does not itself imply causation. "A statistical relationship, however strong and however suggestive, can never establish causal connection: our ideas of causation must come from outside statistics, ultimately from some theory or other" (Kendall and Stuart, 1961, p. 299). Thus, before we engage in regression analysis, we should have a clear idea of the direction of the relationship between the variables being studied. In the examples above, biologists and health-care workers have long documented a connection that runs from smoking cigarettes to the development of various forms of cancer. Similarly, atmospheric scientists increasingly argue that the increase in carbon dioxide in the atmosphere is one of the factors contributing to global warming. In human geography, and the social sciences more generally, the direction of relationships between variables often is more ambiguous. Throughout this section, we assume that dependent and independent variables have been identified correctly.

In Figure 4-8 the regression lines do not pass through all the points in the scatterplots. This is a characteristic of most statistical, or *stochastic,* relationships, between what are typically *random variables* (see Chapter 5). In a statistical relationship, knowledge of the value of one or more explanatory variables will not enable the researcher to predict the value of a dependent variable exactly. Missing variables and errors in measurement, as well as random variation in the relationship between dependent and independent variables, ensures some unpredictability in all regression analysis. In contrast, in a *deterministic* relationship, the value of variable A can be related precisely to the value of variable B with a mathematical expression. For example, Boyle's famous gas law states that $PV = c$: for a given mass, at a constant temperature, the pressure (P) times the volume (V) of an ideal gas is constant (c). Using this deterministic equation, if we know exactly by how much the pressure or the volume is changed, then we can determine precisely the change in the value of the constant. Figures 4-7a–d show examples of a deterministic relationship between two variables.

Although we limit our discussion in this chapter to analysis of the linear regression model, regression can be employed to examine situations where the dependence of Y on X follows many different functional forms. Figure 4-9 illustrates several potential alternatives and corresponding functional forms.

Estimation of a Linear Regression Function

The linear regression function, describing the relationship between a dependent variable (Y) and an independent variable (X), takes the form

$$Y = a + bX \tag{4-4}$$

where a is the *intercept* of the regression line, the point where the line crosses the vertical, or Y-axis of the scatterplot, and b is the *slope* or *gradient* of the line, showing by how much the value of the dependent variable changes when the independent variable increases by one unit. Ordinarily, we do not know the values of the *regression coefficients a* and b, and we must estimate their values from data. Once we know a and b, we can precisely locate our regression line within the scatterplot.

We reproduce in Figure 4-10 the scatterplot of the relationship between cigarette smoking and death rates from lung cancer, to which we have added a regression line. How do we estimate, or fit, the regression line through a scatter of observations as in Figure 4-10? If the points in the scatterplot lie quite close to a straight line we could do a reasonable job of fitting the line "by eye" through the middle of the data points. We could then simply read off the values of a and b from the scatterplot. However, this is rarely the case. In Figure 4-10, for example, different people would probably locate a line representing the scatterplot in different locations. What we need is an

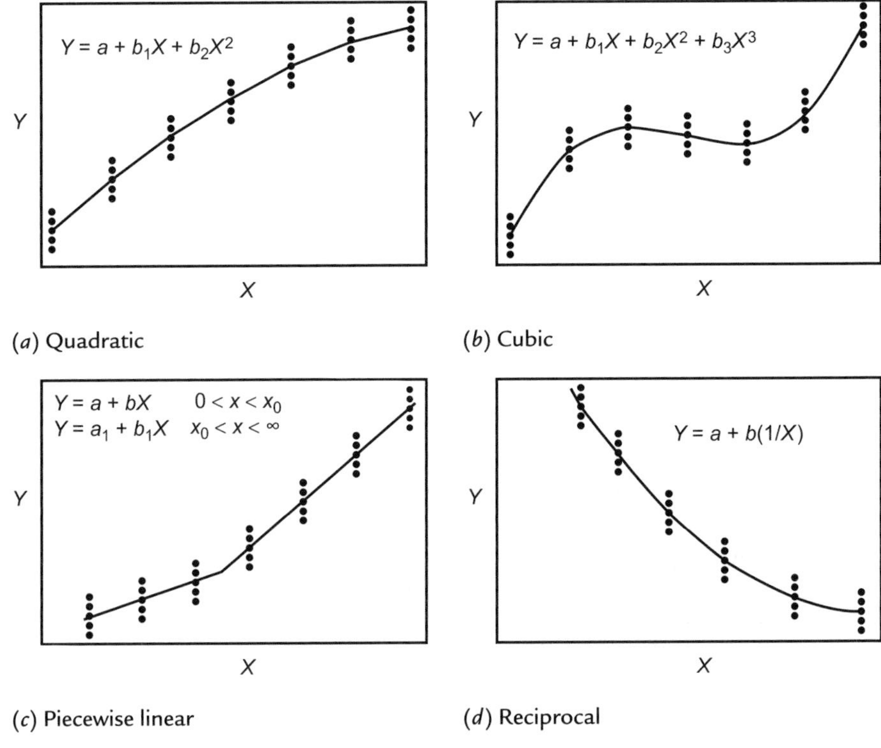

(*a*) Quadratic

(*b*) Cubic

(*c*) Piecewise linear

(*d*) Reciprocal

FIGURE 4-9 Nonlinear regression relationships.

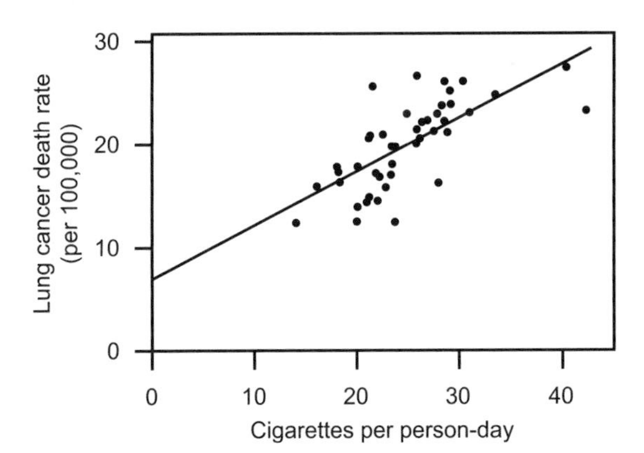

FIGURE 4-10. Regression line linking cigarette use and death rates from lung cancer.

objective way of fitting the regression line that can be easily operationalized. Fortunately, a simple algebraic solution to this problem exists.

Possible Criteria for Fitting a Regression Line

As a starting point, we must choose the criterion to assess the success of our line of best fit. An approach based on the deviations of the points to the line is ideal. Because we are interested in explaining the variation in the dependent variable Y, it seems natural to measure the vertical deviations from the observed values of Y to the regression line. We do this because the regression line should pass through the mean value of Y for any given value of X. Our error is the difference between the mean value of Y predicted by our line and the observed value of Y. This measure of error is illustrated in Figure 4-11.

How do we calculate the value of the error? First, we locate the point on the regression line corresponding to the observed X by defining $\hat{Y}_i$ (read "Y hat i") $= a + bX_i$. Second, we find the error or deviation as $e_i = Y_i - \hat{Y}_i$, the difference between the actual value of Y for a given X and the value predicted from the regression line. Note that when the observed Y_i lies above the line, this error is positive, and when the observed Y_i lies below the line, it is negative. Now we might think that a good criterion for fitting the regression line is to minimize the sum of these errors taking into account all our observations:

$$\min \sum_{i=1}^{n} (Y_i - \hat{Y}_i) \tag{4-5}$$

Unfortunately, the positive deviations offset the negative deviations, and it is possible to draw several lines through the set of points, each with a zero sum of deviations. To overcome this problem, the German scientist Carl Gauss argued that we should fit the line by utilizing the *least squares* criterion:

$$\min \sum_{i=1}^{n} (Y_i - \hat{Y}_i)^2 \tag{4-6}$$

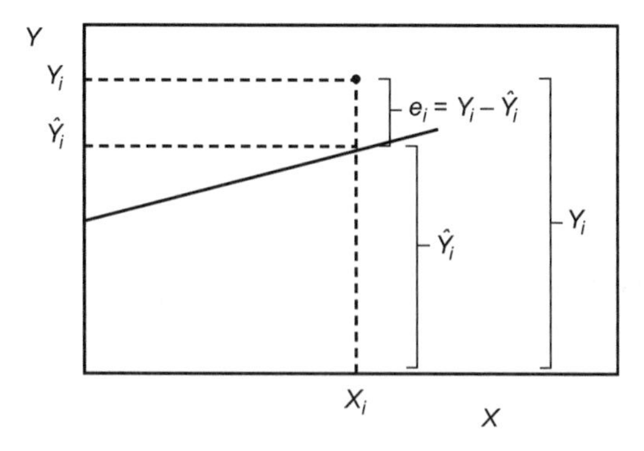

FIGURE 4-11. Error in fitting a regression line to a typical point.

which selects a and b to minimize the sum of squared deviations. If this line is very close to the points, then this sum is small; as the points become increasingly spread about the line, the sum of squared deviations increases. If all the observed points lie on the line, then the sum equals zero. The use of squared deviations overcomes the "sign problem" inherent in the use of Equation 4-5. Moreover, as we will see in Chapter 12, there are important theoretical justifications for using this criterion.

ORDINARY LEAST SQUARES ESTIMATION

How, then, can we find the values of a and b that minimize the sum of the squared deviations? We start by substituting $\hat{Y}_i = a + bX_i$ into Equation 4-6. The problem now becomes one of finding the values of a and b that minimize the expression

$$\min \sum_{i=1}^{n} (Y_i - a - bX_i)^2 \tag{4-7}$$

This is a simple problem of differential calculus, the details of which are reviewed in Appendix 4b. The solution to the problem yields the following pair of simultaneous *normal equations:*

$$na + b\sum_{i=1}^{n} X_i = \sum_{i=1}^{n} Y_i \tag{4-8}$$

$$a\sum_{i=1}^{n} X_i + b\sum_{i=1}^{n} X_i^2 = \sum_{i=1}^{n} X_i Y_i \tag{4-9}$$

From Equation 4-8, if we divide throughout by n and simplify, we can solve for a as

$$a = \frac{\sum_{i=1}^{n} Y_i}{n} - b\frac{\sum_{i=1}^{n} X_i}{n} \tag{4-10}$$

which, from the identities $\bar{X} = \Sigma_{i=1}^{n} X_i / n$ and $\bar{Y} = \Sigma_{i=1}^{n} Y_i / n$ leads to

$$a = \bar{Y} - b\bar{X} \tag{4-11}$$

Since $\bar{X}$ and $\bar{Y}$ are calculated from the data, we can easily determine the intercept a once we know b. There is also one important implication that is derived from Equation 4-11. If we rewrite (4-11) as $\bar{Y} = a + b\bar{X}$, we see that the least squares regression line passes through the point $(\bar{X}, \bar{Y})$. Note that there is no error at this point.

How do we solve for b? Now that we have a, we can substitute (4-11) into our second normal equation (4-9):

$$-\sum_{i=1}^{n} X_i Y_i + \left(\frac{\sum_{i=1}^{n} Y_i}{n} - b\frac{\sum_{i=1}^{n} X_i}{n} \right)\left(\sum_{i=1}^{n} X_i \right) + b\sum_{i=1}^{n} X_i^2 = 0$$

This can be multiplied to yield

$$-\sum_{i=1}^{n} X_i Y_y + \left(\frac{\sum_{i=1}^{n} X_i \sum_{i=1}^{n} Y_i}{n} - b \frac{\left(\sum_{i=1}^{n} X_i \right)^2}{n} \right) + b \sum_{i=1}^{n} X_i^2 = 0$$

$$nb \sum_{i=1}^{n} X_i^2 - b \left(\sum_{i=1}^{n} X_i \right)^2 = n \sum_{i=1}^{n} X_i Y_i - \sum_{i=1}^{n} X_i \sum_{i=1}^{n} Y_i$$

Finally, we solve for b:

$$b = \frac{n \sum_{i=1}^{n} X_i Y_i - \sum_{i=1}^{n} X_i \sum_{i=1}^{n} Y_i}{n \sum_{i=1}^{n} X_i^2 - \left(\sum_{i=1}^{n} X_i \right)^2} \qquad (4\text{-}12)$$

Equations 4-11 and 4-12 are the most straightforward computational formulas for solving linear regression problems by hand. Of course, we use computers for large-scale applications of linear regression analysis (small ones too!). By convention, a is known as the *regression intercept,* or just the *intercept,* and b is the *regression slope coefficient,* sometimes just the *slope.*

EXAMPLE 4-5. Let us now calculate the regression coefficients, a and b, for the smoking and lung cancer data of Figure 4-10. For brevity we only show some observations: the complete dataset may be found on the book's website. Table 4-8 provides an extremely useful table, summarizing the calculations required for the linear regression problem. Strictly speaking, we do not need the last column, that contains $\Sigma_{i=1}^{n} Y_i^2$, but this term is used to calculate the correlation coefficient between X and Y. As we shall see below, the correlation coefficient is a useful measure of the *goodness of fit* of our regression line.

From Table 4-8 we see that the regression of the lung cancer death rate on smoking levels yields the equation

$$\hat{Y}_i = 6.473 + 0.529\, X_i \qquad (4\text{-}13)$$

where $\hat{Y}_i$ is the estimated or predicted mean death rate from lung cancer for a given level of cigarette smoking. It is relatively straightforward to graph the regression line within the scatter of observations. To precisely locate the line we need only two points. Although we can determine any two points by substituting two different values of X into Equation (4-13), it is often simpler to use the points $(\bar{X}, \bar{Y})$ and $(0, a)$. We know that $(\bar{X}, \bar{Y})$ is on the line from Equation 4-11, and by definition the coefficient a is the intercept on the Y-axis. We can connect these two points with a straight line. If the intercept is not close to the range of our data (this is a normal occurrence), we can utilize any other convenient point. In our example of smoking and cancer, we plot the points (0, 6.473) and (24.914, 19.653). The resulting regression line is drawn in Figure 4-12.

TABLE 4-8
Calculations for a Simple Linear Regression Problem

Observation	X	Y	XY	X^2	Y^2
1	18.20	17.05	310.31	331.24	290.70
2	25.82	19.80	511.24	666.67	392.04
3	18.24	15.98	291.48	332.70	255.36
4	28.60	22.07	631.20	817.96	487.08
5	31.10	22.83	710.01	967.21	521.21
6	33.60	24.55	824.88	1128.96	602.70
7	40.46	27.27	1103.34	1637.01	743.65
8	28.27	23.57	666.32	799.19	555.54
9	20.10	13.58	272.96	404.01	184.42
.	.	.	.	.	.
44	30.34	25.88	785.20	920.52	669.77
$n = 44$	$\sum_{i=1}^{n} X_i = 1096.2$	$\sum_{i=1}^{n} Y_i = 864.7$	$\sum_{i=1}^{n} X_i Y_i = 22250.9$	$\sum_{i=1}^{n} X_i^2 = 28646.9$	$\sum_{i=1}^{n} Y_i^2 = 17763.6$

$$\bar{X} = \frac{1096.2}{44} = 24.914 \qquad \bar{Y} = \frac{864.7}{44} = 19.653$$

$$a = 19.653 - 0.529(24.914) \qquad b = \frac{44(22250.9) - (1096.2)(864.7)}{44(28646.9) - (1096.2)^2}$$

$$a = 6.473 \qquad b = 0.529$$

Source: Fraumeni (1968).

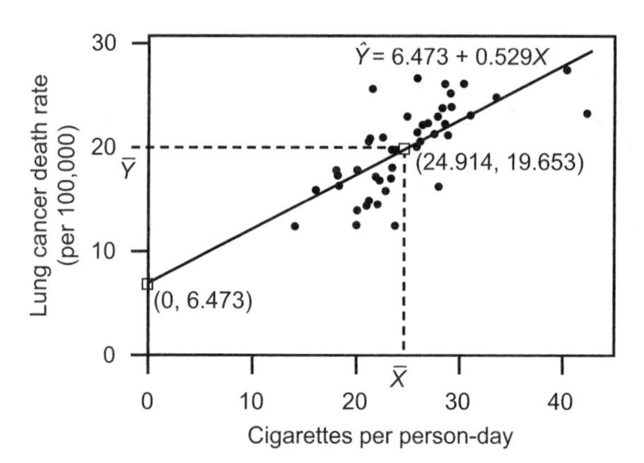

FIGURE 4-12. Drawing the regression line.

Interpreting the Regression Equation

How do we interpret the regression equation? Collecting some of the terms introduced above, and continuing our use of the smoking and lung cancer example, the regression equation tells us how the dependent variable, the lung cancer death rate, varies with changes in the independent variable, in this case, the number of cigarettes smoked each day. What about the individual terms in the regression model?

The intercept, a, of the regression equation predicts the average value of the dependent variable, when the independent variable takes the value zero. When $X = 0$, $\hat{Y}_i = a = 6.473$, and thus, on average across the study sample, people who do not smoke have an expected death rate due to lung cancer of 6.473 people per 100,000. Note that this also implies that the intercept has the same dimension as the dependent variable; both measure the death rate from lung cancer per 100,000 people. The intercept is required to fit the regression equation, but in most instances the intercept cannot be readily interpreted. Only when the independent variable realistically can take the value zero, and when the regression line has been fit over a range of X values that include zero, can we reasonably interpret the value of the intercept.

In our smoking and lung cancer data, the smallest value of the independent variable $X = 14.0$. Thus, when we infer from the regression equation that people who do not smoke have an expected death rate from lung cancer of 6.473 per 100,000, we are *extrapolating,* or predicting Y for a value of X beyond the range of the data. We must be extremely cautious in making such claims, for we do not know whether the scatterplot of Figure 4-12, if extended toward the value $X = 0$, would retain its observed shape. In Chapter 12 we return to this forecasting problem and illustrate the method used to make point and interval estimates for $\hat{Y}$ at a desired level of confidence.

The slope coefficient, b, indicates by how much the value of the dependent variable changes when the independent variable increases by one unit. With $b = 0.529$, the regression analysis argues that for every additional cigarette smoked each day, the expected death rate from lung cancer increases by 0.529 people per 100,000. Thus, if cigarette consumption was 20 per day, the expected value of the lung cancer death rate would be $\hat{Y}_i = 6.473 + 0.529(20) = 17.053$. If $X = 30$, then $\hat{Y}_i = 6.473 + 0.529(30) = 22.343$ and if $X = 31$, then $\hat{Y}_i = 6.473 + 0.529(31) = 22.872$.

While the value of the intercept might not always be meaningful, the value of the regression slope coefficient always is. At present, our interest in regression is in finding the characteristics of a straight line that best represents a scatterplot or that most accurately predicts the value Y given X. For this task, we might not be that concerned with the sign or the value of the slope coefficient. However, as we will see in Chapters 12 and 13, regression is also used to try and establish whether, and how, a particular independent variable is statistically related to the dependent variable. In this task, the slope coefficient plays the lead role. Of course, even when the job at hand is simply to estimate a best-fit line, we do have to pay some attention to the value of the slope coefficient. Theory suggests quite strongly that the relationship between smoking and lung cancer is positive. If our regression line had a negative slope, this would be cause for considerable concern.

Assessing the Goodness of Fit of the Regression

We now know that the ordinary least squares estimates of a and b given by Equations 4-11 and 4-12 provide the best possible fit of a straight line to the data in a scatterplot according to criterion (4-7). Since we can calculate the two regression coefficients from any set of observations having at least two different values of X, the question remains: how well does the line fit the original data? We cannot use the total sum of squares as our index since it clearly depends on the scale of the numbers involved as well as on the number of observations. What we require is an index that is scale independent, or dimensionless, and therefore comparable from one dataset to the next. There are three commonly used measures of the *goodness of fit* of a regression line:

1. The Pearson product–moment correlation coefficient r
2. The coefficient of determination r^2
3. The standard error of the estimate $s_{Y \cdot X}$

CORRELATION COEFFICIENT r

In the last section we found that the Pearson product–moment correlation coefficient, r, is a dimensionless measure of the linear association between two variables. The correlation coefficient can be used to compare the association between two sets of variables drawn from the same set of observations, or of the same two variables drawn from different data sets. The correlation coefficient has a range $-1 \leq r \leq 1$; the closer r is to ± 1 the greater the strength of the linear relationship between two variables. Pearson's r represents a useful summary measure in regression analysis. If the dependent and independent variables are positively related, then $r > 0$ and $b > 0$. If they are negatively related, then r, $b < 0$. We can say that a regression line with a correlation coefficient higher than that of another equation fits the scatterplot or the data better.

In fact, an important algebraic relationship between b and r confirms the close connection between correlation and regression. A quick comparison of Equations 4-3 and 4-12 should convince you of this fact. To illustrate this relationship, let us recall the definition of the correlation coefficient

$$r = \frac{\sum_{i=1}^{n}(X_i - \bar{X})(Y_i - \bar{Y})}{\sqrt{\sum_{i=1}^{n}(X_i - \bar{X})^2}\sqrt{\sum_{i=1}^{n}(Y_i - \bar{Y})^2}} \qquad (4\text{-}14)$$

Define $x_i = X_i - \bar{X}$ and $y_i = Y_i = \bar{Y}$. Substituting these terms into Equation 4-14, we have

$$r = \frac{\sum_{i=1}^{n}(x_i)(y_i)}{\sqrt{\sum_{i=1}^{n}(x_i)^2}\sqrt{\sum_{i=1}^{n}(y_i)^2}} \qquad (4\text{-}15)$$

This is often known as the *deviations form* of the correlation coefficient.

We can express the formula for the regression slope coefficient in deviations form as

$$b = \frac{\sum_{i=1}^{n} x_i y_i}{\sum_{i=1}^{n} x_i^2} \qquad (4\text{-}16)$$

To formalize the relationship between the correlation coefficient and the regression slope coefficient, let us take the ratio b/r

$$\frac{b}{r}=\frac{\sum_{i=1}^{n}x_iy_i/\sum_{i=1}^{n}x_i^2}{\sum_{i=1}^{n}x_iy_i/\left(\sqrt{\sum_{i=1}^{n}x_i^2}\sqrt{\sum_{i=1}^{n}y_i^2}\right)}$$

$$=\frac{\sqrt{\sum_{i=1}^{n}x_i^2}\sqrt{\sum_{i=1}^{n}y_i^2}}{\sum_{i=1}^{n}x_i^2}=\frac{\sqrt{\sum_{i=1}^{n}y_i^2}}{\sqrt{\sum_{i=1}^{n}x_i^2}} \qquad (4\text{-}17)$$

Now, if we divide the numerator and the denominator of the right-hand side of Equation 4-17 by $\sqrt{n-1}$, we obtain

$$\frac{b}{r}=\frac{\sqrt{\sum_{i=1}^{n}y_i^2/(n-1)}}{\sqrt{\sum_{i=1}^{n}x_i^2/(n-1)}}=\frac{s_Y}{s_X} \qquad (4\text{-}18)$$

by using the familiar formula for the sample standard deviation. We can rearrange Equation 4-18 as

$$b=r\frac{s_Y}{s_X} \qquad (4\text{-}19)$$

which shows that the slope of the regression line is equal to the product of the correlation coefficient and the ratio of the standard deviation of Y to the standard deviation of X. Since the standard deviation is always positive, this implies that b and r must have the same sign. In addition, $b = 0$ if $r = 0$, and vice versa. When $r = 1$, the change in the predicted value $\hat{Y}$ is equal (in standard deviation units) to the change in X. As the correlation between X and Y weakens, changes in the predicted value $\hat{Y}$ are less responsive to changes in X. Table 4-9 provides a numerical example of the relation between the correlation coefficient and the regression slope coefficient, using the smoking and lung cancer data.

COEFFICIENT OF DETERMINATION r^2

The coefficient of determination is perhaps the most widely used measure of the goodness of fit of the regression relationship, showing how well the regression line represents the observations comprising a scatterplot of Y on X. If the regression line passed through all the points in the scatterplot, there would be a perfect relationship between Y and X. In this case, for a given value of X, the regression equation would predict the corresponding value of Y without error. Typically, however, the value of *Y predicted* from the regression equation is different from the *observed* value of Y in the data (see Figure 4-11). This difference is known as the *regression error* or *residual.* There is one

TABLE 4-9
Work Table Illustrating Equation 4-19

The simple correlation coefficient for the smoking and lung cancer data, obtained by using computational formula (4-3) with the data from Table 4-8, is

$$r = \frac{22250.9 - [(1096.2)(864.7)]/44}{\sqrt{28646.9 - (1096.2)^2/44}\,\sqrt{17763.6 - (864.7)^2/44}} = 0.698$$

Using the computational formula for the sample standard deviation, we find

$$s_X = \sqrt{\frac{28646.9}{43} - \frac{(1096.2)^2}{44(43)}} = 5.575 \quad s_Y = \sqrt{\frac{17763.6}{43} - \frac{(864.7)^2}{44(43)}} = 4.232$$

Then, $b = 0.698\left(\frac{4.232}{5.575}\right) = 0.529$, which agrees with the value of b in Table 4-8.

residual for each of the n observations. Together the residuals provide extremely useful information on the extent to which the calculated regression line fits the data. Generally, a good regression line is one that has small residuals, and a poor regression line is one with large residuals. Unfortunately, residuals are measured in the same units as the dependent variable, and thus the size of the residuals is dependent on the measure of Y. Ideally, if we are to use the residuals to gauge the goodness of fit of the regression equation, we need an adjustment to make them independent of the units of measurement of Y. The *coefficient of determination* provides this adjustment.

DEFINITION: REGRESSION ERROR OR RESIDUAL

The errors or *residuals* from a regression line are defined by

$$e_i = (Y_i - \hat{Y}_i) \tag{4-20}$$

In order to understand what the coefficient of determination represents, let us rewrite Equation 4-20 as

$$Y_i = \hat{Y}_i + e_i$$

Subtracting $\bar{Y}$ from both sides of this equation yields

$$Y_i - \bar{Y} = \hat{Y}_i - \bar{Y} + e_i$$

which can be rewritten

$$y_i = \hat{y}_i + e_i \tag{4-21}$$

where $y_i = (Y_i - \bar{Y})$ and $\hat{y}_i = (\hat{Y} - \bar{Y})$.

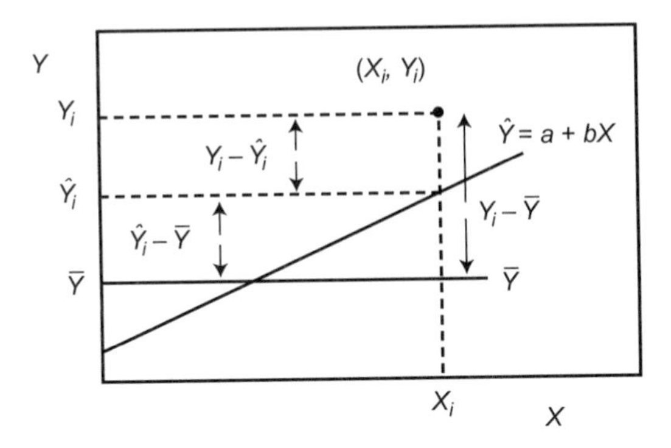

FIGURE 4-13. Decomposition of the total variation in *Y*.

If we now square both sides of Equation 4-21 and sum over the sample, we obtain

$$\sum_{i=1}^{n} y_i^2 = \sum_{i=1}^{n} \hat{y}_i^2 + \sum_{i=1}^{n} e_i^2 + 2\sum_{i=1}^{n} \hat{y}_i e_i = \sum_{i=1}^{n} \hat{y}_i^2 + \sum_{i=1}^{n} e_i^2 \tag{4-22}$$

since $\Sigma_{i=1}^{n} \hat{y}_i e_i = \Sigma_{i=1}^{n} (\hat{Y}_i - \bar{Y})(Y_i - \hat{Y}_i) = 0$.

Expanding Equation 4-22 produces

$$\sum_{i=1}^{n} (Y_i - \bar{Y})^2 = \sum_{i=1}^{n} (\hat{Y}_i - \bar{Y})^2 + \sum_{i=1}^{n} (Y_i - \hat{Y}_i)^2 \tag{4-23}$$

or

$$\text{TSS} = \text{RSS} + \text{ESS} \tag{4-24}$$

where TSS is the total sum of squares, or the total variation in the observed *Y* values about their mean; RSS is the regression sum of squares, or the variation in *Y* explained by the regression; and ESS is the error sum of squares or the variation in *Y* that is not explained by the regression. Equation 4-24 states that the total variation of *Y* can be divided into two parts: that explained by the regression (RSS) and that left unexplained (ESS). Figure 4-13 illustrates the different components of the variation in the dependent variable. These components are discussed in more detail in Chapter 11.

We have still not developed a unit-free measure of goodness of fit, since all our components of variation depend on the units in which the dependent variable is measured. To eliminate this difficulty, let us divide both sides of Equation 4-24 by TSS to get

$$1 = \frac{\text{RSS}}{\text{TSS}} + \frac{\text{ESS}}{\text{TSS}}$$

We can now define the coefficient of determination.

DEFINITION: COEFFICIENT OF DETERMINATION r^2

$$r^2 = 1 - \frac{\text{ESS}}{\text{TSS}} = \frac{\text{RSS}}{\text{TSS}} \tag{4-25}$$

The coefficient of determination, r^2 can be interpreted as the proportion of the total variation in Y that is explained by the regression relationship. Since $0 \le \text{RSS} \le \text{TSS}$, we see that $0 \le r^2 \le 1$. When RSS = 1, the regression line passes through all the points in the scatterplot, $r^2 = 1$ and the regression has explained 100% of the total variation in Y. When RSS = 0, $r^2 = 0$, and variable X is of no use in predicting the value of Y. Most often, r^2 lies between these extremes.

In general, a high value of r^2 means we have a good fit, and a low value means we have a poor fit. It is no accident that we define the coefficient of determination as r^2. This is because it is always numerically equal to the square of the correlation coefficient r. We can use this relationship to aid in the interpretation of the correlation coefficient. An $r^2 = 0.25$ implies that variable X "explains" 25% of the variation in Y. This corresponds to a correlation coefficient of $r = 0.5$. We can thus translate r into values of r^2 and get a better "feel" for the strength of the association. Thus a correlation coefficient of $r = 0.4$ might seem quite weak when we realize that it means only 0.16 or 16% of the variation of Y is explained by the variation in X.

In Table 4-10 we calculate the coefficient of determination for our sample problem relating the death rate from lung cancer to smoking. The total sum of squares or the total variation in the dependent variable is 768.7. (Note that this value is different

TABLE 4-10
Work Table for the Calculation of r^2 and $s_{Y \cdot X}$

Obs. No.	X_i	Y_i	$\hat{Y}_i$	$Y_i - \hat{Y}_i$	$(Y_i - \hat{Y}_i)^2$	$(Y_i - \bar{Y})^2$
1	18.20	17.05	16.101	0.949	0.901	6.776
2	25.82	19.80	20.133	–0.332	0.111	0.022
3	18.24	15.98	16.122	–0.142	0.020	13.491
4	28.60	22.07	21.603	0.467	0.218	5.842
5	31.10	22.83	22.926	–0.096	0.009	10.093
6	33.60	24.55	24.249	0.301	0.091	23.981
7	40.46	27.27	27.878	–0.608	0.370	58.019
8	28.27	23.57	21.429	2.141	4.585	15.343
9	20.10	13.58	17.106	–3.526	12.434	36.881
.	.	.	.	.	.	.
44	30.34	25.88	22.523	3.356	11.263	38.776
		$\bar{Y} = 19.653$		$\sum_{i=1}^{n}(Y_i - \hat{Y}_i) = 0$	$\sum_{i=1}^{n}(Y_i - \hat{Y}_i)^2 = 394.8$	$\sum_{i=1}^{n}(Y_i - \bar{Y})^2 = 768.7$

$$r = \frac{768.7 - 394.8}{768.7} = 0.486 \qquad s_{Y \cdot X} = \sqrt{\frac{394.8}{44 - 2}} = 3.066$$

coeff of var = .13

from the variance of Y, which is equal to TSS/(n – 1). The unexplained variation of Y is 394.8, and so the regression sum of squares is 373.9. To check the calculation r^2 = 0.486, we compare it to the square of the correlation coefficient between the death rate from lung cancer and smoking levels from Table 4-7. Notice that 0.486 = $(0.697)^2$.

The coefficient of determination should be used with caution. Many researchers often place undue emphasis on obtaining a high value of r^2. However, the coefficient of determination alone is not always the best measure of the utility of a regression model. For example, it is relatively easy to obtain high r^2 values using time-series data. One variable growing over time is quite likely to do a good job explaining the growth of another variable over time, whether or not the two variables are causally linked. Similarly, it can be difficult to get even moderately large r^2 values for some variables where the causal link is theoretically clear. This is particularly the case when analysis focuses not on explaining the levels of a dependent variable but on explaining changes in the values of that variable.

Finally, we should emphasize that calling r^2 the coefficient of "determination" does not imply that X necessarily causes Y. The index merely says something about the statistical association found in the data. Likewise, when we speak of the variance in Y "explained" by X, there is no implication of a theoretical connection between the variables. As stated above, questions of causality and explanation are outside the purview of statistics.

STANDARD ERROR OF ESTIMATE $s_{Y \cdot X}$

In many applications of regression analysis, the ultimate goal is to come up with a good estimate or prediction for Y by using variable X. Our success is measured by the degree to which we can predict Y accurately. In this case, we are interested primarily in the numerical value of the error we are likely to make when utilizing X to estimate Y. The best measure of the goodness of fit in this instance is the *standard error of the estimate* $s_{Y \cdot X}$. The standard error of the estimate is the standard deviation of the residuals about

---

DEFINITION: STANDARD ERROR OF ESTIMATE

The standard error of the estimate is defined as

$$s_{Y \cdot X} = \sqrt{\frac{\sum_{i=1}^{n}(Y_i - \hat{Y})^2}{n-2}} \qquad (4\text{-}26)$$

---

the regression line. The subscript $Y \cdot X$ is used in Equation 4-26 to denote the fact that we are not calculating the standard deviation of Y around its mean, but the standard deviation of Y around values $\hat{Y}$ predicted by the regression equation.

The *standard error of the estimate* is sometimes called the *root mean square error,* or root MSE. Notice that it is calculated as the square root of the average or mean squared error. Why do we divide by n – 2 and not n? As we will see in Chapter 12, $s_{Y \cdot X}$ is really an *estimate* of the standard deviation of the error term in our model.

We treat our data of n observations as if they were a sample from a population. As we saw with the calculation of the sample standard deviation and variance, when we make an estimate in inferential statistics, we divide by the appropriate degrees of freedom, not the sample size. The degrees of freedom are equal to $n - 2$ because we lose two degrees of freedom in the data when we estimate the regression coefficients a and b since they require $\bar{X}$ and $\bar{Y}$ for their calculation. We can also use a straightforward geometric argument. because two points are needed to determine a line, any points in excess of two provide degrees of freedom in defining the location of the line. Thus $n - 2$ degrees of freedom are always associated with a simple regression based on n observations.

Table 4-10 shows the calculation of the standard error of estimate for the lung cancer death rate and smoking data. $s_{Y \cdot X}$ is found by taking the square root of the residual variation, or ESS, and dividing by the number of observations minus two. Note that the residual sum of squares is always generated in the computation of r^2. In this example, $s_{Y \cdot X} = 3.066$. Since the standard error of estimate is always measured in units of the dependent variable, we can interpret this index of goodness of fit as 3.066 deaths per 100,000 people. Therefore, on average for a given level of cigarette smoking, the regression predicts the death rate from lung cancer to within 3.066 of the actual rate. But does this represent a "good fit"? We can make this judgment only within the context of the problem at hand, that is, in relation to some predetermined standard of prediction.

If the standard error of estimate is 3.066, then one standard deviation on either side of the regression line roughly corresponds to six deaths per 100,000 people. Two standard deviations would correspond to a death rate of 12. If the errors about the regression line are normally distributed (see Chapter 5), then we might expect approximately 68% of our observations to fall within one standard deviation of the regression line and 95% to fall within two standard deviations. So our predictive accuracy could be very crudely estimated as a range of about 12 people 95% of the time. With a mean death rate of 19.653, our regression equation does not appear to be all that accurate! This only confirms what we know from the coefficient of determination, which indicates the regression of Y on X accounts for just under half the total variation in the death rate.

Finally, let us consider a standardized measure of the goodness of fit known as the *coefficient of variation.* Recall that in Chapter 3 we defined the coefficient of variation CV to be the ratio of the standard deviation to the mean. We use this statistic to compare the variability of two or more variables with different standard deviations and different means. In the context of regression, the appropriate definition is

$$CV = \frac{s_{Y \cdot X}}{\bar{Y}} \times 100$$

where we have multiplied by 100 to express the coefficient as a percentage. If $CV = 0$, this implies that $s_{Y \cdot X} = 0$ and we have a perfect fit between the regression line and the observed values of the dependent variable. When $b = 0$ and the line of best fit reduces to $\hat{Y} = \bar{Y}$, then $CV = 100$. In most cases the CV lies between these extremes. The lower the value of the CV, the better the fit to the regression line.

4.5. Temporal Autocorrelation

In Chapter 2 we introduced temporal data or time series as observations on a variable collected over successive increments in time. In this section, we develop our understanding of correlation to consider the relationships among observations on a single time-series variable. Having just defined correlation as a measure of the linear association between two variables, thinking about correlation in terms of a single variable might sound a little strange. However, time-series data can be organized or ordered by time as well as by the values of a variable of interest. Thus, the correlation of a times-series variable and itself, the *autocorrelation* of the variable over time, concerns whether or not values of the variable at some times are linearly associated with values of the variable measured at other times. The existence of autocorrelation in temporal data has important implications for time-series analysis (Chapter 15) and for inferential aspects of regression (Chapters 12 and 13). Note that some texts refer to temporal autocorrelation as serial correlation.

DEFINITION: TEMPORAL AUTOCORRELATION

Temporal autocorrelation measures the linear association between observations on a single variable gathered over successive units of time.

We can reinforce our understanding of autocorrelation in time-series data by looking at the stylized examples in Figure 4-14. The left-hand column shows time-series plots where the individual observations on variable Y have been indexed by time. The right-hand column presents corresponding scatterplots that relate the value of Y at time t (Y_t) to the value of Y at time $t - 1$ (Y_{t-1}). Autocorrelation between a variable and values of that same variable *lagged* by one period is known as *first-order autocorrelation.* More generally, autocorrelation between a variable and values of that same variable lagged by k periods is known as k-order autocorrelation.

Figure 4-14a shows a time series of the variable Y_t that exhibits a significant upward trend. A scatterplot of Y_t against the first-order lag Y_{t-1} is drawn in Figure 4-14b. Note that the points in this figure fall mostly in the "positive" quadrants I and III. This is because most adjacent observations in the time-series graph are either below the mean or above the mean. That is, if a value is above average, its successor is likely to be above average as well. Similarly, a below-average value is likely to be followed by a below-average value. The time series in Figure 4-14c oscillates above and below the mean in a predictable fashion. Adjacent observations in this plot are on opposite sides of the mean value of the series. This is confirmed by the first-order lag scatterplot of Figure 4-14d, where most observations fall in the "negative" quadrants II and IV. The first-order linear autocorrelation coefficient for the scatterplot in Figure 4-14d is –0.757, indicating a strong negative relationship between Y_{t-1} and Y_t. The time series in Figure 4-14e oscillates above and below the mean apparently at random. Indeed, the first-order lag scatterplot of this time series shows the lack of a strong linear association between Y_t and Y_{t-1}. The autocorrelation coefficient for these variables is 0.071, indicating the absence of a linear relationship between them.

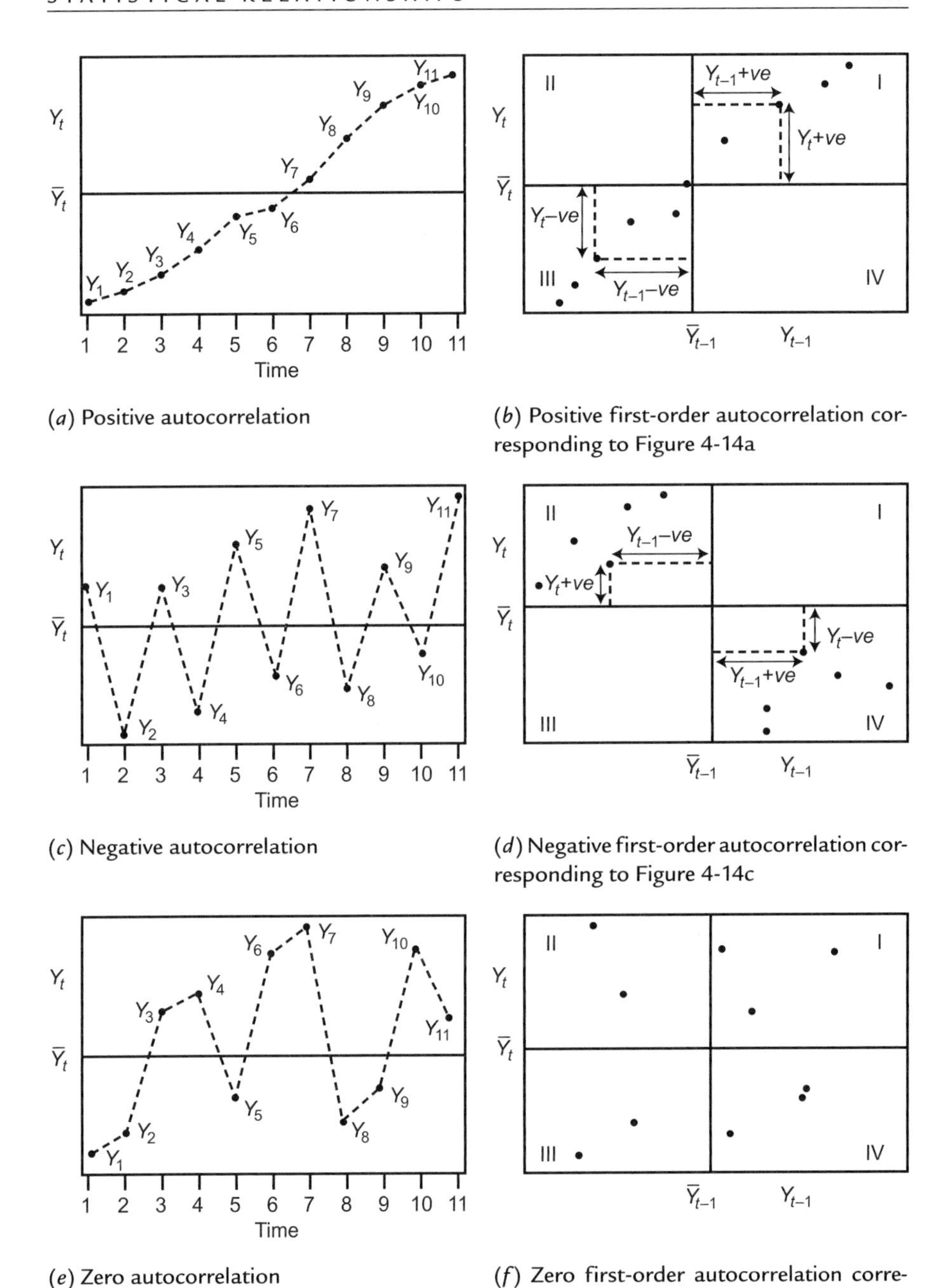

(*a*) Positive autocorrelation

(*b*) Positive first-order autocorrelation corresponding to Figure 4-14a

(*c*) Negative autocorrelation

(*d*) Negative first-order autocorrelation corresponding to Figure 4-14c

(*e*) Zero autocorrelation

(*f*) Zero first-order autocorrelation corresponding to Figure 4-14e

FIGURE 4-14. Autocorrelation in time-series data.

Following Equation 4-2, the (first order) sample autocorrelation coefficient between a variable Y_t and values of that same variable lagged one time period Y_{t-1} is

$$r_{Y_t \cdot Y_{t-1}} = \frac{\sum_{i=2}^{n}(Y_{ti} - \bar{Y}_t)(Y_{t-1i} - \bar{Y}_{t-1})}{\sqrt{\sum_{i=2}^{n}(Y_{ti} - \bar{Y}_t)^2}\sqrt{\sum_{i=1}^{n}(Y_{t-1i} - \bar{Y}_{t-1})^2}} \tag{4-27}$$

EXAMPLE 4-6. Let us now use Equation 4-27 to calculate the first-order autocorrelation coefficient for an actual time series. Figure 4-15a shows the time series of daily maximum temperatures for June 2007 in Madison, Wisconsin. Visually, we see no strong overall trend, but there are heat waves in the second and third weeks and a cool period in the first week. Figure 4-15b contains the first-order scatterplot for the same data. Both the time series and the scatterplot suggest a positive autocorrelation, with warm days followed by warm days and cool days followed by cool days.

To confirm this suggestion, Table 4-11 presents a worksheet for the calculation of the autocorrelation coefficient. The calculated value of $r_{Y_t \cdot Y_{t-1}} = 0.616$. Thus the correlation between the maximum temperature and its first-order lag is indeed positive.

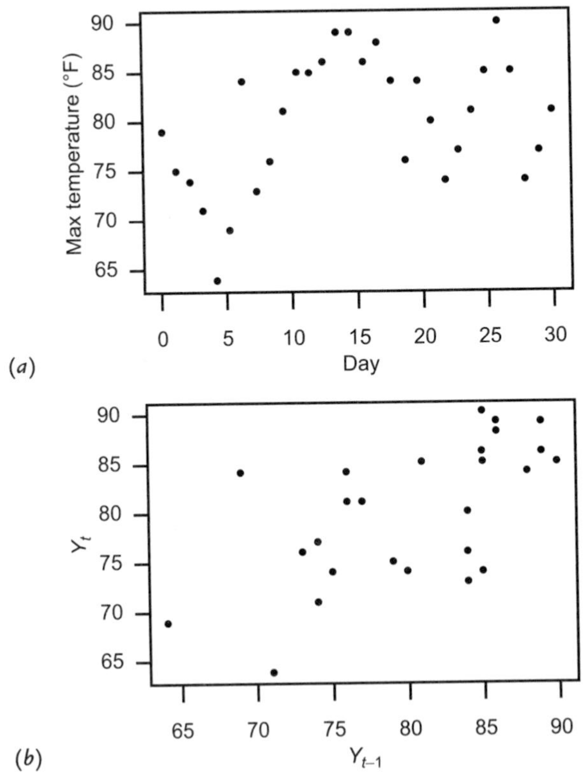

FIGURE 4-15. June 2007 daily maximum temperatures for Madison, Wisconsin.

TABLE 4-11
Worksheet for the First-Order Autocorrelation Coefficient

Day (t)	Y_t	Y_{t-1}	Y_t^2	Y_{t-1}^2	$Y_t Y_{t-1}$
1	79				
2	75	79	5,625	6,241	5,925
3	74	75	5,476	5,625	5,550
4	71	74	5,041	5,476	5,254
...	...	...	...	...	...
28	74	85	5,476	7,225	6,290
29	77	74	5,929	5,476	5,698
30	81	77	6,561	5,929	6,237
Totals	2,323	2,321	187,319	186,999	186,683

Note: Lags are available only from day 2 onward; thus the summation runs from day 2 through day 30 and $n = 29$. We are using the computational formula in Equation 4-3

$$r_{Y_t \cdot Y_{t-1}} = \frac{186683 - (2323)(2321)/29}{\sqrt{187391 - (2323)^2/29}\,\sqrt{186999 - (2321)^2/29}} = \frac{763}{\sqrt{35.195}\,\sqrt{35.199}} = 0.616$$

(The full dataset is available on the book's website.)

The relatively high value of the autocorrelation coefficient reveals that the linear relationship is relatively strong.

We can calculate the sample autocorrelation coefficient for lags greater than one. In general, the kth order sample autocorrelation coefficient between the variables Y_t and Y_{t-k} is

$$r_{Y_t \cdot Y_{t-k}} = \frac{\sum_{i=k+1}^{n} (Y_{ti} - \bar{Y}_t)(Y_{t-ki} - \bar{Y}_{t-k})}{\sqrt{\sum_{i=k+1}^{n} (Y_{ti} - \bar{Y}_t)^2}\,\sqrt{\sum_{i=1}^{n-k} (Y_{t-ki} - \bar{Y}_{t-k})^2}}$$

In Chapter 15, we show how autocorrelation coefficients of different order can be used to identify the characteristics of a time series.

4.6. Summary

When the value taken by one variable affects the value taken by another variable, then the two variables are said to be statistically dependent. We began the exploration of relationships between variables by examining how information on a categorical variable may be used to separate groups of observations on a numeric variable. This separation can be a useful tool for understanding the distribution of the numeric variable. Relationships between two quantitative variables were examined with the scatterplot. The correlation coefficient provides a numerical measure of the linear association between two quantitative variables that are measured across the same set of observations. The concept of correlation is closely related to that of regression. The introduction to regression analysis in this chapter showed how to fit a regression line to a scatterplot and how to assess the goodness of fit of that line. In Chapters 12 and 13 we extend the discussion of regression to questions of inference. The final section of this chapter discussed the concept of autocorrelation in time-series data. Analysis of time-series data is developed further in Chapter 15. Chapter 14 explores autocorrelation in a spatial context as we review statistical techniques for analysis of spatial relationships.

Appendix 4a. Review of the Elementary Geometry of a Line

A straight line is precisely defined by the coordinates of any two points on the line since it continues forever in the same direction. Straight lines have a constant slope. Let us consider the points $P_1(X_1, Y_1)$ and $P_2(X_2, Y_2)$ of Figure 4-16. We define the slope of the line passing through these points to be

$$\frac{\Delta Y}{\Delta X} = \frac{Y_2 - Y_1}{X_2 - X_1} = b \qquad (4\text{-}28)$$

where Δ means change or difference. When $\Delta X = 1$, $\Delta Y = b$, and we can interpret b as the change in Y that results from a unit increase in X.

If we know two points on a line or one point and the slope of the line, we can determine the unique equation for that line. Since any two points on the line will suffice, we choose the two points of Figure 4-17. The coordinates are $(0, a)$ for P_1 and (X_2, Y_2) for point P_2. If the line is to have constant slope, then, by Equation 4-28

$$\frac{Y - a}{X - 0} = a \quad \text{and} \quad Y - a = bX \quad \text{or} \quad Y = a + bX \qquad (4\text{-}29)$$

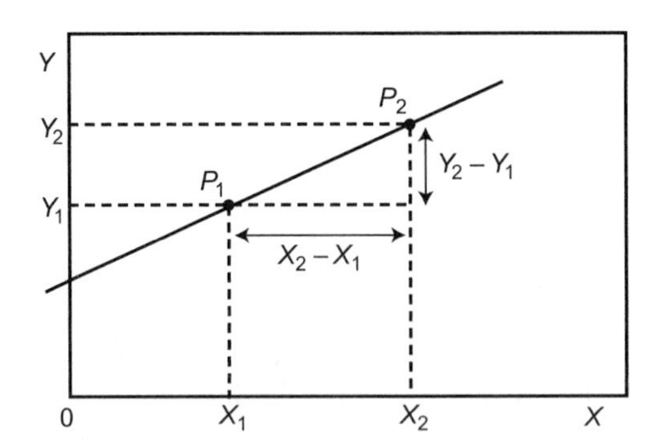

FIGURE 4-16. Slope of a line.

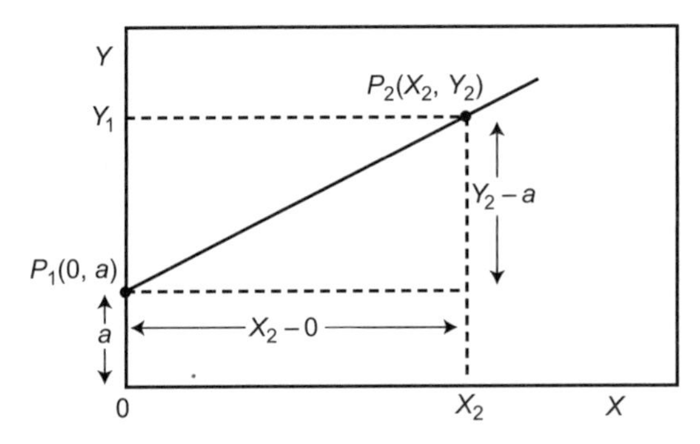

FIGURE 4-17. Derivation of the equation for a line.

This is the slope-intercept form of the equation for a straight line with intercept *a* and slope *b*. The coefficients *a* and *b* are unrestricted in sign or value.

Figure 4-18 illustrates several examples of straight lines and their equations. The line in Figure (b) differs from a in that it has a steeper slope ($1 > 0.5$). The line depicted in (c) is parallel to b because it has the same slope (+1), but differs because

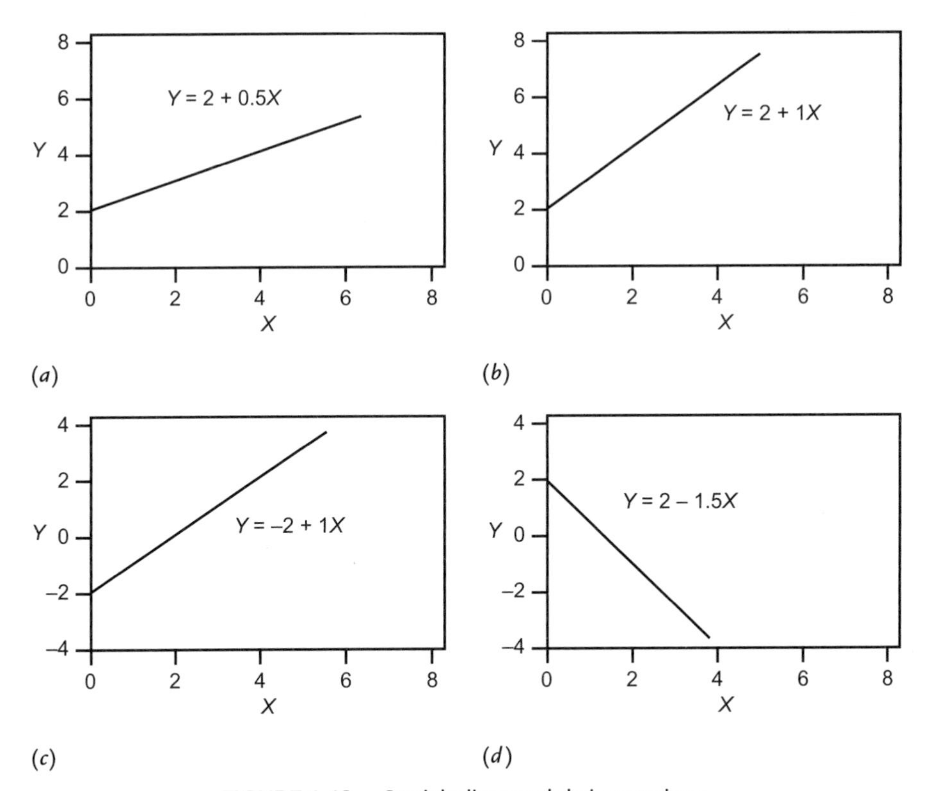

FIGURE 4-18. Straight lines and their equations.

it has a lower intercept (–2 < 2). In (d) we see the graph of a line with a negative slope (–1.5).

Appendix 4b. Least Squares Solution via Elementary Calculus

The normal equations (4-8) and (4-9) that lead directly to the computational formulas for a and b can be easily derived from elementary calculus. Let

$$S(a,b) = \sum_{i=1}^{n}(Y_i - a - bX_i)^2 \tag{4-30}$$

be the total sum of squared deviations from the regression line to the n points. The values of a and b that minimize S can be derived by differentiating Equation (4-39) with respect to a and b, to obtain

$$\frac{\partial S}{\partial a} = -2\sum_{i=1}^{n}(Y_i - a - bX_i)$$

$$\frac{\partial S}{\partial b} = -2\sum_{i=1}^{n}(Y_i - a - bX_i)X_i$$

Setting each derivative equal to zero and simplifying yields

$$\sum_{i=1}^{n}(Y_i - a - bX_i) = 0$$

$$\sum_{i=1}^{n}X_i(Y_i - a - bX_i) = 0$$

These equations can be expanded by using simple rules of summations to produce

$$\sum_{i=1}^{n}Y_i - na - b\sum_{i=1}^{n}X_i = 0$$

$$\sum_{i=1}^{n}X_iY_i - a\sum_{i=1}^{n}X_i - b\sum_{i=1}^{n}X_i^2 = 0$$

These equations are identical to the normal equations (4-8) and (4-9) with a simple rearrangement of terms.

REFERENCES

R. A. Fisher, "The Use of Multiple Measurements in Axonomic Problems," *Annals of Eugenics* 7 (1936), 179–188.

J. F. Fraumeni, "Cigarette Smoking and Cancers of the Urinary Tract: Geographic Variations in the United States," *Journal of the National Cancer Institute* 41 (1968), 1205–1211.

P. D. Jones, T. M. Wigley, and P. B. Wright, "Global Temperature Variations Between 1861 and 1984," *Nature* 322 (1986), 430–434.

C. D. Keeling and T. P. Whorf, "Atmospheric CO2 Records from Sites in the SIO Air Sampling network," in Carbon Dioxide Information Analysis Center, *Trends: A Compendium of Data on*

Global Change, (Oak Ridge National Laboratory, U.S. Department of Energy, Oak Ridge, TN, 2005).

M. G. Kendall and A. Stuart, *The Advanced Theory of Statistics 2: Inference and Relationship.* (London: Griffin, 1961).

R. Summers and A. Heston, "The Penn World Tables (Mark 5): An Expanded Set of International Comparisons, 1950–1988." *Quarterly Journal of Economics* 106 (1991), 327–368.

P. Taylor, "A Pedagogic Application of Multiple Regression Analysis," *Geography* 65 (1980), 203–212.

U.S. Bureau of the Census. *Census of Manufactures, various years.* Washington, DC.

FURTHER READING

Most elementary statistics textbooks cover the concepts of correlation and regression. Good introductions are provided by Moore and McCabe (1993) and Freedman et al. (1978). Stanton (2001) provides an interesting history of the development of these concepts.

D. Freedman, R. Pisani, and R. Purves, *Statistics.* (New York: Norton, 1978).

D. Moore and G. McCabe, *Introduction to the Practice of Statistics,* 2nd ed. (New York: Freeman, 1993).

J. Stanton, "Galton, Pearson, and the Peas: A Brief History of Linear Regression for Statistics Instructors," *Journal of Statistics Education* 9 (2001), 1–12.

DATASETS USED IN THIS CHAPTER

convergence.html
co2.html
iris.html
smoking.html
temperature.html

PROBLEMS

1. Explain the meaning of the following terms or concepts:
   - Scatterplot
   - Correlation coefficient
   - Covariance
   - Coefficient of determination
   - Standard error of estimate
   - Residual
   - Autocorrelation in temporal data

2. What is the difference between correlation and regression?

3. Differentiate between a functional relationship and a statistical relationship.

4. In Figure 4-4 we showed the relationship between levels of GDP and GDP growth for different countries, some members of the OECD and some not. These data are on the website for the book (convergence.html).
   a. Produce separate scatterplots of GDP against GDP growth for OECD nations and for non-OECD nations.
   b. Calculate the correlation coefficients for the data in these two scatterplots.
   c. Interpret your results.

5. The smoking and cancer data are on the website (smoking.html). Using the format shown in Table 4-6, calculate the correlation coefficients between the different forms of cancer and smoking. Check your results against the matrix of correlation coefficients in the text.

6. Measurements of sound levels and distance from the centerline of an urban expressway are as follows:

Observation	Distance (ft.)	Sound level (dB)
1	45	83
2	63	81
3	160	66
4	225	68
5	305	69
6	390	66
7	461	58
8	515	57
9	605	55
10	625	61

   a. Draw a scatterplot of the data with distance as the independent variable on the X-axis.
   b. Try and plot a best-fit line representing the relationship between the two variables.
   c. Using the format of Table 4-8, estimate the slope and intercept of the regression equation.
   d. Now fit the real regression line to the scatterplot. How good was your answer to question b?
   e. Interpret the meaning of the regression slope coefficient and the intercept for this problem.
   f. Suppose the distance variable was measured in hundreds of feet rather than feet. How would the regression equation change? (*Hint:* No new calculations are required.)
   g. Which observation is most closely predicted by the regression line? least closely?
   h. Calculate the coefficient of determination for this problem. What does this value mean?
   i. Calculate the standard deviation of the dependent variable and the standard deviation of the independent variable. Use these values, together with the regression slope coefficient, to estimate the correlation coefficient between your variables. Check this value by finding the square root of the coefficient of determination.

7. Get the carbon dioxide and atmospheric temperature data (*warming.html*) from the website for this book.
   a. Which of these variables should be dependent and which independent?
   b. Calculate the regression equation for these variables.

c. Interpret the meaning of the regression intercept and slope coefficients.
d. Calculate the standard error of estimate and the coefficient of determination.
e. Find the residuals from this problem. Are there any outliers?

8. In deriving the formula for the coefficient of determination, we stated that $\Sigma\hat{y}_i e_i = 0$. Show why this equality holds.

9. Obtain the Dow Jones stock market data from the book website (*dow.html*). Calculate the first-order autocorrelation for these data. (Don't try this by hand!)

III

INFERENTIAL STATISTICS

5

Random Variables and Probability Distributions

Chapters 2, 3, and 4 presented a series of techniques for describing and summarizing data. Looking at data, that is, visualizing it in some way, and exploring data with the aid of summary measures are critical first steps in more sophisticated statistical investigation and inference. In this chapter we develop the link between descriptive and inferential statistics. That link rests squarely on the concepts of random variables and probability distributions. Although most people are familiar with notions of probability in games of chance—calling heads or tails in a coin toss and winning or losing a bet—the fundamental concepts of probability are sometimes difficult to comprehend. Definitions of statistical experiments, statistical independence, and mutual exclusivity seem far removed from the context of these games of chance. However, probability theory lies at the core of statistical inference because all statistical judgments are necessarily probabilistic.

Central to the discussion in this chapter are arguments concerning *probability, random variables,* and *probability distributions.* In Section 5.1 the fundamental concepts of probability are briefly reviewed. In Section 5.2 a random variable is formally defined and related by example to the process of population sampling. We differentiate between two classes of random variables: those that are discrete and those that are continuous. Sections 5.3 and 5.4 are devoted to the specification of probability distribution models for these two classes of random variables. In Section 5.5 we extend our knowledge of random variables and probability distributions to include *bivariate* random variables and bivariate probability distributions. Bivariate probability distributions describe the *joint* distribution of *two* variables. Section 5.6 offers a brief summary of the chapter.

5.1. Elementary Probability Theory

Most people have some idea of what probability means. Everyday conversation contains numerous references to it: The *chance* of rain this holiday weekend is 50%. The *odds* are small that a river will flood in a particular year. The *likelihood* of contracting

West Nile virus in a US state is positively related to the *probability* of contracting the virus in nearby states. All these statements invoke a notion of probability, though a somewhat vague one. In this section, we discuss some fundamentals of probability theory and we show how to rigorously define the term *probability* as it is used in conventional statistical inference. We also briefly explain the methods used to calculate the probability of an event occurring. These tasks are critical to the discussion that follows because the theory of probability provides the logical foundation for *statistical inference,* the process whereby information from a sample dataset is used to estimate the characteristics of a larger population.

Statistical Experiments, Sample Spaces, and Events

The basic concepts of probability theory rest upon statistical experiment, or random trial.

DEFINITION: STATISTICAL EXPERIMENT

A statistical experiment, or random trial, is a process or activity in which one outcome from a set of possible outcomes occurs. Which outcome occurs is not known with certainty before the experiment takes place.

For example, consider the process of sampling from a population. Each time we select a single member of a population and record the value of a variable, we are performing a statistical experiment. The values of the variable for different members of the population represent the possible outcomes in the experiment. Which member of the population is selected, and therefore what value of the variable of interest is recorded, is unknown until the experiment is performed. Repeated trials of this experiment would yield a sample of observations from the population.

DEFINITIONS: ELEMENTARY OUTCOMES AND SAMPLE SPACE

Each different outcome of an experiment is known as an elementary outcome, and the set of all elementary outcomes constitutes the sample space.

Let us examine the statistical experiment of selecting a card from a playing deck. Each of the 52 individual cards in the deck represents an elementary outcome. Together the 52 outcomes comprise the sample space of the experiment.

In some cases the elementary outcomes of a statistical experiment can be defined in different ways. Suppose we define an experiment as the toss of two coins. We could define the elementary outcomes as getting no heads, getting one head, and getting two heads. Alternatively, we might define the elementary outcomes as the complete set of combinations of heads and tails possible from a single toss of the two coins. In this case we might define the outcomes as {HH}, {HT}, {TH}, and {TT}, all of which have an equal probability of occurring.

After the elementary outcomes of an experiment have been defined, it is possible to define collections of elementary outcomes as events.

DEFINITION: EVENT
An event is a subset of the sample space of an experiment, a collection of elementary outcomes.

The particular subset used to define an event can include one or more elementary outcomes. Returning to the card-drawing experiment, we might define event A as drawing a spade, event B as drawing an ace, and event C as drawing the ace of spades. Event A includes 13 elementary outcomes, each of the 13 spades in the playing deck. Similarly, event B includes four elementary outcomes, and event C includes one elementary outcome. A *null event,* denoted $\varnothing$, is an event that contains no elementary outcomes: it cannot occur.

Computing Probabilities in Statistical Experiments

The outcome of a statistical experiment will be one, and only one, of the elementary outcomes of the sample space. To determine the probability that a specific event occurs in a single trial of an experiment, it is necessary to define the likelihood, or the *probability,* of each elementary outcome E_i of the experiment occurring. Leaving aside for the moment the thorny question of how these values are obtained, let us define three postulates that formalize certain properties of probabilities.

Denote the n elementary outcomes of an experiment as the set $S = \{E_1, E_2, \ldots, E_n\}$ that defines the sample space, and let the assigned probabilities of these outcomes be $\{P(E_1), P(E_2), \ldots, P(E_n)\}$.

POSTULATE 1

$$0 \leq P(E_i) \leq 1 \quad \text{for} \quad i = 1, 2, \ldots, n \tag{5-1}$$

For every elementary outcome in the sample space, the assigned probability must be a non-negative number between zero and one inclusive. There are several important arguments in this seemingly simple postulate. First, probabilities are non-negative. Second, if we say an outcome, for example, outcome j, is *impossible,* we are saying that $P(E_j) = 0$. If we say an outcome, for example, outcome k, is *certain* to occur, then we are saying that $P(E_k) = 1$.

POSTULATE 2

$$P(A) = \sum_{i \in A} P(E_i) \tag{5-2}$$

The second postulate states that the probability of any event A occurring is the sum of the probabilities assigned to the individual elementary outcomes that constitute event A.

POSTULATE 3

$$P(S) = 1$$
$$P(\varnothing) = 0 \qquad (5\text{-}3)$$

The third postulate states that the probability associated with the entire sample space must equal one, and the probability of a null event must be zero. Although these two statements are intuitive, it is possible to deduce a number of important results from them.

First, it is easily shown that the sum of the probabilities of all the elementary outcomes must equal one:

$$\sum_{i=1}^{n} P(E_i) = 1 \qquad (5\text{-}4)$$

This follows directly from Postulates 2 and 3. Since the sample space contains all the elementary outcomes, that is, $S = \{E_1, E_2, \ldots, E_n\}$, and from Postulate 2 we know that $P(S) = \Sigma_{i=1}^{n} P(E_i)$, it follows from Postulate 3 that $\Sigma_{i=1}^{n} P(E_i) = 1$.

Second, since any event A must contain a subset of the elementary outcomes of S, and since from Postulate 1 the probability of an elementary outcome occurring is between zero and one, it follows that for any event A

$$0 \leq P(A) \leq 1 \qquad (5\text{-}5)$$

In other words, every event has a probability of occurring that lies between zero and one.

Third, if events A and B are *mutually exclusive,* meaning that no elementary outcomes are common to both events, then from Postulate 3

$$P(A \cap B) = 0 \qquad (5\text{-}6)$$

where $\cap$ denotes the *intersection* of A and B, the set of elementary outcomes that belong to both events. This follows from the fact that the intersection of two mutually exclusive events must be the null set: $A \cap B = \varnothing$. For example, counting the spots after rolling a die gives a number 1, 2, . . . , 6. All of these numbers are mutually excusive events. According to Equation 5-6, there is no chance of observing a 2 and a 5 on the same roll.

Definitions of Probability

The three probability postulates tell us that probability is a measure that, for any elementary outcome or event, is a number between zero and one. But where do probability values come from, and how are they to be interpreted? It turns out that there are different ways of determining probabilities and thus different ways of interpreting what they represent. A simple distinction can be made between *objective* and *subjective* interpretations of probability. According to the objective view, different individuals will use the rules of probability to assign the same probabilities to the elementary outcomes of a particular experiment. According to the subjective view, estimates of the likelihood of a particular event occurring in the trial of an experiment will vary from person to person depending on their individual assessment of the relevant evidence at that particular moment in time. Let us briefly consider these alternative ways of understanding the meaning of probability.

There are two methods of finding probabilities according to the objective interpretation of probability. The first of these, the *classical* view of probability, is usually associated with games of chance. For many such games, each elementary outcome has the same probability of occurrence. In these games, we can use arguments of symmetry or geometry to deduce the probability of a given event occurring. Accordingly, in an experiment with n equally likely outcomes, the probability of occurrence of any one elementary outcome is $1/n$. More generally, if event A is consistent with m elementary outcomes and the sample space comprises n equally likely outcomes,

$$P(A) = m/n \tag{5-7}$$

For example, the sample space S for the card selection experiment consists of $n = 52$ elementary outcomes. If we define event A to be drawing a diamond, then, with $m = 13$, the probability of event A occurring is $13/52 = 1/4$. Similarly, the probability of drawing an ace is $4/52 = 1/13$, since $m = 4$ and $n = 52$.

If Equation 5-7 cannot be used to calculate event probabilities, then other, more time-consuming methods must be used. First, it is necessary to identify and count all of the elementary outcomes in the sample space. Then, the subset of the sample space that comprises the event space must be specified. Finally, the addition of the probabilities of the elementary outcomes in the event space will yield the event probability. The simplification of this operation by using Equation 5-7 is obvious, particularly when there may be thousands or even millions of elementary outcomes in the sample space and/or event space. In fact, it may even be difficult to generate a list of the elementary outcomes in a sample space. In such cases we often use *counting rules* to compute both m and n. These counting rules are shown in Appendix 5a.

The second objective view of probability is the relative frequency interpretation, which is typically associated with prominent statisticians such as Pearson and Fisher. Unlike the deductive logic applied to find probabilities in the classical model, supporters of the relative frequency position adopt an inductive approach to probability determination, whereby the probability of an event is given by the relative number of times that event occurs in a large number of trials. This definition can be understood more formally in the following way.

DEFINITION: RELATIVE FREQUENCY INTERPRETATION OF PROBABILITY

Let event A be defined in some experiment. If the experiment is repeated N times and event A occurs in n of those trials, then the relative frequency of event A is n/N. The probability of event A occurring is $P(A) = n/N$ as $N \to \infty$, that is, where the number of trials of the experiment approaches infinity (becomes extremely large).

Since experiments cannot be conducted indefinitely, the observed relative frequency in an extended set of trials can be used as an estimate of the probability of an event occurring.

The relative frequency interpretation of probability is limited to situations where we have access to a lot of repetitive data or where repeated trials of a single experiment may be performed. Note that we do not need to actually perform those trials, but we must believe that repeated trials are possible. If a trial cannot be repeated, the concept of "relative frequency" obviously has no meaning.

In many situations an experiment cannot be repeated exactly, and we are often forced to rely on subjective probabilities. According to the subjective interpretation, probability refers to an individual's "degree of belief" that a particular event may occur. This allows us to assign probabilities to trials with no prospect of repetition. For example, we could assign a probability to recent global warming being the result of human activity. There is only one earth, and only one history of greenhouse emissions. Either these emissions are implicated in global warming, or they are not. To say humans are "highly likely" to have affected climate invokes the subjective interpretation of probability.

Unlike the objective interpretations given above, the subjective probability of an event occurring may change depending on whom we ask and also on when we ask them to assign a probability. Timing is important because an individual gathers information over time and this information may alter the individual's beliefs about a particular event. As an example, many of us can probably relate to a situation where a doctor is asked to make a prognosis about recovery from a particular injury. This prognosis will of course depend on the nature of the injury, and the age and medical history of the patient. The doctor might suggest that the chance, or the probability, of a complete recovery is 0.8 or 80%. It is common for patients to seek a "second opinion," in part, because different doctors have different views about the likelihood of recovery from the same injury. Thus, a second doctor might put the probability of a complete recovery for the same patient at 60%. Because each person can arrive at his or her own answer for the probability of an event, many parts of probability theory cannot utilize subjective probabilities. In many instances, however, we may have no choice but to rely on probability values estimated in this way.

Basic Probability Theorems

Several useful theorems follow directly from the probability postulates discussed above. These theorems can be used to determine probabilities for different types of events. Because they are derived from the postulates, the theorems apply equally to objective and subjective interpretations of probability. In other words, we are free to interpret probabilities emerging from the theorems from either viewpoint.

ADDITION THEOREM

The addition theorem is used to obtain the probability that at least one of two events occurs.

ADDITION THEOREM

Let A and B be any two events defined over the sample space S. Then

$$P(A \cup B) = P(A) + P(B) - P(A \cap B) \tag{5-8}$$

where $A \cup B$ represents the *union* of events A and B, the set of elementary outcomes that belongs to at least one of the events A and B, and where $A \cap B$ represents the *intersection* of events A and B, the set of elementary outcomes that belongs to both events A and B.

For example, consider the problem of determining the probability of drawing a spade or an ace in a single draw from a deck of shuffled cards. We define event A as the selection of a spade and event B as the selection of an ace. Event A comprises 13 elementary outcomes, the 13 spades; event B comprises four outcomes; and one outcome, the ace of spades, is common to both events A and B. Thus, for these two events we know

$$P(A) = 13/52 \quad P(B) = 4/52 \quad P(A \cap B) = 1/52$$

Substituting these known values into Equation 5-8 yields $P(A \cup B) = 13/52 + 4/52 - 1/52 = 16/52 = 0.308$. Since the elementary outcome, the ace of spades, is common to both events A and B, that outcome is counted twice when $P(A)$ and $P(B)$ are added. To avoid double-counting, $P(A \cap B)$ is subtracted.

The addition theorem can be simplified when the two events A and B are mutually exclusive, because $P(A \cap B) = 0$, from Equation 5-6. In this case the addition theorem reduces to

$$P(A \cup B) = P(A) + P(B) \tag{5-9}$$

If we define event C as drawing a diamond from a playing deck, then events A and C have no elementary outcomes in common. The events are mutually exclusive because a spade and a diamond cannot be simultaneously drawn. Therefore, $P(A \cup C) = P(A) + P(C) = 13/52 + 13/52 = 26/52 = 0.5$: there is a 50% chance of drawing either a spade or a diamond.

The addition theorem for mutually exclusive events can be generalized for any number of events. For three mutually exclusive events, D, E, and F, the addition rule is

$$P(D \cup E \cup F) = P(D) + P(E) + P(F) \tag{5-10}$$

Generalizing the addition rule for more than two nonmutually exclusive events is much more complicated. Equation 5-8 would have to be modified to include the intersections between all pairs of events included in the addition.

COMPLEMENTATION THEOREM

It is sometimes more difficult to find the probability of an event, say A, occurring than it is to find the probability of the complementary event $\bar{A}$ occurring. The set of elementary outcomes not in event space A constitute the complementary event to A. Using the complementation theorem, we can derive $P(A)$ from $P(\bar{A})$.

COMPLEMENTATION THEOREM

For any event A and its complement $\bar{A}$ of a sample space S

$$P(A) = 1 - P(\bar{A}) \quad \text{or} \quad P(\bar{A}) = 1 - P(A) \tag{5-11}$$

Consider event D, defined as the selection of any card but a club in a single draw from a deck of playing cards. Event $\bar{D}$ is thus the selection of a club in a single draw. From Equation 5-11,

$$P(D) = 1 - P(\bar{D}) = 1 - 13/52 = 39/62 = 0.75$$

CONDITIONAL PROBABILITIES AND STATISTICAL INDEPENDENCE

The intersection probability $P(A \cap B)$ is sometimes called the joint probability of A and B. To find the joint probability, the probability that A and B occur, it is necessary to introduce the notion of a *conditional probability*. A conditional probability is the probability of an event, say A, occurring, given that another event, say B, has occurred.

DEFINITION: CONDITIONAL PROBABILITY

If events A and B are defined over a sample space S, then the conditional probability of event A, given B, denoted $P(A \mid B)$, is defined as

$$P(A \mid B) = \frac{P(A \cap B)}{P(B)} \tag{5-12}$$

The vertical line separating the A and B in $P(A \mid B)$ denotes a conditional probability. The term $P(A \mid B)$ is read as, "the probability of A, given B."

To illustrate the concept of conditional probability, let us consider a slightly different experiment. Suppose the experiment is to roll a perfectly balanced single die and record the number of dots appearing on the upward face. The sample space S for this experiment consists of six elementary outcomes, the numbers 1 through 6. Define event A as the upward face containing an odd number of dots and event B as the upward face containing fewer than four dots. What is $P(A \mid B)$? In other words, knowing that fewer than four dots appeared, what are the chances that the number of dots is odd? We know that

$$P(A) = 3/6 = 1/2$$
$$P(B) = 3/6 = 1/2$$
$$P(A \cap B) = 2/6 = 1/3$$

Thus, from Equation 5-12, $P(A \mid B) = 1/3 \div 1/2 = 2/3$. The probability that the upward face of a die contains an odd number of dots is 2/3, given that the number of dots must be less than four.

Two events are statistically independent if the probability that one event occurs is unaffected by whether or not the other event has occurred. (Compare this definition of statistical independence to that given in the last chapter.) This is formalized in the following definition.

DEFINITION: STATISTICALLY INDEPENDENT EVENTS

Two events A and B defined over sample space S are said to be independent if and only if

$$P(A \mid B) = P(A) \text{ or, equivalently, } P(B \mid A) = P(B) \qquad (5\text{-}13)$$

If $P(A \mid B) \neq P(A)$, then events A and B are said to be dependent. For the two events A and B defined for the single roll of a die, we have already shown that $P(A \mid B) = 2/3$ and $P(A) = 1/2$, so these events are statistically dependent. That is, knowing that the die roll is less than four affects the probability that it is odd. For events A and B of the card selection experiment, drawing a spade and drawing an ace, respectively, $P(A) = 13/52$, $P(B) = 4/52$ and $P(A \cap B) = 1/52$. From Equation 5-12, $P(A \mid B) = (1/52)/(4/52) = 1/4 = P(A)$ and so these two events must be independent. This implies that knowing the outcome is an ace does not affect the probability that the card drawn is a spade.

MULTIPLICATION THEOREM

The multiplication theorem is used to find the joint probability that two events occur. The multiplication theorem can be derived by rearranging Equation 5-12, the definition of a conditional probability.

MULTIPLICATION THEOREM

For any two events A and B defined over a sample space S,

$$P(A \cap B) = P(A \mid B) \cdot P(B) \text{ or, equivalently, } P(A \cap B) = P(B \mid A) \cdot P(A) \qquad (5\text{-}14)$$

Here $P(A \cap B)$ is the probability that both events A and B occur. For events A and B of the card selection experiment,

$$P(A \cap B = P(A \mid B) \cdot P(B) = \frac{1}{4} \cdot \frac{1}{13} = \frac{1}{52}$$

The probability of getting an ace and a spade ÷ 1/52. The single elementary outcome that satisfies both event definitions is the ace of spades. For the two events of the experiment calling for the roll of a single die, $P(A \cap B) = P(A \mid B) \cdot P(B) = (2/3)(1/2) = 1/3$. Two of the six elementary outcomes in this experiment are both odd and have fewer than four dots.

For independent events, as in the card selection experiment, the multiplication theorem can be simplified. Since $P(A \mid B) = P(A)$ for statistically independent events, Equation 5-14 can be reduced to

$$P(A \cap B) = P(A) \cdot P(B) \tag{5-15}$$

The multiplication theorem for independent events can be generalized to include n independent events. For three events, the joint probability is given as

$$P(A \cap B \cap C) = P(A) \cdot P(B) \cdot P(C) \tag{5-16}$$

5.2. Concept of a Random Variable

Suppose we consider any measurable characteristic of the observations that constitute a typical population. Because this characteristic can take different values in the population we normally refer to it as a variable. For example, let us assume that we have a population comprising eight different households and that the variable under consideration is the total number of persons in the household. If we were to select one household at random from that population, then the value of our variable would be given by the size of the household chosen. Because the value of the variable is determined through random sampling, we refer to it as a *random variable.*

DEFINITION: RANDOM VARIABLE

A random variable is a numerical function that assigns values to members of a sample space.

The eight different households, and the value of the variable of interest for each, may be considered as elementary outcomes from a statistical experiment, or random trial. The sample space from this experiment is shown in Table 5.1. The values of the random variable called size of household are limited to 2, 3, 4, and 5. The sampling experiment defines one and only one numerical value for each household. If the first household is selected, then the value of the random variable is 3; if the second household is chosen, it is 2; and so on. We must emphasize that the word "random" simply means that we do not know the outcomes with certainty. It does not mean that all values are equally likely. Also, it does not mean that all elements of the sample space have the same probability of being selected, although later we will see that there are good reasons for selecting samples that way. Finally, it certainly does not mean that a household makes a random decision to include a certain number of persons.

Let us define the random variable as X and denote any particular value that the random variable may take as x. A random variable always has numerical values and probabilities associated with these values. The *probability distribution* or *probability function* is a listing or a rule that associates probabilities with all values of a random variable.

TABLE 5-1
The Concept of a Random Variable

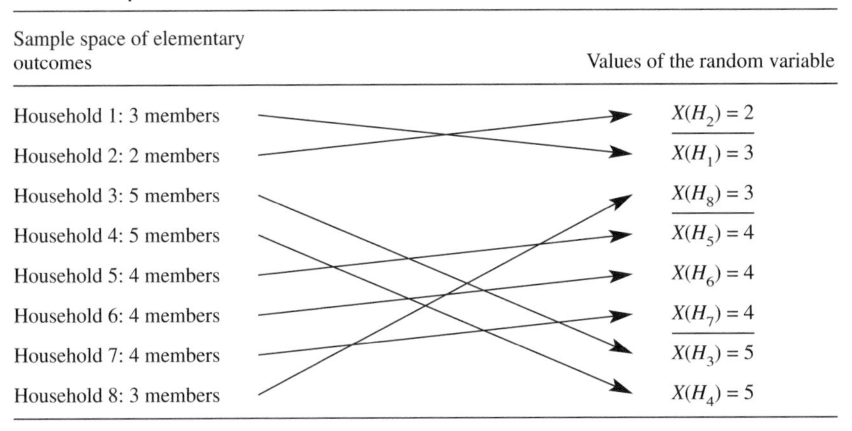

DEFINITION: PROBABILITY DISTRIBUTION OR FUNCTION
A table, graph, or mathematical function that describes the potential values of a random variable X and their corresponding probabilities is a probability function.

As noted in this definition, the probability function can be presented in several ways: (1) in tabular form as in Table 5-2; (2) in graphical form such as in Figure 5-1; and (3) as a mathematical formula such as in Equation 5-23. The probabilities of Table 5-2 are relative frequency probabilities of the random variable size of household for the population described in Table 5-1. The graph in Figure 5-1 is constructed from the same data. Henceforth, we will use the notation $P(X)$ to denote the probability distribution of the random variable X and $P(X = x)$ or $P(x)$ to denote the probability that the random variable X takes on any particular value x. For example, $P(2)$ is the probability that $X = 2$, in this case the probability that the household size is two persons. Because the population values are limited to 2, 3, 4, and 5, the probability distribution

TABLE 5-2
Relative Frequency Probabilities for the Random Variable Household Size of Table 5-1

x_i	$P(x_i)$ or $P(X = x_i)$
2	1/8 = 0.125
3	2/8 = 0.250
4	3/8 = 0.375
5	2/8 = 0.250
	1.0

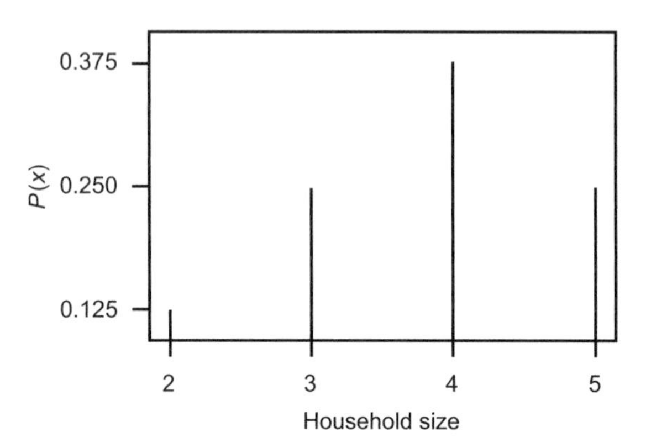

FIGURE 5-1. Graph of a probability function.

is also limited to these values. Of course we know that other household sizes are possible, but these sizes are not represented in this set of households.

Classes of Random Variables

There are two major classes of random variables: discrete and continuous.

DISCRETE RANDOM VARIABLES

There are many different types of discrete random variables, some of which are listed in Table 5-3. The key aspect is that discrete variables are limited to a finite number of values or to a countably infinite number of values. For example, the number of spots on a die is limited to just six values; thus it is clearly a discrete variable. The number of hurricanes in a year is also discrete. Although there is no upper limit, we are always able to count the number of hurricanes in any given year. We might observe 100 or 101 hurricanes, but we know that there are no values between 100 and 101 hurricanes.

DEFINITION: DISCRETE RANDOM VARIABLE
A random variable is said to be discrete if the set of values it can assume is finite or countably infinite.

Values obtained by sampling, such as size of household, gender, and many rating scales can all be considered discrete random variables if the number of values possible is limited. Experimental values, such as the outcome of a coin toss, the roll of a six-sided die, and the selection of a playing card from a shuffled deck are also discrete random variables since the set of possible values that the random variable can take is finite in each case.

In many instances, the discrete outcomes of an experiment are qualitative. To be random variables, each must be assigned a numerical value because a random vari-

TABLE 5-3
Examples of Discrete Random Variables

Random variable X	Values of random variables
From sampling	
Size of household	1, 2, 3, . . .
Gender	0, 1 or 1, 2 or any two integers
Rating scale (strongly disagree, disagree, neutral, agree, strongly agree)	1, 2, 3, 4, 5 or –2, –1, 0, 1, 2
From simple experiments	
Number of heads in N tosses of a coin	0, 1, 2, . . . , N
Number of dots on the upper face of a die on a single throw	1, 2, 3, 4, 5, 6

able is by definition a numerically valued function. The actual numerical assignments to the qualitative classes are purely arbitrary, though very simple assignments are most often utilized. For example, Table 5-3 indicates that we could assign the numerical values male = 0 and female = 1 or male = 1 and female = 2 to the variable called gender. The 0–1 distinction is more common. For rating scales, the scale –2, –1, 0, 1, 2 is useful, since negative attitudes are associated with negative values, neutral attitudes with zero, and positive attitudes with positive values.

One important subgroup of discrete random variables are *counting variables*—those restricted to the value of zero and the positive integers: 1, 2, 3, 4 . . . Counting variables typically arise when we are observing or recording the number of occurrences of some event in a particular time or space interval. The number of shoppers patronizing some store, the number of seedlings germinating in a given experimental plot, and the number of accidents at a specific road intersection are all examples of counting variables.

The probability distribution for a discrete random variable is specified by a *probability mass function.* This function assigns probabilities to the values taken by the discrete random variable. Formally, we say that a probability mass function assigns probabilities to the k discrete values of a random variable X with the following two provisions:

$$0 \leq P(x_i) \leq 1 \qquad i = 1, 2, \ldots, k$$

$$\sum_{i=1}^{k} P(x_i) = 1$$

where x_i, $i = 1, 2, \ldots, k$ are the k different values of the discrete random variable. The first condition simply requires that all probabilities be non-negative and less than one. The second condition requires the sum of all probabilities to be equal to one. The probability mass function for the variable called household size, illustrated in Figure 5-1 and summarized in Table 5-2, satisfies both conditions. Why is the function called

TABLE 5-4
Cumulative Mass Function

x	$P(x_i)$	$F(x_i)$
2	0.125	0.125
3	0.250	0.375
4	0.375	0.750
5	0.250	1.000

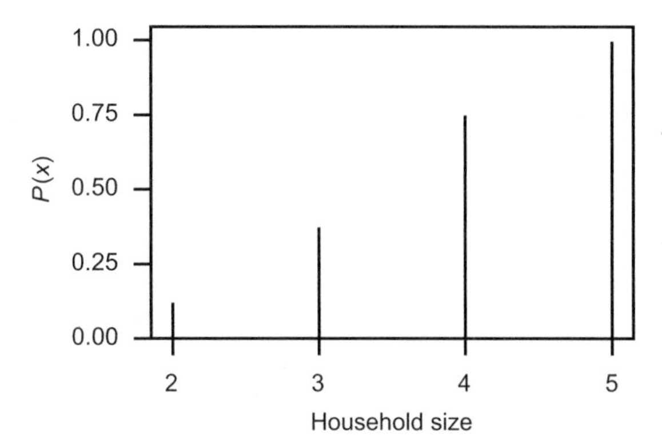

FIGURE 5-2. A cumulative probability mass function.

a mass function? It is called a *mass function* because the probabilities are *massed* at discrete values of the random variable, not spread across an interval.

We can portray probability mass functions in a cumulative form in much the same way that frequency tables and histograms can be cumulated. For a discrete random variable with values ordered from lowest to highest at $x_1 < x_2 < x_3$, the cumulative mass function $F(x_i) = P(X \leq x_i)$ is determined by summing the probabilities that $X = x_1, X = x_2, \ldots, X = x_i$. That is,

$$F(x_i) = \sum_{x=x_1}^{x=x_i} P(x) = P(x_2) + \ldots + P(x_i) \tag{5-17}$$

Consider again the discrete random variable household size of Table 5-2. To cumulate these values, note $F(2) = P(X = 2) = 0.125$, $F(3) = P(X = 2) + P(X = 3) = 0.125 + 0.250 = 0.375$, and so on. The complete mass function is listed in Table 5-4 and illustrated in Figure 5-2. Their construction and display mirror those of the cumulative frequency table and cumulative frequency histogram examined in Chapter 2.

Just as an empirically derived variable has a mean value and variance, so too does a random variable. The average value of a random variable is known as its *expected value.*

DEFINITION: EXPECTED VALUE OF A DISCRETE RANDOM VARIABLE

For a discrete random variable with values $x_1, x_2, \ldots, x_k$, the expected value of X, denoted $E(X)$, is defined as

$$E(X) = \sum_{i=1}^{k} P(x_i)x_i \qquad (5\text{-}18)$$

For the household data we find the following:

x_i	$P(x_i)$	$P(x_i)x_i$
2	0.125	0.250
3	0.250	0.750
4	0.375	1.500
5	0.250	1.250
		3.750

Thus, the mean or expected value of household size is 3.75. Note that this need not be one of the original discrete values of the variable X. Equation 5-18 is a "weighted" mean in which the weights are defined as the relative frequencies or probabilities of each discrete value.

The variability of a discrete random variable is measured by the *variance* $V(X)$.

DEFINITION: VARIANCE OF A DISCRETE RANDOM VARIABLE

For a discrete random variable, the variance of X denoted $V(X)$ is defined by the following equation:

$$V(X) = \sum_{i=1}^{k} [x_i - E(X)]^2 P(x_i) = \sum_{i=1}^{k} x_i^2 P(x_i) - [E(X)]^2 \qquad (5\text{-}19)$$

The second form for the variance given in $V(X)$ is a more efficient computational formula. For the variable household size,

x_i	$P(x_i)$	x_i^2	$x_i^2 P(x_i)$
2	0.125	4	0.500
3	0.250	9	2.250
4	0.375	16	6.000
5	0.250	25	6.250
			15.000

and $V(X) = 15.000 - 14.0625 = 0.9375$. The standard deviation of the random variable may be calculated as the square root of the variance: $\sqrt{0.9375} = 0.9682$.

CONTINUOUS RANDOM VARIABLES

A random variable is said to be continuous if it can assume all the real number values in some interval of the real number line.

DEFINITION: CONTINUOUS RANDOM VARIABLE

A random variable is said to be continuous if the set of values it can assume constitutes the real number line, or one or more intervals on the real line.

The number of different values that a continuous random variable can assume is therefore infinite. Table 5-5 lists several different variables that can be considered to be continuous random variables. For example, although always greater than or equal to zero, there is no limit to the number of values that rainfall can take on. Thus, we consider rainfall to be a continuous random variable. Again, we are not assuming that all values are equally likely, or that mysterious "random" processes generate rainfall. Rather, it is simply that in the act of observing the variable, the outcome is uncertain.

In addition, many discrete variables are often modeled as continuous variables. For example, the population of a city or region or census tract can be treated as if it were a continuous variable even though, in the strictest sense, it is discrete. This is a common simplification for many discrete variables with large ranges.

The probability distribution of a random continuous variable is represented by a *probability density function.* Let X be a continuous random variable defined over some interval of the real number line, say from a to b. A probability density function of X denoted $f(X)$, satisfies three conditions:

1. $f(x)$ is a real, single-valued function. That is, $f(x)$ generates exactly one real value for any x.
2. $f(x) \geq 0$ for $a \leq x \leq b$.
3. The area under $f(x)$from $x = a$ to $x = b$ must be equal to 1.

The latter two conditions are analogous to the two conditions defined for discrete random variables. In the discrete case, the probabilities are said to be *massed* at the

TABLE 5-5
Examples of Continuous Random Variables

Random variable X	Values of random variables
Average rainfall	$X \geq 0$
Distance traveled	$X \geq 0$
Stream discharge	$X \geq 0$
pH value of water sample	Commonly $0 \leq X \leq 14$
Pseudocontinuous variables	
Population of city, region, or census tract	$X \geq 0$
Interaction between two places	$X \geq 0$

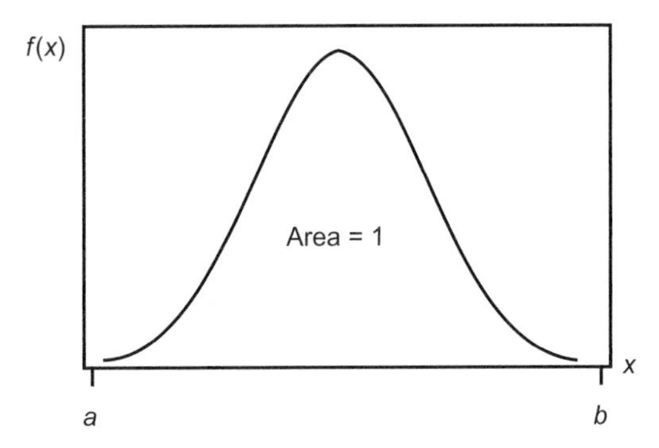

FIGURE 5-3. A probability density function.

discrete values of the random variable. In the continuous case, we say the probability is *spread densely* over the range of the random variable.

The probability density function for a continuous random variable is illustrated in Figure 5-3. Just as the heights of the vertical bars in a probability mass function of a discrete variable sum to one, *the area under a probability density function must also equal one.* It is possible to define continuous random variables over an infinite-length real number line, that is, from $-\infty$ to $+\infty$. If $f(x)$ is asymptotic to the x-axis as it approaches $-\infty$ and $+\infty$, then $f(x)$ may still define a continuous random variable as long as the area under $f(x)$ is 1. The most famous of all continuous distributions, the normal distribution, is defined as a continuous variable with potential values in the range $[-\infty, +\infty]$.

How can we compute probabilities using these probability distributions? Suppose we are interested in computing $P(c \le x \le d)$ for some random variable X. If X is discrete, then this is simply a matter of summing the probabilities of X taking on any discrete value between (and including) c and d:

$$P(c \le x \le d) = \sum_{x=c}^{x=d} P(x) \tag{5-20}$$

For a continuous random variable, the probability that X takes on a value between c and d is equal to the area under the probability density function $f(x)$ between c and d. This is illustrated in Figure 5-4. Depending on the shape of the probability density function, calculating the probability that a continuous random variable assumes a number within some range may not be an easy matter. However, it is not necessary to know advanced calculus to comprehend the concept. For computational purposes, tables of probabilities for commonly used distributions are provided at the end of this text.

A simple example illustrates the procedure for computing probabilities of a continuous variable. Suppose the probability density function is given by

$$f(x) = \begin{Bmatrix} 0.5 & 0 \le x \le 2 \\ 0 & x < 0 \text{ or } x > 2 \end{Bmatrix} \tag{5-21}$$

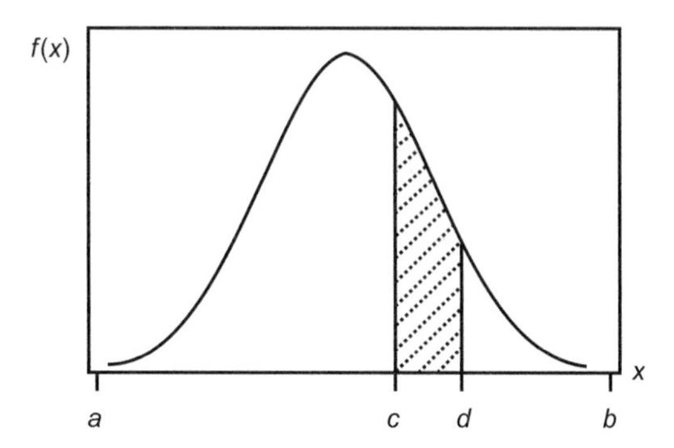

FIGURE 5-4. The probability that $c \leq x \leq d$.

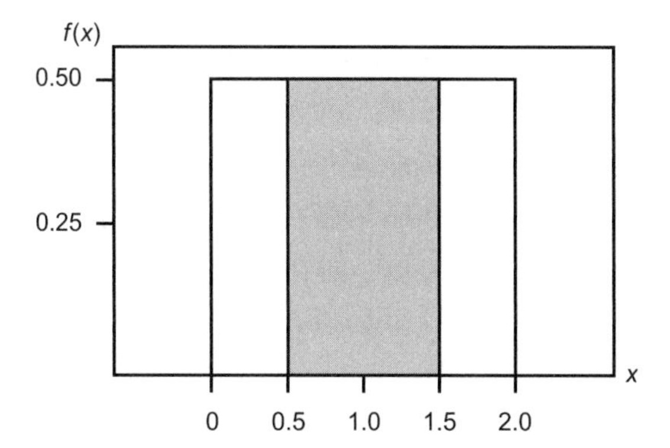

FIGURE 5-5. Computing a probability from a simple random continuous probability distribution.

as illustrated in Figure 5-5. The total area under this function can be calculated as the area of a rectangle with base = 2 and height = 0.5. The area under this function is 0.5 $\times$ 2 = 1. Note also that $f(x) \geq 0$ for all x. To calculate $P(0.5 \leq x \leq 1.5)$, simply find the area under the density function between these two limits. The base is $1.50 - 0.5 = 1.0$, and the height is 0.5, so the total area, and thus the probability, is $0.50 \times 1.0 = 0.50$. As illustrated in Figure 5-5, this is one-half the area under the probability density function. Unfortunately, it is not always possible to use such simple arguments from elementary geometry for more complex probability density functions.

Although there are many similarities between continuous and discrete variables, there is one key difference. Consider the expression $P(X = r)$. For a discrete variable, this is simply the probability that the variable takes on a discrete value r. This probability can be determined directly from the probability mass function as the height of the bar graph for the value $X = r$. The same operation applied to a continuous random variable does not give a probability. In fact, for a continuous random variable, $P(X = r) = 0$. Why? Assume that the variable X is defined on the real number line from $X = a$ to $X = b$. If we were to let $P(X = r) = f(r)$, the height of the density function at $X = r$, then this would have to apply to all other values in the interval $[a,b]$. Since there are an infinite number of such values, say $r_1, r_2, \ldots r_n$, the sum of the probabilities $f(r) + f(r_1) + f(r_2) + \ldots + f(r_n)$ would undoubtedly exceed 1.0.

It is therefore not possible to interpret $f(r)$ directly as a probability. In fact, there is nothing in the definition of a probability density function that prohibits any single value of $f(r)$ from exceeding one. The height of $f(r)$ is only constrained to be non-negative. Consider the probability density function

$$f(x) = \begin{cases} 2 & 0 \le x \le 0.5 \\ 0 & x < 0 \text{ or } x > 0.5 \end{cases} \tag{5-22}$$

Clearly, $f(0) = f(0.25) = f(0.5) = 2$. However, the area under the probability density function, $P(0 \le x \le 0.5) = 0.5 \times 2 = 1$.

There is another explanation of why $P(X = r)$ for any continuous variable must equal zero. Let us suppose we calculate $P(c \le x \le d)$ and derive some non-negative probability. As we move c and d closer together, this probability must get smaller as the area under $f(x)$ declines. In the limit, $c = d$, and there is no area under the curve. The probability that a continuous random variable assumes a particular value must always be zero, regardless of the value of the underlying probability distribution $f(x)$.

Computing the expected value $E(X)$ and variance $V(X)$ of a continuous random variable is not nearly so straightforward as the computation for discrete variables. The derivation of these quantities requires knowledge of introductory calculus beyond the scope of this text. The expected value and variance of the probability distributions described in Section 5.4 are therefore presented without derivations. Students with a background in calculus may wish to consult Appendix 5b at the end of the chapter for a more complete specification of the derivation of $E(X)$ and $V(X)$ for continuous variables.

The properties of continuous and discrete random variables are summarized in Table 5-6. Both types of variables must satisfy an identical set of basic properties,

TABLE 5-6
A Comparison of Continuous and Discrete Random Variables

	Discrete random variable	Continuous random variable
Number of values that can be assumed	Finite (or countably infinite), say k	Infinite, from a to b or from $-\infty$ to ∞
Graphical summary	Probability mass function as linear bar chart	Probability density function $f(x)$, area under curve
$P(X = a)$	$P(a)$	0
$P(c \le x \le d)$	$\sum_{x=c}^{d} P(x)$	$\int_c^d f(x)dx$
Conditions on $P(x)$		
(1)	$0 \le (P(x_i) \le 1 \; i = 1, 2, \ldots, k$	$f(x) \ge 0, a \le x \le b$
(2)	$\sum_{i=1}^{k} P(x_i) = 1$	Area under $f(x) = 1$, that is: $\int_a^b f(x)dx = 1$

including (1) non-negativity and (2) the sum of probabilities equal to one. The major difference arises from the number of different values that the random variable can assume. For a discrete random variable, this is always finite or countably infinite. By definition, a continuous random variable can assume an infinite number of different values. This distinction leads to several important differences in the specification of the properties of these two classes of random variables.

5.3. Discrete Probability Distribution Models

Discrete random variables can be represented by a number of different probability mass functions. If a complete listing of the *population* of interest is available, then the relative frequency distribution is the appropriate probability mass function. Figure 5-1 represents the probability mass distribution for the variable called household size as distributed in the population listed in Table 5-1. For cases in which the available data are a sample from some larger population, this procedure is not feasible. Instead, it is useful to see how well the available data fit certain probability distribution models derived from specific statistical experiments. For discrete variables, three different probability distributions are often used in statistical analyses:

1. Discrete uniform distribution
2. Binomial distribution
3. Poisson distribution

It is not unusual to find close parallels between the assumptions underlying the statistical experiments governing these distributions and the nature of an empirical statistical problem.

Discrete Uniform Distribution

One of the simplest probability distributions is the discrete uniform distribution. It occurs frequently in games of chance. Examine the probability of getting a head or a tail in the single toss of a fair coin. Each outcome has an equal probability of 0.5. In a single draw from a shuffled deck of playing cards, the probability of drawing a card of any suit is 0.25. Each of these experiments defines a discrete uniform random variable. In such a distribution, the probability of occurrence for each outcome, or value of the random variable, is equal. For two outcomes, each has a probability of 0.5; for three outcomes, each has a probability of 0.33; and so on.

The uniform distribution also has applications in statistical decision making in situations of complete uncertainty. Suppose, for example, someone is lost and arrives at an unmarked fork in the road. There is no reason to suspect that either road leads to the desired destination. Which road should be taken? In the absence of any other information, each road could be assigned an equal probability. A coin toss could be used to make the decision. Many other decision-making situations are similar to this example. All involve the notion of complete uncertainty.

To formalize this probability mass function, assign a single integer from 1, 2, . . . , k to each of the different k outcomes that can be assumed by the random variable X.

DEFINITION: DISCRETE UNIFORM DISTRIBUTION

If X is a discrete uniform random variable, then

$$P(X) = \frac{1}{k} \quad x = 1, 2, \ldots, k \tag{5-23}$$

That is, each state or value of the random variable has an equal (hence the name *uniform*) probability of occurrence. Figure 5-6a illustrates the general form of the probability mass function of a discrete uniform variable. Figures 5-6b and 5.6c illustrate the probability mass functions for the card selection and coin toss examples.

What are the mean and variance of a discrete uniform variable? Using Equation 5-18, we can calculate the expected value as

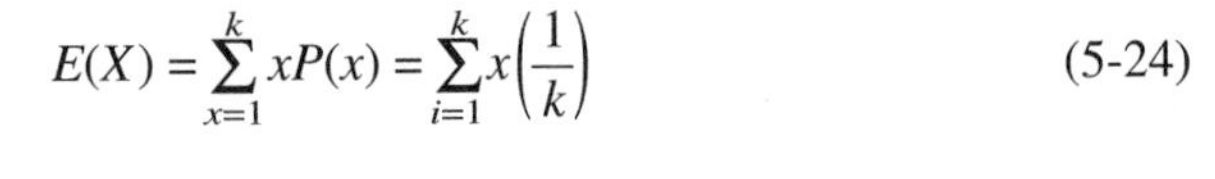

$$E(X) = \sum_{x=1}^{k} xP(x) = \sum_{i=1}^{k} x\left(\frac{1}{k}\right) \tag{5-24}$$

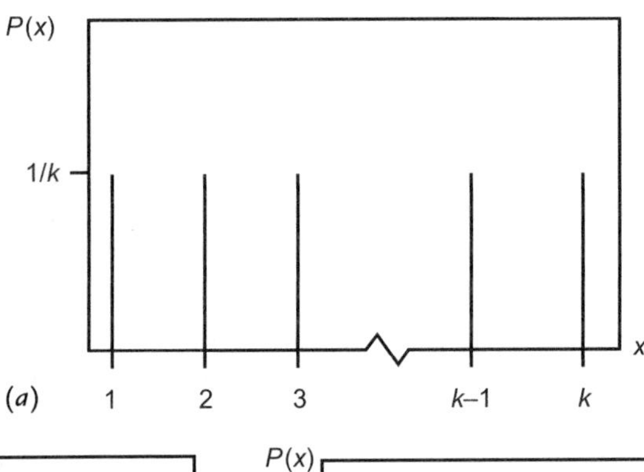

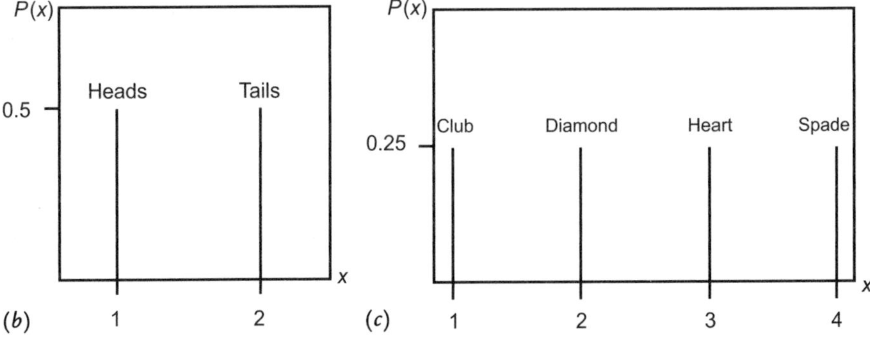

FIGURE 5-6. Discrete uniform random variables.

Now, since $1/k$ is a constant, it can be brought outside the summation (recall Rule 2 of Appendix 3a in Chapter 3). The expression $\Sigma_{x=1}^{k} x$ is the sum of the first k integers and can be simplified to $k(k + 1)/2$. Therefore,

$$E(X) = \frac{1}{k}\left[\frac{k(k+1)}{2}\right] = \frac{k+1}{2} \tag{5-25}$$

Although it is somewhat more difficult to derive, the variance of a discrete uniform variable can be shown to be

$$V(X) = \frac{k^2 - 1}{12} \tag{5-26}$$

As an example, consider again the outcome of a single coin toss. If we let $x = 1$ be the result of getting a head and $x = 2$ the result of getting a tail, then $P(1) = P(2) = 0.5$ for the $k = 2$ potential values of the random variable. The mean or expected value is $(2 + 1)/2 = 3/2 = 1.5$. Note once again that this average is not one of the two values that can be assumed by X. Rather, it is midway between each. From a decision-making point of view, this tells us that neither a head nor a tail is a preferable choice in a coin toss. Using Equation 5-19, we find that $V(X) = (2^2 - 1)/12 = 0.25$. This has no immediate interpretation.

Binomial Distribution

A second important distribution of a discrete random variable is the *binomial distribution.* It is applicable to variables having only two possible outcomes: a person is male or female; a person answers a certain question yes or no; or the coin yields a head or a tail. The simplest example of a binomial variable is the operation of a coin toss. Suppose we toss a coin n times and record the number of heads occurring in the n tosses. If the tosses are all independent events and the random variable X is defined as the number of heads occurring in the n tosses, then X is a binomial random variable.

A binomial random variable is produced by a statistical experiment known as a *Bernoulli* process, or a set of *Bernoulli* trials. A set of Bernoulli trials is defined by the following four conditions:

1. There are n independent trials of the experiment.
2. The same pair of outcomes is possible on all trials.
3. The probability of each outcome is the same on all trials.
4. The random variable is defined to be the number of "successes" in the n trials.

Let us relate this definition to the example of a coin toss experiment. Our interest lies in the number of heads, or "successes," that occur in an experiment in which a coin is tossed n times. The probability of a head, or a success, is labeled π, and the probability

TABLE 5-7
Possible Outcomes of the Coin Toss Experiment

$n = 1$				T		H		
$n = 2$			TT		TH			
					HT		HH	
$n = 3$		TTT		TTH				
				THT				
				HTT		THH		
					HTH			
					HHT		HHH	
$n = 4$	TTTT		TTTH		TTHH		THHH	HHHH
			TTHT		THTH		HTHH	
			THTT		HTTH		HHTH	
			HTTT		THHT		HHHT	
				HTHT				
				HHTT				

of a tail, or failure, is $1 - \pi$. Each trial is represented by one toss of the coin. Table 5-7 lists the possible outcomes from this experiment in which a coin is tossed $n = 1, 2, 3,$ or 4 times.

Consider the first row of the table. There are only two possible outcomes, a head (H) or tail (T). We may get either zero heads or one head, depending on the outcome. The probability of no heads is $1 - \pi$, and the probability of one head is π. Since $\pi + 1 - \pi = 1$, this is a legitimate probability distribution for a random variable. What if the coin is tossed twice? Now, there are three different possible outcomes: zero, one, or two heads. Although there is only one way of getting zero heads or two heads, there are two ways of obtaining one head. To get two heads requires a head on each of the $n = 2$ tosses. This is listed as outcome {HH}. Similarly, only one sequence leads to two tails, {TT}. But to obtain one head, two different sequences are possible: {HT} and {TH}.

For $n = 3$ tosses there are now eight possible outcomes. For example, there are now three ways of obtaining one head. It can appear on the third, second, or first toss of the coin. These three outcomes are listed as {TTH}, {THT}, and {HTT}. Note that these three alternatives can be derived from the sequences for $n = 2$, which are listed immediately above. The sequence {TTH} results from the addition of a head to the sequence {TT}, and {THT} and {HTT} derive from the two sequences {TH} and {HT} with the addition of a tail. Furthermore, the three sequences with two heads—{THH}, {HTH}, and {HHT}—can be generated by the addition of the appropriate result to the sequences listed above as {TH}, {HT}, and {HH}. The entire list of outcomes in Table 5-7 is constructed in such as way that the sequences in any row can be generated from the immediately preceding row with the addition of the appropriate outcomes. The size of this triangle expands quite rapidly as the number of tosses increases.

One way of summarizing these results is to calculate the different number of sequences that lead to any outcome for a given number of tosses. For example, there

TABLE 5-8
Pascal's Triangle of Binomial Coefficients

$n = 1$						1		1					
$n = 2$					1		2		1				
$n = 3$				1		3		3		1			
$n = 4$			1		4		6		4		1		
$n = 5$		1		5		10		10		5		1	
$n = 6$	1		6		15		20		15		6		1

are three ways of getting one head in $n = 3$ tosses. These numbers are termed the binomial coefficients and are displayed in Table 5-8 for $n = 1$ to $n = 6$ Bernoulli trials. This representation is known as *Pascal's triangle* in recognition of the role that French mathematician Blaise Pascal played in the development of these results. Note that any number in this table can be generated by adding the number of sequences above the entry in the immediately preceding row. The six in the fourth row is 3 + 3, the sum of the two numbers directly above it. How are these binomial coefficients to be interpreted? Each entry represents *the number of different ways of obtaining exactly x heads in n tosses of a coin.* There are exactly six ways of obtaining two heads in four tosses of a coin. This is denoted C_2^4 or $C(4,2)$ since this many combinations of heads and tails lead to the desired result. From the counting rules discussed in Appendix 5a, C_2^4 is calculated as $4!/[2!(4–2)!] = 24/4 = 6$ combinations. These six combinations are listed in the third column of the $n = 4$ row of Table 5-7. The five numbers in the $n = 4$ row refer to the appearance of 0, 1, 2, 3, or 4 heads in the $n = 4$ tosses of the coin.

Now, consider the probability that any of these individual sequences occurs. We know that the probability of a head is π, the probability of a tail is $1 - \pi$, and the trials are independent. What is the probability of each of the six sequences {TTHH}, {THTH}, {HTTH}, {THHT}, {HTHT}, and {HHTT}? Each contains $x = 2$ heads in $n = 4$ tosses. It is rather easy to show that the probabilities of each of these sequences are equal and are given by $\pi^2(1 - \pi)^2$. For the first sequence, the independence of trials allows us to rewrite this as

$$P(TTHH) = P(T) \cdot P(T) \cdot P(H) \cdot P(H)$$

Since $P(T) = 1 - \pi$ and $P(H) = \pi$, this simplifies to $(1 - \pi) \cdot (1 - \pi) \cdot \pi \cdot \pi$ or, with rearrangement, $\pi^2 \cdot (1 - \pi)^2$. All sequences can be rearranged in this form.

Our ultimate concern is determining the probability of obtaining exactly x heads in n tosses of a coin. To continue the same example, what is the probability of obtaining exactly two heads in four tosses of a coin? Since there are six different sequences with exactly two heads, each with a probability of $\pi^2(1 - \pi)^2$, the required probability is $6\pi^2(1 - \pi)^2$. The probability of obtaining x heads in n tosses of a coin for up to $n = 4$ trials is listed in Table 5-9. For example, for $n = 4$ and $x = 2$, the five coefficients are 1, 4, 6, 4, and 1. These are the entries of Pascal's triangle, Table 5-8,

TABLE 5-9
Binomial Distributions

Number of trials n	Number of heads	Probability of number of heads
1	0	$(1 - \pi)$
	1	π
2	0	$(1 - \pi)^2$
	1	$2\pi(1 - \pi)$
	2	π^2
3	0	$(1 - \pi)^3$
	1	$3\pi(1 - \pi)^2$
	2	$3\pi^2(1 - \pi)$
	3	π^3
4	0	$(1 - \pi)^4$
	1	$4\pi(1 - \pi)^3$
	2	$6\pi^2(1 - \pi)^2$
	3	$4\pi^3(1 - \pi)$
	4	π^4

for the row $n = 4$. For this reason the entries in Pascal's triangle are known as the binomial coefficients. Also, the exponent of π is the number of heads, and the exponent of $1 - \pi$ is the number of tails. In general, the probability of obtaining x heads in n tosses of a coin is equal to the product of the number of sequences of length n with x heads and the probability that the sequence contains x heads with $n - x$ tails. Let us now formally define the probability mass function for a binomial random variable.

DEFINITION: BINOMAL RANDOM VARIABLE
If X is a binomial random variable, then the probability mass function is given by

$$P(X) = C(n, x)\pi^x(1 - \pi)^{n-x} \quad x = 0, 1, 2, \ldots, n \qquad (5\text{-}27)$$

where n is the number of trials, x is the number of successes and π is the probability of a success in each trial.

The specific form of the binomial distribution depends on the values of the two parameters π and n. Given values of π and n, we can easily compute the probability for any number of successes using Equation 5-27. If $\pi = 0.2$ and $n = 4$ and we wish to compute the probability of three successes, then by Equation 5-27, $P(3) = C(4, 3)(0.2)^3(0.8)^1 = 4(0.008)(0.8) = 0.0256$. Figure 5-7 illustrates the probability mass functions for three members of the *family* of binomial probability distributions. If the coin we are tossing is a fair coin, then $\pi > 0.5$ and the distribution of successes is always symmetric. If $\pi < 0.5$, then the distribution is positively skewed; if $\pi > 0.5$, the distribution is negatively skewed. When n is very small, the probability mass function for a binomial random variable can be used directly to compute any probabilities

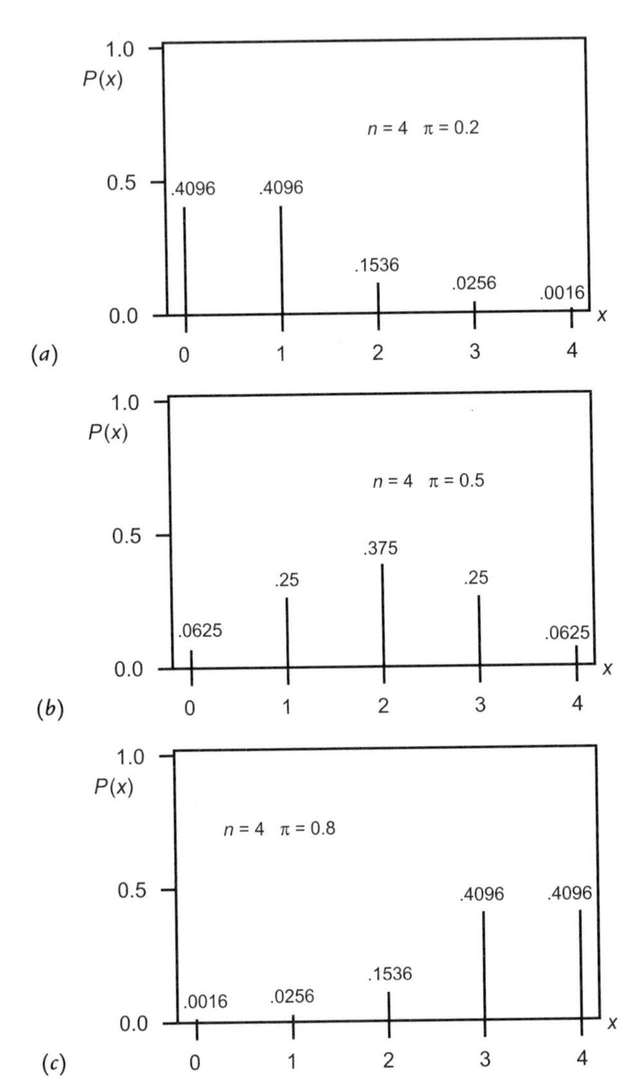

FIGURE 5-7. Some binomial probability distributions.

required for a specific problem. However, when n is large, it is more convenient to consult statistical tables that compile binomial probabilities for many different values of π and n. Statistical tables for a few members of the binomial family are included at the end of this text.

Using Equations 5-18 and 5-19, we can determine the mean and variance of a binomial probability distribution.

DEFINITION: MEAN AND VARIANCE OF A BINOMAL RANDOM VARIABLE

If X is a binomial random variable, then the mean and variance of X are given by

$$E(X) = \sum_{x=0}^{n} xC(n, x)\pi^x(1 - \pi)^{n-}x = n\pi \qquad (5\text{-}28)$$

$$V(X) = \sum_{x=0}^{n} [x - E(X)^2C(n, x)\pi^x(1 - \pi)^{n-}x = n\pi(1 - \pi) \qquad (5\text{-}29)$$

EXAMPLE 5-1. Suppose a student taking a course in statistics is required to take an examination consisting of 10 multiple-choice questions. The questions are designed in such a way that the probability of correctly guessing the answer to any question is 0.2, as they consist of five equally plausible choices. If the student can *only* guess at the answers to each question (never attended lectures, labs, nor even read the textbook), what is the expected number of correct answers? What is the standard deviation of the number of correct answers? Assuming a grade of 50% is required to pass the test, what is the probability that the student passes?

Let the binomial random variable X be the number of correct answers on the test. There are $n = 10$ trials, and it is known that $\pi = 0.2$. The appropriate probability mass function is

$$P(x) = C(10, x)(0.2)^x(0.8)^{10-x} \quad x = 0, 1, 2, \ldots, 10$$

where x is the number of correct answers. The expected number of correct answers is obtained from (5-28) as

$$E(X) = n\pi = 10(0.2) = 2$$

and $V(X)$ is calculated from (5-29):

$$V(X) = n\pi(1 - \pi) = 10(0.2)(0.8) = 1.6$$

Therefore, the standard deviation of the number of correct answers is $\sqrt{1.6} = 1.26$ correct answers. To determine the probability that the student passes the test, it is necessary to calculate $P(5) + P(6) + P(7) + P(8) + P(9) + P(10)$, since all these outcomes lead to passing grades. Expressing these values to three decimal places, we get:

$$P(5) = C(10, 5)(.2)^5(.8)^5 = 0.264$$
$$P(6) = C(10, 6)(.2)^6(.8)^4 = .006$$
$$P(7) = C(10, 7)(.2)^7(.8)^3 = .001$$
$$P(8) = C(10, 8)(.2)^8(.8)^2 = .000$$
$$P(9) = C(10, 9)(.2)^9(.8)^1 = .000$$
$$P(10) = C(10, 10)(.2)^{10}(.8)^0 = .000$$

The sum is 0.033, and thus, as a good approximation, there is a 3.3% chance a student could pass this test just by guessing.

In our discussion of the binomial distribution, we assumed that the values of the two parameters π and n are known. With these values we can specify the appropriate probability mass function, calculate the expected value and variance, and solve many probability problems. What if they are not known? Then it is necessary to estimate these parameters from some sample data. We see how this problem in estimation is solved in Chapter 7. In Chapter 10 we describe how to solve the related problem of testing the fit of the binomial distribution to sample data. Such methods are extremely useful for determining which particular probability mass function best describes empirical data that we wish to analyze.

Poisson Probability Distribution

The final discrete distribution discussed in this chapter, the *Poisson probability distribution,* is very important in the analysis of geographic point patterns. As we saw in the last chapter, point pattern analysis involves a variety of techniques that describe the spatial distribution of phenomena represented by points on the traditional dot map. Figure 5-8 illustrates a typical dot map. Each dot represents the location of one observation of an item of interest. For example, the dots might represent settlements, individual dwelling units, households of some particular ethnic group, or geographic features such as drumlins, erratics, or members of specific plant species. To derive the Poisson probability distribution and to develop the notion of a Poisson random variable, consider the analysis of such a dot map. Let us suppose that the dots represent drumlins and that these drumlins are randomly distributed over the area depicted on the dot map with no location more likely than any other to have a drumlin. Suppose it is also known that there is, on average, one drumlin per square mile.

One way of characterizing this dot map is to record the number of dots appearing in the *quadrats* of the map. Quadrats are simply areal subdivisions of a map. Usually they are regular in size and square. Imagine that the region depicted in Figure 5-8 is divided into 100 quadrats. Initially, let us analyze this map as a Bernoulli process

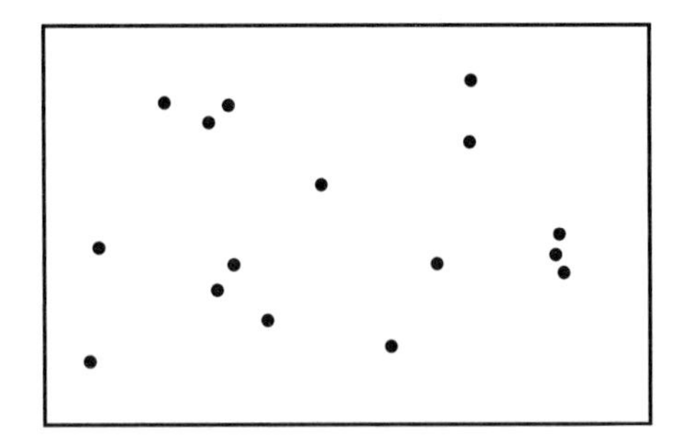

FIGURE 5-8. A typical dot map.

in which the examinations of quadrats are treated as "trials" of a statistical experiment. When examining a quadrat, we define a success as the existence of a drumlin and a failure as the absence of a drumlin. Because there is, on average, one drumlin per square mile, the expected number of drumlins in any quadrat is one.

Now suppose each 1-mi^2 quadrat is divided into four equal-sized 0.25-mi squares. Let us make the additional assumption that no more than one drumlin can occur in any 0.25-mi quadrat. Again, we might examine these four 0.25-mi squares as a set of Bernoulli trials. Each quadrat either contains a drumlin or does not. Because the expected number of drumlins per square mile is one, the probability that there is a drumlin in any of the four quadrats is 1/4. Examining four quadrats as trials of a Bernoulli process leads to five possible outcomes: zero drumlins, one drumlin, or two, three, or four drumlins in these four quadrats. Clearly, this is a binomial random variable. The probability that the four quadrats contains exactly two drumlins is:

$$P(2) = \frac{4!}{2!(4-2)}\left(\frac{1}{4}\right)^2\left(\frac{3}{4}\right)^2 = 0.2109 \qquad (5\text{-}30)$$

This expression is derived by the application of Equation 5-27 with π 1/4, $1 - \pi = 3/4$, $n = 4$, and $x = 2$. Note that $P(2) = 0.2109$ is also the probability of observing two drumlins in a 1-mi^2 quadrat.

To use Equation 5-27, we must make the assumption that the probability of finding more than one drumlin in any 0.25-mi quadrat is zero. This violates the assumption that the distribution of points on the map is random. If the distribution is truly random, then there must be some probability that any quadrat could contain two, three, or even all dots of the map. To get a closer approximation, let us divide these four 0.25-mi quadrats into 16 quadrats of 1/8-mi square. This time we assume that there can be at most one drumlin in each of these 16 quadrats. The possible outcomes of the set of Bernoulli trials based on these 16 quadrats are 0, 1, 2, . . . , 16 drumlins. The probability of observing exactly two drumlins in 16 quadrats is:

$$P(2) = \frac{16!}{2!(16-2)!}\left(\frac{1}{16}\right)^2\left(\frac{15}{16}\right)^{14} = 0.1899 \qquad (5\text{-}31)$$

since for this set of trials $\pi = 1/16$, $1 - \pi = 15/16$, $n = 16$, and $x = 2$. Again note that $P(2) = 0.1899$ is the probability of observing two drumlins in a 1-mi^2 quadrat.

The estimate of $P(2)$ given by Equation 5-31 is more accurate than Equation 5-30 since the probability of more than one drumlin in any quadrat is smaller. This is because the quadrats themselves are smaller. To get an even better estimate, let us divide our 16 quadrats into 64 quadrats. The probability that our 64 quadrats contain exactly two drumlins or, equivalently, that a randomly chosen 1-mi^2 quadrat contains exactly two drumlins is given by:

$$P(2) = \frac{64!}{2!(64-2)!}\left(\frac{1}{64}\right)^2\left(\frac{63}{64}\right)^{62} = 0.1854$$

Because 64 quadrats are a better basis for the approximation than 16 quadrats, we might ask what would happen if we examined this process as the number of quadrats approached infinity, and the probability that any quadrat contained a drumlin approaches zero. To determine the expression for $P(2)$ for an infinite number of trials requires an evaluation of the limit as n approaches infinity of the following expression:

$$P(2) = \frac{n!}{2!(n-2)!}\left(\frac{1}{n}\right)^2\left(\frac{n-1}{n}\right)^{n-2}$$

Although some calculus is required to derive the result, it can be shown that this limit converges to

$$P(2) = \frac{e^{-1}(1)^2}{2!} = 0.1839$$

where e is the base of the natural logarithms, or 2.71828. The limit of $P(2)$ thus converges to 0.1839, and the increasing accuracy of our estimate with the increasing number of quadrats is clearly illustrated.

This result can now be extended to determine the limits for $P(0)$, $P(1)$, $P(2)$, $P(3)$ and the probabilities for the rest of the positive integers. Let λ be the average number of drumlins per unit area. In the example above, $\lambda = 1$ drumlin per square mile. The random variable X is the number of drumlins found in a randomly selected quadrat, and X may take on the values $x = 0$, $x = 1$, $x = 2$, or any other positive integer. The probability that a quadrat contains exactly x drumlins is a Poisson random variable.

DEFINITION: POISSON RANDOM VARIABLE

If X is a Poisson random variable, then the probability mass function is given by:

$$P(x) = \frac{e^{-\lambda}\lambda^x}{x!} \tag{5-32}$$

There is only one parameter for this distribution, λ.

To generalize, the experiment generating a Poisson random variable is described in the following way. Let the random variable X be the number of occurrences $x = 0, 1, 2, \ldots$ of some specific event in a given continuous interval. The interval can be a time interval, a spatial interval (such as a quadrat), or even the length of some physical object such as a rope or a chain. The random variable X is a Poisson random variable if the experiment generating the values of X satisfies the following conditions:

1. The number of occurrences of the event in two mutually exclusive intervals is independent.

2. The probability of an occurrence of the event in a small interval is small and proportional to the length of the interval (i.e., the event is rare).
3. The probability of two or more occurrences of the event in a small interval is near zero.

For the point pattern problem, the event is the existence of a dot in the quadrat. Because the pattern is random, the number of dots in any two quadrats is independent. Finally, the probability of two or more dots in one quadrat is extremely small.

There is a family of Poisson probability mass functions, each member of which is specified by selecting a particular value of λ. Figure 5-9 illustrates the probability

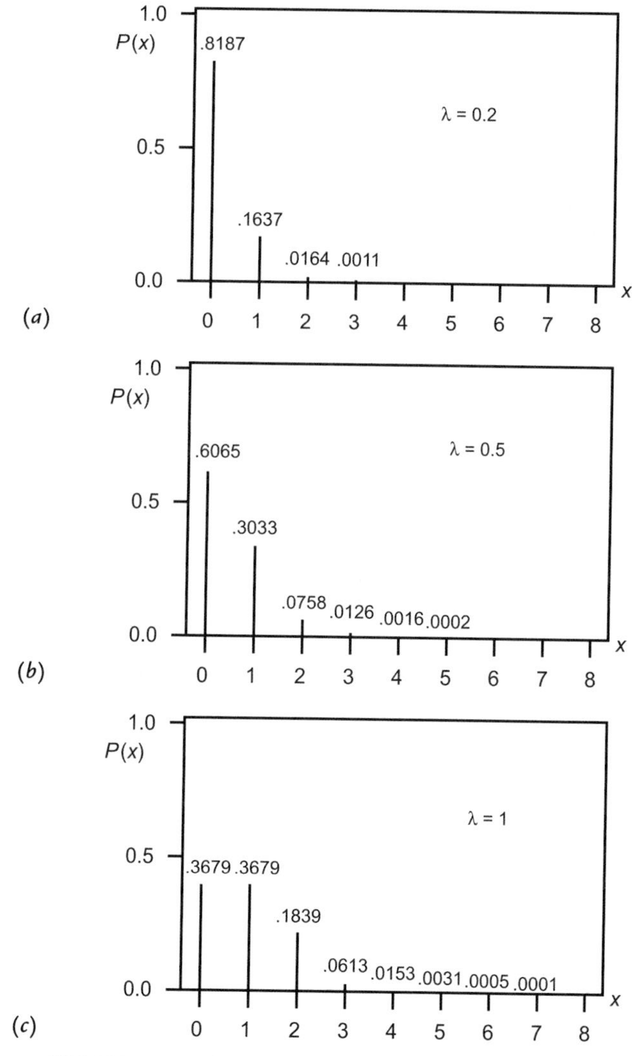

FIGURE 5-9. Three Poisson probability distributions.

mass functions for three members of this family. Note that all distributions are truncated and depict probabilities only for small values of x. But, as indicated in Equation (5-32), the probability mass function of a Poisson random variable assigns probabilities to all positive integers. (The Poisson is an example of a countably infinite, discrete variable.) The probabilities of larger values of x are extremely small and have been omitted for clarity. A comparison of these three probability mass functions indicates that the probabilities of a Poisson random variable become increasingly spread over the values of X as λ increases. Compact tables for a few values of λ are provided at the end of the text.

The mean and variance of a Poisson random variable can be shown to be:

$$E(X) = \sum_{x=0}^{\infty} xP(x) = \sum_{x=0}^{\infty} x \frac{e^{-\lambda}\lambda^x}{x!} = \lambda \tag{5-33}$$

$$V(X) = \sum_{x=0}^{\infty} [x - E(X)]^2 \frac{e^{-\lambda}\lambda^x}{x!} = \lambda \tag{5-34}$$

Thus, a characteristic feature of a Poisson random variable is the equality of the mean and variance. This is a useful check for examining the applicability of the Poisson random variable to some sample data.

EXAMPLE 5-2. The number of arrivals to a wilderness park is found to be Poisson distributed with a mean of 2.5 camping groups per day. On a particular day during the summer period, what is the probability that no groups arrive at the park? What is the probability that between one and three groups arrive? What is the most likely number of arriving camping groups?

To solve this problem, we make use of the appropriate probability mass function

$$P(x) = \frac{e^{-2.5}(2.5)^x}{x!} \quad x = 0, 1, 2, \ldots \tag{5-35}$$

where x is the number of camping groups arriving at the park per summer season day. To find the probability that no groups arrive, we simply compute $P(0)$ from Equation 5-35:

$$P(0) = \frac{e^{-2.5}(2.5)^0}{0!} = 0.821$$

Thus, there is an 8% chance that no groups arrive on a particular day. This computation can be checked by consulting the Poisson statistical table at the end of the text for the value $x = 0$ and $\lambda = 2.5$. To determine the probability that between one and three groups arrive, $P(1 \leq x \leq 3)$, we can either consult the table or calculate $P(1) + P(2) + P(3)$ from Equation 5-35. Using the table, we get $P(1 \leq x \leq 3) = P(1) + P(2) + P(3) = 0.2052 + 0.2565 + 0.2138 = 0.6755$.

Let us examine the Poisson table for the value $\lambda = 2.5$. We note that

$$P(0) = 0.0821 \qquad P(1) = 0.2052$$
$$P(2) = 0.2565 \qquad P(3) = 0.2138$$

The mode is $x = 2$ because it occurs with the greatest possibility. The most likely number of arriving camping groups is therefore two.

The Poisson distribution is applied extensively in problems related to the modeling of the distribution of the number of persons joining a queue, or line. Examples of this situation include arrivals at service stations or traffic facilities such as toll booths or ferries. These types of models are then used to make decisions about how many servicing units (for example, toll booths or ferries with a given capacity) to provide to keep the length of the queue manageable.

The experiment that leads to the value of a random variable should always be compared to the experimental conditions generating a Poisson random variable. In the example of the wilderness park, the random variable X is the number of camping groups arriving in a one-day period. The possible values that X can assume are $x = 0, 1, 2, \ldots, \infty$. If we consider the one-day interval to be composed of many small subintervals, each 10 minutes long, then the conditions for X to be considered a Poisson random variable appear to be satisfied. In a 10-minute interval:

1. The number of arriving camping groups is independent of the number arriving in any other 10-minute interval of the day.
2. The probability that a camping group arrives in a 10-minute interval is small and proportional to the length of the interval.
3. The probability of more than one camping group arriving in a 10-minute interval is extremely small.

The Poisson distribution appears to be a reasonable model for the random variable in this case. A specific member of the family of Poisson distributions is chosen by the estimation of λ from some sample data.

5.4. Continuous Probability Distribution Models

Unlike a discrete random variable, a continuous random variable can assume an infinite number of values. Although the distribution of a continuous random variable can also take on many forms, only one distribution is presented in this section. Our focus is on the bell-shaped curve discovered by Gauss and now known as the normal distribution, perhaps the most important probability distribution in conventional statistical analysis.

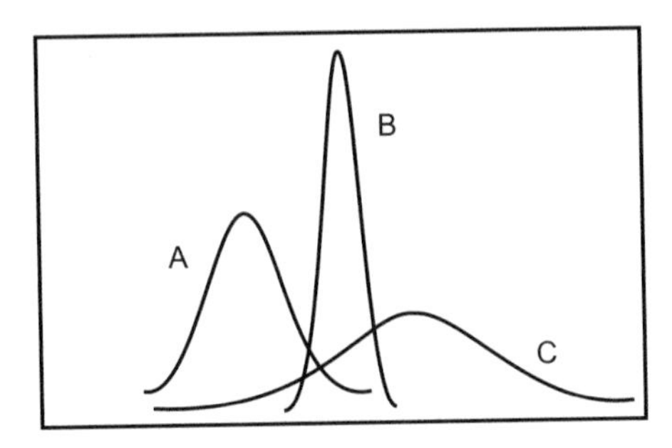

FIGURE 5-10. Three normal distributions.

Normal Probability Distribution

For many continuous random variables, the governing probability distribution is a specific bell-shaped curve known as the *normal distribution.* Originally developed by the German mathematician Carl Gauss, this distribution is sometimes termed the *Gaussian distribution.* The basic shape of the normal distribution is a simple bell. Three different normal distributions are illustrated in Figure 5-10. Why is this distribution so common? One reason is that this basic bell shape is extremely flexible; only a few of its many forms are illustrated in Figure 5-10. As implied by the figure, the center of the distribution can be located at any point along the real number line. In addition, the distribution can be very flat, or it can be very peaked.

DEFINITION: CONTINUOUS NORMAL RANDOM VARIABLE

A continuous random variable X is normally distributed if its probability density function is given by

$$f(X) = \frac{1}{\sigma\sqrt{2\pi}} \cdot e^{\left[\frac{1}{2}\left(\frac{x-\mu}{\sigma}\right)^2\right]} \qquad (5\text{-}38)$$

where $-\infty < x,\ \mu < \infty$, and $\sigma > 0$.

What do all three distributions have in common? All normally distributed random variables follow a particular mathematical expression that defines the characteristic bell shape. The probability density function for this continuous random variable is defined below. The values π and e are the familiar mathematical constants defined as approximately 3.14159 and 2.71828, respectively. There are two parameters of the normal distribution, μ and σ. Although the proof requires knowledge of calculus, it can be shown that the expected value and variance of a continuous normal random variable are given by

$$E(X) = \mu \tag{5-36}$$

$$V(X) = \sigma^2 \tag{5-37}$$

The parameter $\mu$is the mean of the probability distribution and locates the center of the distribution along the real number line. The variance σ^2 controls the dispersion of the values of X around the mean. It should now be clear why the members of the normal family come in such a variety of shapes. Two *independent* parameters can be used to combine any degree of dispersion with a central value located anywhere along the real number line. Although there is great diversity in the appearance of many random normal distributions, all are symmetric about μ. Thus, all normal variables are characterized by the equality of the mode, the median, and the mean. Not only is the highest point on the probably density function located at μ, but it also divides the distribution into two equal parts. Exactly 0.5 of the area under the probability density function lies on either side of μ.

To compute the probability that a normal random variable lies in the range between two values of X, say c and d, we must calculate the area under the probability density function given by Equation 5-36 between $X = c$ and $X = d$. This area is illustrated in Figure 5-11. How do we calculate this area? One option is to integrate Equation 5-36 for values of X between c and d. However, a second option, using the *standard normal distribution* is more appealing. While there exist an infinite number of normal distributions, each with a unique combination of μ and σ that control their precise shape, *any* normally distributed random variable can be converted into a single probability distribution known as the standard normal distribution. Thus, if we transform our normal probability distribution into the standard normal distribution, we can then make use of readily available statistical tables that show the area under the standard normal distribution for different ranges of our variable of interest. This transformation process is outlined in the following discussion.

The standard normal transformation process converts a variable measured in units of x to a standard normal variable measured in units of z.

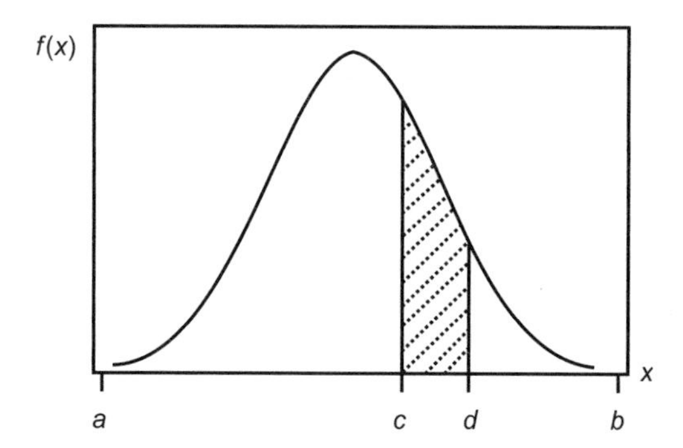

FIGURE 5-11. Probability that a normal random variable assumes a value in the interval $[c, d]$.

DEFINITION: STANDARD NORMAL TRANSFORMATION

$$z = \frac{x - \mu}{\sigma} \tag{5-39}$$

When measured in units of x, the variable has a mean of μ and a standard deviation of σ. When it is converted to a standard normal variable, it has a mean of zero and a standard deviation (and also a variance) of one. For example, consider the two normal random variables X_1 and X_2, the first with a mean $\mu_1 = 100$ and a standard deviation $\sigma_1 = 100$ and the second with a mean $\mu_2 = 10$ and a standard deviation $\sigma_2 = 2$. Both can be converted to standard normal variables by taking the original measurements and coding them according to Equation 5-39. Consider the first variable X_1 illustrated in Figure 5-12a. The value $x_1 = 110$ is converted to a standard z value by calculating $z_{110} = (110 - 100)/10 = 1.0$. In what units is z measured? It is always measured in *standard deviation units* of the variable being considered. Thus, the value $z = 1.0$ implies that a value of $x_1 = 110$ is one standard deviation above the mean $\mu_1 = 100$. Similarly, $x_1 = 90$ is one standard deviation, $\sigma_1 = 10$, below the mean of $\mu_1 = 100$, so that $z_{90} = (90 - 100)/10 = -1.0$.

DEFINITION: THE STANDARD NORMAL DISTRIBUTION

The probability density function for the standard normal distribution is given by

$$f(z) = \frac{1}{\sqrt{2\pi}} \cdot e^{\frac{1}{2}z^2} \tag{5-40}$$

Variable X_2 of Figure 5-12b can be transformed in the same way. Consider first a measurement of $x_2 = 14$. Expressed as a *standard score* or *z-score*, $z_{14} = (14 - 10) / 2 = 2.0$. That is, the value of $x_2 = 14$ lies two standard deviations above the mean $\mu_2 = 10$. As another example, we know that $x_2 = 4$ lies three standard deviations below the mean of $\mu_2 = 10$. To verify this, we calculate the z-score as $z_4 = (4 - 10) / 2 = -3.0$. Any value of x can be converted to an equivalent value of z and vice versa. For variable X_2 what value of x_2 corresponds to $z = -2.0$? From Equation 5-39 we want to find the value of x_2 that satisfies the equation $(x_2 - 10)/2 = -2.0$. The solution is $x_2 = 6$; that is, the value of six lies two standard deviations below the mean of 10. Thus, we can now see that any variables like X_1 and X_2 can be converted to the form of the standard normal distribution illustrated in Figure 5-12c.

When this transformation is made, it is easier to see the characteristics shared by all normally distributed random variables. As illustrated in Figure 5-12c, 68.26% or 0.6826 of the values of a random normal variable lie within one standard deviation of the mean. For the standard normal, this is between the values of $z = 1.0$ and $z = -1.0$. Thus, using the standard normal transformation, we know that for the variable X_1, 68.26% of the area under the probability density function lies between one standard deviation below the mean, 90, and one standard deviation above the mean, 110.

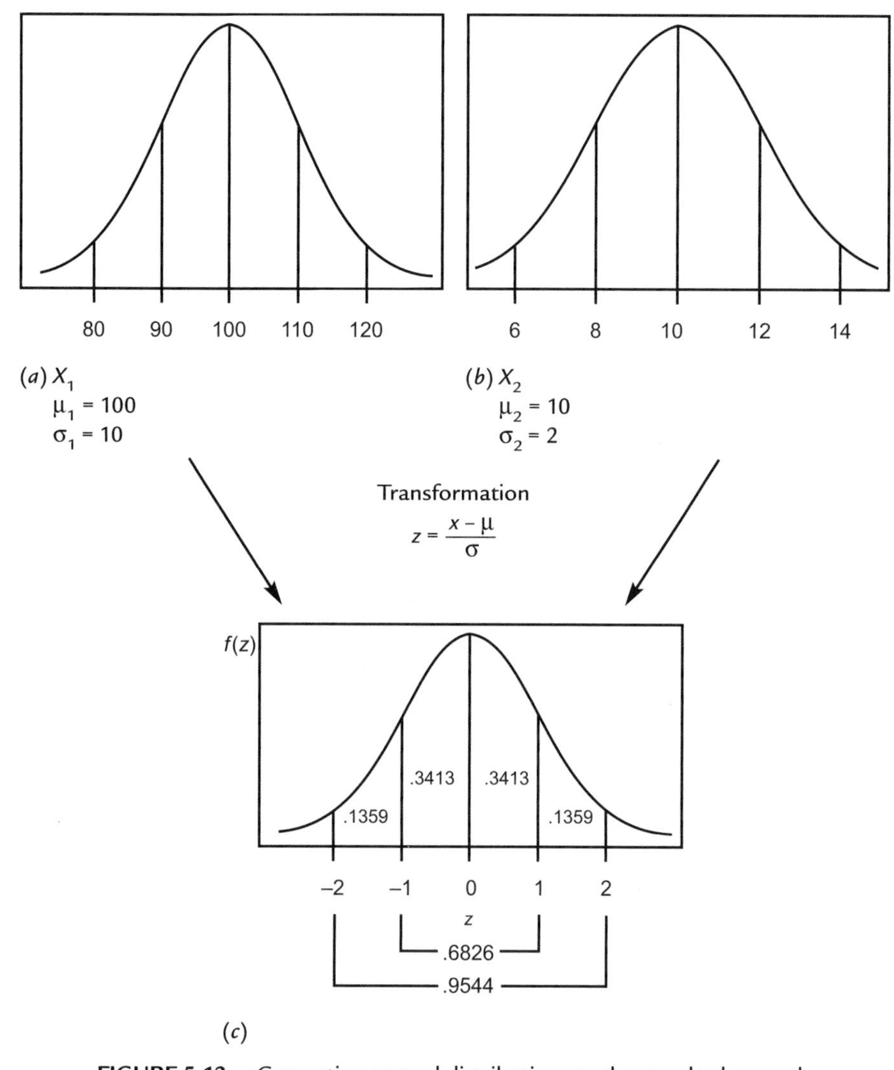

FIGURE 5-12. Converting normal distributions to the standard normal.

For variable X_2 the corresponding limits are $x_2 = 8$ and $x_2 = 12$. Also, 95.44% of the area under the probability density function of the standard normal lies within two standard deviations of the mean. For X_1 these limits are 80 and 120, and for X_2 the limits are 6 and 14. For all normally distributed random variables, the limits to the distribution are $+\infty$ and $-\infty$, but virtually 100% of the area under the curve (in fact, 99.72%) lies within three standard deviations of the mean. Because we can convert any normally distributed variable to the standard normal distribution, it is not necessary to provide an elaborate set of statistical tables that show the areas under the normal probability density curve between different limits for various values of the two parameters μ and σ. Rather, it is usual to provide a single table, the standard normal distribution. This table is included in the Appendix at the end of this text. To access this table, we must convert our data measured in units of x to units of z.

EXAMPLE 5-3. The travel time for the journey to work in a metropolitan area averages 40 minutes with a standard deviation of exactly 12.5 minutes. Assuming these distances are normally distributed, what is the probability that a randomly selected person's journey to work is longer than 1 hour? To solve this problem, we must first convert the value of $x = 60$ minutes (1 hour) to units of z so that we can employ the standard normal probability table. Note that $z = (60 - 40) / 12.5 = 1.6$. Next, we must determine the probability that a value of z in a standard normal distribution exceeds 1.6. Using the third column of the standard normal table (Table A-3 in the Appendix at the end of this text), we find this probability to be 0.055. Note the various ways in which the standard normal table can be expressed. Equivalently, we can say that there is a 94.5% chance that the work trip for our randomly selected commuter is likely to be less than 60 minutes in length.

5.5. Bivariate Random Variables

When two or more random variables are jointly involved in a statistical experiment, their outcomes are generated by a *multivariate probability function. Bivariate probability functions* are a class of multivariate functions in which two random variables are jointly involved in the outcome of a statistical experiment. In turn, bivariate probability functions can be classed as either discrete or continuous. The concepts of bivariate random variables are most easily explained in relation to discrete distributions without resorting to the mathematical complications inherent in continuous models. The relevant features of continuous models can then be explained by analogy to the discrete case.

Bivariate Probability Functions

The *joint* or *bivariate probability mass function* for two discrete variables is a function that assigns probabilities to joint values of X and Y, such that two conditions are satisfied:

1. The probability that random variable X takes on the value $X = x$ and that random variable Y takes on the value $Y = y$, denoted $P(x, y)$, is non-negative and less than or equal to 1. That is, $0 \leq P(x, y) \leq 1$ for all (x, y) pairs of the discrete random variables X and Y.
2. The sum of the probabilities $P(x, y)$ taken over all discrete values of X and Y is 1. That is, $\Sigma_x \Sigma_y P(x, y) = 1$.

To understand these two conditions, consider the relationship between the size of a household and the number of cars owned by the household. Suppose there is a com-

TABLE 5-10
Car Ownership and Household Size

Household size	Cars owned 0	1	2	3	Total
2	10	8	3	2	23
3	7	10	6	3	26
4	4	5	12	6	27
5	1	2	6	15	24
Total	22	25	27	26	100

TABLE 5-11
Probabilities of Car Ownership and Household Size

Size	0	1	2	3	$P(x)$
2	0.10	0.08	0.03	0.02	0.23
3	0.07	0.10	0.06	0.03	0.26
4	0.04	0.05	0.12	0.06	0.27
5	0.01	0.02	0.06	0.15	0.24
$P(y)$	0.22	0.25	0.27	0.26	1.00

munity of 100 households and we take a census noting the number of members in each household and the number of cars owned by that household. The responses of the households are classified in Table 5-10. Of the 100 households in the population, 10 have two members and no cars available, 8 have two members and one car available, and so on. When households are categorized by size, 23 have two members, 26 have three members, 27 have four members, and 24 have five members. Similarly, the totals by number of cars owned are given in the last row of the table.

These data can be converted to a bivariate probability distribution by dividing each entry in Table 5-10 by the size of the population, $N = 100$. The bivariate probabilities, $P(x, y)$, for these two variables are given in Table 5-11. In the last column of the table is the *marginal probability distribution* of random variable X, the probability distribution of random variable X taken alone. The marginal probability distribution of X can be calculated form the relation $P(x) = \Sigma_y P(x, y)$. Note that, for example, the probability of a randomly selected household having exactly three members is $P(x = 3) = 0.07 + 0.10 + 0.06 + 0.03 = 0.26$. Similarly, the marginal probability distribution of Y can be determined from $P(y) = \Sigma_x P(x, y)$. The probability that a household has no cars is obtained by summing the first column $P(Y = 0) = 0.10 + 0.07 + 0.04 + 0.01 = 0.22$. It is thus possible to construct the marginal probability distributions $P(X)$ and $P(Y)$ from a joint or bivariate probability function $P(X, Y)$.

The joint probability distribution $P(X, Y)$ can also be used to determine the *conditional probability functions* for variables X and Y. What is the conditional probability distribution of Y given that X takes on some specific value $X = x$? For example, what is the conditional probability distribution of car ownership, given that a household contains three members? The conditional probability distribution given $X = x$ is calculated from the relation

$$P(y \mid x) = \frac{P(x, y)}{P(x)} \tag{5-41}$$

And the conditional distribution of X given $Y = y$ is

$$P(x \mid y) = \frac{P(x, y)}{P(y)} \tag{5-42}$$

The conditional probability distribution of car ownership for three-person households can be computed from Equation 5-41 with the data of Table 5-11. The conditional probability function $P(y \mid x = 3)$ is

$$P(y = 0 \mid x = 3) = \frac{.07}{.26} = .269$$

$$P(y = 1 \mid x = 3) = \frac{.10}{.26} = .385$$

$$P(y = 2 \mid x = 3) = \frac{.06}{.26} = .231$$

$$P(y = 3 \mid x = 3) = \frac{.03}{.26} = .115$$

Note that this is a valid probability distribution because it sums to one. To obtain this conditional distribution, we simply divide each entry in the row for $x = 3$ by the row sum $P(3)$ given in the last column. It is not possible to use the entries in the row as they stand. Why? Because they do not sum to one. From a purely mechanistic standpoint the division by 0.26 standardizes the distribution so that it sums to one. The logic behind the standardization is as follows. In knowing that $x = 3$, the population of possible values of y has been reduced, so that the probability of any single value is higher than what is given by the joint distribution. Instead of the entire sample space ($P = 1$), the candidate values lie in a region of total probability 0.26. Each of the ratios above is the fraction of the reduced sample space. Similarly, the conditional probability distribution of household size given that the household has three cars is

$$P(x = 2 \mid y = 3) = \frac{.02}{.26} = .077$$

$$P(x = 3 \mid y = 3) = \frac{.03}{.26} = .115$$

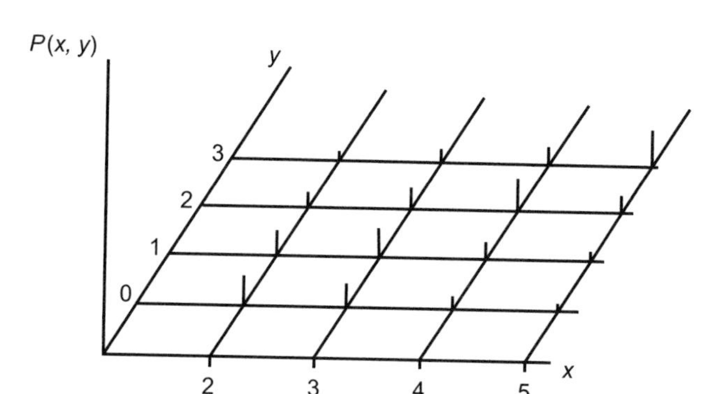

FIGURE 5-13. Graph of bivariate probability distribution of Table 5-11.

$$P(x = 4 \mid y = 3) = \frac{.06}{.26} = .231$$

$$P(x = 5 \mid y = 3) = \frac{.15}{.26} = .577$$

It is also possible to illustrate a bivariate probability function by using a simple three-dimensional graph. The vertical axis is used for $P(x, y)$, and the two random variables are depicted in the XY plane. Where there are only a few discrete values for the variables, such illustrations give a visual impression of the joint distribution of the two random variables involved. Figure 5-13 illustrates the bivariate probability distribution for the data in Table 5-11. It is apparent, for example, that larger households tend to own more cars than smaller households. In fact, there appears to be a direct relation between the size of household and the number of cars owned. How can the strength of this relation be measured? Following the discussion in Chapter 4, we can examine how closely these two random variables are related by estimating their covariance.

Covariance of Two Random Variables

The covariance of two discrete random variables is defined as

$$COV(X, Y) = E\{[X - E(X)][Y - E(Y)]\} = \sum_x \sum_y P(x, y)[X - E(X)][Y - E(Y)] \quad (5\text{-}43)$$

though a more convenient formula for calculation is

$$COV(X, Y) = E(X, Y) - E(X)E(Y) = \sum_x \sum_y xyP(x, y) - E(X)E(Y) \quad (5\text{-}44)$$

Here $COV\ (X, Y)$ is the covariance, and E is the expectation operator. Recall from Equation 5-18 that $E(X) = \Sigma_{i=1}^{k} x_i P(x_i)$. The covariance is a direct statistical measure

TABLE 5-12
Computation of Covariance Data of Table 5-11 Using Equation 5-44

x	$P(x)$	$xP(x)$	Y	$P(y)$	$yP(y)$	x, y	$P(x, y)$	$xyP(x, y)$
2	0.23	0.46	0	0.22	0.00	2, 0	0.10	0.00
3	0.26	0.79	1	0.25	0.25	2, 1	0.08	0.16
4	0.27	1.08	2	0.27	0.54	2, 2	0.03	0.12
5	0.24	1.20	3	0.26	0.78	2, 3	0.02	0.12
		$E(X) = 3.52$			$E(Y) = 1.57$	3, 0	0.07	0.00
						3, 1	0.10	0.30
						3, 2	0.06	0.36
						3, 3	0.03	0.27
						4, 0	0.04	0.00
						4, 1	0.05	0.20
						4, 2	0.12	0.96
						4, 3	0.06	0.72
						5, 0	0.01	0.00
						5, 1	0.02	0.10
						5, 2	0.06	0.60
		$COV(X, Y) = 6.16 - (3.52)(1.57) = 0.6336$				5, 3	0.15	2.25
								$\sum = 6.16$

of the degree to which two random variables X and Y tend to vary together. Whenever large values of X tend to be associated with large values of Y, and small values of X with small values of Y, then $COV(X, Y)$ has a large positive value. When large values of X are associated with small values of Y and small values of X with large values of Y, then $COV(X, Y)$ is a large negative number. Whenever there is no pattern, $COV(X, Y) = 0$ or is close to zero.

Table 5-12 employs Equation 5-44 to calculate the covariance between the household size and car ownership data from Table 5-11.The calculated value of $COV(X, Y)$ indicates that the two variables tend to covary, or move together about their respective expected values. Because $COV(X, Y)$ is positive, household size and car ownership are positively correlated. Unfortunately, the covariance is not an easily interpretable measure of correlation. The numerical value of $COV(X, Y)$ is completely dependent on the magnitudes of the two random variables X and Y. If either of these two variables is measured in different units, then $COV(X, Y)$ must change. Recall that Pearson's product–moment correlation coefficient, introduced in Chapter 4, standardizes the covariance and overcomes this problem. Here, because we are discussing random variables (populations) rather than samples, the formula for the correlation coefficient is slightly different. As will be seen below, the formula involves population values rather than sample values.

Independence of Two Random Variables

Extending the discussion of statistical dependence encountered in Chapter 4, two discrete variables are said to be independent if

$$P(x, y) = P(x, y) = P(x)P(y) \tag{5-45}$$

for all possible combinations of X and Y. To show that two variables are dependent, it is sufficient to show that Equation 5-45 does not hold for any one combination of X and Y. Clearly, household size and car ownership are not independent, for the value assumed by one of the variables depends in part on the value obtained by the other variable. To see this, examine the relation given by (5-45) for the values $x = 3$ and $y = 1$. From Table 5-11

$$P(x = 3) = 0.26 \quad P(y = 1) = 0.25 \quad P(x = 3, y = 1) = 0.1$$

Thus, $P(x = 3)P(y = 1) = (0.26)(0.25) = 0.065 \neq P(x = 3, y = 1) = 0.1$.

Independence and covariance are closely related. If two random variables X and Y are independent, then their covariance must equal zero. However, it is not necessarily true that the two variables are independent if the covariance is zero. It is possible for two variables to have a covariance $COV(X, Y)$ yet still not satisfy the demanding requirements of Equation 5-45.

Bivariate Normal Random Variables

For continuous bivariate random variables, the joint probability distribution is specified by using a density function $f(X, Y)$. The most important distribution is the *bivariate normal distribution.* Figure 5-14 contains a graphical representation of a bivariate normal distribution. For every pair of (x,y) values, there is a density $f(x, y)$ represented by the height of the surface at that point. The surface is continuous, and probability corresponds to the volume under the surface. If two random variables are jointly normally distributed, then

1. The marginal distributions of both X and Y are univariate normal.
2. Any conditional distribution of X or Y is also univariate normal.

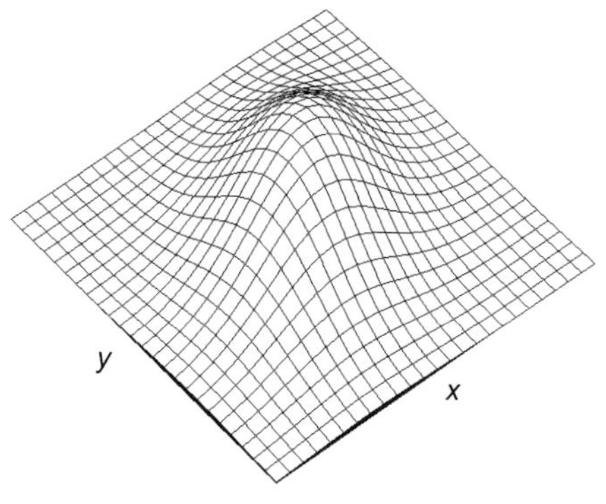

FIGURE 5-14. Bivariate normal probability density function.

The implication of item 2 is that the cross section of a "slice" through the bivariate normal distribution for a given value of X, say $X = x_a$, has the characteristic shape of the normal curve. Thus, in Figure 5-14, each curve parallel to the Y-axis is a normal distribution. A slice at any $Y = y_b$ has the same characteristic.

There are five parameters to the bivariate normal density function: μ_x and μ_y, σ_x and σ_y; and ρ_{xy}. The first four parameters are the means and standard deviations of the marginal distributions for X and Y. The parameter ρ_{xy} is the correlation coefficient between random variables X and Y defined as

$$\rho_{xy} = \frac{COV(X, Y)}{\sigma_x \sigma_y}$$

The characteristic values of the correlation coefficient were discussed in the last chapter.

It is common to portray a bivariate normal distribution by using a contour diagram. A contour diagram is created by taking horizontal slices through the bivariate normal distribution, as in Figure 5-15. A contour is composed of all (x, y) pairs hav-

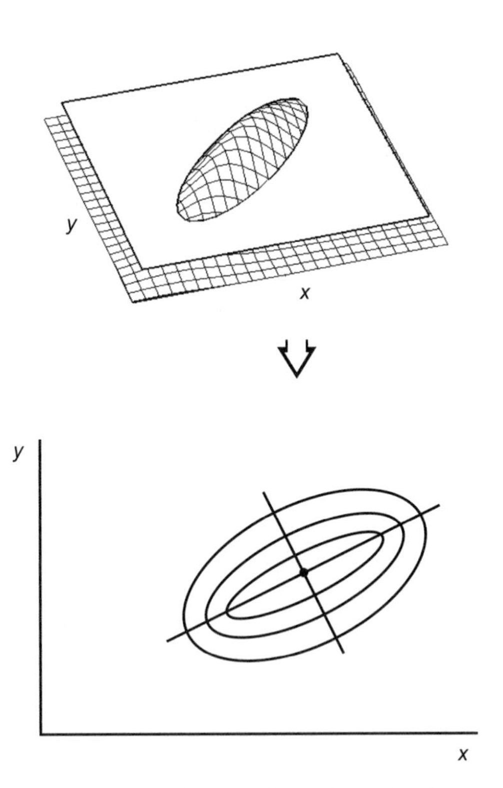

FIGURE 5-15. Three- and two-dimensional representations of the bivariate normal distribution.

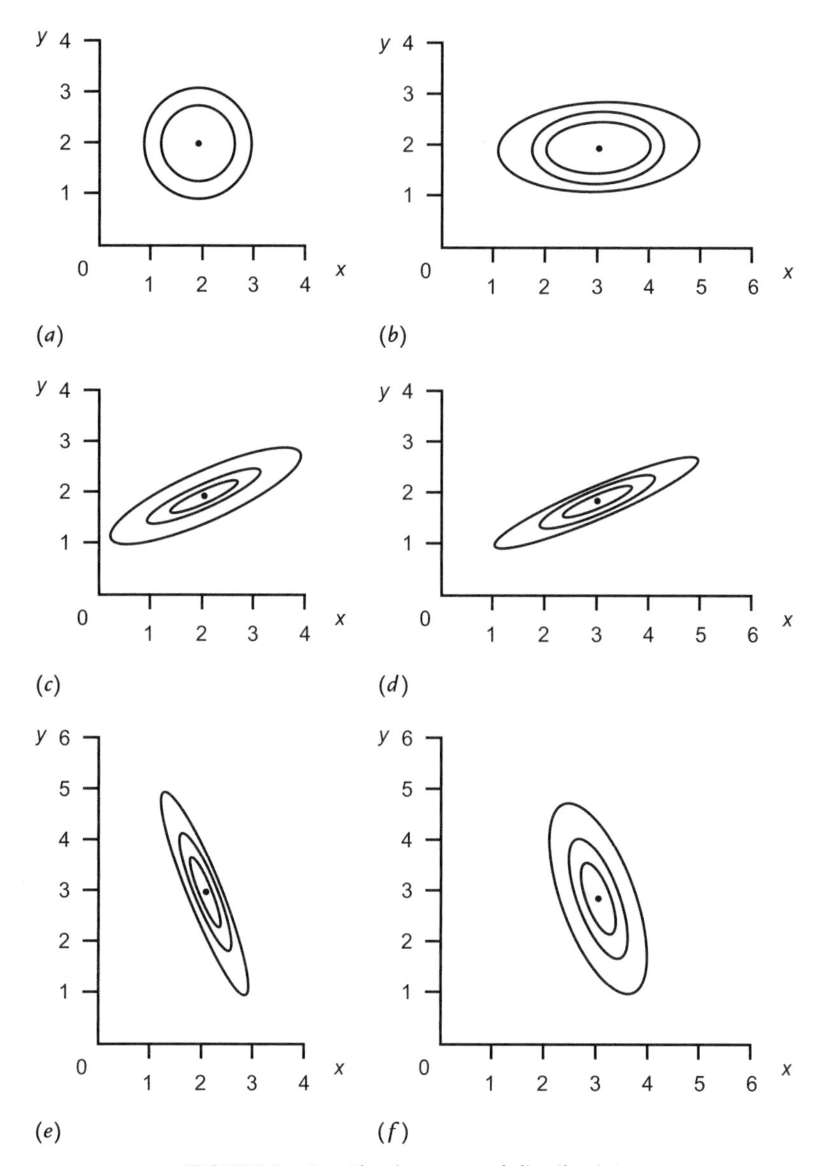

FIGURE 5-16. Bivariate normal distributions.

ing constant density $f(x, y)$. The contour curves of all bivariate normal distributions are elliptical, except where $\rho_{xy} = 0$ and $\sigma_x = \sigma_y$. In this single situation, the contours will appear as concentric circles. Several examples of bivariate normal distributions are shown in Figure 5-16. In (figure 5-16b) note that $\rho_{xy} = 0$, but the contours are elliptical since $\sigma_x > \sigma_y$. In both (c) and (d), variables X and Y are positively related and

$\rho_{xy} > 0$. The principal axis of each ellipse has a positive slope, indicating that the surface tends to run along a line with positive slope. Similarly, when X and Y are negatively related and $\rho_{xy} > 0$, the principal axis has a negative slope. This case is illustrated in (e) and (f). In all cases in Figure 5-16, the centers of the ellipses are at coordinates (μ_x, μ_y). The roles of the parameters are clearly illustrated by these eight cases. The two means μ_x and μ_y control the location of the surface, and the two standard deviations and the correlation coefficient control the shape.

5.6. Summary

This chapter formally introduces the concepts of probability and random variables. When used in statistics, the term *probability* is a measure of the likelihood of an event occurring. That measure must take on values between zero and one. The basic arguments of probability are founded on the notion of a statistical experiment. To compute event probabilities, it is usual to determine the elementary outcomes of the experiment. These outcomes should constitute the set of mutually exclusive and collectively exhaustive outcomes of the experiment. Together, these outcomes represent the sample space of the experiment. If all elementary outcomes are equally likely, the probability that an event occurs in the experiment can be determined by taking the ratio of the number of outcomes in the event space to that of the sample space. The enumeration of the elementary outcomes in either the event or the sample space is often made easier by the application of one or more counting rules that are reviewed in Appendix 5a of this chapter. The probabilities associated with a complex event can be identified once the relationships between each of the individual events that comprise it are known. Whether events are mutually exclusive or independent is the most important consideration in deciding which probability laws should be applied to solve a particular problem.

As we moved toward a consideration of statistical inference, the notion of probability was extended through discussion of random variables. A random variable is any numerically valued function that is defined over a sample space. We identified two different types of random variables: continuous and discrete. A list of all the values that a random variable might assume, together with the probabilities of their occurrence, is known as a probability distribution. A number of probability distribution models for both discrete and continuous random variables were introduced in Sections 5.3 and 5.4. Section 5.5 briefly developed these ideas to consider bivariate random variables and bivariate probability distributions, describing the joint distribution of two variables.

Appendix 5a. Counting Rules for Computing Probabilities

The following four counting rules may be used to quickly enumerate the number of elementary outcomes in a sample or an event space, without specifically listing each one.

Product Rule

RULE 1: THE PRODUCT RULE

Suppose there are r groups of objects. In group 1 there are n_1 objects, in group 2 there are n_2 objects, and so on. If we define the experiment to be the selection of one object from each of the r groups, then there are $n_1 \cdot n_2 \cdot \ldots \cdot n_r$ elementary objects in the experiment.

To illustrate the use of this counting rule, consider the following example.

EXAMPLE 5a.1. A planner is interested in surveying retail patronage at three types of shoe stores: factory outlets, discount stores, and specialty stores. She decides to sample one store of each type from the outlets in the town. If there are 10 factory outlets, 7 discount stores, and 3 specialty stores in the town, how many different combinations of stores are there?

There are $r = 3$ groups of objects, with $n_1 = 10$, $n_2 = 7$, and $n_3 = 3$. If one store must be selected from each category, then there are $(10)(7)(3) = 210$ possible combinations of stores from which the planner may choose.

Combinations Rule

The second and third counting rules are concerned with a slightly different problem. The experiment consists of selecting a subset of r objects from a set of n objects, where $r \leq n$.

EXAMPLE 5a.2. Suppose a city is considering adding two new libraries to its existing system. To minimize expenses, the libraries will be constructed on two of five vacant properties currently owned by the city. How many different locational plans for the system of libraries are there?

Label the five available sites as *A, B, C, D,* and *E.* The possible outcomes of the experiment are:

AB	*BC*	*CD*	*DE*
AC	*BD*	*CE*	
AD	*BE*		
AE			

Note that each of the possible plans contains two different sites; these are combinations. Note also that we list CD as a site combination, but not DC. The reason is that by definition, order does not matter in a combination. From the city's perspective, there is no difference between outcome CD and DC, because buildings will be erected on the same two sites.

DEFINITION: COMBINATION
A combination is a set of r distinguishable objects, regardless of order.

The counting rule for combinations is concerned with the number of different combinations of size r that can be drawn from a master set of n objects.

RULE 2: COMBINATIONS RULE
The number of combinations of r objects taken from n different objects is

$$C_r^n = \frac{n!}{r!(n-r)!}$$

Sometimes the number of combinations C_r^n is written $C(n, r)$ or $\binom{n}{r}$. The symbol $n!$ is read "n factorial" and is defined as $n \cdot (n-1) \cdot (n-2) \cdot \ldots \cdot 2 \cdot 1$. For example, $5! = 5 \cdot 4 \cdot 3 \cdot 2 \cdot 1 = 120$ and $2! = 2 \cdot 1 = 2$. By definition, $0! = 1$.

For the library example, $n = 5$ and $r = 2$; therefore

$$C_2^5 = \frac{5!}{2!(5-2)!} = \frac{5 \cdot 4 \cdot 3!}{2 \cdot 1 \cdot 3!} = \frac{20}{2} = 10$$

These are the 10 combinations listed above. As you can imagine, the combinations rule can save a great deal of time in enumerating the elementary outcomes of an experiment.

Permutations Rule

Let us now consider a slightly different problem. Suppose the city has five sites for two new libraries, one that will house the main library and one that will house digital and other special collections. The sites are labeled *A* to *E* as before. Now consider the outcome *AB*. This is interpreted as meaning site *A* is to be used for the main library (the first letter of the pair) and site *B* (the second letter) is to be used for the special collections. Notice that outcome *AB* is not the same as outcome *BA*, because the two libraries are placed on different sites. In *BA*, the main library is located at site *B* while in *AB* the main library is on site *A*. In this instance, the *order* of the elements in the set is important.

DEFINITION: PERMUTATION
A permutation is a distinct ordering of r objects.

The counting rule for permutations defines the number of different ordered arrangements of size r that can be drawn from a set of n objects.

RULE 3: PERMUTATIONS RULE

The number of permutations of r objects taken from a set of n different objects is

$$P_r^n = \frac{n!}{(n-r)!}$$

Sometimes, P_r^n is written as $P(n, r)$. In the library example, $n = 5$ and $r = 2$, so that

$$P_2^5 = \frac{5!}{(5-2)!} = 20$$

That is, there are twice as many permutations as combinations in a problem of this size. For each combination *AB* there is the reverse arrangement *BA*. This is not a general result. The formal relation between the number of combinations and the number of permutations is given by

$$C_r^n = \frac{1}{r!} \cdot P_r^n$$

Whenever $r = 2$, there are twice as many permutations as combinations.

Hypergeometric Rule

The final rule combines the product rule with the combinations rule.

RULE 4: HYPERGEOMETRIC RULE

A set of n objects consists of n_1 of type 1 and n_2 of type 2, where $n_1 + n_2 = n$. From this group of n objects, define an experiment in which r_1 of the first type and r_2 of the second type are to be chosen. The number of different groups that can be drawn is

$$C_{r1}^{n1} \cdot C_{r2}^{n2}$$

As an example, consider the following problem. A survey is to be undertaken of neighborhoods in some city. There are 10 suburban neighborhoods and 8 inner-city neighborhoods. A decision is to be made to survey three suburban and two inner-city neighborhoods. How many different combinations of neighborhoods are possible for the survey? Clearly, the hypergeometric rule applies. There are two groups with $n_1 =$ 10 and $n_2 = 8$. From these two groups, $r_1 = 3$ and $r_2 = 2$ objects are to be drawn. From the hypergeometric rule, the number of combinations of neighborhoods that could be drawn is

$$C_3^{10} \cdot C_2^8 = \frac{10!}{3!(10-3)!} \cdot \frac{8!}{2!(8-2)!} = 120(28) = 3360$$

Appendix 5b. Expected Value and Variance of a Continuous Random Variable

There is a direct correspondence between the area under a probability density function between any two limits c and d and the probability that a continuous random variable X assumes a value between these two limits. Students familiar with calculus will recognize that the area under $f(x)$ between c and d is given by

$$\int_c^d f(x)dx$$

where the symbol $\int$ represents the integration operator. It is directly analogous to the summation operator Σ used for discrete variables. The requirement that the sum of the probabilities of the values that may be assumed by a random variable equal one can be stated as

$$\int_a^b f(x)dx = 1$$

where a and b are the limits of the values that may be assumed by the random variable. For continuous random variables defined over the infinite length of the real number line, the appropriate equation is

$$\int_{-\infty}^{\infty} f(x)dx = 1$$

A second important property is that the probability of a continuous variable assuming a particular value, say $X = c$, is always zero. That is,

$$\int_c^c f(x)dx = 0$$

because any definite integral over the limits c to c must be zero.

The expected value of a continuous random variable is

$$E(X) = \int_a^b xf(x)dx$$

and the variance is

$$V(X) = \int_a^b [x - E(X)]^2 f(x)dx = \int_a^b x^2 f(x)dx - [E(X)]^2$$

As an example, let us consider the case of the uniform continuous random variable defined by

$$f(x) = \begin{cases} \frac{1}{10} & 0 \le x \le 10 \\ 0 & \text{otherwise} \end{cases}$$

First, to show the area under $f(x)$ equals one, we note

$$\int_0^{10} \frac{1}{10}\, dx = \frac{1}{10}(10 - 0) = 1$$

To calculate the expected value, we write

$$E(X) = \int_0^{10} x\left(\frac{1}{10}\right)dx = \frac{1}{10}\int_0^{10} x\,dx$$

$$= \frac{1}{10}\,\frac{x^2}{2}\Big|_0^{10}$$

$$= \frac{1}{10}\left(\frac{10^2}{2} - \frac{0^2}{2}\right) = 5$$

The variance is

$$V(X) = \int_0^{10} x^2\left(\frac{1}{10}\right)dx - [E(X)]^2$$

$$= \frac{1}{10}\,\frac{x3}{3}\Big|_0^{10} - 5^2$$

$$= \frac{1}{10}\left(\frac{1000}{3}\right) - 25 = 8.33$$

FURTHER READING

Introductory statistics textbooks for geographers seldom discuss the concept of random variables in detail, nor do they give an extended treatment of probability distribution models. Introductory statistics textbooks for business and economics students and some texts for applied statistics courses usually give a more complete presentation of these topics. Three representative examples are Neter et al. (1982); Pfaffenberger and Patterson (1981); and Winkler and Hays (1975). The development of the Poisson probability distribution using the analogy to a dot map only hints at the widespread use of probability models in this situation. Some additional material on quadrat analysis and a related technique known as nearest neighbor analysis is provided in Chapter 14. Extended treatment of these topics can be found in Getis and Boots (1978), Taylor (1977), and Unwin (1981).

A. Getis and B. Boots, *Models of Spatial Processes* (Cambridge, UK: Cambridge University Press, 1978).

J. Neter, W. Wasserman, and G. Whitmore, *Applied Statistics* (Boston: Allyn and Bacon, 1982).

R. Pfaffenberger and J. Patterson, *Statistical Methods for Business and Economics* (Homewood, IL: Richard Irwin, 1981).

P. Taylor, *Quantitative Methods in Geography* (Boston: Houghton Mifflin, 1977).
D. Unwin, *Introductory Spatial Analysis* (London: Methuen, 1981).
R. Winkler, and W. L. Hays, *Statistics: Probability Inference and Decision Making,* 2nd ed. (New York: Holt, 1975).

PROBLEMS

5.1
5.2
5.3
5.4
Apendix

1. Explain the meaning of the following terms:
   - Statistical experiment or random trial
   - Elementary outcome of an experiment
   - Conditional probability
   - Probability distribution
   - Expected value of a random variable
   - Standard normal distribution
   - Bivariate probability function
   - Binomial coefficients
   - Sample space
   - Event
   - Random variable
   - Standard score
   - Variance of a random variable
   - Covariance
   - Bernoulli trial
   - Pascal's triangle

2. Differentiate between an *objective* and a *subjective* interpretation of the concept of probability. Give an example of each. Why are objective interpretations preferred in statistical analysis?

3. Residents of a city are asked to rank the desirability of five different neighborhoods *A, B, C, D,* and *E.* How many different ways can they be ranked, assuming there can be no ties?

4. A traveling salesperson beginning a trip at city *A* must visit (in order) cities *X, Y,* and *Z* before returning to *A.* Several roads connect each pair of cities. There are four different ways of traveling between cities *A* and *X,* three routes between *X* and *Y,* five between *Y* and *Z,* and two between *Z* and *A.* By how many different routes can the salesperson complete the trip?

5. Construct an outcome tree outlining the elementary outcomes for three tosses of a coin. What is the probability that three consecutive tails occur, assuming the coin is fair?

6. The license plates in a certain jurisdiction have six alphanumeric digits. The first digit is *A, B,* or *C.* The second digit is *N, S, E,* or *W.* The last four digits are restricted to the integers 0, 1, 2, . . . , 9. How many different license plates are possible?

7. The probability that it rains on a given day in July is 0.1.
   a. Assuming independent trials, what is the probability that it does not rain for three consecutive days?
   b. Again assuming independent trials, what is the probability that one day of rain is followed by two days without rain?
   c. Is the assumption of independent trials in this experiment reasonable?

8. A retail geographer surveys 200 shoppers after each has visited one of three shopping centers *A, B,* or *C.* She records whether each made a purchase, with a yes (*Y*) or no (*N*). The survey results are as follows:

Center	Y	N	Total
A	25	25	50
B	10	50	60
C	65	25	90
Total	100	100	200

a. Find $P(C)$.
b. Find $P(A \cup B)$.
c. Find $P(B \cup Y)$.
d. Find $P(A \cap B| N)$.

Explain in your own words the meanings of parts (a) to (d).

9. Differentiate between the following:
   a. A discrete and a continuous random variable
   b. A probability mass function and a probability density function
   c. A normal probability distribution and the standard normal distribution
   d. A marginal probability distribution and a conditional probability distribution
   e. A probability mass function and a cumulative mass function

10. Let X be a random variable with the following probability distribution:

X	P(x)
0	0.40
1	0.30
2	0.15
3	0.15

   a. Verify that this is a valid probability distribution model.
   b. Determine $E(X)$.
   c. Determine $V(X)$.
   d. What is the mode of X?

11. Graph the probability mass functions for the following discrete probability distribution models:
   a. Uniform: $k = 3$, 5, and 10
   b. Binomial: $n = 5$, $\pi = 0.20$; $n = 5$, $\pi = 0.50$; $n = 5$, $\pi = 0.70$.
   c. Poisson: $\lambda = 0.20$, 0.40 and 0.60

12. The maximum temperature reached on any day can be classified as above freezing (a success) or below freezing (a failure). In a certain city of eastern North America, January weather statistics indicate the probability a January day will be above freezing is 0.3. Use the binomial distribution to determine the following probabilities:
   a. Exactly 2 of the next 7 January days will be above freezing.
   b. More than 5 of the next 7 days will be above freezing.
   c. There will be at least 1 day above freezing in the next 7 days.
   d. All 7 days in the next week will be above freezing.
   e. Is this a reasonable application of the binomial distribution? Why or why not?

6

Sampling

In Chapter 1, statistical methodology was conveniently divided into *descriptive statistics* and *inferential statistics.* In inferential statistics, a descriptive characteristic of a sample is linked with probability theory, so that a researcher can generalize the results of a study of a few individuals to some larger group. This idea was made more explicit in Chapter 5, where the notions of a *random variable* and its *probability distribution* were defined. At the core of inferential statistics is the distinction between a *population* and a *sample.* From a statistical universe or population, a small subset of individuals is often selected for detailed study. This sample is used to estimate the value of some population characteristic or to answer a question about a particular characteristic of the population. However, to make such inferences, the sample must be collected in a specific way. It is not possible to make statistically reliable inferences from any sample. Whereas street corner interviews, for example, tend to make interesting news items, they may not reflect the views of the population they are supposed to represent.

Ideally, it would be best to have a sample that is a good representation of the population from which it has been drawn. High-quality inferences are made by using high-quality samples. Unfortunately, unless we know everything about the population, say from a census, we have no way of knowing if we do have a representative sample. The very act of sampling thus introduces some uncertainty into our inferences, simply because the sample may not be representative of the population. This is known as *sampling error.* Suppose, for example, we wished to sample the students at a university and determine the number of hours the average student spends studying in any given week. By mere chance, we may just select a sample that includes more industrious students than average students and thus overestimates the amount of time spent studying. We might be led to believe that the average student spent more time studying than is, in fact, the case. Our only safeguard against such sampling error is to select a larger sample. The larger the sample, the more likely it includes a true cross section of the population—that is, the more likely it is to be *representative* of the population. Notice that sampling error is not a "mistake," such as choosing the "wrong" sample or some other methodological failing. All samples deviate from the population in some way; thus sampling error is always present. The associated uncertainty is the price one pays for using a subset of the population rather than the entire population. The appeal of statistics is not that it removes uncertainty, but rather that it permits inference in the presence of uncertainty.

DEFINITION: SAMPLING ERROR

Sampling error is uncertainty that arises from working with a sample rather than the entire population.

Besides sampling error, there is another reason our sample may not be representative of a population: *sampling bias*. This occurs when the way in which the sample was collected is itself biased. In the example of university student study habits, a sample would surely be biased if it were selected on the basis of interviews of students leaving the university library late in the evening!

DEFINITION: SAMPLING BIAS

Sampling bias occurs when the procedures used to select the sample tend to favor the inclusion of individuals in the population with certain population characteristics.

Sampling bias can usually be avoided, or at least minimized, by selecting an appropriate sampling plan. Errors in recording, editing, and processing sample data can also be limited by various checks. When data are collected through mail questionnaires, a form of bias due to *nonresponse* often occurs. The respondents to the questionnaire may not be representative of the overall population. Many studies have found their respondents to be more highly educated, wealthier, and more interested in the subject of the questionnaire than members of the population at large. Since the quality of the inferences from a sample depends so much on the sample itself, any researcher must carefully select a sampling plan capable of minimizing, or at least controlling to acceptable limits, both sampling error and sampling bias.

Sampling techniques with this characteristic are the focus of this chapter. First the advantages of sampling are enumerated in Section 6.1. Why do we favor a sample over a complete census of a population? In Section 6.2 an extremely useful four-step procedure for sampling is outlined. The tasks defined in these steps are encountered in every sampling problem. Section 6.3 identifies various types of samples. Only specific types of samples can be used to generalize to a population—with a *known* degree of risk. The most commonly used type, the simple random sample, is explained in Section 6.4. This sampling strategy is then compared to a few other sample designs. Section 6.5 introduces the concept of a sampling distribution. The sampling distribution of sample statistics such as the mean ($\bar{X}$) and proportion P is central to both the estimation and the hypothesis-testing procedures of statistical inference. Finally, issues arising in geographic sampling are presented in Section 6.6.

6.1. Why Do We Sample?

Seldom must we collect information from all members of a population in order make reliable statements about the characteristics or attributes of that population. Often a sample constituting only a small percentage of the total population is sufficient for such inferences. There are several reasons for choosing a *sample* rather than a census of an entire population.

1. Usually, it is not necessary to take a complete census. Valid, reliable generalizations about the characteristics of a population can be made with a sample of modest size—if the sample is properly taken. The uncertainty inherent in generalizing from the few to the many not only is within acceptable limits, but sometimes is even less than the uncertainties that arise when we try to precisely control the enormous amount of data generated from a complete enumeration of an extremely large population. It is simply far easier to check the data of a small sample than those of a large population.
2. The time, cost, and effort of collecting data from a sample are usually substantially less than are required to collect the same information from a larger population. The workforce or available financial resources usually constrain a researcher from taking a full census.
3. The population of interest may be infinite, and therefore sampling is the only alternative. We could, for example, consider the population to be the water temperature at a certain depth at a given reach of a particular stream. There are an infinite number of times when we could record the water temperature—even in a small time interval. Since space itself can be treated as a continuous variable, there are an infinite number of places in any area where a set of sample measurements could be taken. This issue is explored more fully in Section 6.6.
4. The act of sampling may be destructive. To estimate the mean lifetime of light bulbs, for example, any light bulb in a sample must be tested until it is no longer of use. A census of the light bulbs produced by a manufacturer would destroy that producer's entire stock!
5. The population may be only hypothetical. In the case of the light bulb manufacturer, the real population of interest is the set of bulbs that the manufacturer will produce in the future. At the time any sample is taken, this population is not observable.
6. The population may be empirically definable, but not practically available to a researcher. Not all slopes in a study region may be accessible to a geomorphologist interested in studying the dynamics of freeze-thaw weathering. Even an experienced climber may find only a few suitable sites for study.
7. Information from a population census can be quickly outdated. Given the volatility of political polls, it would be unwise to determine the party supported by each member of a population. *Repeated* censuses of this type would be impossible, and sufficient accuracy can be obtained by using only a small proportion of voters. Repeated polls of this type are a usual occurrence now.
8. When the topic of the study requires an in-depth study of individuals in the population, only a small sample may be possible. By restricting attention to only a

few individuals, extremely comprehensive information can be collected. A study of the residential mobility of people living in a large city might require detailed questions concerning the history of moves, characteristics of the current and past residences, motivations for each move, the search process used to locate new residences, and characteristics of the household itself. Probing for sufficient detail in all these areas precludes the possibility of a complete census of the population. Such a task would certainly be beyond the resources of most institutions. For a variety of reasons, then, many research questions must be answered through the use of a small sample from a population. Providing the sample is collected properly, valid conclusions about key characteristics of the population can be drawn, with only a surprisingly small degree of uncertainty.

6.2. Steps in the Sampling Process

Having decided that no suitable data exist to answer a research question, and concluding that a sample is the only feasible method of collecting the necessary data, the researcher must specify a sampling plan. Rushing out to collect the data as quickly as possible is often the worst thing the researcher could do. It is far better to follow the simple five-step sampling procedure illustrated in Figure 6-1. Many potential problems not anticipated by the researcher can be addressed and successfully solved before a considerable effort has been put into the actual task of data collection. In literally hundreds, if not thousands, of studies, insufficient time and care in devising a sampling plan have led to the collection of large datasets with only limited possibilities for statistical inference. Let us consider each of these five steps in turn.

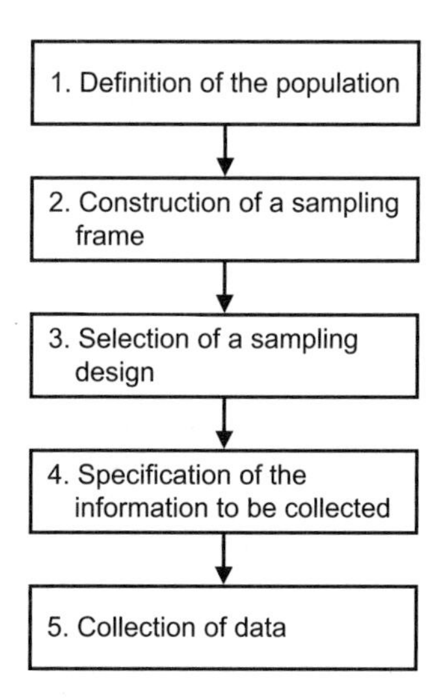

FIGURE 6-1. Steps in the sampling process.

Definition of the Population

The first step is to define the population. What at first glance might appear to be a trivial task often proves to be an extremely difficult chore. It is easy to conceive of a statistical population as a collection of individual elements that may be individual people, objects, or even locations. However, to actually identify which individuals should be included in the population and which should be excluded is not so simple. To see some of the potential issues and difficulties, consider the problem of defining the population for a study of the elderly in a city. A number of practical questions immediately surface:

1. How will we distinguish the elderly from the nonelderly? by age? If so, what age? age 60? age 65?
2. Or will the elderly be defined by occupational categories? Should we restrict ourselves to retired persons? Or should we restrict the population to individuals over 65 years of age *and* retired ?
3. Will we include all elderly, or those living independently, that is, not in a long-term care home?
4. Is the study concerned with elderly individuals or households of the elderly? What about mixed households with both elderly and nonelderly members?

As you can see, even if we can conceptualize the population of interest, arriving at a strict, operationally useful definition may require considerable thought and difficult choices. In the study of the elderly, it is still necessary to define a time frame and a geographical limit to the study region.

Construction of a Sampling Frame

Once we have chosen the specific definition to be used in identifying the individuals of a population, it is necessary to construct a *sampling frame.*

DEFINITION: SAMPLING FRAME

A sampling frame (also called a population frame) is an ordered list of the individuals in a population.

The sampling frame has two key properties. First, it must include all individuals in the population; that is, it must be exhaustive. Second, each individual element of the population must appear once and only once on the list. Obtaining a sampling frame for a particular population may itself be a time-consuming task. It is usually easy to compile a list of all current students at a university in a given academic year from existing academic records or transcripts. But what if the population of interest is not regularly monitored in any way? Where, for example, could a list of all the elderly residents of a city be obtained? It may be possible to extract a fairly complete list of the elderly by examining the list of recipients of Social Security or old-age assistance from a government agency. But would the list include all the elderly? What about noncitizens or residents otherwise ineligible for this type of aid?

As a second example, consider the use of telephone surveys for evaluation of voter preferences for political parties. Although the population of interest is all eligible voters, the population actually sampled is composed of those residents with telephones—or, more accurately, the set of individuals who answer these phones. Clearly, these two groups overlap a great deal, but they are not exactly the same. Restricting ourselves to those with listed telephone numbers may exclude some relatively wealthy residents with unlisted numbers as well as some poorer households without telephones. It is now becoming increasingly common for some households to have only cell phones, and so they would never appear in a conventional telephone book or listing. It is therefore useful to distinguish the *target* population from the *sampled* population.

DEFINITION: TARGET AND SAMPLED POPULATIONS

The target population is the set of all individuals relevant to a particular study. The sampled population consists of all the individuals listed in the sampling frame.

Obviously, it is desirable to have the sampled and target populations as nearly identical as possible. When they do differ, it is extremely important to know the particular way(s) that they vary, since this is a form of sampling bias. It is sometimes necessary to qualify the inferences made by using a sampled population when it differs in significant ways from the target population. This is equivalent to recognizing the limitations imposed on the study by the sampling frame.

Selection of a Sampling Design

Next we must decide how we are going to select individuals from the sampling frame to include in the sample.

DEFINITION: SAMPLE DESIGN

A sample design is a procedure used to select individuals from the sampling frame for the sample.

There are several ways to select a sample. We could simply take the first n individuals listed in the sample frame. Or we could select the last n individuals or every kth individual on the list until we get n members for the sample. There are many types of samples and sample designs. Because of the importance of this step, it is described in depth in Sections 6.3 and 6.4. At this point it is sufficient to note that a *random* sample is an extremely useful design in statistical analysis. Individuals to be included in the sample are chosen by using some procedure incorporating chance. The mechanical devices used in many lotteries are one example. An urn is filled with identical balls, one for each member of the sampled population. The sample is chosen by selecting balls from a well-mixed urn, one at a time.

The important characteristic of this type of sample is that we know the probability of each individual in the population being included in the sample. In this case, each individual has an equal chance of being included. A number of variations of this design are explained in Section 6.4.

Specification of the Information to Be Collected

This step can usually be accomplished at any point prior to beginning data collection. The particular format used to collect data must be rigorously defined and pretested by using a pilot sample.

DEFINITION: PILOT SAMPLE, OR PRETEST

A pilot sample, or pretest, is an extended test of data collection procedures to be used in a study in advance of the main data collection effort.

In a field study, the pretest can be used to check instruments, data loggers, and all other logistics. For surveys—mail, telephone, or personal interview—the pretest can sometimes reveal deficiencies for any of the following reasons: difficulty in locating population members; dealing with an abnormally high percentage of refusals or incomplete questionnaires; problems in questionnaire wording, question sequence, or format; unanticipated responses; or inadequately trained interviewers.

Collection of Data

Once all the problems indicated in the pretest have been successfully solved, the task of data collection can begin. At this stage, careful tabulation and editing are particularly important if we wish to minimize *nonsampling* error.

6.3. Types of Samples

In this section, we expand on the ideas discussed in the third step of the sampling process, the selection of a sampling design. Sampling designs can be conveniently divided into two classes: probability samples and nonprobability samples. Simple random sampling is one type of probability sample.

DEFINITION: PROBABILITY SAMPLE

A probability sample is one in which the probability of any individual member of the population being picked for the sample can be determined.

Because we know only the probability that an individual is included in the sample, an element of chance, or uncertainty, is introduced into any inferences made from the sample. Expressed simply, it could happen that a particular sample is quite unrepresentative of the population it is supposed to reflect. The advantage of a probability sample is not that there is no uncertainty in the results, but rather that we can assign

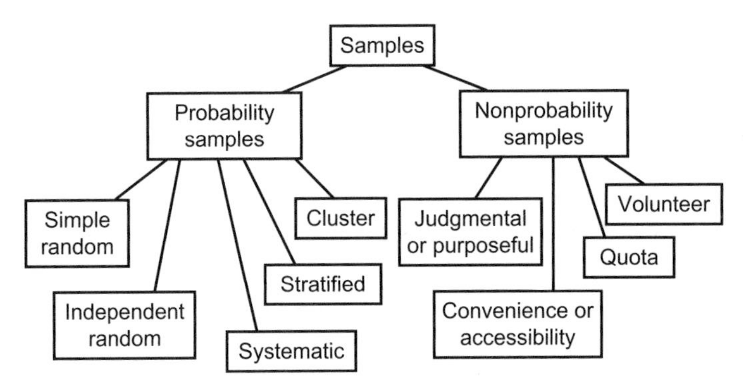

FIGURE 6-2. Types of samples.

a quantitative uncertainty value to the results. The five principal types of probability samples are shown in Figure 6-2. Because of their importance in statistical inference, these samples are discussed in depth in the following section.

Nonprobability samples may also be excellent or poor representations of the population. The difficulty is that whether it is a good or bad sample can never be determined. Four types of nonprobability samples are sometimes used to collect sample data: *judgmental, convenience, quota,* and *volunteer.*

DEFINITION: JUDGMENTAL, OR PURPOSEFUL, SAMPLE

This type of sample, also known as a purposeful sample, is one in which personal judgment is used to decide which individuals of a population are to be included in the sample. These are individuals whom the investigator feels can best serve the purpose of the sample.

Obviously, a very skillful investigator with considerable knowledge of a population can sometimes generate extremely useful data from such a sample. If the deliberate choices turn out to be poor, however, poor inferences are made. The risk is not known. Purposeful samples are sometimes selected in pretest, or pilot, samples. A range of respondents, including both "typical" and "unusual" individuals, is chosen. This sample is used to get an idea of what the full range of questions in the survey should be. In addition, it is not unusual to use these interviews to preview the types of answers respondents are likely to give in the actual survey. Many times, this information can be used to significantly improve the survey instrument. It would be completely erroneous to utilize such a sample to draw conclusions about the whole population. Because they include only easily identifiable members of a population, convenience or accessibility samples are subject to sampling bias. These individuals are almost always special or different from other population members in some way. Street corner interviews rarely reflect overall opinion; only certain individuals will allow themselves to be interviewed by the media.

DEFINITION: CONVENIENCE, OR ACCESSIBILITY, SAMPLE
A convenience, or accessibility, sample is one in which only convenient, or accessible, members of the population are selected.

There are two other types of nonprobability samples. Before probability samples were widely accepted in empirical surveys, a pseudoscientific approach known as quota sampling was common.

DEFINITION: QUOTA SAMPLING
Quota sampling is an attempt to obtain a representative sample by instructing interviewers to acquire data from given subgroups of the population.

The basic idea behind quota sampling is sound. Suppose, for example, that we were attempting a voters' poll and knew that voter preference varied by whether the respondent was male or female, resided in an urban or rural area, and was young or old. If we knew the breakdowns of the population among these categories, we would instruct one of our interviewers to obtain a specific number of young male voters in some given rural area. Similar instructions could be given to other interviewers for other categories. Some judgment is involved since the interviewer must select the actual respondents in that category. A more sophisticated form of quota sampling known as *stratified random sampling* is discussed in the following section.

Another type of sample that fails to take adequate safeguards against sampling bias is the volunteer sample.

DEFINITION: VOLUNTEER SAMPLE
A volunteer sample consists of individuals who self-select from the population.

Such samples are rarely representative. Usually, these individuals are more motivated and have a higher interest in the topic than the general population. Since the respondents to mail questionnaires sometimes have these characteristics, this type of survey is often subject to nonresponse bias.

The sampled population units in a probability sample are always chosen according to some probability plan. This minimizes or even eliminates the sampling bias. Although the resulting sample cannot be guaranteed to be representative of the population, we can determine the probable accuracy of our results. For these reasons, a probability sample is the preferred sampling method.

6.4. Random Sampling and Related Probability Designs

In most social science applications and many physical science applications, the population being sampled is finite. For example, the number of people in a target population is always limited, and the number of trees in a forested area—though it may be very large—is certainly less than infinity. As we will see, the size of the sample relative to the population size has implications for the way samples are constructed. We begin with a basic probability sample, the so-called simple random sample.

DEFINITION: SIMPLE RANDOM SAMPLE

A simple random sample is one in which each possible sample of a given size n has an equal probability of being selected.

To illustrate this definition, consider a population consisting of the elements A, B, C, and D. Suppose we draw a sample of size $n = 2$ from this population by selecting two individuals at a time from the four members of the population. First, how many samples of size $n = 2$ are there? As illustrated in Figure 6-3, there are $C_2^4 = 6$ samples of size $n = 2$. To meet the definition, each must have an equal chance of being selected; that is, the probability of obtaining any of the six possible samples must be 1/6.

How can we ensure that each sample is equally likely? In this example, we could label the samples from 1 to 6 as in Figure 6-3 and "randomly" choose a number from 1 to 6. Obviously, this calls for some way to generate a random number. One possibility is to put six identical balls numbered from 1 to 6 in an urn, mix well, and draw one. If the number 3 resulted, we would use the third sample, consisting of the *AD* pair. This is often done mechanically in lottery draws, although the number of balls is of course much larger. For populations and samples of even modest size, it is not feasible to develop a list of all possible samples as in Figure 6-3. For a population of size $N = 10{,}000$ and a sample of $n = 100$, there are $C_{100}^{10{,}000}$ samples, which is a huge number to list. Instead of attempting that, we could randomly select one observation at a time from the sampling frame and continue until the sample is filled out. To illustrate this procedure, consider the problem of selecting a sample of size $n = 10$ from a population with $N = 87$ members. The procedure described for this case can be easily generalized to larger samples from larger populations.

To perform this procedure, we number each individual in the sampling frame consecutively from 1 to 87. Using some type of randomization device, we select the first sample member from the population. We could employ numbered balls as above, but this time there would be 87 balls, each of which refers to an individual in the

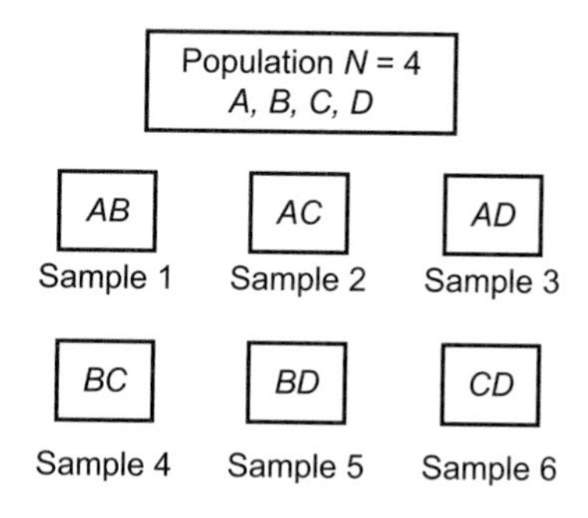

FIGURE 6-3. Samples of size $n = 2$.

population. A total of 10 balls are removed, which together constitute the simple random sample.

Notice that both of the methods we have described require a way to produce random numbers. These days, mathematical rather than physical methods are most commonly used for this purpose.

DEFINITION: MATHEMATICAL RANDOM NUMBER GENERATOR

A mathematical random number generator is an equation or mathematical procedure that produces sequences of random digits.

Mathematical random number generators are commonly encountered in everyday life. They are available on hand calculators, embedded in many video games, and can be found as user functions in most statistical programs and computer spreadsheets, to name just a few examples. From now on, we will refer to them simply as "random number generators." Before describing how a random number generator works, we need to explain what the term *random* means in this context. In other words, what properties do we require "random" numbers to possess? Though simple to phrase, this is a surprisingly difficult question to answer; we will cover just two issues, uniformity and independence. Let us suppose that we have a population of size 2,000, so that we require 4-digit random numbers. A successful random number generator will provide numbers that are:

1. *Uniform.* We want all individuals to have an equal chance of being included in the sample; thus no value produced by a random number generator should be any more likely than any other. This means that the probability distribution of the numbers produced must be uniform. Equivalently, all ten digits from 0 to 9 must have an equal probability of occurring in each of the four positions.
2. *Independent.* As numbers are generated, we want successive numbers to be statistically independent of one another. In other words, there should be no possibility of predicting the next number before it appears. Note that this is not the same as uniform. A generator that produces the integers 1 to 1,000 in numerical order is giving uniform numbers, but not independent numbers. Knowing that the last number was 534, we can predict the next number will be 535 with certainty. Imagine the consequences if winning lottery numbers were chosen in this fashion! Thinking more about independence, it seems clear that independence in the individual digits of the number is required. If the hundreds digit cannot be predicted knowing any of the other digits, then the sequence of 4-digit numbers will be independent.

Thus, if we can generate a long sequence of random single digits, then we can form random numbers of any length. What works for 4-digit numbers will work for 10-digit numbers, and so forth. We will just pull out 4-digit or 10-digit clumps as needed. Can we think of any examples of long random numbers? The digits of π are one such example. There is no pattern (yet discovered) to those digits. One can start at any point in the sequence of digits and will find a random number. Knowing what came before provides no clue about what follows, and no digit appears more often than any other. If we wanted to, we could use π as a source of random numbers. Notice that, though random, the method is deterministic in the sense that it always produces exactly the same sequence from the same starting point. As a result, such generators are more properly called *pseudorandom* number generators.

The most popular mathematical generator works somewhat like our imaginary π-based method. It is called the *linear-congruential method,* and it works by dividing one number by another and using the *remainder* as the random number. If it doesn't sound like this approach could produce a random number, recall that π is the ratio of a circle's circumference to its diameter. The digits in the remainder, the fractional part of π are random. The general procedure is

1. Choose a starting, or seed number X_0.
2. Form a new number $Y = aX_0 + b$, where a and b are constants.
3. Divide Y by another constant c, and let the remainder be the first random number X_1.
4. Return to step 2 above, this time using X_1 to form Y.

Of course, the constants a, b, and c must be carefully selected—dividing 4 by 2 does not give a random remainder—but otherwise the method is just this simple. Notice that the sequence is determined completely by the starting value, so the method gives pseudorandom numbers. Table 6-1 shows a sequence of numbers generated by this technique (similar tables are often printed as appendices in statistics texts).

As an example of how to use such a table, suppose that we want to draw a simple random sample of size 87 from a population consisting of $N = 100$ members. We need to associate one random number with each member of the population listed in the sampling frame. For any population with up to $N = 100$ members, a 2-digit random number will suffice. The random number pair 01 will refer to the first individual listed in the sampling frame, the pair 02 refers to the second individual, and so on. The random number pair 00 is assigned to the last, or 100th, individual of the population. Any 2-digit random number drawn thus leads us to a particular individual in

TABLE 6-1
A Block of 100 Random 2-Digit Numbers

61	56	24	90	10	28	59	01	45	47
28	61	34	93	93	02	83	73	99	29
42	40	63	36	82	63	15	01	28	36
68	02	63	83	41	24	30	10	18	50
41	05	14	37	40	64	10	37	57	18
04	59	34	82	76	56	80	00	53	36
69	16	43	43	46	04	35	64	32	05
74	90	09	63	92	36	59	16	05	88
25	29	81	66	99	22	21	86	79	64
43	12	55	80	46	31	98	69	22	22

the population. For a population with up to $N = 1{,}000$ members, a 3-digit random number can be associated with each individual in the sampling frame.

To select the sample, choose a place in the random number table to begin, for example, by looking away from the page and placing a finger somewhere in the table. Suppose this technique leads to the entry in the fourth row and fourth column, 83. The 83rd individual in the sampling frame is the first element of the chosen sample. Proceeding down the column (moving across columns or even diagonals are also possible), the following individuals will be selected for the first seven members of the sample:

Sample member	Population member identifier
1	83
2	37
3	82
4	43
5	63
6	66
7	80

Then, moving to the next column of the random number table, we see the eighth member of the sample is identified as individual 10 on the sampling frame. The next random number of the list is 93. Since this number is larger than 87, it is discarded and sampling is continued. The last two individuals are then selected by using the random numbers 82 and 41. This simple procedure can be followed in all cases where random sampling from a finite population is required. There is no fundamental difference whether the numbers come from a table or are generated directly in a computer program. In both cases we obtain an identifying number, uniquely tied to an individual in the population.

We see that simple random samples are not terribly difficult to obtain, providing we have a suitable source of random numbers. Moreover, a simple random sample sounds appealing because it does not privilege any particular sample over another—by definition all are equally likely. However, if the sample size is large relative to the population, the outcomes within the sample will not be independent of one another. For example, from Figure 6-3, we see that if that if the sample contains an A, we can be certain it contains a B, a C, or a D as the other member. This lack of independence greatly complicates any ensuing statistical analysis; thus a more restrictive variation known as an *independent random sample* is greatly preferred.

To appreciate what this means, imagine that we draw repeated samples from the same population. Obviously, the individual appearing in the first position will vary from sample to sample, as will the second, and so forth. We see that there is a probability distribution associated with position one, two, and so on. In other words, X_1, X_2, and so on, are random variables. An independent random sample requires that these random variables have the same distribution and that they also be independent. Independence in sampling means that the value of X_1 provides no information about the value of X_2 or any other observation. These considerations are formalized in the following definition:

DEFINITION: INDEPENDENT RANDOM SAMPLE

An independent random sample is one whose n sample probability distributions are identical, independent of one another, and equal to the distribution of the underlying random variable X. This means that the distribution for X_1, X_2, X_3 and so on are the same as that of X. That is, $f(X_1) = f(X_2) = f(X_3) = \ldots = f(X_n) = f(X)$. In addition, the value of X_2 cannot be predicted knowing the value of X_1, and similarly for all other pairs of sample members. Or equivalently, using the concept of a joint probability distribution developed in the previous chapter, $f(X_1, X_2, X_3, \ldots X_n) = f(X_1)\,f(X_2)\,f(X_3) \ldots f(X_n)$.

This is a great advantage because it means that when we operate on observations in an independent random sample, we will be operating on independent and identically distributed variables. As we have seen in Chapter 5, independence greatly simplifies probability theory. Many important theorems in statistics will therefore call for an independent random sample, not because they are so easy to obtain, but because the theory is so much more tractable.

Does a simple random sampling design guarantee an independent random sample? This is sometimes the case, but not always. Notice that the simple random procedure used with Figure 6-3 failed the independence test because once a member appeared in the sample (*A*, *B*, etc.), there was no chance it could be chosen again. In other words, we sampled from the population without replacement, and the effect was to violate independence. As a practical matter, if the population is reasonably large, there is little difference between sampling with replacement or without. That is, for a large population, there would be almost no chance of selecting the same individual twice, and therefore it does not matter if a sampled individual is eligible for resampling. Notice that "large" is relative; in this context, a population of 10,000 would be considered small if a sample of 5,000 were drawn. The point is that whenever the population is much larger than the sample size, we do not need to be concerned about sampling with replacement. The simple random sampling procedure will give samples that are effectively independent random samples. Because in practice populations are usually much larger than samples, in the following sections we will assume that simple random sampling is sufficient.

Systematic Sampling

The *systematic random sample* is a variation on simple random sampling. This method is generally easier to operationalize than simple random sampling in terms of both time and cost.

DEFINITION: SYSTEMATIC RANDOM SAMPLE

To draw a systematic random sample, every *k*th element of the sampling frame is chosen, beginning with a randomly chosen point.

Consider a population with $N = 200$ elements from which we wish to draw a random sample of size $n = 10$. This sample can be drawn by selecting every kth $= N/n = 200/10 = 20$th individual listed on the sampling frame. The next thing to be decided is *where* on the sampling frame selection is to begin. The easiest procedure is to draw a single random number between 0l and 20. The possible samples include the following:

01	21	41	...	181
02	22	42	...	182
03	23	43	...	183
...	...	...	...	...
19	39	59	...	199
20	40	60	...	200

Why is a systematic random sample *not* a simple random sample? Notice there are not C_{10}^{200} different samples that could be selected, but only 20. For example, it is impossible to collect a sample that includes individuals 2, 3, and 27. Only a small fraction of the potential samples of size n may be drawn.

The systematic sample is likely to be as good as a random sample, provided that the arrangement of the individuals in the sampling frame is random. Where there are regularities or periodicities in the sampling frame, these will be picked up in the systematic sample. To see the potential problems this may create, consider a sampling frame that includes all the apartments in some high-rise block. Suppose there are 10 floors with 20 apartment units per floor. As is usual in such cases, apartment units above and below any given apartment are identical in floor plan, size, and other features. A systematic sample of size $n = 10$ would probably include 10 identical units. Furthermore, we might expect them to be occupied by roughly similar-sized households. Our systematic sample might include 10 households, each living in a two-bedroom unit. This would hardly be a representative sample if the building contained, one-, two-, and three-bedroom units. As we shall see in Section 6.6, this potential for bias also exists in spatial sampling. Before a systematic sample is drawn, it is always necessary to check for randomness in the sampling frame itself. If the list has been collected in an orderly manner, then one must be sure that such an ordering will not lead to a biased sample through systematic sampling. Many obvious periodicities exist in data ordered in time sequence or by regular spatial intervals.

Stratified Sampling

One drawback of random sampling is the ever-present possibility that individuals selected for a sample may be unrepresentative of the population. Unlike with random sampling, stratified random sampling is not based entirely on chance.

DEFINITION: STRATIFIED RANDOM SAMPLE

A stratified random sample is obtained by forming classes, or strata, in the population and then selecting a simple random sample from each.

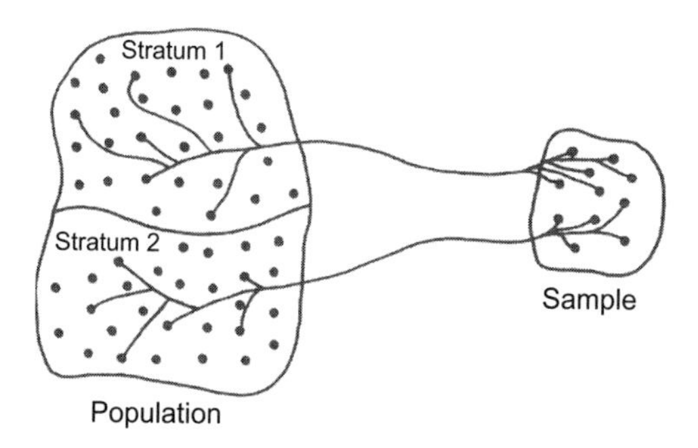

FIGURE 6-4. Stratified random sampling from a population with two strata.

The basic idea behind stratified sampling is illustrated in Figure 6-4. The population is divided into a set of strata before sampling takes place. The divisions must be both exhaustive and mutually exclusive. That is, each member of the population must appear in one and only one class. Simple random samples are drawn independent of class. Population units may be stratified on the basis of one or more characteristics. Social surveys often stratify the population by age, income, and location of residence.

Why do we stratify? If it is properly organized, stratified sampling can use the additional control over the sampling process to reduce sampling error. Put simply, stratified sampling can decrease the likelihood of obtaining an unrepresentative sample. To reduce sampling error, we define homogeneous strata, that is, strata consisting of individuals who are very much alike in terms of the principal characteristic of the study. Suppose we are interested in the trip-making behavior of the households in a metropolitan area. Knowing that households with two or more cars make more total trips, make different types of trips, and are less likely to use public transport means that they are less variable than the population as a whole. By defining subpopulations or substrata with less internal variability, we effectively reduce sampling error.

A stratified sample is preferred whenever two conditions are met: First, we must know something about the relationship between certain characteristics of the population and the problem being analyzed. Second, we must actually be able to achieve stratification, that is, to identify population members in each class. Sometimes we may not be able to develop a sampling frame for this purpose. Where, for example, could we obtain a list of households according to the number of cars owned? Various *post hoc* procedures of classifying households after they are sampled are usually less than satisfactory. Because of the advantages of stratification, it should be undertaken whenever possible. Virtually all political polls employ stratified sampling techniques, particularly by geographic area.

Cluster Sampling

In cluster sampling, illustrated in Figure 6-5, the population is first divided into mutually exclusive and exhaustive classes, as in the initial step in stratified sampling.

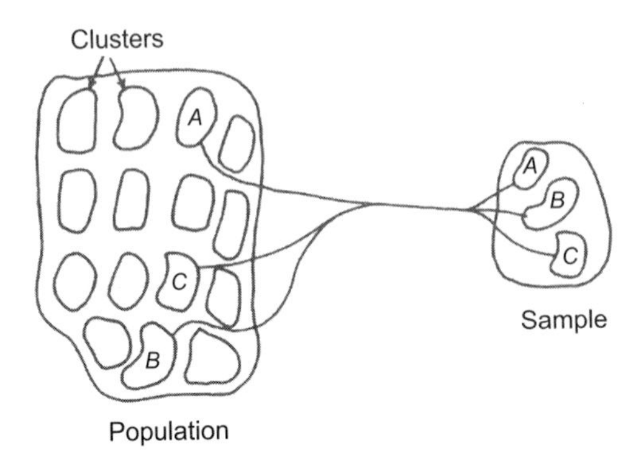

FIGURE 6-5. Cluster sampling.

These classes are usually defined on the basis of convenience rather than some variable thought to be important to the problem under study. Next, certain clusters are selected for detailed study, usually by some random procedure. However, a random sample is *not* drawn from *within* each cluster. Cluster sampling is completed either by taking a complete census of those clusters selected for sampling or by selecting a random sample from these clusters. These units are then combined and constitute the complete cluster sample. Ideally, each cluster should be internally heterogeneous. This is in direct contrast to the situation in stratified sampling where the strata are defined to be as internally homogeneous as possible.

Cluster sampling can often give very poor results. The sampling error from a cluster sample is usually higher than the sampling error from a simple random sample or a stratified sample. A cluster sample is effective only when each class or cluster defined for the problem is representative of the population as a whole. If these clusters are not representative, then the cluster sample is likely to be biased and create considerable sampling error. Randomly or arbitrarily defined clusters are unlikely to be representative. Obviously, great care should be taken in specifying the clusters into which any population is to be divided.

Why are cluster samples taken? Usually, the reason is efficiency in time and cost. These efficiencies are particularly important whenever the population to be sampled is geographically dispersed over a large area. Suppose, for example, that the problem under study dictates that a personal interview be taken from a sample of such a population. If we were to select a random or systematic sample, this would require a considerable amount of traveling and hence would be a very costly collection procedure. But by restricting our sampling to a small number of clusters, each covering only a small area, these costs are significantly reduced.

Choice of Sampling Design

In terms of sampling efficiency, stratified samples are best and cluster samples are worst. Random samples usually lie somewhere between these two extremes. That is, for a given level of precision, stratified samples require the fewest respondents and cluster samples require the most respondents. Unfortunately, sampling efficiency alone does not dictate the type of design chosen in any study. The selection of a stratified design presumes that the relevant information necessary to define the strata and construct the sampling frame is available. Also, the time and cost of collecting the data must be considered. If the sample requires a survey of some human population, then another concern is whether the information is to be collected through personal interview, telephone interview, or mail questionnaire.

Finally, many sampling designs used in practice are hybrids that combine elements of several different designs. A sample of housing in a metropolitan area might use stratification to ensure adequate representation of African American, Asian, and other ethnic groups. Within these strata, cluster sampling might be used to select particular neighborhoods and/or blocks within these neighborhoods. A simple random or systematic random sample might be used to select the individual households for the sample.

6.5. Sampling Distributions

In Chapter 3, we introduced several numerical measures of datasets, such as the mean, median, standard deviation, and variance. We also made a distinction between numerical measures calculated for a dataset that could be considered a *population* and measures for a data set that is a *sample.* Let us now clarify the distinction between such numerical measures. If the measure is computed from a population dataset, it is termed the *value of the population parameter;* if it is calculated from a sample dataset, it is called the *value of the sample statistic.* For a finite population, the *population mean parameter* and the *population standard deviation parameter* are defined in the following way.

DEFINITION: POPULATION MEAN AND STANDARD DEVIATION PARAMETERS (FINITE POPULATION)

Let $x_1, x_2, \ldots x_N$ be the values of some population characteristic X in a finite population of N elements. The value of the population mean parameter, denoted μ is

$$\mu = \frac{\sum_{i=1}^{N} x_i}{N} \tag{6-1}$$

and the value of the population standard deviation parameter, denoted σ, is

$$\sigma = \sqrt{\frac{\sum_{i=1}^{N} (x_i - \mu)^2}{N}} \tag{6-2}$$

There is only one population mean parameter and one population standard deviation parameter in any population. All population parameters are denoted by Greek letters.

The equivalent numerical measures for a sample, the values of the sample statistics for the mean and standard deviation, are defined as follows.

DEFINITION: SAMPLE AND STANDARD DEVIATION STATISTICS

Let $x_1, x_2, \ldots x_n$ be a set of values of the n elements taken from a population. The value of the sample mean statistic, denoted $\bar{x}$, is defined as

$$\bar{x} = \frac{\sum_{i=1}^{n} x_i}{n} \qquad (6\text{-}3)$$

and the value of the sample standard deviation statistic, s, is

$$s = \sqrt{\frac{\sum_{i=1}^{n} (x_i - \bar{x})^2}{n - 1}} \qquad (6\text{-}4)$$

There are two principal differences between these two formulas. First, sample statistics are calculated by using the n values of a sample, and population parameters are calculated by using the N values of the population, $n < N$. Second, the divisor within the radical of σ is N, but for s it is $n - 1$. However, there is another more important difference. Whereas there is only one value for a population parameter such as μ, there are many different possible values for the equivalent sample statistic $\bar{x}$. Why? Each different sample of size n drawn from a finite population of N elements can have a different value for $\bar{x}$. Also note that *if we choose a random sample from the population, then the value of the sample mean must itself be a random variable.*

In Chapter 5, we distinguished a random variable by an uppercase letter X and a specific value of this random variable by the corresponding lowercase symbol x. This distinction is also useful in the sampling context. When we refer to a random sample of size n drawn from a population *before* a value has been drawn, the sample is designated $X_1, X_2, \ldots X_n$. *After* a single sample has been drawn, the values of these random variables are known. These n numbers constitute the sample $x_1, x_2, \ldots x_n$, which is a set of random values. They are random because the value of any of these variables depends on which member of the population is drawn while sampling. We must emphasize that saying X_1, X_2, and so on, are random is not to say that that all values in a population are equally likely to occur. That would imply X is uniformly distributed, something that will be true only in special circumstances.

Sample Statistics

The sample mean is also a random variable since its value is not known before the sample is actually drawn. Prior to taking the sample, the sample mean is a random variable

$$\bar{X} = \frac{\sum_{i=1}^{n} X_i}{n} = \frac{X_1 + X_2 + \ldots + X_n}{n} \tag{6-5}$$

Formally, we say that $\bar{X}$ is a sample statistic.

DEFINITION: SAMPLE STATISTIC

A sample statistic is a random variable based on the sample random variables $X_1, X_2, \ldots X_n$. The *value* of this random variable can be determined once the values of the observations in a specific random sample have been drawn from the population.

Just as there are many numerical measures of a dataset, so there are many sample statistics. In fact, each measure has an associated sample statistic. Thus, for example, the sample mean statistic $\bar{X}$ given by Equation 6-5 is a random variable, and its value is given by Equation 6-3. The notation that differentiates $\bar{X}$ from $\bar{x}$ is necessary so that we appreciate that the statistic $\bar{X}$ is a random variable whose value will vary from sample to sample for a fixed sample size n. Given that $\bar{X}$ is a random variable, the obvious question is how that random variable is distributed.

Sampling Distributions

DEFINITION: SAMPLING DISTRIBUTION OF A STATISTIC

A sampling distribution is the probability distribution of a sample statistic. The sampling distribution of a statistic can be found by taking all possible samples of size n from a population, calculating the value of the statistic for each sample, and drawing the distribution of these values.

To illustrate this concept, let us generate the sampling distribution of the sample means $\bar{X}_i$ for the following population consisting of five elements:

Element	Values of x
A	$x_1 = 6$
B	$x_2 = 6$
C	$x_3 = 5$
D	$x_4 = 4$
E	$x_5 = 4$

First, let us calculate the population mean and standard deviation, using Equations 6-1 and 6-2:

$$\mu = \frac{\sum_{i=1}^{5} x_i}{5} = \frac{6 + 6 + 5 + 4 + 4}{5} = 5 \tag{6-6}$$

$$\sigma = \sqrt{\frac{\sum_{i=1}^{5}(x_i - \mu)^2}{5}} \tag{6-7}$$

$$= \frac{(6-5)^2 + (6-5)^2 + (5-5)^2 + (4-5)^2 + (4-5)^2}{5} = \sqrt{\frac{4}{5}} = 0.89$$

Notice that the divisor in each case is equal to 5, the number of elements in the population.

Suppose a simple random sample of size $n = 3$ is to be taken from this population. How many samples are possible? There are $C_3^5 = 10$ different samples of size 3. These possibilities are listed in Table 6-2, along with the different possible values of the statistic $\bar{X}$. The sampling distribution of the statistic $\bar{X}$, the distribution of sample means, is illustrated in Figure 6-6.

Computing the mean and standard deviation of this distribution yields

$$\mu_{\bar{X}} = \frac{\sum_{i=1}^{10}\bar{x}_i}{10} = \frac{5.67 + 5.33 + \ldots + 4.33}{10} = 5 \tag{6-8}$$

$$\sigma_{\bar{X}} = \sqrt{\frac{\sum_{i=1}^{10}(\bar{x}_i - \mu_{\bar{x}})^2}{10}} \tag{6-9}$$

$$= \sqrt{\frac{(5.67-5)^2 + (5.33-5)^2 + \ldots + (4.33-5)^2}{10}} = 0.365$$

The notation used to denote the mean of $\bar{X}$ is $\mu_{\bar{X}}$. Note that the mean of the sample means for all samples of a given size, $\mu_{\bar{X}}$, is a different variable than the overall population mean, μ. Also note that the divisor for $\mu_{\bar{X}}$ in Equation 6-8 is 10, the number

TABLE 6-2
Possible Samples of Size n = 3 from a Population N = 5

Elements in sample	Values of X	Sample mean $\bar{x}$
ABC	6, 6, 5	5.67
ABD	6, 6, 4	5.33
ABE	6, 6, 4	5.33
ACD	6, 5, 4	5.00
ACE	6, 5, 4	5.00
ADE	6, 4, 4	4.67
BCD	6, 5, 4	5.00
BCE	6, 5, 4	5.00
BDE	6, 4, 4	4.67
CDE	5, 4, 4	4.33

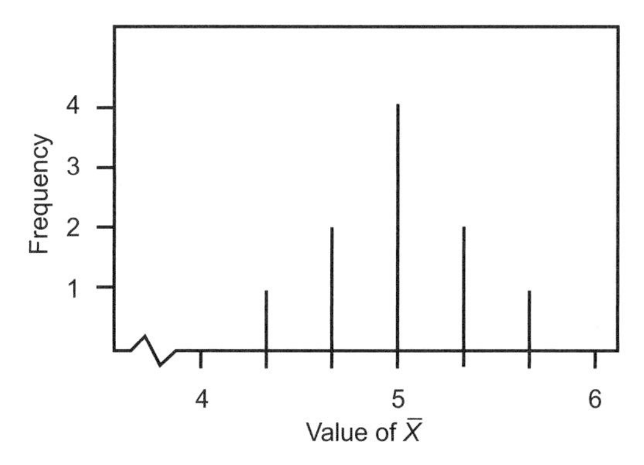

FIGURE 6-6. Sampling distribution of $\bar{X}$.

of samples of size $n = 3$ that can be drawn from a population of size $N = 5$. When we think about how $\bar{X}$ will vary from sample to sample, we would expect it to be distributed about the population mean μ. Indeed, one of the important properties of the sample mean statistic is that its mean is equal to the population mean, or $\mu_{\bar{X}} = \mu$.

To compute the standard deviation of the sample means, the mean of each sample of size 3 (given in the last column of Table 6-2) is subtracted from the mean of the sample means ($\mu_{\bar{X}}$) and squared. The sum is then divided by 10, the number of samples of size 3 that can be drawn from a population with $N = 5$ elements. The fact that $\sigma_{\bar{X}} < \sigma$ should not be surprising. We should expect less variability in the distribution of sample means than in the population itself. Why? Because when we take a sample of items from the population, we tend to select samples in which large values of the variable X are averaged with small values, and the result lies close to the population mean. To distinguish σ from $\sigma_{\bar{X}}$, it is common to call σ the population standard deviation and $\sigma_{\bar{X}}$ the *standard error of the mean.* By convention, the standard deviation of a sampling distribution is known as its standard error.

Inferential statistics would be a time-consuming process if it were necessary to manually generate sampling distributions in this way. Fortunately, this is not the case. A set of theorems can be used to specify the nature of the sampling distributions of many important sample statistics under quite general conditions. We illustrate the principal theorem involved, using as an example the sampling distribution of the sample mean $\bar{X}$.

Central Limit Theorem

DEFINITION: CENTRAL LIMIT THEOREM

Let $X_1, X_2, \ldots, X_n$ be a random sample of size n drawn from a population with mean μ and standard deviation σ. Then for large *n,* the sampling distribution of $\bar{X}$ is approximately normally distributed with mean μ and standard deviation $\sigma/\sqrt{n}$. In the special case where X is normal, the distribution of $\bar{X}$ is exactly normal regardless of sample size.

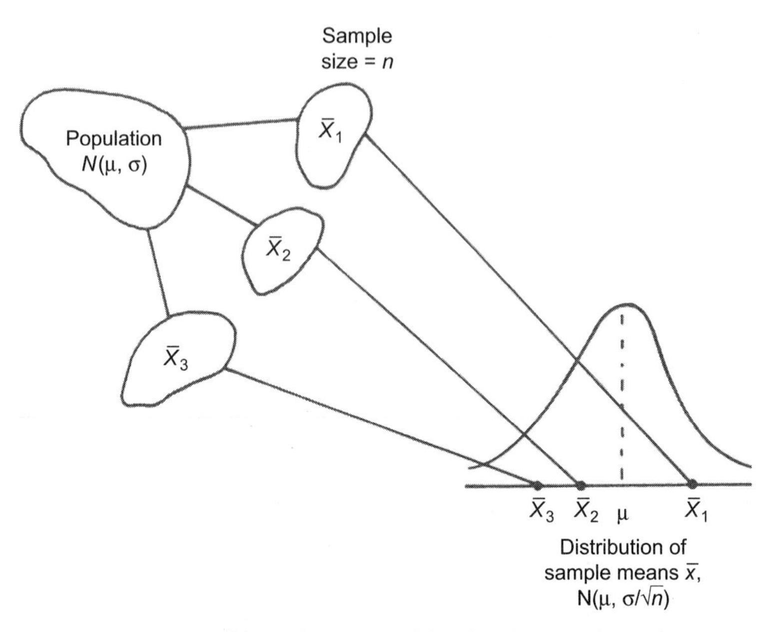

FIGURE 6-7. Central limit theorem and the distribution of sample means.

This is an extremely powerful theorem. It says that no matter what the distribution of X might be, we know that the distribution of $\bar{X}$ is approximately normal. Moreover, as we will see later, the approximation is very good for even moderately sized samples: $\bar{X}$ will be nearly normal for n as small as 10 or 20. The implications of the central limit theorem are illustrated in Figure 6-7 for a normal random variable.

We need to make several important observations from this theorem. The most important observation is the inverse relation between sample size and the variability of the sample statistic $\bar{X}$. As the sample size n increases, the variability of $\bar{X}$ decreases since $\sigma_{\bar{X}} = \sigma/\sqrt{n}$ falls. Suppose the population standard deviation is $\sigma = 100$. Then the following relations always hold:

n	1	4	16	25	100
$\sigma_{\bar{X}}$	100	50	25	20	10

This is intuitively appealing. If we were to randomly select one element of the population, we would expect our choice to exactly mirror the underlying population. Larger sample sizes give smaller standard errors. Notice that as n approaches infinity, the standard error approaches zero. What does this imply? It tells us that as n increases, the sampling distribution of $\bar{X}$ converges to μ, the value of the population mean. This is also intuitively obvious. We would expect the value of $\bar{X}$ to be closer to μ as the sample size becomes larger. This feature is illustrated in Figure 6-8.

It is necessary to distinguish between situations where the random variable X is normally distributed and where it is not. Where X is not normally distributed, the

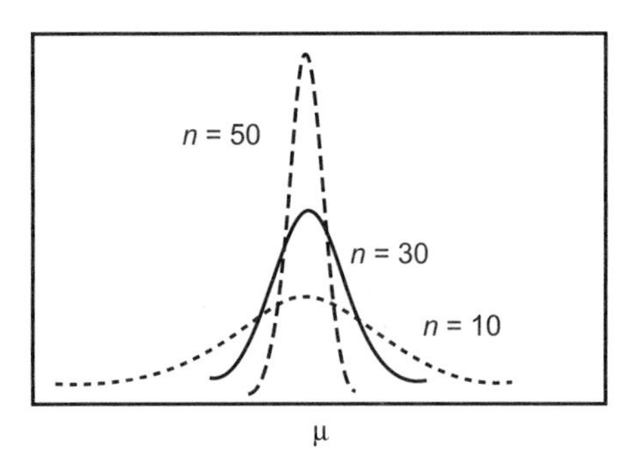

FIGURE 6-8. Sampling distributions of $\bar{X}$ with increasing *n*.

theorem states that $\bar{X}$ is only approximately normal. As we have hinted, the approximation improves very rapidly with increasing sample size. This feature can be illustrated by a simple example. Suppose the population characteristic of interest is the time between the arrival of recreation parties at a wilderness park. Parties are delivered by buses, which arrive at various intervals, but never more than two hours apart. We assume that these intervals are uniformly distributed. Because the interval cannot be negative, the random variable runs from 0 to 2, with a mean of 1 hour.

Using a pseudorandom number generator, hypothetical samples of X were drawn by computer. One thousand random samples were drawn for sample sizes $n =$ 1, 2, 4, 16, and 32. The mean for each sample was computed and accumulated in the frequency distributions plotted in Figure 6-9. Note that for a samples of size $n = 1$, the mean is just the single value contained in the sample. Figure 6-9a shows the sample values more or less uniformly distributed, as would be expected given that X is uniform. For $n = 1$, the sample mean $\bar{X}$ is far from normally distributed, as it cannot be any different than the distribution of X. However, as the sample size increases, the central limit theorem suggests that the frequency distribution of $\bar{X}$ should become progressively closer to normal, no matter what is the distribution of X. We do, in fact, see this behavior in panels (b)–(f). Notice how each of the histograms is centered on the population mean μ. We explained this earlier in terms of small values canceling large values.

Notice also that although X is uniform with no value preferred above any other, $\bar{X}$ is much more likely to be near μ than far away. A little thought reveals why this is so. Think first about what is needed to obtain a large value for $\bar{X}$. Obviously, the only way this can happen is for the sample to have many large values of X. Similarly, the only way to obtain a small $\bar{X}$ is for the sample to contain mainly small values. Both of these events are relatively unlikely. It will be much more common to obtain samples that contain a mixture of high and low values. Clearly, a *mixed* sample will produce a sample mean of *intermediate value.* Because intermediate values of $\bar{X}$ are more common, the distribution of $\bar{X}$ is clustered around its mean. For example, with an n of only 16, none of the 1,000 sample means is outside the interval 0.5 to 1.5 (Figure 6-9e).

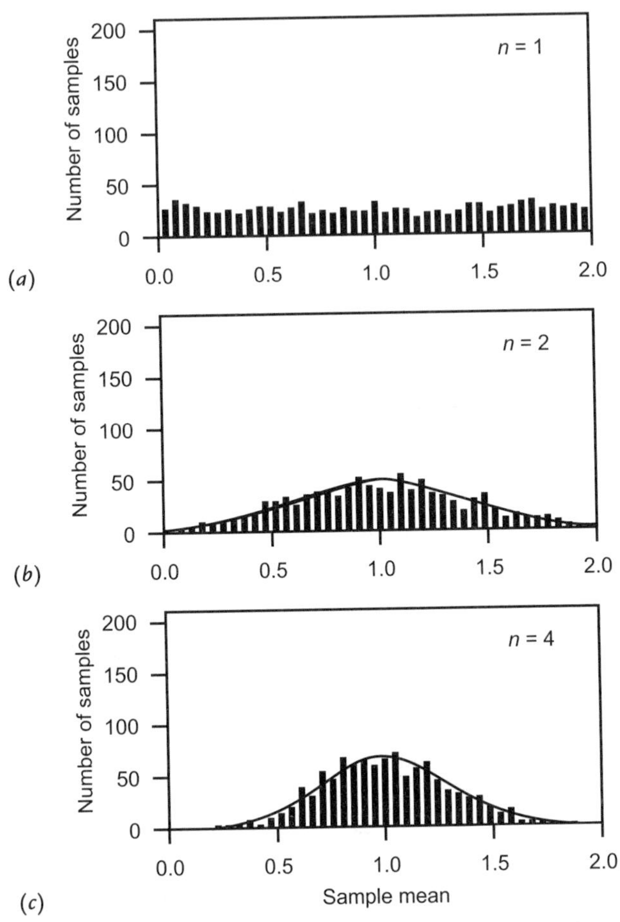

FIGURE 6-9. Distribution of sample means from a uniform distribution. Solid curve is the normal distribution given by the central limit theorem.

We see, therefore, that the sampling distribution of $\bar{X}$ is clustered rather than uniform. Why does the degree of clustering increase with *n?* To answer this question, imagine that one of our samples has a mean of zero. This can happen, but only if all the sample values are zero. This might occur for *n* equal to 4 but would extremely unlikely for *n* equal to 32. In other words, extreme values of $\bar{X}$ become less and less likely as *n* increases. Figure 6-9 shows this behavior clearly for uniform *X,* but we would see the same result regardless of how *X* is distributed. The closer *X* is to normal, the closer to normal will be the distribution of $\bar{X}$. In the extreme case where *X* is normally distributed, $\bar{X}$ will be exactly normal.

As defined earlier, the central limit theorem applies to both continuous and discrete random variables. We might wonder how a discrete random variable can possibly

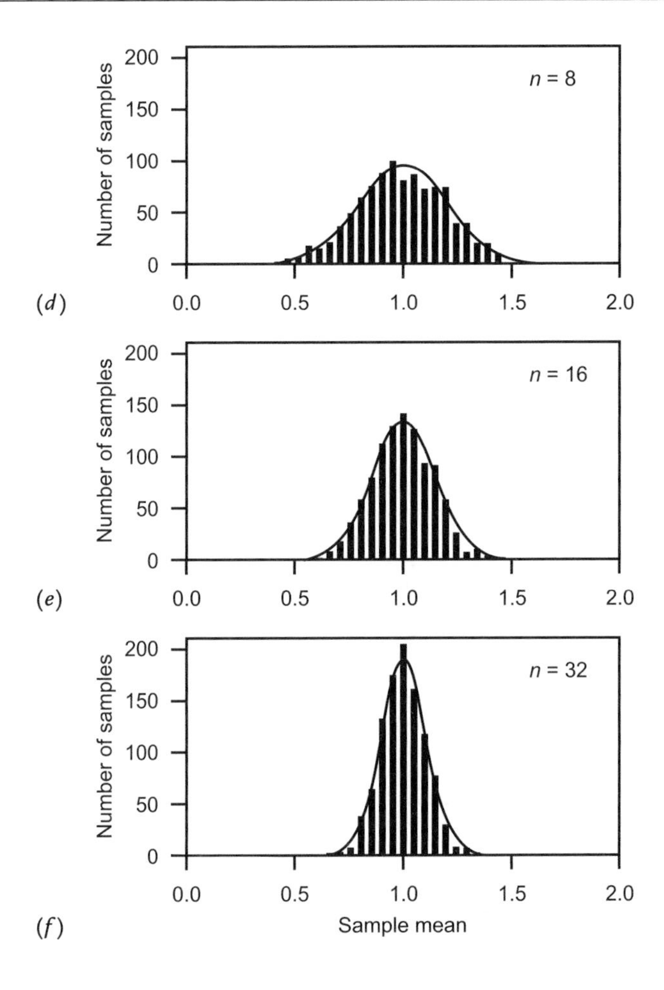

give rise to a continuous (normal) variable. To see how, consider the probability distribution in Figure 6-10a. The random variable takes on only three values (1, 2, or 3), with the central value less likely than the other two. According to the central limit theorem, if we sample from this distribution, the sample means will be approximately normal for large n. In Figure 6-10b, we have drawn 1,000 random samples of size two, and we have plotted a histogram of the resulting sample means. Although X takes on only three values, there are five possible values for the sample mean. Because there are three ways for a sample to have a mean of two, $\bar{x} = 2$ is the most common result. For $n = 4$, there are more outcomes, with intermediate values still more common. Notice how the probability of extreme outcomes decreases for this larger sample (Figure 6-10c). As the sample size increases to eight and beyond, more outcomes are possible, and they are distributed ever more closely to the central value. In the limit (infinite n), the number of outcomes is infinite and the distribution of $\bar{X}$ is continuous, as implied by normality.

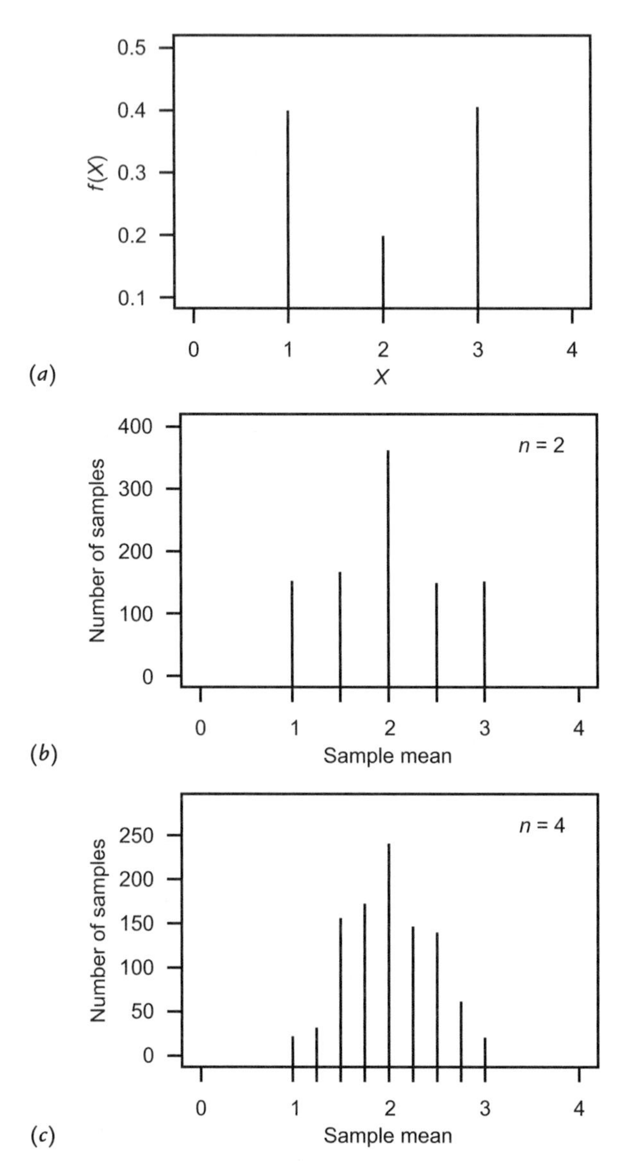

FIGURE 6-10. Sampling from a discrete probability distribution. (As n increases, sample means become more numerous, approaching a continuous normal distribution.)

The great value of the central limit theorem is that it provides a way of deducing results concerning the outcome of a sample, based only on knowledge of the population mean and standard deviation. In particular, the theorem allows us to determine the probability that the sample mean statistic $\bar{X}$ is larger than a given value, or smaller than a given value, or falls within a specified interval. This is best illustrated by a simple

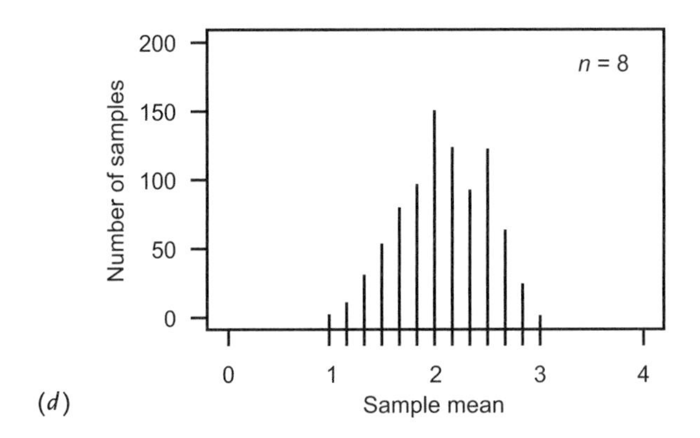

numerical example. Geography students at UCLA well know that the wait times for a coffee at Lu Valle Commons can be quite lengthy. Let us suppose that waiting times are normally distributed with $\mu = 8$ minutes and $\sigma = 4$ minutes. What is the probability that, in a sample of $n = 36$ students waiting for coffee, the mean wait is greater than 10 minutes?

Since the random variable X is normally distributed, we know the distribution of $\bar{X}$ is normal, with a mean of 8 and a standard deviation $\sigma_{\bar{X}} = 4/\sqrt{36} = 0.667$ minute. This sampling distribution is shown in Figure 6-11. We are interested in determining the probability that $\bar{X}$ is greater than 10 minutes, the shaded area of the sampling distribution. To find this probability, we must first standardize the variable $\bar{X}$ using the standard normal transformation,

$$z_{\bar{X}} = \frac{\bar{x} - \mu}{\sigma_{\bar{X}}} = \frac{10 - 8}{0.667} = 3.0$$

Then, using the normal table, we see the desired probability is $P(Z > 3) = 0.001$. Thus, there is a relatively low probability of waiting more than 10 minutes to get a coffee.

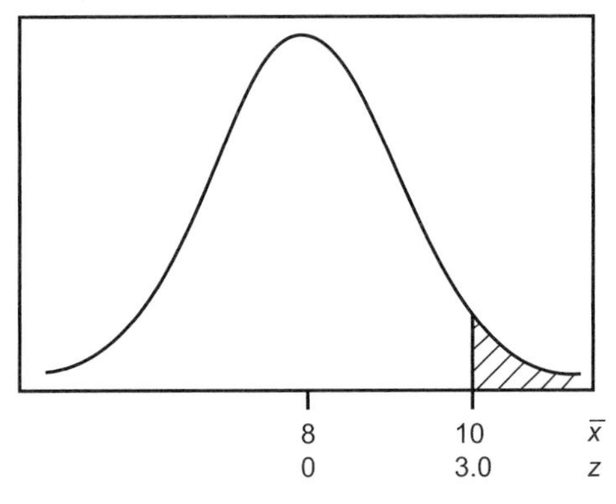

FIGURE 6-11. Sampling distribution of waiting times for $\mu = 8$, $\sigma = 4$, and $n = 36$.

In Chapter 7 we will use this theorem in another way. We will take a characteristic of a sample such as $\bar{X}$ and then determine the probability that it comes from a population with hypothesized population characteristics.

Since the central limit theorem does not apply only to the sample mean, we will be able to apply the theorem to other statistics, such as the sample proportion p. Still, other statistics, such as the sample standard deviation s, cannot be handled by the central limit theorem and have different sampling distributions. In some cases, these sampling distributions are known, but only under certain restrictive assumptions about the distribution of X. By way of contrast, the central limit theorem has the great advantage of requiring almost no assumptions. Indeed, it is precisely because the central limit theorem is so widely applicable that the normal distribution is so important in statistics. The specification of other (non-normal) sampling distributions is left to Chapter 7.

6.6. Geographic Sampling

There are at least two instances in which geographers must take spatially distributed samples. In the first instance, the geographer must go into the field and sample some areally distributed phenomenona. A biogeographer may wish to study the distribution of a particular plant species or even the distribution of seeds from a particular plant or set of plants. As another example, field sampling may be required for the compilation of a soil map. Second, areal sampling is sometimes necessary in direct sampling of locations on a map. This is particularly important when we are sampling from maps over which some phenomenon is continuously distributed. Land-use maps of cities, soil maps, and vegetation maps are typical examples of this form of sampling. For phenomena discretely distributed on a map, conventional sampling procedures can be used. Consider the problem of selecting a set of farms distributed on a map of some rural area. It is sufficient to compile a list of all farms in the area and then sample from this sampling frame in the usual way. We may even sample from all parts of the map by dividing the list into strata based on location. Such a procedure is basically independent of the map itself. For the complex patterns that exist on most maps of continuous variables, such a procedure is inappropriate. Instead, these maps are sampled by using traverse samples (or lines), quadrats, or points. These three alternatives are illustrated in Figure 6-12.

Before examining the sampling designs used with each of these techniques, we specify the sampling frame implicit in spatial sampling. This is equivalent to the or-

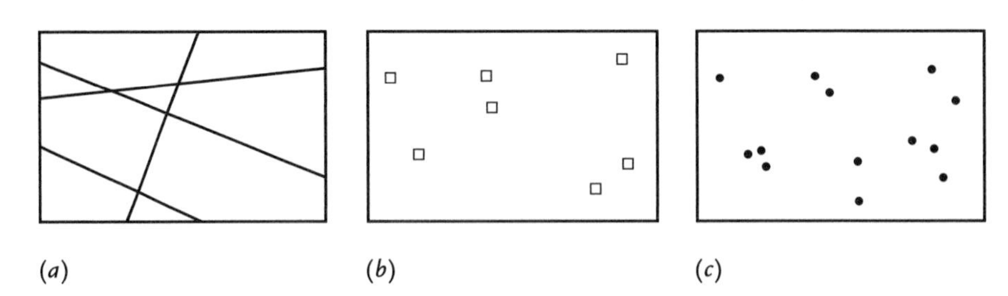

FIGURE 6-12. (a) Traverse samples; (b) quadrat samples; (c) point samples.

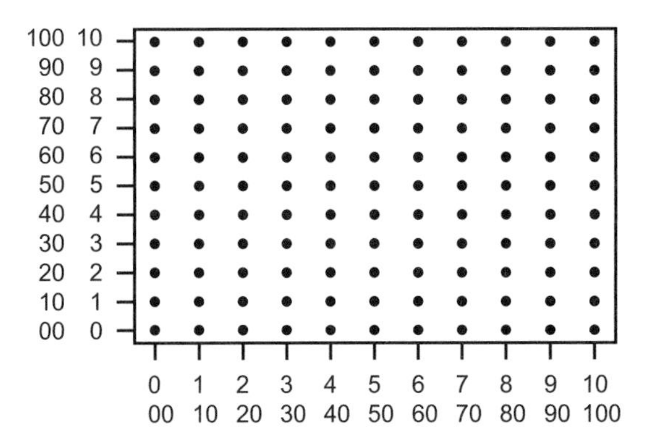

FIGURE 6-13. Coordinate systems and areal sampling.

dered list of the elements in a population used in conventional spatial sampling. For sampling from a map, the Cartesian coordinate system is sufficient to identify any point on the map and hence any element of the population. This coordinate system meets all the requirements of a sampling frame. It is ordered, with a unique address for each element of the population. The (x, y) coordinate pairs are mutually exclusive since no two points on the map can possibly have the same coordinates. Cartographers will recognize that we are ignoring problems of spatial distortion associated with small-scale maps. Finally, it is exhaustive of all the elements in the population since it covers the entire map.

Consider the coordinate system used to sample from the area illustrated in Figure 6-13. The two axes follow the usual north–south and east–west orientations, with the origin placed at the lower left, or southwest, corner of the map. The axes are then subdivided into units at any desired degree of accuracy. Two possibilities are shown in Figure 6-13. The simplest 1-digit grid uses the numbers 0 to 9 to identify particular points on both the x- and y-axes. Then random numbers can be used to locate any point on the map. The first random number drawn is used to define the x coordinate, and the second number drawn, the y coordinate. If 1-digit random numbers are used, then only the 100 points shown as grid intersections can actually be sampled. This *available population* is quite different from the *target,* or *total, population,* that is the entire map. If 2-digit random numbers are used for the grid on each axis, then there are $100 \times 100 = 10{,}000$ possible points to be included in this sample. The grid used to define the sampling frame should be sufficiently fine that features with only small areas are included in the available population. Depending on the map in question, it may be necessary to use 2-, 3-, or even 4-digit grid references so that the set of available points adequately covers the target map.

Quadrat Sampling and Sampling Traverses

One of the more commonly used sampling methods is quadrat sampling. Quadrat sampling differs slightly from the quadrat systems used to analyze point patterns. Variations on three different sampling designs are illustrated in Figure 6-14. First, consider

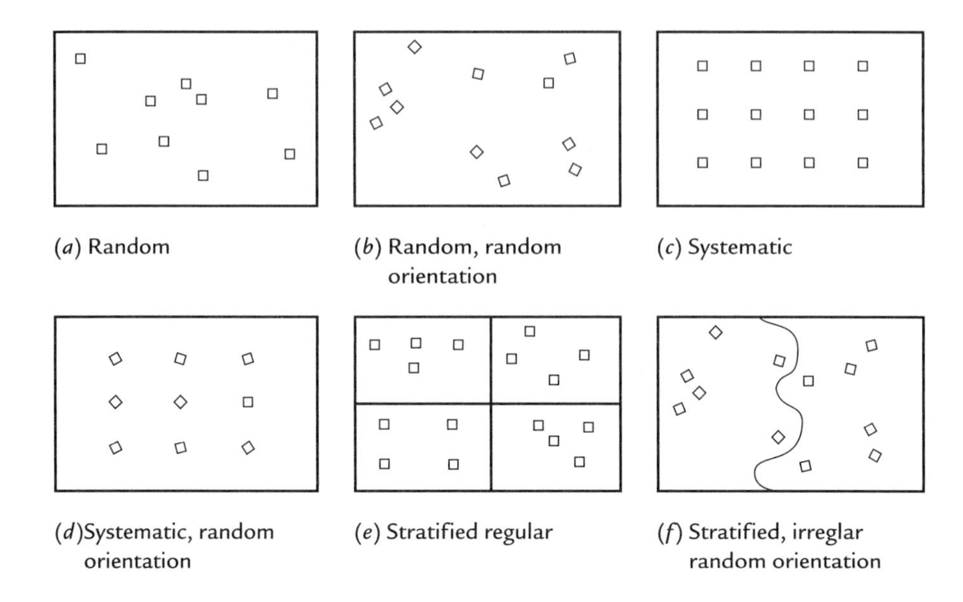

(*a*) Random

(*b*) Random, random orientation

(*c*) Systematic

(*d*)Systematic, random orientation

(*e*) Stratified regular

(*f*) Stratified, irreglar random orientation

FIGURE 6-14. Sampling designs for quadrat sampling.

the random quadrat sampling design in Figure 16.4a. Such a design can be developed by using three simple rules:

1. Select a quadrat size. There is some choice in size, but larger quadrats sometimes become difficult to analyze. If one is forced to calculate, say, the proportion of the quadrat in various land-use categories, then this task itself becomes extremely time-consuming for a large quadrat.
2. Draw two random numbers. Use the first random number to locate the x coordinate of the quadrat centroid and the second to locate the y coordinate of the centroid.
3. Select as many quadrats as necessary to reach a sample size deemed sufficient.

As a variation of this algorithm, it is also possible to randomly orient the quadrats by selecting a third random number to control the orientation of the quadrats. This alternative is shown in Figure 6-14b. For very complex maps, the irregular distribution of the map pattern *within* a quadrat may make analysis very difficult. In such a case, point sampling may be a desirable alternative.

A systematic sampling design is illustrated in Figure 6-14c. A pair of random numbers is used to locate one quadrat on the map, and the remaining quadrats are placed according to a prespecified spacing interval that can yield a sufficiently large sample. Figure 6-14d illustrates a systematic sample with a random orientation. In the stratified sampling designs of Figure 6-14e and f, the number of quadrats in any stratum is proportional to the area of the map in that stratum. These areally defined strata

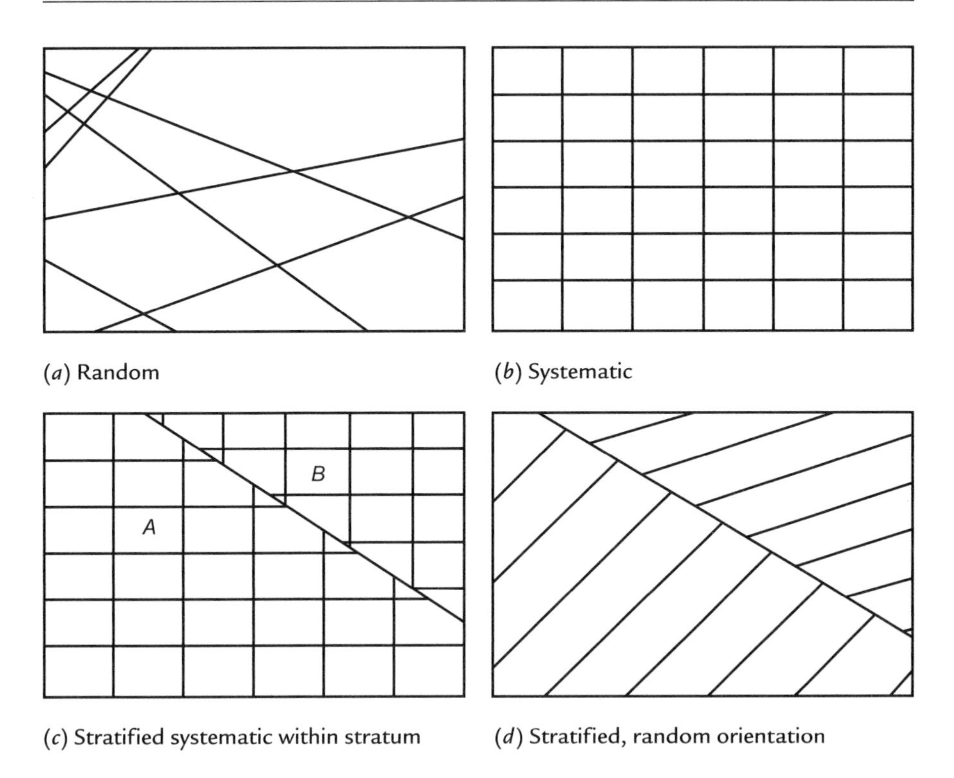

FIGURE 6-15. Sampling designs for traverse sampling.

may be based on either regularly or irregularly shaped portions of the map. In both designs, the quadrats are randomly located within each stratum.

Four designs for sampling traverses are illustrated in Figure 6-15. To locate these traverses, the sampling frame must be slightly modified. As illustrated in Figure 6-15a, the grid references are defined in a clockwise manner from the lower left corner. Again, there is some choice of precision depending on the number of digits used in the grid referencing system. Let us suppose we wish to sample L miles of linear traverse. For the random sampling design, this is accomplished in the following way. First, a single two-digit random number is drawn. This number specifies a point on the map boundary that can be used as one end of the first traverse. A second number is used to locate a point on the boundary for the other end of the traverse. A line connecting these two points determines the first traverse. Next, we determine the length of this line. Let this length be l_1 miles. We now have l_1 miles in our sample and require only $L - l_1$ miles to complete the sample. Two more random numbers are used to locate a second traverse, the length of the sample traverse is updated to $l_1 + l_2$, and the remaining length to be sampled is reduced to $L - l_1 - l_2$ miles. This process is continued until exactly L miles or, alternatively, until at least L miles of traverse has been selected for sampling.

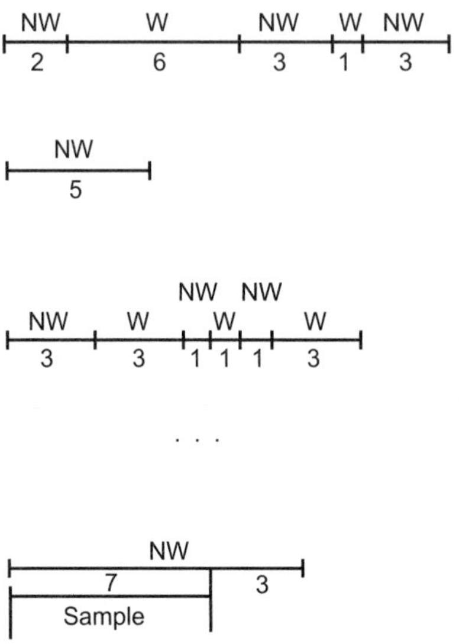

l	l_W	l_{NW}	Cumulative C_l	C_{lW}	C_{lNW}
15	7	8	15	7	8
5	0	5	20	7	13
12	7	5	32	14	18
			. . .		
			193	71	122
10	0	7	200	78	122

$$\text{Proportion of woodland} = \frac{78}{200} = .39$$

FIGURE 6-16. Estimating the proportion of woodland on a map using traverses.

Let us consider the sequence of traverses generated in a study attempting to estimate the proportion of a 100-square-mile map that is woodland, given a traverse of 200 miles. Figure 6-16 illustrates the sequence of traverses used to generate the sample. The first traverse is 15 miles long, of which 7 miles pass through woodland (W) areas and 8 miles through nonwoodland (NW) areas. These figures are recorded in the first three columns. A cumulative, or running, total of the traverse length, in W and NW areas, is provided in the last three columns. Notice that to complete the sample of *exactly* 200 miles, the last traverse is truncated, and only 7 of its 10 miles are included in the sample. Alternatively, a sample of $L = 203$ miles could be used by accepting the complete length of the final traverse in the sample. The estimated proportion of woodland on the map is 78/200 = 0.39.

The systematic design of Figure 6-15b is based on a regular pattern of perpendicular traverses with a prespecified interval. Suppose a sample of $L = 250$ traverse miles is to be taken from a 25-square-mile area. There will be five traverses in each of the north–south and east–west directions, each 25 miles long. A spacing of 5 miles between traverses is used. A single random number can be used to locate the east–west traverses and another to locate the north–south ones. Where there are periodicities on the map, particularly linear or near-linear features, this sampling procedure can lead to biased estimates of certain map characteristics. Streets, highways, stream networks, eskers, and other similar map features can lead to significant sampling error when maps are sampled by using a systematic design.

Randomly selecting the orientation of traverses in a systematic sample is possible, although it is slightly more difficult to operationalize than regular traverses. Additional modifications are also required to sample from an irregularly shaped area. Nevertheless, systematic sampling can yield significant time efficiencies, often making the added complexity well worth the effort.

An areally stratified sampling design for an irregular map is illustrated in Figure 6-15c. The areas or strata may be defined contour intervals, geomorphic or geologic features, or even municipal boundaries in some urban area. Two strata, *A* and *B,* are depicted in Figure 6-15c, in the proportion 60:40. Thus, 60% of the traverse length should be drawn from area *A* and 40% from *B*. As usual, random numbers are drawn to locate individual traverses.

As each traverse is selected and enumerated, a running total is kept of the total traverse length in each stratum. Once the length of the traverses in area *A* reaches 60% of the desired sample traverse length, only the parts of subsequent traverses in stratum *B* are utilized to complete the sample. If the traverses first exhaust the desired length in area *B,* then subsequent traverse lengths in *B* are ignored and only lengths in *A* are included in the sample. This procedure can be extended to handle any number of areas or strata, of regular or irregular shape. For regular shapes, however, it may be easier to sample each stratum individually. Linear traverses for areally stratified maps can also be taken in a systematic manner and with a randomly chosen orientation, as in Figure 6-15d.

Point Sampling

Perhaps the simplest form of spatial sampling is point sampling. Selecting a *simple random point sample* from a map is not much more difficult than selecting a random sample from a conventional sampling frame. To identify each point in the sample, two random numbers are necessary. One is used for the *X* coordinate and the other for the *Y* coordinate. Two random numbers are drawn for each point in the sample until *n* points have been located. Figure 6-17a illustrates a typical random point pattern. Usually in a few areas on the map no points are sampled, and a few areas have relatively high point density. Since the design is purely random, there is always the possibility that a nonrepresentative sample could be drawn.

To ensure that all parts of the map are sampled and there are no overconcentrations of sampled points, a systematic random point sample can be drawn. A typical systematic sample is illustrated in Figure 6-17b. The 16 points for this particular map are regularly spaced at equal intervals in terms of both the *X* and *Y* coordinates. This design is exceptionally easy to operationalize. The interval is selected so that the number of points that fall on the map equals the desired sample size. First, the point in the southwest corner of the map is randomly chosen within the small area delimited by the dashed lines; then others are chosen at fixed intervals away. Only two random numbers are required for a systematic point sample of any size. Once the

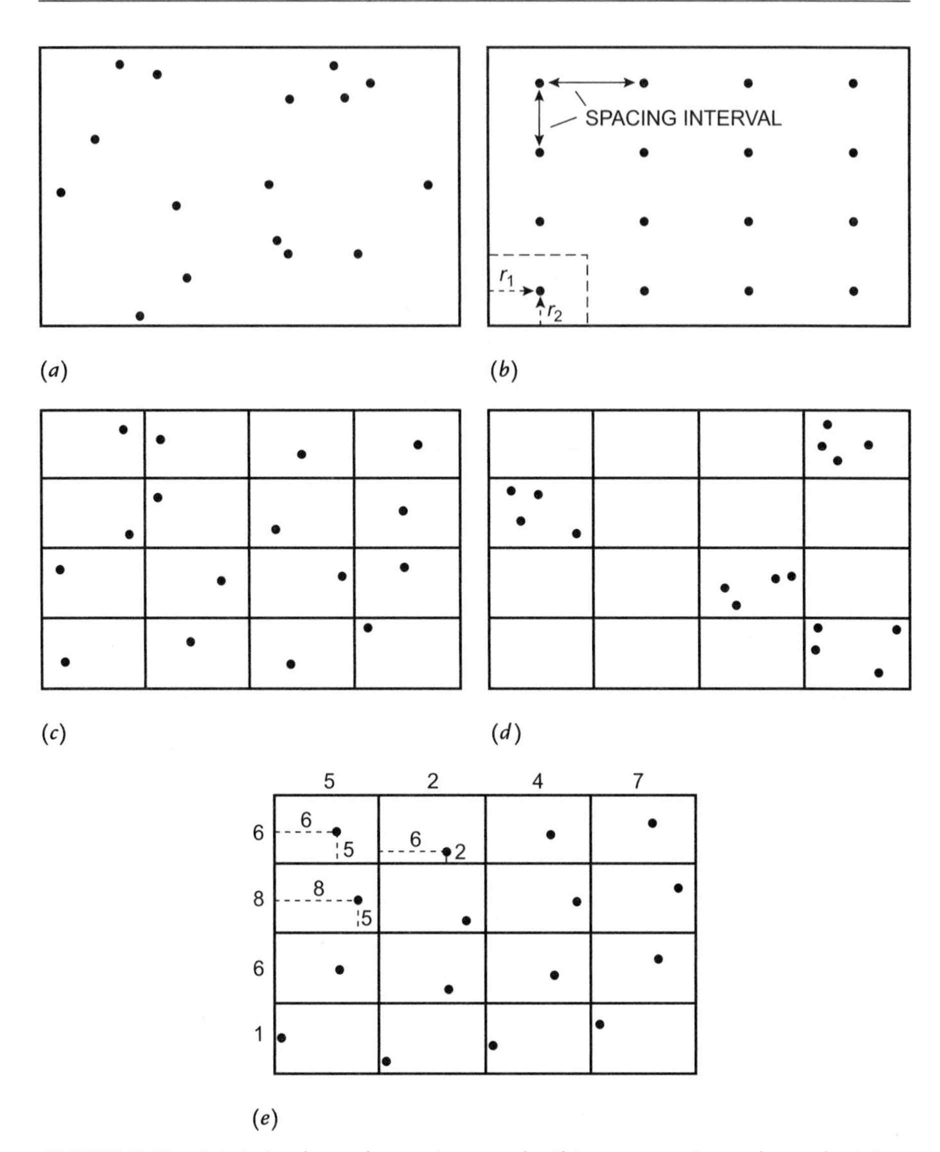

FIGURE 6-17. (a) A simple random point sample; (b) a systematic areal sample; (c) an areally stratified random sample (d) a cluster sample; (e) a stratified systematic unaligned areal sample.

location of a single point in the design is known, the location of all the other points is given.

Because this design selects points at regular intervals, the systematic design can lead to a biased estimate of the proportion of a map covered by some phenomenon. Consider the land-use map of a large city. The land-use map of the typical North American city shows many structural regularities. Streets tend to follow a rectangular grid with a standard distance between street intersections. Successive points taken at regular intervals might all lie on streets. Or perhaps only somewhat more of the points lie on a street. In either case, an estimate of the fraction of land devoted to streets would be biased. The bias admitted with a systematic sample is usually small, except when the map has a systematic linear or near-linear trend. Most often, the ease with which this sample can be drawn outweighs the slight bias introduced by systematization. Systematic sampling is widely used in studies of soils and vegetation.

When *stratified random sampling* is used in spatial sampling, it has the same advantage as systematic sampling, namely, more or less complete coverage of the map. As illustrated in Figure 6-17c, however, the procedure is less likely to pick up any regularities in the map. Stratified designs have smaller sampling errors than either simple random or systematic designs. A *cluster sample* can also be taken from a map that has been areally stratified. As shown in Figure 6-17d, the usual procedure is to randomly select certain areas on the map and then intensively sample randomly within the clusters. The advantage of sampling ease, particularly in field sampling, must be weighed against the possible disadvantage of omitting large areas of the map.

A hybrid design incorporating elements of these simpler designs is the stratified systematic unaligned sample shown in Figure 6-17e. The map is stratified because it is divided into cells prior to sampling. It is systematic because it is unnecessary to draw two random numbers for each point in the sample. However, unlike the case of a simple systematic sample, the pattern produced by this procedure is not aligned. To develop this axis, we define a simple (x, y) grid within each cell. A single random number is drawn for each row and each column. The row random number defines the X coordinate within each cell in that row, and the column random number locates the Y coordinate for all cells in that column. The locations of the points in the three most northwesterly cells are illustrated in Figure 6-17e.

The choice of sampling design in spatial sampling depends very much on the nature of the phenomenon being studied. Each spatial design shares the basic characteristics of its nonspatial counterpart. If the underlying map has systematic features, then a systematic design is a poor choice. If there is no obvious pattern on the map, then all designs will give fairly similar results. Typically, stratified samples are more precise than random samples, which in turn are more precise than cluster samples. Research experiments to test the precision of all sample designs have generally found the stratified systematic unaligned sample design to be decidedly superior to all other designs.

6.7. Summary

Samples can be drawn from a population in a number of ways. They are usually taken to minimize both the time and the cost of collecting data. Four steps usually precede the actual collection of data: specification of the population, delineation of a sampling frame, selection of a sampling design, and pretest of the data collection procedures. For purposes of statistical inference, a probability sample is the required sampling procedure. Although a probability sample still can lead to the selection of an unrepresentative sample, it is always possible to specify the probable accuracy of the results. Simple random samples are the most commonly used method of data collection, although other variations sometimes prove necessary and desirable in practice. *For the remainder of this text, all inferential procedures are based on the assumption that the sample has been collected in this manner.*

Sample statistics are random variables. The value of any sample statistic varies from sample to sample for all samples of a fixed size drawn from the same population. For sufficiently large samples, the sampling distribution of many statistics such as $\bar{X}$ or P is known to be normal or approximately normal. By knowing the characteristics of sampling distributions, it is possible to answer probability questions about one sample. That is, we can determine the likely sampling error in our sampling experiment. In Chapter 7, we show how these sampling distributions can be used to either *estimate* population characteristics or make judgments concerning their hypothesized value.

In this chapter, we briefly examined the nature and problems of geographic sampling. An exceedingly rich variety of sample designs can be devised based on linear, point, or quadrat sampling. All methods find considerable application in the geographical literature.

FURTHER READING

Textbooks devoted to advanced topics in sampling theory include Mendenhall, Ott, and Schaeffer (1971) and Cochran (1977). Many textbooks of this type, and in particular the text by Cochran, require considerable background. A more practical approach to sampling questions is taken by Sudman (1976). There are literally hundreds of journal articles on the practical techniques of sampling. Two very useful journals to consult are the *Journal of Marketing Research* and *Public Opinion Quarterly.* For an analysis of sampling techniques in geography, see Dixon and Leach (1978) and Berry and Baker (1968).

B. J. L. Berry and A. Baker, "Geographic Sampling," in B. J. L. Berry and D. F. Marble (eds.), *Spatial Analysis,* pp. 91–100 (Englewood Cliffs, NJ: Prentice-Hall, 1968).

W. Cochran, *Sampling Techniques* (New York: Wiley, 1977).

C. J. Dixon and B. Leach, *Sampling Methods for Geographical Research.* (Norwich, England: Geo Abstracts, 1978).

R. M. Groves, F. J. Fowler, M. P. Cooper, J. M. Lipkowski, E. Singer, and R. Tourangeau. *Survey Methodology.* (New York: Wiley, 2004).

W. Mendenhall, L. Ott, and R. L. Schaeffer, *Elementary Survey Sampling.* (Belmont. CA: Duxbury, 1971).

S. Sudman, *Applied Sampling.* (New York: Academic, 1976).

PROBLEMS

1. Explain the meaning of the following terms:
   - Sample
   - Sampling error
   - Non-response bias
   - Target population
   - Pilot sample or pre-test
   - Simple random sample
   - Stratified random sample
   - Sampling distribution
   - Finite population correction factor
   - Standard error of a sampling distribution
   - Stratified, systematic, unaligned point sample
   - Population
   - Sampling bias
   - Sampling frame
   - Sample design
   - Random number table
   - Systematic sample
   - Sample statistic
   - Sampling distribution

2. Explain why, in general, it is advisable to sample rather than to attempt a complete census of a population.

3. Give three examples of finite populations and three examples of infinite populations.

4. Differentiate among (a) probability samples, (b) quota samples, (c) convenience samples, (d) judgmental samples, and (e) volunteer samples.

5. Why does a simple random sample from an infinite population differ from one from a finite population?

6. In stratified sampling, it is desirable to define strata that are internally homogeneous but different from other strata. In cluster sampling, both internal heterogeneity and between-cluster homogeneity are desirable. Why?

7. Why is a sample statistic a random variable?

8. Select a hypothetical population of 100 members. Draw the frequency distribution of some variable of interest. Illustrate the operation of the central limit theorem by drawing repeated samples of sizes $n = 4$ and $n = 10$ and drawing the sampling distributions of sample means. How do these sampling distributions illustrate the implications of the central limit theorem?

9. Distinguish among (a) the population mean of a variable, (b) the sample mean, and (c) the mean of the sampling distribution of the random variable.

10. A population consists of the following seven observations

Observation	A	B	C	D	E	F	G
Value	4.0	3.0	5.0	7.0	4.0	8.0	11.0

   a. Find the population mean and standard deviation.
   b. Create a table listing all samples of size $n = 2$. For each sample, find the sample mean $\bar{x}$. Draw a histogram of the sampling distribution.

c. Find the mean and standard deviation of the sampling distribution of means generated in part (b).

d. Repeat parts (b) and (c) for samples of size $n = 3$.

11. The following map classifies the land use in a square city into four categories: residential, commercial, industrial, and open space. The percentage of the city land use in each class is as follows: residential (R) = 69%; commercial (C) = 8%; industrial (I), = 10%; open space (0) = 13%. Estimate the proportion of the map in each land-use category, using a point sample of 25 observations. Draw samples using the following designs: simple random, stratified random, systematic, and systematic stratified unaligned.

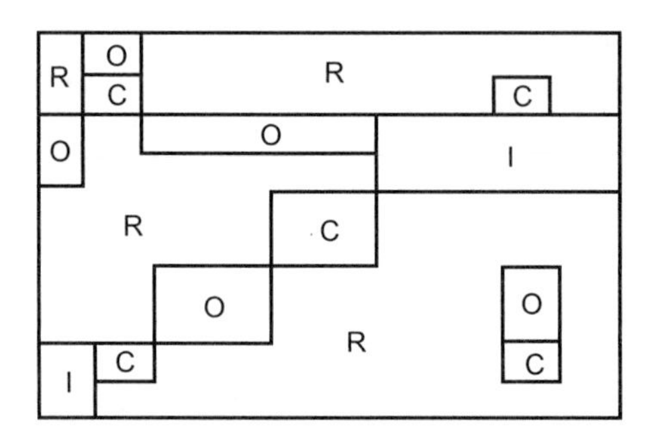

7

Point and Interval Estimation

The principal objective of statistical inference is to draw conclusions about a population based on a subset of that population, the sample. To be perfectly correct, statistical inference includes all procedures based on using a *sample statistic* to draw inferences about a *population parameter.* In Chapter 1, we distinguished between two ways of drawing inferences about population parameters. The first is *estimation,* in which we guess the value of a population parameter based on sample information. This process is illustrated in Figure 7-1 for the population parameter μ. From some statistical population, we draw a sample of size n and elect to use the sample statistic $\bar{X}$ in making inferences about μ. Once the sample is taken, the observations of the sample variable X are known to be $x_1, x_2, \ldots, x_n$, and the *value* of the test statistic $\bar{x}$ is used to estimate the population parameter μ.

The second way of drawing inferences about the value of a population parameter, you will recall, is called *hypothesis testing.* When we test a hypothesis, we make a reasonable assumption about the value of μ (or some other population parameter), and then we use sample information to decide whether or not this hypothesis is supported by the data. For example, we may be interested in a population parameter π, the proportion of residents of a city who moved in the last year. In particular, we wish to determine whether the city has some abnormally high rate of residential mobility, say greater than $\pi = 0.20$. A sample of residents is taken, and the percentage of movers within the last year is determined as p. We then use this value of p to test whether $\pi > 0.20$ or $\pi \leq 0.20$ is the more reasonable conclusion.

Both *estimation* and *hypothesis testing* are based on statistical relationships between samples and populations. It should not be surprising, therefore, that these two methods of statistical inference are intimately related. The relationship is fully explained in Chapter 8. In this chapter, we focus solely on estimation. Section 7.1 explains the distinction between two main categories of statistical estimation, *point* and *interval* estimation. This section also develops criteria differentiating good estimators from bad ones. In Section 7.2, point estimators for the population parameters μ, π, and σ are considered. Section 7.3 discusses interval estimation of the first two parameters and introduces the notion of a *confidence interval.* Finally, in Section 7.4 we

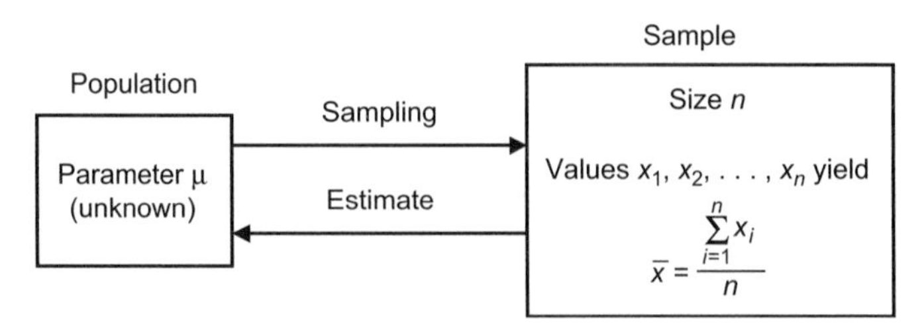

FIGURE 7-1. Estimating the value of the population parameter μ.

examine how the estimation methods can be applied "in reverse" to determine the sample size needed to estimate a parameter with a given degree of confidence.

7.1. Statistical Estimation Procedures

The procedures used in statistical estimation can be conveniently divided into two major classes: point estimation and interval estimation.

DEFINITION: POINT ESTIMATION

In making a point estimate, a single number is calculated from the sample and is used as the best estimate of some unknown population parameter.

This number is called a *point estimate* because the population parameter is estimated by one point on the number line. For example, we might say that based on our sample of some size n, the best estimate of the proportion of movers in a city (π) is 0.26. Note that we do not get the actual value of π from the sample, but rather we compute some estimate of the true value.

DEFINITION: INTERVAL ESTIMATION

In interval estimation, the sample is used to identify a range or interval within which the population parameter is thought to lie, with a certain probability.

We might say, for example, that we are 95% certain that the proportion of people who moved in the city last year is in the interval [0.20, 0.32]. Let us consider each type of estimation in more detail.

Point Estimation

Point estimation of some population parameter is illustrated in Figure 7-2. The population parameter of interest might be the mean, variance, standard deviation, proportion, or *any* other characteristic of the population. A random sample gathered to

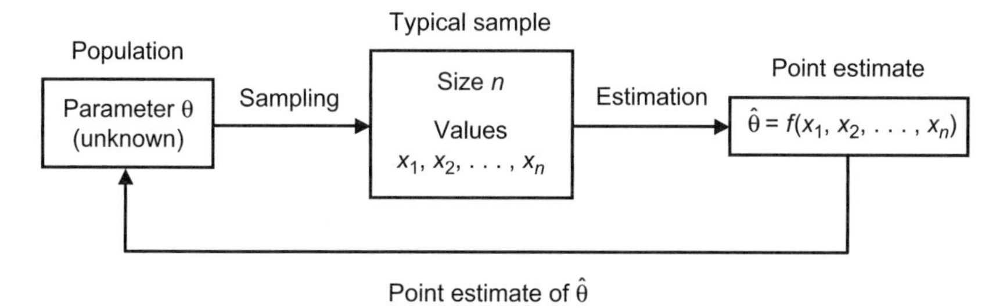

FIGURE 7-2. Point estimation of a population parameter.

estimate the value of an unknown population parameter will typically comprise n observations of the variable of interest. The estimator of the population parameter is some function of these sample observations.

DEFINITION: STATISTICAL ESTIMATOR AND STATISTICAL ESTIMATE
A statistical estimator is a function of the n observed values, $x_1, x_2, \ldots, x_n$, sampled from a random variable X. An estimator is, therefore, also a random variable. The *value* of the estimator, $\hat{\theta}$, can thus be calculated. It is known as the *statistical estimate* of the unknown population parameter, θ. (We follow statistical convention and denote the sample estimate of a population parameter with the "hat" notation ^.)

As an example of θ consider the population mean parameter. There are several possible point estimators for μ. The sample mean is the obvious choice, but why should it be chosen over the sample median, or even the sample mode? All are simple functions of values $x_1, x_2, \ldots, x_n$, and can therefore serve as estimators.

There are two principal criteria for selecting an estimator for a population parameter θ: *bias* and *efficiency*. These criteria can be understood most easily by thinking about the error associated with an estimate. Obviously, when we use a sample statistic to guess at the value of a parameter, we do not expect the sample estimate to equal the population value—we expect some estimation error $\hat{\theta} - \theta$. If we overestimate θ and $\hat{\theta} > \theta$, then we have a positive error, whereas if we underestimate θ and $\hat{\theta} < \theta$ we have a negative error. Moreover, we know different samples will give different estimates, so the error will vary from sample to sample. Both bias and efficiency relate to error of estimation $\hat{\theta} - \theta$. In the case of bias, we are identifying the mean, or expected error. If the estimator *on average* gives values that are too high, we would say it is biased high. Similarly, if the estimator tends to underestimate θ, it is biased low. To arrive at the bias, we need to know something about the sampling distribution of the estimator; in particular, we need to know its mean value. Figure 7-3 shows the sampling distributions for two estimators, $\hat{\theta}_1$ and $\hat{\theta}_2$. The first is biased low, as its mean is less than the true value θ. We would very likely prefer $\hat{\theta}_2$, because on average it neither underestimates nor overestimates θ. In other words, $\hat{\theta}_2$ is *unbiased*.

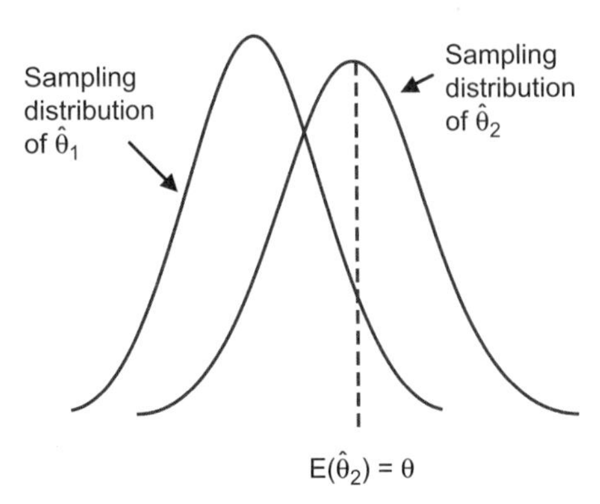

FIGURE 7-3. Sampling distributions for a biased and an unbiased estimator of θ.

Bias is related to error in the following way:

$$\text{Bias} = \text{Mean error} = E(\hat{\theta} - \theta) = E(\hat{\theta}) - \theta$$

DEFINITION: UNBIASED ESTIMATOR
An estimator $\hat{\theta}$ of a population parameter is said to be unbiased if the expected value of the estimator is equal to the population parameter. That is, $\hat{\theta}$ is unbiased if $E(\hat{\theta}) = \theta$.

The sample mean $\bar{X}$ is a good example of an unbiased estimator. Given a random sample, the expected value of $\bar{X}$ is μ, the same value we are trying to estimate. In other words, we would compute the bias of $\bar{X}$ as

$$\text{Bias of } \bar{X} = E(\bar{X}) - \mu = \mu - \mu = 0$$

Notice that using an unbiased estimator does not imply that an *individual* estimate will equal the true value of the population parameter, or even that it will be close to the true value. We know only that the errors are zero *on average*. Small individual errors are expected only if the distribution of $\hat{\theta}$ is tightly clustered around θ. Figure 7-4 shows this graphically for two estimators, both of which are unbiased. In both cases the mean estimation error is zero, but $\hat{\theta}_3$ is more tightly clustered. If we use $\hat{\theta}_3$, we are more likely to obtain an estimate that is near the true value. We would say that $\hat{\theta}_3$ is more *efficient* than $\hat{\theta}_4$.

To make the concept of efficiency more precise, we need to consider estimation errors more carefully. We have seen that bias contributes to error and is clearly undesirable. Does this mean we always choose unbiased estimators? The answer is no, because bias is not the only source of estimation error. Rather than minimize mean error (bias), we probably want individual errors to be small. After all, we have only

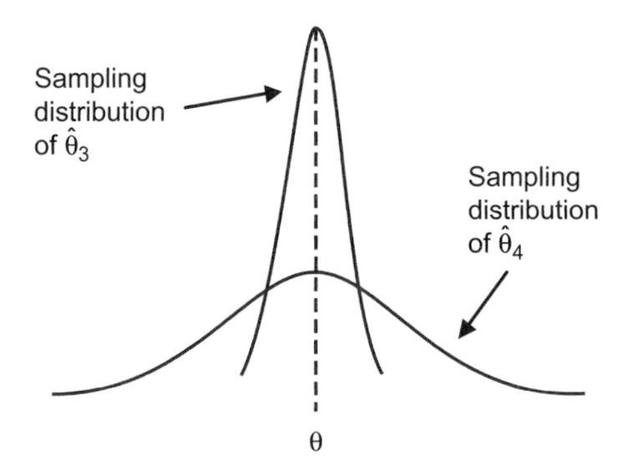

FIGURE 7-4. Sampling distributions for two unbiased estimators of θ.

a single sample in hand; thus it isn't necessarily helpful to know that the mean error is small. We would prefer to know that the error expected from a single (individual) sample is small. Consider, for example, Figure 7-5. Even though it is biased, estimator $\hat{\theta}_5$ would probably be preferred to $\hat{\theta}_6$. The degree of bias introduced by using this estimator is more than offset by its high efficiency. Clearly, to evaluate an estimator, we need a measure that tracks more than just bias. Bias is inadequate because positive and negative errors cancel one another, giving low mean error even when individual errors are large. What we are after is an index that measures the departure between $\hat{\theta}$ and θ without regard to sign.

One approach is to use squared differences. Rather than calculate the expected error, we will calculate the expected *squared* error. This seems like a subtle difference but can have profound effects on the choice of estimator. The expected squared error is known as the *mean squared error,* or MSE, and is given by

$$\text{MSE} = E(\hat{\theta} - \theta)^2$$

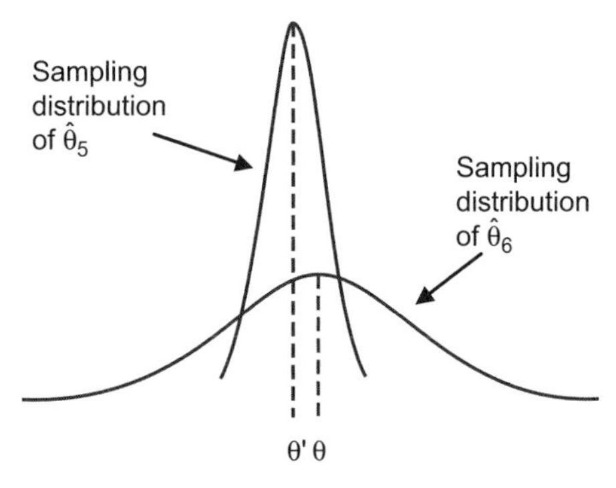

FIGURE 7-5. Difficulties in choosing a potential estimator.

This is appealing because although individual errors can be positive or negative, all are squared and thus there is no cancellation. After a little algebra, it is easy to arrive at an alternative, more useful expression for MSE:

$$\text{MSE} = E(\hat{\theta} - \theta)^2 = [E(\hat{\theta}) - \theta^2 + E[\hat{\theta} - E(\hat{\theta})]^2 = \text{BIAS}^2 + \text{VARIANCE}$$

In other words, MSE is the sum of bias squared plus the variance of $\hat{\theta}$. Bias tells us where the sampling distribution is centered, while variance indicates the amount of dispersion around the central (mean) value. If both are small, the sampling distribution is both centered near the true value *and* tightly clustered around the true value. In short, the estimator will be efficient when MSE is small. This suggests a definition of efficiency in terms of MSE:

DEFINITION: EFFICIENCY OF AN ESTIMATOR

The efficiency of an estimator is inversely proportional to its mean square error, MSE. As MSE increases, efficiency decreases. For an estimator to be efficient, both its bias and variance must be small.

For example, consider the sample mean. We know that it is unbiased and that its variance is σ^2/n. Thus we can say MSE = 0 + σ^2/n. As the sample size increases, MSE decreases, and efficiency improves. (This is expected, of course, as larger samples contain more information, and thus they ought to give better estimates.) Looking again at Figure 7-4, we might think of the curves as representing $\bar{X}$ based on different sample sizes, with $\hat{\theta}_3$ coming from the larger sample. Alternatively, the curves might represent two completely different estimators. For example, $\hat{\theta}_4$ could be the sample median, and $\hat{\theta}_3$ the sample mean. Both of these are unbiased; thus, we do not prefer one over the other on the basis of bias. However, if X has a normal distribution, it can be shown that the sample mean has a smaller variance than any other unbiased estimator of μ, including the median. It is therefore more efficient than the median, and it would be preferred over the median.

Notice that bias and sampling variability both affect MSE. If we choose an estimator with small MSE, we are saying we don't care how the errors arise, we simply want them to be small. If this means using a biased estimator, so be it. Looking again at Figure 7-5, we see that the preferred estimator has smaller MSE, in spite of nonzero bias. We therefore expect that an individual estimate based on $\hat{\theta}_5$ will be closer to the true value than one based on $\hat{\theta}_6$.

MSE obviously provides very useful information about an estimator, but it also suffers from an important shortcoming. Because it is the average squared error, MSE has a dimension of θ^2 and is sometimes hard to interpret. For example, if we are estimating mean age in years, MSE has units of "years squared," hardly a meaningful dimension. To overcome the interpretation difficulty, a related efficiency measure is more commonly used, the *root mean square error,* or RMSE. This is simply the square root of MSE:

$$\text{RMSE} = \sqrt{\text{MSE}} = \sqrt{\text{BIAS}^2 + \text{VARIANCE}}$$

Now we have two measures of overall error, MSE and RMSE. Should we choose an estimator that minimizes the former or the latter? Happily, this is never an issue. Because RMSE is a monotonic function of MSE, minimizing MSE is equivalent to minimizing RMSE. In other words, if an estimator is the minimum MSE estimator, it is also the minimum RMSE estimator. Choosing to minimize RMSE instead of MSE does not give a different estimator; it simply gives a different numerical measure of error, one that happens to be easier to interpret than MSE. We emphasize that RMSE always has the same dimension as the parameter being estimated. It is interpreted as the average estimation error, where the averaging process involves squared, not absolute, errors. We will see RMSE again in Chapter 12, where it will be used as a way of summarizing errors of prediction.

As a final general comment about point estimation, we should mention that a number of commonplace statistical methods are based on minimum-variance unbiased estimates. That is, we first restrict the search to unbiased estimators and then choose the one with the smallest MSE. Such estimators are often called *best unbiased estimators,* where the word "best" indicates minimum variance. For example, assuming X is normal, the sample mean is the best unbiased estimator of μ. No other unbiased function of the sample observations has a smaller variance, thus no other unbiased estimator has a smaller MSE.

Interval Estimation

Interval estimates improve upon point estimates by providing a range of values for θ. Although we might know that in general some point estimator gives the best guess as to the true value of the population parameter, we have no idea about how good a particular estimate is. In statistical estimation, we obviously want to know how far the point estimate is likely to be from the actual value of the population parameter. This is the province of interval estimation and is illustrated in Figure 7-6. Sampling from a population, we utilize our observed sample values $x_1, x_2, \ldots, x_n$ to arrive at two points that together define an interval, or range of values for the parameter θ. The two points represent a lower limit $\hat{\theta}_L$ and an upper limit $\hat{\theta}_U$ for the interval. In making an interval estimate we claim *with a known degree of confidence* that the interval contains the unknown value of the population parameter. For this reason, interval estimates are often termed *confidence intervals.* A typical conclusion might be, "Based on a sample of 100 light bulbs, we estimate the population mean lifetime of bulbs to be between 1,000 and 1,200 hours, with 95% confidence." The word "confidence," when it is used in this context, has a particular meaning, thoroughly explained in Section 7.3.

The usefulness of our inference is based on the *probability* that the specified confidence interval contains the actual value of the population parameter. In the example cited above, we would attach a probability of 0.95 to the interval [1,000, 1,200] based on a sample size of $n = 100$. In doing so, we recognize that because of sampling variability, the interval constructed might be in error. That is, it might not contain the actual mean lifetime. The confidence level, 0.95, indicates that the probability of error is known to be 0.05, so that the chances of error are 5 out of 100.

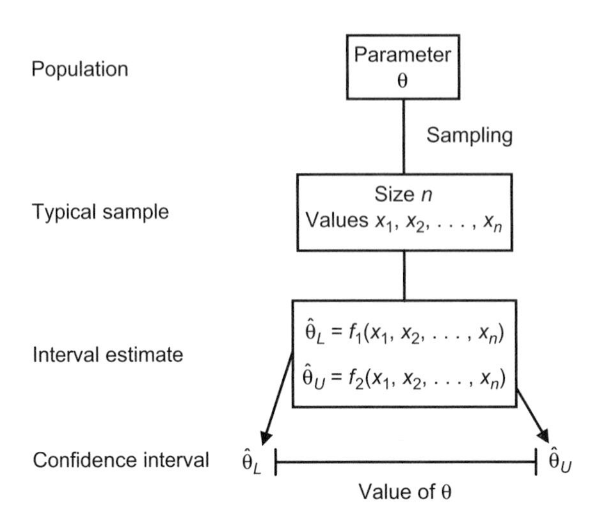

FIGURE 7-6. Interval estimation of a population parameter.

Just as there are good point estimators, so are there good interval estimators. The principal requirement of a good interval estimator is that the specified interval $[\hat{\theta}_L, \hat{\theta}_U]$ be as small as possible for a given sample size and confidence level. Interval estimators for π and μ are covered in Section 7.3.

7.2. Point Estimation

The point estimators for the population parameters μ, π, and σ^2 are given in Table 7-1. Let us consider each in turn.

Population Mean

Without question the most common estimator of the population mean μ is the sample mean $\bar{X}$. Its popularity is in large part explained by the fact that it combines high efficiency with no bias. In fact, one can show that if X is normal, the sample mean is more efficient than all other unbiased estimators. For example, its variance is about 56% smaller than the variance of the sample median. Though both are unbiased estimates of μ, the sample mean is preferred because the expected estimation error (RMSE) is smaller. For distributions that depart from normality, even symmetric distributions, the sample mean is not necessarily best. Consider, for example, a distribution that has more probability mass in its tails than the normal distribution. A sample from such a distribution is more likely to contain very high or very low values than would be true for the normal distribution. In Chapter 3 we noted that $\bar{X}$ is sensitive to extreme values; thus it can be pulled away from μ by a single unusually large or small observation. Here then, is a case where the sample mean is not very satisfactory, and we seek a more robust estimator.

TABLE 7-1
Point Estimators of μ, π, and σ^2

Population parameters	Point estimator	Formula for point estimate
μ	$\bar{X}$	$\bar{x} = \sum_{i=1}^{n} x_i/n$
	Median	Middle value in sample (50%-ile)
	25% trimmed mean	Mean of middle 50% of values in sample
	10% trimmed mean	Mean of middle 80% of values in sample
π	P	$p = x/n$ where x = number of successes in n trials
σ^2	S^2	$s^2 = \sum_{i=1}^{n}(x_i - \bar{x})^2/n - 1$

An obvious alternative is the sample median. Because it responds only to the central value in the sample, it is not affected by the extremes. Compared to $\bar{X}$, the median therefore has a smaller RMSE for some non-normal distributions. Yet another alternative is to use a *trimmed mean.* The sample values are placed in rank order, and an equal number of observations are discarded from the bottom and top of the sample. The trimmed mean is just the mean of the sample values that remain. For example, the 25% trimmed mean discards one-quarter from both ends and is the average of the middle half of the sample. The procedure effectively removes (trims) the extreme values in the sample, so that they cannot influence the estimate. The amount of data discarded can be as much or little as seems appropriate, though typically the trim fraction is between 20 and 50%.

In the limit, when all observations except one have been trimmed away, the trimmed mean converges to the sample median. In other words, we can think of the median as an extreme example of a trimmed mean. Likewise, we can think of $\bar{X}$ as a trimmed mean with a trim fraction of zero. In addition to the median and trimmed mean, there are other robust estimators that can outperform the sample mean. Regardless of which is used, if we want to claim that an estimator is better than $\bar{X}$, we must know the probability distribution of X. Given that we do not even know the mean of X, it's unlikely that we will have detailed knowledge about the distribution of that variable. In actual practice therefore, it will be hard to claim with certainty that a particular estimator is better than the sample mean. Nevertheless, one should not blindly assume $\bar{X}$ is always the appropriate choice. Prudence suggests that several estimators be computed and compared with one another. When they give widely divergent values, one should be cautious in using the sample mean.

We note in passing that there is another, and very compelling, reason to use $\bar{X}$ as an estimator—its sampling distribution is known! Although we might have no information about the underlying distribution of X, we know from the central limit theorem that $\bar{X}$ is approximately normal. This allows us to assign probabilities to various values of $\bar{X}$ and, as will be seen below, provides a way of attaching uncertainty to the estimate of μ. That is, by using $\bar{X}$, we are able to say something about how good is the estimate we have just made. If some other estimator is used, we will have obtained a point estimate but will have no idea about how close it is to the true value of μ.

Population Proportion π

The recommended estimator of the population proportion is the sample statistic $P = X/n$ where X is the number of "successes" found in the sample of n observations. The parameter π is one of the two parameters in the binomial distribution and corresponds to the proportion of "successes" in the population at large. To estimate π, we simply calculate the observed number of successes x in a sample of n trials and estimate π as x/n. Suppose, for example, we are interested in the proportion of residents in a metropolitan area who choose public transit for their journey to work. A sample of 100 residents finds that 26 took public transit. Our best estimate of π is $p = 26/100 = 0.26$.

Population Variance σ^2 and Standard Deviation σ

For the population variance σ^2, the suggested estimator is S^2. Given a sample, $x_1, x_2, \ldots, x_n$, the sample value of this random variable is

$$S^2 = \frac{\sum_{i=1}^{n} (x_i - \bar{x})^2}{n - 1}$$

Recall that $\sigma^2 = E(X - \mu)^2$, the mean square departure from the population mean. It therefore makes perfect sense that our estimate of σ^2 is found by averaging the squared departures from the sample mean. Not so obvious is why we should divide by $n - 1$ rather than n. The answer is that doing so gives us an unbiased estimator. If we took many samples of size n from a population, the values of S^2 computed for each sample would average to σ^2. On the other hand, if we were to divide by n, the estimator would be biased low and tend to underestimate σ^2. The formal proof of this is beyond an introductory text (see, e.g., Pindyck & Rubinfeld, 1981), but it is not hard to understand the problem in general terms. In estimating σ^2, we are replacing the unknown population mean μ with the sample mean $\bar{X}$. That is, we are computing squared deviations around the sample mean, not the population mean. From Chapter 3, we know that the squared deviations around the sample mean are smaller than the deviations around any other number. This implies that the deviations around $\bar{x}$ are *smaller* than the deviations around μ. However, it is deviations around μ that we are trying to estimate! Clearly, if the mean deviation around $\bar{x}$ is smaller than the mean deviation around μ, it must underestimate σ^2. In order to compensate for this bias, we inflate the sum of squares by dividing by a number smaller than n, namely, $n - 1$. In addition to being unbiased, S^2 is also the most efficient (minimum MSE) unbiased estimator of σ^2, providing X is normally distributed.

The estimator for σ, the population standard deviation, is the square root of S^2, or S. Although it is most commonly used, it is actually a biased estimator of σ. However, the degree of bias is very small for reasonable sample sizes, and S is much more convenient than alternative unbiased estimators. We will use it exclusively.

7.3. Interval Estimation

Statistical inference typically includes both an inference and some measure about the reliability of the inference. In point estimation, we select a single *value* that is used as an estimate of the population parameter in question. Ideally, along with the estimate, we should provide information about how far the estimate is likely to be from the true value. What would be useful is some measure of the probability that the population parameter of interest lies within a certain range. Specifying these ranges, and the probabilities attached to them, is the goal of interval estimation. In this section we show how to develop interval estimates for two population parameters, μ and π, based on knowledge of the sampling distributions of the estimators, $\bar{X}$ and P.

Interval Estimator for μ When X Is Normally Distributed

Interval estimates for μ are based on $\bar{X}$ and its sampling distribution. With regard to the sampling distribution of $\bar{X}$, remember that it is normal with mean μ and standard deviation $\sigma/\sqrt{n}$. Knowing this, we can find the probability of $\bar{X}$ falling into an interval, or range of values. This is most easily done if we convert to the standard normal distribution using the Z transformation. Consider the sampling distribution shown in Figure 7-7. Since 95% of the area under the standard normal curve is between –1.96 and +1.96, we know 95% of the values of $\bar{X}$ lie between –1.96 and 1.96 standard deviations of μ. That is, we can say

$$P\left(-1.96 \le \frac{\bar{X}-\mu}{\sigma_{\bar{X}}} \le 1.96\right) = 0.95 \tag{7-1}$$

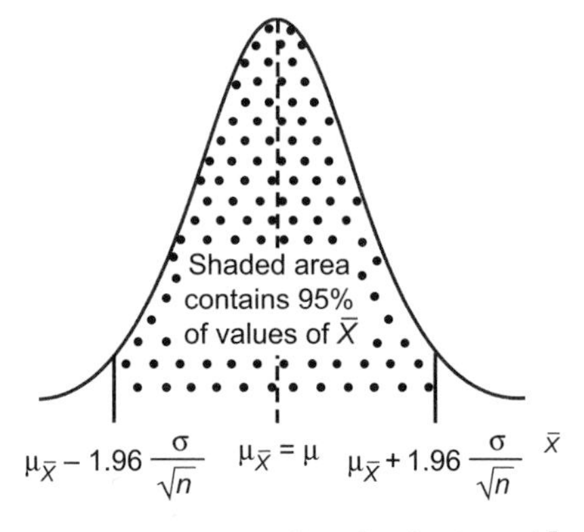

FIGURE 7-7. Sampling distribution of $\bar{X}$.

Multiplying all parts of the inequality by $\sigma_{\bar{X}}$ gives

$$P(-1.96\sigma_{\bar{X}} \leq \bar{X} - \mu \leq 1.96\sigma_{\bar{X}}) = 0.95$$

Now substitute for $\sigma_{\bar{X}}$ to get

$$P\left(-1.96\frac{\sigma}{\sqrt{n}} \leq \bar{X} - \mu \leq 1.96\frac{\sigma}{\sqrt{n}}\right) = 0.95$$

Subtracting $\bar{X}$ and multiplying by –1 expresses Equation 7-1 in terms of the unknown μ as

$$P\left(\bar{X} - 1.96\frac{\sigma}{\sqrt{n}} \leq \mu \leq \bar{X} + 1.96\frac{\sigma}{\sqrt{n}}\right) = 0.95 \tag{7-2}$$

How can this probability statement be interpreted? Let us suppose repeated samples of size n are taken from a population with a mean of μ and a standard deviation σ. See Figure 7-8. Using the data in each of these samples, we calculate a value for $\bar{X}$ using

$$\bar{x} = \frac{x_1 + x_2 + \ldots + x_n}{n}$$

Each time we find a new interval with upper and lower bounds $\bar{x} + 1.96\sigma/\sqrt{n}$ and $\bar{x} - 1.96\sigma/\sqrt{n}$. Then, as illustrated in Figure 7-8, 95% of the samples yield an interval con-

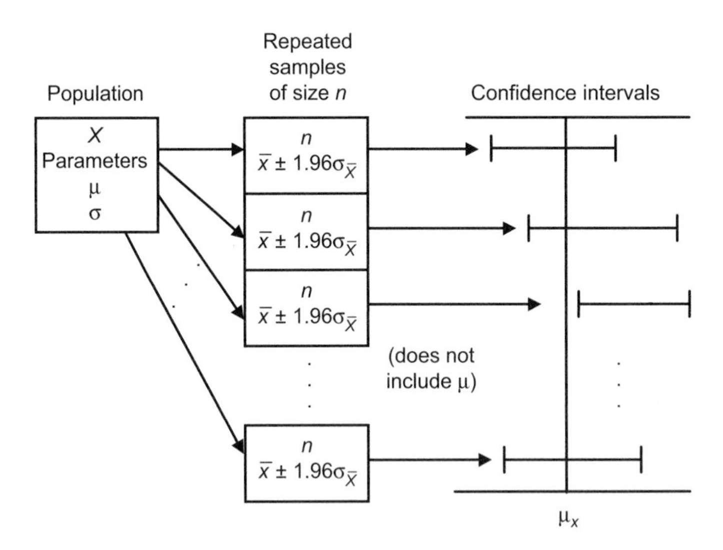

FIGURE 7-8. Interval estimates constructed from repeated samples of size *n*.

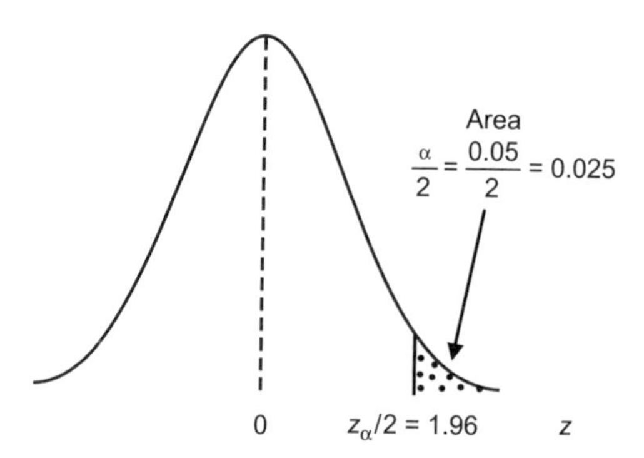

FIGURE 7-9. Calculation of $z_{\alpha/2}$ for a 95% confidence interval.

taining the true value of μ. The other 5% of the samples give intervals that do not contain μ.

Intervals specified in this way are known as *confidence intervals*. In practice, a confidence interval is constructed from a single sample. Obviously, this interval either contains μ or it does not. Because of the randomness inherent in sampling, we cannot be completely sure which of these two possibilities occurs. However, we do know from the central limit theorem that there is a 0.95 chance of selecting *a single sample* that produces an interval containing μ; thus we are 95% confident in the interval obtained from our particular sample. This is a classical interpretation of a confidence interval, which states that the interval has a 95% chance of containing μ. In other words, the classical interpretation is that 95 times out of 100, intervals constructed in this manner will contain the true population mean.

The subjectivist interpretation of a confidence interval is different and somewhat less cumbersome. It states that there is a 95% chance that μ is within the interval. Equivalently, we could say that the probability of μ being within the interval is 0.95. Note that for an objectivist, this is nonsense. For them μ is a constant, not a random variable, and there are no probabilities that can be assigned to μ.

Confidence intervals can be constructed for any desired degree of confidence. Depending on the problem at hand, we may wish to be 90, 95, or 99% or even more confident in our result. Let $1 - \alpha$ be the desired degree of confidence specified as a proportion. For example, if α is 0.05, the degree of confidence is 0.95, representing 95% confidence. Similarly, α values of 0.1 and 0.01 correspond to 90 and 99% confidence. Let us define $z_{\alpha/2}$ as the value of the standard normal table such that the area to the *right* of $z_{\alpha/2}$ is $\alpha/2$. For a 95% confidence interval we wish $\alpha/2 = 0.05/2 = 0.25$ of the area of the curve to be right of $z_{\alpha/2}$ (see Figure 7-9). From the standard normal table, $z_{.025} = 1.96$. Different confidence levels imply different values for $z_{\alpha/2}$ and therefore for the upper and lower bounds of the confidence interval. In general, confidence intervals for μ are written $\bar{x} \pm z_{\alpha/2}\sigma/\sqrt{n}$. A simple example will explain this procedure.

EXAMPLE 7-1. A random sample of $n = 25$ shoppers at a supermarket reveals that patrons travel, on the average, 16 minutes from their homes to the supermarket. Furthermore, we are told that the population standard deviation for this variable is $\sigma = 4$ and that these travel times are normally distributed. The problem is to construct the 99% confidence interval for the average travel time to this supermarket. The desired interval for 0.99 confidence is

$$\bar{x} - z_{\alpha/2}\frac{\sigma}{\sqrt{n}} \le \mu \le \bar{x} + z_{\alpha/2}\frac{\sigma}{\sqrt{n}}$$

which, by substitution of $\bar{x} = 16$, $\sigma = 4$, $n = 25$, and $\alpha = 0.01$, can be rewritten as

$$16 - z_{.005}\frac{4}{\sqrt{25}} \le \mu \le 16 + z_{.005}\frac{4}{\sqrt{25}}$$

From the normal table, $z_{.005} = 2.58$, so that the 99% confidence interval is given by

$$16 - 2.58\frac{4}{5} \le \mu \le 16 + 2.58\frac{4}{5} = 16 \pm 2.064 = [13.936, 18.064]$$

The value of μ may or may not be in this range. Our hope is that the random sample used to construct the interval is one of the 99% of all samples of size $n = 25$ that do contain the true value for μ.

The width of a confidence interval is influenced by two principal factors under control of the researcher: the desired degree of confidence $(1 - \alpha)$ and sample size n. Let us examine the impact of both factors on confidence intervals for travel time. First, note how varying degrees of confidence affect interval width (Table 7-2). As our confidence level increases, $z_{\alpha/2}$ increases, and the width of the interval increases. Figure 7-10 shows the same behavior in graphical form over a wider range of confidence levels. Notice that as the level of confidence approaches 100%, the width approaches infinity, representing the entire number line. This is because we can *never* be 100% sure that an interval contains μ unless the interval covers the entire range of possible values of the population variable X. If X is normal, the range of possible values runs from $-\infty$ to $+\infty$. Notice also that the interval width increases very rapidly for confidence levels above 95%. Increasing the confidence level from 0.7 to 0.95 costs relatively little in terms of increasing width, but diminishing returns set in at higher levels. We can demand any arbitrarily high level of certainty, but only at the expense of greatly increasing interval width. This behavior is consistent with the idea that a sample of fixed size contains a finite amount of information. By increasing the confidence level, we are demanding greater certainty that the interval contains μ. We have not added any new information; thus this increase in certainty can only come at the expense of a wider interval. That is, we become increasingly certain but make progressively less precise statements regarding the value of μ.

TABLE 7-2

Confidence Intervals for μ with Varying Degrees of Confidence

				Lower bound	Upper bound	Values of bound for travel time problem		
Confidence level	α	$\alpha/2$	$z_{\alpha/2}$	$\bar{x} - z_{\alpha/2}\sigma/\sqrt{n}$	$\bar{x} + z_{\alpha/2}\sigma/\sqrt{n}$	Lower	Upper	Interval width
0.90	0.10	0.050	1.65	$\bar{x} - 1.65\sigma/\sqrt{n}$	$\bar{x} - 1.65\sigma/\sqrt{n}$	14.688	17.312	2.624
0.95	0.05	0.025	1.96	$\bar{x} - 1.96\sigma/\sqrt{n}$	$\bar{x} + 1.65\sigma/\sqrt{n}$	14.432	17.568	3.136
0.99	0.01	0.005	2.58	$\bar{x} - 2.58\sigma/\sqrt{n}$	$\bar{x} + 2.58\sigma/\sqrt{n}$	13.936	18.064	4.128

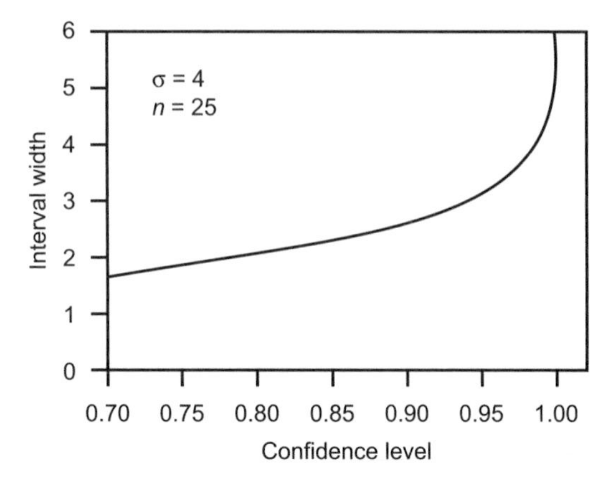

FIGURE 7-10. Effect of confidence level on interval width.

TABLE 7-3
Confidence intervals for μ with varying sample sizes

Sample size n	Confidence level	$z_{\alpha/2}$	Lower bound $\bar{x} - z_{\alpha/2}\frac{\sigma}{\sqrt{n}}$	Upper bound $\bar{x} + z_{\alpha/2}\frac{\sigma}{\sqrt{n}}$	Interval width
25	0.95	1.96	14.432	17.568	3.136
100	0.95	1.96	15.216	16.784	1.568
400	0.95	1.96	15.608	16.392	0.784

Table 7-3 summarizes the impact of sample size on interval width. Interval widths are calculated for sample sizes of 25, 100, and 400. As the sample size increases, the width of the confidence interval decreases, everything else being equal. The same relation is shown in Figure 7-11, where it can be seen that the width approaches zero as the sample size approaches infinity. The center of the interval is always the sample mean $\bar{X}$, but this approaches μ as n approaches infinity. With the interval narrowing and at the same time becoming more nearly centered on μ, we see the *confidence interval* for μ converging on the *value* of μ with larger sample sizes. This has great intuitive appeal. As our sample includes more and more members of the population, we become more and more knowledgeable about the population. If the sample includes the entire population, there should be no doubt about the value of the population parameter; it can simply be calculated from known information. The figure shows how gathering more information (increasing n) permits increasingly precise statements about μ. However, once again we see diminishing returns. For example, doubling the sample size decreases the width of the confidence interval, but by less than a factor of two.

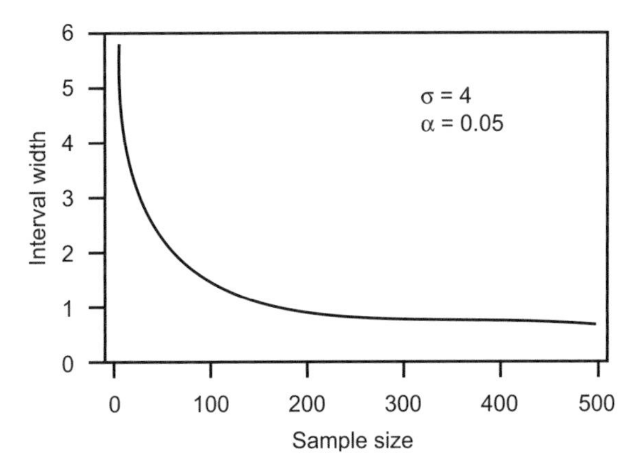

FIGURE 7-11. Effect of sample size on confidence interval width.

There are thus two ways of decreasing the width of confidence intervals, that is, of making more precise statements about μ. First, we can decrease the level of confidence. Second, and a usually more worthwhile alternative, we can increase the sample size. The formula for the confidence interval of the population mean can now be summarized.

DEFINITION: INTERVAL ESTIMATE OF POPULATION MEAN μ

Let $x_1, x_2, \ldots, x_n$ be a representative sample of size n randomly drawn from a normally distributed random variable X. The confidence interval with confidence level $1 - \alpha$ is

$$\bar{x} \pm z_{\alpha/2} \frac{\sigma}{\sqrt{n}} \tag{7-3}$$

where $\bar{x}$ is the sample mean, σ is the population standard deviation, and $z_{\alpha/2}$ is the value of the standard normal distribution for which the area to the right of $z_{\alpha/2}$ is $\alpha/2$.

Interval Estimation of μ with X Not Normally Distributed, σ Known

If the population random variable X is not normally distributed, then we cannot appeal to the strict form of the central limit theorem to justify the normality of the sampling distribution of $\bar{X}$. All we can say is that the sampling distribution is *approximately* normal. It follows that the confidence interval given in Equation 7-3 can also be considered *approximately* correct. As a rule of thumb, the approximation can be considered sufficiently close when the sample size (n) exceeds 30.

Interval Estimation of μ with σ Unknown

An extremely common problem in developing a confidence interval for μ is that the population standard deviation, σ is unknown. After all, it does seem that μ would be known along with σ, given that it is needed to compute σ. When σ is unknown, there is no way to find the standard error of the mean $\sigma_{\bar{X}} = \sigma/\sqrt{n}$; thus we cannot use (7-3) to find the confidence interval. Luckily, there is a simple way around this problem. First, the values of the sample $x_1, x_2, \ldots, x_n$ are used to estimate σ by

$$s = \sqrt{\frac{\sum_{i=1}^{n}(x_i - \bar{x})^2}{n-1}} \tag{7-4}$$

Recall that this is the most commonly used estimate of σ identified in Section 7.2. With s, we can replace $\sigma_{\bar{X}} = \sigma/\sqrt{n}$ by the estimate $s/\sqrt{n}$. However, when this is done, we can no longer use the normal distribution to calculate the confidence interval. To see why, recall that confidence intervals were based on probability statements involving the random variable $\bar{X}$ having the form

$$P\left(-k \leq \frac{\bar{X} - \mu}{\sigma/\sqrt{n}} \leq k\right) = 1 - \alpha$$

(Equation 7-1). Knowing the middle quantity is Z, we could find k for any confidence level desired. For example, for $\alpha = 0.05$, k is 1.96. When σ is unknown, the analogous statement is

$$P\left(-k \leq \frac{\bar{X} - \mu}{s/\sqrt{n}} \leq k\right) = 1 - \alpha \tag{7-5}$$

The middle quantity now contains two random variables ($\bar{X}$ and s) and is not a standard normal variable. Fortunately, the distribution is known, *providing X is normal.* In particular, assuming X is normal, the ratio

$$T = \frac{\bar{X} - \mu}{s/\sqrt{n}}$$

follows the so-called t-distribution. This is also called the *Student's t-distribution,* because the inventor wrote under the pseudonym *Student.* The t-distribution is symmetric and somewhat similar to the standard normal distribution. Not surprisingly, its variance is larger than Z. Because the population parameter σ is replaced by an estimate s, more uncertainty is introduced, and there is greater variability in the distribution.

Obviously, the bigger the sample, the better the estimate s will be. It follows that the bigger the sample, the more the t-distribution resembles the standard normal.

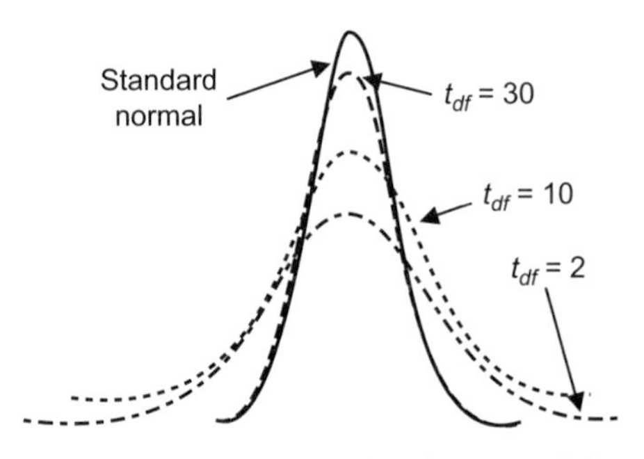

FIGURE 7-12. Normal distribution and the *t*-distribution for 2, 10, and 30 *df.*

We see that the form of the t distribution depends on the sample size n. Because we utilize $n - 1$ in the denominator for s in Equation 7-4, the distribution of t is tabulated according to its *degrees of freedom* (*df*) rather than the sample size n. The degrees of freedom are always equal to the sample size minus 1, or $n - 1$. Figure 7-12 compares the standard normal with the t-distributions for degrees of freedom 2, 10, and 30. At $df = 30$, note the similarity between the standard normal and t-distributions.

One way of quantifying the convergence of the standard normal and the t-distributions with increasing degrees of freedom is to examine the value of t-corresponding to a certain tail area and compare it to the standard normal. For example, we know that a value of $z = 1.96$ for the standard normal leaves 0.025 of the area under the curve in each tail of the distribution. Now, if we compare this to the values of t for several members of the family of t-distributions, we find

Degrees of freedom (*df*)	2.5% value of *t*
1	12.806
2	4.303
3	3.182
10	2.228
20	2.086
30	2.042
40	2.021
60	2.000
120	1.980
Infinity	1.960

As the sample size and degrees of freedom increase, the value of t converges to 1.96. Moreover, by the time $df = 30$ or so, there is a close correspondence between z and t. Table A-4 in the Appendix has *t-values* for several other confidence levels. In all cases, t converges to z with increasing *df.* Where before we used $z_{\alpha/2}$ to help find the boundary of the confidence level, now we use $t_{\alpha/2,n-1}$. The notation is similar to $z_{\alpha/2}$, in that we seek an upper-tail probability of $\alpha/2$. The notation for t includes the degrees of freedom $n - 1$, indicating that the appropriate value of t depends on sample size.

DEFINITION: CONFIDENCE INTERVAL ESTIMATION FOR μ WITH UNKNOWN POPULATION STANDARD DEVIATION σ

Let $x_1, x_2, \ldots, x_n$ be a sample of size n drawn from a normal population. The confidence interval for μ at level of confidence $1 - \alpha$ is

$$\bar{x} \pm t_{\alpha/2,n-1} \frac{s}{\sqrt{n}} \tag{7-6}$$

where $\bar{x}$ is the sample mean, s is the sample standard deviation estimated using Equation 7-4, and $t_{\alpha/2,n-1}$ is the value of the t-distribution with degrees of freedom equal to $n - 1$, such that the area to the right of t is $\alpha/2$.

The confidence interval specified by Equation 7-6 differs from that of Equation 7-4 in two ways. First, because σ is unknown, it is replaced by the estimate s. Second, a value from the t-distribution is used rather one from the z-distribution. As an example of confidence interval formula (7-6), let us return to the travel time example. The sample size is $n = 25$, and the mean is $\bar{x} = 16$. Now, let us change the situation slightly by assuming that the population standard deviation is unknown but has been estimated as $s = 4$. The confidence interval formula from the t-distribution in Equation 7-6 is

$$\bar{x} \pm t_{\alpha/2,n-1} \frac{s}{\sqrt{n}}$$

which for a $1 - \alpha = 0.95$ or 95% confidence is equal to

$$16 \pm 2.064 \frac{4}{\sqrt{25}} = 16 \pm 1.651 = [14.349,\ 17.651]$$

The value of $t = 2.064$ is found in the t table for $t_{.025,\ 24}$ by using the column labeled .025 and the row (for degrees of freedom) corresponding to 24. This interval is slightly wider than the interval found when σ was known, [14.432, 17.568]. Because we are in a situation of greater uncertainty with σ unknown, this is expected. Because the t- and the z-values give very similar confidence intervals in large samples, the normal distribution with z is often used, even in cases where σ is unknown. For small samples, t should always be used. It is important to remember that Equation 7-6 requires that X be normally distributed. If this is not true, the confidence intervals will be approximate, with the approximation improving as sample size increases.

Interval Estimation for π

For binomial random variables, the population proportion of successes can be estimated by using

$$P = \frac{X}{n} \tag{7-7}$$

where X represents the number of successes in n trials. To define the confidence interval for π, it is necessary to specify the sampling distribution of the statistic P. This is not too difficult, because P is an average of sorts and therefore falls under the central limit theorem. To see why P is an average, think of the sample as a collection of 0's and 1's, where 1 corresponds to a "success." The number of successes, [X in Equation 7-7] is just the sum of the sample values Σx_i, because only the 1's contribute to the sum. Dividing by n gives the proportion of successes P, which also happens to be the mean of the sample. As a result, the sample mean P is approximately normally distributed for large sample sizes. To determine the parameters of the sampling distribution μ_p and σ_p, we can use what we know about the parameters of the binomial random variable X. First, note the mean of X, $E(X) = n\pi$ [see Equation 5-28],

$$\mu_P = E(P) = E\left(\frac{X}{n}\right) = \frac{1}{n}E(X)$$

$$= \frac{1}{n}(n\pi) = \pi$$

As expected, the distribution of P is centered on π, the population proportion. This simply proves a point made in Section 7.2: P is an unbiased estimator of π. The standard deviation of the sampling distribution of P can be found in a similar way by using the formula for the variance (V) of a binomial random variable. Recall that if X is binomial, its variance is $V(X) = n\pi(1 - \pi)$ [see Equation 5-29]. With this, the variance of P can be found as:

$$\sigma_P^2 = V\left(\frac{X}{n}\right) = V\left(\frac{1}{n}X\right)$$

$$= \frac{1}{n^2}V(X) = \frac{n\pi(1-\pi)}{n^2}$$

$$= \frac{\pi(1-\pi)}{n}$$

and thus

$$\sigma_P = \sqrt{\frac{\pi(1-\pi)}{n}}$$

The principal difficulty here is that the standard deviation of P, or σ_P, itself requires a knowledge of π for estimation. This problem is solved by substituting our sample proportion p for π. The confidence interval for π can now be summarized.

DEFINITION: CONFIDENCE INTERVAL FOR π

Let X be a binomial random variable representing the number of successes in n trials where the probability of success in each trial is π. The confidence interval for π at a level of confidence $1 - \alpha$ is

$$p - z_{\alpha/2}\sqrt{\frac{p(1-p)}{n}} \leq \pi \leq p + z_{\alpha/2}\sqrt{\frac{p(1-p)}{n}} \qquad (7\text{-}8)$$

where p is the proportion of successes in a sample of size n and $z_{\alpha/2}$ is the value of the standard normal distribution with an upper-tail probability of $\alpha/2$.

The confidence interval for π in Equation 7-8 should be used only for large samples, say $n > 100$. For smaller sample sizes, the exact sampling distribution of the statistic P is the discrete binomial distribution. The binomial mass function (5-27) can be used to generate the exact sampling distribution, or else tables can be consulted.

EXAMPLE 7-2. An urban geographer interested in estimating the proportion of residents in a city who moved within the last year conducts a survey. Summary statistics reveal that 27 of the 270 people surveyed changed residences. What are the 95% limits on the true proportion of people who moved in the city?

Because the sample size of 270 is greater than 100, we can utilize formula (7-8) with the estimate $p = 27/270 = .10$. The limits are therefore

$$.10 \pm 1.96\sqrt{\frac{(.10)(.90)}{270}}$$

$$.10 \pm .036$$

$$[.064, .136]$$

EXAMPLE 7-3. Prior to a referendum on a bylaw requiring mandatory recycling in a city, a random sample of $n = 100$ voters is taken and their voting intentions are obtained. Of those surveyed, 46 favor the bylaw, and 54 are opposed. What are the 90% confidence limits on the population proportion in favor of this bylaw?

Once again the sample is large, so formula (7-8) can be used. With $p = .46$, the interval is

$$.46 \pm 1.65\sqrt{\frac{(.46)(.54)}{100}}$$

$$.46 \pm .082$$

$$[.378, .542]$$

Notice that the interval includes $\pi = 0.5$. Despite the fact that only 46% in the sample favor the bylaw, it is still possible that the true proportion in favor is a majority.

7.4. Sample Size Determination

One common question asked of statisticians is, "How large a sample will I need in my research?" This is an extremely complicated question involving many different issues. However, in certain restricted conditions, we can use our knowledge of sampling distributions and confidence intervals to determine appropriate sampling sizes. To determine the sample size necessary in a study, we must know three things:

1. Which population parameter is under study? Is it a mean, proportion, or some other parameter?
2. How close do we wish the estimate made from our sample to be to the actual (or true) value of the population parameter?
3. How certain (or, in statistical jargon, confident) do we wish to be that our estimate is within the tolerance specified in item 2?

The answer to item 1 tells us which sampling distribution governs the statistic of interest. In this chapter we limit ourselves to the population mean μ and proportion π.

We illustrate the procedure for sample size determination in the context of estimating μ. Suppose we wish to take a sample that is capable of yielding a point estimate of μ that is no more than E units away from its actual value. Short for *error,* E is a statement of the level of precision we desire. If E is small, we are saying we can tolerate only a small error in the estimate; if it is large, our required precision is less. Let us also assume that we wish to be 95% confident. In other words, a 5% chance of drawing a sample of size n with an estimate $\bar{x}$ for μ that is more than E units away from the true value of the population mean is acceptable.

To see how we can determine the sample size n meeting these requirements, let us examine the sampling distribution of $\bar{X}$ and the 95% confidence interval illustrated in Figure 7-13. We know that the interval $\mu \pm 1.96\sigma/\sqrt{n}$ contains exactly 95% of the values of the statistic $\bar{X}$. On this basis, the 95% confidence interval for μ with an estimate $\bar{x}$ is $\bar{x} \pm 1.96\ \sigma/\sqrt{n}$. If we wish to be no more than E units from μ with the estimate $\bar{x}$, then we can set

$$E = 1.96 \frac{\sigma}{\sqrt{n}}$$

which is one-half the interval width. In general, for a level of confidence $1 - \alpha$, we can set

$$E = z_{\alpha/2} \frac{\sigma}{\sqrt{n}} \tag{7-9}$$

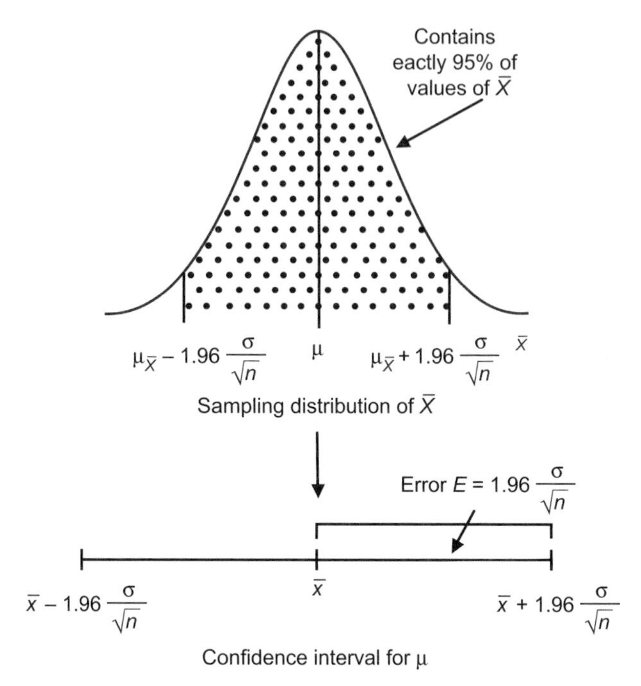

FIGURE 7-13. Relationship between sampling distribution, confidence limit and error.

We now must solve Equation 7-9 for n. With a few straightforward algebraic manipulations, we find

$$n = \left(\frac{z_{\alpha/2}\sigma}{E} \right)^2 \tag{7-10}$$

This formula indicates that the required sample size depends on the desired level of confidence (the higher the confidence, the larger $z_{\alpha/2}$, and hence the larger n), the desired precision (the higher the precision, the smaller E, and hence the larger n), and the variability of the underlying population (the greater the variability in the population, the larger σ, hence the larger n). All these relations have great appeal. It makes perfect sense that increased precision implies a greater sample size.

The single difficulty with Equation 7-10 is that seldom will σ be known beforehand, except where a population census has been completed. If this were the case, μ would be known and would not require estimation! Usually, then, we must use an estimate of σ to determine n. The obvious candidate is s. If a rough estimate of s is not available from a recent survey, then a small pilot study can be undertaken to generate s. This is a quite common procedure in social science research. If no estimate of s exists, then the following rough rule of thumb can be used: It is very unlikely to observe a value more than 2 standard deviations above or below the mean. So, if we could guess at the range of X, we could take that as 4 σ. For example, if we estimate the range of values for the random variable X to be 100, we could use 25 as an estimate for σ. Usually, it is unnecessary to resort to this crude procedure.

EXAMPLE 7-4. A soils scientist is interested in determining the number of soil samples in a specific soil horizon necessary to estimate the extractable P_2O_5 (measured in milligrams per 100 g). Previous research indicates the standard deviation of P_2O_5 is about 0.7 mg. The desired precision is $E = 0.2$, and a confidence level for the study is set at $\alpha = 0.05$ or 95%. Using formula (7-10) gives

$$n = \left[\frac{1.96(0.7)}{0.2}\right]^2 = \left(\frac{1.372}{0.2}\right)^2 = 47.06$$

So, roughly 47 samples should be taken. Note that the entire procedure is based on the assumption that the random variable X is normally distributed, or that the required sample size is large enough ($n > 30$), and the central limit theorem applies.

An exactly analogous argument can be used to develop the following sample size formula for the population proportion π:

$$n = \left[\frac{z_{\alpha/2}\sqrt{p(1-p)}}{E}\right]^2 \tag{7-11}$$

This formula utilizes the estimate p for π in the confidence interval formula. Evidently, some knowledge of the likely value of π, say from a previous study, can be used to make the necessary sample size decision. If no estimate of p is available, then a useful upper bound on n can be found by assuming $p = 0.5$. With $p = 0.5$, the value of $p(1-p)$ is maximized, as shown in the following table:

p	0.30	0.40	0.50	0.60	0.70
$\sqrt{p(1-p)}$	0.46	0.49	0.50	0.49	0.46

Because $p\sqrt{(1-p)}$ appears in the numerator of (7-11), the value of n is maximized by maximizing this expression. We see that using 0.5 as an estimate is conservative in the sense that the value of n will be, if anything, larger than actually needed.

EXAMPLE 7-5. An urban geographer is interested in determining the sample size necessary to estimate the proportion of households in a large metropolitan area who moved in the last year. Past experience indicates that approximately 20% of all residents in North American cities move each year. How many households should be sampled? Assume a confidence level of 95% and that the geographer will tolerate an error of $E = 0.03$. Using Equation 7-11 with an estimate of $p = .2$ yields

$$n = \left[\frac{1.96\sqrt{(0.2)(0.8)}}{0.03}\right]^2 = 682.95$$

TABLE 7-4
Summary of Point Estimators and Confidence Intervals for π and μ

Population parameter	Point estimator	Formula for confidence level	Approximate conditions
μ	$\bar{X}$	$\bar{x} \pm z_{\alpha/2}(\sigma/\sqrt{n})$	Exact for any sample size when σ is known and X normally distributed. Approximate when X is not normally distributed, but $n > 30$.
μ	$\bar{X}$	$\bar{x} \pm t_{\alpha/2,n-1}(s/\sqrt{n})$	Exact when σ is unknown and X normally distributed. Approximate when X is not normal, but $n > 30$.
π	P	$p \pm z_{\alpha/2}\sqrt{p(1-p)/n}$	Approximate when $n > 100$.

or approximately 683 households. It might also be interesting to determine an upper bound on the sample size by assuming $p = .5$:

$$n = \left[\frac{1.96\sqrt{(0.5)(0.5)}}{0.03}\right]^2 = 1067.11$$

Under these conditions, a very large sample of 1067 is required. If the researcher feels the percentage of people in the city who moved is likely to be larger than the 20% characteristic of North American cities, it would be wise to enlarge the sample.

7.5. Summary

This chapter has developed the ideas in statistical inference used to estimate the value of a population parameter. We distinguished between two types of statistical estimation: point estimation and interval estimation. As is suggested by its name, a point estimate is a single value or point used to predict a population parameter, and it is based on some function of the sample values. To indicate the reliability of an estimate, it is common to construct a confidence interval—a range of values known to contain the true value of the population parameter with a prescribed level of confidence. The more commonly used confidence levels are 90, 95, and 99%, but the technique is general enough to produce intervals for any desired confidence level. One can increase the confidence level to an arbitrarily high value, but only at the expense of a wider interval. That is, every confidence interval represents a trade-off between precision (interval width) and degree of certainty (confidence level). The formulas for point and interval estimates for μ and π are summarized in Table 7-4. Although other point estimates are available, their sampling distributions are in general unknown; thus they do not easily lead to confidence intervals.

REFERENCE

R. S. Pindyck and D. L. Rubinfeld, *Econometric Models and Economic Forecasts,* 2nd ed. (New York: McGraw-Hill, 1981).

FURTHER READING

Most introductory textbooks on statistical methods for geographers also contain the material covered in this chapter. Usually, these methods are discussed in the context of representative applications in various fields. The problems at the end of this chapter also indicate several possible applications of these techniques in geographical research.

PROBLEMS

1. Explain the meaning of the following concepts:
   - Point estimation
   - Interval estimation
   - Unbiased estimator
   - Statistical estimator
   - Efficient estimator
   - Statistical estimate
   - Confidence interval

2. Discuss the two principal types of statistical estimation.

3. When should the t-distribution be used in place of the z-distribution in the construction of confidence intervals for the population mean?

4. The yields in metric tons per hectare of potatoes in a randomly selected sample of 10 farms in a small region are 32.1, 34.4, 34.9, 30.6, 38.4, 29.4, 28.9, 32.6, 32.9, and 44.9. Assuming that these yields are normally distributed, determine a 99% confidence interval on the population mean yield.

5. Past experience shows that the standard deviation of the distances traveled by consumers to patronize a "big-box" retail store is 4 km. Adopting an error probability of 0.05, how large a sample is needed to estimate the population mean distance traveled to within 0.5 km? 1 km? 5 km?

6. The proportion of automobile commuters in a given neighborhood of a large city is unknown. A random sample of 50 households is taken, and 38 automobile commuters are found. Determine the 95% confidence interval of the proportion of commuters by automobile in the neighborhood.

7. A historical geographer is interested in the average number of children of households in a certain city in 1800. Rather than spend the time analyzing each entry in the city directory, she decides to sample randomly from the directory and estimate the size from this sample. In a sample of 56 households, she finds the average number of children to be 4.46 with a standard deviation of 2.06.
   a. Find the 95% confidence interval, using the t-distribution.
   b. Find the 95% confidence interval, using the z-distribution.
   c. Which interval is more appropriate? Why?

8. A geographer is asked to determine the sample size necessary to estimate the proportion of residents of a city who are in favor of declaring the city a nuclear-free zone. The estimate must not differ from the true proportion by more than 0.05 with a 95% confidence level. How large a sample should be taken? at 99%?

9. Suppose we double the size of a sample taken when trying to estimate the confidence interval for a mean. What will the effect be on the width of the confidence interval, assuming all other parameters (significance level and standard deviation) are held constant?

8

One-Sample Hypothesis Testing

Chapter 7 presented the form of statistical inference known as estimation and covered the techniques used for point and interval estimation for a few representative statistics. This chapter introduces the form of statistical inference known as *hypothesis testing.* The methods used in hypothesis testing are closely related to confidence interval estimation. They are simply two ways of looking at the same problem, and both are based on the same theory. In Section 8.1, a highly structured method for testing hypotheses known as the *classical test of hypothesis* is outlined. Section 8.2 presents a newer and widely used variant of this procedure, the PROB-VALUE, or *p*-VALUE method. Hypothesis tests for population parameters μ and π are detailed in Sections 8.3 and 8.4. The link between hypothesis testing and confidence intervals is explained in Section 8.5. In many practical problems, the principal concern is not to test a hypothesis about the value of some parameter for a *single* population. Rather, we may want to compare *two* populations with regard to some quantitative characteristic or parameter. For example, suppose we have measures of carbon monoxide concentration from different residential neighborhoods in some city. We wish to compare the two samples. The statistical question being asked is whether these two samples come from a single population or from two different populations. So-called two-sample tests of hypotheses for means and proportions are explained in Chapter 9. These two-sample tests also utilize the basic testing method developed in Sections 8.1 and 8.2.

8.1. Key Steps in Classical Hypothesis Testing

A general conceptual framework for hypothesis testing is illustrated in Figure 8-1. We are interested in making inferences about the value of a population parameter θ. We hypothesize a value for this unknown population parameter, say $\theta = \theta_0$. A random sample of a given size n is collected having values $x_1, x_2, \ldots, x_n$. From these sample data a point estimator $\hat{\theta}$ is calculated. On the basis of $\hat{\theta}$, we evaluate our hypothesis by determining whether the sample value $\hat{\theta}$ does or does not support the contention that $\theta = \theta_0$. If $\hat{\theta}$ is close to θ_0, we might be led to believe the hypothesis is true. If $\hat{\theta}$

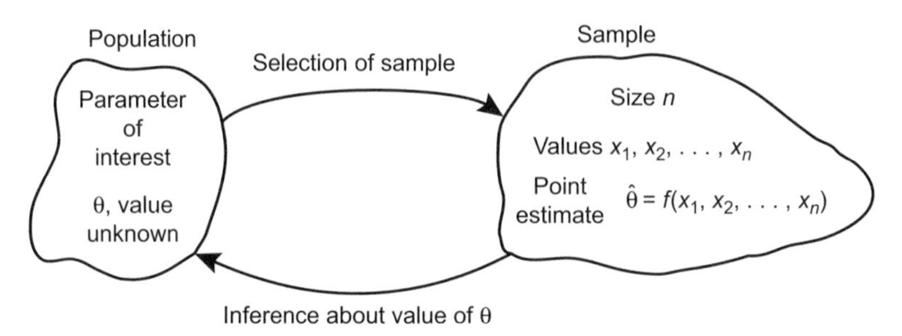

FIGURE 8-1. Sampling and statistical hypothesis testing.

is very much different from θ_0, it is less likely that the hypothesis $\theta = \theta_0$ can be sustained. Recall that two reasons explain why $\hat{\theta}$ might differ from $\theta = \theta_0$. First, the hypothesis $\theta = \theta$ may actually be untrue. Second, $\hat{\theta}$ might differ from θ simply because of sampling error. The framework shown in Figure 8-1 can be illustrated with a simple example involving residential mobility.

EXAMPLE 8-1. The average residential mobility in a North American city can be measured by the probability that a typical household in the city changes residence in a given year. Most studies have found that about 20% of the population of the city moves in any one year, or $\pi = 0.2$. On this basis, we would expect a typical *neighborhood* to have this overall rate of residential mobility. Let us now consider a specific neighborhood in a specific city. How could we determine whether this neighborhood has the citywide mobility rate? We first *hypothesize* a value of $\pi = \pi_0 = 0.2$. From a random sample of households in the neighborhood, we then determine the proportion who have moved into the neighborhood in the past year. Since this is a sample proportion, we call this value *p*, a specific value of the estimator *P*. Now, on the basis of *p*, we must decide whether this is a typical neighborhood (in terms of residential mobility). If *p* were close to our hypothesized value of 0.2, we would probably decide that the hypothesis is true and conclude it is a typical neighborhood. If, however, the sample proportion were $p = 0.8$, we would be less likely to believe the hypothesis $\pi = 0.2$. In rejecting our hypothesis, we must always consider the possibility that this is an incorrect judgment. Our judgments are always based on samples and are therefore subject to sampling error.

This section outlines a rigorous procedure for testing any hypothesis of this type. A classical test of hypothesis follows the six steps shown in Table 8-1. We now consider each of these six steps, in turn, and then illustrate the procedure using a straightforward example.

Step 1: Formulation of Hypotheses

Statements of the hypothesis to be evaluated in a classical test *always* can be expressed in one of these three ways:

(*A*)	(*B*)	(*C*)
$H_0 : \theta = \theta_0$	$H_0 : \theta \geq \theta_0$	$H_0 : \theta \leq \theta_0$
$H_A : \theta \neq \theta_0$	$H_A : \theta < \theta_0$	$H_A : \theta > \theta_0$

The symbol H stands for hypothesis. There are two parts to any hypothesis, labeled H_0 and H_A. In H_0 a statement is made about the value of some population parameter θ. The value θ_0 is the hypothesized, or conjectured, value for θ. In the first form (*A*), the statement $H_0 : \theta = \theta_0$ means we are interested in deciding whether the value of the population parameter θ is equal to θ_0. In H_0, we assert that θ does equal θ_0. This is called the *null hypothesis*. We can either reject this null hypothesis or accept it, depending on the information we collect from the sample. If H_0 is not true, something else must be. As an alternative to H_0, we offer the statement H_A: $\theta \neq \theta_0$. Obviously, if we reject the statement $\theta = \theta_0$, then the only remaining possibility is $\theta \neq \theta_0$. Thus H_A is called the *alternate hypothesis*. In our residential mobility example, the parameter of interest is π, the population proportion, so the null and alternate hypotheses are

$$H_0 : \pi = 0.2 \qquad H_A : \pi \neq 0.2 \tag{8-1}$$

This short form is an especially useful way of summarizing the hypotheses in a classical test. The population parameter of interest is clearly identified as π along with the hypothesized value of $\theta_0 = 0.2$.

There are two other possible formats for H_0 and H_A, shown as hypothesis sets *B* and *C*. Sometimes we are interested in asserting *range* of values rather than an *exact* value for the population parameter θ. In format *B*, for example, we wish to test whether our sample is consistent with the statement $\theta \geq \theta_0$. That is, can we say the true population parameter is *at least* as large as θ_0? The alternative is that θ is definitely *less than* θ_0, as is expressed in H_A. In format *C*, we are interested in whether our sample supports the statement $\theta \leq \theta_0$; that is, the true value of the population parameter is *no larger than* θ_0. In this case, the alternate hypothesis must be that θ is *larger than* θ_0.

Which of these three forms should we choose? The correct format depends on the question we wish to answer. To see this, let us return to Example 8-1 and rates of residential mobility in urban neighborhoods. If we express our hypothesis concerning residential mobility in the form of Equation 8-1, we are implicitly saying that we are interested in determining whether residential mobility in our study neighborhood

TABLE 8-1.
The Six Steps of Classical Hypothesis Testing

Step 1:	Formulation of hypothesis.
Step 2:	Specification of sample statistic and its sampling distribution
Step 3:	Selection of a level of significance.
Step 4:	Construction of a decision rule.
Step 5:	Compute the value of the test statistic.
Step 6:	Decision.

differs from the typical rate of 0.20. We are *not* trying to determine whether mobility is either lower or higher in this particular neighborhood. Were we to express the hypotheses as in format *B*

$$H_0 : \pi \geq 0.2 \qquad H_A : \pi < 0.2 \tag{8-2}$$

we would be testing a hypothesis about whether the neighborhood has a lower than average residential mobility. If we can reject H_0, then our alternate hypothesis tells us that the neighborhood is more stable than the average. If the neighborhood under study is an older suburb where there is predominantly single-family housing, this may be a hypothesis well worth evaluating. Notice that by rejecting the null hypothesis H_0, we are accepting H_A, that the neighborhood is less mobile than the city at large. Unlike format (*A*), this form of H_A does not allow for the neighborhood to be more mobile.

By casting our hypotheses in format *C*

$$H_0 : \pi \leq 0.2 \qquad H_A : \pi > 0.2 \tag{8-3}$$

we could test the hypothesis of whether the neighborhood has a higher turnover than average. If we were studying the residential neighborhood around a major university, a high rate of mobility might be expected. In all three formats *A, B,* and *C,* what we are really interested in saying is stated in the alternate hypothesis H_A, which we prove by evaluating H_0 and then rejecting it, if possible. This apparently peculiar way of operating is actually the only way of proceeding. The reasons will become clear when we discuss the second step in the hypothesis-testing procedure.

Hypothesis format *A* is sometimes referred to as a *two-sided,* or *nondirectional,* test. This is because the alternate hypothesis does not say on which *side,* or *direction,* the true population parameter θ lies in relation to the hypothesized value θ_0. Formats *B* and *C* are usually called *directional hypotheses* because the alternate hypothesis H_A does specify on which side of θ_0 the true parameter lies. Sometimes format *B* is termed a *lower-tail test* and format *C* an *upper-tail test,* on the basis of the direction specified in H_A. The reasons for these names and the choice of formats among A, B, and C are explained further in the examples that follow.

In setting up H_0 and H_A, it is essential that they be *mutually exclusive* and *exhaustive.* We are trying to choose between the two hypotheses, so there must be no chance that both are true. H_0 must exclude the possibility of H_A, and vice versa. In addition, we want to be able to say that if H_0 is false, H_A is definitely true. There must be no possibility that some other (unspecified) hypothesis is true instead of H_A. In order to meet this requirement, H_0 and H_A must contain (exhaust) all possible values of θ.

Whenever sample data are used to choose between H_0 and H_A, there are risks of making an incorrect decision. These errors are known as *inferential errors,* because they arise when we make an incorrect inference about the value of the population parameter. Table 8-2 summarizes the way errors arise in hypothesis testing. Note that any decision about the null hypothesis may have four possible outcomes. In two cases,

TABLE 8-2
Decisions and Outcomes in a Classical Test of Hypothesis

	True state of nature	
Decision	H_0 is true	H_0 is false
Reject H_0	Type I error	No error
Accept H_0	No error	Type II error

the decision is the correct one. This occurs when we rightfully reject a false H_0 or accept a true H_0. In the other two cases we commit either a Type I or a Type II error. If we reject H_0 n it is true, we commit a Type I error; if we fail to reject H_0 when it is false, we commit a Type II error.

DEFINITION: TYPE I ERROR

A Type I error occurs when one rejects a null hypothesis that is actually true. The probability of committing a Type I error is denoted α (alpha).

DEFINITION: TYPE II ERROR

A Type II error occurs when one accepts a null hypothesis that is actually false. The probability of making a Type II error is denoted β (beta).

In adopting a method for choosing between H_0 and H_A, it is important to evaluate the risks of making both types of error. The chances of committing Type I and Type II errors are denoted α and β, respectively; that is, $\alpha = P$ (Type I error) and $\beta = P$ (Type II error). The relationship of α and β to the various choices about H_0 s indicated in Table 8-3. Moving down a column in the table, we select one of the two alternatives—either we reject or we accept H_0. Our two choices are complementary; thus the total probability of each column sums to 1. Notice that regardless of our decision, there is some risk that it is wrong. After all, there is no way to be completely certain of our conclusion short of taking a complete census of the population. Although there is no

TABLE 8-3
Probabilities of Making a Correct or Incorrect Decision in a Test of Hypothesis

	True state of nature	
Decision	H_0 is true	H_0 is false
Reject H_0	α	$1 - \beta$
Accept H_0	$1 - \alpha$	β
Total probability	1	1

way of avoiding risk completely, we can *limit* the level of risk associated with our decision. The way this is done is explained in step 3 of the classical test of hypothesis.

Step 2: Selection of Sample Statistic and Its Sampling Distribution

The second step of the classical procedure is to select the appropriate *sample statistic* on which we base our decision to reject or not to reject H_0. From Chapters 6 and 7 we know that the appropriate statistic to use is the minimum error estimator of the population parameter under study. When used in this way, the sample statistic is called the *test statistic*. Test statistics for μ, π and σ are summarized in Table 7-1. By now it should make sense that tests concerning μ are based on the sample value of the estimator $\bar{X}$, and, similarly, tests of π are based on the estimator P.

Returning to the example of residential mobility, we see that the decision about the *proportion* of households in the neighborhood changing residence should be based on the sample proportion statistic P. What is the sampling distribution of P? Recall that if the sample is large, we can appeal to the central limit theorem for the approximate sampling distribution of P: for large n, the sampling distribution of P is approximately normal with mean $\mu_p = \pi$ and $\sigma_P = \sqrt{\pi(1-\pi)/n}$. For π, we use the value given in the null hypothesis. In this example, therefore, we can say the sampling distribution of P is approximately normal with a mean of 0.2 and a standard deviation of $\sigma_P = \sqrt{0.2(0.8)/100} = 0.04$.

The essential point is that we use a test statistic whose sampling distribution can be specified. By using parameter values given in H_0, we obtain a sampling distribution consistent with H_0. This means that in hypothesis testing we *always* use a sampling distribution that assumes H_0 is true.

Step 3: Selection of a Level of Significance

In this third step of the classical hypothesis-testing procedure, we specify the probability of error associated with our ultimate decision. Obviously, we want to be confident about whatever decision we reach, and so we would prefer to have both α and β be small. Unfortunately, it is almost always impossible to simultaneously minimize the probability of both types of error; we can limit α, or β, but not both. Faced with this problem, classical hypothesis testing adopts the strategy of controlling only α. That is, the usual procedure is to select a very low value for α, the probability of making a Type I error. Typically, α is chosen to be 0.10, 0.05, or 0.01. The α value selected is known as its significance level.

This posture has some very important implications for the way we frame H_0 and H. In making α small, we are saying that if we reject H_0, we will do so with only a small probability of error. In other words, if we reject H_0 we will be confident in our decision. On the other hand, because β is not controlled, we cannot be confident that a decision to accept H_0 is correct. In light of this consideration, *the null hypothesis should be something we want to reject, rather than something we want to confirm.* Rejection of H_0 needs to be the scientifically interesting result, because rejecting H_0 is the only action we can take with low probability of error.

DEFINITION: LEVEL OF SIGNIFICANCE

The level of significance of a classical test of hypothesis is the value chosen for α, the probability of making a Type I error.

Whenever we report a decision about the null hypothesis, we always state the level of significance of our result. We say, for example, that a null hypothesis is rejected at the 0.05 level or that it is *statistically significant* at the 0.05 level. This means that the probability we are making an error is low, no higher than 5%.

The usual analogy in explaining the concept of a level of significance is based on the nature of justice in a court of law. In North American law, a defendant is innocent until proven guilty beyond a reasonable doubt. The burden is placed entirely on the prosecution to prove the defendant guilty. Why? Because society prefers to free a guilty person as opposed to imprisoning an innocent one. This is tantamount to saying that a Type I error is more serious than a Type II error. The reasonable doubt in a statistical test is the α level. If we are going to reject H_0, we must be very sure we are correct. This is why the usual levels of significance are so low. When we report a result as statistically significant, we are actually saying that it is *probably* correct. Using the word "probably" reflects our knowledge that we may be incorrect owing to random sampling.

The level of significance chosen for a particular problem depends entirely on the nature of the problem and the uses to which the results will be put. We are likely to be very demanding if we are testing toxicity levels of drugs, given the consequences of marketing a drug with serious side effects. In that instance, an α level of 0.0001 may be appropriate. For most social and physical science purposes, α is likely to be set equal to 0.10, 0.05, or 0.01. The consequences of an error, for example, in the study of neighborhood mobility, cannot be considered to be very serious. For our example of residential mobility, we choose a significance level $\alpha = 0.05$.

To this point, we can summarize the test of hypothesis as follows:

- Step 1: $H_0 : \pi = 0.2$ and $H_A : \pi \neq 0.2$.
- Step 2: P is chosen as the sample or test statistic. Its sampling distribution is approximately normal with mean 0.2 and standard deviation 0.04.
- Step 3: $\alpha = 0.05$.

Step 4: Construction of a Decision Rule

To construct a decision rule for H_0, it is now necessary to specify which values of the test statistic will lead us to reject H_0 in favor of H_A and which values will not. First, we assume $H_0 : \pi = 0.2$ is true and generate the sampling distribution of P, as in Figure 8-2. Note that it is centered over a value of $P = 0.2 = \pi$. The actual value p generated in our sample will most likely be near 0.2 if H_0 is true and far away from 0.2 if H_0 is false. For a given level of significance, we can divide the sampling distribution into two parts—those outcomes consistent with H_0 and those leading to a rejection of H_0. Since we have specified a two-tailed test, we reject H_0 only if the sample

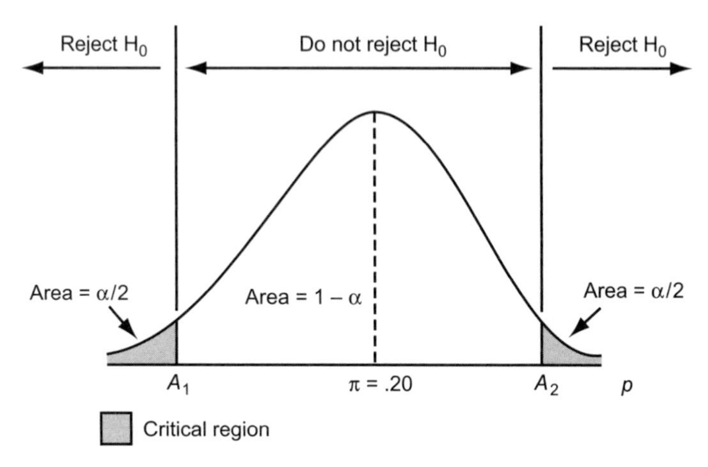

FIGURE 8-2. Sampling distribution of *P*, centered on hypothesized value $\pi = 0.2$.

outcome is in the most extreme parts of the sampling distribution. The most extreme outcomes occur in either tail of the distribution, so we place $\alpha/2$ of our possible outcomes in each tail. If the value of p is in the remaining $1 - \alpha$ fraction of outcomes, we cannot reject the null hypothesis. The values of p that lead us to reject H_0 are known as the *critical region* of the test. When we construct a decision rule for our hypothesis, we are specifying the values for A_1 and A_2 in Figure 8-2. These are sometimes termed the *action limits,* or *critical values,* for the hypothesis.

DEFINITION: CRITICAL REGION AND CRITICAL VALUES

The critical region corresponds to those values of the test statistic for which the null hypothesis is rejected. The critical values are at the less extreme (more central) edges of the critical region.

It is easy to see how the decision rule is constructed by considering the residential mobility example. Suppose we randomly sample n = 100 households in the neighborhood and calculate p, the proportion of people who moved in the last year. What values of p will lead us to conclude the neighborhood is less mobile than the city average (i.e., what is A_1?), and what values of p will lead us to conclude the neighborhood is more mobile than the city average (i.e., what is A_2)? Let us solve this problem by generating the sampling distribution of P for $n = 100$, $\pi = 0.2$, and $\alpha = 0.05$. The appropriate sampling distribution is illustrated in Figure 8-3. First, note that the sampling distribution in this figure is drawn to be normal (or approximately so) since we know from step 2 that this is the appropriate sampling distribution for the test statistic *P*. Also, note that we have placed $\alpha/2 = 0.025$ of the outcomes for P in each tail of the distribution.

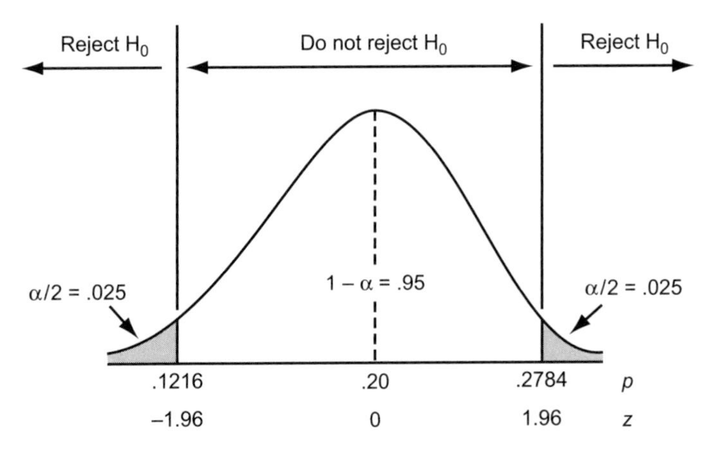

FIGURE 8-3. Determining critical region limits for residential mobility example, α = .05.

To determine the values of the critical limits A_1 and A_2, we use our knowledge of the normal distribution. First, we know this sampling distribution has a mean centered on $\pi = 0.2$. The standard deviation of this sampling distribution is

$$\sigma_P = \sqrt{\frac{\pi(1-\pi)}{n}} = \sqrt{\frac{0.2(0.8)}{100}} = 0.04$$

The critical limits on the sampling distribution are determined by the conditions

$$P(P < A_1) = \frac{\alpha}{2} \text{ and } \quad P(-\mathrm{P} > A_2) = \frac{\alpha}{2} \tag{8-4}$$

According to Equation 8-4, we must choose A_1 such that the probability of the sample statistic P being less than A_1 is $\alpha/2$. Because the sampling distribution of P is normal, A_1 and A_2 are symmetric about $\pi = 0.2$. To find A, we use the standardizing formula for the normal distribution:

$$z_{A_1} = \frac{A_1 - \pi}{\sigma_P} = z_{\alpha/2}$$

From the significance level α, we can say z_{A_1} is the value of z, which has $\alpha/2$ in the lower tail of the distribution. In other words, $z_{A_1} = z_{\alpha/2}$. Similarly,

$$z_{A_2} = \frac{A_2 - \pi}{\sigma_P} = z_{1-\alpha/2}$$

can be used to determine z_{A_2}. Substituting the known values of π, α, and σ_P yields

$$z_{.025} = \frac{A_1 - 0.2}{.04} \quad \text{and} \quad z_{1-.025} = \frac{A_2 - 0.2}{.04} = z_{.975}$$

We find $z_{.025}$ by looking in the body of the normal table to find the z-score that has 0.025 of the area under the curve in the left tail. Using column 6 of Table A-3 in the Appendix gives a z-value of –1.96. The other z-score, $z_{.975}$, calls for 0.025 of the area in the right tail, corresponding to $z = 1.96$ (see Table A-3, column 3). Substituting the z-values into our equations, we find the critical limits for our decision rule:

$$\begin{aligned} \frac{A_1 - 0.2}{0.04} &= -1.96 & \qquad \frac{A_2 - 0.2}{0.04} &= 1.96 \\ A_1 - 0.2 &= 0.0784 & A_2 - 0.2 &= 0.0784 \\ A_1 &= 0.1216 & A_2 &= 0.2784 \end{aligned} \tag{8-5}$$

The decision rule can now be summarized as follows:

Reject H_0 if $p < 0.1216$ or $p > 0.2784$. Otherwise, do not reject H_0.

This is a very compact way of expressing what action we will take depending on the value of p generated in the sample. There is an equivalent, widely used way of expressing the decision rule. Instead of making the decision rule based on the value of the sample statistic p, we express it in terms of z.

Reject H_0 if $z < -1.96$ or $z > 1.96$. Otherwise, do not reject H_0.

In this case, we determine the value of z for the decision rule on the basis of the standardizing formula

$$z = \frac{p - \pi}{\sigma_P} \tag{8-6}$$

What we are doing in this transformation is locating the observed value of sample statistic p within the standard normal distribution. In using Equation 8-5 to find the critical limits A_1 and A_2, we are actually reworking Equation 8-6 to solve for p.

Let us now explain the reasoning behind the use of these decision rules in greater detail. If the null hypothesis $\pi = 0.2$ were true, we would expect the sample value p to be fairly close to .2. For $\alpha = 0.05$, we would expect p to be in the range [.1216, .2784] with probability $1 - 0.05 = 0.95$. If the sample gives a value for p outside that range, we reject H_0. This decision rule is thus based on the sampling distribution of the test statistic P, obtained under the assumption that H_0 is true.

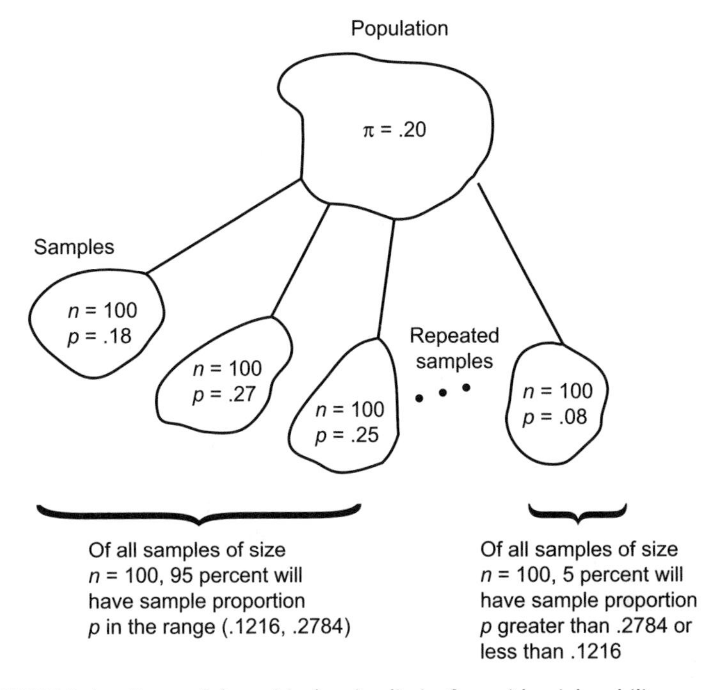

FIGURE 8-4. Determining critical region limits for residential mobility example, $\alpha = 0.05$.

At first thought, this procedure seems wrong: we have said H_0 should be a hypothesis we want to reject, yet we assume it is true in performing the test. Actually, this is not strange at all. Consider the courtroom example. As a prosecutor, we want to reject the assumption of innocence. In order to do so, we evaluate evidence under the assumption of innocence. If things have occurred that are unlikely under that assumption, we reject innocence in favor of the alternative hypothesis. This is expressed in a slightly different way in Figure 8-4.

Step 5: Compute the Value of the Test Statistic

At this point, we collect our sample and calculate the value, p, of the sample statistic P. Suppose that 26 of 100 randomly selected households in the neighborhood have moved in the past year. Then the value of the test statistic is $p = 26/100 = 0.26$. We can use this value directly with the decision rule constructed in the previous step. Alternatively, if we wish to express our result in terms of z, we calculate

$$z = \frac{0.26 - 0.2}{0.04} = \frac{0.06}{0.04} = 1.50$$

and use this value in conjunction with the decision rule based on z.

Step 6: Decision

The last step is to formally evaluate the null hypothesis based on the sample. This is simply a matter of examining the calculated value of the test statistic in relation to the decision rule. Since $p = 0.26$ is between 0.1216 and 0.2784, it *does not* lie in the critical region. We therefore cannot reject H_0. If we were to reject H_0 on the basis of this sample, the probability of making a Type I error would exceed 0.05. In adopting a significance level of $\alpha = 0.05$, we have said a 5% error is the largest we can tolerate, so we must not reject H_0. Notice that even if the computed value of p were 0.2783, we still could not reject H_0. As a matter of fact, this rigidity in our decision rule is one problem with the classical test of hypothesis. Because the decision to reject or not is binary, it does not admit shades of certainty regarding H_0. But surely, would not a sample with $p = 0.26$ and a sample with $p = 0.2$ give us varying degrees of belief in H_0? In Section 8.2 we introduce the PROB-VALUE method, which gets around this problem. Instead of simply rejecting H_0 in favor of H_A, we determine our degree of belief in H_A; that is, we compute the value of α implicit in our sample.

What if $p = 0.31$ is the calculated sample value of p? In this instance, the decision is different. We reject H_0 because p falls in the critical region ($0.31 > 0.2784$). Rejecting H_0 leads to two possibilities. Either we made a correct decision and $\pi \neq 0.2$, or we obtained a sample that did come from a population with $\pi = 0.2$. In the latter case, we made a wrong decision and committed a Type I error. However, we should feel relatively secure, knowing the probability of such an occurrence is necessarily less than 0.05.

To evaluate the hypothesis in terms of a decision rule based on z, we note that $z = 1.5$ calculated in the previous step is in the range [–1.96, 1.96], and thus we must not reject H_0. Of course, any decision based on z is equivalent to one made by using p, and vice versa. Some texts in statistics always work with z, others with the value of sample statistic p, and still others both. The advantage of using p (or $\bar{x}$ or s, as the case may be) is that the decision is being made using the units in which the variable of interest is measured. In the current example, we could say that we will reject the hypothesis that the neighborhood has a typical level of residential mobility if the proportion of people who moved recently is less than 0.1216 or greater than 0.2784. To say the same thing in terms of some standardized normal variable removes the decision from the context of the problem at hand. Nevertheless, it is the more common approach. The complete text of hypothesis testing for the example is summarized in Table 8-4. It provides a convenient way for beginning students to organize classical tests in inferential statistics, and we will use it throughout the remainder of the text. In subsequent examples in this chapter, many of the variations in the classical test are introduced. In particular, we will cover the directional tests depicted in formats *B* and *C*, also called one-tailed tests.

There is one final, very important issue regarding decisions in hypothesis testing. In our example, the sample value did not lie in the critical region. It would seem that we should therefore accept H_0. Strangely enough, we did not do so, saying instead, "We therefore cannot reject H_0" In other words, rather than accept H_0, we make no decision. The reason for this has to do with our choice to make α small. If we were

TABLE 8-4
Summary of Test of Hypotheses for Residential Mobility Example

Step 1:	$H_0 : \pi = 0.2$ and $H_A : \pi \neq 0.2$.
Step 2:	P is chosen as the sample statistic.
Step 3:	$\alpha = 0.05$.
Step 4:	Reject H_0 if $p < 0.1216$ or $p > 0.2784$. (See Figure 8-3.)
Step 5:	From random sample, $p = 0.26$.
Step 6:	Because $0.1216 < p < 0.2784$, do not reject H_0.

to accept H_0, we would do so with an error probability of β, a value we have made no attempt to control. How can we calculate β? It seems clear that because β applies when H_A is true, we need to know how likely it is to obtain $p = 0.26$ when H_A is true. Here we have a problem, because H_A is vague, simply saying $\pi \neq 0.2$. H_A admits many values for π, actually an infinite number. If we are going to accept H_0, we must be convinced that *none* of these possibilities is true. On the basis of $p = 0.26$, are we prepared to say π is definitely 0.2, rather than 0.25 or 0.27? Clearly, if we conclude H_0 is true, we run a large, unknown, risk of error. Though we could make this decision, we could not attach an uncertainty value to our decision; thus we would accomplish nothing.

It is often said that anything can be proved with statistics. Our experience with hypothesis testing thus far suggests nothing could be more wrong! Instead of proving things with statistics, we find that we are only sometimes able to *disprove* some things (H_0), and we can do that only with some chance of error (α). Fortunately, the level of error is known and is under our control.

8.2. PROB-VALUE Method of Hypothesis Testing

In Section 8.1 we outlined a strict procedure for hypothesis testing known as the classical method. This procedure is useful as a means of understanding issues that lie at the heart of all hypothesis testing, such as the distinction between a Type I and a Type II error. However, the classical procedure is less often used in actual practice than an alternative procedure known as the PROB-VALUE, or p-VALUE, method. (We will call it ***PROB-VALUE testing*** rather than p-VALUE since the letter p is used in so many other contexts.) What is wrong with hypothesis testing done in the classical way? First, it requires us to choose a value for α, the level of significance for the test. Unfortunately, there are few instances where this value can be rationally chosen. No doubt it should always be small, but exactly how small? We can easily think of cases where we would reject H_0 at $\alpha = 0.05$ but not at $\alpha = 0.01$. If two researchers happen to choose different α levels, they might reach different conclusions from the same data. This is unsettling, given that convention rather than theory drives the choice of α.

A second shortcoming is that our findings are dichotomous. We choose α and report only whether or not H_0 is rejected for that α. Another researcher, who prefers a different α, cannot tell from our report if she or he would arrive at the same decision. The classical method can therefore be criticized as not providing enough information. Along these same lines, suppose we subscribe to the subjectivist view of probability. If so, we might be willing to think of H_0 as having some probability of falsehood. (The objectivist interpretation doesn't allow this—H_0 is either true or not true, and no probability can be assigned.) The classical method of testing offers no information regarding how confident we are about a decision to reject. But surely, if the test statistic is far in the tail of the sampling distribution, we will be more confident about rejecting H_0 than if the statistic is barely within the critical region. From a subjectivist viewpoint, we have various degrees of certainty about H_0's falsehood depending on where in the critical region the statistic lands. If the degree of certainty is available, it should be reported along with our decision to reject H_0.

The PROB-VALUE method is preferred because it addresses both of the deficiencies mentioned above. As it is best explained by example, we will reconsider the test about residential mobility developed in Section 8.1. After using the example to show how the PROB-VALUE method differs from the classical method, we will turn to more general applications.

PROB-VALUE Method for a Two-Sided Test

Figure 8-3 illustrates the decision rule developed for the hypothesis set

$$H_0 : \pi = 0.2 \quad H_A : \pi \neq 0.2$$

for the residential mobility problem. Selecting a significance level of $\alpha = 0.05$ led to this decision rule:

1. Reject H_0 if $p < 0.1216$ or $p > 0.2784$ (or $z < -1.96$ or $z > 1.96$).
2. Do not reject H_0 if $0.1216 \leq p \leq 0.2784$ (or $-1.96 \leq z \leq 1.96$).

Here p is the value of the sample proportion statistic P, and z is the value of the standardized variable

$$Z = \frac{P - \pi_0}{\sqrt{\pi_0(1 - \pi_0)/n}}$$

Since our sample proportion is 0.26, we do not reject H_0. Now, the principal difficulty with the classical method is that it provides no information as to *exactly* how likely it is that the null hypothesis is false. *Any* value of p in the critical region leads us to reject H_0. But a sample proportion $p = 0.9$ suggests that H_0 is less likely to be true than a sample proportion $p = 0.28$ (just on the edge of the critical region). It is just this problem that the PROB-VALUE method is designed to solve.

In the example, the actual sample proportion $p = 0.26$ leads us not to reject H_0. But since it is close to the critical limit of $p = 0.2784$, we might wonder, "If we were

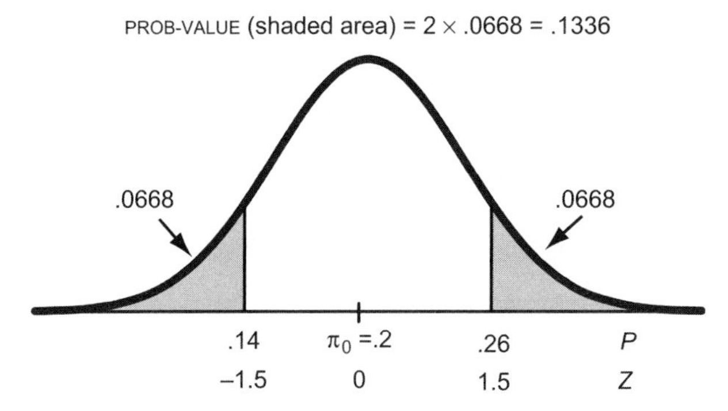

FIGURE 8-5. Determining the PROB-VALUE in a two-sided hypothesis test for the residential mobility example.

to reject H_0 on the basis of $p = 0.26$, how much faith would we have in this decision?" To answer this question, consider Figure 8-5. On this diagram we have established the critical region at the value of $p = 0.26$ on the upper side of $\pi = 0.2$ and at $p = 0.14$ on the lower side (0.2 – 0.06). Together, these two values define another critical region. Associated with this new critical region is a new, different, α-value. If we happened to have used the new α, the test result would have been just on the margin of rejection. The probability of a Type I error would have been *exactly* equal to the new α.

Simply put, this new α is the "probability" in the PROB-VALUE method. The method calls for us to find and report the value of α for which the sample would just barely permit us to reject H_0. The phrase "just barely" means that the sample statistic is on the edge of the critical region. Notice how the PROB-VALUE method differs from the classical method. There we chose the probability α, which determined the critical region. Here we are using the test statistic to determine a critical region, and we are finding a probability from the area of that region. How do we interpret this probability? It should be clear that the PROB-VALUE is the probability of making a Type I error. In other words, should we elect to reject H_0, the PROB-VALUE tells us how likely it is that we are wrong.

Returning to Figure 8-5, we see that the PROB-VALUE is equal to the sum of the shaded areas in the tails of the sampling distribution. This area can be easily determined knowing that P is normally distributed. First, convert the sample value of $p = 0.26$ to a standardized value z, using

$$z = \frac{p - \pi_0}{\sqrt{\pi_0(1 - \pi_0)/n}} = \frac{0.26 - 0.2}{\sqrt{0.2(1 - 0.2)/100}} = 1.5$$

Next, determine the area to the right of $z = 1.5$. Using the normal table gives 0.0668. Since the distribution is symmetric, the area in both tails is 2(0.0668) = 0.1336. This is the PROB-VALUE attached to H_0 for this problem. The probability of 0.0668 is doubled because we are evaluating a two-sided alternative. Recall that in the classical test

TABLE 8-5
Summary of PROB-VALUE Hypothesis Testing

Step 1:	State the null and alternative hypotheses.
Step 2:	Identify a test statistic with known sampling distribution.
Step 3:	Choose a level of error α.
Step 4:	Compute (or observe) the test statistic.
Step 5:	Compute the PROB-VALUE associated with the test statistic.
Step 6:	Reject H_0 if PROB-VALUE < α; otherwise do not reject H_0.

of a two-sided hypothesis the significance level α is halved, and $\alpha/2$ is placed in the two rejection areas at the tails of the sampling distribution.

The PROB-VALUE of 0.1336 can be interpreted in the following way: *it is the lowest value at which we could set the significance of the test and still reject* H_0. If we reject H_0 on the basis of $p = 0.26$, we are saying that the probability of committing a Type I error of $\alpha = 0.1336$ is acceptable to us. In short, the PROB-VALUE indicates the *degree of belief* that H_0 should be rejected. As the PROB-VALUE approaches 0, we are ever more sure that rejecting H_0 is the correct decision. Calculating a PROB-VALUE is clearly a superior alternative to the classical test. It avoids the thorny problem of needing to dismiss an alternate hypothesis simply because the sample statistic lands just outside the critical region. It also allows other researchers to interpret our sample results using whatever level of uncertainty (α) they choose.

DEFINITION: PROB-VALUE

The PROB-VALUE associated with a null hypothesis is equal to the probability of obtaining a value of the sample statistic as extreme as the value observed, if the null hypothesis is true. If a decision is made to reject H_0, the PROB-VALUE gives the associated probability of a Type I error.

Virtually all statistical packages compute PROB-VALUES. These values report what the sample data tell us about the credibility of a null hypothesis, rather than forcing a decision about H_0 on the basis of a possibly arbitrarily defined level of significance. Of course, determining a PROB-VALUE does not preclude performing a classical test. Rejecting a null hypothesis with a PROB-VALUE of 0.05 or less is equivalent to a classical test of hypothesis with $\alpha = .05$. Table 8-5 summarizes PROB-VALUE testing in general terms.

PROB-VALUE Method for a One-Sided Test

Determining the PROB-VALUE in a one-sided test of hypothesis, either lower or upper tail, differs in only one respect from the method discussed above. Suppose the hypothesis being evaluated in the residential mobility example is

$$H_0: \pi \le 0.2 \ H_A: \quad \pi > 0.2$$

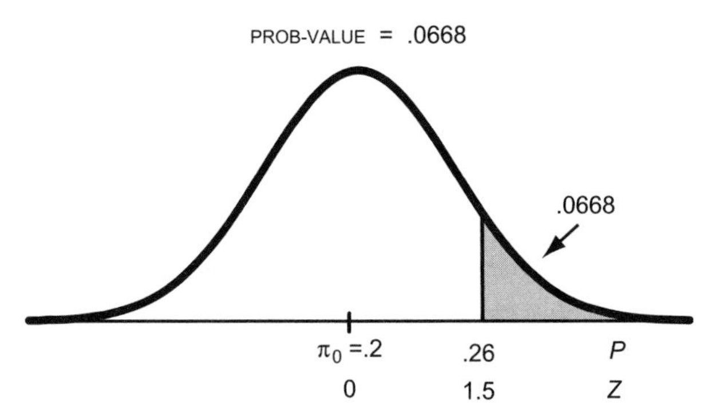

FIGURE 8-6. Determining the PROB-VALUE in a one-sided hypothesis test for the residential mobility example.

This is test format *C*, in which we are examining whether the neighborhood has a *higher* mobility rate than average. The PROB-VALUE associated with the sample value $p = 0.26$ is 0.0668, as illustrated in Figure 8-6. There is no need to double the probability in a one-sided test since *the critical region is always placed in one tail or the other, depending on whether it is a lower-or an upper-tail test.*

Does it make sense that the PROB-VALUE is lower for the one-tailed test than for the two-tailed test? Consider the fact that in doing a one-tailed test, we are saying the only alternative to H_0 is $\pi > 0.2$. In the two-sided test we allowed for the additional possibility $\pi < 0.2$. In other words, we come to the one-sided test with more knowledge, having restricted H_A by one-half. Because we ask a narrower question, the sample is able to give a more precise answer to the question we ask. Here the value of $p = 0.26$ indicates the neighborhood is experiencing higher mobility than average. A PROB-VALUE of 0.0668 (as opposed to 0.1336) reflects our improved ability to evaluate H_0. The lesson is that in a one-tailed test, a less extreme sample result is needed to reject H_0. Whenever one has good reason for using a one-tailed test, it should be used. Doing so will maximize use of sample information.

In Figure 8-6, we have drawn the sampling distribution of P assuming $\pi = 0.2$. Likewise, we computed the PROB-VALUE assuming $\pi = 0.2$. Looking at the null hypothesis, we see that it allows for many values for π. In fact, *any* value is allowed as long as it is less than or equal to 0.2! Clearly, had we used some other value, say $\pi = 0.1$, the sampling distribution and resulting PROB-VALUE would be different. Why, then, did we choose $\pi = 0.2$?

Imagine that we use one of the allowable (smaller) values. Obviously, the sampling distribution in Figure 8-6 will move to the left. The sample result $p = 0.26$ stays the same, so that the shaded area and PROB-VALUE must *decrease*. This shows that the PROB-VALUE we obtained using $\pi = 0.2$ is the *largest* that can be obtained under this null hypothesis. We can be sure, therefore, that no matter which value of π is true, our PROB-VALUE is, if anything, too large. If the computed PROB-VALUE leads us to reject H_0, we can be positive that H_0 would be rejected for all the other values of π admitted under H_0. In other words, using $\pi = 0.2$ is a conservative approach.

8.3. Hypothesis Tests Concerning the Population Mean μ and π

In the remainder of this chapter, a series of hypothesis tests are described for the population parameters μ and π. Examples illustrate typical situations where these tests may be used, and revisit many of the theoretical issues discussed in Sections 8.1 and 8.2. Because it is so common, all examples employ the PROB-VALUE method. However, readers are reminded that the PROB-VALUE method permits classical testing. In particular, if the PROB-VALUE obtained is lower than some α level chosen *a priori,* the null hypothesis should be rejected.

The first test concerns hypotheses related to the population mean μ. The sample mean statistic $\bar{X}$ is the appropriate test statistic in this case, and its sampling distribution is given by the central limit theorem. Just as we distinguished between two cases for specifying a confidence interval for μ, hypothesis tests of μ must be based on what is known about the population standard deviation σ:

- Case 1: Population standard deviation σ is known.
- Case 2: Population standard deviation σ is unknown.

Case 1: Population Standard Deviation σ Known

In Chapter 7 we showed that the sample mean statistic $\bar{X}$ is approximately normally distributed with mean $E(\bar{X}) = \mu_{\bar{X}} = \mu$ and standard deviation $\sigma = \sigma/\sqrt{n}$. Tests of hypothesis concerning μ can therefore be evaluated by using the standard normal statistic

$$Z = \frac{\bar{X} - \mu}{\sigma_{\bar{X}}} = \frac{\bar{X} - \mu}{\sigma/\sqrt{n}} \tag{8-7}$$

This standard normal statistic finds extensive application in hypothesis testing. As we will see below, it can be used to generate decision rules for tests of sample proportions. In fact, for many different sample statistics $\hat{\theta}$, the general form of the standard normal statistic

$$Z = \frac{\hat{\theta} - \theta_0}{\sigma_{\hat{\theta}}} \tag{8-8}$$

can be used to construct confidence intervals or test hypotheses. In Equation 8-8, θ_0 is the hypothesized value of the population parameter and the mean of the sampling distribution of the sample statistic $\hat{\theta}$. The denominator or $\sigma_{\hat{\theta}}$ is the standard deviation of the sampling distribution of $\hat{\theta}$, sometimes called the *standard error* of the sample statistic. The purpose of the standardized statistic Z is to locate the value of the sample statistic within the standard normal distribution.

The value of z obtained from Equation 8-8 for a particular sample tells us how many standard deviations (or standard errors) $\hat{\theta}$ is from the mean of the sampling distribution. The higher the absolute value of z, the more unusual the sample is, and the less likely that H_0 is true. To test H_0, we simply need to assign a probability to the observed value under the assumption that H_0 is true. The PROB-VALUE method for any sample statistic that is normally distributed will always be as follows for a *two-sided test:*

1. Compute the sample value z from Equation 8-8 using the observed $\hat{\theta}$ and hypothesized value θ_0.
2. Find the PROB-VALUE: $P(Z < -z) + P(Z > z) = 2\,P(Z > z)$

Let us return to hypothesis testing for μ by considering the following example.

EXAMPLE 8-2. The mean household size in a certain city is 3.2 persons with a standard deviation of $\sigma = 1.6$. A firm interested in estimating weekly household expenditures on food takes a random sample of $n = 100$ households. To check whether the sample is truly representative, the firm calculates the mean household size of the sample to be 3.6 persons. Test the hypothesis that the firm's sample is representative of the city with respect to household size. What PROB-VALUE can be attached to this hypothesis?

Solution. The values of $\mu = 3.2$ and $\sigma = 1.6$ are given, and we know $\bar{X}$ is approximately normal for $n = 100$. The null and alternate hypotheses are

$$H_0 : \mu = 3.2 \quad H_A : \mu \neq 3.2$$

A two-sided test is needed because we want to know if the sample families are different from the general population, either significantly smaller or significantly larger. The sample has mean $\bar{x}$ of 3.6, from which we compute z:

$$z = \frac{3.6 - 3.2}{1.6/\sqrt{100}} = \frac{0.4}{0.16} = 2.5$$

Using column 7 from Appendix Table A-3, we find a PROB-VALUE of 0.012. We therefore conclude it is highly unlikely that the sample is representative of the larger population. Figure 8-7 shows the results graphically. Note that if we were testing this

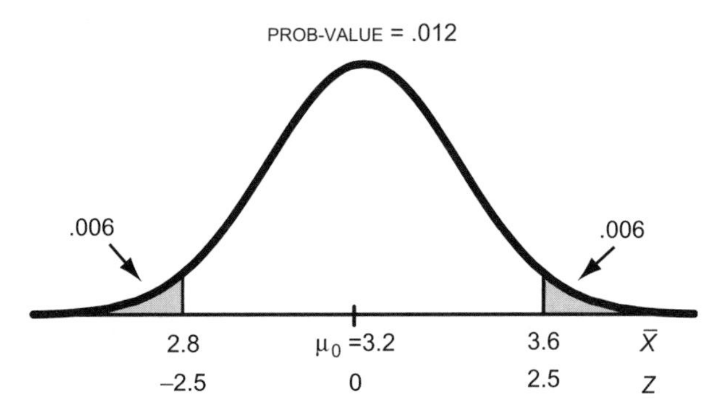

FIGURE 8-7. Sampling distribution for Example 8-2.

hypothesis at the 0.05 level, we would reject H_0. Testing at the 0.01 level, we would not reject H_0.

This example illustrates an important point about tests for μ. Household size is a discrete variable, and one with a small range of values (few families have more than five children, none less than zero). We can therefore be sure that X is not normally distributed. However, with the large sample size used here, $\bar{X}$ will be very close to normal—so close that the error is negligible. Note that if we were to repeat the test with a small *n,* we would have to consider the PROB-VALUE approximate, knowing that we are working with a crude approximation to the sampling distribution of $\bar{X}$.

Case 2: Population Standard Deviation σ Unknown

If σ is unknown, it must be estimated from the sample using the estimator *S*. From Chapter 8, we know the standardized random variable

$$T = \frac{\bar{X} - \mu}{S/\sqrt{n}} \tag{8-9}$$

is *t*-distributed with $n - 1$ degrees of freedom, providing *X* is itself normal. Thus for a normally distributed random variable *X,* Equation 8-9 can be used to test hypotheses and determine PROB-VALUES. When X is not normally distributed but *n* is large, the use of expression (8-9) gives approximate results.

When we used the t-distribution for confidence intervals, we chose a probability (α) and used Appendix Table A-4 to find the corresponding value of *t.* Here we need to do just the reverse. In the PROB-VALUE method we compute a value of *t* and use a table to arrive at the corresponding probability. Tables A-5 and A-6 give one-and two-tailed probabilities for the *t*-distribution. Notice that the rightmost columns (infinite *df*) are equal to standard normal probabilities.

EXAMPLE 8-3. A commuter train advertisement asserts that the time on the train between a certain station and the downtown terminal is 23 min. A random sample of $n = 30$ trains in a single month yields an average time of 26 min with a standard deviation of 6 min. Test the assertion that the trains are actually slower than advertised, using $\alpha = 0.05$.

Solution. Viewing this as a hypothesis-testing problem, we want to know if the mean is greater than 23. This gives

$$H_0 : \mu \leq 23 \quad H_A : \mu > 23$$

We choose a one-sided test because we are asking only if the trains are slower than advertised. If we reject H_0, we can make the assertion we would like: Average times are longer than advertised. Based on a sample of size $n = 30$, and the estimates $\bar{x} = 26$, $s = 6$, we compute the observed *t*-value:

$$t = \frac{26 - 23}{6/\sqrt{30}} = 2.74$$

From Table A-6 in the Appendix, we obtain a PROB-VALUE of about 0.005, 10 times smaller than the 0.05 level we have chosen for α. We therefore reject H_0.

Notice that we used 23 for the hypothesized value of μ. Once again, this is a conservative approach. Although H_0 admits many values for μ, each with a different PROB-VALUE, we can be sure that none is larger than the PROB-VALUE we get using $\mu = 23$.

The tests of hypothesis concerning the population parameter μ are summarized in Table 8-6. The table distinguishes between two cases depending on whether σ is known beforehand. Both cases give exact PROB-VALUES when X is normal, and approximate values otherwise. The approximations improve with increasing sample size and will give excellent results for $n > 30$ or so.

Hypothesis Tests Concerning the Population Proportion π

This test was used as the example in presenting the classical test of hypothesis in Section 8.1. We cover it again here, describing in more detail how the sampling distribution is obtained. As we shall see, there are two variations depending on the sample size.

Consider first the situation when n is large. As was seen in Chapter 6, the random variable X is a Bernoulli variable, taking on only two values, zero or unity. To compute the sample proportion p, we divide the number of 1's in the sample by n. This is equivalent to computing the sample mean $\bar{X}$, which we know to be approximately normally distributed for large n. We can say, therefore, that the random variable P is approximately normal. As in tests for μ, we will compare the sample value for the test statistic with the value hypothesized under H_0. If the sample value diverges greatly from the hypothesis, we will reject H_0 in favor of H_A.

EXAMPLE 8-4. Household surveys of residents in a suburban neighborhood have consistently found the proportion of residents taking public transit to be 0.16. The city council, at the urging of neighborhood residents, increased the bus service to this neighborhood. However, the council vowed to discontinue the service if there was not an increase in patronage. Three months after the introduction of the new service, the council hired a consultant to survey $n = 200$ residents and determine their current modes of transportation. Forty-two residents indicated they took public transit. Should the council continue the service?

Solution. Since the proportion of residents taking public transit has consistently been 0.16 and the council has specified that the proportion must *increase,* we can specify the two hypotheses as follows:

$$H_0 : \pi \leq 0.16 \quad H_A : \pi > 0.16$$

TABLE 8-6
Single-Sample Tests for μ

Background

A single random sample of size n is drawn from a population, and the sample mean $\bar{x}$ is calculated. The sample value will be compared to the hypothesized value to determine if H_0 should be rejected. The sampling distribution of $\bar{X}$ is known exactly if X is normal, which permits one to attach an exact PROB-VALUE to the sample result. If X is approximately normal, the PROB-VALUE is approximate.

Hypotheses

$H_0: \mu = \mu_0$	$H_0: \mu \geq \mu_0$	$H_0: \mu \leq \mu_0$
$H_A: \mu \neq \mu_0$	$H_A: \mu < \mu_0$	$H_A: \mu > \mu_0$
(*A*)	(*B*)	(*C*)
(two-tailed)	(one-tailed)	(one-tailed)

Test statistic

Case 1: σ known. The test statistic Z is normally distributed. Using the sample mean $\bar{x}$, we see that the observed value is

$$z = \frac{\bar{x} - \mu_0}{\sigma/\sqrt{n}}$$

Case 2: σ unknown. The test statistic T is t-distributed with $n - 1$ degrees of freedom. Using the sample mean and standard deviation ($\bar{x}$, s), we find that the observed value is

$$t = \frac{\bar{x} - \mu_0}{s/\sqrt{n}}$$

PROB-VALUES (PV) and decision rules

Two-tailed tests (format A):

$$PV = 2P(Z > |z|) = P(Z < -|z|) + P(Z > |z|) \quad PV = 2P(T > |t|) = P(T < -|t|) + P(T > |t|)$$

Reject H_0 if PROB-VALUE $< \alpha$

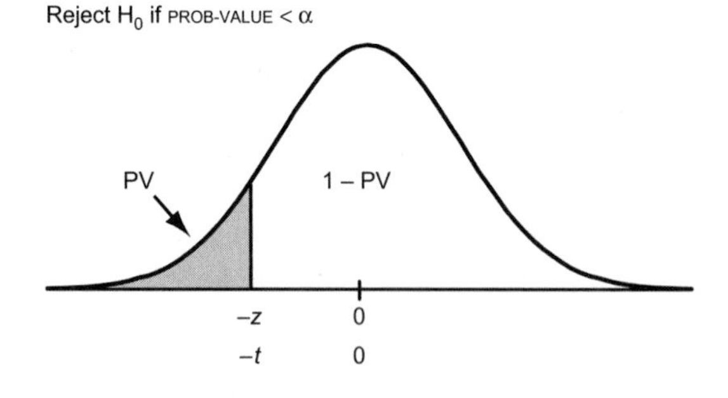

One-tailed tests (format *B*):

$$PV = P(Z < z) \quad PV = P(T < t)$$

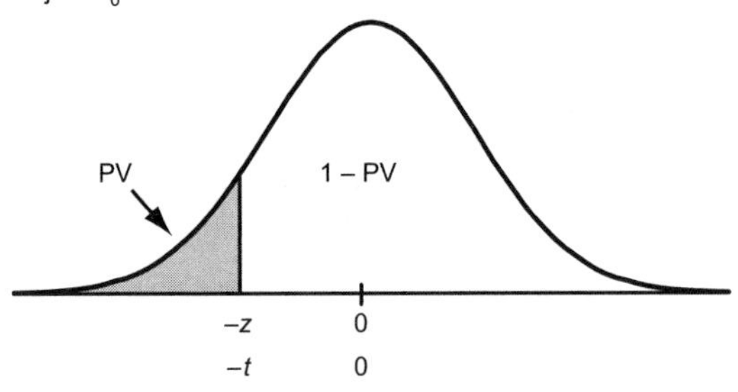

One-tailed tests (format *C*):

$$PV = P(Z > z) \quad PV = P(T > t)$$

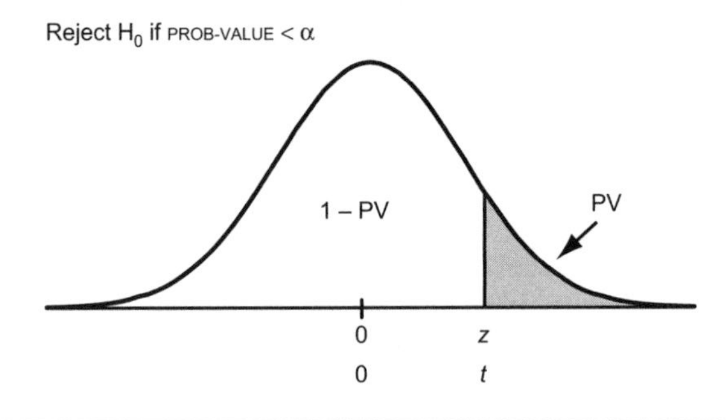

This is an upper-tail test. The corresponding sampling distribution of P is shown in Figure 8-8. With $\pi = 0.16$, the standard deviation of the sampling distribution for a sample size of $n = 200$ is $\sigma_P = \sqrt{0.16(0.84)/200} = 0.026$. The sample value for p is 42/200= 0.21. Using the hypothesized value for π, we compute z as

$$z = \frac{0.21 - 0.16}{0.026} = 1.9$$

According to Table A-3, the PROB-VALUE is $P(Z > 1.9) = 0.029$. The null hypothesis can therefore be rejected with an error probability just less than 3%. If the city council deems this an acceptable level of error, it will continue the new bus schedule. Alternatively, a council requiring stronger evidence might authorize another survey, or perhaps even pronounce the new service a failure.

The procedure we have described above will not work for small n, because the sample proportion P will not be close to a normal distribution. However, the tools

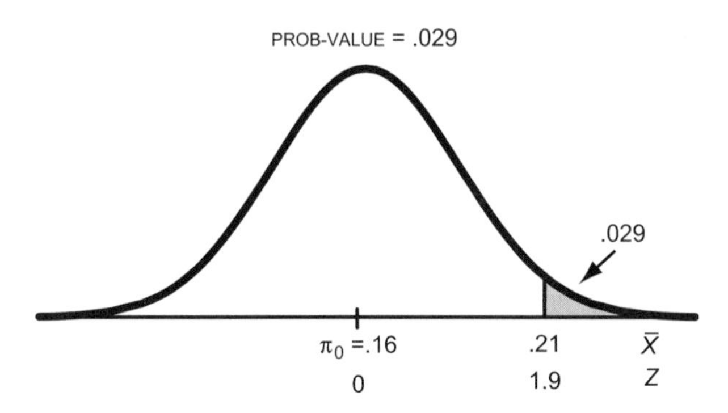

FIGURE 8-8. Sampling distribution for Example 8-4.

needed for small n have already been presented. Instead of working with the *proportion* of the sample taking on a value of unity, we can work with the *number* of 1's in the sample. That is, rather than use a fraction of the sample that are unity, we will count the number that are unity. If we let X be the number of 1's in the sample, we know that X follows the binomial distribution. Knowing the distribution of X lets us find the probability associated with any observed value.

EXAMPLE 8-5. In unpolluted waters, abnormal growths are known to occur in 15% of all sturgeon. A sample of 10 fish obtained near a chemical plant yields 3 fish with tumors. Is there reason to believe fish living in waters near the plant have an unusually high tumor rate? Use a 10% confidence level for α.

Solution. The question calls for a one-tailed test, with null and alternate hypotheses:

$$H_0 : \pi \leq 0.15 \quad H_A : \pi > 0.15$$

We let X be the number of successes in 10 trials, where an abnormal growth is called a "success." The sample has a 30% success rate, which seems quite large if the population rate π is only 15%. To assess the probability, we make use of the fact that X is binomial, assuming the value for π indicated in H_0. We need to find the probability of obtaining 3 or more successes in 10 trials. If 3 or more successes occur often purely by chance, we will not reject H_0. One the other hand, if the sample value turns out to be unusual, we will embrace the alternate hypothesis.

We need $P(X \geq 3)$. Binomial probabilities are given in Appendix Table A-2 for $\pi = 0.15$ and $n = 10$. Rather than sum the probabilities over $X = 3, 4, \ldots, 10$, we can use

$$\begin{aligned} P(X \geq 3) &= 1 - P(X = 0) - P(X = 1) - P(X = 2) \\ &= 1 - 0.1969 - 0.3474 - 0.2759 = 0.1970 \end{aligned}$$

If we reject H_0, we run a 20% chance of a Type I error. Because this is greater than α, we do not reject H_0. This sample does not allow us to conclude fish living near the chemical plant are experiencing an unusually high rate of abnormal growths.

One-sample tests for π are summarized in Table 8-7.

8.4. Relationship between Hypothesis Testing and Confidence Interval Estimation

Hypothesis testing and confidence interval estimation are closely related methods of statistical inference. The *significance level* of a test is actually the complement of the *confidence* in a confidence interval. The fact that we denote the level of significance as α and the confidence level as $1 - \alpha$ is no coincidence. The relationship between these two procedures for statistical inference is best explained by a simple example. For this purpose, let us reconsider Example 8-1, which dealt with household sizes. Based on the sample value for z, we showed that the PROB-VALUE for the two-tail test is .0124.

Now, for comparative purposes, let us construct both the 95 and the 99% confidence intervals for μ, using the sample $\bar{x} = 3.6$ and $\sigma = 1.6$:

$$\text{95\% interval:} \quad \bar{x} \pm z_{.025}\frac{\sigma}{\sqrt{n}} = 3.6 \pm 1.96\frac{1.6}{\sqrt{100}} = [3.28, 3.91]$$

$$\text{99\% interval:} \quad \bar{x} \pm z_{.005}\frac{\sigma}{\sqrt{n}} = 3.6 \pm 2.58\frac{1.6}{\sqrt{100}} = [3.19, 4.01]$$

Notice that the 95% confidence interval does not contain the hypothesized value of $\mu_0 = 3.2$, but the 99% interval does. Since the PROB-VALUE was found to be 0.0124, this is the expected result. That is, the PROB-VALUE says we are between 95 and 99% sure that $\mu \neq 3.2$, as do the two confidence intervals.

This interpretation can now be extended as follows. If the $(1 - \alpha)$ confidence interval *does not contain* the hypothesized value θ_0, we can *reject* the hypothesis $\theta = \theta_0$ at the α level of significance. On the other hand, if the $(1 - \alpha)$ confidence interval includes θ_0, testing the hypothesis does not allow H_0 to be rejected. We see that associated with every confidence interval is an implicit two-sided hypothesis test. The α level of the test is equal to (1 – confidence level).

8.5. Statistical Significance versus Practical Significance

Hypothesis testing provides a formal, objective means of evaluating a statistical hypothesis. When we obtain a statistically significant result, we know that H_0 should be rejected with no more than a small (α) chance of error. Does this mean that all statistically significant results are important from a practical, or scientific standpoint?

TABLE 8-7
Single-Sample Tests for π

Background

A single random sample of size n is drawn from a population, and we observe the number of successes in the sample, x. The associated random variable X follows the binomial distribution. If n is small, we will assign a probability to the sample value directly using the binomial distribution. For large n, we will divide by the sample size to obtain the sample proportion $p = x/n$. Using the central limit theorem, we know its associated random variable P is approximately normal.

Hypotheses

$H_0: \pi = \pi_0$	$H_0: \pi \geq \pi_0$	$H_0: \pi \leq \pi_0$
$H_A: \pi \neq \pi_0$	$H_A: \pi < \pi_0$	$H_A: \pi > \pi_0$
(A)	(B)	(B)
(two-tailed)	(one-tailed)	(one-tailed)

Test Statistic

Case 1: Small n. The test statistic is the number of successes in the sample X, which follows the binomial distribution:

$$P(X = x) = C(n, x)\, \pi_0^x (1 - \pi_0)^{n-x} \quad x = 0, 1, \ldots, n$$

Case 2: Large n. The observed value of the test statistic is the proportion of successes in the sample $p = x/n$, which is approximately normal. Converting to standard normal gives

$$z = \frac{p - \pi_0}{\sqrt{\pi_0(1 - \pi_0)/n}}$$

PROB-VALUE (PV) and decision rules

Two-tailed test (format A):

$$\text{PV} = 2P(Z > |z|) = P(Z < -|z|) + P(Z > |z|)$$

Reject H_0 if PROB-VALUE $< \alpha$

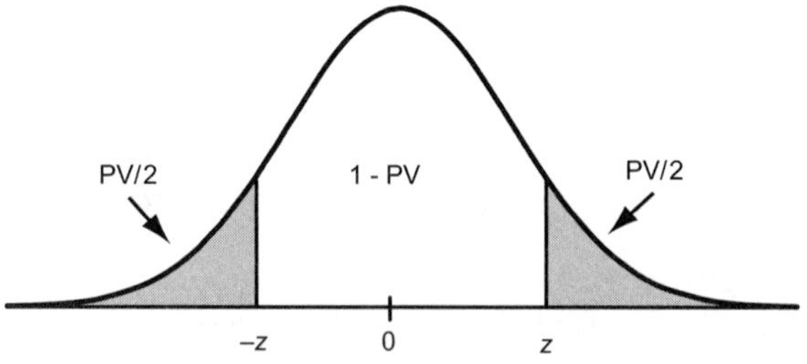

PV/2
1 - PV
PV/2
$-z$
0
z

One-tailed test (format *B*):

$$PV = P(Z < z) \quad PV = P(X \leq x)$$

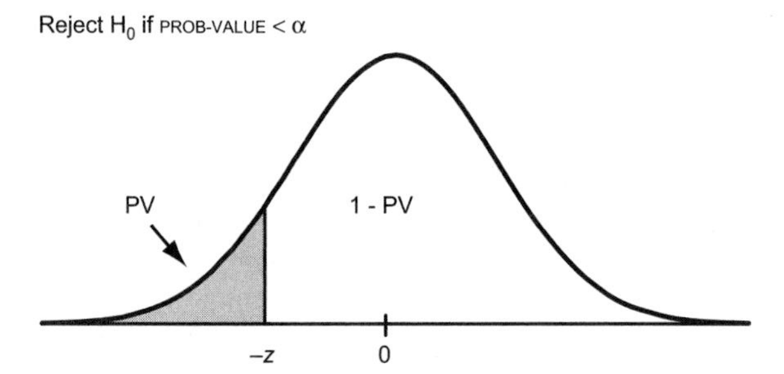

One-tailed test (format *C*):

$$PV = P(Z > z) \quad PV = P(X \geq x)$$

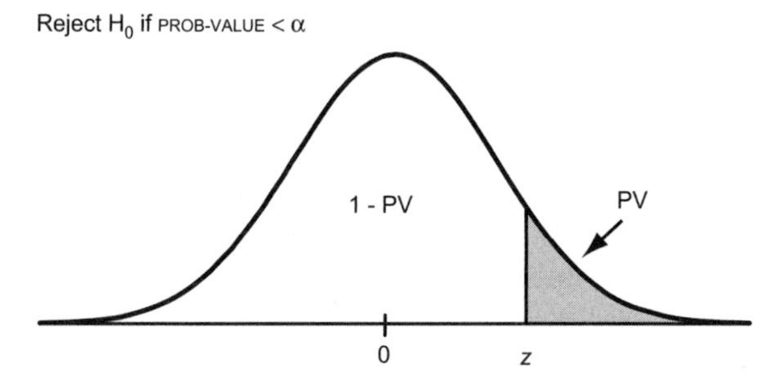

The large sample probabilities are approximate, whereas the binomial PROB-VALUES are exact.

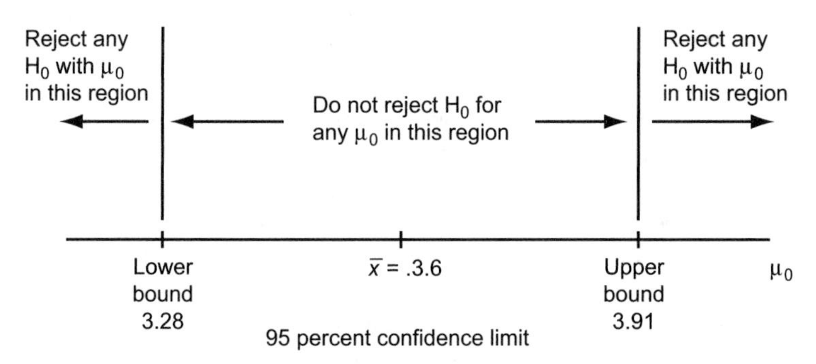

FIGURE 8-9. Relationship between confidence interval estimation and hypothesis testing for Example 8-1.

Certainly not! Whether or not we obtain important results depends on whether or not we test important hypotheses. Although this truth seems obvious, and is universally acknowledged, it is often overlooked.

To see how easy it is to go wrong on this point, think back to the public transit example. Let us suppose that everyone concerned is satisfied with the 95% confidence level employed in the test. If that is so, then we certainly proved that ridership increased following the bus-schedule change. The question is, did we obtain a result of any practical significance? Well, what are we prepared to say about the increase? Can we tell the city council that ridership is now 0.21 (the sample p)? No, we did not test that hypothesis—we tested for an unspecified increase, however small. Thus, all we can say from the test is that ridership is more than the old value of 0.16. It might be 0.21 or even higher, or it might be much lower, say 0.1600001. Is it likely the council would want to fund the new bus schedule knowing the increase might be only 0.0000001?

Unfortunately, situations similar to the above occur all too often, where a null hypothesis of no difference is adopted. There are instances where rejecting such a hypothesis is important, but not many. One example might be testing the efficacy of a drug, where even a marginal mortality decrease translates into lives saved and is therefore important. Most often, however, we are not interested in marginal differences. Showing that state income is below the national average is not important, unless the difference is large enough to result in a different standard of living, or has other economic importance. Likewise, showing that a new agricultural practice decreases erosion is not worthwhile, unless erosion declines by a scientifically or environmentally significant amount. Practical and scientific importance usually requires more than just the existence of a difference.

This problem of scientific significance has two solutions. One approach is to test only important hypotheses. Instead of settling for any difference, members of the city council should have asked whether or not ridership increased by some practically significant amount. For example, they might have looked for an increase large enough to recover the marginal cost of the improved service. Finding that, they could be confident that the neighborhood is paying for the extra service. Of course, the council needn't look for complete cost recovery; they might ask for only a 50% payback. For that matter, they could ask that the service more than pay for itself in order to be continued. The point is, we should have been looking for a difference large enough to be of practical importance. Having defined scientific importance, we could adjust H_0 and H_A accordingly. For example, if the council deems 20% ridership as the minimum needed to justify the new service, we could test $\pi \leq 0.2$ against $\pi > 0.2$.

The other alternative is to use a confidence interval rather than a hypothesis test. If the confidence interval includes only scientifically significant values, then we have obtained an important result.

Notice that both alternatives call for us to define what is "scientifically significant" or "practically significant." Just what these terms mean is not a statistical issue, but rather one that depends on the subject being investigated. However difficult it is to arrive at these definitions, it is necessary to do so. Although it might seem that testing for no difference avoids the problem, in point of fact it simply defines "significant" to include immeasurably small differences. Hardly ever is this an acceptable definition.

8.6. Summary

In the form of statistical inference known as hypothesis testing, a value of a population parameter is assumed or hypothesized, and sample information is used to see whether or not this hypothesis is tenable. One of three forms of a null hypothesis is chosen, depending on the nature of the problem under study. A six-step procedure known as the classical test of hypothesis has been outlined and applied to several single-sample tests for μ and π. A significance level α is chosen by the researcher and represents the acceptable level of risk for a Type I error. The probability of committing a Type II error, β, is usually unknown and not controlled. As a result, one will normally either reject H_0 r make no decision. The risks of both types of error decrease with increasing sample size.

A variant of the classical test known as PROB-VALUE method has been explained, as well as the relationships between the classical test, the PROB-VALUE method, and confidence interval estimation. By using these procedures, it is possible to evaluate hypotheses and determine their statistical significance. Any evaluation must include a statement of this significance level, although it can be expressed as a PROB-VALUE or made implicit in a confidence interval.

In any application of hypothesis testing, one must be careful to differentiate between statistical significance and practical significance. Showing some result to be statistically significant does not necessarily impart any practical or scientific significance. Proving the statistical significance of a meaningless hypothesis is of no value. Only a thorough understanding of the problem under study can lead to the specification of hypotheses worthy of statistical evaluation.

FURTHER READING

The material covered in this chapter appears in most elementary textbooks of statistical methods. Additional examples of situations in which hypothesis testing can be undertaken are described in the texts listed at the end of Chapter 2 and in the problems.

PROBLEMS

1. Explain the meaning of the following terms or concepts:
   - Classical test of hypothesis
   - Null hypothesis
   - Alternate hypothesis
   - Directional hypothesis
   - Significance level
   - Test statistic
   - Critical, or rejection, region
   - Action limits (or critical values)
   - Decision rule
   - PROB-VALUE method

2. Differentiate between (a) a one-sided (or one-tail) and a two-sided test, (b) a Type I and a Type II error, and (c) statistical and practical significance.

3. List the steps in the classical test of hypothesis.

4. The mayor and city council in a certain jurisdiction want to reduce the average waste of their residential households. Results in other jurisdictions suggest the average weight of garbage per household, where there is a weekly pickup cycle, is 22 kg with a standard deviation of 7 kg. They want to run an experiment in which they monitor the weight of the refuse from approximately 100 randomly selected homes following the introduction of a program designed to divert waste to recycling alternatives. If the program is a success, they would like to broaden the experiment for the entire city. Discuss the specific hypotheses they might wish to test if their goal is to come up with conclusions having practical implications. What other information might be useful?

5. When should one use the t-distribution rather than the z-distribution in a hypothesis test of population mean?

6. For a given t, the probabilities in Appendix Table A-6 are half those in Table A-5. Why? In each of the following questions, follow the steps in the classical test of hypothesis outlined in this chapter. Where necessary, modify this procedure to incorporate the requirements of the PROB-VALUE method.

7. Department of Agriculture and Livestock Development researchers in Kenya estimate that yields in a certain district should approach the following amounts, in metric tons per hectare: groundnuts, 0.50; cassava, 3.70; beans, 0.30. A survey undertaken by a district agricultural officer on farm holdings headed by women reveals the following results:

	$\bar{x}$	s
Groundnuts	0.40	0.03
Cassava	2.10	0.83
Beans	0.04	0.30

There are 100 farm holdings in the sample.

a. Test the hypothesis that these farm holdings are producing at the target levels of this district for each crop.
b. Test the hypothesis that the farm holdings are producing at lower than target levels.
c. Why would you want to be extremely careful in interpreting the results of this study, even if great care were taken in selecting a truly random sample?

8. A variety of popular health books suggest that the readers assess their health, at least in part, by estimating their percentage of body fat. Then, they can compare their body fat to a standard set by the average body fat based on gender and age. The following measurements are percentages of body fat from a random sample of men aged 20 to 29. The data come from a larger dataset in the StatLib collection: *lib.stat.cmu.edu/datasets/bodyfat.*

12.6	6.9	24.6	10.9	27.8	20.6	19	12.8	5.1
12	7.5	8.5	16.1	19	15.3	14.2	4.6	4.7
9.4	6.5	13.4	9.9	10.8	14.4	19	28.6	6.1
24.5	9.9	19.1	10.6	16.5	20.5	17.2	30.1	10.5

a. Test the hypothesis that the mean body fat for males in this category is 12 at $\alpha = 0.05$ level.
b. Calculate the 95% confidence limits on the body fat percentage of males in this age group.
c. How do the results of (b) relate to your decision in part (a)?

9. A stream has been monitored weekly for a number of years, and the total dissolved solids in the stream averages 40 parts per million and is constant throughout the year. Following a recent change in land use in the drainage basins of the stream, a fluvial geomorphologist finds that the mean parts per million of dissolved solids in a 25-week sample to be 52 with a standard deviation of 32. Has there been a change in the average level of dissolved solids in this stream?

10. Mean annual water consumption per household in a certain city is 6,800 L. The variance is 1,440,000. A random sample of 40 households in one neighborhood reveals a mean of 8,000.
    a. Test the hypothesis that these households have (i) different and (ii) higher consumption levels than the average city household.
    b. Comment on the practical significance of this result.
    c. What is the PROB-VALUE for each hypothesis in part (a)?

11. An exhaustive survey of all users of a wilderness park taken in 1960 revealed that the average number of persons per party was 2.6. In a random sample of 25 parties in 1985, the average was 3.2 persons with a standard deviation of 1.08.
    a. Test the hypothesis that the number of persons per party has changed in the intervening years.
    b. Determine the corresponding PROB-VALUE.

12. A manufacturer of barometers believes that the production process yields 2.2% defectives. In a random sample of 250 barometers, 7 are defective.
    a. To test whether the manufacturer's belief is confirmed by this sample, calculate the PROB-VALUE for a test of hypothesis of $H_0 : \pi = .022$.
    b. At which of the following significance levels will the null hypothesis be rejected? (i) 0.20, (ii) 0.10, (iii) 0.05, (iv) 0.01, (v) 0.001.

13. The average score of geography majors on a standard graduate school entrance examination is hypothesized to be 600; that is, $H_0 : \mu = 600$. A random sample of $n = 75$ students is selected. The decision rule used to reject H_0 is

    Reject H_0 : $\bar{x} < 575$ or $\bar{x} > 625$
    Do not reject H_0 : $575 \leq \bar{x} \leq 625$

    Assume that the standard deviation is 100 and X is normally distributed. Find α for this test.

14. A shipper declares that the probability that one of the shipments is delayed is .05. A customer notes that in her first $n = 200$ shipments, 12 have been delayed.
    a. Test the assertion of the shipper that $\pi = 0.05$ at $\alpha = 0.05$.
    b. Calculate the corresponding PROB-VALUE.

9

Two-Sample Hypothesis Testing

In a large number of inferential problems, the goal is to compare two populations with one another. We may want to *estimate* the mean difference between two populations, or we may wish to *test* whether or not two populations have the same mean. For example, suppose we want to compare the average distances traveled for major clothing purchases by males and by females. To do this, we would take a random sample of male shoppers and an independent random sample of female shoppers and try to draw conclusions about the population means based on the sample means of each gender. Because we are comparing statistics from two samples, these problems are known as *two-sample tests.* They could equally well be labeled *two-population tests,* since the fundamental question asked is whether or not the two samples come from the same population. For the shopping behavior example, the question we are trying to answer is, "Do women and men travel, on average, the same distance for purchases [are they from the same population?], or do they travel different distances [are they from two populations]?" Note that when we test whether the means of two populations are equal, we actually test whether the *difference* between them is zero. For this reason it is most often termed the *difference-of-means* test.

In Section 9.1 we consider the widely used two-sample difference-of-means test and describe a methodology for creating confidence intervals for the difference in population means. We introduce a difference-of-means test for paired observations in Section 9.2. This is a special form of the difference-of-means test where the observations are *matched* or *paired.* In this test both samples consist of the same observations. We might survey individuals on their shopping behavior, and then survey them again after some significant change in the environment has occurred—for example, the opening of a major new shopping mall. Has their behavior changed? This test is commonly used in before and after situations. The two-sample difference-of-proportions test is detailed in Section 9.3. It is used when our data are counts and we can calculate the percentage or proportion of the counts having a particular characteristic. For example, we might wish to test whether the proportion of voters supporting a certain school board issue differs by city neighborhood.

As we shall see in Section 9.1, difference-of-means tests require that we know something about the variances of the two populations we are testing. Specifically, we need to know whether or not they are equal. Obviously, a statistical test for this is necessary. Testing the equality of variances is covered in Section 9.4. In addition to being useful in the difference-of-means test, the test can be used in many other situations where we are interested in the dispersion rather the average values of our samples. As shall see, much of the theory behind the procedures has already been presented in Chapter 7. What follows then is mainly an extension of the ideas developed for one-sample hypothesis testing.

9.1. Difference of Means

For two-sample tests, it is necessary to use additional subscripts to distinguish between the two samples and populations. Thus, μ_1 and μ_2 are the means of population 1 and population 2, respectively. Similarly, there are n_1 units sampled from the first population and n_2 units drawn from the second population. $\bar{X}_1$, and $\bar{X}_2$, are the random variables corresponding to the sample mean statistics for sample 1 and sample 2, respectively, and $\bar{x}_1$, and $\bar{x}_2$, are the values of these random variables calculated from the two samples. To complete the notation, we use a double subscript to label the sample values x_{ij}. The first subscript, *i*, is the population from which the sample value has been drawn ($i = 1, 2$), and subscript j designates the *j*th sample unit drawn from the *i*th population. The necessary data for a two-sample difference-of-means test are of the following form:

Sample from population 1	Sample from population 2
x_{11}	x_{21}
x_{12}	x_{22}
...	...
x_{1n_1}	
	x_{2n_2}

The two samples do not have to be of equal size; and n_1 need not equal n_2. From these samples we calculate the values of sample mean statistics:

$$\bar{x}_1 = \frac{\sum_{j=1}^{n_1} x_{1j}}{n_1} \qquad \bar{x}_2 = \frac{\sum_{j=1}^{n_2} x_{2j}}{n_2}$$

The methods described in this section assume that the populations are normally distributed. If they are not normal, then the tests will still be approximately correct as long as n_1 and n_2 are *large*, say both $n_1, n_2 \geq 30$.

Hypotheses about μ_1 and μ_2

We can ask a number of questions about the relationship between the two population means μ_1 and μ_2. The simplest hypothesis is whether or not the two means are identical. In this test we are looking for *any* difference between the two population means, however small. This is the most common test, and the one most computer programs report by default. There are one-and two-tailed versions of this test, in the following formats:

	Two-tailed	One-tailed
H_0 :	$\mu_1 = \mu_2$	$\mu_1 \leq \mu_2$
	or	or
	$\mu_1 - \mu_2 = 0$	$\mu_1 - \mu_2 \leq 0$
H_A :	$\mu_1 \neq \mu_2$	$\mu_1 > \mu_2$
	or	or
	$\mu_1 - \mu_2 \neq 0$	$\mu_1 - \mu_2 > 0$

As was true in one-sample testing, the two cases differ primarily with respect to the format of the alternative hypothesis. The two-tailed version is used when the hypothesis being evaluated contains nothing about the direction or sign of the difference. In other words, it is used when we want to show the means are different, but we are not hypothesizing which one is larger. The one-tailed test, on the other hand, is used to show that μ_1 is the larger of the two. Note that we always label population 1 the larger of the two values in our hypothesis, so that this specification corresponds to the case where we are seeking to show that μ_1 is the larger of the two means. As the null hypothesis suggests, we do this by *disproving* that μ_1 is less than or equal to μ_2. The subscripts are arbitrary, and we should feel free to label them however we like. So, for example, in testing for the difference of-means for distances traveled by men and women for major clothing purchases, if we are interested in testing whether women travel farther, we would label women (1) and men (2) and use the one-tailed version.

When we wish to show that the difference-of-means exceeds some given value D_0, we use slightly different versions of these basic hypotheses:

	Two-tailed	One-tailed
H_0 :	$\mu_1 - \mu_2 = D_0$	$\mu_1 - \mu_2 \leq D_0$
H_A :	$\mu_1 - \mu_2 \neq D_0$	$\mu_1 - \mu_2 > D_0$

Again, the one-sided version assumes that the mean from sample 1 is the larger of the two means. If it were not the largest, then the *difference* $\mu_1 - \mu_2$ would be *negative* and we would replace D_0 with $-D_0$. Since we are focusing on the *magnitude* of the difference between the two means, it makes no difference which of the two means we label 1 or 2. For the one-tailed test, we are testing an hypothesis that one of the population means exceeds the other by some given amount, D_0. When $D_0 = 0$, these hypotheses reduce to those described above. If $D_0 = 0$, we are saying *any* difference is significant, but if we select some other value for D_0, we are saying that we are interested in knowing whether or not the magnitude of the difference meets some given standard.

Suppose we gathered two random samples of calls for emergency ambulance services from a certain part of a city in consecutive years recording the elapsed time from the time of the original call to the arrival of the emergency vehicle at the accident site. If we are interested in knowing whether times are *increasing,* we would use the simpler form of the hypotheses. However, we may wish to test whether response times have increased by, say, at least 60 seconds, using the argument that any difference below this amount may have no practical significance. In this case, we would use the more complete specification and choose $D_0 = 60$.

Sampling Distributions

The basic inferential framework for comparing two population means is illustrated in Figure 9-1. We have two populations with means μ_1 and μ_2 and variances σ_1^2 and σ_2^2, respectively. From each of these populations an independent random sample is drawn and the values of the sample mean and variance are calculated using the formulas presented in Chapter 3. For sample 1, these values are labeled $\bar{x}_1$ and s_1^2, and for sample 2 the values are $\bar{x}_2$ and s_2^2. Naturally, we use these sample values to make inferences about the two population means. To decide about the difference in population means $\mu_1 - \mu_2$, we will examine the difference in sample means $\bar{x}_1 - \bar{x}_2$. If this difference is

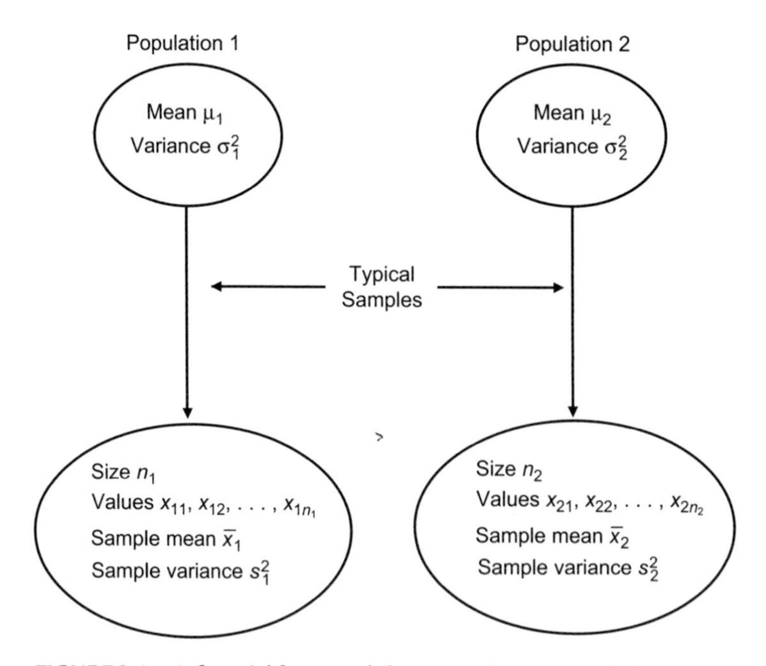

FIGURE 9-1. Inferential framework for comparing two population means.

large, we are likely to conclude that the two population means are different. But what difference is large enough? To answer this question, we need information about the sampling distribution of the random variable $\bar{X}_1 - \bar{X}_2$. Knowing how this difference varies due to sampling error, we will be in a position to evaluate the significance of our observed difference $\bar{x}_1 - \bar{x}_2$.

To arrive at the sampling distribution of $\bar{X}_1 - \bar{X}_2$, we need to know something about the population variances σ_1^2 and σ_2^2. Are we likely to know these values? Well, probably not. If we did, we would most certainly know the values of the two population means since they are used in the calculation. Let us consider the more usual situation when they are unknown. There are two possibilities: we can either assume (or prove) (see Section 9.4) that the two variances are equal or assume that they are different. In either case, the distribution of the difference of sample means follows a T-statistic with the following form:

$$T = \frac{(\bar{X}_1 - \bar{X}_2) - D_0}{\hat{\sigma}_{\bar{X}_1 - \bar{X}_2}} \tag{9-1}$$

Although it looks complicated, this formula is similar to Equation 8-7. The numerator compares the difference in the sample means to the hypothesized difference D_0, and the denominator is an estimate of the standard deviation of the difference in sample means, that is, the standard error. Put simply, the numerator calculates the size of the difference, and the denominator standardizes it relative to the sampling distribution. When the difference is near zero, we tend to accept the null hypothesis; as it gets larger and larger, we are more likely to reject it in favor of the alternative hypothesis. If we meet certain assumptions and choose the correct estimator for the standard error in the denominator, then the value of T calculated in Equation 9-1 can be shown to follow the t-distribution. The standard error to be used depends on the population variances of X_1 and X_2.

CASE 1: POPULATION VARIANCES EQUAL: $\sigma_1^2 = \sigma_1^2$

In this case there is an unknown value for the population variance σ^2, which is assumed to be identical in the two populations, that is, $\sigma_1^2 = \sigma_2^2$. Our sample data contain two independent estimates of this variance, s_1^2 and s_2^2. Rather than just use one of these estimates, it makes sense to *combine* them together into a *pooled variance estimate*

$$s_p^2 = \frac{(n_1 - 1)s_1^2 + (n_2 - 1)s_2^2}{(n_1 - 1) + (n_2 - 1)} = \frac{(n_1 - 1)s_1^2 + (n_2 - 1)s_2^2}{(n_1 + n_2 - 2)}$$

since they are both estimates of the same unknown σ^2. The p subscript in this formula for the estimate of the variance refers to *pooled.* If we examine this equation we can see that it is a weighted average of the two sample variances. The weights chosen are not the sample sizes but the *degrees of freedom,* which in this case is 1 less than the sample size of each of the two samples. If the sample sizes are equal, then the formula weights each of the variance estimates equally. But if, for example, $n_1 > n_2$, the estimate from the first sample will be more important in determining the value of the pooled estimate s_p^2 than the estimate from the second sample. The appropriate estimator for the standard error $\hat{\sigma}_{\bar{X}_1-\bar{X}_2}$ is

$$\hat{\sigma}_{\bar{X}_1-\bar{X}_2} = s_p\sqrt{\frac{1}{n_1}+\frac{1}{n_2}} \tag{9-2}$$

Note that this formula contains a value for s_p that is not the pooled estimate of the variance itself, which is s_p^2.

With this as background we can formalize our results for the sampling distribution of $\bar{X}_1 - \bar{X}_2$ in the following theorem:

DEFINITION: SAMPLING DISTRIBUTION OF $\bar{X}_1 - \bar{X}_2$, ASSUMING EQUAL POPULATION VARIANCES

Assume X_1 and X_2 are normal with a difference in means $\mu_1 - \mu_2 = D_0$. Assuming the variance σ^2 is the same for both populations, then the following has a *t*-distribution:

$$T = \frac{\bar{X}_1 - \bar{X}_2 - D_0}{\hat{\sigma}_{\bar{X}_1-\bar{X}_2}} = \frac{\bar{X}_1 - \bar{X}_2 - D_0}{s_p\sqrt{\frac{1}{n_1}+\frac{1}{n_2}}} \tag{9-3}$$

with degrees of freedom

$$df = n_1 + n_2 - 2$$

To use this result in hypothesis testing, we simply substitute our sample values of $\bar{x}_1$, $\bar{x}_2$, and s_p for the random variables $\bar{X}_1$, $\bar{X}_2$, and S_p.

CASE 2: POPULATION VARIANCES NOT EQUAL: $\sigma_1^2 \neq \sigma_1^2$

The test outlined in the previous section rests on the assumption that the variances of the two random variables X_1 and X_2 are equal. This can be assumed, but if there are large differences in the sample variances, this assumption is worth testing using the test for homogeneity of variances described in Section 9.4. If it turns out that the two samples can be shown to have different variances, then we can still use an approximate test for the difference of means. The specifics are expressed in the following definition.

DEFINITION: SAMPLING DISTRIBUTION OF $\bar{X}_1 - \bar{X}_2$, POPULATION VARIANCES UNEQUAL

Assume X_1 and X_2 are normal with a difference in means $\mu_1 - \mu_2 = D_0$ and variances $\sigma_1^2 \neq \sigma_2^2$. The following has an approximate *t*-distribution:

$$T = \frac{\bar{X}_1 - \bar{X}_2 - D_0}{\hat{\sigma}_{\bar{X}_1 - \bar{X}_2}} = \frac{\bar{X}_1 - \bar{X}_2 - D_0}{\sqrt{S_1^2/n_1 + S_2^2/n_2}} \tag{9-4}$$

with degrees of freedom given by

$$df = \frac{(S_1^2/n_1 + S_2^2/n_2)^2}{(S_1^2/n_1)^2/(n_1 - 1) + (S_2^2/n_2)^2/(n_2 - 1)} \tag{9-5}$$

Alternatively, the approximate degrees of freedom can be found using

$$df = \min(n_1 - 1, n_2 - 1)$$

There are two differences from Case 1. First, we do not pool the variances together to form a single estimate of the variance. The standard error for the sampling statistic of the difference of means changes from Equation 9-2 to

$$\hat{\sigma}_{\bar{X}_1 - \bar{X}_2} = \sqrt{s_1^2/n_1 + s_2^2/n_2} \tag{9-6}$$

Second, the degrees of freedom are not the sum of the two sample sizes minus 2 but are estimated using the rather complicated formula given as Equation 9-5. A simpler and more conservative rule is to set the degrees of freedom to one less than the smaller of the two sample sizes, that is, min $(n_1 - 1, n_2 - 1)$. For example, suppose our two variance estimates were $s_1^2 = 50$ and $s_2^2 = 400$ from samples of size $n_1 = 50$ and $n_2 = 50$, respectively. We could estimate our degrees of freedom as

$$df = \frac{(50/50 + 400/50)^2}{(50/50)^2/49 + (400/50)^2/49} = \frac{(1 + 8)^2}{1/49 + 64/49} = \frac{81}{1.3265} \approx 61$$

Note that this is well below what we would have used for the equal variance test of $n_1 + n_2 - 2 = 50 + 50 - 2 = 98$. The correction increases with the disparity between the two variances. If $S_2^2 = 200$ and is only four times the variance in the first sample (note it is eight times as large when $S_2^2 = 400$), the $df \approx 72$. The alternative rule of using one less than the smallest sample size would suggest approximately 49 degrees of freedom, an even more conservative result. We will be less likely to reject an H_0 using this approximation.

Confidence Intervals for $\mu_1 - \mu_2$

From an *estimation* point of view, we can use the results from Cases 1 and 2 to derive confidence intervals for the difference in population means $\mu_1 - \mu_2$. Not unexpectedly, the best point estimator for the difference is simply $\bar{X}_1 - \bar{X}_2$, which we can determine from our sample values as $\bar{x}_1 - \bar{x}_2$. To establish confidence intervals around this estimate, we simply use the theory outlined above, and, as usual, we will need to assume normality for the two random variables X_1 and X_2. Two-sided confidence intervals will rely on the t-distribution and have the form

$$\bar{x}_1 - \bar{x}_2 \pm t_{\alpha/2}\hat{\sigma}_{\bar{X}_1 - \bar{X}_2} \quad (9\text{-}7)$$

where the confidence level is $1 - \alpha$. The value of t leaving $\alpha/2$ in each tail of the distribution is $t_{\alpha/2}$. This t-value is then multiplied by the standard error of the difference of means $\sigma_{\bar{X} - \bar{X}2}$. If we are assuming equal variances for X_1 and X_2, we use Equation 9-2 to estimate this standard error, otherwise Equation 9-6.

EXAMPLE 9-1. Considerable evidence has been gathered pointing to significant differences for climate outcomes such as temperature and precipitation between El Niño and non-El Niño years. In Southern Africa, interest lies in the role of El Niño in delivering lower than average rainfall in key months of the growing season and thus lower than average agricultural yields and ultimately food supply shortages. The timing of rainfall is also critical. Is there a relationship between early- and late-season rains? Would it be possible to predict crop conditions in the late season based on early season rains? There is also some evidence that overabundant early season rains in October due to El Niño seem to lead to lower than normal January rains. Research in this area is currently being undertaken by the Famine Early Warning System (FEWS) Network of the United States Geological Survey. This is one potential application of a rainfall model for Africa described by Funk et al. (2003).

Let us consider a simple prototype problem that can be expressed as a difference-of means test. There are two samples: El Niño and non-El-Niño years. El Niño years are defined as years in which the sea-surface temperature was greater than 0.6°C above the mean sea surface temperature. On this basis, a number of years in the historical record are classified as being either El Niño or non-El Niño years. There appeared to be no pattern in the time record as to which years were categorized in either group. The probability any year was that El Niño does not appear to be related to whether or not the previous year was classified as being El Niño; the same is true for non-El Niño years.

Our hypothesis to be tested is whether or not January rainfall at this climatic station was significantly drier during El Niño years. The following table contains the summary statistics for January rainfall from the historical record for an individual climatic station in southern Zambia.

	Non-El Niño years	El Niño years
Mean January precipitation (mm)	225	165
Standard deviation (mm)	80	60
Number of years (sample sizes)	35	13

Solution. Examining the data, we note that the two standard deviations lead to variances of $s_1^2 = 6400$ and $s_2^2 = 3600$ for the two samples. We will begin by using the assumption that the variances are equal and make a *pooled estimate* of the variance:

$$s_p^2 = \frac{(35-1)6400 + (13-1)3600}{(35-1)+(13-1)} = \frac{217600 + 43200}{(48-2)} = 5669.565$$

leading to an estimate of $s_p = 75.2965$. Due to the difference in sample sizes, the pooled estimate is closer to $s_1^2 = 6400$ than to $s_2^2 = 3600$. With our samples labeled in this fashion, we wish to show that El Niño years are drier; therefore, we expect a positive value for the difference-of-means. The sample t value is

$$t = \frac{225 - 165}{75.2965\sqrt{1/35 + 1/13}} = \frac{60}{24.456} = 2.453$$

The sampling distribution for the difference of means in this case is illustrated in Figure 9-2. Note that it is centered on the value of 0, and the critical region is in the upper tail of the distribution. The value of t isolating 5% of the distribution in the upper tail is 1.679 for 46 *df.* Table A-5 in the Appendix does not contain a row for 46 *df,* as most t-tables usually provide values for between 1 and 30 degrees of freedom and a few other values. This is because we know the t-distribution is extremely close to the normal once we have 30 *df* or more and we can use the normal table to give an approximately correct result. Using this t-table and the row for 40 degrees of freedom (we always select *fewer* degrees of freedom when we cannot find our exact value), we find the PROB-VALUE for this hypothesis to be in the range [.005, .01]. It turns out that the exact PROB-VALUE is .009.

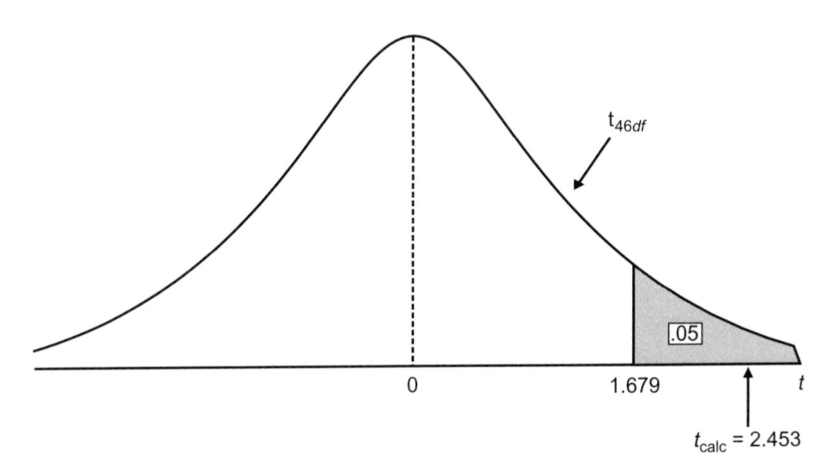

FIGURE 9-2. Sampling distribution for difference of means (Example 9-1).

So, we can now say that January precipitation in El Niño years appears to be significantly drier. But by how much? First, let us see if we can show that it is 10 mm drier. Assuming nothing else changes, calculate the t-value as

$$t = \frac{225 - 165 - 10}{75.2965\sqrt{1/35 + 1/13}} = \frac{50}{24.456} = 2.044$$

which can be shown to yield a PROB-VALUE of .023. If we were to use $D_0 = 20$, we would find our calculated t-value to be 60/24.456 = 1.636 and just outside the critical region.

To estimate the magnitude of the difference in means, we could also determine a confidence interval using Equation (9-7). Using this approach, we provide a range of values within which we think the precipitation difference lies. Ignoring the fact that we wish to show monthly precipitation in January is drier, let us use a two-tailed version of the 95% confidence interval for the mean difference in precipitation:

$$(225 - 165) - 2.013(24.456) \le \mu_1 - \mu_2 \le (225 - 165) + 2.013(24.456)$$
$$10.770 \le \mu_1 - \mu_2 \le 109.230$$

The value of $t_{\alpha/2}$ used in this equation is the exact value for 46 *df* and has not been approximated from the t-table in Appendix Table A-4. This confidence interval suggests that January rainfall in El Niño years is between 11 mm and 109 mm drier than that of non-El Niño years.

Now, suppose we do not wish to rely on the assumption that the variances of rainfall in El Niño years and non-El Niño years are equal. In this case, we can follow the procedure for Case 2. Our t-value would be calculated as

$$t = \frac{225 - 165}{\sqrt{6400/35 + 3600/13}} = \frac{60}{21.442} = 2.798$$

Our estimated degrees of freedom using Equation 9-5 are

$$df = \frac{(6400/35 + 3600/13)^2}{(6400/35)^2/34 + (3600/13)^2/12} = 211397.851/7373.966 = 28.668$$

so our degrees of freedom decline slightly. Using the t-table in Appendix Table A-5 for a one-tailed test and 28 *df*, we find the PROB-VALUE to be < .005. No matter which case we apply, the results appear to be highly statistically significant, though we must temper this result with a knowledge that the decrease in rainfall could be rather minimal and as little as 11 mm over the entire month. In Section 9.4 we will see which of these two cases applies when we test for the assumption that the variances in the two samples are the same.

9.2. Difference of Means for Paired Observations

The difference-of-means test outlined in Section 9.1 assumes that the two samples being tested are completely independent of one another. In other words, it is assumed that no relation exists between the individuals in one sample and the individuals in the second sample. They have been drawn independently of one another. So, if we found our sample 2 data was somehow defective—say it was contaminated with measurement error—we could simply collect a new sample 2 without considering sample 1.

In many instances, particularly in experimental situations, the observations do not have this type of independence. For example, many *before and after* studies use the same observations in both samples. Consider a pharmaceutical experiment in which we record patient characteristics such as blood pressure in a control situation and then again after the participants have taken a course of drugs. We would say we have a sample of *paired observations,* often called *matched pairs,* because the two values across samples are paired together. It does not make sense to compare the blood pressure of Patient A after treatment to that of Patient B before the experiment. Special methods are needed to analyze these data because the observations violate the assumption of independence. Knowing that observations are paired together, however, actually improves our ability to detect differences in the populations.

EXAMPLE 9-2. Urban planning officials are interested in measuring the impact of various traffic-calming schemes on traffic volumes in various residential neighborhoods in a large city. Twenty different locations in the city were randomly selected and then monitored using traffic count rollovers for a one-month period. Vehicle counts were calculated on a daily basis, and then a measure was taken of the *average daily traffic volume* (ADT). In order to lower traffic, a number of *traffic-calming* procedures were introduced. Barriers were installed at some intersections to prohibit turns, stop signs were introduced on major collector roads, and rumble strips were installed midway along some streets. For four weeks no data were collected while traffic patterns changed to accommodate the new situation. Then, traffic counters were replaced at the identical 20 locations, and new measurements of traffic volumes were taken. The data are summarized in Table 9-1. Have traffic volumes at these test locations decreased as a consequence of the traffic diversion scheme?

In the two-sample difference-of-means test, we used the difference in sample means $\bar{X}_1 - \bar{X}_2$ to draw inferences about the population means. With paired observations, the random variable used is the difference in X measured for each pair of values. Denote the pairs in values (x_{1j}, x_{2j}) where x_{1j} is the value of the jth observation in sample 1, and x_{2j} is the value of the jth observation in sample 2. In paired samples, we calculate the differences between the paired values

$$d_j = x_{1j} - x_{2j}$$

These differences can be positive or negative and also depend on which sample is labeled sample 1 and which sample 2. A positive value means the observation

TABLE 9-1
Traffic Counts for Example 9-2

Location	(Before) Initial volume	(After) Postcalming volume	Location	(Before) Initial volume	(After) Postcalming volume
1	1,005	509	11	559	666
2	646	476	12	892	758
3	583	888	13	794	610
4	1,064	433	14	979	582
5	817	993	15	757	722
6	703	665	16	1,079	822
7	788	531	17	854	519
8	1,025	540	18	902	494
9	546	964	19	767	402
10	527	867	20	953	728

in sample 1 is larger, whereas a negative value means that the observation in sample 2 is larger. In Example 9-2, if we label the "before" sample 1 and the "after" sample 2, then a positive result implies less traffic and a successful intervention. If we labeled them in the opposite way, negative values of d_j would imply traffic reductions.

We begin by calculating the mean and standard deviation of our differences:

$$\bar{d} = \frac{\sum_{j=1}^{n} d_j}{n} \tag{9-8}$$

$$s_d = \sqrt{\frac{\sum_{i=1}^{n} (d_j - \bar{d})^2}{n-1}} \tag{9-9}$$

If there has been significant decline in average daily traffic, then most of the differences will be positive, and $\bar{d}$ will also be positive. If some traffic volumes decline but others increase, then $\bar{d}$ will be near zero and we will not have any evidence for a difference in population means. This is exactly the strategy in the matched-pairs difference of means test.

Equation 9-9 estimates $\bar{d}$, a *sample* mean difference. Following the convention of using an uppercase letter for a random variable, we would say that $\bar{d}$ is a particular value of the random variable $\bar{D}$. Once we know the sampling distribution of $\bar{D}$ we can undertake hypothesis tests and construct confidence intervals. The sampling distribution is defined by the following theorem:

DEFINITION: SAMPLING DISTRIBUTION OF PAIRED-OBSERVATION MEAN $\bar{D}$

Assume X_1 and X_2 are normally distributed with a difference in means given by $\mu_1 - \mu_2 = D_0$. Given a random sample of n paired observations, the following quantity has an approximate t-distribution:

$$T = \frac{\bar{D} - D_0}{S_d/\sqrt{n}} \qquad (9\text{-}10)$$

with $n - 1$ degrees of freedom

Equation 9-10 has a form similar to all test statistics. The numerator is the difference between a mean value and some hypothesized value (normally but not necessarily zero), and the denominator estimates the standard error of the sampling distribution. We use our sample values for $\bar{d}$ and s_d given by expressions (9-8) and (9-9) for the random variables $\bar{D}$ and S_d. The mean of the sampling distribution is the value for our hypothesized difference D_0. This may or may not be zero. For Example 9-2, we may wish to test the hypothesis that there has been a *decline* in traffic volumes and assume $D_0 = 0$, or else test an hypothesis concerning the magnitude of this traffic reduction—perhaps that the decline is at least $D_0 = 100$ vehicles per day.

Returning to our example, we calculate the differences in average daily traffic volumes and find that average volumes have declined by 153 vehicles, that is, $\bar{d} = 153$. First, let us determine whether we can show that traffic volumes have declined and $\bar{D} > 0$. Our null hypothesis is that the before-traffic volumes are either lower or at most equal to after volumes:

$$H_0 : \bar{D} = \mu_1 - \mu_2 \leq 0$$

versus the alternative

$$H_A : \bar{D} = \mu_1 - \mu_2 < 0$$

Using the calculations of Table 9-2 with an estimated value for $s_d = 295.74$, we find that our sample test statistic is

$$t = \frac{\bar{d} - D_0}{S_d/\sqrt{n}} = \frac{153 - 0}{295.74/\sqrt{20}} = 2.3136$$

The sampling distribution for this problem is illustrated in Figure 9-3. The relevant t-distribution has 19 *df,* one less than the sample size of $n = 20$. It is a one-tailed test since we are trying to prove that traffic has *declined* and that $\bar{D} > 0$, so the critical region appears on the right-hand side of the distribution. If we accept a significance level of $\alpha = 0.05$, the t-table in Appendix Table A-4 reveals that we require $t > 1.729$

TABLE 9-2
Calculations for Example 9-2

Location	(Before)	(After)	Difference
1	1,005	520	485
2	646	476	170
3	583	888	–305
4	1,064	433	631
5	817	993	–176
6	703	665	38
7	788	531	257
8	1,025	540	485
9	546	964	–418
10	527	867	–340
11	559	666	–107
12	892	758	134
13	794	610	184
14	979	582	397
15	757	722	35
16	1,079	822	257
17	854	519	335
18	902	494	408
19	767	402	365
20	953	728	225

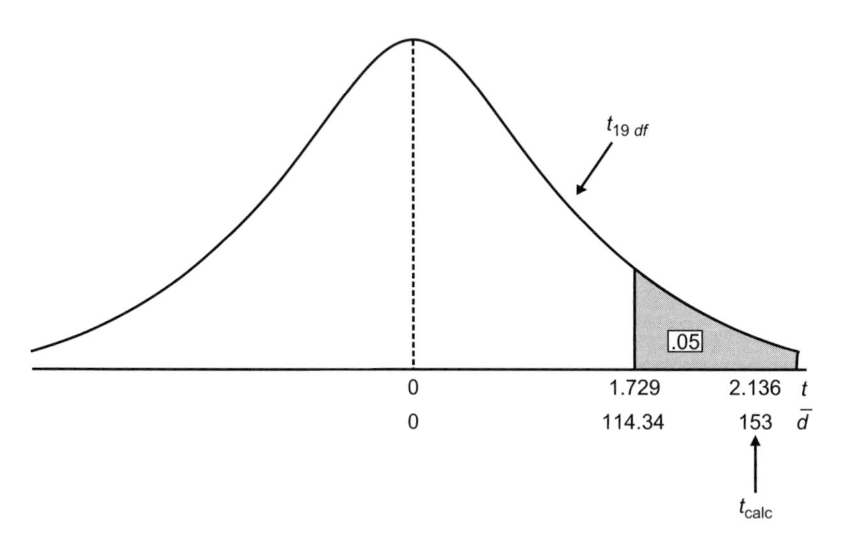

FIGURE 9-3. Sampling distribution for matched pairs test of Example 9-2.

in order to reject H_0 and shows that traffic *is* declining. To express our decision rule in terms of $\bar{d}$, we require $\bar{d} > 0 + 1.729(295.74/\sqrt{20})$ or $\bar{d} > 114.34$. It is clear that the observed difference of 153 is well above this threshold. If we take our value of 2.3136 to the t-table of Appendix Table A-5, we find PROB-VALUE of about 0.016 for this test. In sum, the evidence points to a decline in traffic, and we reject the hypothesis that there is no difference in traffic volumes.

9.3. Difference of Proportions

Just as the two-sample difference-of-means test is used to evaluate hypotheses about two population means μ_1 and μ_2, the difference-of-proportions test is used for hypotheses involving two population proportions π_1 and π_2. Inferences concerning the population difference are based on the difference in sample proportions, the two random variables P_1 and P_2. As usual, it is common to assume under the null hypothesis that the difference in proportions is 0, and hence the two samples come from a common population. In the most general case, we test whether the proportions differ by any given amount D_0, and we use a simple two-sided formulation:

$$H_0 : \pi_1 - \pi_2 = D_0 \quad \text{or} \quad (\pi_1 - \pi_2 - D_0) = 0$$
$$H_A : \pi_1 - \pi_2 \neq D_0 \quad \text{or} \quad (\pi_1 - \pi_2 - D_0) \neq 0$$

Tests of whether one of the proportions *is significantly larger* than the other $\pi_1 > \pi_2$ or that it is larger by a given proportion D_0, $\pi_1 - \pi_2 > D_0$ are based on one-tailed versions of the hypotheses:

$$H_0 : \pi_1 - \pi_2 \leq D_0 \quad \text{or} \quad (\pi_1 - \pi_2 - D_0) \leq 0$$
$$H_A : \pi_1 - \pi_2 > D_0 \quad \text{or} \quad (\pi_1 - \pi_2 - D_0) > 0$$

For all one-tailed tests, it is assumed that π_1 is the larger of the two proportions, so no version of the hypotheses for $\pi_2 - \pi_1$ is given.

Hypothesis tests and confidence intervals are based on a version of the central limit theorem that specifies the sampling distribution of the difference in sample proportions. This is summarized in the following theorem:

DEFINITION: SAMPLING DISTRIBUTION OF $P_1 - P_2$

Assume we have two population proportions that differ by some finite amount D_0 so that $\pi_1 - \pi_2 = D_0$. Given two random samples with $n_1, n_2 > 100$, the following expression has an approximately normal distribution:

$$Z = \frac{P_1 - P_2 - D_0}{\hat{\sigma}_{P_1 - P_2}} = \frac{P_1 - P_2 - D_0}{\sqrt{P_1(1 - P_1)/n_1 + P_2(1 - P_2)/n_2}} \qquad \text{(9-11)}$$

From the theorem summarized in Equation 9-11, we know that under certain sample size conditions the distribution of the difference of sample proportions is approximately normal with a standard error of

$$\sqrt{P_1(1 - P_1)/n_1 + P_2(1 - P_2)/n_2}$$

so that we can construct the following confidence interval for $(\pi_1 - \pi_2)$:

$$p_1 - p_2 \pm z_{\alpha/2}\sqrt{p_1(1 - p_1)/n_1 + p_2(1 - p_2)/n_2} \qquad (9\text{-}12)$$

where, as usual, the limits are centered on the observed difference $p_1 - p_2$ and are placed $z_{\alpha/2}$ standard errors on each side.

EXAMPLE 9-3. Researchers studying in the effects of air pollution on forest vitality are interested in the differential impact of pollution on different tree species. They centered their research in an area where ozone is known to be a significant problem. In particular, they focused on whether Jeffrey pine is more susceptible to air pollution than white fir. To analyze this problem, 100 different 50 m by 50 m quadrats were randomly selected where Jeffrey pine is dominant, and the proportion of dead trees in each area was noted. The study was repeated in 100 different quadrats of equal size in a nearby forest where white fir is dominant and the proportion of dead trees was calculated. The results are summarized in the following table:

	Proportion of dead trees	Sample size
Jeffrey pine	.37	100
White fir	.13	100

Can we conclude that Jeffrey pine are more susceptible to the pollution than the white fir?

Solution. We have two samples in excess of $n = 100$, so we can use the normal approximation based on the central limit theorem. Since we are interested in whether Jeffrey pine are *more* susceptible, we must use a one-tailed test. We are given $p_1 = 0.37$, $p_2 = 0.13$, and $n_1 = n_2 = 100$. The sample test statistic is

$$z = \frac{0.37 - 0.13 - 0}{\sqrt{0.37(0.63)/100 + 0.13(0.87)/100}}$$

$$z = \frac{0.24}{0.0588} = 4.079$$

which is highly significant. The decision rule for this hypothesis is illustrated in Figure 9-4. Note that the value for the difference needed for significance at $\alpha = 0.05$ is $z_{.05}(0.0588) = 1.645(0.0588) = 0.097$. The observed difference of 0.24 is well above this value.

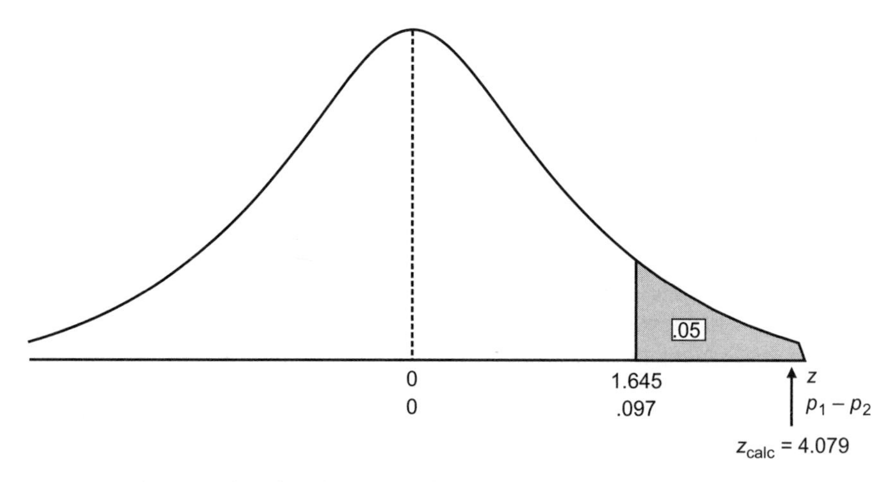

FIGURE 9-4. Sampling distribution for difference-of-proportions test of Example 9-3.

How *large* is the likely difference in proportions? Using the formula for a confidence interval and arbitrarily selecting $\alpha = 0.05$, we find

$$0.24 \pm z_{.025}(0.0588) = 0.24 \pm 1.96(0.0588) = [0.125, 0.355]$$

On the basis of this research, we can say that the proportion of Jeffrey pine that die in this forest is certainly higher than that of white fir. At the minimum, the difference in proportion is 0.125. Since the proportion of white fir dying is 0.13, individual Jeffrey pine are almost twice as likely to die than white fir.

9.4. The Equality of Variances

The two-sample difference-of-means test requires us to make a decision about whether we can assume that the variances of two populations are equal. If we can assume they are equal, we pool the information from the two samples in order to estimate the population variance. If they are not assumed to be equal, we should not use the pooled estimate. In this section, we describe a test that can be used to help us decide whether or not to pool our data. This test can be used to tell us whether our data are *homoscedastic* or *heteroscedastic.*

DEFINITION: HOMOSCEDASTICITY

If the variance about the means of two samples is equal, this condition is called *homoscedasticity.* If the variances are unequal, the two samples are termed *heteroscedastic.*

We shall see in Chapters 11 and 12 that homoscedasticity is an important assumption in both the *analysis of variance* and *regression analysis.*

Although we motivate this test as an aid for deciding the proper formulas to use in a difference-of-means test, this test is useful whenever we need to compare population variances or, by extension, population standard deviations. Is the variation or dispersion greater in one sample than in another sample? Variance tests also have broad applications in geographical research. For example, a climatologist may wish to know whether January rainfall exhibits more variability than rainfall in another month.

The null hypothesis for this test is that the two variances are equal and $\sigma_1^2 = \sigma_2^2$. As usual, this statement is evaluated based on the values of our sample variances. If they are very different, then H_0 is rejected. The test differs from all those that we have discussed so far in that it is based on the *ratio* of the sample variances s_1^2/s_2^2, rather than the differences $s_1^2 - s_2^2$. This is because the sampling distribution of the estimator for σ_1^2/σ_2^2 is well known when the samples are randomly and independently selected from two normal populations. A test statistic close to the value of 1.0 thus supports the null hypothesis, but one close to 0 or much greater than 1.0 suggests that the null hypothesis will be rejected.

To arrive at a firm decision rule, we need to know the sampling distribution of the test statistic. It so happens that if we can assume the random variables X_1 and X_2 are normally distributed, then the ratio of the sample variances follows the F-distribution. It takes on values in the range $[0, +\infty]$. It can never be negative since by definition variances must be non-negative, and therefore so is their ratio. Unlike the normal and t-distribution, an F-distribution can be skewed to the right, or to the left, or can even be symmetric about its mean. Its exact shape depends on the *degrees of freedom* associated with the two estimates S_1^2 and S_2^2, which are always $v_1 = n_1 - 1$ and $v_2 = n_2 - 1$, respectively. Like the t-distribution, there is a family of F-distributions, one for each pair of values (v_1, v_2).

DEFINITION: SAMPLING DISTRIBUTION OF THE VARIANCE RATIO

Assume X_1 and X_2 are both normally distributed from populations with variances σ_1^2 and σ_2^2. Given independent random samples of size n_1 and n_2, the statistic

$$F = \frac{S_1^2/\sigma_1^2}{S_2^2/\sigma_2^2}$$

will follow an F-distribution with $n_1 - 1$ and $n_2 - 1$ degrees of freedom.

Now when our null hypothesis of the equality of the variances is true and $\sigma_1^2 = \sigma_2^2$, we can write

$$F = \frac{S_1^2/\sigma_1^2}{S_2^2/\sigma_2^2} = \frac{S_1^2}{S_2^2}\frac{\sigma_2^2}{\sigma_1^2} = \frac{S_1^2}{S_2^2} \tag{9-13}$$

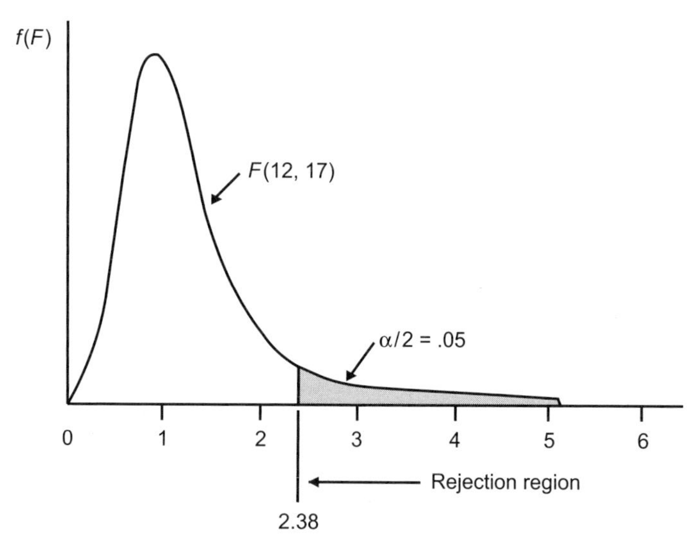

FIGURE 9-5. The F-distribution for $df = (12, 17)$.

Thus, our test statistic contains only known values, the observed ratio of the sample variances.

Figure 9-5 illustrates an F-distribution for $df = (12, 17)$. Notice that the mode of the distribution is 1 since under the null hypothesis we expect the variances to be equal and their ratio to be one. The critical value of F that leaves 0.05 in the tail is 2.38 for this pair of degrees of freedom. This means that 95% of the time we would expect the ratio of the variances to be less than or equal to 2.38. Knowing the F-distribution lets us assign a PROB-VALUE to any observed ratio, and like all of our distributions we can determine the exact proportion or area under the curve for any observed value of f. Most of the commonly used statistical software packages report the PROB-VALUE directly, but if we are working by hand we will need to use tabled alues of F. We will illustrate the use of the tables provided in Appendix Table A-7 in an example at the end of this section.

As usual, there are one- and two-tailed variants of this test. The one-tailed test is summarized in Table 9-3. As we can see, there are two options in the one-tailed version of this test, depending on which sample we are suggesting has the larger variance, sample 1 or 2. In either case, the critical value of F is the same, the value in the F-distribution for which the area in the upper tail equals α.

For the two-tailed version of this test, we need to show that the variances are different from one another. This format for the test is summarized in Table 9-4. Notice that by always taking the ratio with the *largest* variance in the numerator, we only need to look at the upper tail of the distribution, and look for the value of F that leaves $\alpha/2$ in this tail. We need not be concerned with values in the lower tail, even though this is a two-tailed test.

TABLE 9-3
One-Tailed Variance Ratio Test

H_0 :	$\sigma_1^2/\sigma_2^2 = 1$	(i.e., $\sigma_1^2 = \sigma_2^2$)
H_A :	$\sigma_1^2/\sigma_2^2 > 1$	(i.e., $\sigma_1^2 > \sigma_2^2$)
	or	
H_A :	$\sigma_2^2/\sigma_1^2 > 1$	(i.e., $\sigma_2^2 > \sigma_1^2$

Test statistic

$$F = \frac{s_1^2}{s_2^2} \quad \text{or} \quad F = \frac{s_2^2}{s_1^2}$$

Rejection region

$$F > F_{\alpha}$$

TABLE 9-4
Two-Tailed Variance Ratio Test

H_0 :	$\sigma_1^2/\sigma_2^2 = 1$	(i.e., $\sigma_1^2 = \sigma_2^2$)
H_A :	$\sigma_1^2/\sigma_2^2 \neq 1$	(i.e., $\sigma_1^2 \neq \sigma_2^2$)

Test statistic

$$F = \frac{\text{Larger sample variance}}{\text{Smaller sample variance}}$$

$$F = \frac{s_1^2}{s_2^2} \quad \text{when} \quad s_1^2 > s_2^2$$

$$F = \frac{s_2^2}{s_1^2} \quad \text{when} \quad s_2^2 > s_1^2$$

Rejection region

$$F > F_{\alpha/2}$$

EXAMPLE 9-1 (REVISITED). We are now in a position to test whether or not the variances for El Niño and non-El Niño years in Example 9-1 can be assumed to be equal. We are given the two variance estimates of $s_1^2 = 6{,}400$ and $s_2^2 = 3{,}600$ based on sample sizes of $n_1 = 35$ and $n_2 = 13$, respectively. This is a two-tailed test since we are only testing whether one of the variances is significantly larger than the other. Computing the test statistic using Equation 9-14

$$f = \frac{6{,}400}{3{,}600} = 1.778$$

following the convention of putting the larger variance estimate in the numerator. We are now in a position to compare this value to the values in the *F*-table of Appendix Table A-7. Notice that this table contains degrees of freedom for the numerator v_1 along the top of the table to determine the correct column, and the degrees of freedom for the denominator v_2 is used to select the correct row. Like many *F*-tables, this one does not contain a column for $df = (34, 12)$, so we are forced to improvise and use the column for 30 degrees of freedom. Also, this table contains critical values for four different α levels: 0.10, 0.05, 0.025, and 0.01. Since we are employing a two-tailed test and using, say, $\alpha = 0.05$, we seek the critical value within the table having 0.025 in the upper tail. The critical value is 2.963. As we are well below this value, we accept the null hypothesis of equal variances. The actual value for F(34, 12) is 2.937 for $\alpha = 0.025$. The PROB-VALUE for our calculated value of 1.778 is roughly 0.288 since this value isolates 0.144 in the upper tail. We are now in a position to say that the best difference-of-means test to employ in this example would be Case 1 using a pooled estimate of the variance.

9.5. Summary

In this chapter, hypothesis-testing procedures have been outlined for the difference in population means $\mu_1 - \mu_2$ (Section 9.1), the difference in proportions $\pi_1 - \pi_2$ (Section 9.3), and the ratio of two population variances σ_1^2/σ_2^2 (Section 9.4). For the difference-of-means test, two cases are identified, based on an assumption regarding the variances of the two samples. The suitability of the assumption can be tested using the variance-ratio test. A special test for comparing means where the observations are paired or matched has been discussed in Section 9.2. This test is used when the two samples are not independent of one another. This is most commonly characteristic of so-called *before and after* studies.

It should be apparent by now that the statistical significance of the results alone does not guarantee that the hypothesis being evaluated is of any practical significance. Statistical and practical significance don't necessarily occur together. With a large enough sample, we can often find a very minuscule absolute difference between two samples to be statistically significant. Suppose, for example, we sampled 10,000 ambulance response times in two consecutive years and were able to conclude through a one-tailed difference-of-means test that response times were increasing with a very low PROB-VALUE. Does this result have any practical significance? Maybe, and maybe not. Quite clearly, this depends on the *actual magnitude of the difference in response times.*

There are two simple ways of assessing the practical significance of our result. First, we can determine the confidence interval within which the actual time difference must lie. If it is to between 3 and 13 seconds, we may find this to be of no practical value. But if the confidence interval suggested the difference to be between 120 and 130 seconds, we might feel differently. If we are going to evaluate an hypothesis of the difference of means and we know that any time difference less than 60 seconds has no practical impact, we should test the hypothesis for the difference of means (using Equation 9-1, for example) with a value of $D_0 = 60$. Using either of these two approaches, we are sure to improve the *validity* of our statistical analysis.

FURTHER READING

W. L. Carlson and B. Thorne, *Applied Statistical Methods* (Upper Saddle River, NJ: Prentice Hall, 1997).

P-S. Chu and J. Wang, "Tropical Cyclone Occurrences in the Vicinity of Hawaii: Are the Differences between El Niño and Non-El Niño Years Significant?," *Journal of Climate* 10 (1997), 2683–2689.

C. Funk, et al., "The Collaborative Historical African Rainfall Model: Description and Evaluation," *International Journal of Climatology* 23 (2003), 47–66.

PROBLEMS

1. Explain the meaning of the following terms:
   - Pooled variance
   - Paired samples
   - Homoscedasticity
   - Practical significance

2. Under what conditions is a matched-pairs test undertaken instead of a simple difference-of-means test?

3. Why does it make sense to undertake a test for the equality of variances in our samples before completing a difference-of-means test?

4. A developing country is implementing a number of innovations to increase the efficiency of food production and thereby reduce its reliance on food imports. You have been asked to determine if maize production under a new trickle irrigation system is as high as the productivity under the old spray irrigation system. The new system uses less water because of lower evaporation losses. A decision has been made to implement this new system if it can be shown that maize yields *are not lower* than those under the old system. An experimental design is developed using 25 fields for each system.

   In the fields using spray irrigation, maize production averages 1,092 kg/ha with a variance of 3,600. In the fields irrigated by the trickle system, the maize yields averaged is 993 with a variance of 900.

   a. Does Case 1 or Case 2 apply to this problem? Why? Justify your choice.
   b. Should the new trickle system be implemented?
   c. What PROB-VALUE can you assign to this conclusion?
   d. Suppose the decision to adopt the new trickle system would be made only if it were proven superior to the existing spray system.
   e. Other than the difference in yield, what other considerations would be important in evaluating the decision to convert to the new irrigation technology?

5. A waste management company is interested in knowing whether the proportion of households participating in a recycling program is affected by the frequency of material pickups. They devise a simple experiment. In one neighborhood, residents were invited to participate in a weekly pickup service over the subsequent six months, provided with recycling boxes and given reminder letters about the availability of the service and service details. In a second neighborhood, the same procedures were used but residents were informed that their service would be bi-weekly. In the last month of the experiment, operators determined the number of participants in the recycling program by counting the number of houses where they made pickups. The results are summarized in the following table:

	Frequency of pickup	
	Weekly	Biweekly
Eligible households	131	121
Participants	64	39

a. Does the frequency of service matter?
b. If you were redesigning this study, what other data do you think it would have been useful to collect? What other changes would you make in the research design?

6. A fluvial geomorphologist is interested in whether there is a change in the dissolved oxygen levels in a river before and after weed removal. Dissolved oxygen levels were measured at 10 different sites, before and after the removal of vegetation from the stream bed.

	Dissolved oxygen levels, mg/L	
Site	Before	After
1	10.6	10.2
2	9.8	9.4
3	12.3	11.8
4	9.7	9.1
5	8.8	8.2
6	8.8	8.2
7	8.8	8.2
8	10.6	10.5
9	9.3	9.4
10	9.5	9.1

a. Conduct a matched-pairs test to determine whether there is a change in dissolved oxygen levels following weed removal.
b. Suppose the hypothesis being tested is that dissolved oxygen levels have declined. How would this change your results?
c. If you hadn't noticed that the data were matched pairs, what might you have concluded in a simple difference-of-means test?

7. A company producing digital thermometers believe its new design significantly reduces the variation in reported temperatures at very low temperatures compared to its existing design. Twenty-five thermometers of each type were used in an experiment that recorded reported temperatures in a controlled environment. The results are summarized as follows:

	Reported temperatures	
	Old design	New design
Number of samples	25	25
s^2	1.04	0.51

Does their new design give better results?

10

Nonparametric Methods

All the methods of statistical inference discussed so far require specific assumptions about the nature of the population from which the samples are obtained. We select a population parameter of interest (μ or σ or π), collect a random sample, calculate a point estimator ($\bar{x}$ or s or p), construct a sampling distribution, and then employ hypothesis-testing decision rules or confidence interval formulas. Assumptions about the population probability distribution are needed to obtain the sampling distribution, which in turn permits assignment of probabilities to sample outcomes. As we have seen, the most common assumption is that the population probability distribution is normal with mean μ and variance σ^2. Because these methods require a number of assumptions about one or more population parameters, they are termed *parametric* statistical methods. The tests described in this chapter are referred to as *nonparametric,* or *distribution-free,* tests. Compared with parametric tests, nonparametric tests require less stringent assumptions about the nature of the underlying population.

Nonparametric methods are advantageous in three situations. First, the underlying probability distribution may be unknown, or is known to be different from what a parametric method requires. For example, because rainfall amounts are never negative, the probability distribution is truncated at zero, and therefore rainfall cannot truly be normally distributed. For a dry climate with mean rainfall near zero, the departure from normality will be quite pronounced. In this situation it would be a mistake to rely on a normality assumption in testing hypotheses about σ. Notice that we do not need to know the exact form of the probability distribution to say that X is nonnormal. Generally speaking, nonparametric methods are preferred unless we have good reason to believe the population meets the required assumptions, or unless we know the test is insensitive to violations of the assumptions. Later in the chapter we will present methods to evaluate normality and other assumptions about the form of the population probability distribution. In these so-called *goodness-of-fit tests,* an observed frequency distribution can be compared to the distribution we might expect according to some theory or by some assumption. Quite frequently, one of the first steps in parametric testing is to use a goodness-of-fit test to see whether the assumption of normality is warranted. When the assumption of normality is rejected by a goodness-of-fit test, the alternative is to employ a nonparametric procedure.

The second situation calling for nonparametric methods occurs when the level of measurement falls below what is required for a parametric technique. For example, we might believe that a random variable is normally distributed, but perhaps we have only ordinal measurements on the variable. A t-test is impossible, and we cannot even compute the necessary sample statistics (mean and standard deviation). Clearly, meeting the assumptions of a parametric technique is not sufficient—we must also satisfy data requirements.

Finally, there are many situations for which no suitable parametric technique exists. For example, we might be interested in the median of a continuous random variable. Although the underlying random variable is measured at a level appropriate for parametric analysis, there is no parametric technique for constructing confidence intervals for the population median. In this case we would turn to a nonparametric approach. In other words, nonparametric techniques are also useful because they extend the family of statistical methods to situations not handled by a parametric approach.

Section 10.1 presents comparisons between parametric and nonparametric procedures and describes the advantages and disadvantages of each method. The type of test to be used in a specific case is shown to depend on a number of conditions. Sections 10.2 to 10.7 describe several nonparametric procedures useful in five different situations. Section 10.2 discusses the nonparametric equivalents to several one-and two-sample parametric tests presented in Chapters 8 and 9. Section 10.3 is devoted to a test for comparing more than two samples, a topic that is discussed in a parametric framework in the next chapter. Goodness-of-fit tests are presented in Section 10.4. In Section 10.5, tests dealing with contingency tables or cross-classified count data are described. A nonparametric approach to estimating a probability distribution is presented in Section 10.6, somewhat similar to a histogram but requiring much more calculation. Finally, in Section 10.7 we discuss *bootstrapping,* a relatively new technique that is applicable to a wide variety of problems. The chapter is summarized in Section 10.8.

10.1. Comparison of Parametric and Nonparametric Tests

If nonparametric methods demand little in the way of assumptions and level of measurement, why aren't they always used? Or equivalently, why would we ever prefer a parametric technique given the possibility of violating an assumption and other associated liabilities? Actually, both parametric and nonparametric tests have points of superiority. The choice of statistical method depends on balancing several competing factors. The following paragraphs outline some of the considerations bearing on this decision.

Power and Statistical Efficiency

In hypothesis testing we frame the null hypothesis in a way that rejection is the scientifically interesting result. Thus a good test is one that will successfully detect a false null hypothesis. Equivalently, a good test will have a low probability of Type II error. Such a test would be called a powerful test because it has good discriminating ability. Formally, *power* is defined as a $1 - P(\text{Type II error}) = 1 - \beta$. It is nearly always the case that when equivalent parametric and nonparametric tests are available, the parametric test is more powerful. Thus if all of its assumptions are satisfied, the parametric test should be applied. Using the nonparametric alternative would be more likely to result in "no decision" when in fact H_0 is false.

Note that the power of a test can always be increased by collecting more information. Obviously, if one measured the entire population, one could detect a false H_0 without fail. Thus if one test is more powerful than another, using that test with a smaller sample will provide the same discriminating ability as would the less powerful test with a larger sample. We would say that the more powerful test is more *efficient,* because it produces the same result with less information. The bottom line is that whether one thinks in terms of power or the related concept of efficiency, a parametric test is preferred as long as the assumptions are met.

Robustness and Resistance

When performing a statistical test or establishing a confidence interval, we state results with some level of uncertainty (α) that is under our control. Thus, for example, we say that H_0 is rejected at the 5% level, or we claim that a confidence interval includes the unknown parameter with 95% certainty (5% uncertainty). Clearly, for the result to be of any value, the stated level of uncertainty needs to be close to the actual uncertainty. It might not matter very much if the probability of a Type I error was 4% rather than 5%, but it would certainly matter if the actual probability was 40%. A test is said to be robust if it is relatively insensitive to violation of assumptions. That is, a robust test is one in which the actual value of α is unaffected by failure to meet assumptions. Similarily, a robust test is one whose power (or efficiency) is not degraded by violation of assumptions. The robustness of a test depends on the extent to which violation of its assumptions alters the sampling distribution of the test statistic. This varies with the test being considered, the degree to which the assumptions are violated, and the size of the sample.

For example, the one-sample t-test described in Chapter 8 is considered a robust test—the actual value of α will be close to the nominal (stated) value even when the population is non-normal. Moreover, the test will be affected less by departures from non-normality if the population distribution is symmetric than if the distribution is asymmetric. In addition, because of the central limit theorem, the larger the sample, the less critical the assumption of normality. To cite a counterexample, the F-test for equality of variance is notoriously nonrobust to departures from normality. Even symmetric populations that appear normal can yield very different α-values than expected —so much so that some authors recommend never using the F-test to test for variances unless one is absolutely certain the underlying variables are normal. Clearly, robustness is a matter of degree. Violations of some assumptions, such as normality, may be less critical than a violation of, say, the homogeneity of variance. When a specific assumption is violated, it is important to check, first, the magnitude of the the violation and, second, the sensitivity of the test to the departure. Where the violation is significant, it is often advantageous to revert to a nonparametric test that does not require that particular assumption.

Robustness is actually a rather general concept. It also includes insensitivity to data errors, a property formally known as *resistance*. Treatment of resistance is beyond the level of this book; hence we will simply note that nonparametric techniques are often more resistant than parametric equivalents. For example, the sample mean (a parametric estimator) could be drastically changed by a single very large error. The sample median, on the other hand, is totally resistant to such an error. In recent years, a subfield called robust statistics has developed a considerable body of theory and techniques surrounding these issues. Detailed treatments can be found in Maronna et al. (2006) and Rousseeuw and Leroy (2003).

Scope of Application

Since they are generally based on fewer or less restrictive conditions than parametric tests, nonparametric tests can be legitimately applied to a much larger set of populations. Even when the populations are known to be skewed or bimodal or rectangular, most nonparametric tests can still be applied. Also, statistics calculated from variables measured at the nominal or ordinal scale can be tested by using nonparametric methods. This is a distinct advantage in survey design. It can be difficult to extract interval-scale responses from mail questionnaires or personal interviews. Preference ratings or approve/disapprove questions typically yield ordinal and/or nominal responses. As an example, consider the problem of getting respondents in a mail questionnaire to accurately answer a question about their annual income. For reasons of confidentiality, many respondents are willing to answer a question concerning income only if it allows them to check a box corresponding to a range of income, such as "From $50,000 to $75,000."

Simplicity of Derivation

The derivation of the sampling distribution of the test statistics utilized in parametric tests is based on quite complicated theorems such as the central limit theorem. Usually, the form of the sampling distribution is a function of the form of the population probability distribution. Given the mathematical complexity of the normal distribution (recall the equation for the normal density function, Equation 5-36), many users of parametric statistical methods cannot comprehend the derivation of the key results on which the tests are based. Some proponents of nonparametric statistics suggest that many who employ parametric statistics must therefore apply them in a "cookbook" manner.

In contrast, the sampling distributions of the test statistics used in nonparametric tests are derived from quite basic combinatorial arguments and are easily understood. Several examples that illustrate the simplicity of the derivations are given in this chapter. For this reason, nonparametric tests are an inviting alternative when writing for a statistically unsophisticated audience. The claim of superiority based on simplicity must be tempered by the fact that nonparametric tests are also misused in some social science applications. At times, they, too, are inappropriately applied or else interpreted incorrectly.

Sample Size

The size of the available sample is an important consideration in determining whether to use a parametric or nonparametric test. When sample size is extremely small, say n is less than 10 or 15, violations of the assumptions of parametric tests are particularly hard to detect. Yet a violation of an assumption can have extremely deleterious effects under these conditions, as there is no help from the central limit theorem or other large-sample advantages. The use of nonparametric tests in these situations is often advisable. In general, the smaller the sample, the more attractive (and conservative) is the choice of a nonparametric method. The implications of the central limit theorem support this assertion. Where sample size is large, the presence of non-normality is often insignificant.

As a final point concerning the relation between parametric and nonparametric methods, one should realize that "nonparametric" means only that one does not make parametric assumptions about the random variable being observed. This does not mean that parametric distributions have no place in the application of those methods. As a matter of fact, familiar distributions (especially the normal distribution) appear frequently in nonparametric methods. Although this might seem counterintuitive, the explanation is really quite simple.

Typically, one makes very weak (nonparametric) assumptions about a random variable. The random variable is sampled, and a test statistic is computed. Very often the test statistic will follow a normal distribution, or some other distribution we think of as "parametric." This is especially common for large sample sizes. When sample sizes are small, the sampling distribution typically follows a less familiar, but nevertheless known, distribution. In the small-sample case, most tests yield PROB-VALUES directly, and a classical test of hypothesis can only be approximated. In the large-sample case, classical and PROB-VALUE tests can be performed with equal ease.

In sum, parametric tests have a significant advantage over nonparametric tests when their assumptions are met. Violation of these assumptions introduces inexactness into the test. The seriousness of the violation depends on the size of the sample, the specific assumption violated, and the robustness of the test to the particular violation.

10.2. One- and Two-Sample Tests

The first set of nonparametric procedures that we discuss are the one-and two-sample nonparametric equivalents to the t-tests described in Chapter 8. They are used whenever an interval-scaled dataset violates an assumption of the appropriate parametric test or when the variables of the dataset are at an ordinal level of measurement.

Sign Test for the Median

The simplest parametric test discussed in Chapter 8 is the one-sample t-test, which compares the sample mean $\bar{x}$ and some hypothesized (or known) value of the population mean μ specified in the null hypothesis. As such, the test is one for the middle of the distribution, which is of course described by the mean for a normally distributed variable. (Recall that normality is assumed in the t-test). For a normal distribution the mean is also the median; thus a t-test for μ is also a test for the median. The sign test is a nonparametric equivalent, used to test the value of the sample median against some hypothesized value of the population median η (Greek letter eta) specified in H_0. The sample median M is an unbiased consistent estimator of η, just as the sample mean estimates μ. The null hypothesis in the sign test for the median is that the true median η equals some hypothesized value η_0, that is, H_0: $\eta = \eta_0$. If H_0 is true, then in any sample we are just as likely to get an observation above the value η_0 as below it. If an observation x_i lies above η_0, then the difference $x_i - \eta_0$ has a positive sign. If it lies below η_0, then the difference has a negative sign. Thus, the null hypothesis suggests we should get roughly an equal number of plus and minus signs. Put another way, if H_0 is true, the probability of getting a negative sign is $\pi = P(X < \eta_0) =$ 0.5. In any random sample we will not necessarily get an equal number of positive and negative signs. The question boils down to this: How few minus signs or how few plus signs must there be before we can reject the null hypothesis that η_0 is the true median?

The test statistic is simply the number of observations in the sample with values greater than the value η_0 specified in the null hypothesis. Call this number B, where B is a binomial random variable with $\pi = 0.5$. There are only two outcomes for each observation (above or below η_0), and there are n observations or trials. Thus, the sign test turns out to be equivalent to a test of the population proportion $\pi = 0.5$. Here π stands for the proportion of the population greater than (or less than) the hypothesized value η_0. The test is summarized in Table 10-1.

EXAMPLE 10-1. The travel time of an individual's journey to work is randomly sampled on 20 different workdays over three months. Each day, the total door-to-door travel time is recorded, and the following values are obtained: 22, 23, 25, 27, 27, 28, 28, 29, 29, 31, 33, 37, 37, 38, 39, 42, 43, 43, 43, 58. For purposes of simplicity, the times have been sorted in ascending order. Inspecting the sample values, we can see that the 10th and 11th values are 31 and 33, respectively; thus the sample median is 32 minutes. Previously, before a number of changes in street signage, studies found a value of 30 minutes. Do these data suggest the median is different than 30 minutes? That is, at the 0.05 level of significance, can we claim the median travel time has changed?

Solution. First, it is necessary to calculate the number of minus signs, or observations above the hypothesized median of $\eta = 30$. Since the observations are in order, a simple count reveals $b = 9$ observations above the hypothesized median of 30. Or, equivalently, the probability of a negative sign in this sample is $p = b/n = 11/20 = 0.55$.

TABLE 10-1
Sign Test for the Population Median

Background

The sign test is used to examine hypotheses regarding η, the population median. There are two versions of the test—one for small samples and one for large samples (accurate for $n > 10$). Both versions begin by counting the sample values that are above η_0, the hypothesized median. Note that if η_0 is the true median, half of the population values are larger than η_0. That is, stating that $\eta = \eta_0$ is equivalent to stating $\pi = 0.5$.

Hypotheses

For small samples, the hypotheses are framed in terms of η:

$H_0: \eta \leq \eta_0$	$H_0: \eta \geq \eta_0$	$H_0: \eta = \eta_0$
$H_A: \eta > \eta_0$	$H_A: \eta < \eta_0$	$H_A: \eta \neq = \eta_0$
(one-tailed)	(one-tailed)	(two-tailed)
(Format A)	(Format B)	(Format C)

For large samples, the hypotheses are framed in terms of π:

$H_0: \pi \leq 0.5$	$H_0: \pi \geq 0.5$	$H_0: \pi = 0.5$
$H_A: \pi > 0.5$	$H_A: \pi < 0.5$	$H_A: \pi \neq = 0.5$
(one-tailed)	(one-tailed)	(two-tailed)
(Format A)	(Format B)	(Format C)

Test statistic

Small samples:
Let b be the number in the sample greater than η_0. The random variable B has a binomial distribution with $\pi = 0.5$. For small samples the test is based on b, using the binomial distribution directly.

Large samples:
The test is based on the *proportion* of the sample larger than the hypothesized median, $p = b/n$. For large samples, the random variable P is approximately normal with mean $\pi = 0.5$ and standard deviation $\hat{\sigma}_P = \sqrt{\pi(1-\pi)/n}$. The test statistic is

$$z = \frac{p - 0.5}{\sqrt{0.25/n}}$$

PROB-VALUES (PV) and decision rules

Small samples:
- Format A: $PV = P(B \geq b)$, where B is a binomial variable.
- Format B: $PV = P(B \leq b)$.
- Format C: We need the probability of landing in either tail of the binomial distribution. If fewer than half the sample values are below η_0 (that is, if $b < n/2$), use $PV = 2\,P(B \leq b)$. For $b > n/2$, use $PV = 2P(B \geq b)$.

Large samples:
- Format A: $PV = P(Z \geq z)$, where Z is a standard normal variable.
- Format B: $PV = P(Z \leq z)$.
- Format C: We need the probability of landing in either tail of the Z-distribution:

$$PV = P(Z \leq -|z|) + P(Z \geq |z|) = 2P(Z \geq |z|).$$

In all cases, reject H_0 when $PV < \alpha$.

TABLE 10-2
Sign Test for Example 10-1

Hypotheses
H_0: $\eta = 30$ H_A: $\eta \neq 30$
Test statistic and test to be employed
Because the sample size is greater than 10, we use the large-sample version of the sign test. We evaluate H_0 using the proportion of the sample larger than 30, or equivalently, the proportion of positive signs in the sample, P. (Note that the observed value of the random variable P is denoted p.) A Z-test is used.
Level of significance
$\alpha = 0.05$
Computation of test statistic
There are 11 positive signs in the sample of 20 observations, thus $p = 11/20 = 0.55$ Under H_0, the test statistic is approximately normal with a mean of 0.5 and a standard deviation of 0.112. We therefore convert p to a standard normal variate: $z = (0.55 - 0.5)/0.112 = 0.446$
Computation of PROB-VALUE
We need the two-tailed probability of Z being smaller than –0.446 or larger than 0.446. Thus the PROB-VALUE is $PV = P(Z < -0.446) + P(Z > 0.446) = 2P(Z > 0.446) \approx 0.65$
Decision
Do not reject H_0.

Because $n\pi = 20(0.5) = 10$ is greater than 5, the sampling distribution of P is closely approximated by the normal. Since the null hypothesis makes the assumption that $\pi = 0.5$, we estimate the standard deviation of the sampling distribution of P as $\hat{\sigma}_P = \sqrt{(0.5)(0.5)/20} = 0.112$. The standard hypothesis-testing procedure for a test of p versus π as described in Chapter 8 is followed in Table 10-2. The null hypothesis cannot be rejected.

EXAMPLE 10-2. A recreation geographer wants to show that maintenance of a neighborhood park has degraded to the point that users are no longer satisfied with the park. A random sample of $n = 9$ visitors is sampled, and each visitor is asked to rate satisfaction with the park on a 5-point scale with labels *completely dissatisfied. dissatisfied, neutral, satisfied, and completely satisfied.* The responses are then scored by assigning a value of 1 to *completely dissatisfied,* 2 to *dissatisfied,* and so on. The nine responses are 2, 3, 1, 2, 1, 5, 3, 1, and 2. Using the $\alpha = 0.05$ significance level, can we claim that most people in the population of park users are dissatisfied with the park?

Solution. A test of hypothesis for this question uses H_0: $\eta \geq 3$ versus the alternate hypothesis H_A: $\eta < 3$. Because we are interested only in knowing if park users are unsatisfied, we use a one-tail test. The sequence of signs is (–,0,–,–,–,+,0,–,–). There are two observations with a value of zero. By convention, these observations are omitted, and the sample size is reduced to $n = 7$ observations with $b = 1$ plus sign. Note that if the alternative hypothesis is true ($\eta < 3$), we would expect to see a small number of positive signs. In this sample only one of the signs is positive. Is the number observed (one) small enough to conclude H_A is true? To answer, we need the probability of observing one or fewer positive signs in a sample of size 7.

Given the small sample size, we use the binomial tables directly. For $\pi = 0.5$ and $n = 7$, the probabilities for various values of B are as follows:

b	0	1	2	3	4	5	6	7
$P(b)$	.0078	.0547	.1641	.2734	.2374	.1641	.0547	.0078

As indicated in Table 10.1, the required PROB-VALUE is given by $P(B \leq 1) = 0.0547 + 0.0078 = 0.0625$. If we were to conclude H_0 is false, we would do so with a 6.25% chance of error. This exceeds the 5% limit we can tolerate; thus we fail to reject H_0. This sample does not support the supposition that the majority of park users are dissatisfied.

Sign Test for Other Percentiles

Hypothesizing a value for the population median is of course equivalent to hypothesizing a value for the 50th percentile. Thus the sign test outlined above amounts to a test for the 50th percentile. Suppose that instead of testing a value for η, we need to test for the 20th percentile, or the 75th percentile, or any other percentile. For example, imagine that you administer a social program targeted at citizens earning less than \$17,200, which is supposed to include the bottom 20% of wage earners. Over time you have come to question the \$17,200 eligibility cutoff, thinking that with inflation the 20th percentile has shifted to a larger value. In this case you might want to test a hypothesis about the 20th percentile. In particular, letting $X_{.2}$ denote the 20th percentile of the random variable X, you would test the following one-sided hypothesis:

$$H_0: X_{.2} \leq \$17{,}200$$
$$H_A: X_{.2} > \$17{,}200$$

Rejection of H_0 would show that the 20th percentile had increased. The sign test for η is easily extended to this situation. We first specify a value for whatever percentile is of interest in H_0, and then we count the number in the sample above the hypothesized value. The test statistic is either this sample count B or, if n is large, the sample proportion P. To assign a PROB-VALUE we use $\pi = 0.2$, or $\pi = 0.75$, or whatever value is appropriate for the percentile being tested. Thus we can follow the procedure in Table 10-1, by replacing $\pi = 0.5$ with some other value.

EXAMPLE 10-3. Referring to Example 10-2, suppose we want to show that at least 40% of all park users are dissatisfied, or equivalently, that the 40th percentile is smaller than 3. That is, we want to test H_0: $X_{.4} \geq 3$ against H_A: $X_{.4} < 3$ at the $\alpha = 0.05$ level.

Solution. We count the number in the sample greater than 3, obtaining signs of (–, 0, –, –, –, +, 0, –, –), which is of course no different than in Example 10-2. But this time we use $\pi = 0.4$ in the binomial distribution, giving the following table of probabilities:

b	0	1	2	3	4	5	6	7
$P(b)$	.0280	.1306	.2613	.2903	.1935	.0774	.0172	.0016

The probability of obtaining one or fewer positive signs is now PV = $P(B \leq 1)$ = 0.1306 + 0.0280 = 0.1586. At the 5% level we therefore cannot reject H_0 and conclude that less than 40% of the population is satisfied with the park.

Large-sample tests for percentiles follow large-sample tests for η, with two differences. First, the normal approximation requires somewhat larger samples, because both $n\pi$ and $n(1 - \pi)$ must exceed than 5. Second, the estimate for the standard deviation of P is now $\hat{\sigma}_P = \sqrt{\pi(1 - \pi)/n}$, where π is something other than 0.5. Both of these modifications follow from the single-sample test for π outlined in Chapter 8.

Mann–Whitney Test

The Mann–Whitney test is analogous to a two-sample t-test. Recall that the t-test assumes two populations are normally distributed and evaluates the question of whether or not they have the same mean. Because the normal distribution is symmetrical, the mean is also the median, which implies that we are also asking if the two have the same 50th percentile. If so, an observation randomly selected from one population has a 50% chance of being larger than a random observation from the other population. Thus we can also think of the t-test as asking this question about two randomly selected observations. The Mann–Whitney is similar to the t-test in that it assumes the two populations have the same shape, and it tests the same hypothesis about two randomly selected observations. However, it does not make a normality assumption and requires only an ordinal level of measurement. Although not strictly true, the Mann–Whitney test in effect tests the hypothesis that the two populations have the same median. Indeed, the distinction between this and the actual hypothesis is so slight that we will present the test as a test for equality of medians. Although we assume the two populations have the same distribution, we need not state (or know) the distribution in order to use the Mann–Whitney test.

Accepting the technical caveat mentioned above, the null hypothesis is that the two populations have the same median. If we think of the populations as *X* and *Y*, the null hypothesis is that the median of population *X* equals the median of *Y*. That is H_0: $\eta_X = \eta_Y$. As with the *t*-test, there are two-tailed and one-tailed alternatives, giving the following possibilities:

(A)	(B)	(C)
$H_0 : \eta_X = \eta_Y$	$H_0 : \eta_X \geq \eta_Y$	$H_0 : \eta_X \leq \eta_Y$
$H_A : \eta_X \neq \eta_Y$	$H_A : \eta_X < \eta_Y$	$H_A : \eta_X > \eta_Y$

As it uses only ordinal data, the Mann–Whitney test is based on ranks. Let x_1, $x_2, \ldots x_{n_X}$ be a sample of size n_x from population *X*. Likewise, let $y_1, y_2, \ldots y_{n_Y}$ be a sample of size n_y population *Y*. The sample sizes do not need to match. The first step in the Mann–Whitney test is to rank the observations without regard to population. To do this, we pool the two samples and sort the values from lowest to highest. Assign a rank of 1 to the lowest, a rank of 2 to the next, and so forth. This results in ranks that range from 1 to $n_x + n_y$ corresponding to the smallest and largest values in the combined sample.

The test is based on how the ranks are distributed across the two populations. In particular, we sum the ranks for observations taken from population *X*. To see how the test works, imagine an extreme case in which *all* of the *X* ranks are smaller than those from population *Y*. This might happen if η_X is much smaller than η_Y. In that case the sum of the ranks of *X* would be much smaller than the sum of the ranks from *Y*. At the other extreme, if η_X is much greater than η_Y, all of the *X* ranks would be greater than the *Y* ranks, and the sum of the ranks of *X* would be larger than the corresponding sum from *Y*. The following table shows the two extremes for samples of size 4.

Ranks when η_x much less than η_Y		Ranks when η_x much greater than η_Y	
Sample *X*	Sample *Y*	Sample *X*	Sample *Y*
1			1
2			2
3			3
4			4
	5	5	
	6	6	
	7	7	
	8	8	
10	26	26	10

Let *S* be the sum of the ranks from the first sample. A little thought shows that if *S* is very small (close to its minimum) or very large (close to its maximum), we would have good reason to doubt that the medians are equal. Knowing the sampling distribution of *S* lets us attach probabilities to observed values. In particular, if both samples are larger than 10 or so, the distribution of S is approximately normal with mean and variance

$$\mu_S = \frac{n_X(n_X + n_Y + 1)}{2} \tag{10-1}$$

$$\sigma_S^2 = \frac{n_X n_Y(n_X + n_Y + 1)}{12} \tag{10-2}$$

Note that the value obtained for μ_s depends on which sample is treated as coming from population X. Equation 10-1 assumes that sample X is the smaller of the two samples. If $n_X = n_Y$ it makes no difference how the samples are labeled. Equations 10-1 and 10-2 can be used in a z-test for large samples. The distribution of S is also known for small samples, but there is no simple formula. Of course, computer programs can readily compute a small-sample PROB-VALUE, or one can turn to tables found in some of the texts listed at the end of this chapter. The full test is outlined in Table 10-3.

EXAMPLE 10-4. Carbon monoxide concentrations are sampled at 20 intersections in heavy-traffic areas of a city. All intersections have approximately equal average daily traffic volumes. Ten of the intersections are controlled by yield signs, and 10 are controlled by stop signs. The concentrations of carbon monoxide (in parts per million) are as follows:

Carbon monoxide levels, ppm	
Intersections with yield signs	Intersections with stop signs
10	31
15	9
16	21
17	28
22	14
27	12
11	58
12	29
18	22
20	29

Because the stop signs cause more cars to travel at lower speeds, encourage periods of engine idling, and create more acceleration–deceleration phases, it is hypothesized that intersections with yield signs will have lower carbon monoxide levels than those with stop signs. Is this hypothesis supported by the data at the $\alpha = 0.1$ level?

Solution. Letting X be the population with yield signs, we need a one-tailed test with the hypotheses

$$H_0: \eta_X \geq \eta_Y$$

$$H_A: \eta_X < h_Y$$

TABLE 10-3
Two-Sample Mann–Whitney Test

Background

This test compares samples of size n_x and n_y drawn from populations X and Y, respectively. The test evaluates whether or not the medians of the two populations are equal. Both X and Y must be measured at the ordinal level or higher, and it is assumed they have the same probability distribution (perhaps with different medians). Ranks are assigned to the combined sample: the lowest observation is given a rank of 1, and the highest observation is given a rank of $n_X + n_Y$.

Hypotheses

The hypotheses are framed in terms of the population medians η_X and η_Y:

H_0: $\eta_X \leq \eta_Y$	H_0: $\eta_X \geq \eta_Y$	H_0: $\eta_X = \eta_Y$
H_A: $\eta_X > \eta_Y$	H_A: $\eta_X < \eta_Y$	H_A: $\eta_X \neq \eta_Y$
(one-tailed)	(one-tailed)	(two-tailed)
(Format A)	(Format B)	(Format C)

Test statistic

The test statistic is based on S, the sum of the ranks from population X. For small samples (n_X or $n_Y < 10$), exact probabilities can be computed. For large samples, S is approximately normal with mean and standard deviation

$$\mu_S = \frac{n_X(n_X + n_Y + 1)}{2}$$

$$\sigma_S = \sqrt{\frac{n_X n_Y(n_X + n_Y + 1)}{12}}$$

where X is the smaller of the two samples ($n_x \leq n_y$). In this case the test is based on z:

$$z = \frac{S - \mu_S}{\sigma_S}$$

PROB-VALUE (PV) and decision rules

Small samples:
Use tabled values of S or a statistical computer package.

Large samples:
- Format A: PV = $P(Z \geq z)$, where Z is a standard normal variable.
- Format B: PV = $P(Z \leq z)$.
- Format C: We need the probability of landing in either tail of the Z-distribution:

$$\text{PV} = P(Z \leq -|z|) + P(Z \geq |z|) = 2P(Z \geq |z|).$$

In all cases, reject H_0 when PV < α.

TABLE 10-4
Mann–Whitney Test for Example 10-3

Carbon monoxide			
Intersections with yield signs		Intersections with stop signs	
Value (ppm)	Rank	Value (ppm)	Rank
10	2	31	19
15	7	9	1
16	8	21	12
17	9	28	16
22	13.5	14	6
27	15	12	4.5
11	3	58	20
12	4.5	29	17.5
18	10	22	13.5
20	11	29	17.5

$$S = 83$$

$$\mu_S = \frac{10(10 + 10 + 1)}{2} = 105$$

$$\sigma_S^2 = \frac{10(10)(10 + 10 + 1)}{12} = 175$$

$$\sigma_S = \sqrt{175} = 13.23$$

The calculations required for the Mann–Whitney test of the carbon monoxide pollution levels are shown in Table 10-4. For each observation we show the CO level (in parts per million) and its rank taken over both samples. Thus the lowest value is 9, the second observation in the Y-sample. The next lowest value is 10 (rank 2) obtained as the first observation in the X-sample. Notice the convention used when ties are encountered. Two observations have carbon monoxide concentrations of 12 parts per million. These two observations are ranked 4th and 5th. Each is assigned the average rank of $(4 + 5)/2 = 4.5$. The convention of assigning the average rank to tied observations can be used for two, three, or more tied observations. For example, if three observations were tied for what would be the 5th, 6th, and 7th ranks, all three would be assigned a rank of $(5 + 6 + 7)/3 = 6$. However, the test assumes there are no ties in the data, and the test is really only approximately correct when ties exist. The decision rule for a hypothesis test must be modified if there are numerous ties. These modifications are discussed in the nonparametric statistics texts cited under Further Reading.

The sum of the ranks of X yields $S = 83$. Since n_X and n_Y are greater than or equal to 10, the normal approximation to the sampling distribution of S is utilized: $\mu_S = 105$ and $\sigma_S = 13.23$.

Converting S to a standard normal, we get

$$z = \frac{83 - 105}{13.23} = -1.66$$

The one-tailed PROB-VALUE is PV = $P(Z < -1.66) \approx 0.0485$. Because PV < 0.1, we reject the null hypothesis that the medians are equal and conclude that intersections with yield signs have generally lower carbon monoxide levels.

Two-Sample Number-of-Runs Test

The Mann–Whitney test is a test for equality of two population medians, under the assumption that both populations have the same form of probability distribution. Recall that we do not need to identify a particular distribution, but we do assume it is common to both X and Y. The *number-of-runs* test relaxes that assumption and lets us ask if there are any differences in the two distributions. That is, this test will detect differences in centrality and/or the shape of the distributions.

The null hypothesis in the runs test is that the two populations have the same probability distribution. Again, that distribution need not be known, and once again only ordinal data are required. The alternative hypothesis is (obviously) that the underlying distributions are different.

As in the Mann–Whitney test, we begin by ranking the observations without regard to the population from which they were obtained. Then the number of runs (R) is noted, where a run is defined as a sequence of one or more consecutive ranks from the same population. For example, suppose we have two samples with $n_X = 4$, $n_Y = 5$. With a total of nine observations, ranks for the combined sample will run from 1 to 9. Suppose the ranks are as follows:

Rank	1	2	3	4	5	6	7	8	9
Source population	x	y	x	x	y	y	x	x	y

The second row of the table indicates whether the corresponding observation was drawn from population X or Y. Notice that in this table there are six runs comprised of the following ranks: (1), (2), (3–4), (5–6), (7–8), and (9). Thus in this example we find $R = 6$ and would use that value in testing H_0.

How does the number of runs (R) relate to the probability distributions of X and Y? To see this, imagine that the probability distributions are identical, as called for by H_0. In that case we would expect high and low ranks to be distributed randomly across the two samples, which would give rise to a relatively large value for R. On the other hand, suppose values in one population are much smaller. In that case, we might see an arrangement like the following:

1	2	3	4	5	6	7	8	9
x	x	x	x	x	y	y	y	y

In this case we find $R = 2$, which is the smallest possible value. Alternatively, suppose that there is no tendency for X to be smaller or larger than Y, but X is characterized by very high and low values, without many intermediate values compared to Y. We would describe X as more dispersed, regardless of its typical rank. In this case sampling from each population might give rise to something like the following:

1	2	3	4	5	6	7	8	9
x	x	x	y	y	y	y	x	x

Again there is little mixing of the ranks between samples, and R is small (3 in this example). Thus we again see that if R is small, the null hypothesis is in doubt. Note that in the last example a small value for R arises because X is more dispersed than Y, not because X and Y have different medians. When applying the runs test, we will reject H_0 when R is small. In particular, H_0 is rejected when R is so small that its probability of occurrence is less than α. The sampling distribution of R is known, and once again a normal approximation is available for large samples. However, the approximation calls for somewhat larger samples (n_X, $n_Y > 20$) than for the Mann–Whitney test. The details are shown in Table 10-5.

EXAMPLE 10-5. Random samples of rainfall for two desert locations are obtained. The research question is whether or not processes responsible for rain at the two places (convective activity, etc.) are sufficiently similar to give rise to the same probability distribution. A t-test is not advised because both distributions are highly skewed. The sample sizes from X and Y are 25 and 30, respectively, with values shown in Table 10-6. After sorting the data, it is seen that most of the observations from station Y have low rank, suggesting that station Y has smaller values. Adopting an α-level of 0.01, we will use the number-of-runs test to confirm this subjective visual impression.

Solution. A total of 24 runs is observed ($R = 24$) in the sorted dataset. With large sample sizes the normal approximation for R can be used. From Equations 10-3 and 10-4, the expected number of runs and its standard deviation are

$$\mu_R = \frac{2(25)(30)}{25 + 30} + 1 = 28.3$$

$$\sigma_R = \sqrt{\frac{2(25)(30)(2(25)(30) - 25 - 30)}{(25 + 30)^2(25 + 30 + 1)}} = 3.64$$

Converting to z, we get $z = (24 - 28.3)/3.64 = -1.181$. The associated PROB-VALUE is $PV = P(Z < -1.181) = 0.119$. Because $PV > \alpha$, we cannot reject H_0. At the 1% level, we cannot conclude that the probability distributions are different. In other words, hypothesis testing does not confirm our subjective impression.

At this point a word about which of these two-sample tests to use is in order. Given that the number-of-runs test detects differences in dispersion as well as centrality, is it always preferred to the Mann–Whitney test? The answer is no, because it

TABLE 10-5
Two-Sample Number-of-Runs Test

Background

This test evaluates the question of whether or not two populations have the same underlying probability distribution. Letting f(x)and g(y)be probability distributions for X and Y, we want to know if f and g are identical.

Hypotheses

The hypotheses are framed in terms of unspecified distributions f and g. For every t in each population, we want to know if the probability functions are the same.

$$H_0: f(t) = g(t)$$
$$H_A: f(t) \neq g(t)$$

Test statistic

The samples are combined, and observations are ranked from lowest to highest without regard to the originating population. Then the number of runs R is computed, where a run is defined as a sequence of one or more ranks from the same sample. If the null hypothesis is false, a relatively small value of R is expected. The sampling distribution of R is computed directly for small samples; alternatively, a normal approximation is used with large samples. In particular, if n_X, $n_Y > 20$, R is approximately normal with mean and standard deviation given by

$$\mu_R = \frac{2n_X n_Y}{n_X + n_Y} + 1 \tag{10-3}$$

$$\sigma_R = \sqrt{\frac{2n_X n_Y(2n_X n_Y - n_X - n_Y)}{(n_X + n_Y)^2(n_X + n_Y + 1)}} \tag{10-4}$$

In that case, the test statistic is an approximate standard normal variate z:

$$z = \frac{R - \mu_R}{\sigma_R}$$

PROB-VALUE (PV) and decision rule

H_0 is rejected only if R is smaller than μR, meaning $z < 0$. The PROB-VALUE is

$$PV = P(Z < z)$$

H_0 is rejected when PV < α, and we conclude that the underlying population distributions are different. Otherwise (when PV $\geq \alpha$), we fail to reject H_0 and draw no conclusion about the equality of f and g.

is less efficient than the Mann–Whitney in detecting differences in centrality. Thus, if one is only interested in testing for the population medians, one should use the Mann–Whitney test—it is less likely that a difference in medians will go unnoticed. On the other hand, if one also wants to interrogate the spread of the populations around their central values, the (less powerful) runs test is preferred.

TABLE 10-6
Data for Example 10-5

Value	12.1	11.3	11.1	10.7	10.2	10.0	9.8	9.5	9.1	8.7	8.5	8.4	8.3	8.0	7.8
Rank	1	2	3	4	5	6	7	8	9	10	11	12	13	14	15
Source	x	x	x	x	y	x	y	x	x	x	y	x	y	y	y
Value	7.7	7.3	7	6.8	6.6	6.5	6.1	6.0	5.9	5.6	5.5	5.4	5.3	5.2	5
Rank	16	17	18	19	20	21	22	23	24	25	26	27	28	29	30
Source	x	x	y	y	y	y	x	x	x	x	x	y	y	y	x
Value	4.9	4.7	4.5	4.3	4.2	3.9	3.7	3.5	3.3	3.2	3.1	3.0	2.9	2.8	2.6
Rank	31	32	33	34	35	36	37	38	39	40	41	42	43	44	45
Source	x	x	y	y	y	x	y	x	y	y	x	x	y	y	y
Value	2.4	2.3	2.0	1.9	1.7	1.6	1.5	1.4	1.3	0.8					
Rank	46	47	48	49	50	51	52	53	54	55					
Source	y	x	y	y	x	y	y	y	y	y					

10.3. Multisample Kruskal–Wallis Test

The *Kruskal–Wallis test* extends the Mann–Whitney test to the case when there are more than two populations. It is the nonparametric equivalent to one-way analysis of variance (ANOVA), which we cover in the following chapter. Like ANOVA, it tests whether or not multiple populations have the same central location. Unlike ANOVA, it does not assume the populations are normal, nor does it assume they have equal variances. In ANOVA, violations of these assumptions are especially serious when the sample sizes are different; thus the Kruskal–Wallis test is a good choice in such instances.

Like the Mann–Whitney test, the Kruskal–Wallis test uses ranked observations, where once again the ranks are assigned without regard to population. The basic idea is that if the populations have the same central location, the average rank should be nearly the same for all samples. Suppose there are k populations. A random sample is taken from each population, with sample sizes denoted $n_1, n_2, \ldots, n_k$. As in the Mann–Whitney test, the observations are ranked across all k samples. Each sample has its own distribution of ranks and its own mean rank. The mean rank for sample k is simply $\bar{R}_k = S_k / n_k$, where S_k is the sum of ranks for the kth sample. Table 10-7 shows this schematically.

In addition to the sample averages, there is a grand mean equal to the average rank for the entire suite of observations. If there are a total of $n = n_1 + n_2 + \ldots + n_k$ observations, the largest rank is obviously n. The smallest rank is 1; thus the grand mean is simply $\bar{R}_{\text{tot}} = (n + 1)/2$.

The null hypothesis is that the populations have identical locations. Under this assumption the departures of each sample average $\bar{R}_k$ rank from the grand mean should be small. The Kruskal–Wallis statistic exploits this and is given by

$$H = \frac{\sum_{j=1}^{k} [n_j (\bar{R}_j - \bar{R}_{tot})^2]}{n(n+1)/12} \qquad (10\text{-}5)$$

TABLE 10-7
Setup for Kruskal–Wallis Test

	Sample 1	Sample 2	Sample 3	. . .	Sample k
	r_{11}	r_{12}	r_{13}	. . .	r_{1k}
	r_{21}	r_{22}	r_{23}	. . .	r_{2k}
	r_{31}	r_{32}	r_{33}	. . .	r_{3k}
	. . .	r_{42}	r_{43}	. . .	r_{4k}
	$r_{n_1 1}$	r_{52}	. . .	. . .	. . .
		r_{62}	$r_{n_3 3}$	. . .	$r_{n_k k}$
		. . .		. . .	
		$r_{n_2 2}$		. . .	
Sums:	S_1	S_2	S_3	. . .	S_k
Means:	$\bar{R}_1 = S_1/n_1$	$\bar{R}_2 = S_2/n_2$	$\bar{R}_3 = S_3/n_3$	. . .	$\bar{R}_k = S_k/n_k$

Note: Ranks r_{jk} are taken over all samples, not within samples.

We see that H is proportional to the sum of squared departures weighted by sample size. If H is small, the average rank for all the samples is near the grand mean. A small H is therefore consistent with H_0. Conversely, if H is large, H_0 is in doubt. This suggests that we should reject the null hypothesis for large values of H. To say exactly how large a value rules out H_0, we need to know the sampling distribution of H.

For large samples—those with at least five observations from each population—H follows a distribution known as the "chi-square" distribution, symbolized by χ^2 (Greek letter chi). The general form of this distribution is shown in Figure 10-1. Note that the x-axis for χ^2 begins at zero, which means that the random variable cannot be negative. This is consistent with our test statistic H, which likewise can never be less than zero. As was true for the T-distribution, the exact shape of χ^2 is controlled by its degrees of freedom. In the case of H, the appropriate degrees of freedom is $k - 1$. To find the PROB-VALUE for any H, we find the area to the right of H under the χ^2 curve.

For small samples (fewer than five observations from each population,), the PROB-VALUE can be computed exactly by formulas we will not reproduce here.

It should be noted that although Equation 10-5 makes clear how H is related to the null hypothesis, it is not the best choice for computation. An alternative formula less sensitive to rounding is

$$H = \frac{12}{n(n+1)} \sum_{j=1}^{k} \frac{S_j^2}{n_j} - 3(n+1)$$

Table 10-8 presents the Kruskal–Wallis test in detail.

EXAMPLE 10-6. Suppose the carbon monoxide concentrations used in the Mann–Whitney test described earlier are augmented by another sample of 10 observations from intersections with no traffic controls. The unregulated intersections obviously constitute another population. We want to know if there are any differences in central

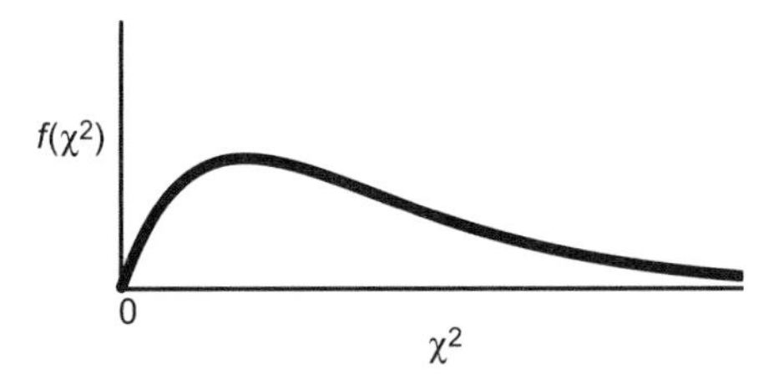

FIGURE 10-1. The χ^2 distribution.

tendency between the three populations, and we choose 5% as an acceptable level of error ($\alpha = 0.05$). The original observations are shown in "Raw Data" columns of Table 10-9.

Solution. There are 30 observations in all. A rank from 1 to 30 is attached to every sample value. Notice how ties are handled. For example, all samples have an observation whose CO value is 12 parts per million. All get the same rank of 9, equal to the average rank for the three $(8 + 9 + 10)/3$. For each column of ranks the average is computed. Here the averages for the three samples are 15.3, 21.0, and 10.2. The overall average rank is $(1 + 30)/2 = 16.5$. The sample average ranks are compared to the overall average in the calculation of H:

$$H = \frac{10(15.3 - 16.5)^2 + 10(21.0 - 16.5)^2 + 10(10.2 - 16.5)^2}{30(31)/12} = 7.5$$

With two degrees of freedom, the corresponding PROB-VALUE is about.02. This is below the stated α-level; thus we reject the null hypothesis and conclude the populations are not identical.

In closing this section on the Kruskal–Wallis test, we should note what it is not. First, it is *not* a test for equality of the means, nor (as often stated) is it a test for the medians. Indeed, it does not assume that the underlying populations necessarily have a mean, and one can construct samples whose medians are identical and yet give significant values of H. The only assumption is that elements of the populations can be ranked with respect to one another. Assuming that can be done, the test examines whether or not the average ranks in the populations are identical. If not, the populations have different "locations," which is what the test is designed to detect.

10.4. Goodness-of-Fit Tests

It is frequently necessary in empirical research to test the hypothesis that a random variable has a specified theoretical probability distribution. This hypothesis may be of interest for two reasons: to confirm some theory about the population or to determine whether the assumptions of a particular statistical inference procedure are satisfied.

TABLE 10-8
Multisample Kruskal–Wallis Test

Background

This test evaluates whether or not three or more populations have the same central value or "location." The test is based on samples of size $n_1, n_2, \ldots, n_k$ drawn from k populations. Ranks are assigned to the observations without regard to population. That is, the lowest observation of the all samples is given a rank of 1 and the highest observation is given a rank of $n = n_1 + n_2 + \ldots + n_k$. The concept behind the test is that if the populations have the same location, the average ranks in the samples should be similar.

Hypotheses

The hypotheses are framed in terms of the mean ranks.

H_0: The mean rank is identical for all populations
H_A: Two or more populations differ in mean rank

Test statistic

The test statistic is based on differences between the average rank of the samples and the overall average rank. Each sample mean rank $\bar{R}_k$ is compared to the overall mean rank $\bar{R}_{tot} = (1 + n)/2$. The test statistic is

$$H = \frac{\sum_{j=1}^{k}[n_j(\bar{R}_j - \bar{R}_{tot})^2]}{n(n+1)/12}$$

An equivalent formula better suited for calculation uses the sum of the ranks for each sample (S_j):

$$H = \frac{12}{n(n+1)} \sum_{j=1}^{k} \frac{S_j^2}{n_j} - 3(n+1)$$

PROB-VALUE (PV) and decision rules

Large samples (at least five observations in each sample):
H has a χ^2 distribution with $k - 1$ *df*. Therefore, one finds the probability of landing in the right-hand tail of a χ 2 distribution:

$$PV = P(\chi^2 > H)$$

Reject H_0 if PROB-VALUE $< \alpha$

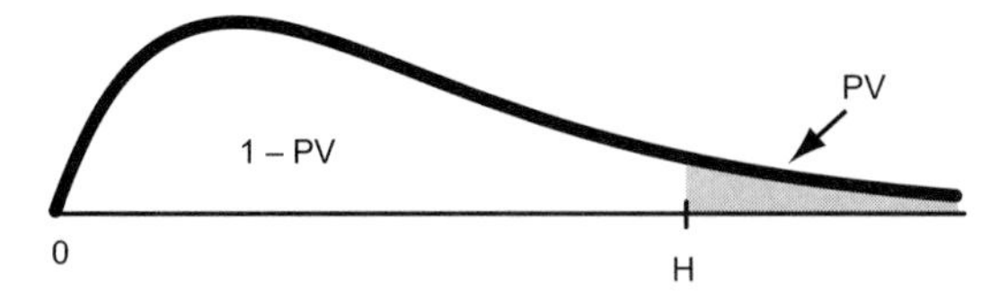

Small samples:
Use tabled values of H or a statistical computer package.

In both cases, reject H_0 when $PV < \alpha$.

TABLE 10-9
Data and Calculations for the Kruskal–Wallis for Example 10-6

Intersections with yield signs		Intersections with stop signs		Intersections with no controls	
Raw value	Rank	Raw value	Rank	Raw value	Rank
				5	1
				6	2
				7	3
				8	4
		9	5		
10	6				
11	7				
12	9	12	9	12	9
		14	12.5	14	12.5
15	14				
16	15				
17	16.5			17	16.5
18	18				
				19	19
20	20				
		21	21		
22	22.5	22	22.5	22	22.5
				23	24
27	25	28	26		
		29	27.5		
		29	27.5		
		31	29		
		58	30		

As an example of the first case, we may wish to test the randomness of a particular point pattern by using the Poisson probability distribution. Or perhaps we want to evaluate a claim about the probability distribution of annual floods. As an example of the second case, recall that most parametric techniques require a normality assumption. If the test or technique is sensitive to that assumption, it is advisable to check the assumption before proceeding. Notice that both cases (confirming a theory, checking an assumption) call for an assessment of the population probability distribution.

As always in empirical work, we will use a sample to draw an inference about the population. Here the question is whether or not the distribution of sample values is consistent with some idea about the population. Goodness-of-fit tests are employed for this purpose. The label "goodness of fit" is used because the goal of the test is to see whether the sample "matches," or agrees with, a particular probability distribution. This judgment is made according to whether the probability distribution specified in H_0 provides a good fit for the sample data. The two most commonly used tests are the chi-square and Kolmogorov–Smirnov tests.

Chi-Square (χ^2) Test

The chi-square test is a simple procedure for comparing an observed distribution to a theoretical distribution of nominal values. That is, this test is appropriate for discrete random variables that take on only categorical values (e.g., land-use type, ethnicity, vegetation type, etc.). As you know from Chapter 4, the probability distribution for such a variable simply assigns probabilities to the various classes. Equivalently, it gives the relative frequency of occurrence for each category. Thus, in this case, comparing an observed distribution with a theoretical distribution amounts to comparing observed frequencies with expected frequencies on a category-by-category basis. The chi-square test is so named because the test statistic follows a χ^2 distribution. As we will see, the test statistic tallies the correspondence between expected and observed frequencies.

Suppose a nominal random variable Y has been sampled, giving n observations spread over k categories. (We use Y to avoid confusion with X^2, the test statistic.) Of course, the observations on Y are not necessarily uniformly distributed, so that some categories will have more observations than others. We will use O_j to stand for the number of observations in category j ($j = 1, 2, \ldots k$). Consider, for example, Table 10-10, which shows field cropping patterns after incentives for erosion control were discontinued. There are four crop categories ($k = 4$), with hay and oats more popular than other crops.

The incentive program compensated farmers choosing hay over the other, more lucrative alternatives. The question is, did loss of the program cause a shift away from the hay? To answer, we need to compare the observed frequencies with what would be expected had the program continued. It is known that with the program in place, 50% of all fields were devoted to hay. Thus in a group of 50 fields, one would expect to find (0.5) (50) = 25 in hay. The percentages for corn, soybeans, and oats are 25%, 15%, and 10%, respectively. Using these percentages, we can compute expected frequencies for all categories using the formula $E_j = n\pi_j$. Thus each expected frequency is simply the sample size times the probability of occurrence for the category (π_j), as seen in the last column of Table 10-10. None of the observed counts agrees with the expected—but are the differences large enough for us to conclude that discontinuing the incentive program had an effect on crop choice?

To answer, we pose as H_0 that the true probability distribution agrees with the theoretical distribution. With that we can use the chi-square statistic, which is based on the squared differences between the observed and expected frequencies:

$$X^2 = \sum_{j=1}^{k} \frac{(O_j - E_j)^2}{E_j} \tag{10-6}$$

Inspection of Equation 10-6 reveals much about the test. First, we see that X^2 can be zero only if all k of the observed and expected frequencies agree. As the differences between O_j and E_j increase, X^2 increases. However, because the squared differences $(O_j - E_j)^2$ are divided by the expected frequencies, a given difference is less significant when the expected frequency is large. Suppose, for example, we expect a count of 40 in some class and we observe 38. This difference leads to a $(38 - 40)^2/40$

TABLE 10-10
Observed and Expected Crop Frequencies

	Observed	Expected	
Crop type	O_j	Percent	E_j
Hay	20	50%	25.0
Oats	13	25%	12.5
Corn	9	15%	7.5
Soy	8	10%	5.0
	50	100%	50

= .1 contribution to X^2. However, the same difference of 2 between an observed count of 6 and an expected count of 4 leads to a $(6-4)^2/4 = 1$ contribution to X^2.

Note also that X^2 is based on expected frequencies, but computation of expected frequencies is not addressed by the test. That is, the test does not tell us how to compute E_j; rather, the test simply compares observed counts with expected counts, however they were derived. Finally, notice that the test is based entirely on *counts* of the random variable, not *values* of the random variable. Even supposing we had assigned numbers to Y (hay = 1, etc.), those numbers would not be used in the test.

When H_0 is true, the sampling distribution of X^2 approximately follows a chi-square (χ^2) distribution. For the approximation to be reliable, every category should have an expected frequency of 2 or higher, most categories (80%) should have an expected frequency greater than 5. If these requirements are not met, categories may be merged to increase E_j, or categories can be dropped from the analysis. Recall from Chapter 8 that the shape of χ^2 is controlled by the degrees of freedom *df*. Here the degrees of freedom is

$$df = k - m - 1 \tag{10-7}$$

where k is the number of categories and m is the number of population attributes that are estimated from the sample. In our crop example no sample information was used; thus m is zero. We will return to the question of m later; for now we simply note that in this example the degrees of freedom is $df = 4 - 0 - 1 = 3$. Large differences between observed and expected frequencies lead to a large value of X^2. Thus H_0 is rejected when X^2 is large. The complete procedure is outlined in Table 10-11.

EXAMPLE 10-7. Using the field crop data, we want to test the hypothesis that the observed frequencies are consistent with the distribution of crops that would be obtained if the incentive program were in force. Thus the null hypothesis is:

$$H_0 : f(y) = \begin{cases} 0.50, & y = \text{hay} \\ 0.25, & y = \text{oats} \\ 0.15, & y = \text{corn} \\ 0.10, & y = \text{soybeans} \end{cases}$$

TABLE 10-11
Chi-Square Goodness-of-Fit Test

Background

The χ^2 test is used to check if a sample distribution agrees with a theoretical probability distribution. The underlying random variable can be measured at the nominal level or higher. The sample consists of n observations distributed over k categories. Each category has O_j observations, thus $\Sigma O_j = n$. The theoretical distribution is used to generate expected frequencies for each category in the sample. For example, in testing a sample against the uniform distribution, each sample category would have the same expected frequency, equal to the sample size divided by the number of categories. The expected count for each category is denoted E_j. The test is based on the differences between O_j and E_j. For this test to be reliable, all E_j should exceed 2, and 80% of the E_j should exceed 5.

Hypotheses

Letting Y be the random variable, and $f(Y)$ be the theoretical probability distribution of Y, the null and alternative hypotheses are:

H_0: The sample was drawn from a population $f(Y)$
H_A: The sample was drawn from some distribution other than $f(Y)$

Test statistic

Under H_0 the following has an approximate χ^2 distribution

$$X^2 = \sum_{j=1}^{k} \frac{(O_j - E_j)^2}{E_j}$$

with degrees of freedom $df = k - m - 1$ where m is the number of parameters in $f(Y)$ estimated from the sample.

PROB-VALUE (PV) and decision rule

Large values of χ^2 cast doubt on the truth of H_0, because they arise rarely when H_0 is true. The PROB-VALUE is

$$PV = P(\chi^2 > X^2)$$

Reject H_0 if PROB-VALUE $< \alpha$

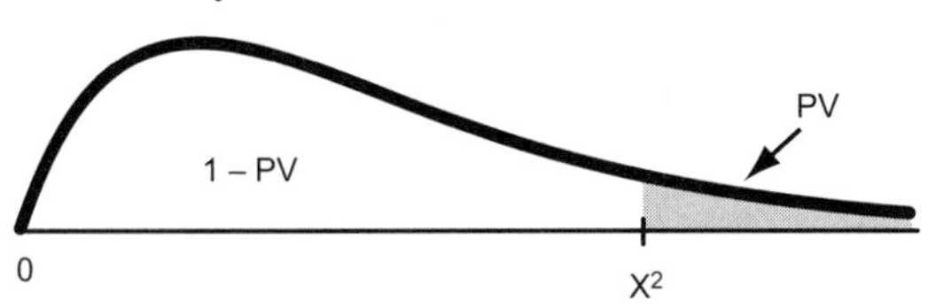

H_0 is rejected when $PV < \alpha$, and we conclude that the sample did not arise from the distribution $f(Y)$. Otherwise (if $PV \geq \alpha$), we fail to reject H_0 and draw no conclusion about $f(Y)$.

TABLE 10-12
Calculations for the Crop χ^2 Example

Crop type	O_j	E_j	$O_j - E_j$	$(O_j - E_j)^2$	$(O_j - E_j)^2/E_j$
Hay	20	25.0	–5.0	25.00	1.00
Oats	13	12.5	.5	0.25	0.02
Corn	9	7.5	1.5	2.25	0.30
Soy	8	5.0	3.0	9.00	1.80
	50				3.12

$X^2 = 1 + 0.02 + 0.30 + 1.80 = 3.12$

Solution. A sample of size 50 is obtained, and the theoretical distribution $f(Y)$ is used to assign expected frequencies: $E_j = nf(Y)$. We then compute the χ^2 statistic as in Table 10-12. The sample chi-square value is 3.12. From Appendix Table A-8, we find $P(\chi^2 > 3.12) \approx .373$; thus we cannot reject H_0.

EXAMPLE 10-8. A plant geographer in interested in testing whether the distribution of plants over space is random. She has divided the study area in to 100 equal-sized quadrats and counted the number of plants found in each quadrat. She finds a total of 150 plants distributed as shown in Figure 10-2. The number of plants per cell ranges from 0 to 6 in the sample, as summarized in the middle column of the following table:

Points per quadrat	Observed number of quadrats	Probability under H_0	Expected number of quadrats
0	27	0.2231	22.31
1	27	0.3347	33.47
2	26	0.2510	25.10
3	15	0.1255	12.55
4	3	0.0471	4.71
5	1	0.0141	1.41
6+	1	0.0045	0.45

At the 0.05 level, is there reason to think that the plants are not randomly distributed?

Solution. If plants are randomly assigned to quadrats, the distribution of number of plants per quadrat will follow the Poisson distribution. Recall that the Poisson distribution has a single parameter λ, which is both its mean and variance. From the sample we estimate the mean to be $\hat{\lambda} = 150/100 = 1.5$, and the theoretical distribution is

$$f(y) = \frac{e^{-1.5}1.5^y}{y!}, \quad y = 0, 1, 2, \ldots \tag{10-8}$$

0	1	2	2	0	1	2	0	1	1
3	2	2	0	0	2	1	2	0	1
0	1	0	0	0	2	3	1	3	1
4	2	3	1	3	1	0	2	3	0
1	1	0	1	2	2	2	0	2	1
1	4	2	3	1	5	2	2	0	3
1	2	0	2	0	0	3	1	4	0
0	2	1	2	1	2	3	3	3	1
1	3	6	0	3	3	0	0	0	0
1	0	2	1	2	2	0	1	1	2

FIGURE 10-2. Data for Example 10-8.

Equation 10-8 gives the probability of obtaining y plants in a cell. For example, the probability of zero plants in a cell is

$$f(0) = \frac{e^{-1,5}1.5^0}{0!} = e^{-1.5} \approx 0.2231$$

We have in effect performed 100 trials by counting plants in each quadrat, and we want to know if the distribution of counts agrees with Equation 10-8. Multiplying the sample size by $f(y)$ gives the expected number of quadrats having y plants. The last column in the table above reflects this computation, giving E_j for the seven values of y observed. Note that the last row pools probabilities for $y \geq 6$, which means the expected value is the number of cells containing six or more plants.

The χ^2 test cannot be applied to the table above, because the bottom two categories have $E_j < 3$. We therefore collapse the bottom three rows and complete the test:

Points per quadrat	Observed number of quadrats O_j	Probability under H_0	Expected number E_j	$(O_j - E_j)^2/E_j$
0	27	.2231	22.31	0.99
1	27	.3347	33.47	1.25
2	26	.2510	25.10	0.03
3	15	.1255	12.55	0.48
4+	5	.0657	6.57	0.38
	100	1.00	100	$\chi^2 = 3.13$

This example has 5 categories ($k = 5$), but we have used the sample to estimate one parameter in $f(y)$, namely ($\hat{\lambda}$). Using sample data like this inflates the agreement between observed and expected; thus we reduce the degrees of freedom by $m = 1$. The result is $df = 5 - 1 - 1 = 3$. Consulting Appendix Table A-8, we obtain PROB-VALUE = 0.373. Because this is well above any reasonable level of α, we cannot reject H_0. This sample does not suggest that the distribution of plants is anything but random.

Kolmogorov–Smirnov Goodness-of-Fit Test

The Kolmogorov–Smirnov test also compares a sample distribution with a theoretical distribution, and—as with the χ^2 test—the null hypothesis is that the sample has been drawn from the theoretical distribution. There are, however, three principal differences. First, the Kolmogorov–Smirnov test assumes that the random variable being evaluated is continuous rather than nominal. Second, the test is based on observed and expected cumulative frequency distributions, rather than on simple frequency distributions as we saw in the χ^2 test. Third, the Kolmogorov–Smirnov test assumes that no sample information has been used to obtain the theoretical distribution $F(x)$. Recall that in the χ^2 test we can use the sample to estimate parameters needed to calculate expected frequencies. The degrees of freedom is reduced by one for every parameter estimated. In the Kolmogorov–Smirnov test there is no such adjustment, because $F(x)$ is assumed to be known completely, without reference to the sample.

By way of review, recall that for a random variable X, the cumulative probability distribution is $F(x) = P(X \leq x)$. For a continuous random variable, $F(x)$ is the probability of finding a value less than or equal to x, and it is the area under the probability distribution $f(x)$ to the left of x. Remember also that every sample of a continuous variable has a cumulative distribution giving the proportion of the sample less than or equal to each *sample* value x_i. Here we will denote the sample distribution as $S(x_i)$. If the sample values are sorted in ascending order ($x_1 < x_2 < \ldots < x_n$), the sample distribution is very easily calculated as $S(x_i) = i/n$. For example, in any sorted sample of size 20, half the sample (10 out of 20) is less than or equal to x_{10}, and 100% of the sample (20 out of 20) is less than or equal to x_{20}.

The Kolmogorov–Smirnov test is based on the differences between $F(x_i)$ and $S(x_i)$. In particular, the test statistic is the largest absolute difference

$$D = \max_i |S(x_i) - F(x_i)| \tag{10-9}$$

Equation 10-9 simply means that we search over all x_i for the largest discrepancy between observed and expected cumulative frequencies. This is shown graphically in Figure 10-3. Notice that S consists of a set of dots, because its values are known only at the sample points x_i. On the other hand, $F(x)$ can be computed for any x; thus it appears as a continuous curve. Of course, the test compares S and F only at the sample locations. D is the largest deviation between the curve and the points.

As usual, under the assumption that H_0 is true, the test statistic D has a known sampling distribution which can be used to assign PROB-VALUES to observed values. (Critical values of D are shown in Appendix Table A-9.) The Kolmogorov–Smirnov test is summarized in Table 10-13.

EXAMPLE 10-9. A cartographer records the times taken by 10 subjects to complete an experiment in which subjects are asked to estimate data values illustrated as proportional circles. After sorting, the times in seconds are (50, 61, 62, 64, 65, 66, 67, 68, 80, 83). Prior research indicates that such estimation times are normally distributed

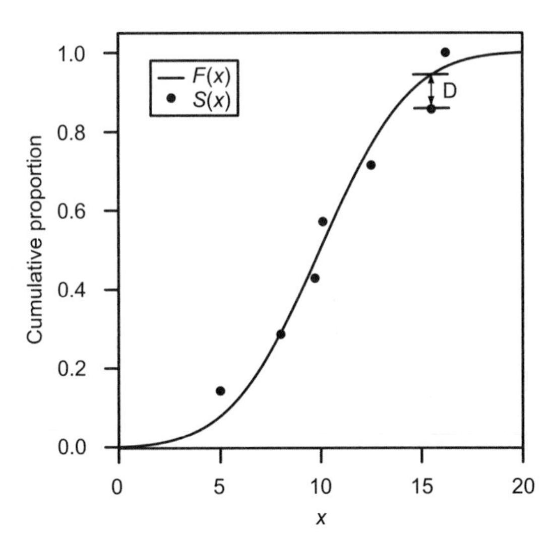

FIGURE 10-3. $F(x)$ and $S(x_i)$ as used in the Kolmogorov–Smirnov test.

TABLE 10-13
Kolmogorov–Smirnov Goodness-of-Fit Test

Background

The Kolmogorov–Smirnov test compares a theoretical cumulative probability function $F(x)$ with a sample cumulative distribution function $S(x)$. We assume the sample has been sorted in ascending order, so $x_1 < x_2 < \ldots < x_n$. The sample function is simply the proportion of the sample less than or equal to each sample value x_i. Thus S is known at the sample points only, with $S(x_i) = i/n$. Accordingly, the test compares $F(x_i)$ to $S(x_i)$ at $i = 1, 2, \ldots, n$.

Hypotheses

H_0: The sample is from the population $F(x)$.
H_A: The sample not from the population $F(x)$.

Test statistic

S and F are compared at all sample values x_i, $i = 1, 2, \ldots, n$. The test statistic is the largest absolute difference:

$$D = \max_i |S(x_i) - F(x_i)|$$

Large values of D cast doubt on the null hypothesis.

PROB-VALUES (PV) and decision rule

Probabilities for various values of D and n are shown in Appendix Table A-9. In all cases, reject H_0 when $PV < \alpha$.

with a mean of 60 seconds and a standard deviation of 10 seconds. Has this sample been drawn from such a distribution? In other words, is the sample consistent with a normal distribution where $\mu = 60$ and $\sigma = 10$?

Solution. We will first convert each sample value to a standard normal using $z_i = (x_i - \mu)/\sigma$. We then find the expected (theoretical) cumulative distribution $F(z_i)$ using the normal Appendix Table A-3. The observed distribution is found using $S(z_i) = i/n$ as described above, and the differences $|F(z_i) - S(z_i)|$ are computed. The following table shows the results:

x	z	$F(z)$	$S(z)$	$\lvert F(z) - S(z)\rvert$
50	−1.0	0.159	0.1	0.059
61	0.1	0.540	0.2	0.340
62	0.2	0.579	0.3	0.279
64	0.4	0.655	0.4	0.225
65	0.5	0.692	0.5	0.192
66	0.6	0.726	0.6	0.126
67	0.7	0.758	0.7	0.058
68	0.8	0.788	0.8	0.012
80	2.0	0.977	0.9	0.077
83	2.3	0.989	1.0	0.011

The largest difference is $D = 0.340$. From Appendix Table A-9 we see that for $n = 10$, the PROB-VALUE $P(D > 0.34) \approx 0.2$. Thus we fail to reject H_0 at the $\alpha = 0.05$ level.

It might be noted that a χ^2 test can also be used with a continuous random variable. We would divide the range of x into classes such that each class has an expected frequency greater than 5, and then proceed with a χ^2 test in the usual way. Although possible (and frequently employed), we do not recommend this course of action. First, in merging the data to create the classes, we discard information about different values of x within classes. Obviously, the χ^2 test will be based on just the k categories, whereas the Kolmogorov–Smirnov test will use all n values. Second, because the χ^2 test does not use the fact that the categories are ordered, it ignores useful information in the classed values. Thus the χ^2 test will use less information than the Kolmogorov–Smirnov test for the same sample. Finally, the χ^2 test is only approximate, whereas the Kolmogorov–Smirnov probabilities are exact. All these factors argue for the Kolmogorov–Smirnov test whenever it can be used.

10.5. Contingency Tables

Suppose we are studying two nominal variables X and Y. Very often we want to know if there is any relationship between X and Y, or conversely, if the two variables are independent of one another. Note that this is a perfectly reasonable question despite the categorical nature of the variables. For example, we might wonder if unemployed people are more likely to be homeless, or if years with major volcanic eruptions are

TABLE 10-14
Example of a Contingency Table

	Crop			
Soil	Corn	Oats	Hay	Total
Elevasil	10	8	3	21
Doreton	7	10	6	23
Lamoile	4	5	12	21
Valton	4	6	8	18
Total	25	29	29	83

also years when El Niño events are more likely. Because the variables are nominal, we are restricted to analysis involving counts of various combinations of X and Y. These counts are conveniently arranged in a *contingency table,* so named because it is used to investigate whether or not values of X are contingent on Y (and vice versa). As an example of a contingency table, suppose X is soil type and Y is crop type. A sample of 83 fields could be displayed in contingency table as in Table 10-14.

Note that the table simply shows sample counts for the various values of X (soil type) and Y (crop). The question posed by by the table is whether or not there is any association between the type of crop grown in a field and the dominant soil in the field. Put differently, we wonder if crop and soil type are independent random variables. Recall that for independent variables the joint probability is simply the product of the individual probabilities. Thus for example, if crop and soil types are independent, the probability of an oat crop grown on Doreton soil is

$$P(X = \text{Doreton and } Y = \text{oats}) = \text{P}(X = \text{Doreton})\ P(Y = \text{oats})$$

The equation shows that to get the probability of a row–column combination (Doreton and oats), we would multiply the probability associated with the row (Doreton) times the probability for the column (oats). More generally, let π_{ij} denote the population proportion in row i, column j of a contingency table, and let π_i^{row} and π_j^{col} be the probabilities associated with row i and column j, respectively. Then, if the rows and columns are independent, we can write

$$\pi_{ij} = \pi_i^{\text{row}} \pi_j^{\text{col}}; \quad i = 1, 2, \ldots, r; \quad j = 1, 2, \ldots, c \tag{10-10}$$

where the random variable X has r categories and Y has c categories. To test for the independence of X and Y, Equation 10-10 is used to generate expected frequencies for all row–column combinations. Then we use the χ^2 test to compare expected frequencies with those observed as in Table 10-10. Note that this assumes we have the row and column probabilities (π_i^{row} and π_j^{col}). In the preceding example, we would need the unconditional probabilities for each of the four soil types and each of the three crop types. If that information is not available, it must be estimated from the sample. Thus, for example, because 23 of the 83 fields have Doreton soil, we could estimate the probability $P(X = \text{Doreton})$ as $p = 23/83 = .278$. In general, we divide the row count by the sample size to estimate the row probability. A similar approach for the columns divides each column count by the sample size. Letting R_i be the count for row i, and C_j be the count for column j, we have

$$p_{ij} = \frac{R_i}{n}\frac{C_j}{n}$$

The expected frequency for any cell is then the sample size times the cell probability:

$$E_{ij} = np_{ij} = n\frac{R_i}{n}\frac{C_j}{n} = \frac{R_i C_j}{n}$$

Having observed and expected frequencies for every cell, let us proceed to a χ^2 test. That is, we compute the chi-square statistic over all cells:

$$X^2 = \sum_{i=1}^{r}\sum_{j=1}^{c}\frac{(O_{ij} - E_{ij})^2}{E_{ij}}$$

The double sum is necessary to capture all row and column combinations. Note that in computing X^2, we have used sample information to estimate the row and column probabilities. There are r rows, but we only need sample information for $r - 1$ rows. Because the probabilities sum to unity, we know the value of the last row probability from the previous $r - 1$ probabilities. Similarly, we have estimated $c - 1$ column probabilities from the sample. Thus to fill out the table of rc cells, we have estimated a total of $m = (r - 1) + (c - 1)$ parameters. In applying the χ^2 test, the degrees of freedom is therefore

$$df = k - m - 1 = rc - (r - 1) - (c - 1) - 1 = (r - 1)(c - 1)$$

Because the expected frequencies are computed under an assumption that X and Y are independent, the contingency table test takes independence as the null hypothesis. If that hypothesis is rejected, we will accept the alternative hypothesis that there is some association between X and Y. The complete test is outlined in Table 10-15.

EXAMPLE 10-10. Use the sample shown in Table 10-14 to test the hypothesis of independence between soil and crop type at the $\alpha = 0.1$ level.

Solution. We first compute expected frequencies for all cells using Equation 10-11. The calculations are as follows:

Soil	Crop Corn	Crop Oats	Hay	Total
Elevasil	$\frac{(21)(25)}{83} = 6.33$	$\frac{(21)(29)}{83} = 7.34$	$\frac{(21)(29)}{83} = 7.34$	21
Doreton	$\frac{(23)(25)}{83} = 6.93$	$\frac{(23)(29)}{83} = 8.04$	$\frac{(23)(29)}{83} = 8.0$	23
Lamoile	$\frac{(21)(25)}{83} = 6.33$	$\frac{(18)(25)}{83} = 5.42$	$\frac{(21)(29)}{83} = 7.34$	21
Valton	$\frac{(18)(29)}{83} = 6.29$	$\frac{(21)(29)}{83} = 7.34$	$\frac{(18)(29)}{83} = 6.29$	18
Total	25	29	29	83

All expected counts are larger than 5; thus we proceed with the test. (If the any of the E_{ij} were less than 2, or if there were a substantial number less than 5, we would face the need to pool or drop cells as discussed earlier.) From Equation 10-12, we obtain $X^2 = 11.169$ with degrees of freedom $df = 12 - 3 - 2 - 1 = 6$. The corresponding is PROB-VALUE = 0.083, which is smaller than the stated α-value of 0.1. Thus we reject H_0 and conclude that soil and crop type are not independent. Notice that if we had adopted a value of $\alpha = 0.05$, we would not reject the null hypothesis.

10.6. Estimating a Probability Distribution: Kernel Estimates

This section and the next (10.7) cover two members of a larger family of methods known as *compute-intensive* methods (sometimes also called *resampling* methods). Though varied in purpose, such techniques are alike in their reliance on considerably more computing power than the traditional methods described so far. Generally speaking, computer-intensive methods replace knowledge (or assumptions) about random variables with sheer volume of calculation. We emphasize, however, that although the amount of calculation may be enormous compared to traditional methods, the burden is not excessive by present computing standards. Most of them, including those discussed here, can easily be performed on personal computers given datasets of moderate size (hundreds to thousands of cases).

This section considers the problem of obtaining probability density estimates from a sample. We assume that a random sample $(x_1, x_2 \ldots x_n)$ has been obtained from an unknown continuous probability distribution $f(x)$. The goal is to construct an estimate of this probability distribution, which we denote $\hat{f}(x)$. There are numerous uses for such estimates. For example, we may be interested in knowing something about the shape of $f(x)$—for example, is it reasonably close to normal, or perhaps bimodal; how different is it from another distribution? Such uses might be satisfied by just a graph of $\hat{f}(x)$. Alternatively, $\hat{f}(x)$ might be used to estimate probabilities. For example, what proportion of the population is below the poverty level, or what are the chances of more than 10 cm of precipitation falling in a month? Such uses call for

TABLE 10-15
Chi-Square Test of Independence in Contingency Tables

Background

A total of n observations of two nominal (or higher) variables are cross-classified in a contingency table of dimension $r \times c$. The observed frequency in each cell is $O_{ij} (i = 1, \ldots, r; j = 1, \ldots, c)$. Expected frequencies for these cells are calculated from

$$E_{ij} = \frac{R_i C_j}{n} \tag{10-11}$$

where R_i and C_j are the observed counts for row i and column j, respectively.

Hypotheses

H_0: $\pi_{ij} = \pi_i^{\text{row}} \pi_j^{\text{col}}$; $i = 1, \ldots, r; j = 1, \ldots, c$

The variables are statistically independent.

H_A: $\pi_{ij} \neq \pi_i^{\text{row}} \pi_j^{\text{col}}$ for at least one ij pair

The variables are not statistically independent.

Test statistic

The following has an approximate χ^2 distribution with $(r - 1)(c - 1)$ degrees of freedom:

$$X^2 = \sum_{i=1}^{r} \sum_{j=1}^{c} \frac{(O_{ij} - E_{ij})^2}{E_{ij}} \tag{10-12}$$

PROB-VALUE (PV) and decision rule

The PROB-VALUE is found as

$$\text{PV} = P(\chi^2 > X^2)$$

The null hypothesis is rejected if $\text{PV} < \alpha$, and we conclude that X and Y are not independent. Otherwise we draw no conclusion about the independence of the random variables.

numerical operations on $\hat{f}(x)$. The operations might be as simple as finding the area between two values of X (a probability), or they could be very complicated. Regardless of whether the use is graphical or numerical, what is needed is a way of estimating $f(x)$ for any value of x.

Consider, for example, Figure 10-4, which shows tropical cyclone activity in the Atlantic area for the period 1950–1990. A value near 100% is a "normal" year, whereas lower and higher values correspond to years with abnormally light or heavy cyclone activity. With these data in hand, we might want to estimate the probability of a year with less than 50% of normal activity, or more than 200% of normal, or some other probability.

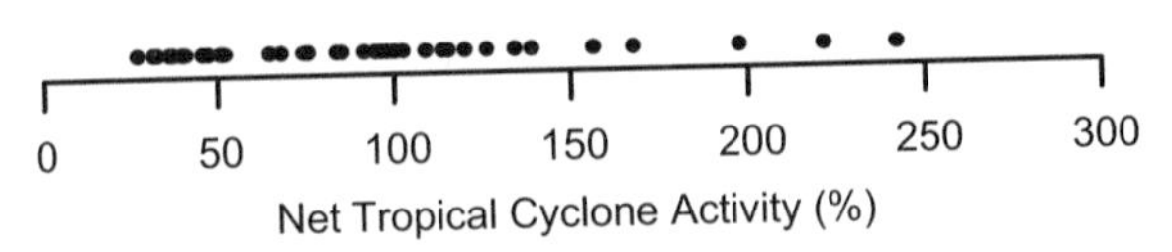

FIGURE 10-4. Atlantic region tropical cyclone activity.

Traditional Approaches

Of course, traditional methods can be employed. For example, one could proceed parametrically by assuming X is normally distributed and use the sample to estimate its mean and standard deviation. Because these are the only parameters of $f(x)$, this is tantamount to estimating $f(x)$ itself. If we use $\bar{x}$ and s as estimates, we get

$$\hat{f}(x) = \frac{1}{s\sqrt{2\pi}} \exp\left[-\frac{1}{2}\left(\frac{x-\bar{x}}{s}\right)^2\right] \tag{10-13}$$

One could evaluate this function to produce a graph, or integrate the function to assign probabilities. (Tables of the standard normal could be used for the latter purpose.) Though very easy to apply, the problem with this approach is that if $f(x)$ is far from normal, the graph and associated probabilities will be misleading. Using the cyclone data as an example, the normal curve gives the probability of an extremely heavy year, $P(X > 200)$, of 0.02. With 3 of the 41 values near or above 200 (more than 7% of the sample), the normal estimate seems too low.

Another option is to proceed nonparametrically by using the sample to construct a relative frequency histogram as in Figure 2-5. If each histogram value is divided by the interval width, we get a probability density estimate, one for each interval. Though no assumptions have been made about the form of $f(x)$, we do have rather arbitrary decisions about the number and position of class intervals. Depending on the data, small changes can have large effects, both in visual appearance and computed probabilities. As mentioned in Chapter 2, another disadvantage is the roughness of the histogram, with the result that $\hat{f}(x)$ changes abruptly at the class limits.

Kernel Methods

Kernel estimation is a simple approach that avoids the difficulties we have discussed. The method provides smooth estimates without requiring explicit assumptions about the form of $f(x)$. Recall that in the histogram approach, each observation is treated as a point. If the point lies in a probability interval, the entire mass of the point, $1/n$, is added to the probability. Kernel methods, on the other hand, spread the mass of each observation around the observed value. The amount of spread or smearing is governed by a function called the kernel. Normally, the kernel function will be symmetric, so that the point is smeared equally toward higher and lower values. The width of the kernel is adjustable—narrow kernels concentrate mass tightly around central (observed) values, whereas a wide kernel gives more smearing. To estimate probability density, we add the kernel values for all the data points. This operation is shown in Figure 10-5,

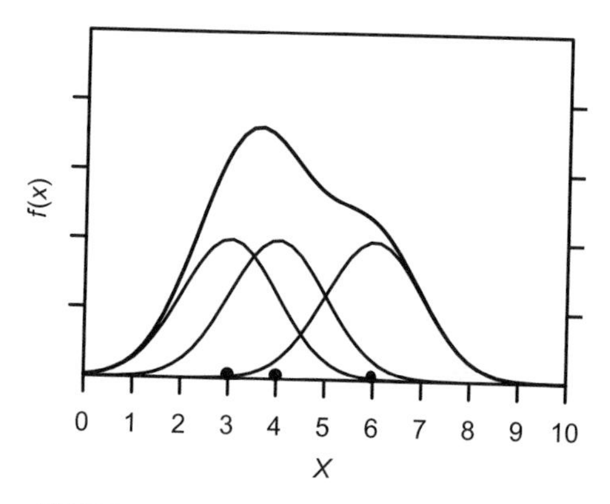

FIGURE 10-5. Kernel probability density.

where the dataset consists of just three points, located at $X = 3$, 4,and 6. Centered over each data point is its kernel function, in this case a standard normal curve. That is, we have placed a Z-distribution over each data point; hence we are using a kernel width of unity. The top curve is the density estimate, $\hat{f}(x)$, obtained by summing the three kernels for 100 different x-values. Notice that at intermediate values of X all three kernels contribute substantially to $\hat{f}(x)$, but the tails are determined by just the two extreme points. To express the operation mathematically, let $K(\bullet)$ represent the kernel function and h the degree of smearing, or bandwidth. With this notation, the kernel density estimate is simply

$$\hat{f}(x) = \frac{1}{nh} \sum_{i=1}^{n} K\left(\frac{x - x_i}{h}\right) \tag{10-14}$$

Looking first at the argument to K, we see that the distance between a point x and each data value is expressed as a multiple of h. Because h has the same units as x, the ratio is dimensionless. This scaled distance is used in the kernel function to find the contribution of each observation. With K a decreasing function of distance, the contribution of an observation far from x will be small, whereas nearby observations will contribute more. The sum of all contributions divided by nh becomes the density estimate. Division by h is required to make a density, with dimensions of probability per unit X.

For example, with K the normal curve and $h = 1$, the density estimate at $x = 5$ based on the three sample points is:

$$\hat{f}(5) = \frac{1}{3(1)}\left[K\left(\frac{5-3}{1}\right) + K\left(\frac{5-4}{1}\right) + K\left(\frac{5-6}{1}\right)\right]$$

$$= \frac{1}{3}[K(2) + K(1) + K(-1)]$$

$$= \frac{1}{3}(0.054 + 0.242 + 0.242) = 0.179$$

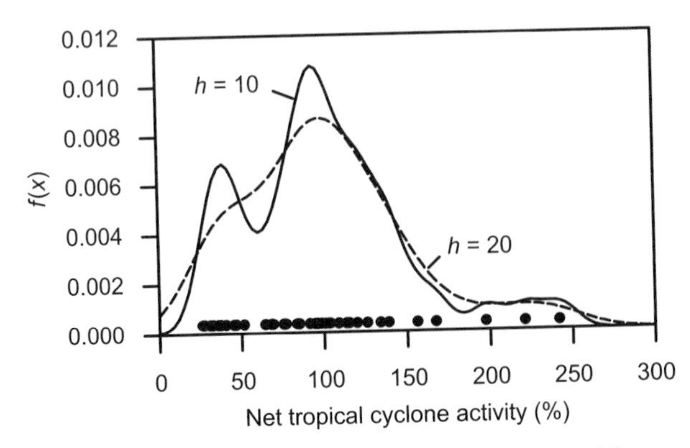

FIGURE 10-6. Kernel probability density estimated for cyclone data.

Because x_1 is farther from $x = 5$, it contributes less than x_2 and x_3 to the density estimate (0.054 vs. 0.242). The other two points share equally in $\hat{f}(5)$ because they are equally close to $x = 5$.

Several general points can be made about this approach. First, $\hat{f}(x)$ inherits many of its properties from the kernel function. In particular, if K is a probability function, $\hat{f}(x)$ will likewise be a probability function (non-negative everywhere, with total area equal to unity). Similarly, if K is smooth, $\hat{f}(x)$ will also be smooth. A second general point is that the amount of detail present in $\hat{f}(x)$ varies directly with the bandwidth h. A large value leads to much smearing of each data point, which in turn results in a broad, smooth density function without much fine structure. This behavior is seen in Figure 10-6, which shows kernel estimates for the cyclone data.

Using the wider kernel ($h = 20\%$) results in a very smooth curve without the density minimum near 60% activity. Notice that both curves reflect the positive skew in the data, illustrating the general principle that a symmetric kernel does not force symmetry in $\hat{f}(x)$. In other words, symmetry is one attribute *not* inherited from the kernel. In this example, the presence of skewness confirms our suspicion regarding the inadequacy of the normal curve. Using either curve, we get a value of 0.06 for $P(X > 200)$, about three times larger than the normal estimate. As a final general point about the kernel method, notice that each density estimate requires n evaluations of K. Given that a large number of estimates might be needed to produce a graph or calculate a probability, this method is obviously not suited to hand calculation. For example, the curves in Figure 10-6 needed 4,100 evaluations of K (100 estimates $\hat{f}(x)$, each based on the 41 data points).

CHOICE OF KERNEL AND BANDWIDTH

Having described the method in general terms, we now need to discuss several practical matters, foremost of which is the choice of kernel function. There are no hard rules about this, though generally one will choose a symmetric probability distribution.

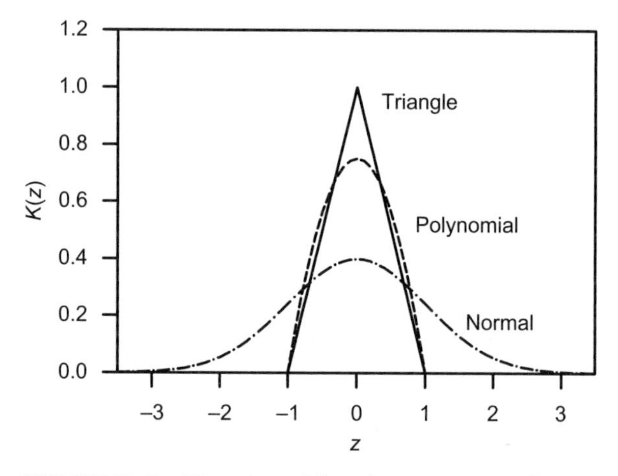

FIGURE 10-7. Three kernel functions, corresponding to Equations 10-15, 10-16, and 10-17. The bandwidth (h) is unity for all curves.

Within the family of symmetric kernels, selection will often be made by balancing computational demands against smoothness. We present three alternatives in Figure 10-7, each of which occupies a different point along the trade-off continuum.

Perhaps the simplest choice is a triangular kernel given by

$$K(z) = \begin{cases} 1 - |z|, & |z| \leq 1 \\ 0, & |z| > 1 \end{cases} \tag{10-15}$$

where

$$z = \left(\frac{x - x_i}{h}\right)$$

This kernel is just a triangle one unit high and two units wide, and therefore has an area of unity. Its chief advantage is ease of evaluation; only division and subtraction are required, and there is no need to consider data points more than h units away from any x. On the other hand, it does not provide as smooth an estimate as do other kernels. (Although the kernel is continuous, it has breaks in slope at the center and edges. These carry over to slope discontinuities in $\hat{f}$). Roughness can be reduced by using a higher degree function in z. For example, a second-degree polynomial will avoid the central slope discontinuity:

$$K(z) = \begin{cases} \frac{3}{4}(1 - z^2), & |z| \leq 1 \\ 0, & |z| > 1 \end{cases} \tag{10-16}$$

Now K varies smoothly across its midpoint; hence $\hat{f}$ is smoother. Once again K becomes zero a finite distance away from the observations, and so only nearby observations need be considered. There are, however, slope discontinuities at the edges of this kernel. Still smoother estimates are produced using the normal kernel:

$$K(z) = \frac{1}{\sqrt{2\pi}} \exp\left(-\frac{1}{2}z^2\right) \tag{10-17}$$

This yields a very smooth density function, but at the cost of more arithmetic. There is the added burden of an exponential, and K is technically nonzero everywhere. As a practical matter, however, because the normal curve approaches zero so quickly, an observation more than a few multiples of h distant cannot contribute much to $\hat{f}$. The calculations can therefore be restricted to nearby points if necessary.

Having selected a kernel function, there remains the question of bandwidth, h. The general principles are easy to state: too large a value prevents small-scale variations in f from emerging in $\hat{f}$, whereas too small a value gives estimates that are too variable. For example, looking at the cyclone data, it may be that the local minimum in $\hat{f}$ is fictitious and would not appear in a larger sample. If so, the lower value of h is too small, because it invents a signal where none exists. In that case, the 20% bandwidth would be preferred. On the other hand, an alternative interpretation is that the minimum is real, and the larger bandwidth obscures an important feature of cyclone frequencies! The lesson learned from this example is that one should experiment with a range of bandwidths, comparing the resulting density estimates with one another. Results that turn out to be sensitive to h should be reported only with qualification.

This said, how does one choose reasonable values of h for experimentation? Intuition suggests that if the data are highly dispersed in X, a large bandwidth is called for. Intuition also suggests that because large samples contain more information, bandwidth should decrease with increasing n, everything else being equal. It so happens that theory and experience confirm these notions. As a first step in choosing the optimum, Silverman (1986, p. 48) recommends first calculating the sample standard deviation s and the interquartile range R. His suggested bandwidth is then the smaller of the following two values:

$$\begin{aligned} h_1 &= 0.9\ s/n^{1/5} \\ h^2 &= 0.67\ R/n^{1/5} \end{aligned} \tag{10-18}$$

Notice because n is raised to a fractional exponent, dispersion is much more important than sample size. If we think of h as a measure of how much information can be extracted from the sample, we see that a very large n is needed to compensate for widely spaced data points. In the case of the cyclone data, h_1 and h_2 are 21% and 18%, respectively.

The equations above provide an easy way to choose h for any dataset and give values that work well for a wide range of density functions. However, they depend only on bulk sample statistics (s, R), not on the details of how the observations are arranged. At the expense of more calculations, two refinements are possible that make greater use of sample information. First, it is possible to optimize the choice of bandwidth by selecting a value that minimizes predictive error in $\hat{f}$. The technique is called cross-validation and uses the data themselves to identify the optimum h. The basic idea is that a good choice for h will give a good estimate for $\hat{f}$. Cross-validation uses the data in hand to evaluate how good is a particular h. That value of h that gives the lowest estimated error in $\hat{f}$ is selected. Cross-validation can be used in many other contexts as well, as has been seen elsewhere in the text.

A second refinement is to employ a so-called adaptive kernel. In this case, the bandwidth is allowed to vary with X; small values are used where the data are closely packed, and large values where observations are sparse. In this way the kernel adapts to local variations in density, providing high resolution where possible without creating spurious features in low-density regions of X. The approach requires two passes through the data, the first of which uses a fixed kernel, giving preliminary density estimates. These pilot estimates are then used to determine a unique bandwidth for each observation. Though considerably more work, adaptive estimates are especially useful if there are large-density variations in the data. Details can be found in the Silverman monograph.

KERNEL METHODS FOR BIVARIATE DATA

Thus far we have been using kernel methods to estimate a function $\hat{f}(x)$, where the observations lie on the number line. The domain of f is obviously one-dimensional—location within the domain is determined by a single coordinate value. Here we consider generalizing the kernel method for use on bivariate data. The domain now is now two-dimensional, and we are after a bivariate function $f(X, Y)$ in which the data are points lying in a plane.

Instead of considering probability density functions, we will first describe how the method works for estimating point density. That is, we are given a set of points in space, and we want to know how the density of points varies throughout the study area. In Chapter 2 we mentioned that Thiessen polygons can be used, but they are not very satisfactory as they give density estimates only at the data points. Kernel methods are attractive because they can provide estimates at any location. In addition, if a smooth kernel is used, the resulting density surface will be smooth. Let d_i be the distance between data point i and some location (x, y):

$$d_i = \sqrt{(x - x_i)^2 + (y - y_i)^2}$$

As before, we need to choose a kernel width h and use distances scaled by h in the kernel. For example, the normal kernel is

$$K\left(\frac{d_i}{h}\right) = \frac{1}{2\pi} \exp\left[-\frac{1}{2}\left(\frac{d_i}{h}\right)^2\right]$$

With this notation, the density estimate is simply

$$\hat{g}(x, y) = \frac{1}{h^2} \sum_{i=1}^{n} K\left(\frac{d_i}{h}\right) \tag{10-19}$$

We have used $\hat{g}$ rather than $\hat{f}$ to indicate that the equation provides point density, not probability density. Point density is appropriate if we want information about spatial variations in the density of observations. For example, we might be studying earthquake epicenters and want to map epicenter density. In that case we would very likely contour the density function. As another example, perhaps we are mapping some climatic variable on the basis of values observed at a network of weather stations. The density of stations plays a key role in determining the resolving power of the network. Hence we will likely need a map of station density in order to know how well the network is going to perform. Once again the kernel method can provide an estimate.

Notice how we divide by h^2 in Equation 10-19. Because h has units of length (miles, kilometers, etc.), the density estimate $\hat{g}$ will have units of points per unit area (e.g., points per square mile, points per square kilometer). Equation 10-19 can be evaluated for any (x, y) location; thus the method provides a continuous density surface.

EXAMPLE 10-11. Figure 10-8a shows a sample of 200 points within a 100 km^2 area. Visual inspection reveals that the northeast is sparsely sampled but that the southwest has relatively high point density. To supplement this impression, we seek numerical information about the variations in sampling density.

Though we could divide the region into cells and count points within each cell, to do so would involve all the disadvantages associated with histograms. We prefer a method that gives a smooth density estimate; thus we will employ the bivariate kernel method. We choose the bandwidth h to be 1 km, or 10% of the range of x and y coordinates. As already mentioned, this will constrain the detail present in the resulting density surface. Given the rather large data gaps in the northeastern part of the study area, a smaller value does not seem justified. Kernel functions are placed over all 200 points and summed at a large number of interior locations. When the resulting values are contoured, Figure 10-8b is the result.

The contour map clearly shows the density maximum in the southwest, along with a secondary maximum to the north of the primary maximum. If we had to rely solely on the point data (Figure 10-8a), we would probably miss the secondary maximum. Note that with 200 points in a 100 km^2 area, the mean density is 2 points/km^2. Looking at the map, we see that below-average densities are confined to the eastern and extreme northern portions of the study area. Notice that density appears to decrease rapidly near the southern and western edges of the map. This is undoubtedly an edge effect, produced by not sampling outside the map boundaries. Had we done so, those points would contribute to the density estimate within the boundaries, so that estimates near the edges would be larger. Because the normal curve falls off rapidly, there is very little contribution from points farther away than $2h$. In other words, we might have wished for points in a band of width $2h$ surrounding the study area, but no more

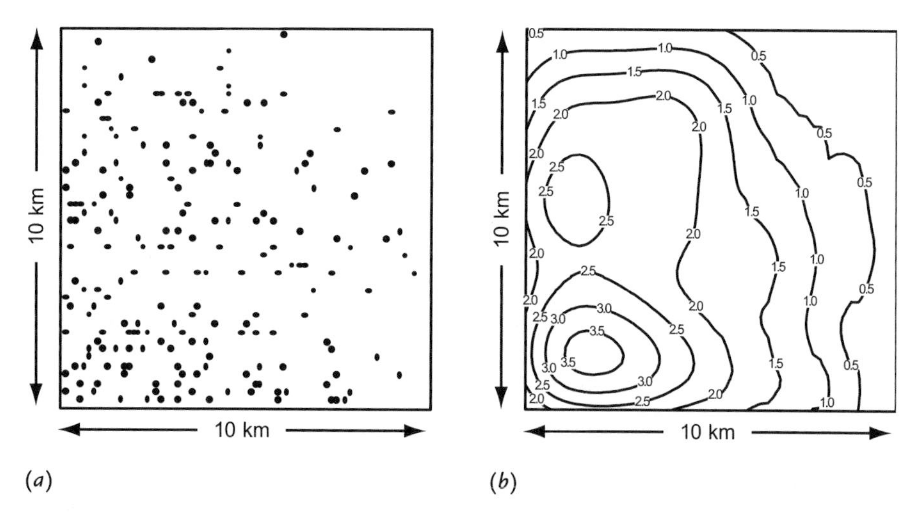

FIGURE 10-8. (a) Sample of 200 points; (b) density contours for (a) in points/km².

than that. By the same logic, edge effects extend no more than $2h$ units into the study area. (We do not see an edge effect along the eastern boundary, because densities are so low that missing points have almost nothing to contribute.) As a final point, we remark that the 0.5 contour has small "wiggles" here and there, smaller even than the bandwidth of 0.1 km. These wiggles are artifacts of the method, an indication that h is too small for these low-density regions. We could use a larger bandwidth but would lose detail in the high-density areas. A better choice is an adaptive kernel method, which automatically adjusts to variations in density, as described earlier.

A related problem is that of estimating a bivariate probability density function. The domain is still two-dimensional, but this time x and y are not necessarily spatial variables. More importantly, the function we seek, $\hat{f}(x, y)$, is a surface that gives probability density, not point density. To convert point density to probability density, we need only divide by the number of data points:

$$\hat{f}(x, y) = \frac{1}{n}\hat{g}(x, y) = \frac{1}{nh^2}\sum_{i=1}^{n} K\left(\frac{d_i}{h}\right)$$

As long as the kernel function is a probability distribution, the estimate $\hat{f}$ will be a probability distribution. If x and y have different units, it will be necessary to standardize them in order for the distance computation to have meaning. A natural choice is to express both as z-scores when computing d. Doing so lets h be interpreted as the width of the kernel in standard deviation units, rather than in raw units like miles or kilometers. In all other regards the bivariate method is a direct extension of the one-dimensional technique.

10.7. Bootstrapping

We will now consider *bootstrapping,* a very general technique for obtaining information about the distribution of a sample statistic. As you know, knowledge of sampling distributions is necessary for hypothesis testing and establishing confidence intervals, which are the essence of inferential statistics. All of the techniques discussed thus far have used statistical theory to obtain sampling distributions. Note that this is true for both parametric and nonparametric methods described above. In the latter case we do not specify a probability distribution for the random variable being studied, but we certainly use a sampling distribution in performing nonparametric tests. For example, in the sign test for the median, we relied on theory to obtain the sampling distribution of the test statistic (the number of positive signs). You might have noticed that, although we presented a test for the median, we did not offer a way to set up a confidence interval for η. The reason is that without knowing the probability distribution of X, we can't know the sampling distribution of the sample median.

Bootstrapping is a way around this difficulty. The bootstrap is very flexible and can be applied to a wide variety of problems. We will restrict our discussion to its use in (1) estimating the sampling distribution of a sample statistic and (2) obtaining confidence intervals.

Bootstrap Samples

Following the notation of Chapter 7, we will use $\hat{\theta}$ to denote a sample estimate of some unknown population characteristic θ. As before, inferential statistics requires that we have information about how $\hat{\theta}$ varies from sample to sample. Bootstrapping is based on the simple idea that if we had a large number of samples, we could obtain information about sampling variability by observing sample-to-sample differences. In actual practice only one sample is available, of course; thus the bootstrap is forced to create these samples from the original sample. These artificial samples are called *bootstrap samples,* each of which is obtained by sampling with replacement from the original observations. Variability across the bootstrap samples serves as a substitute for variability across truly independent (but unavailable) samples.

Figure 10-9 shows the bootstrap procedure for a sample of size 4, much smaller than would be used in actual practice. The original sample is resampled B times, giving B bootstrap samples of size $n = 4$. To construct a bootstrap sample, we generate uniform random numbers between 1 and n inclusive. The random numbers are used to choose which observations appear in the bootstrap sample. Using a uniform random number generator ensures that all observations have an equal probability of appearing in the sample. Because the random numbers are uncorrelated, an observation can appear more than once in the sample and is no less likely to appear again once it has been chosen. Obviously, this procedure amounts to sampling with replacement.

For this example with four observations, there are only 256 different bootstrap samples that might be drawn. However, the number of different bootstrap samples grows rapidly with sample size (see Problem 17); thus for even moderate n the range of outcomes is huge. We calculate $\hat{\theta}$ for each bootstrap sample, and use the resulting

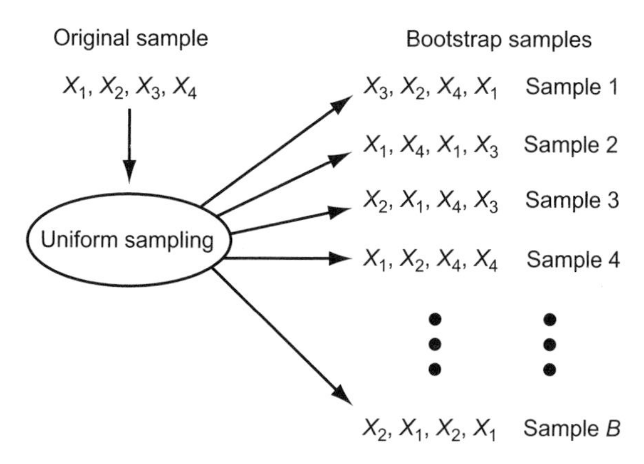

FIGURE 10-9. The bootstrap procedure for a sample of size 4.

distribution of $\hat{\theta}$ as if it had come from the population rather than from a single sample. We are pretending that the B bootstrap samples can be used instead of B real samples. The rationale behind this seemingly extreme approach is actually quite straightforward. Obviously, the probability distribution of $\hat{\theta}$ depends on the unknown probability distribution $f(x)$. The bootstrap replaces $f(x)$ with an empirical estimate developed from the sample. In particular, it uses an estimate $\hat{f}(x)$ hat assumes all observations are equally likely, as is consistent with random sampling. After all, if every observation has an equal chance of appearing in a bootstrap sample, we are saying that there is nothing special about any particular observation that would cause us to weight it more heavily than another. We emphasize that assuming all *observations* are equally likely is not at all the same as assuming all *values* of X are equally likely. Rather, the distribution of values appearing in a bootstrap sample reflects the distribution of values in the original sample. If the sample is representative of the population, the bootstrap samples will vary one from another much like real samples.

At first thought, the idea of sampling with replacement might seem wrong, as it permits a single observation to appear more than once in a bootstrap sample. However, a particular observation is no more likely than any other to appear multiple times; thus no inherent bias is introduced. Although a single bootstrap sample might contain some observation an usually large number of times, many bootstrap samples are used, not just one. It would be very unlikely indeed for the same observation to repeatedly appear a disproportionate number of times in the ensemble of bootstrap samples. As a final point about sampling with replacement, notice that if one samples without replacement, only one outcome is possible, and, except for order, all bootstrap samples will be identical to the original sample!

Bootstrap Estimate of a Sampling Distribution

Suppose we have a single sample giving a single estimate $\hat{\theta}$ and would like to know something about the probability distribution of $\hat{\theta}$. For example, we might want to know the chances of obtaining a value twice as large, or half as large, or perhaps we want to know something about the shape of the sampling distribution. We seek the probability distribution of $\hat{\theta}$, denoted $f(\hat{\theta})$. Notice that this is different from estimating $f(x)$, the probability distribution of the random variable underlying $\hat{\theta}$. The procedure to estimate $f(\hat{\theta})$ is simple:

1. Draw a bootstrap sample of size n from the original observations and compute $\hat{\theta}_1^*$, the sample statistic for bootstrap sample number 1. Repeat for a total of B samples, giving the B bootstrap values $\hat{\theta}_i^*$, $i = 1, 2, \ldots, B$.
2. Use the relative frequencies of $\hat{\theta}^*$ to assign probability density estimates for the sampling distribution $f(\hat{\theta})$.

In step 2, we might calculate the relative frequency over some range of $\hat{\theta}^*$, or build a histogram of $\hat{\theta}^*$. The number of bootstrap samples needed (B) depends on how $\hat{f}(\hat{\theta})$ is used. For example, estimating a probability near the tail of the distribution demands more samples than creating a rough histogram. Though there are no hard rules, experience shows that the number of samples needed is at most on the order of thousands. If there is doubt about whether or not B is large enough, one can always increase B and note any change in the results.

EXAMPLE 10-12. El-Niño/Southern Oscillation Events (atmosphere–ocean changes in the tropical Pacific) are thought to influence climatic conditions at far-away locations. Figure 10-10 shows the relationship between rainfall in sub-Saharan Africa and the southern oscillation index, a measure of surface pressure differences between Darwin, Australia, and Tahiti. The scatterplot indicates that high-index years are associated with higher than normal rainfall, the correlation coefficient (r) being 0.451. We have a point estimate r and want to know something about its sampling variability. The problem is to estimate the probability distribution of r, $f(r)$.

Solution. The classical method is Fisher's Z-transformation, but this requires an assumption that both variables are normally distributed. Because the rainfall values are positively skewed, we will use a bootstrap instead of the Z-transform.

There are 41 sample pairs (x_i, y_i), corresponding to the years 1950–1990. The first step is to construct bootstrap samples of size 41, where each member is an (x, y) pair for one of the years. In this example, a total of 5,000 samples were created, each giving a correlation coefficient. The bootstrap correlations range from a low of –0.027 to a maximum value of 0.750. Dividing this range into 50 bins gives the histogram shown in Figure 10-11. The histogram shows the frequency of occurrence bin by bin—to assign a probability, the count for a bin would be divided by the number of bootstrap samples B. Notice that the histogram is not perfectly symmetric, but rather is negatively skewed. This is in qualitative agreement with normal theory using Fisher's transformation, but there is less skew evident in the bootstrap histogram. Notice that the bootstrap distribution is more tightly clustered around the sample estimate $\hat{\theta}$, suggesting that the variance in $\hat{\theta}$ is smaller than what is indicated by normal theory.

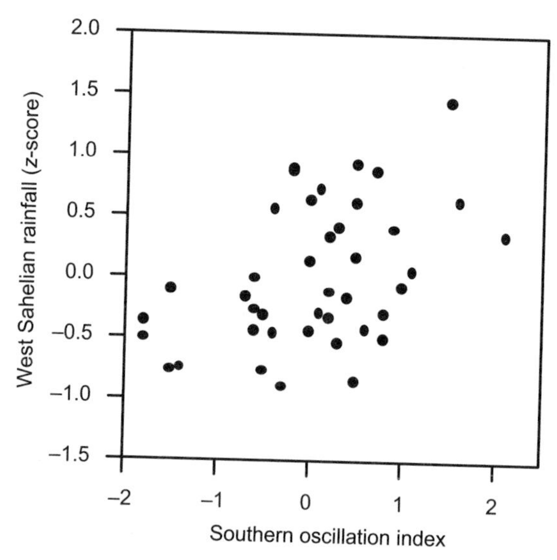

FIGURE 10-10. Scatter plot of Sahelian rainfall versus the southern oscillation index.

Bootstrap Confidence Intervals

A very important use of the bootstrap is to construct confidence intervals for θ. When we faced this problem in a parametric context, we relied on theory regarding the distribution of the sample statistic. By knowing the sampling distribution $f(\hat{\theta})$, we could find an interval likely to contain θ at any desired level of confidence. The bootstrap accomplishes the same thing but uses the bootstrap distribution instead of a theoretical sampling distribution. It is therefore a nonparametric method based on an *empirical* sampling distribution. We will describe two variants—the percentile and bias-corrected percentile method. Both versions require thousands of samples and should probably not be used with B less than 1,000.

PERCENTILE METHOD

Given a set of bootstrap values, this method is trivially easy to carry out. For example, to construct a 90% confidence interval, we simply use the central 90% of the bootstrap distribution. That is, the lower limit will be the 5th percentile value of $\hat{\theta}^*$ and the upper limit will be the 95th percentile. Similarly, the 95% confidence interval is based on the 2.5th and 97.5th percentiles.

Recall that the 90% confidence level corresponds to an α level of 0.10, and a 95% confidence level corresponds to an α level of 0.05. In other words, the confidence *level* is $1 - \alpha$. To find the confidence *interval,* we look for the lower and upper $\alpha/2$ percentiles of $\hat{\theta}^*$. This is easily done by sorting the bootstrap values in ascending order. Given any integer j, the proportion of bootstrap values less than or equal to

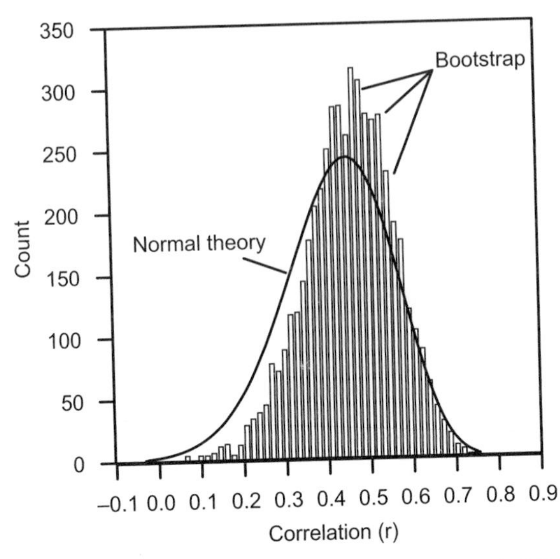

FIGURE 10-11. Bootstrap distribution for the rainfall–southern oscillation index correlation.

the *j*th element in the sorted list is j/B. For example, using 2,000 samples, 5% of the values are less than or equal to the 100th element in the list. More generally, the method is:

1. Choose a confidence level $1 - \alpha$.
2. Draw a bootstrap sample of size n from the original observations and compute $\hat{\theta}_1^*$, the sample statistic for bootstrap sample number 1. Repeat for a total of B samples, giving the bootstrap values $\hat{\theta}_i^*$, $i = 1, 2, \ldots, B$.
3. Sort the bootstrap samples in ascending order of $\hat{\theta}^*$.
4. Use the lower $\alpha/2$ percentile as the lower limit of the confidence interval θ_{Low}. This can be found as the *j*th element in the sorted list, where $j = B\alpha/2$.
5. Use the upper $\alpha/2$ percentile as the upper limit of the confidence interval θ_{High}. This can be found from the *k*th element in the sorted list, where $k = B(1 - \alpha/2)$.

EXAMPLE 10-13. Returning to the travel times of Example 10-1, the sample suggests a skewed distribution for the population. The sample distribution in Figure 10-12 shows tight clustering of values below the sample median of 32 and a long tail to the right indicating some very bad travel days. We would like to construct a 90% confidence interval for the population median.

Solution. We draw 10,000 bootstrap samples of size 20 from the original sample. Each bootstrap sample yields a bootstrap median $\hat{\theta}^*$. After sorting the medians, we search for the 500th value as the 5th percentile, and the 9,500th as the 95th percentile of the bootstrap distribution. The 5th and 95th percentiles are 28 and 38, respectively;

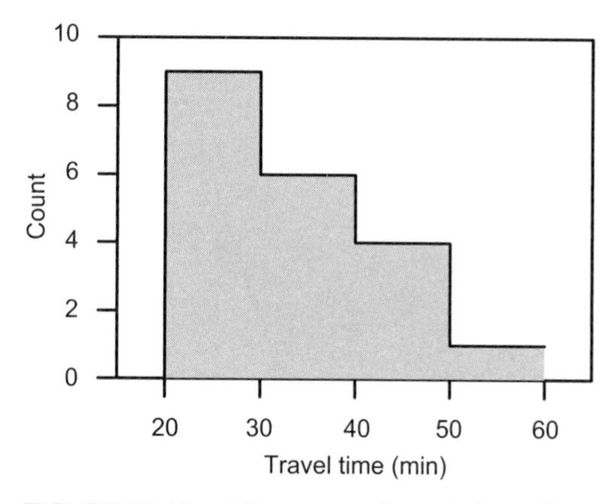

FIGURE 10-12. Histogram of travel times from Example 10-1.

thus the 90% confidence interval runs from 28 minutes to 38 minutes. Note that this interval is *not* symmetric around the sample median (32 minutes). This result is consistent with the positive skewness seen in Figure 10-12.

EXAMPLE 10-14. Use the bootstrap sampling distribution from Example 10-10 to find a 90% confidence interval for the correlation between sub-Saharan rainfall and the southern oscillation index.

Solution. With 5,000 samples, the 5th percentile is at position 250 (5% of 5,000). The 95th percentile is at position 4,750. After sorting the bootstrap values, we obtain $\theta_{\text{Low}} = \hat{\theta}^*_{250} = 0.262$ and $\theta_{\text{High}} = \hat{\theta}^*_{4750} = 0.610$. The interval given by normal theory is wider, running from 0.216 to 0.637.

The percentile method is even easier to describe by considering the cumulative distribution of the bootstrap samples. Recall that the cumulative distribution $F(x)$ is simply the probability that X takes on a value less than or equal to x. That is, $F(x) = P(X \leq x)$. Given a probability distribution $f(x)$, there is obviously a corresponding cumulative distribution $F(x)$. Similarly, given the bootstrap distribution $\hat{f}(\hat{\theta}^*)$, we can define the corresponding cumulative bootstrap distribution $\hat{F}(\theta^*)$. This function gives the proportion of bootstrap samples less than or equal to any value of $\hat{\theta}^*$.

For example, Figure 10-13 shows the cumulative distribution for bootstrap correlations plotted in Figure 10-11. As shown in the figure, if we want an 80% confidence interval, we will look up the values of $\hat{\theta}^*$ corresponding to $\hat{F} = 0.1$ and $\hat{F} = 0.9$. Given a probability, we find the associated $\hat{\theta}^*$. In other words, we need the inverse of $\hat{F}$, which we write as $\hat{F}^{-1}(p)$. Notice that $\hat{F}^{-1}(p)$ gives a value of $\hat{\theta}^*$ for any probability p. With this notation the percentile method is simply:

1. Choose a confidence level $1 - \alpha$.
2. Draw a bootstrap sample of size n from the original observations and compute $\hat{\theta}_1^*$, the sample statistic for bootstrap sample number 1. Repeat for a

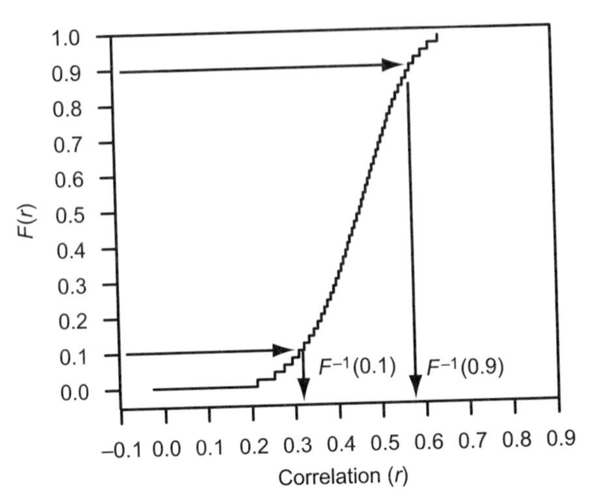

FIGURE 10-13. Cumulative distribution of bootstrap samples.

total of B samples, giving the bootstrap values $\hat{\theta}_i^*$, = 1, 2, . . . , B and the corresponding cumulative distribution $\hat{F}(\theta^*)$.

3. Let $\hat{F}^{-1}(p)$ be the inverse of $\hat{F}$. The lower limit of the confidence interval is $\theta_{\text{Low}} = \hat{F}^{-1}(\alpha/2)$.
4. The upper limit of the confidence interval is $\theta_{\text{High}} = \hat{F}^{-1}\,(1 - \alpha/2)$.

EXAMPLE 10-15. Use $\hat{F}$ to construct an 80% confidence interval for southern oscillation index–rainfall correlation.

Solution. For the lower limit, we want $\hat{\theta}^*$ corresponding the 10% probability (10th percentile). Looking at Figure 10-13, the corresponding correlation is $\theta_{\text{Low}} = \hat{F}^{-1}(0.1) \approx 0.31$. For the upper limit, we need $\hat{F}^{-1}(0.9)$, which is approximately 0.58. Thus our 80% interval runs from 0.31 to 0.58.

BIAS-CORRECTED PERCENTILE METHOD

The percentile method is easy to use, but it is not always unbiased. At the cost of additional complexity, the method can be improved. It can be shown that when the median is being bootstrapped, the bootstrap distribution $f(\hat{\theta}^*)$ is unbiased, or "centered" on the true value. That is, the expected value of the bootstrap median equals the true population median. In the absence of sampling error, we would have the estimate $\hat{\theta}$ equal to the true median, and we would also have the median of the bootstrap distribution ($\hat{\theta}^*_{B/2}$) equal to the true median. In other words, without sampling error, $\theta = \hat{\theta} = \hat{\theta}^*_{B/2}$. Of course, when we generate a particular bootstrap distribution from a real sample, its 50th percentile will not equal $\hat{\theta}$. Knowing that the difference between $\hat{\theta}$ and $\hat{\theta}^*_{B/2}$ arises from sampling error, we are led to a bias correction based on the difference. The theory behind the correction is beyond an introductory text, but correction itself is easily accomplished.

The adjustment for sample bias does not involve any assumptions of normality, but it does require that we use the normal distribution function. Let $\varphi(z)$ be the area to the left of z under the normal curve. That is, $\varphi(z)$ gives $P(Z < z)$, where Z is a standard normal variable. This is recognizable as the cumulative normal curve and should be familiar. Note that column 2 of Appendix Table A-3 gives these probabilities; hence it is actually a table of φ. The inverse of $\varphi(z)$ is denoted $\varphi^{-1}(p)$, and gives a z-value corresponding to a probability value p. The inverse function is like using Table A-3 in reverse; given a probability value in column 2, we find the corresponding z from column 1. Notice that $z_\alpha = \varphi^{-1}(\alpha)$.

With this as background, we can describe the bias-corrected percentile method:

1. Choose a confidence level $1 - \alpha$. Find the lower and upper z-values corresponding to $\alpha/2$. That is, find $z_{\alpha/2} = \varphi^{-1}(\alpha/2)$ and $z_{1-\alpha/2} = \varphi^{-1}(1 - \alpha/2)$.
2. Draw a bootstrap sample of size B from the original observations and compute $\hat{\theta}_1^*$, the sample statistic for bootstrap sample number 1. Repeat for a total of B samples, giving the bootstrap values $\hat{\theta}_i^*$, $i = 1, 2, \ldots, B$ and the corresponding cumulative distribution $\hat{F}(\theta^*)$.
3. Compute $\hat{\theta}$ from the original sample and find its percentile in the bootstrap distribution, $\hat{F}(\hat{\theta}^*)$. Call this p_0, and find $z_0 = \varphi^{-1}(p_0)$.
4. Compute two probabilities, $p_{\text{Low}} = \varphi(z_{\alpha/2} + 2z_0)$ and $p_{\text{High}} = \varphi(z_{1-\alpha/2} + 2z_0)$. Use $\theta_{\text{Low}} = \hat{F}^{-1}(p_{\text{Low}})$ and $\theta_{\text{High}} = \hat{F}^{-1}(p_{\text{High}})$ as the lower and upper limits of the confidence interval.

This looks complicated at first glance but is actually not too different from the percentile method. Before discussion, we present a numerical example. Continuing with the correlation example,

1. We choose a 90% confidence level, giving $\alpha = 0.1$. The corresponding 5% and 95% z-values are -1.645 and 1.645, respectively.
2. We use the 5,000 bootstrap samples from Example 10-11 whose cumulative distribution appears in Figure 10-13.
3. The sample correlation $\hat{\theta}$ is 0.451, giving $p_0 = \hat{F}^{-1}(0.451) = .486$. That is, the sample estimate is approximately the 49th percentile of the bootstrap distribution. Converting this to a z-value yields $z_0 = \varphi^{-1}(.486) = -.033$.
4. With $2z_0 = -.066$, we find the two probabilities

$$p_{\text{Low}} = \varphi(-1.645 - .066) = .044 \quad p_{\text{High}} = \varphi(1.645 + .066) = .934$$

and the confidence limits

$$\theta_{\text{Low}} = \hat{F}^{-1}(.044) = .256 \quad \theta_{\text{High}} = \hat{F}^{-1}(.934) = .606$$

Notice that in the last step p_{Low} and p_{High} call for percentiles slightly lower than the 5th and 95th percentiles of $\hat{\theta}^*$. In this example, bias correction reduces θ_{Low} and θ_{High}

by almost the same amount, effectively shifting the confidence interval to the left. The change is not large, however, because the bootstrap distribution is so nearly centered on $\hat{\theta}$. In other situations, bias correction can lead to a significant change from the simpler percentile method.

Although the bias-corrected method seems much more complicated, in fact z_0 is the only term not present in the simple percentile method. Moreover, the two methods are identical when bias correction is not needed. To see this, suppose that the bootstrap median happens to equal the sample value $\hat{\theta}$. In this case $\hat{\theta}$ is obviously at the 50th percentile of the bootstrap distribution, and p_0 is.5. The corresponding z-value (z_0) is zero. Under these circumstances step 4 above reduces to

$$\theta_{\text{Low}} = \hat{F}^{-1}(p_{\text{Low}}) = \hat{F}^{-1}(\varphi(z_{\alpha/2})) = \hat{F}^{-1}(\alpha/2)$$

and

$$\theta_{\text{High}} = \hat{F}^{-1}(p_{\text{High}}) = \hat{F}^{-1}(\varphi(z_{1\text{-}\alpha/2})) = \hat{F}^{-1}(1 - \alpha/2)$$

The right-hand members of the equations above call for the lower and upper $\alpha/2$ tails of the bootstrap distribution, which is exactly the same as the percentile method. We see that bias correction only appears when the bootstrap distribution is not centered on $\hat{\theta}$, as is consistent with the motivation for the correction. Because of the way it adjusts to the bias apparent in the data, we think of the bias-corrected method as an adaptive technique.

It should be mentioned that both the percentile and bias-corrected percentile methods assume the existence of some function that transforms $\hat{\theta}$ to a normal variable. We do not have to know the form of the function, only that it exists. Actually, reliance on these so-called normal pivotal functions is not unique to bootstrapping. For example, the Fisher Z transformation is a pivotal function for bivariate normal variables. So far as the bootstrap is concerned, we prefer the bias-corrected method to the percentile method because it admits a larger class of pivotal functions and thus applies in a larger number of situations. There are, however, some instances for which no appropriate pivotal function exists, and neither the percentile nor bias-corrected intervals will be correct. Unfortunately, these situations are likely to be complicated, offering little hope of verifying the assumption regarding a pivotal quantity. This is a little like having to assume bivariate normality but not having data available to check on the assumption. However, in the case of bootstrap confidence intervals, the assumption we make is much less restrictive than bivariate normality. Our point, therefore, is not that bootstrap confidence intervals are unreliable, but rather that they are not foolproof.

Discussion

It should be clear that the bootstrap is a very general method and can be used in many contexts other than those described here. For example, we could regress Y on X using some method other than least squares, and we could bootstrap for confidence intervals of the slope and intercept. Common to all applications is the requirement that the values being bootstrapped form a random sample. It is worth repeating that this means the observations must be identically distributed and uncorrelated. This restriction does rule out a number of situations where one might be tempted to bootstrap. For example, if the data exhibit systematic temporal variation, one cannot bootstrap a time series directly. If values in one year are related to those of nearby years, we want bootstrap samples that preserve the temporal correlation. Simple random sampling of the original data will not suffice because it generates uncorrelated sequences. Whereas the real world might not admit a very high value followed by a low value, the bootstrap sample would contain no similar restriction and would therefore be a poor surrogate for repeated sampling of the real world.

Note that this requirement for a random sample is not unusual, and in fact random sampling is nearly always assumed. Bootstrapping is therefore no more restricted by this requirement than are the traditional techniques we have described in other chapters. Given its simplicity and wide applicability, it is not surprising that the bootstrap is becoming ever more common in geographic research. It is particularly useful in modeling spatial processes. One has a single expression of the process, and will want to know if the observations are consistent with some hypothesized theoretical model. That is, could the observations have arisen from a hypothesized model? In the same way that classical methods allow one to test simple hypotheses about parameters, the bootstrap provides a way to examine more complicated hypotheses.

10.8. Summary

When standard parametric procedures are not appropriate, nonparametric techniques are an appealing alternative. They are most useful when (a) assumptions required by a parametric technique are untenable, or (b) when the sample level of measurement does not permit application of a parametric technique, or (c) when no parametric technique can be found to address the research question at hand.

Nonparametric methods are quite popular in geographic research, because they are often much easier to apply than parametric tests, and they are applicable to a wide variety of data encountered in geographic research (nominal variables, ordinal survey data, etc.). Nonparametric methods are not without disadvantages, of course. Chief among these is the general rule that nonparametric methods are less powerful than parametric equivalents. Thus they often require a larger sample to achieve the same discriminating ability. For this reason a parametric test is preferred when its assumptions and data requirements are met.

This chapter has considered a variety of nonparametric methods: analogs to one-and two-sample tests presented as the sign test, the Mann–Whitney test, and the number-of-runs test. The chi-square and Kolmogorov–Smirnov goodness-of-fit tests were presented as a way to compare a sample distribution with a hypothesized theoretical distribution. We also covered contingency problems as a special case of a goodness-of-fit test, again based on the χ^2 distribution. The last two sections of the chapter described computer-intensive methods for estimating probability density functions (kernel estimates) and distributions of sample statistics (bootstrapping).

This brief overview has covered only a few of the nonparametric methods used by geographers. As can be seen in the list of references, entire books have been written on nonparametric approaches—readers are advised to consult the books mentioned to learn about additional methods and for details on the techniques presented here.

FURTHER READING

General texts on nonparametric methods include:

W. J. Conover, *Practical Nonparametric Statistics* (New York: Wiley, 2001).

M. Hollander and D. A. Wolfe, *Nonparametric Statistical Methods* (New York: Wiley, 1999).

E. L. Lehman, *Nonparametrics: Statistical Methods Based on Ranks* (New York: Springer, 2006).

L. Wasserman, *All of Nonparametric Statistics* (New York: Springer, 2005).

More specialized works:

P. Barbe and P. Bertai, *The Weighted Bootstrap* (New York: Springer Verlag, 1995).

A. C. Davison and D. V. Hinckley, *Bootstrap Methods and Their Application* (Cambridge, UK: Cambridge University Press, 1997).

P. Diaconis, and B. Efron, "Computer-Intensive Methods in Statistics," *Scientific American* 248 (1983), 110–130.

B. Efron and R. Tibshirani, "Bootstrap Methods for Standard Errors, Confidence Intervals, and Other Measures of Statistical Accuracy," *Statistical Science* 1 (1986), 54–77.

B. Efron, *The Jackknife, the Bootstrap, and Other Resampling Plans* (Philadelphia: Society for Industrial and Applied Mathematics, 1982).

P. Good, *Permutation Tests: A Practical Guide to Resampling Methods for Testing Hypotheses* (New York: Springer Verlag, 1992).

W. M. Gray, C. W. Landsea, P. W. Mielke, and J. J. Berry, "Predicting Attlantic Seasonal Hurricane Activity 6-11 Months in Advance," *Weather and Forecasting* 7 (1992), 440–455.

R. Maronna, D. Martin and V. Yohai, *Robust Statistics: Theory and Methods* (New York: Wiley, 2006).

C. Z. Mooney and R. D. Duv, *Bootstrapping: A Nonparametric Approach to Statistical Inference* (Beverly Hills, CA: Sage, 1994).

P. Rousseeuw and A. M. Leroy, *Robust Regression and Outlier Detection* (New York: Wiley, 2003).
Resampling Stats Inc., *Resampling Stats v. 5.0* (612 N. Jackson St., Arlington, VA).

J. Shao and D. Tu, *The Jackknife and Bootstrap* (New York: Springer Verlag, 1995).

B. W. W Silverman, *Density Estimation for Statistics and Data Analysis* (New York: Chapman and Hall, 1986).

PROBLEMS

1. Explain the meaning of the following:
   - Nonparametric statistics
   - Robustness
   - Goodness-of-fit test
   - Contingency table

2. Briefly compare the advantages and disadvantages of parametric and nonparametric tests.

3. An urban geographer randomly samples 20 new residents of a neighborhood to determine their ratings of local bus service. The scale used it as follows: 0–very dissatisfied, 1–dissatisfied, 2–neutral, 3–satisfied, 4–very satisfied. The 20 responses are 0,4,3,2,2,1,1, 2,1,0,01,2,1,3,4,2,0,4,1. Use the median test to see whether the population median is 2.

4. A course in statistical methods was team-taught by two instructors, Professor Jovita Fontanez and Professor Clarence Old. Professor Fontanzez used many active learning techniques, whereas Old employed traditional methods. As part of the course evaluation, students were asked to indicate their instructor preference. There was reason to think students would prefer Fontanez, and the sample obtained was consistent with that idea: of the 13 students surveyed, 8 preferred Professor Fontanez and 2 preferred Professor Old. The remaining students were unable to express a preference. Test the hypothesis that the students prefer Fontanez. (*Hint:* Use the median test.)

5. Use the data in Table 10-8 to perform two Mann–Whitney tests: (a) compare uncontrolled in tersections and intersections with yield signs, and (b) compare uncontrolled intersections and intersections with stop signs.

6. Solid-waste generation rates measured in metric tons per household per year are collected randomly in selected areas of a township. The areas are classified as high-density, low-density, or sparsely settled. It is thought that generation rates probably differ because of differences in waste collection and opportunities for on-site storage. Do the following data support this hypothesis?

High density	Low density	Sparsely settled
1.84	2.04	1.07
3.06	2.28	2.31
3.62	4.01	0.91
4.91	1.86	3.28
3.49	1.42	1.31

7. The distances travelled to work by a random sample of 12 people to their places of work in 1996 and again in 2006 are shown in the following table.

	Distance (km)			Distance (km)	
Person	1996	2006	Person	1996	2006
1	8.6	8.8	7	7.7	6.5
2	7.7	7.1	8	9.1	9.0
3	7.7	7.6	9	8.0	7.1
4	6.8	6.4	10	8.1	8.8
5	9.6	9.1	11	8.7	7.2
6	7.2	7.2	12	7.3	6.4

Has the length of the journey to work changed over the decade?

8. One hundred randomly sampled residents of a city subject to periodic flooding are classified according to whether they are on the floodplain of the major river bisecting the city or off the floodplain. These households are then surveyed to determine whether they currently have flood insurance of any kind. The survey results are as follows:

	On the floodplain	Off the floodplain
Insured	50	10
No insurance	15	25

Test a relevant hypothesis.

9. The occurrence of sunshine over a 30-day period was calculated as the percentage of time the sun was visible (i.e., not obscured by clouds). The daily percentages were:

Day	Percentage of sunshine	Day	Percentage of sunshine	Day	Percentage of sunshine
1	75	11	21	21	77
2	95	12	96	22	100
3	89	13	90	23	90
4	80	14	10	24	98
5	7	15	100	25	60
6	84	16	90	26	90
7	90	17	6	27	100
8	18	18	0	28	90
9	90	19	22	29	58
10	100	20	44	30	0

If we define a sunny day as one with over 50% sunshine, determine whether the pattern of occurrence of sunny days is random.

10. Test the normality of the DO data (a) using the Kolmogorov–Smirnov test with the ungrouped data of Table 2-4 and (b) using the χ^2 test with $k = 6$ classes of Table 2-6.

11. Explain the workings of (a) kernel probability density estimates and (b) bootstrapping.

12. How does the choice of kernel bandwidth affect the estimate?

13. Why is the normal distribution a good kernel function?

14. Use the kernel method to estimate the probability distribution for the DO data of Table 2-4. Graph the resulting function and compare with Figure 2-1.

15. Use the kernel method to map station density for the Kansas dataset available from the text website.

16. Explain the difference between a kernel estimate $\hat{f}(x)$ and a bootstrap estimate $\hat{f}(\hat{\theta})$.

17. Given a sample of size 12, how many different bootstrap samples are possible?

18. What is the fundamental assumption behind bootstrapping?

19. Explain the percentile method for constructing bootstrap confidence intervals.

20. Use a straight edge with Figure 10-13 to find an approximate 60% confidence interval for ρ.

21. Randomly select a sample of 25 stations from the Kansas precipitation dataset. Use the sample to bootstrap a 90% confidence interval for the population median.

11

Analysis of Variance

In this chapter, we present a test that may be used to compare two or more sample means. Using this test, known as the ANalysis Of VAriance (or ANOVA for short), we make inferences about whether or not the population means corresponding to two or more samples are equal. ANOVA is an extension of the two-sample difference-of-means test described in Chapter 9, though there are important differences between the formats of these two approaches. Analysis of variance procedures were originally developed to help assist researchers interested in agricultural experiments. Typically, agricultural experiments require an entire growing season and utilize considerable input resources to prepare fields, apply different treatments, harvest the crops, and measure output. Based on these results, the objective is to evaluate different seeds, fertilizers, tillage methods, and irrigation levels to determine which levels and combinations of factor levels lead to the highest crop productivity.

EXAMPLE 11-1. Scientists at an agricultural research station are interested in determining which of two commercial fertilizers should be used to increase corn yield. A statistical consultant suggests that a controlled experiment be undertaken. Eighteen 5-acre plots are *randomly* assigned to be (1) treated with fertilizer XLC, (2) unfertilized, or (3) treated with fertilizer CG100. There are six plots in each of these three treatments. The unfertilized plots constitute a control group. The difference between the yields with XLC or CG100 and these plots gives us an estimate of the effectiveness of each fertilizer. The 18 plots to be used in the experiment are homogeneous with respect to soil type and condition as well as drainage. Irrigation levels and rainfall are monitored so that all plots receive equal amounts of water. The plots selected for fertilizer application are treated with equal amounts of either XLC or CG100. At the end of the season, the total yield from each plot is measured and recorded. The sample means for the control XLC and CG100 plots are to be compared to test the effectiveness of the fertilizers. Scientists wish to know whether or not the corresponding population mean yields for the three *treatments* are equal. In other words, are the differences in sample means across the three treatment groups sufficiently large as to

TABLE 11-1
Corn Yields (bushels per acre) from Experimental Plots

Experimental treatment	Observations						Total	Mean
Fertilizer XLC	84	80	81	84	82	81	492	82
Control (unfertilized)	74	66	70	70	72	74	426	71
Fertilizer CG100	96	84	92	90	80	80	522	87

support the hypothesis that the samples are likely drawn from populations whose mean yields are different?

The sample data collected during the experiment are given in Table 11-1. For each plot the recorded figure is the total corn yield in bushels per acre. The sample corn yields suggest that both fertilizers raise corn yield above those attainable from unfertilized control plots. Also, it is possible that fertilizer CG100 is superior to XLC. The true relationship among the three population means may be as illustrated in Figure 11-1. In this example, corn yield is highest for CG100 and lowest for the control group, with fertilizer XLC having an intermediate yield. The analysis of variance

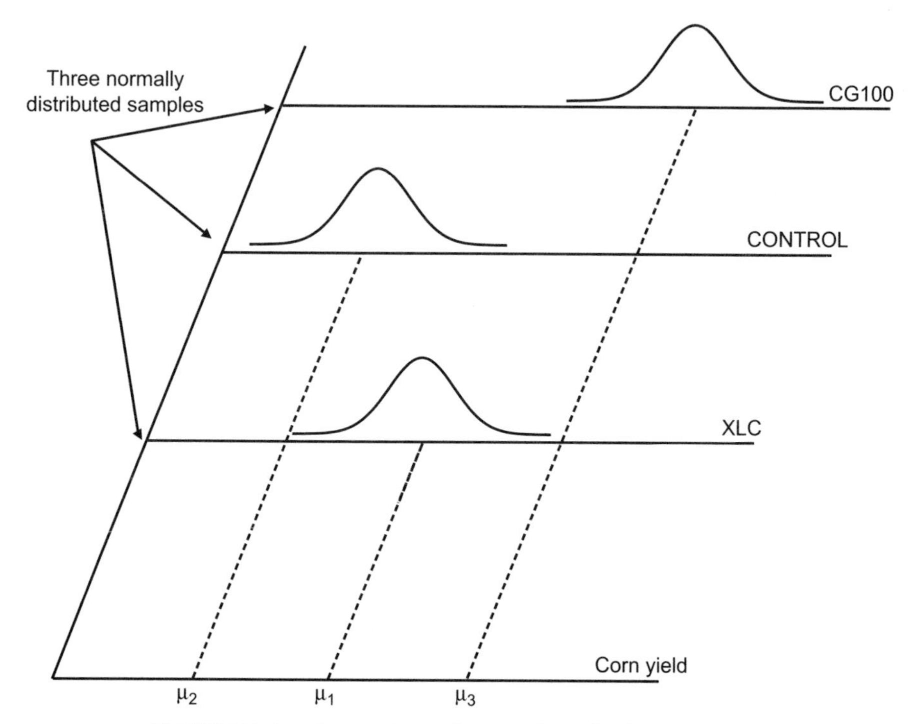

FIGURE 11-1. Comparison of means from the three samples.

test is used to determine whether we should reject the null hypothesis that all three treatments have equal means in favor of the alternative hypothesis that at least one mean is different.

The name of this test seems inappropriate because we are comparing the *means* of the samples and not the *variances*. As we shall see, however, the name is not drawn from the statistic being tested but from the procedure used to test for the difference of the means.

Section 11.1 presents the rationale and procedure for the one-factor or *one-way completely randomized* analysis of variance. This is the appropriate model for the corn yield experiment. It is termed one-factor because we are isolating the effects of a single factor, fertilizer type, on the yield of corn from a set of experimental plots. In Section 11.2, the analysis of variance is extended to study the simultaneous effects of two different factors in a single experiment. Such models are termed *two-factor* (or *two-way)* ANOVAs. Two-factor experiments are used to test the individual effects of the two different factors, as well as possible *interaction* effects between them. For example, suppose we wish to test the effects of these two fertilizers using two different seed hybrids. Both fertilizers may well produce the same corn yield with one seed hybrid, but produce markedly different yields with another seed hybrid. In this case, the two factors of fertilizer and seed hybrid are said to *interact.*

The ANOVA procedure provides a hypothesis test to determine if *any* of the subgroup means are different. In terms of the agricultural example we have described, we would find whether any of the three subgroups, either the control, XLC, or CG100, was different. While this is important, we are also interested in which of the subgroup means are different in a statistical sense. Are average yields for all three treatments the same? Can both fertilizers be distinguished from the control? Can they be distinguished from each other? The answer to such questions can be found using the method of *multiple comparisons.* A number of different procedures can be used. In Section 11.3 we outline the Scheffé method for multiple comparisons, one of several procedures that can be used to answer this type of question. Section 11.4 explains the four key assumptions of ANOVA models and describes appropriate statistical tests. The chapter is summarized in Section 11.5.

11.1. The One-Factor, Completely Randomized Design

The data generated by this agricultural experiment conform to a model of ANOVA known as the *one-factor, completely randomized design.* It is a one-way or one-factor design because we are isolating the effects of a single categorical variable, fertilizer type, on corn yield. The experiment is designed so that all other factors known to effect corn yield are held constant and can therefore have no effect on the observed yields for the experimental plots. It is a completely randomized design because we have assigned each 5-acre plot to one of the three experimental groups in a random manner. The statistical hypothesis to be tested is

$$H_0: \mu_1 = \mu_2 = \mu_3$$
$$H_A: \text{At least one } \mu_i \text{ differs from the rest}$$

If the null hypothesis is true, then the population mean yields are equivalent and fertilizer application is ineffective in increasing corn yield. Why? Simply because it cannot be shown they differ from the *control* plots that have been untreated.

Denote μ as the single population mean *under the null hypothesis.* If the three random samples of size $n = 6$ are drawn from this population, then the appropriate model is

$$Y_{ij} = \mu + e_{ij}, \; i = 1, 2, 3, \quad \text{and} \quad j = 1, 2, \ldots, 6 \qquad (11\text{-}1)$$

where μ is the common population mean, Y_{ij} is the random variable corn yield on plot j with fertilizer treatment i, and e_{ij} is the random error component of the jth plot assigned to the ith fertilizer.

Since μ is common to all plots and fertilizers, it is a constant. Corn yield, Y_{ij}, is a random variable because it depends on the random error component e_{ij}. Certainly, we cannot expect to be able to control all of the influences on corn yield, but we expect to control all major *systematic* ones. This error component can be thought of as the combined effect of many omitted variables, each of which has a small effect on corn yield, but when combined together has no systematic component and is random in nature. For example, it is impossible to ensure that the seeds used on each plot are exactly the same. We would thus not expect an *exact* duplication of our experimental results if we were to repeat the experiment several more times. It is also common to term the Y_{ij} as the *dependent* variable in an ANOVA problem. If we can reject the null hypothesis that the population means are equal, then we have shown that the value of Y_{ij} depends on the sample or treatment to which the experimental plot was assigned. Put simply, we say that corn yield depends on fertilizer type.

The general format of the input data for a one-factor completely randomized ANOVA model is presented in Table 11-2. There are t-samples or treatments, $i = 1, 2, \ldots, t$. The term *treatments* used in ANOVA reflects the development of the technique in the context of controlled experiments of this type. A plot is said to be "treated"

TABLE 11-2
Format for Input Data to ANOVA

	Sample values j =					
Treatment i =	1	2	3	. . .	n	Mean
1	Y_{11}	Y_{12}	Y_{13}	. . .	Y_{1n}	$\bar{Y}_1$
2	Y_{21}	Y_{22}	Y_{23}	. . .	Y_{2n}	$\bar{Y}_2$
. . .	. . .	. . .	. . .	. . .	. . .	. . .
t	Y_{t1}	Y_{t2}	Y_{t3}	. . .	Y_{tn}	$\bar{Y}_t$

with fertilizer XLC or CG100. Table 11-2 presents a *balanced* experimental design with an equal number of sample values for each treatment. The number of samples in each of the t treatments is n, so there are a total of nt observations in the problem. The modifications for handling *unbalanced* designs are discussed at the end of this section. The mean values for each treatment are labeled $\bar{Y}_i$, $i = 1, 2, \ldots, t$.

A quick glance at Table 11-1 suggests that the values of Y_{ij} do not randomly fluctuate about a common mean μ. It appears that each random sample may be drawn from a different population as is illustrated in Figure 11-1. If the *null hypothesis is false,* then we can rewrite Equation 11-1 as

$$Y_{ij} = \mu_i + e_{ij} = \mu + (\mu_i - \mu) + e_{ij}, \quad i = 1, 2, \ldots, t \quad \text{and} \quad j = 1, 2, \ldots, n \quad (11\text{-}2)$$

This indicates that the values of the dependent variable are drawn from two or more populations with different means. The dependent variable Y_{ij} is now reexpressed as the sum of three components, the common mean μ, the difference between the mean value of the ith treatment and μ, and the random error. If τ_i (read "tau i") is defined to be the effect of the ith treatment, then Equation 11-2 can be written as

$$Y_{ij} = \mu_i + e_{ij} = \mu + \tau_i + e_{ij}, \quad i = 1, 2, \ldots, t \quad \text{and} \quad j = 1, 2, \ldots, n \quad (11\text{-}3)$$

where $\tau_i = \mu_i - \mu$. When the null hypothesis is true, it is easy to see that Equation 11-3 reduces to Equation 11-1 since all of the τ_i are equal to 0. An equivalent statement of the null hypothesis is therefore

$$H_0: \tau_1 = \tau_2 = \ldots \tau_t = 0$$

$$H_A: \text{At least one } \tau_i \text{ differs from zero}$$

If H_0 is true, then the t-sample means are drawn from a single population with mean μ and the variation in $\bar{Y}_i$ can be entirely attributed to sampling error. There are a number of ways in which the null hypothesis may not be true. This is because the alternative hypothesis only states that *at least one* of the τ_i is nonzero. It is possible that any one of the t-treatments or populations could be different. Or, two populations could be different from the rest, or three, or so on. It is even possible that all of the t-samples could come from different populations. This last alternative is illustrated in Figure 11-1. How do we know which of these cases applies to a particular ANOVA problem? Once we have rejected the null hypothesis, we can use the method of multiple comparisons to determine the appropriate alternative hypothesis to be selected. To summarize, the one-factor completely randomized, balanced ANOVA design is fully specified in the following definition.

DEFINITION: ONE-FACTOR, COMPLETELY RANDOMIZED, BALANCED ANOVA

The one-factor, completely randomized, balanced ANOVA model can be expressed by the linear equation

$$Y_{ij} = \mu_i + e_{ij} = \mu + \tau_i + e_{ij}, \quad i = 1, 2, \ldots, t \quad \text{and} \quad j = 1, 2, \ldots, n \qquad (11\text{-}4)$$

where Y_{ij} is the dependent response random variable corresponding to the jth experimental unit being subjected to the ith treatment, μ is the common population mean, μ_i is the *mean* population value of the dependent variable corresponding to the ith treatment, τ_i is the effect of the ith treatment, the difference between μ and μ_i, and e_{ij} is the random error component associated with the random variable Y_{ij}.

The Analysis of Variance Procedure

To use the analysis of variance properly, we need to make a few assumptions about the random variable Y_{ij}. First, it is assumed that Y_{ij} is normally distributed in each population being compared. Relating this to the agricultural experiment, the assumption requires that corn yield is normally distributed for each of the three experimental conditions, XLC, control, and CG100. This is illustrated in Figure 11.1 Second, the distributions of Y_{ij} must have the same variance. This is termed *homogeneity of variance* or *homoscedasticity.* The three normal distributions depicted in Figure 11-1 appear to be roughly homoscedastic. This assumption is critical to the ANOVA procedure. (The assumptions for the various ANOVA models are fully explained in Section 11.4.) The rationale for the ANOVA test is as follows. If H_0 is true, then the t-samples must provide unbiased estimates of the common population mean μ and population variance σ_Y^2. Within each sample, we can make an estimate of this variance using the values within that sample. The variance in the first sample or treatment $i = 1$ is

$$s_1^2 = \frac{\sum_{j=1}^{n} (Y_{1j} - \bar{Y}_1)^2}{n - 1}$$

for the second treatment $i = 2$ is

$$s_2^2 = \frac{\sum_{j=1}^{n} (Y_{2j} - \bar{Y}_2)^2}{n - 1}$$

and so on. In general, the variance estimate of the ith sample is

$$s_i^2 = \frac{\sum_{j=1}^{n} (Y_{ij} - \bar{Y}_i)^2}{n - 1} \qquad (11\text{-}5)$$

The *average* of these t estimates is called the *pooled* variance or the *pooled estimate of the variance*. Thus, an estimate of the true population σ_Y^2, denoted $\hat{\sigma}_Y^2$ is

$$\hat{\sigma}_Y^2 = \frac{1}{t}\sum_{i=1}^{t} s_i^2 = \frac{\sum_{i=1}^{t}\sum_{j=1}^{n}(Y_{ij} - \bar{Y})^2}{t(n-1)} \tag{11-6}$$

There is a second important relationship within the data that is also utilized in the ANOVA procedure. Under the null hypothesis, the t-samples are drawn from the same population. Therefore, we can calculate an estimate of the variance of the sample means, $\sigma_{\bar{Y}}^2$, from these t-samples. First, calculate the overall or *grand* mean in the matrix as

$$\bar{Y} = \frac{\sum_{i=1}^{t}\sum_{j=1}^{n} Y_{ij}}{nt} \tag{11-7}$$

The total sample size is nt so that $\bar{Y}$ is simply the average value of all the observations in the table. The variance of the sample means is calculated using the sample means $\bar{Y}_1, \bar{Y}_2, \ldots, \bar{Y}_t$ and the grand mean $\bar{Y}$ in the following way:

$$\sigma_{\bar{Y}}^2 = \frac{\sum_{i=1}^{t}(\bar{Y}_i - \bar{Y})^2}{t-1} \tag{11-8}$$

Now, the key relationship exploited in the analysis of variance is the relation between the two variances given by (11-6) and (11-8). From the central limit theorem for means (see Section 6.5 in Chapter 6) we know that

$$\sigma_{\bar{Y}}^2 = \frac{\sigma_Y^2}{n}$$

or, if both sides are multiplied by n

$$n\sigma_{\bar{Y}}^2 = \sigma_Y^2$$

Of course, this relationship will not hold *exactly* in any set of samples, and we must assess the degree to which the variability in meeting this equation can be attributed to chance or sampling fluctuations. The key question becomes "Is $n\sigma_{\bar{Y}}^2$ large relative to σ_Y^2?" It is customary to calculate

$$f = \frac{n\sigma_{\bar{Y}}^2}{\sigma_Y^2} \tag{11-9}$$

as the test statistic. Whenever H_0 is true, this F-ratio is approximately 1. Whenever it exceeds 1 by a significantly large amount, we can conclude that the t-samples do not come from the same population, at least at a certain α level.

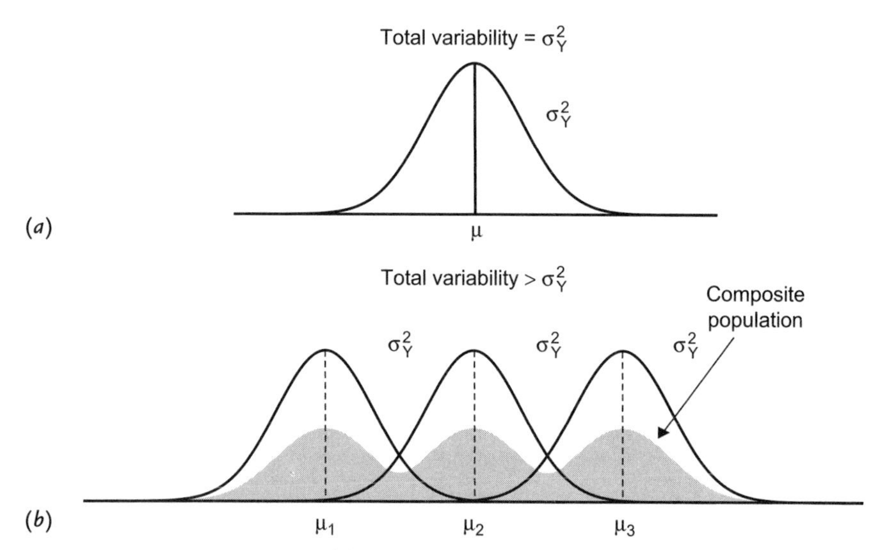

FIGURE 11-2. Comparison of the population variability when the treatment effects are equal and unequal.

It should now be apparent why this test is known as the analysis of variance. The method utilizes variances in order to compare means. The rationale for this test can also be given a straightforward graphical interpretation. The key relationship is depicted in Figure 11-2. When H_0 is true and the "treatment" effects are zero, the samples are all drawn from a single population with variance σ_Y^2 as in panel (a). The variance σ_Y^2 should equal n times the variance of the sample means. If, on the other hand, the treatment effects are not zero (the samples are not drawn from a single population), then the situation is more like that in panel (b). If repeated samples of size n are drawn from the "composite" population indicated by the shaded area in panel (b), then the total variability in this population necessarily exceeds the one in which the means are equal. Why? This is because the population itself is more variable. In this case it is easy to see that $n\sigma_{\bar{Y}}^2 > \sigma_Y^2$.

To illustrate the numerical procedures involved in the analysis of variance, let us reconsider the data generated in the corn fertilizer experiment of Table 11-1. The three sample means are $\bar{Y}_1 = 82$, $\bar{Y}_2 = 71$, and $\bar{Y}_3 = 87$, and the overall grand mean is $\bar{Y} = 80$. The variance estimate in the first sample is

$$s_1^2 = \frac{(84-82)^2 + (80-82)^2 + (81-82)^2 + (84-82)^2 + (82-82)^2 + (81-82)^2}{6-1}$$

$$s_1^2 = 2.8$$

Similarly, Equation 11-5 can be used to calculate $s_2^2 = 9.2$ and $s_3^2 = 44.4$. Using 11-6, the pooled estimate of the variance is calculated as $\hat{\sigma}_Y^2 = 1/3(2.8 + 9.2 + 44.4) =$

1/3(56.4) = 18.8. To determine $\hat{\sigma}^2_{\bar{Y}}$ based on the three sample means, we use Equation 11-8:

$$\hat{\sigma}^2_{\bar{Y}} = \frac{(82-80)^2 + (71-80)^2 + (87-80)^2}{3-1}$$

$$\hat{\sigma}^2_{\bar{Y}} = 134/2 = 67$$

and therefore $\hat{\sigma}^2_{Y} = 6(67) = 402$. The F-statistic

$$f = 402/18.8 = 21.38$$

can be tested for significance. As usual, the degrees of freedom are determined by the number of independent values used to make the estimates. In the numerator there are $t-1$ such values, one less than the number of treatments or samples. For each of the t-samples there are $n-1$ degrees of freedom, one less than the number of observations. In total, then, there are $t(n-1)$degrees of freedom in the denominator. For the corn yield experiment, the critical F for $\alpha = .05$ is $F(0.95, 2, 15) = 3.68$. The calculated value of $f = 21.38$ exceeds this critical value, and therefore we can reject H_0: $\tau_1 = \tau_2 = \tau_3 = 0$.

To determine the PROB-VALUE for this problem, we can use Table A-7 so long as we recognize that the α-values shown are 2-tailed probabilities. For ANOVA, we need 1-tailed probabilities, which are obviously just half of those indicated in Table A-7. Using $df = (2, 15)$, the smallest PROB-VALUE in Table A-7 is 0.001/2 = 0.0005. The corresponding F-value is 13.16, thus we know that the observed F of 21.38 has an even smaller probability. We cannot get an exact PROB-VALUE from the table, but we know it is less than 0.0005. We therefore reject the null hypothesis and conclude that fertilizer applications affect corn yield.

The Variance Decomposition Approach

An alternative way of generating the estimates required for the ANOVA test is based on the concept of the *decomposition of the total variation* in the t-samples of the input data. Consider any observation Y_{ij}. The deviation of this observation from the grand mean $\bar{Y}$ can be divided into two parts:

$$(Y_{ij} - \bar{Y}) = (\bar{Y}_i - \bar{Y}) + (Y_{ij} - \bar{Y}_i) \qquad (11\text{-}10)$$

The first term on the right-hand-side of Equation 11-10 is the difference *between* the sample mean $\bar{Y}_i$ and the grand mean $\bar{Y}$. The second term is the deviation *within* the sample or treatment, the difference between the value of an observation and its sample or treatment mean $\bar{Y}_i$. As an example, consider the first observation in the XLC treatment of Table 11-1, $Y_{11} = 84$. Substituting into Equation 11-10:

$$(Y_{11} - \bar{Y}) = (\bar{Y}_1 - \bar{Y}) + (Y_{11} - \bar{Y}_1)$$

$$(84 - 80) = (82 - 80) + (84 - 82)$$

$$4 = 2 + 2$$

A graphical interpretation of the decomposition of variation is illustrated in Figure 11-3. In this figure, each observation is represented by a ⊠ symbol, and the vertical axis has been arranged to differentiate the three treatments, Control, XLC, and CG100, and the X-axis measures corn yield. The solid vertical line at $\bar{Y} = 80$ represents the grand mean. There are three dashed vertical lines representing the three group means. The deviations around these four lines are represented as horizontal lines. Three solid horizontal lines running from the group means to the grand mean correspond to *between-group* variation. The dashed line from each observation to its group or treatment mean represents *within* variation, and the dotted line from each

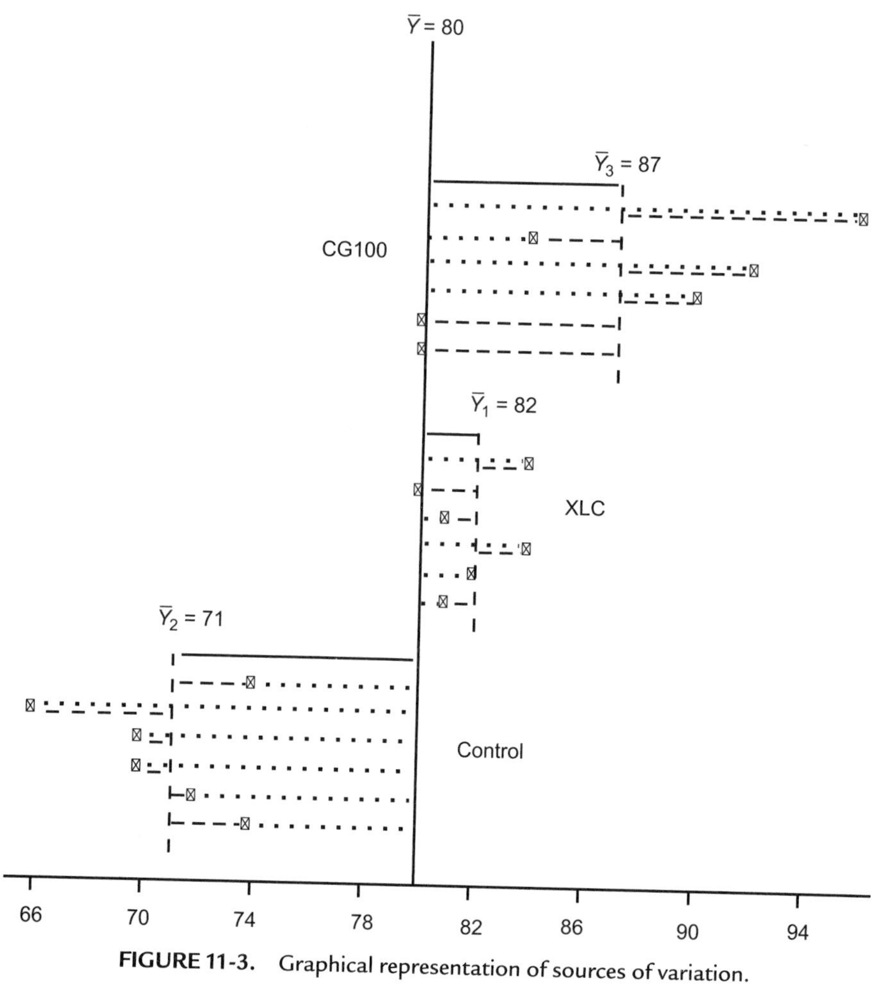

FIGURE 11-3. Graphical representation of sources of variation.

TABLE 11-3
Work Table to Illustrate Equation 11-11 for Corn Yield Data

$$\sum_{i=1}^{t}\sum_{j=1}^{n}(Y_{ij} - \bar{Y})^2$$
$$= (84-80)^2 + (80-80)^2 + (81-80)^2 + (84-80)^2 + (82-80)^2 + (81-80)^2 + (74-80)^2 + (66-80)^2$$
$$+ (70-80)^2 + (70-80)^2 + (72-80)^2 + (74-80)^2 + (96-80)^2 + (84-80)^2 + (92-80)^2$$
$$+ (90-80)^2 + (80-80)^2 + (80-80)^2$$
$$= 38 + 532 + 516$$
$$= 1086$$

$$n\sum_{i=1}^{t}(\bar{Y}_i - \bar{Y})^2$$
$$= 6[(82-80)^2 + (71-80)^2 + (87-80)^2]$$
$$= 6[134]$$
$$= 804$$

$$\sum_{i=1}^{t}\sum_{j=1}^{n}(Y_{ij} - \bar{Y}_i)^2$$
$$= (84-82)^2 + (80-82)^2 + (81-82)^2 + (84-82)^2 + (82-82)^2 + (81-82)^2 + (74-71)^2 + (66-71)^2$$
$$+ (70-71)^2 + (70-71)^2 + (72-71)^2 + (74-71)^2 + (96-87)^2 + (84-87)^2 + (92-87)^2$$
$$+ (90-87)^2 + (80-87)^2 + (80-87)^2$$
$$= 14 + 46 + 222$$

observation to the grand mean represents the observation's contribution to *total* variation. It is easy to see the additivity of the *within* and *between* components to the *total* deviation.

Now, if both sides of Equation 11-10 are squared and we sum over all treatments i and observations j, the following fundamental relationship is derived (this is fully detailed in Appendix 11a):

$$\sum_{i=1}^{t}\sum_{j=1}^{n}(Y_{ij} - \bar{Y})^2 = n\sum_{i=1}^{t}(\bar{Y}_i - \bar{Y})^2 + \sum_{i=1}^{t}\sum_{j=1}^{n}(Y_{ij} - \bar{Y}_i)^2 \qquad (11\text{-}11)$$

Equation 11-11 states that the total *variation* or *sum of squares* can be decomposed into two separate components: *between* and *within* (or *error*) sum of squares. The terms in this equation are called variations because they represent the sum of squared differences from observations to means. The equation is often written compactly as

$$\text{Total} = \text{Between} + \text{Error}$$

$$\text{TSS} = \text{BSS} + \text{ESS}$$

where the abbreviation TSS, for example, refers to *total sum of squares.*

Table 11-3 illustrates the calculations required to verify (11-11) for the corn fertilizer experiment. The total variation or sum of squares TSS is computed by summing the square of each observation to the grand mean. The *between* sum of squares BSS sums the squares of the differences between the sample means and the grand mean. This total is then multiplied by the number of observations in each sample, $n = 6$. The *error* sum of squares ESS sums the squared differences from each observation to its treatment or sample mean. In the corn yield experiment we calculate TSS = 1086,

BSS = 804, and ESS = 282. Since 1086 = 804 + 282, Equation 11-11 is verified for these data.

These variations can be converted to variances by dividing by the appropriate degrees of freedom in each case. These variances can be shown to be exactly equivalent to those presented as Equations 11-6 and 11-8. Note that

$$\sigma_Y^2 = \frac{\sum_{i=1}^{t}\sum_{j=1}^{n}(Y_{ij} - \bar{Y}_i)^2}{t(n-1)} \quad (11\text{-}12)$$

$$n\sigma_{\bar{Y}}^2 = \frac{n\sum_{i=1}^{t}(\bar{Y}_i - \bar{Y})^2}{t-1} \quad (11\text{-}13)$$

utilize the two components of total variation in their numerators and the appropriate degrees of freedom in the denominator. These are the same degrees of freedom used in (11-6) and (11-8).

Analysis of Variance Table

A particularly convenient format for displaying the results of an ANOVA is the *analysis of variance table.* The typical structure of this table is displayed in Table 11-4. There are three rows, each corresponding to a different source of variation: between, error, and total. Sometimes these three categories are given other designations in statistical software packages. The *between* row is often labeled "explained," "treatments," or "model." These terms are used since we are trying to *explain* variations in the dependent variable Y by various *treatments* in a *model.* We might think of Equation 11-4 as an explanatory model for the dependent variable Y_{ij}. In the case of the corn yield experiment, the plots are treated with various fertilizers. ANOVA is used as a model to explain variations in corn yield. The term *treatments* is frequently used in purely experimental applications of ANOVA, most notably in the discipline of psychology. Similarly, the row labeled *Error* is sometimes labeled "unexplained" or "within" or even "experimental error."

TABLE 11-4
The Analysis of Variance Table

Source of variation	Sum of squares	Degrees of freedom	Mean square	*f*-ratio
Between (explained, model, treatments)	BSS	$t-1$	$\frac{BSS}{t-1}$	$f = \frac{\frac{BSS}{t-1}}{\frac{ESS}{t(n-1)}}$
Error (within, unexplained)	ESS	$t(n-1)$	$\frac{ESS}{t(n-1)}$	
Total	TSS	$tn-1$		

TABLE 11-5
An Edited Version of the SPLUS-2000 ANOVA Output

```
                *** Analysis of Variance Model ***
Terms:
                 Fertilizer    Residuals
Sum of Squares      804           282
Deg. of Freedom       2            15

                  Estimated effects are balanced
                 Df    Sum of Sq    Mean Sq    F Value      Pr(F)
Fertilizer        2       804        402.0     21.38298   0.00004056436
Residuals        15       282         18.8
Tables of means
Grand mean
80
Fertilizer
XLC              CONTROL    CG100
82                 71         87
```

The *between* row includes the sum of squares BSS according to Equation 11-11, $t - 1$ degrees of freedom, and the mean square or variance estimate of BSS/$t - 1$. The mean square for the *between* row is the variance estimate for $n\sigma_{\bar{Y}}^2$. The *error* row has $t(n - 1)$ degrees of freedom. For each of the t-samples there is one degree of freedom less than the common number of observations n. In total, there are thus $t(n - 1)$ independent values in this calculation. The *error* sum of squares given in (11-11) is labeled ESS. The variance estimate for σ_Y^2 given by (11-12) is in the fourth column of this row and is equal to ESS/$t(n - 1)$. The last column of the table is the F-statistic, the ratio of the two mean squares. Note that the table satisfies the two known identities. First, we see the total sum of squares identity TSS = BSS + ESS appears as column 2. Second, we see that the total degrees of freedom in column 3 satisfy the identity $(t - 1) + t(n - 1) = tn - 1$.

We can now verify the results of the F-test for the corn yield experiment using the ANOVA table format. Table 11-5 illustrates a portion of the output from the SPLUS-2000 software package for this ANOVA. Note that S-PLUS reports the exact PROB-VALUE. We conclude that, on the basis of this result and an examination of the treatment means, fertilizer application does improve corn yield,.

Unequal Sample Sizes

Data used in an ANOVA can be in either *balanced* or *unbalanced* designs.

DEFINITION: BALANCED DESIGN
A balanced ANOVA design is one that has an equal number of observations in each treatment.

TABLE 11-6
The Analysis of Variance Table for Unbalanced Designs

Source of variation	Sum of squares	Degrees of freedom	Mean square	f-ratio
Between	BSS	$t-1$	$\frac{\text{BSS}}{t-1}$	$f=\frac{\frac{\text{BSS}}{t-1}}{\frac{\text{ESS}}{\left(\sum_{i=1}^{t} n_i\right)-t}}$
Error	ESS	$\left(\sum_{i=1}^{t} n_i\right)-t$	$\frac{\text{ESS}}{\left(\sum_{i=1}^{t} n_i\right)-t}$	
Total	TSS	$\left(\sum_{i=1}^{t} n_i\right)-1$		

While a balanced design is characteristic of many experiments, at other times the number of cases or observations in any sample or treatment is unequal. Such designs are said to be unbalanced. Unbalanced designs can also occur in controlled experiments. Suppose one of the plots in the corn yield experiment becomes infested with a pest. Before the pest is eradicated, a substantial number of plants in the plot have to be destroyed. The net result will be a reduction in yield from this plot, which should now be excluded from the ANOVA experiment. Many geographical applications of ANOVA tend to have unbalanced designs since the data are often generated from census bulletins, other archival sources, or survey questionnaires. In such situations, fewer controls are available to ensure balanced designs. Fortunately, unbalanced designs can be accommodated with only a few modifications to the ANOVA formulas. Let n_i be the number of observations or cases in sample or treatment i. The one-factor, unbalanced, completely randomized design is now expressed as

$$Y_{ij} = \mu + \tau_i + e_{ij}, \quad i = 1, 2, \ldots, t, \quad \text{and} \quad j = 1, 2, \ldots, n_i \qquad (11\text{-}14)$$

which is identical to (11-4) except that the summations extend over the unequal sample sizes. The appropriate ANOVA table is given in Table 11-6. The principal differences between this table and Table 11-4 are in the specification of the degrees of freedom for the various sources and in the formulas for BSS, ESS, and TSS, which are defined with slightly different summation limits. In all other respects, the testing procedure is identical to that employed in the balanced case.

In the unbalanced design, we use the following equations for calculating BSS, ESS, and TSS:

$$\text{BSS} = \sum_{i=1}^{t} n_i(\bar{Y}_i - \bar{Y})^2$$

$$\text{ESS} = \sum_{i=1}^{t} \sum_{j=1}^{n_i} (Y_{ij} - \bar{Y}_i)^2$$

$$\text{TSS} = \sum_{i=1}^{t} \sum_{j=1}^{n_i} (Y_{ij} - \bar{Y})^2$$

11.2. The Two-Factor, Completely Randomized Design

In many applications of ANOVA, we are interested in explaining the variations in some dependent variable as a function of *several* different independent or explanatory variables. For example, we may wish to explain variations in corn yield as a function of both fertilizer application *and* seed type. The analysis of variance is easily generalized to such multifactor or *factorial experiments*. Let us consider the simplest of such models, the two-factor, completely randomized design.

Suppose, for example, that the corn yields given in Table 11-1 were actually produced using two different seed hybrids: Hybrid 1 and Hybrid 2. Each plot was treated with only one of the two seed hybrids. These data, when reorganized to take into account the two-way classification (by fertilizer and seed type), are shown in Table 11-7. Note that the design is still balanced and there are now $n = 3$ observations in each treatment. A "treatment" now corresponds to a *combination* of one *level* of each *factor* in the experiment. For example, the use of fertilizer XLC with Hybrid 1 seeds represents one of the six possible treatments of plots in the corn yield experiment.

From Table 11-7 we can see that there are differences in productivity by fertilizer type. It *appears* that both CG100 and XLC improve yield, a fact confirmed by our one-factor ANOVA results. However, if we examine the column means for the two seed types, it also *appears* that Hybrid 2 *may* be superior to Hybrid 1. The motive for introducing additional factors in an ANOVA model is to reduce the error or unexplained sum of squares.

To introduce the appropriate model and ANOVA for the two-factor, completely randomized design, we must introduce some new notation. First, let us label the two factors A and B. There are $i = 1, 2, \ldots, a$ levels of factor A, and $j = 1, 2, \ldots, b$ levels of factor B. Let Y_{ijk} be the value of an individual observation k, and we will assume

TABLE 11-7
Sample Data for Two-factor Corn Yield Experiment

	Seed type		
Fertilizer	Hybrid 1	Hybrid 2	Row mean
XLC	80 81 82	84 84 81	82
Control	74 66 70	70 72 74	71
CG100	92 80 80	96 84 90	87
Column mean	78.33	81.67	
Grand mean			80

a balanced design with m observations in each treatment. Notice that a treatment is a combination of a particular fertilizer type *and* a particular seed hybrid. There are therefore a total of $n = (a \times b \times m)$ observations. In the corn yield experiment we have $m = 3$, $a = 3$, and $b = 2$ and a total of 18 observations.

The two-factor, completely randomized, balanced ANOVA design is fully specified in the following definition.

DEFINITION: TWO-FACTOR, COMPLETELY RANDOMIZED, BALANCED ANOVA

The two-factor, completely randomized, balanced ANOVA model is specified by the linear equation

$$Y_{ijk} = \mu + \alpha_i + \beta_j + \gamma_{ij} + e_{ijk}$$
$$i = 1, 2, \ldots, a \text{ and } j = 1, 2, \ldots, b \text{ and } k = 1, 2, \ldots, m \qquad (11\text{-}15)$$

where Y_{ijk} is the dependent response random variable corresponding to the kth experimental unit being subjected to the ith treatment of factor A and the jth treatment of factor B, μ is the common population mean, α_i is the effect of the ith level of factor A, β_j is the effect of the jth level of factor B, γ_{ij} is the *interaction* effect of the ith level of factor A at the jth level of factor B, and e_{ijk} is the random error component.

The terms α_i and β_j have interpretations much the same as τ_i in the one-factor ANOVA model. What do we mean by the term *interaction* effect? This concept is easiest to explain in the context of our corn yield experiment. If there is no interaction effect between the two factors of fertilizer and seed type, then corn yield should increase (or decrease) by the same amount for each fertilizer if we change from seed Hybrid 1 to Hybrid 2. This situation is depicted in Figure 11-4. Note that the gains in yield in moving from Hybrid 1 to Hybrid 2 are approximately equal, no matter which fertilizer treatment is applied. In Figure 11-5, on the other hand, there is an apparent interaction effect. Corn yield increases more with CG100 than with either Control or XLC when the seed is changed to Hybrid 2. As we shall see, one of the important inferential questions in factorial experiments concerns the significance of these interaction effects.

Analysis of Variance Table

The structure of a two-factor ANOVA table differs from the simple one-factor table in several respects. First, four sources of variation identified in the two-factor ANOVA structure are shown in Table 11-8. There is one row for each of the principal factors A and B, one row for the interaction effect, and one row for the error variation. With the exception of the interaction sum of squares, the computational formulas for the various sums of squares represent simple generalizations of the one-factor model.

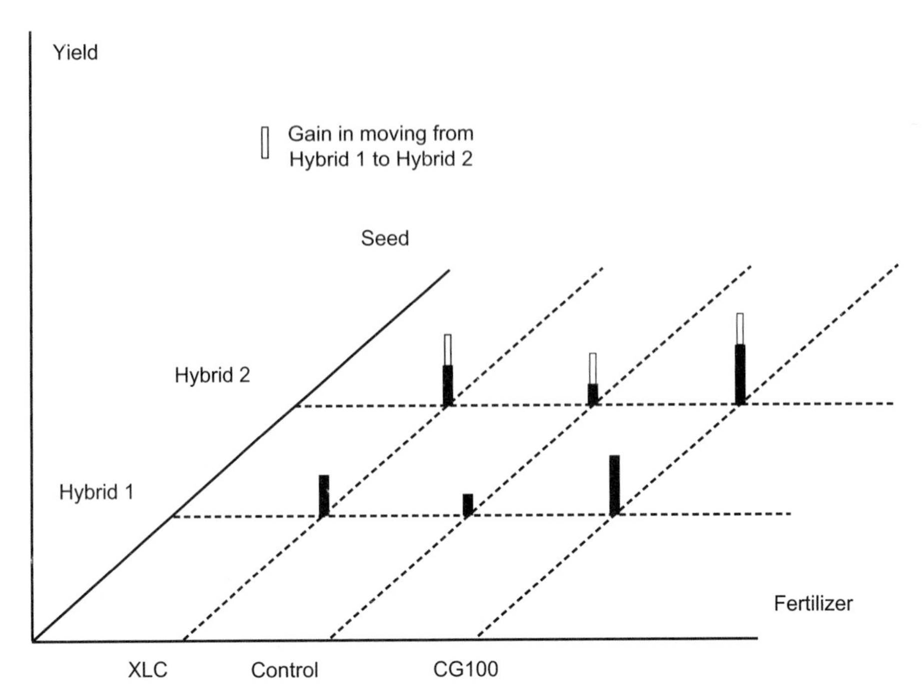

FIGURE 11-4. Graphical interpretation of interaction effects: The case of no interaction.

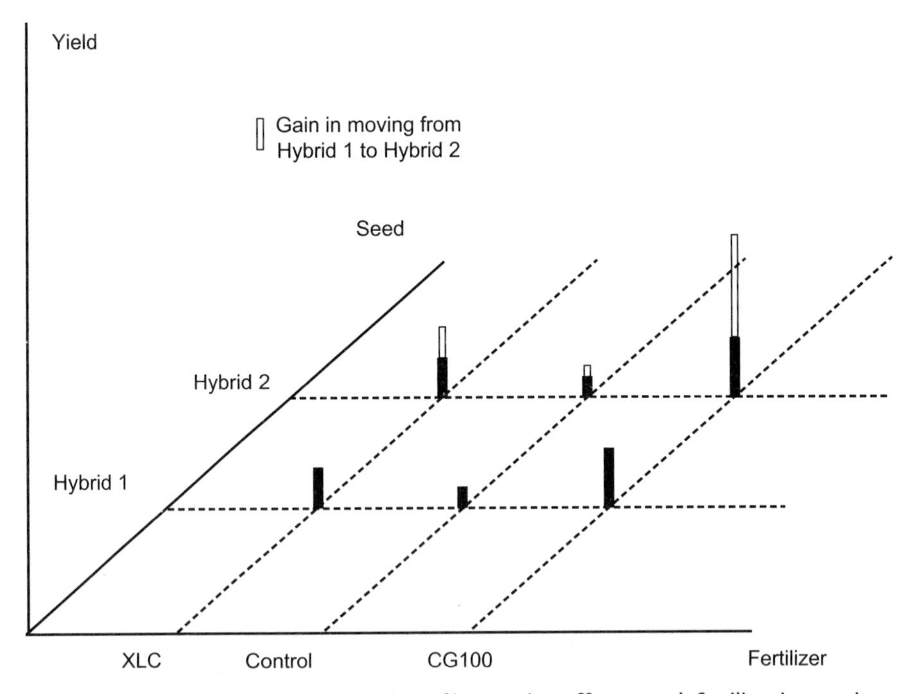

FIGURE 11-5. Graphical interpretation of interaction effects: seed–fertilizer interaction.

TABLE 11-8
The Analysis of Variance Table.

Source of variation	Sum of squares	Degrees of freedom	Mean square	f-ratio
Factor A	ASS	$a-1$	$\frac{\text{ASS}}{a-1}$	$f=\frac{\frac{\text{ASS}}{a-1}}{\frac{\text{ESS}}{ab(m-1)}}$
Factor B	BSS	$b-1$	$\frac{\text{BSS}}{b-1}$	$f=\frac{\frac{\text{BSS}}{b-1}}{\frac{\text{ESS}}{ab(m-1)}}$
Interaction	ISS	$(a-1)(b-1)$	$\frac{\text{ISS}}{(a-1)(b-1)}$	$f=\frac{\frac{\text{ISS}}{(a-1)(b-1)}}{\frac{\text{ESS}}{ab(m-1)}}$
Error	ESS	$ab(m-1)$	$\frac{\text{ESS}}{ab(m-1)}$	
Total	TSS	$abm-1$		

where

$$\text{ASS} = bm\sum_{i=1}^{a}(\bar{Y}_i - \bar{Y})^2$$

$$\text{BSS} = am\sum_{j=1}^{b}(\bar{Y}_j - \bar{Y})^2$$

$$\text{ISS} = m\sum_{i=1}^{a}\sum_{j=1}^{b}(\bar{Y}_{ij} - \bar{Y}_i - \bar{Y}_j + \bar{Y})^2$$

$$\text{ESS} = \sum_{i=1}^{a}\sum_{j=1}^{b}\sum_{k=1}^{m}(Y_{ijk} - \bar{Y}_{ij})^2$$

and

$$\text{TSS} = \sum_{i=1}^{a}\sum_{j=1}^{b}\sum_{k=1}^{m}(Y_{ijk} - \bar{Y})^2$$

Second, note that there are now three F-ratios to be calculated and interpreted. The first F-ratio is used to test the significance of the first factor labeled A, the second is used to test the second factor labeled B and the last F-ratio tests the interaction effect between A and B. In the corn yield experiment described in Example 11-1, A represents fertilizer treatment, B represents the seed hybrid treatment, and the interaction effect (sometimes labeled A * B) represents the potential interaction effects between these two "main factors." What we now need to have is a strategy to evaluate the various F-ratios. Do we test the main factors first, or the interaction effects?

Normally, the first F-ratio to be tested is the interaction effect:

$$H_0: \gamma_{ij} = 0, \quad i = 1, 2, \ldots, \quad a, \text{ and} \quad j = 1, 2, \ldots, b$$

$$H_1: \gamma_{ij} \neq 0, \quad \text{for one } ij \text{ pair}$$

If the interaction proves to be significant, then it is not really possible to undertake significance tests on the *individual* main factors A and B. If it turns out that the interaction is not significant, then the usual F-ratios on the individual factors may be examined; that is,

$$H_0 : \alpha_1 = \alpha_2 = \ldots = \alpha_a = 0$$

$$H_1 : \text{At least one } \alpha_i \text{ is nonzero}$$

or

$$H_0 : \beta_1 = \beta_2 = \ldots = \beta_b = 0$$

$$H_1 : \text{At least one } \beta_j \text{ is nonzero}$$

To generate the appropriate sums of squares for the ANOVA table, it is useful to begin with the summary means, by factor and treatment.

Grand mean		
80		
Fertilizer means		
XLC	*Control*	*CG100*
82	71	87
Seed type means		
Hybrid 1	*Hybrid 2*	
78.333	81.667	
Fertilizer × seed means		
	Hybrid 1	*Hybrid 2*
XLC	81	83
Control	70	72
CG100	84	90

Let us begin by calculating ASS. Using

$$\text{ASS} = bm\sum_{i=1}^{a}(\bar{Y}_i - \bar{Y})^2$$

with $b = 2$ and $m = 3$, and the means by fertilizer type given in the table above, we see that

$$\text{ASS} = 2 \cdot 3[(82 - 80)^2 + (71 - 80)^2 + (87 - 80)^2]$$
$$\text{ASS} = 6[4 + 81 + 49]$$
$$\text{ASS} = 804$$

Similarly, we obtain BSS as

$$\text{BSS} = 3 \cdot 3[(78.33 - 80)^2 + (81.67 - 80)^2]$$
$$\text{BSS} = 9[5.5556]$$
$$\text{BSS} = 50$$

The interaction sum of squares, ISS, is calculated using

$$\begin{aligned}\text{ISS} &= m\sum_{i=1}^{a}\sum_{j=1}^{b}(\bar{Y}_{ij} - \bar{Y}_i - \bar{Y}_j + \bar{Y})^2 \\ &= 3[(81 - 82 - 78.33 + 80)^2 + (83 - 82 - 81.67 + 80)^2 \\ &\quad + (70 - 71 - 78.33 + 80)^2 + (72 - 71 - 81.67 + 80)^2 \\ &\quad + (84 - 87 - 78.33 + 80)^2 + (90 - 87 - 81.67 + 80)^2] \\ &= 3[16/3] \\ &= 16\end{aligned}$$

The appropriate row and column means are subtracted from each treatment mean and then adjusted by the grand mean in this calculation. The total sum of squares is calculated by subtracting each observation from the grand mean, in this case 80, and then squaring this difference. Note that we still have 18 observations in our study, the same grand mean, and therefore the same total sum of squares, TSS = 1086.

Finally, for the error sum of squares, we must subtract each observation from its treatment mean and then square the result:

$$\begin{aligned}\text{ESS} &= \sum_{i=1}^{a}\sum_{j=1}^{b}\sum_{k=1}^{m}(Y_{ijk} - \bar{Y}_{ij})^2 \\ &= [(80 - 80)^2 + (81 - 81)^2 + (82 - 81)^2 + \ldots \\ &\quad + (96 - 90)^2 + (84 - 90)^2 + (90 - 90)^2] \\ &= 216\end{aligned}$$

The two-way ANOVA table for the corn yield experiment generated by the SPLUS-2000 software package is shown in Table 11-9. When compared to the one-way table, Table 11-5 , we first note a considerable decline in the error sum of squares

TABLE 11-9
An Edited Version of the SPLUS-2000 Two-Factor ANOVA Output

```
               *** Analysis of Variance Model ***
Terms:
                     Fertilizer    Seed Type    Fert x Seed    Residuals
Sum of Squares          804           50            16            216
Deg. of Freedom           2            1             2             12

                     Estimated effects are balanced
                  Df     Sum of Sq     Mean Sq      F Value       Pr(F)
Fertilizer         2         804         402       22.33333     0.0000902
Seed Type          1          50          50        2.77778     0.1214472
Fert x Seed        2          16          50        0.44444     0.6513203
Residuals         12         216          18
```

ESS from 282 to 216 due to the introduction of seed hybrid. Has the introduction of seed hybrid improved our model?

First, we test the interaction term. It turns out the F-value of 0.44444 has an associated probability of 0.65, and we conclude that there is *no* interaction effect between fertilizer type and seed hybrid. We are now in a position to test the significance of the factors individually. We find fertilizer type to be highly significant with a P-value less than .0001, but seed hybrid is not significant and has a *P*-value of approximately .12. We conclude that: (1) fertilizer application has a significant impact on corn yield and (2) the two seed hybrids provide approximately equal corn yield, independent of the fertilizer applied.

Caution should always be exercised when drawing inferences about the main or primary factors in a factorial experiment where there are known to be significant interaction effects. For example, consider the enhanced 4×4 factorial design of a corn yield study with three fertilizer types plus a control and four seed hybrids summarized in Table 11-10. We find that there is a significant interaction effect, but neither fertilizer application nor choice of seed hybrid is significant. Note that the row and column means show very little variation, and reflect the lack of significance of these two factors. However, if we examine Table 11-10 closely we can see two anomalies. We

TABLE 11-10
Mean Corn Yields in a 4×4 Factorial Experiment

	Seed type				
Fertilizer	1	2	3	4	Mean
Control	70	72	73	72	71.75
1	68	85	73	69	73.75
2	73	68	82	74	74.25
3	72	72	68	74	71.50
Mean	70.75	74.25	74	72.25	72.8125

see that when fertilizer 1 is applied in *conjunction* with seed Hybrid 2 and when fertilizer 2 is applied in *conjunction* with seed Hybrid 3 there are abnormally high yields. These two fertilizer-seed hybrid *combinations* appear to have significantly higher yields than all other combinations. Such conclusions are consistent with both the intent and nature of multifactor experiments. We should not be dismayed at the lack of significance of our primary factors—we merely have to dig deeper to find important results. Three-factor or even more complex ANOVA designs are simple extensions of the two-factor model. With three factors A, B and C, there are three main effects (A, B, and C), and four interaction effects ($A * B$, $A * C$, $B * C$, and $A * B * C$). Problems of analysis and interpretation increase with the number of different interactions.

11.3. Multiple Comparisons Using the Scheffé Contrast

Once we have determined that the levels of a factor are significantly different, the next step in an ANOVA is to determine *which* sample means differ from the rest of the levels. To see how these pairwise experiments are made, consider the simple one-factor, completely randomized design of the corn yield experiment. From the one-factor ANOVA summarized in Table 11-5, it is known that fertilizer type has a significant impact on corn yield. The question remains—do both fertilizers provide different increases in yield, both about the same increase, or does only one fertilizer improve yield over the control mean? There are four possibilities:

- Case 1: Both fertilizers increase yield from the control plots but by different amounts.
- Case 2: Only fertilizer XLC significantly improves yields over control values.
- Case 3: Both fertilizers increase yield about equally from the control plots.
- Case 4: Only fertilizer CG100 significantly improves yield.

If we examine the sample means of $\bar{Y}_1 = 82$, $\bar{Y}_2 = 71$, and $\bar{Y}_3 = 87$, it is unclear which case applies. We can dismiss Case 2 since the yield of CG100 exceeds that of XLC. From the sample means, we can conjecture that $\mu_3 > \mu_1 > \mu_2$. To test this conjecture we use *Scheffé's contrast method.*

DEFINITION: SCHEFFÉ'S CONTRAST METHOD

If $c_1, c_2, \ldots, c_t$ are known constants such that $\Sigma_{i=1}^{t} c_i = 0$, then the sum $L = c_1\mu_1 + c_2\mu_2 + \ldots, c_t\mu_t = \Sigma_{i=1}^{t} c_i\mu_i$ is called a *contrast.*

We can test for each of the four cases described above by selecting appropriate constants c for each suggested contrast. For example, the null hypothesis $H_0{:}\mu_1 = \mu_2$ can be rewritten as $H_0{:}\mu_1 - \mu_2 = 0$, and the difference $\mu_1 - \mu_2$ is a contrast with $c_1 = 1$ and $c_2 = -1$. Note that $c_1 + c_2 = 0$ and $L = \mu_1 - \mu_2$ is the specified contrast. Depending on which values are inserted for the constants, it is possible to specify an infinite number of contrasts. The three contrasts of interest in the corn yield experiment are

Contrast	Coefficients c_1	c_2	c_3
$\mu_1 - \mu_2$	1	–1	0
$\mu_1 - \mu_3$	1	0	–1
$\mu_2 - \mu_3$	0	1	–1

If all three contrasts are found to be significant, then Case 1 applies and $\mu_1 \neq \mu_2 \neq \mu 3$. If only contrasts 1 and 2 are found to be significant, then Case 2 applies. Case 3 applies if contrasts 1 and 3 are found significant but not 2. If contrasts 2 and 3 are found to be significant, then Case 4 applies.

For each contrast, the Scheffé method produces a confidence interval estimate for L based on

$$\hat{L} - s\hat{\sigma}_L \leq \mathrm{L} \leq \hat{L} + s\hat{\sigma}_L \tag{11-16}$$

where

$$\hat{L} = c_1\bar{Y}_1 + c_2\bar{Y}_2 + \ldots + c_t\bar{Y}_t$$

$$s^2 = (t-1)\,F[1-\alpha, t-1, t(n-1)]$$

and

$$\hat{\sigma}_L^2 = \frac{\text{ESS}}{t(n-1)} \cdot \frac{1}{n}\sum_{i=1}^{t} c_i^2$$

for a balanced ANOVA. Returning to the one-way corn yield experiment ANOVA, we have $t = 3$, $(t - 1) = 2$, ESS/$t(n - 1) = 18.8$, $t(n - 1) = 15$, and $F[0.95, 2, 15] = 3.68$. Therefore, $s^2 = 2(3.68) = 7.36$ and $s = 2.71$. For the contrast $L = \mu_1 - \mu_2$,

$$\hat{L} = (1)\bar{Y}_1 + (\text{-}1)\bar{Y}_2 + (0)\bar{Y}_3 = 82 - 71 = 11$$

$$\sigma_L^2 = 18.8 \cdot \frac{1}{6}\sum_{i=1}^{t}[(1)^2 + (-1)^2 + (0)^2] = 6.27$$

$$\hat{\sigma}_L = 2.50 \quad s\hat{\sigma}_L = (2.71)(2.50) = 6.78$$

From Equation 11-16, the confidence interval for $L = \mu_1 - \mu_2$ is therefore

$$11 - 6.78 \leq L \leq 11 + 6.78 \quad \text{or} \quad [4.22, 17.78]$$

For the contrast, $L = \mu_1 - \mu_3$, we find the interval to be [–11.78, 1.78]. Note that the width of the confidence interval for both contrasts is 2(6.78) = 13.56, which is characteristic of all balanced designs. The final contrast $L = \mu_2 - \mu_3$ also has a width of 13.56 units, [–22.78, –9.22].

The two significant contrasts are $L = \mu_1 - \mu_2$ and $L = \mu_2 - \mu_3$. Neither of these contrasts contains the value of 0 in their intervals. This indicates that they both differ by some nonzero amount. Since the contrast between $L = \mu_1 - \mu_3$ includes 0, we cannot say these two treatments differ. In sum, both fertilizers increase yield significantly, since both can be distinguished from the *control* or untreated plots. However, neither is clearly superior since they cannot be distinguished from each other at $\alpha = 0.05$. Case 3 applies to this data.

Two important considerations should be kept in mind when applying Scheffé's method. First, it should only be applied *after* it has been demonstrated that significant differences between the means exist using ANOVA. If we fail to reject the hypothesis of equal means in ANOVA, Scheffé's method should not be applied as spurious contrasts may be indicated. Second, one might wonder why three simple *t*-tests for differences of means could not be used to arrive at the same result. The reason is that Scheffé's test is applicable to the *entire* set of contrasts with a confidence level of $100(1 - \alpha)$. For three separate *t*-tests, the confidence in the entire set would be $(1 - \alpha)^3$, since they are independent. In this instance, note that $(1 - 0.05)^3 = (0.95)^3 = .857$. The Scheffé method is clearly superior.

Scheffé's method can also be extended to handle two-factor or even more complex factorial designs. Each factor is analyzed separately. For the two-factor randomized design based on fertilizer and seed type, for example, we compare the set of means first for the fertilizer type and then for the seed hybrid. The only modifications to the formulas used in constructing the confidence interval are the degrees of freedom of the F-value in the computation of s^2. For a two-factor design with a levels for the first factor and b levels for the second factor, we use

$$s^2 = (a - 1)F[(1 - \alpha), (a - 1), ab(n - 1)] \quad \text{for factor A} \qquad (11\text{-}17)$$

$$s^2 = (b - 1)F[(1 - \alpha), (b - 1), ab(n - 1)] \quad \text{for factor B} \qquad (11\text{-}18)$$

11.4. Assumptions of the Analysis of Variance

The analysis of variance can be applied when four key assumptions are satisfied by the experimental data: normality, additivity, homogeneity of variances, and independence. An appropriate test can be used to verify whether or not the data input to an ANOVA procedure satisfies each of these assumptions.

Normality

The normality assumption requires that the dependent variable for the population corresponding to each treatment is normally distributed. Nonparametric goodness-of-fit tests described in Chapter 10 (both the Kolmogorov–Smirnov and the chi-square) can be used to rigorously test the normality of Y for each treatment. Given the small sample sizes in each treatment of the corn yield problem, it is virtually impossible to undertake a goodness-of-fit test. For data generated in larger experiments or for survey data with many responses, it is usual to check this assumption. If the data prove to be decidedly non-normal, various transformations can be used to try and achieve a normal dependent variable. Sometimes a logarithmic transformation can be used to ensure normality; the ANOVA test can then be undertaken on this transformed data. The normality assumption for the dependent variable Y_{ij} (or Y_{ijk} for the two-factor model) can also be expressed in terms of the error term e_{ij}. Since μ and τ_i are fixed, we can ensure the normality of Y_{ij} by assuming the error term is a random normal variable with a mean 0 and variance σ^2.

Additivity

This second assumption requires that the fixed effects in the ANOVA model, τ_i in the one-factor model and A_i and B_j, and $(AB)_{ij}$ in the two-factor model are additive. The assumption could be violated, for example, by omitting a significant interaction term in a factorial experiment, or by the presence of an outlier in the data.

Homogeneity of Variances

Each treatment in the ANOVA model must have an equal population variance. As outlined in Section 11.2, this assumption forms the basis of the relationship between the two variance estimates under the null hypothesis. The assumption of homoscedasticity or homogeneity of variance can be checked using an F-test. For example, any pair of treatments i and j in a one-factor ANOVA can be used in a test of the hypothesis:

$$H_0: \sigma_i^2 = \sigma_j^2$$
$$H_A: \sigma_i^2 \neq \sigma_j^2$$

Although all pairs of treatments could be tested in this same way, it is sufficient to test the ratio of the highest and lowest estimated treatment variances. For the one-factor corn yield experiment, the three treatment variances are $s_1^2 = 2.8$, $s_2^2 = 9.2$, and $s_3^2 = 44.4$. All have five degrees of freedom. The ratio 44.4/2.8 =15.86 is greater than F [0.95, 5, 5] = 5.05 and F [0.05, 5, 5] = 1/5.05 = 0.20. Clearly, the ratio 15.86 leads to a rejection of the assumption of homogeneity of variances in this case.

Independence

This assumption requires that the error terms e_{ij} (or e_{ijk} in the two-factor model) are statistically independent. One way in which experiments are designed to avoid violations of this assumption is through *randomization*. In the corn yield experiment, for example, the random assignment of plots to specific treatments is used to avoid error term dependence. For data gathered in uncontrolled experiments, through surveys, or taken from archival sources, it is more difficult to impose randomization procedures. An appropriate test for independence is the nonparametric runs test.

If the assumptions of the ANOVA model are not satisfied by a set of data, the interpretation of the results should be made with caution. However, it should be emphasized that the ANOVA test is quite *robust* with respect to modest departures from these four assumptions. By robust, it is meant that the analysis of variance generally makes accurate judgements about the significance of factor means even when the assumptions are not completely satisfied. If the data depart markedly from one of the assumptions, it may be possible to undertake some transformation of Y_{ij} to correct the problem. Or, as a last resort, it is possible to use the Kruskal–Wallis nonparametric ANOVA described in Chapter 10.

11.5. Summary

The analysis of variance is a parametric inferential test useful for testing hypotheses concerning the equality of a set of two or more population means. Four critical assumptions must be satisfied in order to use this test: (1) *normality* of the dependent variable *Y*, (2) *additivity* of the factor effects, (3) *homogeneity of variances,* and (4) *independence* or *randomness* of the error term. The simplest formulation is the one-factor completely randomized design. The model can be extended to include more than one factor in a *factorial experiment.* Once it has been determined that the means for the levels of a factor differ significantly, then Scheffé's contrast method can be used to identify which levels of the factor differ from one another. Although the analysis of variance has been presented as an extension of the two-sample difference of means test, it also bears a close relationship to several other linear statistical models. These models are used to explore the relationships among several different variables. When the corn yield experiment is seen in this way, the purpose is to examine the relationship between corn yield and fertilizer type and seed hybrid. Though corn yield is an interval variable, both other variables are measured at the *nominal* level. Regression analysis, introduced in the following two chapters, is a useful technique for relating the behavior of one interval variable to one or more variables measured at the *interval* scale.

Appendix 11a. Derivation of Equation 11-11 from Equation 11-10

From Equation 11-10 we know that

$$(Y_{ij} - \bar{Y}) = (\bar{Y}_i - \bar{Y}) + (Y_{ij} - \bar{Y}_i)$$

Squaring both sides of this identity and summing over i and j [note we use the identity $(a + b)^2 = a^2 + 2ab + b^2$] for the right-hand side:

$$\sum_{i=1}^{t}\sum_{j=1}^{n}(Y_{ij} - \bar{Y})^2 = \sum_{i=1}^{t}\sum_{j=1}^{n}(\bar{Y}_i - \bar{Y})^2 + 2\sum_{i=1}^{t}\sum_{j=1}^{n}(\bar{Y}_i - \bar{Y})(Y_{ij} - \bar{Y}_i) + \sum_{i=1}^{t}\sum_{j=1}^{n}(Y_{ij} - \bar{Y}_i)^2$$

The middle or cross-product term on the right-hand-side of this equation can be rewritten as

$$2\sum_{i=1}^{t}[(\bar{Y}_i - \bar{Y})\sum_{j=1}^{n}(Y_{ij} - \bar{Y}_i)]$$

since the first term does not enter the summation over j. One of the properties of the mean is that the sum of the deviations about the mean is always 0. Therefore, the cross-product term must be 0 since the last part of this cross-product term is the sum of the deviations of Y_{ij} around their mean $\bar{Y}_i$. This means we can now reexpress (11-10) as

$$\sum_{i=1}^{t}\sum_{j=1}^{n}(Y_{ij}-\bar{Y})^2 = \sum_{i=1}^{t}\sum_{j=1}^{n}(\bar{Y}_i-\bar{Y})^2 + \sum_{i=1}^{t}\sum_{j=1}^{n}(Y_{ij}-\bar{Y}_i)^2$$

Now let us examine the first term on the right-hand side of this new equation. Since the expression $\Sigma_{i=1}^{t}(\bar{Y}_i - \bar{Y})$ is independent of the summation over j, the summation causes the addition of the quantity n times. This means we can rewrite (11-10) as

$$\sum_{i=1}^{t}\sum_{j=1}^{n}(Y_{ij}-\bar{Y})^2 = n\sum_{i=1}^{t}(\bar{Y}_i-\bar{Y})^2 + \sum_{i=1}^{t}\sum_{j=1}^{n}(Y_{ij}-\bar{Y})^2$$

FURTHER READING

For another presentation of the basic one- and two-way ANOVA, students may wish to consult the textbook on elementary statistics by Allan Bluman (2007). Advanced topics in this area, including the development of MANOVA or Multiple ANalysis of VAriance, are clearly explained in Hair et al. (2004). In MANOVA, there are two or more dependent variables to be explained by two or more independent treatments. Three examples of the use of ANOVA in contemporary research include a study of forestation, stream sediment, and fish abundance by Sutherland et al. (2002); river channel type and in-channel river characteristics by Burge (2004); and seed size and seed depth on germination and seedling emergence by Chen and Maun (1999).

A. Bluman, *Elementary Statistics: A Step by Step Approach, 6th ed.* (New York: McGraw-Hill, 2007).

L. M. Burge, "Testing Links between River Patterns and In-Channel Characteristics Using MRPP and ANOVA," *Geomorphology* 63 (2004), 115–130.

H. Chen and M. A. Maun, "Effects of Sand Burial Depth on Seedling Germination and Seedling Emergence of *Cirsiu pitcheri*," *Plant Ecology* 140 (1999), 53–60.

J. F. Hair, Jr., Willeam C. Black, Barry J. Babin, Rolph E. Anderson and Ronald L. Tatham *Multivariate Data Analysis, 6th Edition* (Upper Saddle River: Prentice Hall, 2006).

A. B. Sutherland, Judy L. Meyer and Edward P. Gardiner (2002) "Effects of Land Cover on Sediment Regime and Fish Assemblage Structure in Four Southern Appalachian Streams," *Freshwater Biology* 47 (2002),1791–1805.

PROBLEMS

1. Explain the meaning of the following terms:
   - One-factor, completely randomized design
   - Homogeneity of variance
   - Mean Square
   - Scheffé's contrast method
   - Interaction
   - Population census
   - Treatments
   - Pooled estimate of the variance
   - Contrast
   - Factorial experiment
   - ANOVA table
   - Secondary data

2. Differentiate between the following:
   a. A one-way (or one-factor) and a two-way ANOVA
   b. A balanced and an unbalanced design
   c. Between sum of squares and error sum of squares

3. Explain the meaning of the assumptions of the ANOVA model.

4. An urban geographer interested in the relationship between crime rates and city size randomly selects eight cities in each of three city size categories and notes the number of assaults per month per 1,000 population. The data are summarized in the following table:

Small cities (<99,999)	Medium cities (100,000–499,999)	Large cities (>500,000)
27	74	146
58	86	92
64	71	42
31	118	166
74	91	108
63	76	100
93	72	118
21	28	73

   a. Calculate the ANOVA table for this problem
   b. Test the hypothesis of equal crime rates at $\alpha = 0.05$
   c. Use Scheffé's method to determine which sample means are different.

5. Noise levels in decibels are taken 100 meters behind noise barriers that have been placed parallel to a high-volume urban expressway. In order to test the effectiveness of the height of the barrier and the material used in its construction, a two-factor experiment is undertaken. The 90th percentile decibel levels for a representative day and are shown in the following table:

	Construction material		
Barrier height	Earth Berm	Wood	Concrete
Low (2 meters)	82 86 90	92 96 100	94 90 86
High (4 meters)	72 76 80	82 86 90	78 82 86

Note that there are three observations for each treatment.

   a. Calculate the 2-factor ANOVA table for this problem.
   b. Calculate the two one-factor ANOVA table for this problem.
   c. Are there significant interaction effects?

6. A cartographer wishes to test the ability of different individuals to use maps by determining the amount of time it takes them to extract information from a standard topographic map. She then classifies them by the discipline of their academic major into four categories:

Geography, Engineering, Humanities, or Physical Sciences. The results of the test expressed in seconds are as follows:

Major			
Geography	Engineering	Humanities	Sciences
32	80	34	90
27	72	24	78
65	30	58	83
60	42	74	49

a. Test the hypothesis of equal test times at $\alpha = .05$.
b. Use the Scheffé method to determine which majors have significantly different times.

7. Suppose that the city observations in Problem 4 come from four different regions of the country. Within each city-size category, the first two observations come from the North, the next pair from the South, the next two from the East, and the last two from the West. Use a two-way ANOVA to test for differences in crime rates in regions and by city size.

8. A scaling technique is used to estimate the familiarity of a set of neighborhood residents with a set of randomly selected retail stores in a city. Familiarity is given a score from 1 to 5. The retail stores are classified according to the sector in which they lie in relation to the neighborhood. The goal is to ascertain the possible directional bias in familiarity of the residents. The scores are:

North	South	East	West
3.2	2.7	2.4	4.0
3.0	2.6	2.9	4.0
3.4	2.5	2.8	3.6
3.2	2.6	2.3	3.6

a. Test the hypothesis that store familiarity is not directionally biased using ANOVA.
b. Use the Scheffé method to determine which directions have significantly different levels of familiarity.

12

Inferential Aspects of Linear Regression

In Chapter 4, we introduced the notions of *correlation* and *regression* in the description of the relationship between two variables. We showed how it was possible to measure the correlation or association between any two variables X and Y using Pearson's product–moment correlation coefficient r based on the concept of covariance. Moreover, we found we could fit linear functions known as regression lines to two-variable scatter plots using the technique of least squares analysis. In the next two chapters we discuss two important extensions to this framework. In this chapter, we explain the inferential aspects of these two tools, and in Chapter 13 we discuss the construction of regression models with multiple independent variables. We first present an overview of the model-building process, outlining the context in which to view the logical steps of regression and correlation analysis.

12.1. Overview of the Steps in a Regression Analysis

Figure 12-1 illustrates the overall strategy for building a regression model. It is convenient to divide the model-building process into three phases:

1. Data collection and preparation
2. Model construction and refinement
3. Model validation and prediction

Let us consider each of these phases in turn.

Data Collection and Preparation

Normally, we undertake our correlation and regression analysis in the context of some overall research task that begins with the collection of data. The particular data collection requirements depend on the nature of the study we are undertaking. Our

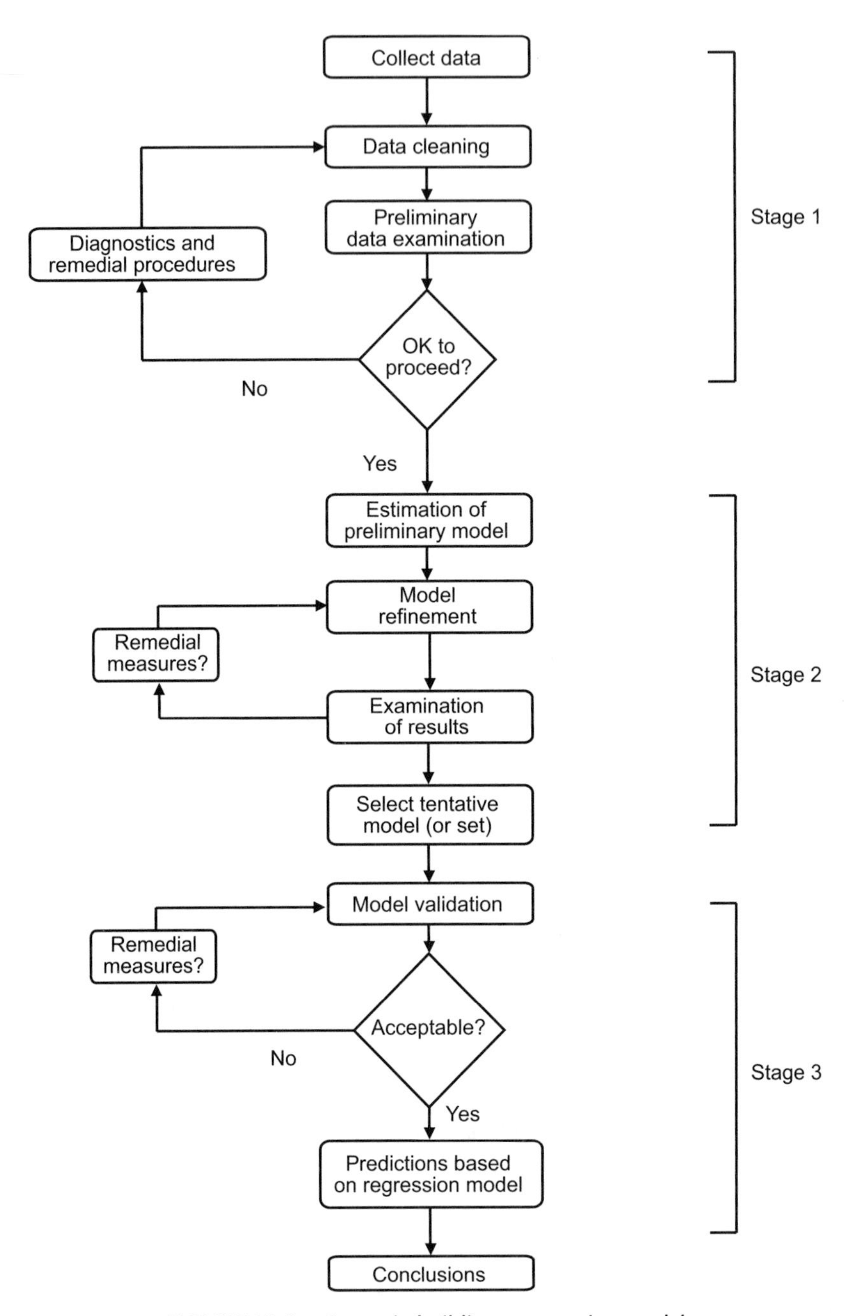

FIGURE 12-1. Stages in building a regression model.

research topic might be described as a *controlled experiment* in which we manipulate a number of explanatory variables and study their effects on a response or dependent variable. In our study of crop yields in Chapter 11, for example, we were interested in examining the role of seed hybrid and amount of fertilizer on crop yields. Data collection for such statistical experiments is generally straightforward, though by no means simple. There may be difficult measurement problems as well as the issue of determining the exact levels and combination of levels of our independent variables to be used in the experiment. If our study is based on *observational* or *survey* data, we must collect data for explanatory variables based on prior knowledge and previous studies as well as new variables that we now feel may be applicable or important to the study. Normally, these studies are undertaken to confirm (or not to confirm as the case may be) hypotheses derived from theoretical models. The variables are not controlled as in an experimental study, though sometimes we are able to ensure that our data include sufficient numbers of observations on particularly important explanatory variables. For example, if we are studying shopping patterns, we may wish to include gender if it is shown to be an important variable explaining individual shopping trip behavior.

Sometimes there is inadequate knowledge about a particular research topic, and the data collection phase of the study might be described as *exploratory* in nature. That is, we may be searching for explanatory variables and include a large number of potential variables in our study. Or one of the variables of interest might be loosely defined, and we may use several different measures of the general concept in our search for a useful variable for our model. For example, we may have different measures of income such as family or household, individual, before tax, or income net of tax. We may be able to obtain data on these variables from different sources, and each may have particular advantages and disadvantages. At other times we may be seeking out a *proxy* variable for a variable that is part of a theoretical model but not directly measurable. For example, any of our measures of income might be used as a proxy for the theoretical income stream of a household in the next 10 years. In a sense, we are prospecting. It is clear that the set of variables can be large, and we may be interested in ensuring that we also obtain data on many different combinations of these variables. It could be important to obtain observations not only on households with low, middle, or high incomes, but also households of varying sizes and household composition in each income class.

Once our data have been collected, many different editing checks and plots can be analyzed to identify pure errors but also data outliers. The investigator should always try to minimize data errors. This often requires the investigator to closely examine an extremely large number of histograms, boxplots, and scatterplots. If we are interested in undertaking inferential tests on our regression equation or our correlation coefficients, we may also find it useful to see whether or not we satisfy the assumptions of the model. We will return to this topic later in this chapter.

Notice in Figure 12-1 that we sometimes find results that force us to collect new or improved data (if this is possible), and we don't proceed to the next major phase of model building until these tasks are complete.

Model Construction and Refinement

At this stage in the model-building process, we may evaluate one or more tentative regression models and perform any number of diagnostic tests on each equation. We may find that some of these models violate key assumptions of regression or correlation analysis and we must take remedial steps to correct these problems. The key assumptions of regression analysis, are detailed in Section 12.2, and various graphical diagnostics are described in Section 12.4. In some cases, *more than one* independent variable is related to the dependent variable, or a nonlinear function is more appropriate for the systematic part of the relationship. These two issues are explored in Chapter 13. The model itself may not be completely specified until the data have been collected. If we are unsure of the nature of the function linking the two variables, we might explore different functional forms using the available empirical data. To do this we undertake *transformations* of our variables.

To *estimate* the parameters of the regression model, we use the least squares procedure. This procedure has been fully explained in Chapter 4 for the case of simple regression, and it is explored further in Chapter 13 for the case of multiple regression. Throughout the model-building process, we may be continually fitting different regression models, using the results of one equation to make subsequent improvements in our model. We may find that we can identify a single best model, or we might find several candidate models that are more or less equally useful. In this phase of model building, we are often faced with the task of making inferences about the parameters of our regression model. In parametric statistics, we used the sample mean $\bar{X}$ to estimate the population mean μ, and we developed a statistical test to determine whether or not a particular sample could have come from a population with a particular parameter value. Later in this chapter we will see that we can make similar inferences about the slope and intercept of our calibrated regression model.

Figure 12-1 shows that this component of model building is also not necessarily a simple set of tasks that unambiguously lead to a single model. At times, the results of fitting one model may suggest we need more data or that we have to undertake some remedial actions based on the diagnostic tests we apply. For example, we may find we have outliers that are extremely influential and may be providing suspicious results. This stage in model building and refinement may be a lengthy exercise in which many options are explored, some discarded, and we use large doses of pragmatic judgment as we assess our results. At the end of this phase of model building, we hope to identify a single model or a series of candidate models that we can explore further in the final phase of our analysis: *validation.*

Model Validation and Prediction

Model *validity* refers to the stability and reasonableness of the regression models we have created. Are the regression coefficients plausible? Is it possible to generalize the results of our analysis to other situations or places? Sometimes we can compare our results to theoretical arguments or to previous empirical studies. If our results tend to support theoretical expectations, or if they are in agreement with similar studies, this tends to support the notion that we have created a valid regression model. If not, we are less likely to believe our model is a valid model for the data.

Sometimes there may be no studies that exactly parallel ours, and it is difficult to validate our model in this way. We can validate the model in two other ways:

1. Collect a new set of data and check the model and its predictive ability on this new set.
2. Use a *holdout* sample to check the model and its predictive ability.

By far the best technique to validate our model is to collect new data and test the ability of the model to predict the values of our dependent variable for this data. To do this, take the simple regression equation, substitute in the value of X from an observation in the new dataset, calculate $\hat{Y}$, and then compare it to the actual observed value of Y for the new data. If our model is reasonable, our predicted values of $\hat{Y}$ should be close to the observed values of Y in the new set.

Unfortunately, this technique is rarely feasible because it would involve another lengthy sampling and data collection exercise. An alternative approach is to split our data beforehand into two parts. The first portion, or *training* set, is used to develop our regression model. The second portion of the dataset is called the *validation* or *prediction* set and is used to evaluate the reasonableness of the developed model. This type of analysis is often called *cross-validation.* We can also reverse the process by developing a model using the second dataset and validating it on the first dataset. When both procedures are used, we are performing a double cross-validation. In Section 13.3 we discuss the specific ways in which this comparison can be made.

Once it is developed, we often wish to use the regression equation to *predict* values of the dependent variable. Normally, confidence intervals are used to establish a range within which the dependent variable can be predicted at any desired level of confidence. This issue is explored in detail in Section 12.3 and again in Chapter 13.

Whenever a regression analysis is undertaken, it may not necessarily include all of these steps. Sometimes we may not be interested in prediction at all, and the ultimate goal is only to evaluate the strength of the empirical evidence in support of some prespecified theoretical model. Other times, the overriding goal may be to use the model for prediction, and interest in hypothesis testing may be secondary. For example, we might wish to develop a regression equation using data collected for one period, and then evaluate its predictive capabilities using data from another period. Quite often both issues are important. The remainder of this chapter explores the computational techniques and issues involved in both hypothesis testing and prediction. First, however, it is necessary to enumerate the assumptions that are necessary in order to make valid inferences in regression analysis.

12.2. Assumptions of the Simple Linear Regression Model

In order to fully understand the inferential framework for regression analysis, it is essential to consider the specification of the simple linear regression function in a more formal way.

Linear Regression Model

A linear regression model formally expresses the two components of a statistical relationship. First, the dependent variable varies systematically with the independent variable or variables. Second, there is a scattering of the observations around this systematic component. Although the systematic component can take on any form and we can have as many variables as is necessary, we restrict ourselves to the simple two-variable regression model. We conceptualize the problem in the following way.

FIXED-*X* MODEL

Suppose we are able to control the value of variable X at some particular value, say at $X = X_1$. At $X = X_1$, we have a distribution of Y values and a mean of Y for this value of X. There are similar distributions of Y and mean responses at each other fixed value of X, $X = X_2$, $X = X_3$, . . . , $X = X_M$. The regression relationship of Y on X is derived by tracing the path of the mean value of Y for these fixed values of X. Figure 12-2 illustrates a regression relation for such a fixed-X model. Although the regression function that describes the systematic part of this statistical relation may be of any form whatsoever, we will restrict ourselves to situations where the path of the means is linear. More complicated nonlinear functions are described in Section 13.2. The function that passes through the means has the equation

$$Y_i = a + bX_i \tag{12-1}$$

which is the familiar slope-intercept form of a straight line.

EXAMPLE 12-1. To examine the rate of noise attenuation with distance from an urban expressway, sound recorders are placed at intervals of 50 meters over a range of 50 to 300 meters from the centerline of an expressway. Sound recordings are then analyzed, and the sound level in decibels, which is exceeded 10% of the day, is calculated. This measure, known as L10, is a general indicator of the highest typical sound levels at a location. Recordings are then repeated for five consecutive weekdays using the same sound recorders placed at the identical locations. A plot of the data generated from this experiment is shown in Figure 12-3.

The graph reveals the typical pattern of distance decay. The variation in decibel levels at any fixed location either is due to the effects of other variables—perhaps variations in traffic volumes or relative elevation—or can be thought of as random error. Let us presume that all other variables that might have caused sound-level variations are more or less constant and that the variations do represent random error. Note that the path of the mean sound levels with distance seems to follow the linear function superimposed on the scattergram. One would feel quite justified in fitting a linear regression equation to this scatter of points.

RANDOM-*X* MODEL

If situations in which we can control the values of X so that they are set at predetermined levels were the only conditions in which regression analysis could be applied,

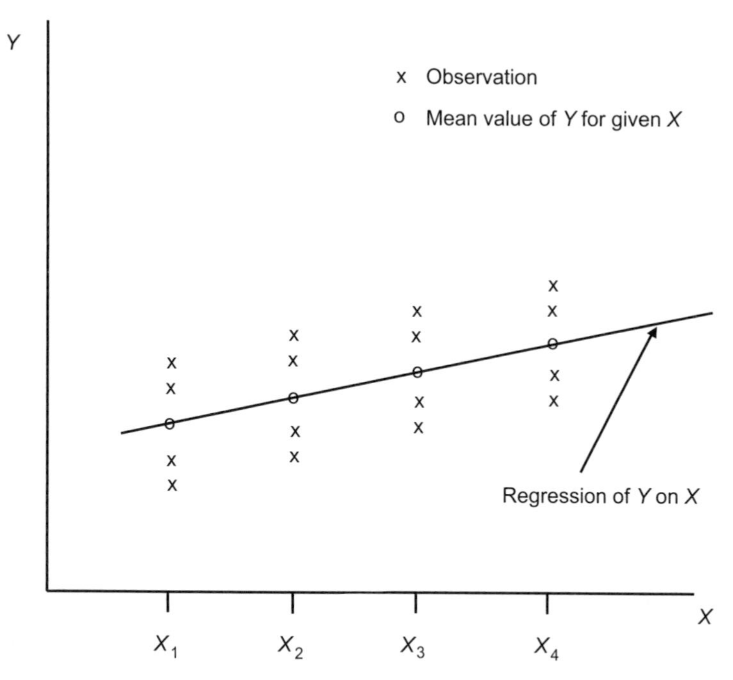

FIGURE 12-2. The fixed-X regression model.

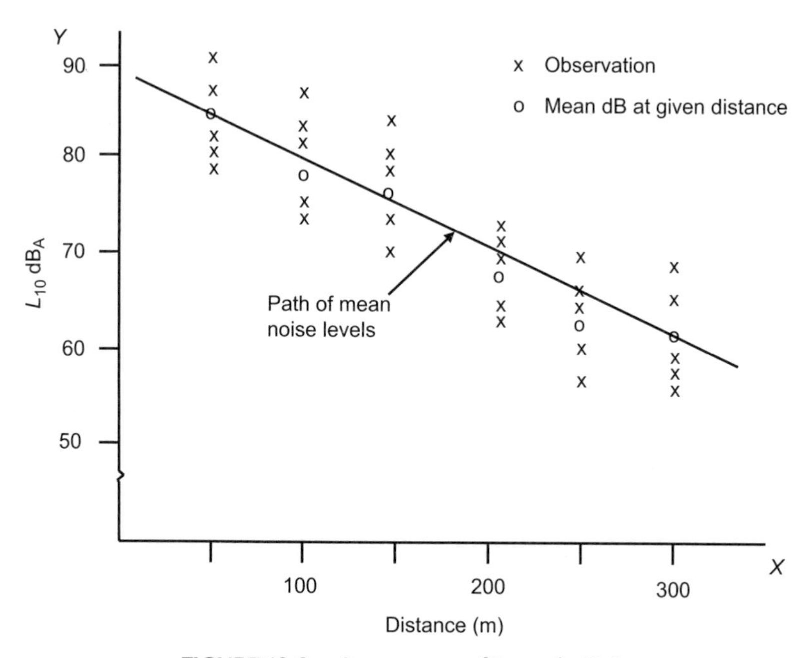

FIGURE 12-3. Scattergram of Example 12-1.

it would be of limited value in geographical research. Only rarely do geographers gather data of this form. They are more commonly collected in experimental situations. All the mechanical operations presented in this chapter can be performed on any set of paired observations of two variables X and Y. Virtually all the inferential results of regression analysis apply to data characteristics of the fixed-X model and to data in which both X and Y can be considered random variables. This greatly generalizes the applicability of the regression model. Let us consider an example of a random-X model.

EXAMPLE 12-2. One of the key tasks in metropolitan area transportation planning is to estimate the total number of trips made by households during a typical day. Data collected from individual households are usually aggregated to the level of the traffic zone. Traffic zones are small subareas of the city with relatively homogeneous characteristics. The dependent variable is the number of trips made by a household per day, known as the *trip generation rate.* One of the most important determinants of variations in household trip generation rates is household income. We would expect the total number of trips made by a household in a day to be positively related to the income of the household.

Consider the hypothetical city of Figure 12-4 which has been divided into 12 traffic zones. Values of the average trip generation rates and household income for each of 12 zones are listed in Table 12-1. The data are typical for a North American city. Inner-city traffic zones 1 to 4 are composed of lower-income households that make few trips per day. Suburban zones 5 to 12 are populated with higher-income households with higher trip generation rates. The scattergram of Figure 12-5 verifies the positive relation between trip generation and household income. Notice how these data differ from the data of a fixed-X model. First, we do not have more than one observation of trip generation rates for any single level of household income. Second, we do not have observations of trip generation rates for systematic values of household income, say every $1,000 or $5,000 of income. Nevertheless, it appears that the simple linear regression of household trip generation on income represents a useful statistical relation. We shall use this example throughout this chapter as well as in Chapter 13.

But first let us examine the nature of these data. In what sense can we think of variable X as being a random variable? We might think of household income as a random variable since it is derived from one particular spatial aggregation of households into traffic zones. The traffic zones of Figure 12-4 can be thought of as one random choice among all the possible ways in which we might build traffic zones from the household data. Other aggregations to different traffic zones would yield different average incomes and different trip generation rates and therefore potentially different regression equations. This is not a very persuasive argument. We would think of income as being a random variable if the original observations from which we generated Table 12-1 were a random sample taken from all the households in the city. If this were the case, then we might have some justification in treating income as a random variable. But what if the aggregated data represent all the households in the city? Can we make a case for treating income as a random variable? As you see, many

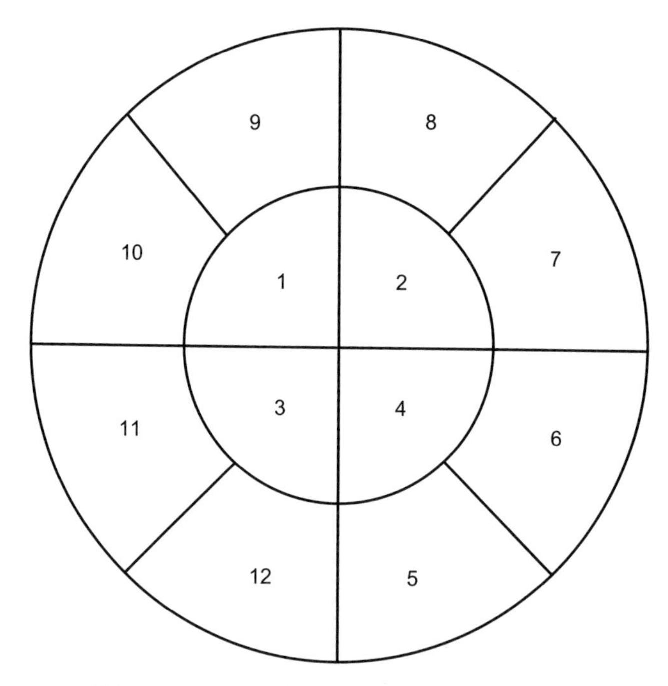

FIGURE 12-4. Traffic zones in a hypothetical city.

TABLE 12-1
Trip Generation Rates and Household Income in a Hypothetical City

Traffic zone	Trips per household per day	Average household income ($000)
1	3	30
2	5	36
3	5	33
4	4	27
5	6	42
6	7	48
7	9	63
8	7	54
9	8	66
10	8	72
11	6	51
12	5	45

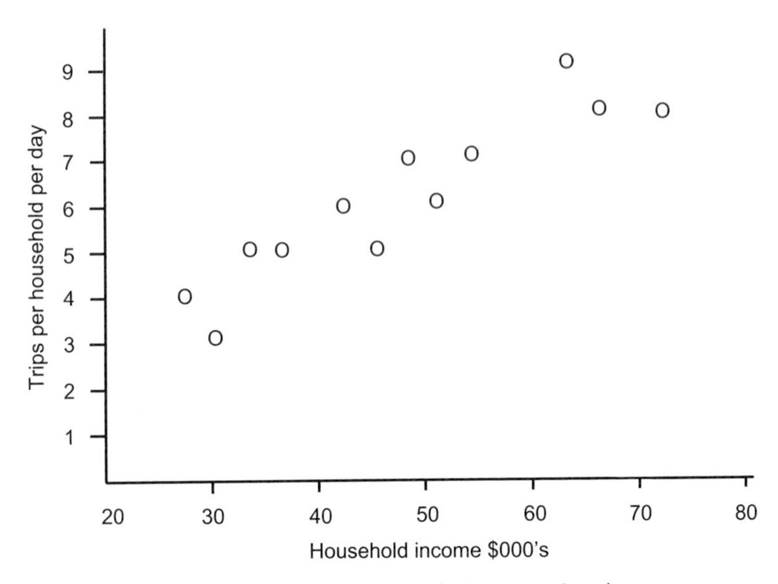

FIGURE 12-5. Scattergram of trip generation data.

applications of regression analysis in geography raise thorny theoretical issues. Most of these issues arise in situations in which we utilize regression analysis in an inferential mode and are discussed in more detail in Section 12.3.

No matter which model is used to derive the sample data for a regression problem, the method of least squares is used to derive the two parameters of the linear equation. To extend the analysis to include inferential statistics, it is necessary to make several assumptions about the parent population from which the sample data are drawn. To clarify this notion, it is useful to examine Figure 12-6. The true (population) regression line of Y as a linear function of X is the heavy line. The equation of this line is

$$Y = \alpha + \beta X \tag{12-2}$$

where the convention of utilizing Greek letters for population parameters has been followed. Depending on the particular sample drawn from the population of X and Y values, we may estimate this true regression by any of a possibly infinite number of sample regression lines. Each of these equations is expressed as $Y = a + bX$, where a and b estimate (in the statistical sense) the two population parameters α and β, respectively. If we wish to make inferences about this true population regression line, then additional assumptions are required. These assumptions are identical, with one exception, for both the fixed-X and random-X models. For present purposes, the discussion of these assumptions is couched in the context of the fixed-X model.

Formally, the simple linear regression model for the fixed-X case

$$Y_i = \alpha + \beta X_i + \varepsilon_i \quad i = 1, 2, \ldots, M \tag{12-3}$$

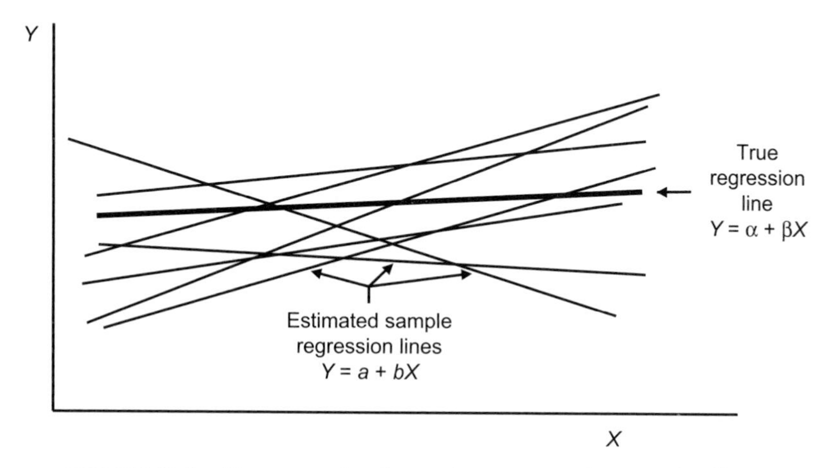

FIGURE 12-6. True regression line and several sample regression lines.

where Y_i is the value of the dependent random variable, X_i is the value of the ith level of the fixed variable X, and ε is the random error or disturbance. In this specification, the subscript i denotes not a particular observation but one of the levels at which the independent variable is set in an experiment. The graphical form of the fixed-X model is seen in Figure 12-7. For each level of X, $X_1, X_2, \ldots, X_M$, there is a probability distribution for the random variable Y. The mean value of the dependent variable Y for $X = X_1$ is given by $Y_I = \alpha + \beta X_1$. To this systematic component of Y we must add the value of the random component ε_i. It is this random error component that makes

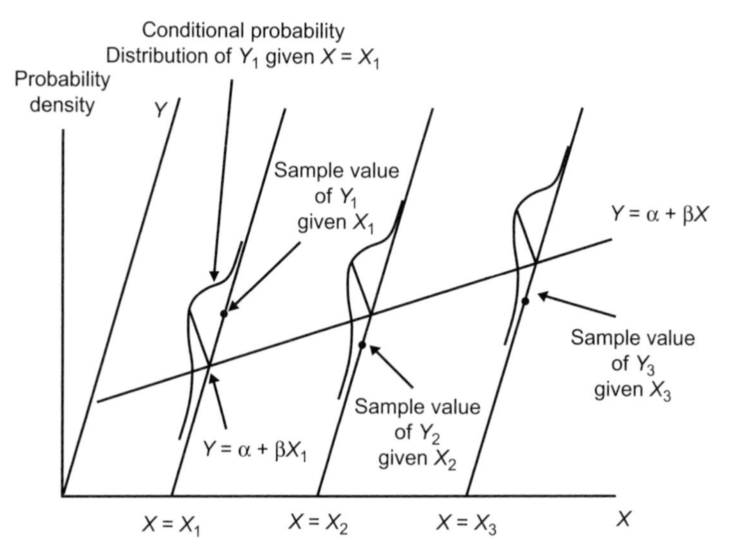

FIGURE 12-7. The fixed-X model.

the dependent variable Y a random variable—even when the independent variable X is fixed.

What is the source of this random error component? The error may be conceived as being derived from two sources. First, the relationship may be imperfect owing to *measurement* error. This source of error should be minimized. The other source of error is termed *stochastic* error. What might this stochastic error represent? Well, the number of trips a household makes is not affected just by household income. For example, it might also be related to the composition of the household, including both the ages and genders of all household members. Individuals might also be less inclined to make vehicular trips for personal reasons, say because of an avid interest in environmental concerns. However, there may also be household members who enjoy making vehicular trips, even short trips that others might make as a pedestrian. If several cars are available in the household, the ability to make vehicular trips is also enhanced. The accessibility of public transit might also encourage or deter trips. Taken together, these effects plus a host of other factors might each have a small influence on household trip making. When these individual factors are aggregated together, the net effect might actually appear to be random. As we shall see in Chapter 13, we can incorporate several independent variables into the regression equation. However, some may have relatively small effects or be otherwise impossible to include in our model and may be best treated as part of the random or stochastic error.

Figure 12-7 represents a crucial diagram for understanding the inferential aspects of regression analysis. Notice that the vertical axis is labeled *probability density* since the dependent variable Y is a random variable. Moreover, associated with each level of X is a conditional probability distribution of the random variable Y. It is important to distinguish between this unobserved population and the observed sample observations on which the regression equation $Y_i = \alpha + \beta X_i$ is estimated. One such sample value is illustrated for each value of X. In fact, there may be more than one sample observation for each value of X. Normally, the experiment would be designed so that a number of different observations of Y are obtained for each value of X. However, even if there are a large number of observations for each value of X and even if the values of X are closely spaced, it would be very fortunate if the estimated regression line $Y = a + bX$ exactly coincided with this true regression equation. By making a few assumptions about the regularity of the distribution of the random error ε_i (or equivalently Y_i), it is possible to make statistical inferences about $Y = \alpha + \beta X$ on the basis of the sample information $Y = a + bX$. There are four essential assumptions.

ASSUMPTION 1

For the ith level of the independent variable X_i the expected value of the error component ε_i is equal to zero. This is usually expressed as $E(\varepsilon_i) = 0$ for i = 1, 2, . . . , M.

This first assumption states that, on the average, the random error component of Y_i is zero. If there are a series of sample observations of Y_i at some particular value of X, say $X = X_i$, then some of these observations should lie above the regression line and

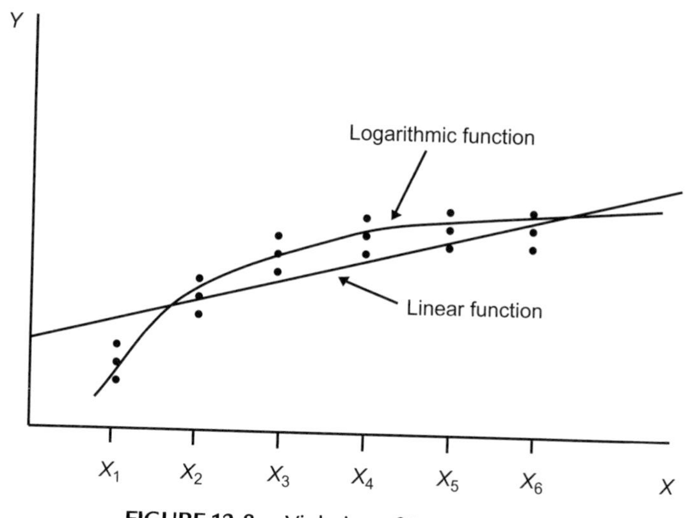

FIGURE 12-8. Violation of Assumption 1.

others below it. Recall that the regression line passes through the mean value of Y_i. To show this, note that $E(Y_i) = E(\alpha + \beta X_i + \varepsilon_i)$. Since α and β and X_i are constants, $E(Y_i) = \alpha + \beta X_i + E(\varepsilon_i)$. By this first assumption $E(\varepsilon_i) = 0$ and thus $E(Y_i) = \alpha + \beta X_i$.

How can this assumption be violated? Suppose we were to fit a linear regression model to two variables that are actually related by a nonlinear function. In Figure 12-8, for example, a linear equation $Y = \alpha + \beta X$ is fit to a set of data. Note that Assumption 1 is not satisfied since the equation does not appear to follow the path of the means. In this instance, if a logarithmic equation of the form $Y = \alpha + \beta \log X$ is fit to the data, it appears that Assumption 1 could be satisfied. But what if the exact form of the true regression equation that links the two variables is unknown? It may not be immediately apparent that the assumption has been violated. An examination of the residual errors from the estimated equation $Y = a + bX$ may uncover certain violations. In Figure 12-8, for example, the residuals from the line $Y = a + bX$ show a definite pattern and do not appear random at all. On the average, the Y-values for $X = X_1$ are below the line, but the Y-values for all intermediate values of X, X_3, X_4, and X_5, are above the line. This is clearly not random and is especially apparent when we examine the residuals from the linear equation. Virtually all residuals for intermediate values of X are positive. The examination of residuals is discussed in detail in Section 12.4, and nonlinear models are briefly introduced in Chapter 13.

For Assumption 1 to hold, the regression model must be correct; that is, it must represent the true underlying process. Once the correct specification is formally assumed, estimating the model parameters becomes a relatively straightforward mechanical procedure. In reality, however, we can never be sure that the model is correctly specified. Two types of problems exist. First, we may have omitted relevant variables from the regression equation. If so, the random error term ε_i cannot be considered random since it contains a systematic component. Second, the wrong functional form may have been chosen. Applied researchers usually examine more than one possible specification in an attempt to find the single function that best describes the process. A detailed examination of the residuals is one of the best ways to make sound judgments concerning model specification. Some simple graphical diagnostics are discussed in Section 12.4.

ASSUMPTION 2

The variance of the error component ε_i is constant for all levels of X ; that is, $V(\varepsilon_i) = \sigma^2_{Y \cdot X}$ for $i = 1, 2, \ldots, M$.

The second assumption also concerns the error term ε_i. The variance of the probability distribution of ε_i is constant for all values of X, each X_i in the fixed-X model. Note that this is not the variance of Y, but rather is the variance of Y at some X. That is, $\sigma^2_{Y \cdot X}$ measures the variability of Y around the regression line, not the variability around the overall mean. This variance is unknown, but it can be estimated from the sample residuals. This is known as the assumption of *homoscedasticity*, or simply *equal variance*. Since $E(\varepsilon_i = 0)$ and $V(\varepsilon_i) = \sigma^2_{Y \cdot X}$, it is possible to specify the variance of the conditional probability distribution of Y_i or $V(Y_i)$. Note that $V(Y_i) = V(\alpha + \beta X_i + \varepsilon_i)$. Since $\alpha + \beta X_i$ is a constant, it has no variance and can be omitted. Simplifying, we see that $V(Y_i) = V(\varepsilon_i) = \sigma^2_{Y \cdot X}$. The probability distributions of Y_i and ε_i differ only by their means. Violations of the assumption of homoscedasticity are also usually uncovered through the analysis of residuals (see Section 12.4).

ASSUMPTION 3

The values of the error component for any two ε_i and ε_j are pairwise uncorrelated.

By this assumption, the outcome Y_i for any given value of X, say $X = X_i$ neither affects nor is affected by the outcome Y_i for any other value of X, say $X = X_j$. If this assumption is not satisfied, then *autocorrelation* is present in the error term. Autocorrelation can exist in many forms. If our data represent geographic areas and can be mapped, we would expect no patterns in the errors to exist. If all of our positive errors were in the western part of a region and all the negative residuals in the eastern part of the study region, we would be in violation of this assumption. Our errors would reveal *spatial autocorrelation*. If our observations represented data from different times, and the pattern of errors showed a trend or else were regularly cyclical, our errors would exhibit *serial autocorrelation*.

ASSUMPTION 4

The error components ε_i are normally distributed.

This last assumption completes the specification of the error term in the fixed-X regression model. It *is not required* for calculating point estimates of α, β, and σ, nor is it required to make point estimates of the mean value of Y for any given X_i, $\hat{Y}_i$. It *is required* for the construction of confidence intervals and/or tests of hypotheses concerning these parameters. This assumption should not be surprising since virtually all the confidence interval and hypothesis-testing methods described in Chapters 7 and 8 are based on the assumption of the normality of the population distribution. The verification of this assumption is also based on an examination of the residuals from the sample regression line $Y = a + bx$. A simple plot to verify this assumption is presented in Section 12.4.

For a random-X model, this assumption must be augmented with an additional constraint governing the probability distribution of the random variable X. One of the most important members of the class of random-X models assumes that the distribution of X and Y is bivariate normal. Data conforming to the random-X model might be realized by a sampling procedure in which individuals, households, industrial firms, zones, regions, or cities are randomly selected as observations from a statistical population. If the underlying probability distribution of the two variables is bivariate normal (see Figure 5-14), then the standard inferential tests described in Section 12-3 are still valid. Often, however, the data used by geographers in a regression model do not conform to either the fixed-X or the random-X model. The data for trip generation Example 12-2 illustrates this point. The validity of the use of the standard inferential tests in this situation is unclear. Sometimes other justifications for their use are given.

Gauss–Markov Theorem

The principal analytical results of regression theory derive from the Gauss–Markov theorem. This theorem justifies use of the least squares procedure for estimating the two parameters of the linear regression equation. Also, it establishes the principal properties of the least squares estimators a and b.

GAUSS–MARKOV THEOREM

Given the first three assumptions of the simple linear regression model, the least squares estimators a and b are unbiased and have the minimum variance among all the linear unbiased estimators of α and β.

To examine the implications of this theorem, consider the sampling procedure depicted in Figure 12-9. From some statistical population, repeated random samples of some given size, say $n = 50$, are drawn, and least squares estimators a and b are calculated. If the four assumptions are satisfied, then several characteristics of the sampling distributions of a and b follow. First, from the Gauss–Markov theorem estimators a and b are unbiased estimators of α and β. In Figure 12-9 the sampling distributions of a and b shown at the bottom of the figure are centered on $E(a) = \alpha$ and $E(b) = \beta$ respectively. Second, these sampling distributions are themselves normal. Finally, and most important, a and b are the most efficient estimators in the sense that the sampling distributions of a and b have less variability (smaller variances) than any other linear and unbiased estimators of α and β. As we can now see, the framework for statistical testing in regression and correlation analysis closely parallels the framework introduced for one- and two-sample parametric tests. Statistical tests on various aspects of the fitted regression equation $Y = a + bX$ follow directly from the Gauss–Markov theorem.

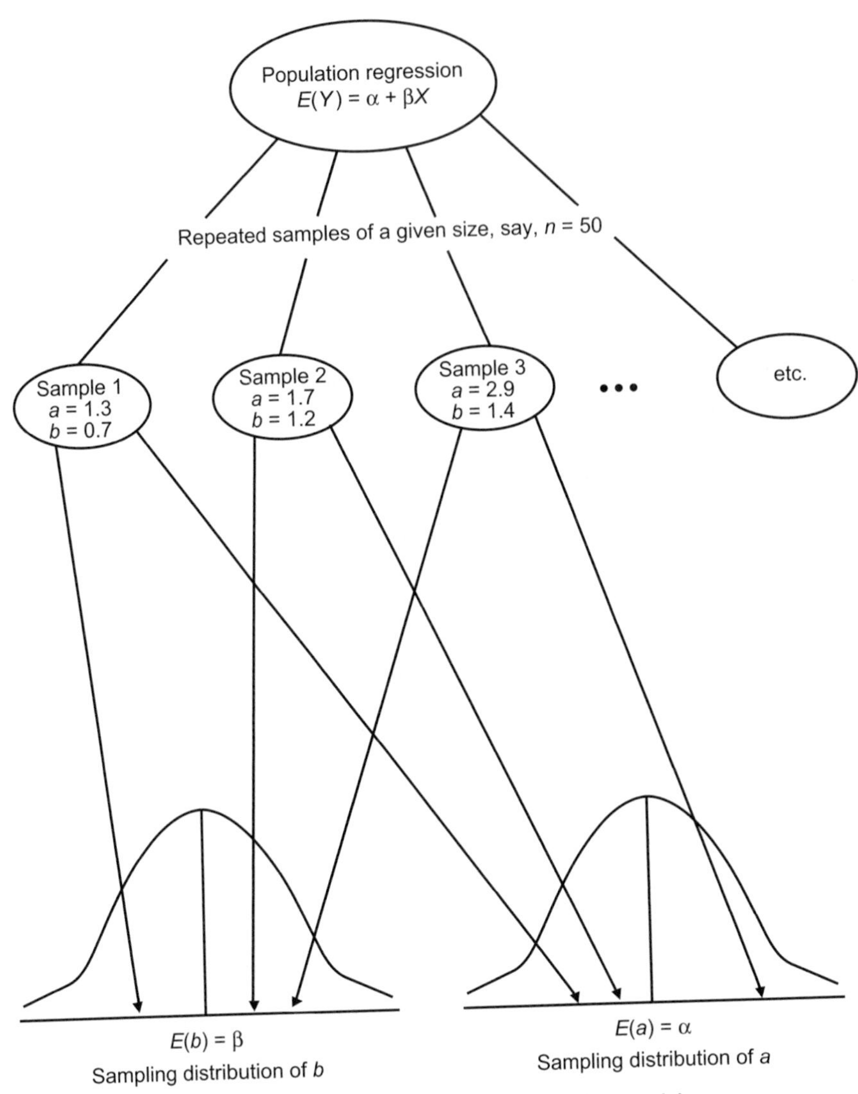

FIGURE 12-9. Sampling for a regression model.

12.3. Inferences in Regression Analysis

Point Estimates

In many applications of regression analysis, it is important to *estimate* the value of some parameter in the regression equation. Statistical estimation in the context of regression is based on the process of statistical estimation described in Chapter 7. In Chapter 7 two types of statistical estimation were distinguished: *point* estimation and *interval* estimation. In point estimation, a single number is computed from the sample

TABLE 12-2
Estimators in Regression Analysis

Regression Model Parameter	Point Estimator	Equation
α	a	(4-11)
β	b	(4-12)
$\sigma_{Y \cdot X}$	$S_{Y \cdot X}$	(4-26)
ρ	r	(4-14)
ρ^2	r^2	(4-25)
$\mu_{Y \cdot X}$	$a + bX$	(12-1)
Y_{new}	$a + bX_{new}$	(12-1)

information and is used as an estimate of some population parameter. Note that all the point estimates in regression analysis *require only that the first three assumptions explained in Section 12.2 be satisfied.* No assumption of normality for the error term is required. However, the assumption of normality *is essential* for both hypothesis testing and the construction of confidence intervals. Table 12-2 summarizes the principal parameters and their point estimators in regression analysis.

From the Gauss–Markov theorem, we know that a and b are the best estimators of α and β. Similarly, $S_{Y \cdot X}$, r, and r^2 provide point estimates of $\sigma_{Y \cdot X}$, ρ, and ρ^2, respectively. If we wish to predict the mean value of Y for any level of X in the range of the sample data or $E(Y)$, the best point estimate is clearly $E(Y_i) = a + bX_i$. This should not be surprising since the regression equation is defined as the path of the means if the first three assumptions outlined in Section 12.2 hold. The expressions $E(Y)$ and $\mu_{Y \cdot X}$ are interchangeable. Here $E(Y)$ is used since Y_i is a random variable, and $\mu_{Y \cdot X}$ is used to reflect the fact that it is a population parameter, which clearly depends on the value of variable X. Finally, suppose we have a single new value of X, denoted X_{new}, and wish to predict the most likely value of Y for X_{new}, or Y_{new}. The best estimator for this is again $Y_{\text{new}} = a + bX_{\text{new}}$. This is the same as $\mu_{Y \cdot X}$. However, as we shall see, the interval estimates of these two parameters are much different. This should be expected. We will be more confident making a prediction of the mean of several numbers than a single number. In the same sense we found the sampling distribution of X to be much more compact than the variable X itself. The standard error of the mean is equal to the standard deviation of variable X only when the sample size is 1!

Inferences Concerning the Slope of the Regression Line

The slope of the regression line is particularly important in regression analysis. Sometimes the interest in b derives from the fact that it measures the sensitivity of variable Y to changes in variable X. For example, the slope of the trip generation equation tells a transportation planner the *magnitude* of the increase in the number of trips expected from increasing incomes. Estimating β is thus extremely important. For this purpose a confidence interval for β can be constructed by using the least squares estimator b. There is a second reason for examining β. If it is possible to infer from a sample regression line that $\beta = 0$, then it can be concluded that the equation is of no use as a predictor since $E(Y) = a + (0)X = a$. For every value of X, the best estimate of Y is $\bar{Y}$. Whether it is possible to infer that the slope of the population regression is nonzero therefore has important ramifications for the use of the regression equation for predictive purposes. To establish confidence interval and hypothesis testing formulas for β, it is necessary to describe the sampling distribution of the least squares estimator b.

The sampling distribution of b is fully characterized given the four regression assumptions (see Section 12.2 and especially Figure 12-9). It has the following features.

1. It is normal.
2. It has a mean value centered on β; that is, it is unbiased and $E(b) = \beta$.
3. It has a standard error, denoted σ_b,

$$\sigma_b = \frac{\sigma_{Y \cdot X}}{\sqrt{\Sigma X_i^2 - (\Sigma X_i)^2/n}}$$

Note that the formula for the standard error of b contains one unknown, $\sigma_{Y \cdot X}$. However, this unknown can be estimated from the sample data by $S_{Y \cdot X}$, and an estimate of σ_b, denoted S_b, is obtained:

$$S_b = \frac{S_{Y \cdot X}}{\sqrt{\Sigma S_i^2 - (\Sigma X_i)^2/n}} \qquad (12\text{-}4)$$

It turns out that the sampling distribution of b is no longer normal once this substitution has been made. Now the sampling distribution of the statistic is t-distributed with $n - 2$ degrees of freedom. There are only $n - 2$ degrees of freedom since 2 degrees of freedom are lost in the estimation of $\sigma_{Y \cdot X}$.

The value of the standard error, S_b, depends on two separate factors. First, S_b will increase with increases in the standard error of estimate. This is expected. If our residuals are large and the fit of the line is poor, one is bound to be less confident about estimating the slope of the regression line. However, the *greater* the sum of squared deviations of X (that is, the greater the variability in X), the *lower* S_b. To see why, compare Figure 12-10a with (b). Suppose all the values of X used to fit the regression line are concentrated at a few values of X quite close to $\bar{X}$. This is the situation in Figure 12-10a. The estimated slopes of various sample regression lines are likely to be quite variable. By comparison, the estimated slopes of possible regression lines in situations in which there is considerable variance in the values of X show less variability, as is illustrated in Figure 12-10b. In general, then, we are more confident about predicting the true value for β when the prediction is based on observations over a wide range of the independent variable X.

The test statistic is

$$t = \frac{b - \beta}{S_b} \qquad (12\text{-}5)$$

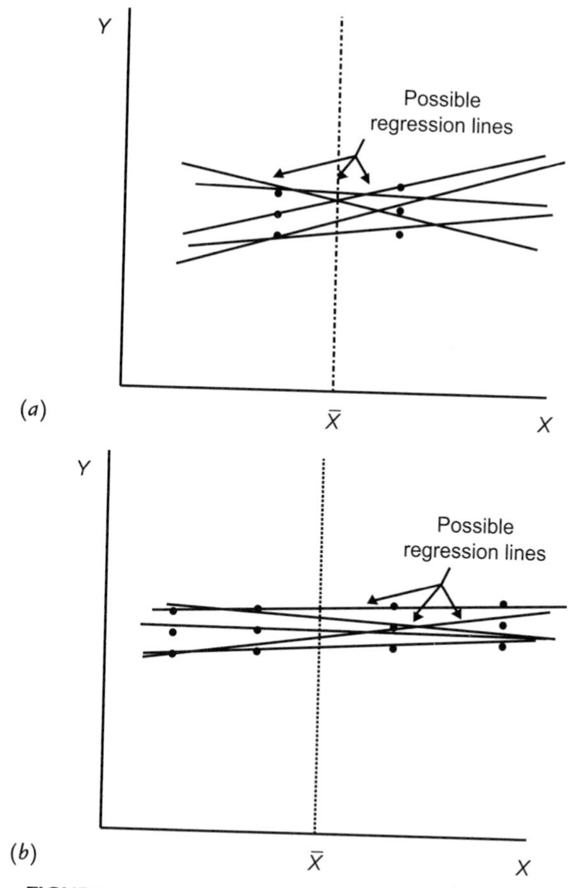

FIGURE 12-10. Variability in the standard error of *b*.

This reduces to $t = b/S_b$ for the most common null hypothesis: $H_0 : \beta = 0$. As an alternative to a formal test of hypothesis, it is possible to construct a confidence interval for β from the point estimate *b:*

$$b - t_{\alpha/2,n-2}S_b \leq \beta \leq b + t_{\alpha/2,n-2}S_b \tag{12-6}$$

If this interval contains zero, then we will not reject the null hypothesis that $\beta = 0$ and conclude that the regression line is of no use in prediction.

It is now possible to test the usefulness of household income variations in household trip generation rates for the regression equation of Example 12-2. A summary of the output from conventional statistical packages is illustrated in Table 12-3. First, notice that the equation is estimated as $Y = .8031 + .1118X$ and the overall equation has a correlation coefficient of .9153 and a standard error of estimate $S_{Y \cdot X}$ of .7527.

TABLE 12-3
Summary of Regression Output for Example 12-2

r	0.9153
r^2	0.8377
$S_{Y \cdot X}$	0.7527

ANOVA

	df	*SS*	*MS*	*F*	Sign.
Regression	1	29.2510	29.2510	51.6281	0.0000
Residual	10	5.6657	0.5666		
Total	11	34.9167			

Equation

	Coeff.	Std. Error	*t*-Stat	*P*-value	Lower 95	Upper 95
Intercept	0.8031	0.7663	1.0480	0.3193	–0.9044	2.5106
Income	0.1118	0.0156	7.1853	0.0000	0.0771	0.1464

We calculate the standard error for b using the values for ΣX_i^2 and $(\Sigma X_i)^2$ from our original data:

$$S_b = \frac{.7527}{\sqrt{29133 - ((567)^2/12)}} = 0.0156$$

and therefore

$$t = \frac{0.1118}{0.0156} = 7.185$$

Note how these agree with the values reported by the statistical package. The value of the t-statistic of 7.1853 for 10 degrees of freedom exceeds the highest value in the t-table in the Appendix so that we know the PROB-VALUE < .001. Normally, statistical packages can give even more precise results, and we see it is reported as 0.0000. The lower and upper 95% confidence bounds are also provided in the output. Given the *estimate* of $b = .1118$, and the standard error $S_b = 0.0156$, the 95% confidence interval for β is calculated using Equation 12-6 as

$$.1118 - 2.228(0.0156) \leq \beta \leq .1118 + 2.228(0.0156)$$

$$0.0770 \leq \beta \leq 0.1466$$

This is in close agreement with the output generated by the statistical software that uses more decimal places in its calculations. With 95% confidence, then, we can say the true slope β lies in the range [0.0771, 0.1464]. This interval does not contain zero, and the sample regression line $Y = 0.8031 + 0.1118X$ is a useful equation for predicting household trip generation rates for traffic zones with average household incomes between $27,000 and $72,000.

Inferences Concerning the Regression Intercept α

Confidence intervals and hypothesis tests for the intercept α are rarely used in regression model building. In many practical applications the Y-intercept does not have an important meaning. Often the range of observations of the independent variable X does not envelop the value $X = 0$. For example, the trip generation rate for a household with an annual income of zero is not particularly meaningful. When we consider that the independent variable in this case is really the average income of all households in a traffic zone, the intercept is even less meaningful! In some cases there is an important interpretation of the intercept, and it does merit inferential consideration. Consider again the example of predicting crop yield Y as a function of the amount of fertilizer applied X to various test plots in an agricultural experiment. The intercept α represents the expected yield of the crop in the absence of fertilizer application. This has meaning as a "control" value.

The sampling distribution of a, under the four regression assumptions, has the following characteristics:

1. It is normal.
2. It has a mean value centered on α; that is, it is unbiased and $E(a) = \alpha$.
3. It has a standard error σ_a of

$$\sigma_a = \sigma_{Y \cdot X} \sqrt{\frac{1}{n} + \frac{\bar{X}^2}{\Sigma X_i^2 - (\Sigma X_i)^2/n}}$$

Again, since σ is unknown, it is estimated by $S_{Y \cdot X}$, and the estimated standard error of a becomes

$$S_a = S_{Y \cdot X} \sqrt{\frac{1}{n} + \frac{\bar{X}^2}{\Sigma X_i^2 - (\Sigma X_i)^2/n}} \qquad (12\text{-}7)$$

The appropriate test statistic is t-distributed with $n - 2$ degrees of freedom. It is most common to test the hypothesis that $\alpha = 0$, but this test statistic can be used to test a null concerning any value of α.

$$t = \frac{a - \alpha}{S_a} \qquad (12\text{-}8)$$

By using the standard t-statistic, the $100(1 - \alpha)$ confidence interval for α from a given least squares estimate a is

$$\mathrm{a} - t_{\alpha/2,n-2} S_a \le \alpha \le a + t_{\alpha/2,n-2} S_a \qquad (12\text{-}9)$$

As we have noted, the intercept of the household trip generation equation is not particularly meaningful. Nevertheless, a test of this intercept is useful to illustrate the necessary calculations of this test. The calculated t-value is

$$t = \frac{0.8031}{0.7527\sqrt{\dfrac{1}{12} + \dfrac{(47.25)^2}{[29133 - (567)^2/12]}}} = \frac{0.8031}{0.7663} = 1.0480$$

leading to an exact PROB-VALUE of 0.3193. Using the t table of Appendix Table A-6, we could estimate this as between 0.297 and 0.341. At a significance level of 0.05, we would thus not reject the null hypothesis that $\alpha = 0$. This seems to be a reasonable conclusion. Without any income, households in a traffic zone might be expected to make zero trips. Since the intercept is well out of the range of the available data, it is of minor concern in any case. If the data were to suggest a rejection of the null hypothesis, this would also be of no real significance since we are examining a value of X beyond the range of our data. Given the estimated value $a = 0.8031$, the 95% confidence limits from Equation 12-9 are

$$0.8031 - 2.228(.7663) \le \alpha \le 0.8031 + 2.228(0.7663)$$

$$-0.9042 \le \alpha \le 2.5104$$

These results differ from those of Table 12-3 purely as a result of small rounding errors. The fact that this range includes the value of 0 agrees with our assessment of the t-test for the hypothesis $\alpha = 0$.

Analysis of Variance for Regression

It is also possible to test the null hypothesis $\beta = 0$ by using the F-distribution. The analysis of variance for regression always produces results equivalent to those obtained from the t-test for the null hypothesis $\beta = 0$. In fact, the value of the F-statistic computed in the analysis of variance is the exact square of the value of the sample t-statistic given by Equation 12-5. This will be demonstrated for the household trip generation equation. Although these two tests are exactly equivalent for simple linear regression, they are not equivalent for cases in which the regression equation includes more than one independent variable. In multiple regression analysis, the t-test is used to test the significance of a single independent variable given the null hypothesis $\beta_j = 0$ for any independent variable j. The analysis of variance is used to test the *joint* hypothesis that all the independent variables have an insignificant effect on the dependent variable Y, that is, $\beta_1 = \beta_2 = \ldots \beta_j = 0$.

The general form of the analysis of variance table for simple linear regression is given in Table 12-4. The total sum of squares (TSS) is equal to $\Sigma(Y_i - \bar{Y})^2$. Associated with this total variation are $n - 1$ degrees of freedom, 1 less than the number of observations. The two mean squares are calculated by apportioning these $n - 1$ degrees of freedom to two sources: the regression model and the error, or residual variation.

TABLE 12-4
Analysis of Variance Table in Simple Regression

Source of variation	Sum of squares	Degrees of freedom	Mean square	F-ratio
Regression (model)	RSS	1	$\frac{\text{RSS}}{1}$	$F = \frac{\text{RSS}/1}{\text{ESS}/n-2}$
Error (residual)	ESS	$n-2$	$\frac{\text{ESS}}{n-2}$	
Total	TSS	$n-1$		

The residual or error sum of squares (ESS) is calculated as $\Sigma(Y_i - \hat{Y}_i)^2$. The degrees of freedom for the residual error (ESS) are equal to the number of observations minus the number of independent variables minus 1. In simple regression this is always equal to $n - 2$. The remaining sum of squares are those attributable to the regression model (RSS) and defined as $\Sigma(\hat{Y}_i - \bar{Y})^2$. The degrees of freedom attributable to the model are always equal to 1. The test statistic F is simply the ratio of the two mean squares. When RSS is high and ESS is low, the model explains a substantial portion of the variance in Y and F will be high. When the reverse is true, the independent variable will be of little use in explaining variations in the dependent variable Y and F will be low.

Table 12-3 contains the ANOVA results for the trip generation equation. Based on the calculated F-value of 51.63, we can say the PROB-VALUE for this test is 0.0000 to four decimal places (this actually makes it less than 0.00005; otherwise it would be rounded up!) and conclude that household income is very useful in predicting household trip generation rates. The calculated value of $F = 51.68$ is almost the exact square of the t-value calculated above for the same null hypothesis:

$$(7.1853)^2 \simeq 51.6281$$

Any difference between the figures is due solely to the number of decimal places used in the calculation of the individual components.

Confidence Interval for $\mu_{Y \cdot X}$ of Given X

One of the most important applications of regression analysis is to use the equation $Y = a + bX$ to estimate a value for the conditional mean of Y, denoted $\mu_{Y \cdot X}$, for a given value of X. We have already noted that the point estimate of $\mu_{Y \cdot X}$, is determined by substituting the appropriate value of X, say X_0, into the equation so that $Y_0 = a + bX_0$. Of course, there is some error involved in such an estimate, and it is usual to construct a confidence interval around this point estimate. To develop this confidence interval, it is necessary to specify the sampling distribution of $\hat{Y}_0$. Again using the four assumptions, the characteristics of this sampling distribution are as follows:

1. It is normal.
2. It has a mean value of $\mu_{Y \cdot X_0}$, and it is unbiased.
3. It has a standard error of $\sigma_{\hat{Y}_0}$ of

$$\sigma_{\hat{Y}_0} = \sigma_{Y \cdot X} \sqrt{\frac{1}{n} + \frac{(X_0 - \bar{X})^2}{\Sigma X_i^2 - (\Sigma X_i)^2/n}}$$

Estimating σ by $S_{Y \cdot X_0}$ leads to

$$S_{\hat{Y}_0} = S_{Y \cdot X} \sqrt{\frac{1}{n} + \frac{(X_0 - \bar{X})^2}{\Sigma X_i^2 - (\Sigma X_i)^2/n}} \qquad (12\text{-}10)$$

The standard error of $S_{\hat{Y}0}$ increases with (1) the standard error of estimate $S_{Y \cdot X}$, (2) the reciprocal of the sum of squared deviations, and (3) the difference between the value X_0, and the mean $\bar{X}$. It decreases with the sample size *n*. As *n* increases, the term $1/n$ decreases, and therefore the standard error must decline. The standard error of estimate and the sum of squared deviations of *X* are important for the same reasons outlined for S_b, and a larger sample size is always better for estimating a mean such as $\hat{Y}_0$.

The implications of the final factor, the difference between X_0, and $\bar{X}$, are illustrated in Figure 12-11. The confidence interval is narrowest at $\bar{X}$. The farther X_0 lies from $\bar{X}$, the wider the interval. There is a simple explanation for the peculiar shape of these confidence intervals. Suppose that the bands were parallel to the regression line and not shaped as the interval in Figure 12-11. This would imply that the *only* error in predicting *Y* derives from mistakes in estimating *a*. Once it is recognized that both *a* and *b* can be in error, then the shape of these limits is easier to understand. The farther the point is from $\bar{X}$, the greater the impact of an error in estimating β on the estimated value $\hat{Y}_0$. The dangers of extrapolation are clear.

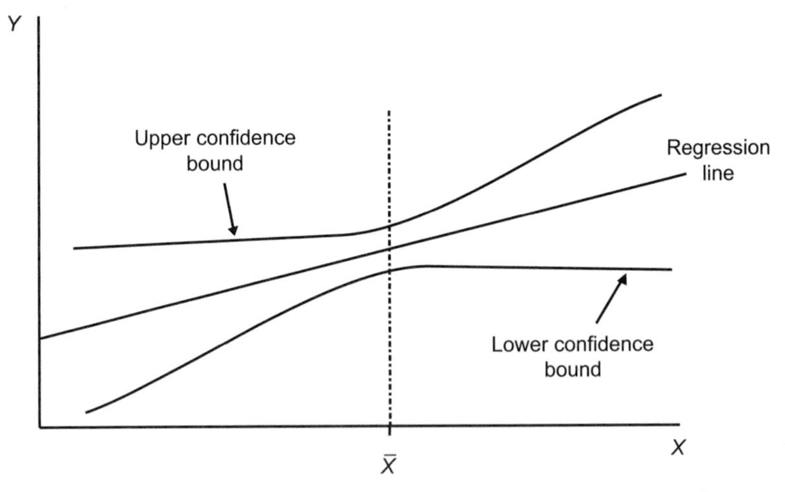

FIGURE 12-11. Characteristic shape of confidence intervals for $\hat{Y}$.

Because of the importance of household trip generation equations in making urban area trip forecasts, the confidence interval about $Y = 0.8031 + 0.1118$ holds special interest. Consider first the confidence interval about the line at the point $X = X_0$. For the average income household of 47.25 (thousands of dollars), the estimated number of trips per day is $Y_0 = \hat{Y} = 6.083$ [recall the regression line always passes through $(\bar{X}, \bar{Y})$]. To develop the confidence interval, we use the results of our regression equation summarized in Table 12-3:

$$S_{\hat{Y}_0} = S_{Y \cdot X}\sqrt{\frac{1}{n} + \frac{(X_0 - \bar{X})^2}{\Sigma X_i^2 - (\Sigma X_i)^2/n}}$$

$$S_{\hat{Y}_0} = 0.7527\sqrt{\frac{1}{12} + \frac{(0)^2}{29133 - (567)^2/12}}$$

$$= 0.7527(0.288) = .217$$

Now, setting $\alpha = 0.05$ we use the value of $t_{.025,10df} = 2.228$ in our confidence interval:

$$\hat{Y}_0 - t_{\alpha/2,n-2df}(S_{\hat{Y}0}) \le \mu_{\hat{Y}_0} \le \hat{Y}_0 + t_{\alpha/2,n-2df}(S_{\hat{Y}_0})$$

$$6.083 - 2.228(0.217) \le \mu_{\hat{Y}_0} \le 6.083 + 2.228(0.217)$$

$$5.559 \le \mu_{\hat{Y}_0} \le 6.567$$

Table 12-5 summarizes the estimates for $\hat{Y}_0$ at several other values of X within the observed range of the independent variable X. The interval is narrowest at $X = \bar{X}$ and wider for values of X greater than or less than $X = 47.25$ and therefore has the characteristic shape illustrated in Figure 12-11.

Prediction Interval for Y_{new}

Often we are not interested in making inferences about the mean of the conditional distribution of Y_0 at any given level of $X = X_0$. Instead, we wish to make inferences about the value of a *single* new observation of X, say X_{new}. Suppose we must predict

TABLE 12-5
95% Confidence Intervals for $\hat{Y}$

Annual household income ($000)	Estimated mean trips per day	Lower bound	Upper bound	Upper–lower estimate
30	4.156	2.310	6.001	3.690
36	4.826	3.038	6.615	3.577
42	5.491	3.742	7.252	3.510
47.25	6.083	4.337	7.829	3.491
54	6.838	5.076	8.599	3.522
63	7.843	6.014	9.672	3.658
66	8.179	6.316	10.041	3.725

the number of trips per household per day made by households in a traffic zone with a forecasted income of $X = 60{,}000$. We cannot use the confidence interval for this purpose since it is based on the expected mean number of trips made by households in several traffic zones with an income of 60,000. It is much more difficult to know a *single* value than a mean value, so we expect that this *prediction interval* will be wider than the corresponding *confidence interval* at any value of X.

The construction of a prediction interval is virtually identical to the procedure followed for a confidence interval. The only difference is the standard error of $\hat{Y}_{\text{new}}$ is

$$S_{\hat{Y}_{\text{new}}} = S_{Y \cdot X}\sqrt{1 + \frac{1}{n} + \frac{(X_{\text{new}} - \bar{X})^2}{\Sigma X_i^2 - (\Sigma X_i)^2/n}} \tag{12-11}$$

This standard error is almost identical to $S_{\hat{Y}0}$. In fact, it is larger by an additional $S_{Y \cdot X}$. To distinguish Equation 12-10 from 12-11, the quantity $S_{\hat{Y}_{\text{new}}}$ is often called the *standard error of the forecast.* The $100(1 - \alpha)$ percent *prediction or forecast interval* is

$$\hat{Y}_{\text{new}} - t_{\alpha/2,n-2df}(S_{\hat{Y}_{\text{new}}}) \le \mu_{\hat{Y}_{\text{new}}} \le \hat{Y}_{\text{new}} + t_{\alpha/2,n-2df}(S_{\hat{Y}_{\text{new}}}) \tag{12-12}$$

Consider now the specification of prediction intervals for the trip generation equation. We begin the specification of the prediction interval for $X_{\text{new}} = \bar{X}$ by calculating

$$S_{\hat{Y}_{\text{new}}} = S_{Y \cdot X}\sqrt{1 + \frac{1}{n} + \frac{(X_{\text{new}} - \bar{X})^2}{\Sigma X_i^2 - (\Sigma X_i)^2/n}}$$

$$S_{\hat{Y}_{\text{new}}} = 0.7527\sqrt{1 + \frac{1}{12} + \frac{(0)^2}{29133 - (567)^2/12}} = 0.7527(1.04) = 0.783$$

Now, setting $\alpha = .05$, we use the value of $t_{\alpha/2,10df} = 2.228$ in our prediction interval:

$$\hat{Y}_{\text{new}} - t_{\alpha/2,n-2df}(S_{\hat{Y}_{\text{new}}}) \le \mu_{\hat{Y}_{\text{new}}} \le \hat{Y}_{\text{new}} + t_{\alpha/2,n-2df}(S_{\hat{Y}_{\text{new}}})$$
$$6.083 - 2.228(0.783) \le \mu_{\hat{Y}_{\text{new}}} \le 6.083 + 2.228(0.783)$$
$$4.337 \le \mu_{\hat{Y}_{\text{new}}} \le 7.829$$

Prediction intervals for several values of X are summarized in Table 12-6. In comparison with the confidence intervals of Table 12-5, we can see that both intervals have the same characteristic shape, but that the prediction interval is wider than the confidence interval for all values of X.

Significance Testing for Correlation Coefficients

We have seen in Chapter 4 that Pearson's correlation coefficient r is a useful summary of the strength of the linear association between two variables. Not surprisingly, this measure can also be used in an inferential mode to estimate the strength of association between variables X and Y in a population where the correlation coefficient is

TABLE 12-6
95% Prediction Intervals for $\hat{Y}_{new}$

Annual household income ($000)	Estimated mean trips per day	Lower bound	Upper bound	Upper–lower estimate
30	4.156	2.310	6.001	3.690
36	4.826	3.038	6.615	3.577
42	5.491	3.742	7.252	3.510
47.25	6.083	4.337	7.829	3.491
54	6.838	5.076	8.599	3.522
63	7.843	6.014	9.672	3.658
66	8.179	6.316	10.041	3.725

specified as ρ (rho). As usual, the sample correlation coefficient r is the best unbiased estimator of ρ and can also be used in an hypothesis test where the normal null hypothesis is that no correlation exists between the two variables: H_0: $\rho = 0$.

The sampling distribution of r under the null hypothesis that $\rho = 0$ is t-distributed with $n - 2$ degrees of freedom and has a standard error of

$$S_r = \sqrt{\frac{1 - r^2}{n - 2}}$$

and the required test statistic is

$$t = \frac{r}{S_r} = \frac{r}{\sqrt{1 - r^2/n - 2}} = \frac{r\sqrt{n - 2}}{\sqrt{1 - r^2}} \qquad (12\text{-}13)$$

Testing the null hypothesis where $\rho \neq 0$ is more complicated as the sampling distribution of r becomes skewed as ρ approaches the limits ± 1. This test is also normally undertaken only for observational data that conforms to the random-X regression model. Where there are multiple observations for any single value of X, as in the fixed-X model, the upper limit of r is always less than 1 (or greater than –1) since it is impossible to have a perfect fit of a line in this case.

Testing the results of Example 12-2, we find that for the observed correlation coefficient of $r = 0.9153$ and $n - 2 = 10$ degrees of freedom, we have

$$t = \frac{0.9153\sqrt{10}}{\sqrt{1 - 0.9153^2}} = 7.186$$

a value identical to the test on the slope of the regression equation b. In the case of simple linear regression, these two tests are equivalent. Thus, if a test on the correlation coefficient r is significant, so too will be tests on b and the ANOVA test on the entire regression equation. The ANOVA is essentially a test on the coefficient of determination r^2, and it can be shown that the F-ratio in Table 12-4 is again the square of the t-test of equation 12-13. As is now even more obvious, correlation and regression analysis are intimately related.

12.4. Graphical Diagnostics for the Linear Regression Model

Specific statistical tests are available for checking whether the assumptions underlying the use of the linear regression model are satisfied by the data utilized in any empirical study. These tests employ the residuals of the sample regression equation $Y_i = a + bX_i + e_i$ as estimates of the error term of the true population regression model $Y_i = \alpha + \beta X_i + \varepsilon_i$. These tests are outlined in detail in Kutner et al. (2004). In many instances, however, several simple diagnostic plots can reveal clear or possible violations of these assumptions.

1. Dot diagram or histogram or boxplot of absolute or standardized residuals
2. Normal probability plot
3. Plot of residuals against $\hat{Y}$
4. Plot of residuals against omitted variables

Plots of Simple Residuals

Table 12-7 summarizes the results of the regression analysis for the 12 observations in the data. The first type of plot graphs the simple residuals or the standardized residuals of the last column as a dot diagram, a histogram, or a boxplot. Standardizing the sample residuals by dividing by the standard error of the estimate, $S_{Y \cdot X}$ converts the fifth column to the sixth column. For the first observation, for example, –1.535 = –1.156/0.7527. Now under Assumption 4 of the linear regression model, the residuals from the sample regression line should be normally distributed with mean 0 and variance $\sigma^2_{Y \cdot X}$. The *standardized* residuals are thus also normally distributed with a zero mean, but have a unit variance and standard deviation. Notice that the shape of the distribution is not altered by creating standardized residuals; all we have done is divide our residuals by a constant, the standard error of estimate.

What should we look for in these residual plots? Since our errors should be normally distributed, roughly 95% of the observations in the standardized residuals should

TABLE 12-7
Residuals for the Regression Model of Example 12-2

i	Y_i	X_I	$\hat{Y}_i$	$e_i = Y_i - \hat{Y}_i$	$z_i = (Y_i - \hat{Y}_i)/S_{Y \cdot X}$
1	30	3	4.156	–1.156	–1.535
2	36	5	4.826	0.174	0.231
3	33	5	4.491	0.509	0.676
4	27	4	3.820	0.180	0.239
5	42	6	5.497	0.503	0.669
6	48	7	6.167	0.833	1.106
7	63	9	7.843	1.157	1.537
8	54	7	6.838	0.162	0.216
9	66	8	8.179	–0.179	–0.237
10	72	8	8.849	–0.849	–1.128
11	51	6	6.502	–0.502	–0.667
12	45	5	5.832	–0.832	–1.105

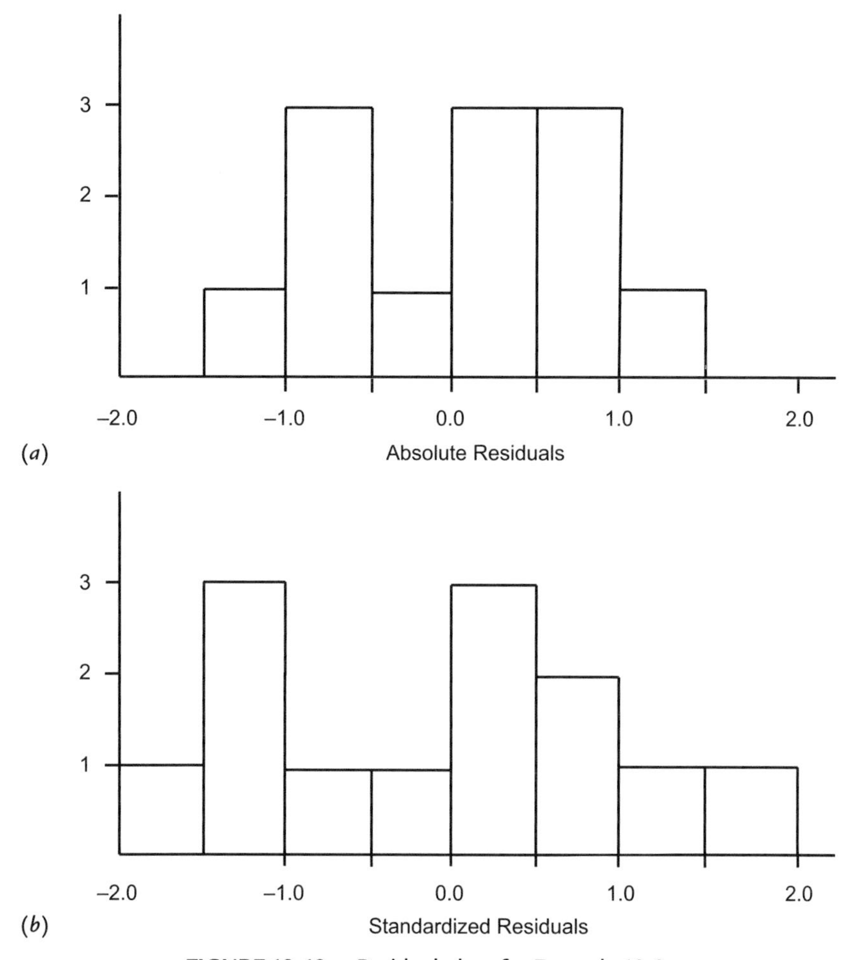

FIGURE 12-12. Residual plots for Example 12-2.

fall in the range [–2, +2], or more exactly [–1.96, +1.96]. Figure 12-12 illustrates plots of absolute and standardized residuals for Example 12-2. Both plots will usually have the same shape, and the standardization has altered only the scale. In our case, there are so few observations that this is not readily apparent. Notice that 100% of the standardized residuals lie within the range [–2, +2]. Given the small sample size of $n = 12$, this is not unexpected. Checking for *outliers* is also convenient when the residuals are expressed in this form. Any observation with a residual outside the range [–2, +2] can be defined as an *outlier* in this context and is worth examining in more detail. Statistical software packages often routinely identify such observations in the course of their results. It is important to examine such observations in detail and try to determine why they are so poorly predicted. Such an analysis might point to other factors not currently accommodated within a simple linear regression model.

Normal Probability Plots

These simple plots can be augmented by a more useful plot termed a normal probability plot, which is commonly available in statistical software packages. In this plot, each residual is plotted against its expected value under the assumption of normality. If the plot is linear, we conclude that the distribution agrees with the assumption of normality of the regression model. Otherwise, the plot suggests that the residuals and therefore the error term may not be normally distributed as required.

To find the expected values of the residuals under normality, we make use of the fact that the expected value of the error terms for a regression model is 0, and the standard deviation can be estimated by the standard error of estimate, $S_{Y \cdot X}$. Statistical theory has shown that a reasonable approximation for the expected value of the kth smallest observation in a random sample of n is

$$S_{Y \cdot X}\left[z\left(\frac{k - .375}{n + 0.25}\right)\right]$$

where $z(A)$ denotes the $(A)/100$ percentile of the standard normal distribution. Using this approximation, let us calculate the expected value of observation 2, which is the sixth smallest residual. Using $S_{Y \cdot X} = 0.753$, $n = 12$, and $k = 6$, we find

$$\frac{k - 0.375}{n + 0.25} = \frac{6 - 0.375}{12 + 0.25} = \frac{5.625}{12.25} = .4592$$

so that the expected value of this residual under normality is

$$0.7527[z(0.4592)] = 0.7527(-0.10245) = -0.077$$

and we see that the actual value of this residual of .174 is slightly over what we would expect. However, given our small sample size, this is not particularly worrisome. The normal plot of the residuals illustrated in Figure 12-13 reveals that the plot is near linear, and there appears to be no significant departure from normality. Were the plot to reveal a scatter of points with a significant concave upward trend or convex downward trend, we would conclude there was significant skewness in the residuals, and it may be necessary to undertake some remedial action. Departures from normality are quite difficult to assess and will be significantly affected by other assumption violations such as using the wrong functional form for the equation.

Plots of Residuals against $\hat{Y}_i$

An extremely useful residual plot graphs the sample residuals e_i (or standardized residuals) versus the predicted value of Y, $\hat{Y}$. How should such a plot appear? If the first three assumptions of the linear regression model hold, the shape of the pattern of all the individual observations should be roughly rectangular, as in Figure 12-14a.

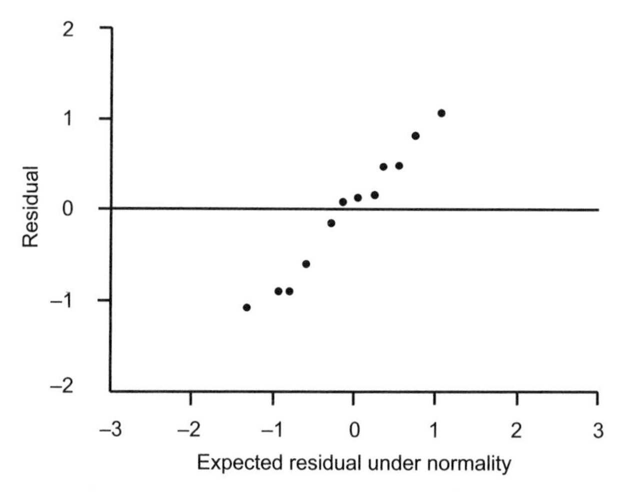

FIGURE 12-13. Normal probability plot for residuals of Example 12-2.

Any significant departure from this pattern is usually evidence of some violation in one of these three assumptions. There are several common patterns indicative of specific problems.

1. A plot in which the width of the residual band increases with $\hat{Y}$ indicates that the assumption of homoscedasticity has been violated. In Figure 12-14b the residuals appear to have a greater variance for large values of $\hat{Y}$ than for smaller values. The least squares estimators a and b do not have the properties specified by the Gauss–Markov theorem but corrective action can often be taken to correct this problem (see Kutner et al., 2004) Although it is less common, the variance of the residuals may also decrease with $\hat{Y}$ or even increase at both extremes.
2. Unsatisfactory residual plots can also be obtained when the wrong functional form is used for the systematic component of the relationship. In Figure 12-14c, for example, we notice that positive residuals occur at moderate values of $\hat{Y}$ and negative residuals for high and low values of $\hat{Y}$. In Chapter 13 we discuss using transformations to fit appropriate nonlinear models.
3. If the data have been generated by an experiment that took place over time, it is often instructive to plot the residuals in time order. In Figure 12-14d, the residuals appear to increase with time. The appropriate corrective action is to introduce a linear function of time into the regression equation by developing a multiple regression model. A multiple regression equation incorporating a time variable would solve the technical violation illustrated in Figure 12-14d, but the real task of the researcher is to find some assignable cause for this trend.

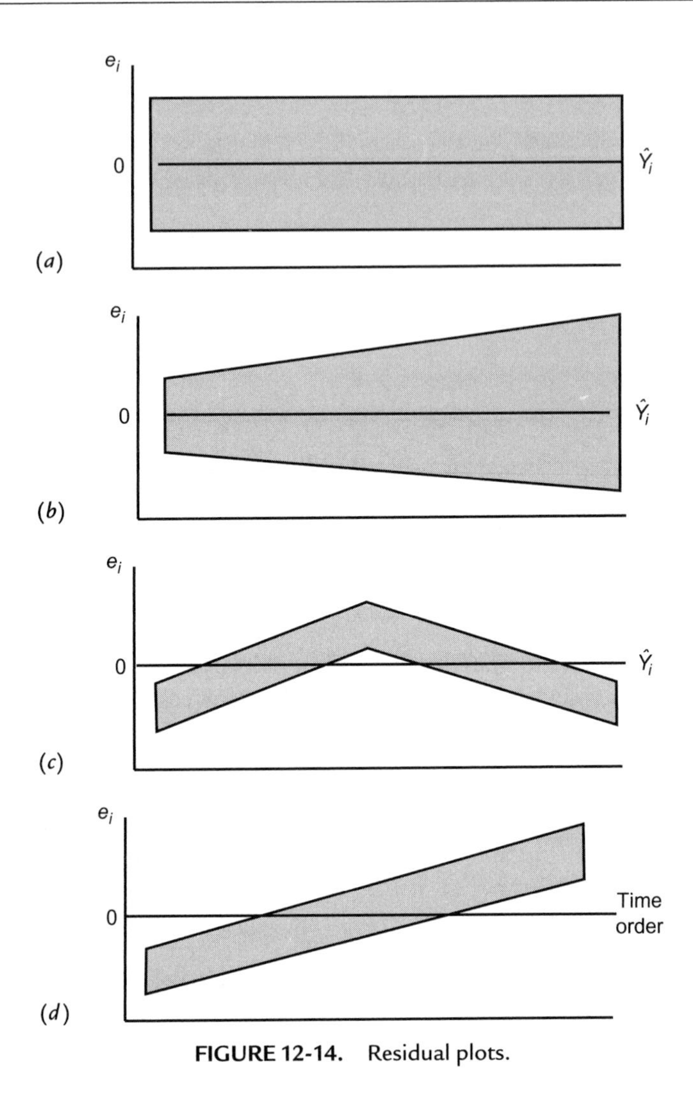

FIGURE 12-14. Residual plots.

The residuals e_i and the fitted values $\hat{Y}_i$ of the trip generation data are plotted in Figure 12-15. There are only 12 residuals, and it is difficult to find any serious deficiency of the simple linear regression model. Since we know that other variables besides income can affect household trip generation rates, for example, household size and car availability, these residuals can hardly be thought of as random. As we shall see in the next section, what appears to be random on the surface may turn out to be a false impression when analyzed in another way.

Plots of Residuals against Omitted Variables

The final type of residual plot to consider is a plot of the residuals from the sample regression equation versus other independent variables. For example, a plot of the

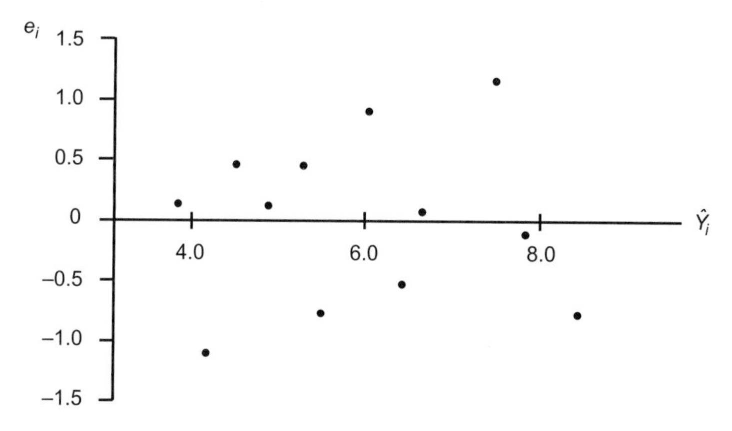

FIGURE 12-15. Residual plot of trip generation equation.

residuals from the household trip generation equation versus the variable household size reveals the essential *nonrandomness* of the error component in the simple linear regression model. The residuals and the household sizes for the 12 traffic zones are listed in Table 12-8 and plotted in Figure 12-16. The residuals are strongly linearly related to average household size. Traffic zones with small average household sizes are *overpredicted* by the regression equation, and those traffic zones with large average household sizes are *underpredicted.* There is a clear indication that a systematic relationship exists between the error component and household size. It is not unreasonable to expect larger households to make more trips simply because more trips are required to maintain larger households. Or there may be more than one or even two working members.

In any case, we have two problems. First, our error is not truly random, and it violates a key assumption of regression analysis. Second, we have found another variable

TABLE 12-8
Residuals and Household Size

Traffic zone	Residual	Average household size
1	−1.156	2.2
2	0.174	3.2
3	0.509	3.3
4	0.180	3.4
5	0.503	3.8
6	0.833	3.4
7	1.157	4.0
8	0.162	2.8
9	−0.179	2.9
10	−0.849	2.3
11	−0.502	2.6
12	−0.832	2.5

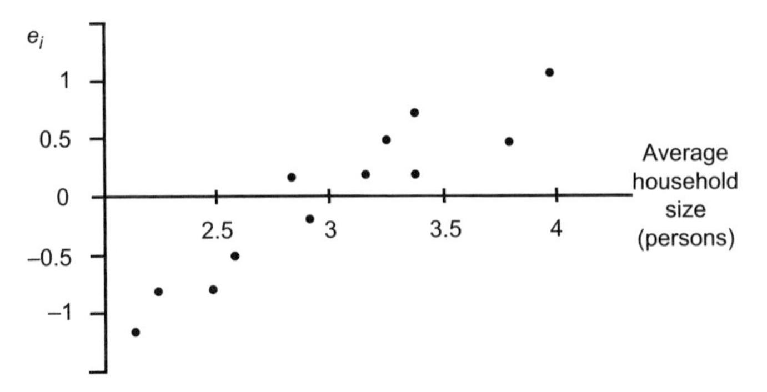

FIGURE 12-16. Plot of residuals and average household size for 12 traffic zones.

that *should* be part of our regression equation. Both of these problems can be solved simultaneously by developing a multiple regression equation in which household size is introduced into the equation. By introducing it into the equation, we remove it from our error term, which now has a greater chance of being truly random error. Second, we will probably be able to make better predictions with a more complete model than the simple one-variable regression model that only contains an income variable. These plots are extremely useful for identifying new variables to add to an equation. A plot of household size against trip generation may not reveal anything. The simple correlation coefficient between household size and trip generation is fairly low $r = 0.28$. The residual plot of Figure 12-16 reveals a much stronger relationship. Once the effect of income is removed, the relationship becomes clearer. This is one of the pitfalls of using simple bivariate models and techniques: we sometimes fail to see important relationships among groups of variables. In the next chapter, we extend our discussion to include the estimation of multiple regression equations and complete the analysis of the trip generation data.

Mapping Residuals

Finally, where the observations used in a regression problem represent contiguous areas of a map, the residuals can be plotted on a map and examined for randomness. A simple two-color map can be obtained by dividing our residuals into two nominal categories: positive and negative residuals. The map is developed by coloring the areas corresponding to positive residuals black and the areas corresponding to negative residuals white.

Referring once more to the household trip generation equation, we see that this procedure generates the two-color map of Figure 12-17. Note that all seven positive residuals are contiguous as are the five negative residuals. Apparently there is *strong positive autocorrelation* since adjacent traffic zones tend to have identically signed residuals. Maps generated from the residuals of a regression analysis can be used to see if there are important spatial considerations in the data. For example, the map

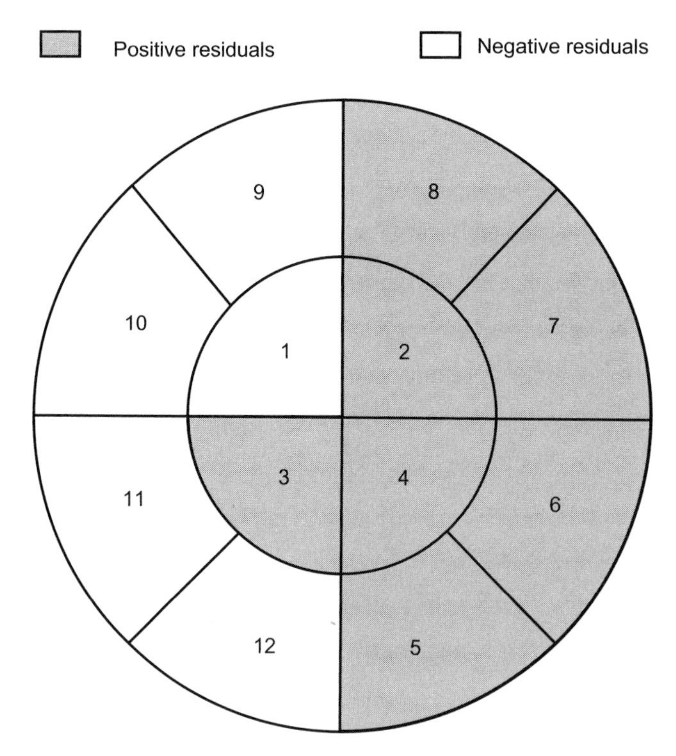

FIGURE 12-17. Two-color map of residuals for 12 traffic zones.

might reveal one or more areas of similarly valued residuals suggesting the existence of a regional effect in the data. Working only with the sign of the residual is the simplest procedure for examining spatial patterns. More sophisticated measures and tests of autocorrelation using the actual values of the residuals are discussed in Griffith (1987) and are discussed in Chapter 14.

12.5. Summary

In this chapter we have extended the role of regression analysis beyond the simple descriptive function introduced in Chapter 4. First, the descriptive role of regression analysis is identified as but one function undertaken in a more comprehensive model-building procedure. Testing the statistical significance of the equation or using the derived model for prediction purposes requires four specific assumptions. With these assumptions it is possible to test the hypothesis that the slope of a regression line equals zero. If this hypothesis can be rejected at a given level of significance, then the model is necessarily useful for purposes of prediction. For this test (as well as for the test of the regression intercept) to be valid, certain key assumptions must be met. These assumptions, including normality, homoscedasticity, and randomness of the error term, have also been discussed. In addition to these tests of the two parameters of the regression model, the procedures that can be used to develop both confidence and prediction intervals for the dependent variable Y have been outlined. We have seen that the overall fit of the model can be tested using a variation of the ANOVA procedure discussed in Chapter 11. A number of simple graphical plots of our residuals can be used as a first evaluation of the suitability of the data in meeting the assumptions of the regression model.

FURTHER READING

C. Daniel and F. S. Wood, *Fitting Equations to Data* (New York: Wiley Interscience, 1980).

D. A. Griffith, *Spatial Autocorrelation: A Primer* (Washington DC: American Association of Geographers, 1987).

T. R. Lange, H. E. Royals and L. L. Connor "Influence of Water Chemistry on Mercury Concentration in Largemouth Bass from Florida Lakes," *Transactions of the American Fisheries Society* 122 (1993), 74–84.

M. H. Kutner, J. Neter, C. J. Nachtsheim, and W. Li, *Applied Linear Regression Models,* 5th edition (Chicago: Irwin, 2004).

DATASETS USED IN THIS CHAPTER

bass.html

PROBLEMS

1. Explain the meaning of the following terms:
   - Measurement error
   - Stochastic error
   - Observational or survey data
   - Experimental data
   - Cross-validation
   - Holdout sample
   - Fixed-X model
   - Random-X model
   - Homoscedasticity assumption
   - Point estimates
   - Model validation
   - Residual plot
   - Spatial autocorrelation

2. Briefly explain the assumptions of the linear regression model and the implications of the Gauss–Markov theorem.

3. Differentiate between a *confidence* and *prediction* interval in the context of a regression equation.

4. Lange et al. (1993) studied the level of mercury contamination in largemouth bass in 53 different lakes in the state of Florida. Using water samples collected from each of these lakes, they measured the pH level, as well as the amounts of chlorophyll, calcium, and alkalinity. The mercury concentration in the muscle tissue of fish caught in each lake was taken and then standardized to take into account the different ages of the fish caught in each lake. From this dataset we have extracted 26 lakes for analysis and report a subset of the variables collected in this study:

Lake	Alkalinity	pH	Calcium	Chlorophyll	Mercury
Alligator	5.9	6.1	3	0.7	1.53
Apopka	116	9.1	44.1	128.3	0.04
Brick	2.5	4.6	2.9	1.8	1.33
Cherry	5.2	5.4	2.8	3.4	0.45
Deer Point	26.4	5.8	9.2	1.6	0.72
Dorr	6.6	5.4	2.7	14.9	0.71
Eaton	25.4	7.2	25.2	11.6	0.54
George	83.7	8.2	66.5	78.6	0.15
Harney	61.3	7.8	57.4	13.9	0.49
Hatchineha	31	6.7	15	17	0.7
Istokpoga	17.3	6.7	10.7	9.5	0.59
Josephine	7	6.9	6.3	32.1	0.81
Kissimmee	30	6.9	13.9	21.5	0.53
Louisa	3.9	4.5	3.3	7	0.87
Minneola	6.3	5.8	3.3	0.7	0.47
Newmans	28.8	7.4	10.2	32.7	0.41
Ocheese Pond	4.5	4.4	1.1	3.2	0.56
Orange	25.4	7.1	8.8	45.2	0.16
Parker	53	8.4	45.6	152.4	0.04
Puzzle	87.6	7.5	85.5	20.1	0.89
Rousseau	97.5	6.8	45.5	6.2	0.19
Shipp	66.5	8.3	26	68.2	0.16
Tarpon	5	6.2	23.6	9.6	0.55
Trafford	81.5	8.9	20.5	9.6	0.27
Tsala Apopka	34	7	13.1	4.6	0.31
Wildcat	17.3	5.2	3	2.6	0.28

These data can be obtained from the Data and Story Library of STATLIB, the online data repository that can be found at *lib.stat.cmu.edu/DASL/Stories/MercuryContaminationinBass.html.*

a. Develop a regression equation using Part I of the dataset in which mercury concentration is made a function of alkalinity using any available statistical software package or spreadsheet program.
b. Test the slope and intercept of the equations for significance. What can you conclude?
c. What PROB-VALUE can you associate with each of these hypotheses?
d. Test the simple correlation coefficient for significance.
e. What does the ANOVA reveal? Show that the results for the test of the slope b, the correlation coefficient r, and the ANOVA are all numerically equivalent.
f. Generate 99% confidence intervals around each parameter.
g. Construct 95% confidence intervals for the mean mercury concentration for alkalinity levels of 20 mg/L and 100 mg/L.
h. Construct 95% prediction intervals for mercury concentration for these same two levels of alkalinity.
i. Generate and interpret a plot of the simple residuals.
j. Create a normal probability plot for the residuals. What does the plot reveal?
k. Create a plot of the residuals against the predicted mercury level. Are there any potential violations of the assumptions of the regression model?
l. Plot the residuals from the regression line against the variables pH, chlorophyll, and calcium. What do these plots reveal? Does it appear that any of these variables might also be a good predictor of mercury concentration?

13

Extending Regression Analysis

In the previous chapter, we developed the inferential framework for regression analysis. At the same time, we outlined a number of different ways in which the simple linear regression model might be improved, and we briefly introduced a number of different extensions to the model. In this chapter, we will explore three of these extensions in more detail. Seldom in the social or physical sciences is it possible to satisfactorily explain the variation in a dependent variable by using a single independent variable. Fortunately, it is possible to extend regression analysis to situations in which several independent variables are used to account for the variability of a single response variable. This is termed *multiple regression analysis* and is explored in Section 13.1. If regression analysis were limited to purely *linear* functions, it would be of limited value. Fortunately, this is not the case. In Section 13.2 we describe a number of different variable *transformations* that can be used within the context of normal least squares analysis. These variable transformations allow us to explore a number of nonlinear functional forms that have great applicability in applied work. Finally, in Section 13.3 we will show how we can validate an existing regression model using a *holdout* sample, a portion of our data set aside specifically for test purposes.

13.1. Multiple Regression Analysis

Multiple regression is one of the most widely used tools in statistical analysis. In all of our examples to this point, we have used a simplified model in which only one independent or predictor variable is used to explain variations in a response or dependent variable. Frequently, such models are of limited value since the predictions they make are too imprecise to be useful. From a *prediction* point of view, it is therefore an obvious advantage to be able to use the information in two or more variables in an explanatory framework. We *should* be able to provide better predictions by using multiple regression analysis.

In Section 12.2, we analyzed a regression equation linking household trip generation rates to household income. We found that income had a significantly positive effect on trip generation rates. Later, however, in Section 12.4, we also found that the error term from this equation *appeared* to be positively related to household size. Our motivation for constructing a multiple regression model for these trip generation rates is obvious: common sense tells us that larger households probably make more trips per day then smaller households. Put simply, more trips are necessary to sustain the household. Let us now develop the format for a multiple regression equation.

When there are two predictor variables X_1 and X_2, the regression model

$$Y_i = \alpha + \beta_1 X_{1i} + \beta_2 X_{2i} + \varepsilon_i \tag{13-1}$$

is now the appropriate specification. Note that the subscript i still refers to the observation and we use a second subscript to distinguish between variables 1 and 2. There are now three parameters to solve for α, β_1, and β_2. How do we interpret these three parameters? The intercept α is the predicted value of Y or the mean response when $X_1 = X_2 = 0$. Otherwise, it has no special meaning as a separate term in the regression model. The parameter β_1 indicates the change in the mean predicted value of Y (or $\hat{Y}$) per unit increase in variable X_1, *when variable X_2 is held constant.* Likewise, β_2 measures the change in the mean predicted value of $\hat{Y}$ per unit increase in variable X_2, when variable X_1 is held constant. When the effect of X_1 on the mean response does not depend on the value or level of X_2, the two variables are said to have *additive* effects or not to interact.

The parameters β_1 and β_2 are normally termed *partial regression coefficients* because they reflect the individual or partial effect of one variable when the other variable is included in the model. By analogy the regression coefficient for variable X in the simple regression model is sometimes called the *gross* regression coefficient. As we shall see, there is no guarantee that these two coefficients will be equal except in a very limiting case.

How can appropriate variables be found? At times, a theory might suggest the proper specification of a relationship between several variables to have the format of a multiple regression equation. In other instances, previous empirical research might suggest logical variables to be included. For example, a great deal of research into household trip generation equations has found a number of different variables that affect trip generation rates. In addition to household size, the number of cars available to the household is known to be associated with higher trip generation rates. Larger households with two or more cars tend to make more trips than smaller households with one or no cars available. The age composition of the household and sometimes the availability of public transport also can affect these trip rates. As research proceeds, replications of studies tend to reveal additional variables or perhaps refined definitions of the variables used in a regression model. Sometimes the exact variables to use and the exact formulation are unknown. We *explore* the available data and look for multiple regression equations that seem to have some predictive or explanatory value.

The Geometry of Multiple Regression

The data for a multiple regression equation with two independent variables takes the form of a simple table:

Observation (i)	Y_i	X_{1i}	X_{2i}
1	Y_1	X_{11}	X_{21}
2	Y_2	X_{12}	X_{22}
...	...	...	...
n	Y_n	X_{1n}	X_{2n}

For each observation *i*, we have observed values for the dependent or response variable Y_i and values for the two independent variables X_{1i} and X_{2i}. Datasets with more variables will have additional columns for each of these variables. Most of the principles involved in evaluating multiple regression equations are valid for models of all sizes, but as we shall see, the geometry of least squares analysis for the two-independent-variable Equation 13-1 is much easier to display.

Figure 13-1 displays the geometry of a two-independent-variable equation using as an example a trip generation equation with two independent variables, household size and household income. We have a three-dimensional space in which the vertical axis is the dependent variable *Y*, in this case the number of trips per day per household. The two horizontally drawn axes represent the two independent variables. Each observation can be located as a single point in this three-dimensional space, with co-

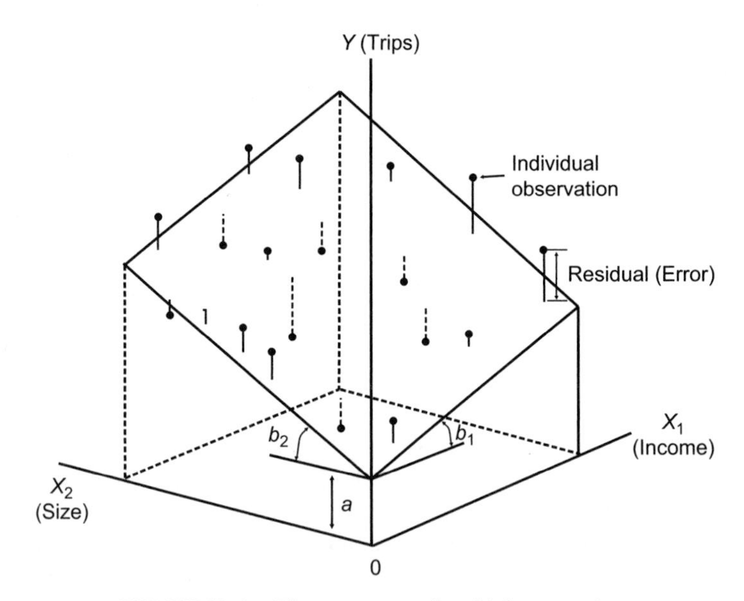

FIGURE 13-1. The geometry of multiple regression.

ordinates $[X_{1i}, X_{2i}, Y_i]$. Our goal is to use least squares analysis to estimate the three parameters a, b_1, and b_2 of the *sample* regression equation:

$$Y_i = a + b_1 X_{1i} + b_2 X_{2i} + e_i \tag{13-2}$$

The regression model given as Equation 13-2 defines a plane in this space. It is common to speak of the regression function as a *response surface* in the context of multiple regression, recognizing that it is not a line as is the case in simple bivariate regression. In Figure 13-1 the surface is a plane, but in other cases the response surface may be curved and have peaks, valleys, and other complex features. The exact orientation of this plane is determined by the values of the three parameters a, b_1, and b_2. Parameter a is the intercept on the Y-axis, where $X_1 = X_2 = 0$. The regression coefficient b_1 defines the *slope* of the plane with respect to the X_1-axis, and the regression coefficient b_2 controls the orientation with respect to the X_2-axis.

The values of these three parameters are chosen so that the sum of squared deviations from all points to this plane is minimized.

$$\min \sum_{i=1}^{n} (Y_i - \hat{Y}_i)^2 = \min \sum_{i=1}^{n} (Y_i - a - b_1 X_{1i} - b_2 X_{2i})^2 \tag{13-3}$$

These deviations represent the vertical distances from the observations to the regression plane. As usual, observations above the plane appear as positive residuals (note the solid lines in Figure 13-1) and those below the plane as negative residuals (dashed lines). So, in many respects, the solution to a multiple linear regression equation parallels simple regression algorithms. In the two-independent-variable case, we have three unknowns and three normal equations to provide us with values for the unknown coefficients a, b_1, and b_2. Each increment in the number of variables is matched by another normal equation used to provide the unique solution to the least squares problem and values for all model parameters.

The general linear regression model with $p - 1$ independent variables and a normal error term is

$$Y_i = \alpha + \beta_1 X_{1i} + \beta_2 X_{2i} + \ldots + \beta_{p-1} X_{p-1i} + \varepsilon_i \tag{13-4}$$

where there are p parameters $(\alpha, \beta_1, \beta_2, \ldots, \beta_{p-1})$ whose values are determined through the technique of least squares. Such models are normally expressed in matrix terms and solved using linear algebra algorithms (e.g., see Kutner et al., 2004).

Let us return to the household trip generation model introduced as Example 12-2. Table 13-1 contains the input data for this problem, combining information from Tables 12-1 and 12-8. At the outset, three aspects of the multiple regression version of this equation are worth examining. First, has the introduction of the new variable, household size, improved the model? To answer this question, we will need to examine our measures of goodness of fit, including the root mean square error (RMSE) and the coefficient of determination. In addition, does the variable have the correct sign for the regression coefficient? Obviously, we would expect the sign to be positive,

TABLE 13-1
Trip Generation Rates Data for Multiple Regression Equation

Traffic zone	Trips per household per day	Average household income ($000)	Household size
1	3	30	2.2
2	5	36	3.2
3	5	33	3.3
4	4	27	3.4
5	6	42	3.8
6	7	48	3.4
7	9	63	4.0
8	7	54	2.8
9	8	66	2.9
10	8	72	2.3
11	6	51	2.6
12	5	45	2.5

reflecting the empirical observation that larger households tend to make more trips per day. Second, what has been the impact of introducing the variable on the parameters of the model, particularly the regression coefficient for the income variable? Has the value of this parameter changed? Why? What does this mean? Third, has this formulation remedied any of the problems with the error term?

Let us begin by analyzing the regression output for this new multiple regression equation given in Table 13-2, comparing it to the results of the simple regression equation of Table 12-3. The first section of our output contains measures of goodness of fit. Notice that the new equation is superior to the simple equation by any measure. The standard error of the estimate, or RMSE, has declined significantly and is about one-third as large. This means that, in general, our residuals are smaller and the multiple regression equation will provide improved predictions.

Notice that the multiple regression equation contains two new measures of goodness of fit: the *multiple correlation coefficient* R and the *coefficient of multiple determination* R^2. These measures are directly related to the simple correlation coefficient r and coefficient of determination r^2. It is usual to employ uppercase letters for the multiple regression equation and lowercase for the simple regression. Recall that the simple correlation coefficient r measures the strength of the association between variables X and Y in a simple regression equation. In a multiple regression equation, we still have a single dependent variable Y, but at least two independents X_1 and X_2. The multiple correlation coefficient measures the degree of linear association between Y and the pair of independent variables *jointly*. This is easiest to imagine if we understand that it is the correlation coefficient between Y_i and $\hat{Y}_i$. Since $\hat{Y}_i = a + b_1X_{1i} + b_2X_{2i}$, it contains the overall effect of both X_1 and X_2. As we see in our results, the correlation has increased significantly from 0.9153 to 0.9910.

Now, we already know that the simple coefficient of determination represents the proportion of the total variation in Y that can be explained by variable X, or

TABLE 13-2
Summary of Multiple Regression Output for Example 12-2

Multiple R	0.9910
Multiple R^2	0.9820
$S_{Y \cdot X}$	0.2643

ANOVA

	df	*SS*	*MS*	*F*	Sign.
Regression	2	34.2881	17.1440	245.4572	0.0000
Residual	9	0.6286	0.6985		
Total	11	34.9167			

Equation

	Coeff.	Std. error	*t*-stat.	*P*-value	Lower 95	Upper 95
Intercept	–3.0081	0.5233	–5.7488	0.0003	–4.1918	–1.8244
Income	0.1167	0.0055	21.2546	0.0000	0.1043	0.1292
Household size	1.1788	0.1388	8.4922	0.0000	0.8648	1.4928

RSS/TSS. The multiple coefficient of determination is computed using the exact same formula, but we now know that RSS, our regression sum of squares, contains the effects of both of our independent variables. To see this, let us examine the ANOVA table in the middle portion of Table 13-2. First, we note that the total sum of squares or TSS in the equation, which represents the total variation in Y, is identical in both equations: 34.9167. This is the total variation available to be explained no matter how many independent variables we have! We would expect a significant increase in RSS when we add a new variable to our equation, and our intuition is correct. Our error ESS declines from 5.6657 to 0.6286, or by 5.0371. Clearly, RSS must increase by this amount, and it does, from 29.2510 to 34.2881.

The last portion of our output contains the actual multiple regression equation. The intercept has changed dramatically from +0.0081 to –3.0081. Should we be concerned with this change of sign? Well, it is almost meaningless to suggest that households with no members and no income make –3 trips per day! This is out of the range of the data for both of our independent variables and thus is of no practical significance. Notice that the coefficient for income has changed marginally from 0.1118 to 0.1167. Why is this coefficient not the same? The reason is that the two independent variables, household income and household size, are not completely independent and are slightly correlated. If we introduce a new variable that is completely independent of all existing variables, the partial regression coefficients of the existing variables will not change. The effects of the new variable are truly *additive* as the equation implies. Most often, the variables will have some correlation, and *gross* and *partial* regression coefficients will not be identical. The relationship among the independent variables

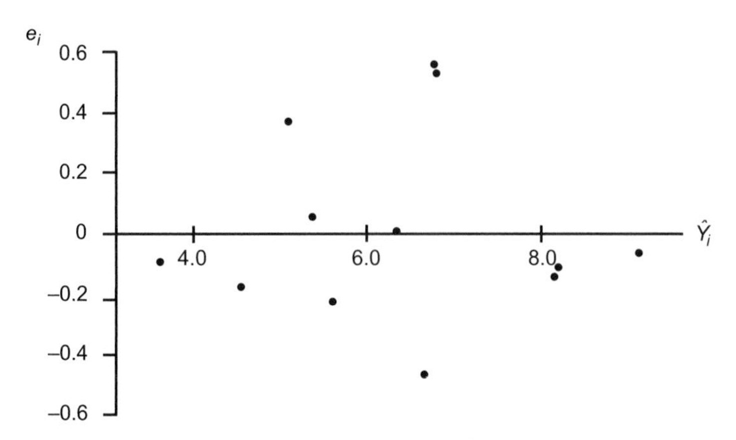

FIGURE 13-2. Residual plot for the multiple regression equation of Example 12-2 (cont.).

is one of the more difficult problems in regression analysis. We will deal with it later in this chapter in the section on *multicollinearity.*

Next, we notice that there are several different tests for significance in a multiple regression equation. First, there is the overall test of the equation expressed in the ANOVA portion of the output. The calculated F-value of 245.4572 is significant with a PROB-VALUE < 0.00005 (otherwise it would have been rounded up to 0.0001 for four digits), and we conclude that our two variables together are significant explanatory variables of household trip generation rates. Unlike simple regression there is no *simple* direct relationship with this F-value and the value of the t-statistic for our individual variables. Our standard error of the estimate, or RMSE seems low at 0.2643 but we must examine the practical side of this value. How many households are there in the city? If we are, say, 95% sure that we can predict within, say, 0.5 trip per day, this amounts to 50,000 trips in a city with 100,000 households.

Notice that we also have tests of significance on the individual partial regression coefficients. In this case, it appears both variables are highly significant as well. One of the perplexing things about multiple regression analysis is that it is possible to have a highly significant equation, but only one of the variables in the equation may be responsible for this. Here, we can conclude that both coefficients are nonzero and have independent effects on trip generation rates.

Does this new equation satisfy the assumptions of regression analysis? Figure 13-2 shows the residual plot of this new equation. Though, of course, there are too few points to make any definitive judgement about patterns in the plot, for some reason it appears that traffic zones with low trip rates and those with high trip rates are better predicted than those with intermediate trip rates of between 5 and 7 per day. Also, it appears that the spatial pattern of our residuals is not as problematic as the east–west pattern of positive and negative residuals apparent in Figure 12-17.

Assessing the Impact of Individual Variables

The derived multiple regression equation for Example 12-2 is

$$Y = -3.0081 + 0.1167X_1 + 1.1788X_2 \quad (13\text{-}5)$$

where X_1 is household income in thousands of dollars and X_2 is household size measured by the number of persons. We are tempted to say that household size is the most important of the two variables since it has the most dramatic effect on total trips. In Equation 13-5 we see that the partial regression coefficient for household size of 1.1788 is almost 10 times as large as the coefficient for household income. This means that each person added to a household seems to add about 1 trip per day to the trip total.

Unfortunately, it is not so simple. The values for the regression coefficients are completely dependent on the *scale* of the numbers. Suppose, for example, we measured household income in tens of thousands of dollars rather than thousands. A household with income of $30,000 would be coded as 3 instead of 30. The partial regression coefficient for income would change to 1.167 (10 times .1167). So for a household with an income of $30,000, trips could be estimated as .1167(30)or 1.167(3). In either case, the answer is the same: the income effect is 3.501 trips. The implications of this are that we must use a measure that is independent of scale in order to compare the relative impact of two independent variables on the dependent variable. A measure commonly termed the *beta weight* or *standardized regression coefficient* is best used to answer questions concerning the relative importance of our dependent variable to the set of independent variables.

DEFINITION: STANDARDIZED REGRESSION COEFFICIENT

If $X_1, X_2, \ldots, X_{p-1}$ are a set of independent variables in a multiple regression with partial regression coefficients $b_1, b_2, \ldots, b_{p-1}$, respectively, then the standardized regression coefficient for each variable is defined as

$$b_i' = b_i \frac{s_{X_i}}{s_Y} \quad (13\text{-}6)$$

where s_Y is the standard deviation of the dependent variable and s_{X_i} is the standard deviation of independent variable X_i.

If we calculate these standardized coefficients for household income and household size

$$b_1' = 0.1167 \frac{14.592}{1.782} = 0.956 \qquad b_2' = 1.1788 \frac{0.577}{1.782} = 0.382$$

and it now appears that, even though the partial regression coefficient for household size is much higher than the coefficient for household income, the standardized effect of income is over twice that of household size. Caution should be undertaken when interpreting these coefficients since they will differ depending on which other variables are included in the equation. They are particularly sensitive to the problem of *multicollinearity* among variables described later in this section.

Selecting Independent Variables

To this point, we have focused on the interpretation and typical output of a single multiple regression equation. In terms of the overall framework for regression analysis of Figure 12-1, this is but one part of a process of fitting potential regression models, interpreting their results, addressing model inadequacies, and deriving a final model or set of models. Typically, a researcher has available a potential set of independent variables. For k different variables there are 2^k different equations that can be constructed from the set of variables. For example, for $k = 3$ variables, Table 13-3 lists the eight different equations that must be evaluated. There is one with no independent variables, one with three variables, three with one variable, and three with two variables. With only $k = 10$ variables there are 1,024 possible regression equations. Since the set may also include different variables measuring the same concept, the task of the research is made even more complicated. For example, education might be measured by the average number of school years completed, the proportion with university degrees, the proportion with advanced degrees, or the proportion completing high school. Methods to reduce the complexity of the task are clearly required.

Two different general approaches to this problem exist. First, it is possible to take a combinatorial approach and complete the *all-possible-subsets regression.* As the name suggests, this approach is based on examining (albeit in the most superficial way) all potential combinations of independent variables. With computerized procedures, this method is now feasible even for fairly large problems with thousands of observations and large sets of independent variables. However, it is usual to limit the output from this procedure to a fairly small set of summary statistics such as R^2, root-mean-square error, and other such measures of the goodness of fit. It is common to ignore many other aspects of individual equations—for example, the appropriate-

TABLE 13-3
Possible Regression Equations with Three Independent Variables

Equation	Number of independents	Variables included
1	0	–
2	1	X_1
3	1	X_2
4	1	X_3
5	2	X_1, X_2
6	2	X_1, X_3
7	2	X_2, X_3
8	3	X_1, X_2, X_3

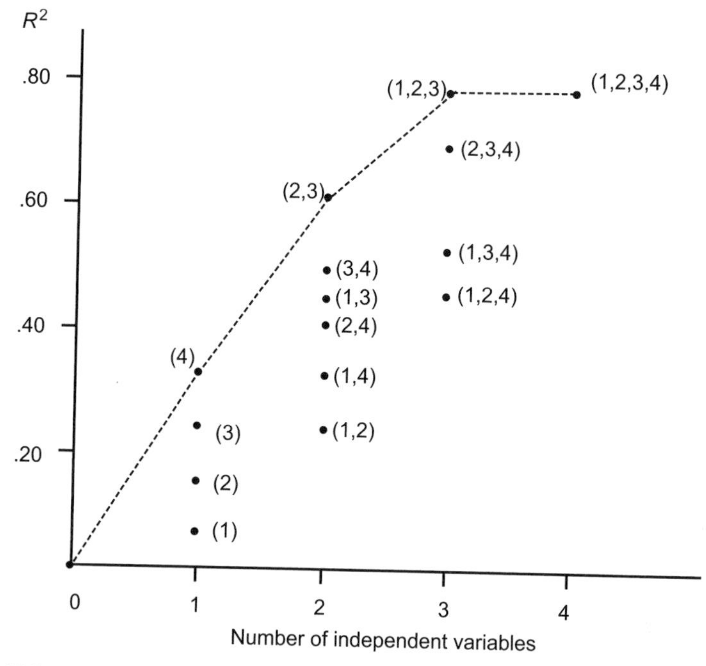

FIGURE 13-3. Hypothetical plot of R^2 for an all possible subsets regression. *Source:* Kutner et al. (2004).

ness of the model in meeting the assumptions of regression. Nevertheless, this approach seems to be very useful in identifying a small number of equations worth exploring in more detail. Consider Figure 13-3. On the *Y*-axis is a measure of goodness of fit, in this case R^2. Where there are four independent variables there are 16 different solutions, and all are ordered by the number of variables in the equation. The dashed line connects the highest R^2 values for the best equation found with a given number of variables. Variable X_4 is the best single variable to include in the model. The best two-variable pair is X_2 and X_3, and the best three-variable combination consists of variables X_1, X_2, and X_3. One advantage of this approach is that it may identify a small number of superior equations that are worth examining in greater detail.

The second major approach to equation development is to use automatic search procedures. *Stepwise* estimation is probably the most popular of these sequential methods of equation development. Using this method, the researcher examines the contribution of each independent variable to the equation and adds the variable with greatest incremental contribution first. Then, the remaining variables are examined to see which new variable can be added to the equation with the most improvement in fit. The process continues until there is no longer any benefit to adding individual variables. In contrast to this forward selection method, a backward stepwise selection method begins with a model that includes all independent variables and then progressively eliminates variables that are not useful until a suitable equation is developed. *Forward addition* and *backward elimination* are largely trial-and-error processes.

In a fully stepwise procedure, variables can be added or deleted at each stage of the process. *By considering only a single variable at a time,* however, a potential bias is introduced. It could be that two variables could *together* explain a significant proportion of the variance, but neither may be significant on their own. It is possible that neither would be considered in any final model. In Figure 13-3, for example, the stepwise procedure would add variable X_4 in the first step since it appears to give the best fit, and then the procedure would add variable X_3 in the second step. This equation would be inferior to the equation consisting of variables X_2 and X_3.

In reality, the analyst often uses both of these approaches as a first step in model identification. Both approaches often provide a small subset of potential equations, which are then subjected to a more detailed analysis. Sometimes we are interested in developing a purely *predictive* model, and we have two primary goals in model development. First, we want to develop a model with the best possible predictions. A second and sometimes conflicting goal is to develop a *parsimonious* model, one with few parameters to be estimated. Smaller models may be preferable to larger models. One with fewer variables is probably easier to maintain than a larger model since the cost of data collection, checking, and maintenance may represent a significant cost. If we are building an *explanatory* model, we know that judgment plays an important role. It may be that we have strong *a priori* grounds for including a particular variable in our final equation. The evidence of previous research and currently accepted theory cannot be ignored. It would be unwise to exclude a variable that is known to be essential. Instead, the researcher is more likely to wonder what it is about her data that has caused her results to differ from previous work or accepted theory.

Regression analysis assumes that our model is correctly specified. Typically, when adding or deleting variables in model development, we may omit a relevant variable, or we may use a superfluous one. Which is a more serious problem? First, incorporating superfluous variables does not bias the model but makes the estimates less precise. If we omit a variable, on the other hand, our regression equation is biased since our error term cannot possible contain purely random error. The error includes the effect of the missing variable. This bias is considered the more serious error of the two. *Specification* error in regression analysis refers to the errors that arise when the model we fit is incorrect and is not a true model for the problem at hand. In addition to the problems of omitted or superfluous variables, the model may be incorrectly specified because we are fitting a linear equation to what is really a nonlinear relationship between the variables. Potential solutions to this problem are discussed in Section 13.2.

Finally, it should be noted that many applications of regression analysis are closely tied to theory, and the choice of independent variables is often specified by the theory itself. In such cases, more interest is often placed on examining the values of the partial regression coefficients. Do they have the correct signs? Are the magnitudes plausible? Using the regression model for prediction may be of minor interest.

Multicollinearity

Perhaps the most vexing of all problems in multiple regression analysis is the problem of *multicollinearity.*

DEFINITION: MULTICOLLINEARITY

Multicollinearity in a multiple regression equation occurs when the independent variables $X_1, X_2, \ldots, X_{p-1}$ are intercorrelated.

Multicollinearity is a matter of degree. There is virtually always *some* intercorrelation among the independent variables of a multiple regression equation. At a certain point, however, the degree of intercorrelation causes problems in terms of both explanation and estimation. The conventional multiple regression equation assumes that all included variables are independent of one another, and therefore it is possible to simply add their effects to determine the value of the dependent variable. In the extreme case of multicollinearity when two of the independent predictor variables are perfectly correlated and $r = 1$, the multiple regression equation cannot even be estimated! Even when it is not perfect, high degrees of multicollinearity can result in regression coefficients having the wrong signs or implausible magnitudes. Why? Because when we are examining the impact of one of our independent variables we are not holding the value of our other independent variables constant! The existence of intercorrelation among our variables means that the two variables move together and not independently.

In addition to the problems related to estimation, there is also considerable difficulty interpreting the effects of individual variables on the dependent variable when a multiple regression equation is characterized by moderate to high multicollinearity. Consider Figure 13-4, which illustrates the possible outcomes in a multiple regression problem with only two variables. If our two variables X_1 and X_2 are perfectly uncorrelated, then the situation is simple and illustrated in panel (a). The variation in our dependent variable Y can be divided into three parts: the part explained by variable X_1, the part explained by X_2, and the unexplained portion. There is no overlap. In panel (b), the case in which there is only a modest amount of correlation between the two variables, the independent effects of the two individual variables will each be apparent. But when the shared component of the explained variation is quite high as in panel (c), it becomes extremely difficult to sort out the effects of the individual variables. The common interpretation of the partial regression coefficient measuring the change in the value of the dependent or response variable to a unit change in the independent variable is no longer applicable. It may be theoretically possible to think of varying X_1 and not X_2 (i.e., holding it constant), but it is unrealistic in the context of the available data.

On the other hand, the existence of multicollinearity does not inhibit us from obtaining good fits or from making predictions of mean response for our dependent variable. In fact, it is possible to envisage a situation in which only one of several variables can be shown to be statistically significant using their t-values, but the ANOVA for the equation as a whole suggests a highly statistically significant result! There is one solution to multicollinearity that is not advisable. Never omit a variable simply because it appears to be highly correlated to another variable and its significance appears to decline when the new variable is added. Remember, omitting a variable is often far worse than having an extra variable in an equation since it can *bias* our

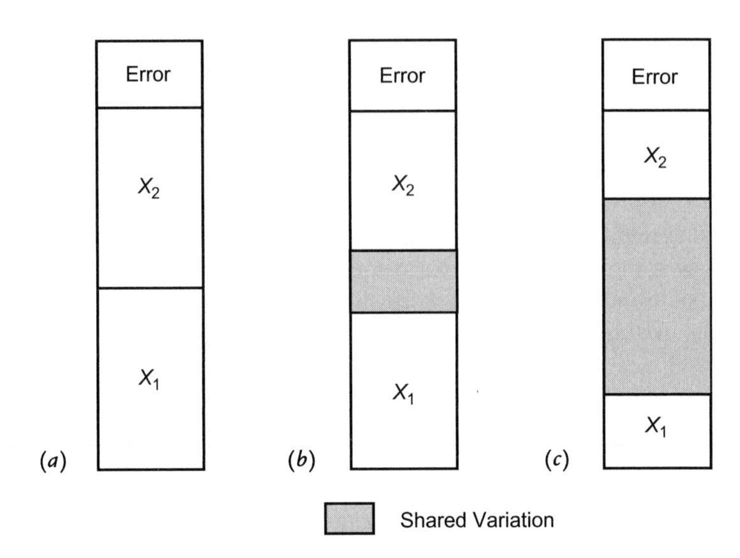

FIGURE 13-4. Decomposition of the variation in dependent variable Y to independent variables X_1 and X_2.

results. Remedies for multicollinearity are described in both Kutner et al. (2004) and Hair et al. (2006).

Regression Diagnostics

One of the key steps in a regression analysis is to examine whether or not a particular equation satisfies the assumptions on which least squares analysis is based. Even when the results of a multiple regression analysis seem particularly promising—a high R^2, low root mean square error (RMSE), statistically significant regressors, and so on—unless we are sure that we have met all the assumptions of regression analysis, our interpretation of these results may be misleading. For this reason, it is common to augment the analysis of the equation itself with a series of diagnostic measures and graphics. This is especially important when data from previous studies have pointed to the potential existence of such problems. In Section 12.4 we introduced several graphical diagnostics for examining normality of the residuals, randomness in the error term, and homoscedasticity. Simple residual plots were found to have significant advantages in assessing the extent of potential violations. In this section we shift our attention to individual observations and focus on two issues—identifying *outlying* observations and *influential* observations. No regression analysis is complete without also assessing the importance of these issues to the results.

DEFINITION: OUTLIERS

Outliers are observations in a set of data that have unique characteristics identifying them as different from other observations in the data.

It has been common in regression analysis to restrict the term *outliers* to those observations having large residuals, that is, observations located far away from the regression line, plane, or multidimensional least squares surface. This is unfortunate. In a broader sense, as our definition suggests, outliers are observations or cases that may be viewed as different from other observations whether or not they have a large residual. Certainly, we must identify and closely examine any observation with a large residual. It is common for most statistical software packages to identify all observations greater than two standard errors (two times RMSE) from the regression line and bring them to the attention of the user. Typically, we would want to know why these observations are poorly predicted. Is the cause simply measurement error, perhaps due to incorrect data entry? More often than not, such observations are found to represent real data points, and it is necessary to determine whether they have a significant impact on the results so that any conclusions we may draw remain valid.

In a more general sense, outliers can be identified by their actual X- and Y-values as in the simple regression scatter of Figure 13-5. Three different outliers exist. Observations 1 and 2 are outlying because of their X-values since they have larger X-values than those of other cases. Observation 3 is not an outlier for this reason, but because of its Y-value; it lies far above the scatter of points representing the other observations. Observation 1 is also an outlier because of its Y-value. Not all of these observations will be found if we only examine the residuals from a regression equation since observation 2 is likely to be fairly closely predicted by the regression line. If we were to eliminate any one of these observations, it is likely that our results would change the most if it were observation 1 and least if it were observation 3. When our results can be dramatically altered by the elimination of a single observation, we must be

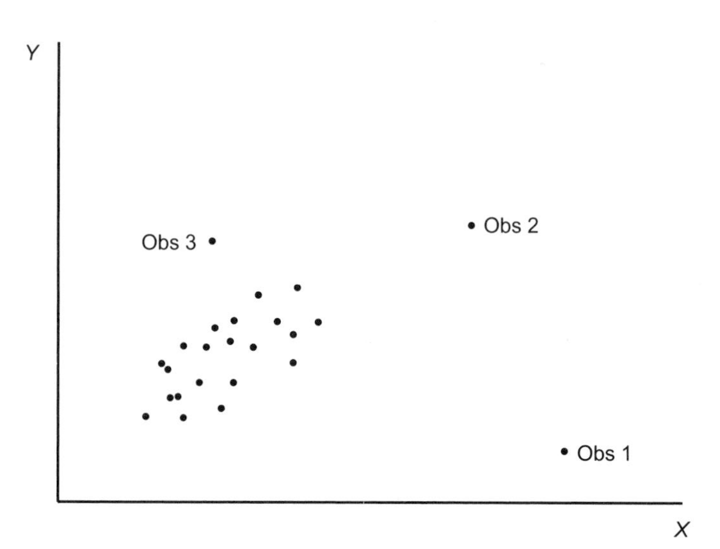

FIGURE 13-5. Identification of outliers in a simple regression framework.

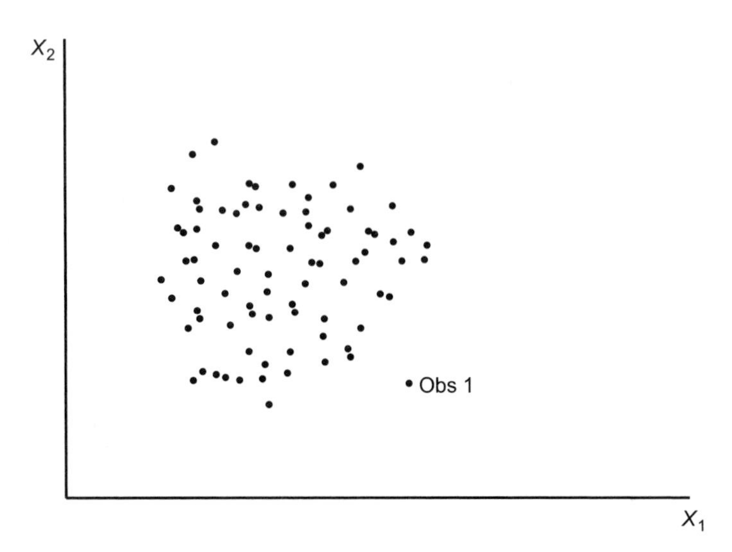

FIGURE 13-6. Identification of dataspace and outliers for two independent variables.

very careful in assessing both the explanatory power of the equation and its utility in prediction.

In the case of a multiple regression equation with several independent variables, the situation is even more complex. An observation can be considered to be outlying not because it has an outlying value of Y for *one* of the independent variables X_1, X_2, $\ldots X_{p-1}$ but because it has a *unique* combination of values for some pair of these variables. Consider Figure 13-6. The outlying observation identified in this figure is not outlying with respect to either variable X_1 or to variable X_2. There are observations with higher and lower values for both of these independent variables, but no other observation has the *combination* of values of this observation. If this observation were to be found to have a high residual from our regression function, this may be the cause. We simply might have insufficient information about the nature of the relationship between our dependent variable Y and these variables at this combination of X_1, X_2 values. Simple bivariate plots such as Figure 13-6 also are useful in defining the *dataspace* within which our equation is valid. Making predictions outside this dataspace is equivalent to *extrapolation* and is much more suspect than making predictions for values of our independent variables for which we have adequate data.

When we have three explanatory variables, it becomes more difficult to display both the dataspace and outliers graphically, though three-dimensional viewers available in many statistical packages have some limited capabilities. Algorithms to find outliers based on distances from the scatter of the other observations exist but are not typically used in most software packages. However, several tools have been developed to find *influential* observations, observations that have a disproportionate impact on the results of our regression analysis. Some of these influential points may be identified as outliers, and others may not.

DEFINITION: INFLUENTIAL OBSERVATIONS

Any observation in a regression analysis whose exclusion causes major changes in the results of the regression analysis is an *influential* observation.

There are two commonly used indicators for identifying influential observations. First, an observation can be shown to be influential if its omission or exclusion causes significant changes in the partial regression coefficients of the model. Consider the influential observation identified in Figure 13-7. When it is included as a data point, the regression line is drawn away from the scatter toward the observation, identified as the solid line. When the influential observation is removed, the slope of the regression line changes dramatically as shown by the location of the new dashed regression line. The simplest measure of this is simply based on the change in the partial regression coefficients:

$$b_k - b_{k(i)} \tag{13-7}$$

where b_k is the regression coefficient for variable k with all observations in the estimation and $b_{k(i)}$ is the value of the coefficient with observation i excluded. These can be calculated for each observation i and each variable in our multiple regression equation.

A second way of identifying influential observations is to calculate the *deleted residual* for each observation. The procedure is quite simple. Delete the ith case, fit the regression function to the remaining $n - 1$ cases, and then obtain a point estimate

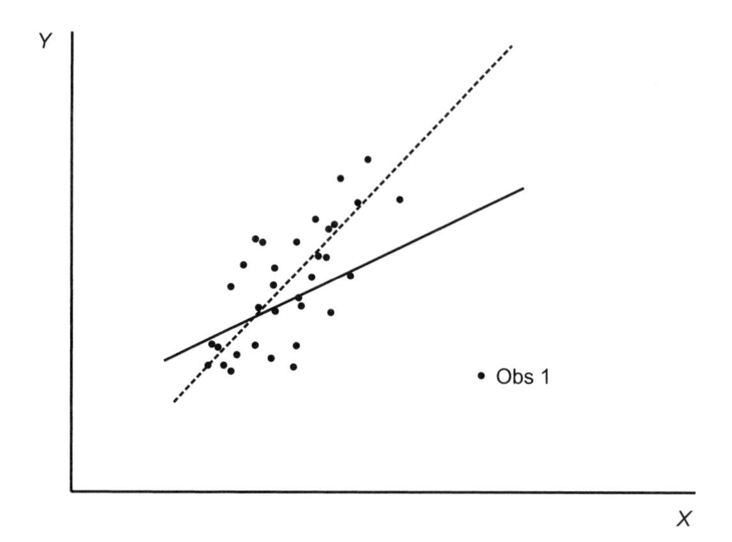

FIGURE 13-7. Identification of an influential observation in a simple regression context.

of the expected value for Y when the X levels are those of the ith case. Denote this expected value $\hat{Y}_{i(i)}$. The difference

$$d_i = Y_i - \hat{Y}_{i(i)} \tag{13-8}$$

is known as the deleted residual for the ith case. Why is this effective? Sometimes an outlier does not have a large residual since it draws the regression equation toward it. However, when the point is deleted and the regression line is reestimated, the residual is likely to be larger. This is clearly illustrated by the residuals from the two regression lines in Figure 13-7.

Methods for standardizing these two measures (known as DFBETAs and DFFITS) and identifying influential observations are discussed in Kutner et al. (2004). There is one drawback to both of these measures. If there were two outlying points in Figure 13-7 quite close together, then neither would necessarily be found using these techniques. The removal of one would not necessarily cause a dramatic change in the regression equation; only if both were eliminated would the location of the regression line shift dramatically. Obviously, more sophisticated measures can be derived by extending the logic underlying these measures to the simultaneous deletion of pairs or larger sets of observations. While it may seem onerous to undertake the calculations for these measures, this is not the case. The calculation of these two measures of influence can be accomplished using only the results from fitting the entire dataset.

13.2. Variable Transformations and the Shape of the Regression Function

We have already noticed that simple scatterplots can often reveal the nature of a regression relationship. It is not at all uncommon to see a scatter of points with an apparently nonlinear pattern. For example, Figure 13-8 illustrates the relationship between mercury concentration and lake alkalinity from the data for the mercury contamination in bass in Problem 4 at the end of Chapter 12. The scatter reveals a general decrease in mercury contamination with increasing lake alkalinity, but at a decreasing rate. If regression analysis were to be limited to purely linear functions and could not be applied to nonlinear relationships such as shown in Figure 13-8, it would not be such a valuable method for analyzing the relationships between variables. Fortunately, this is not the case.

Nonlinear models can be divided into two types: *intrinsically linear* models and *intrinsically nonlinear* models.

DEFINITION: INTRINSICALLY LINEAR MODEL

A nonlinear model is said to be *intrinsically linear* if it can be expressed in the standard linear form given by Equation 12-1 by using a suitable transformation of one or both variables in the model. If such a transformation cannot be found, then the model is *intrinsically nonlinear.*

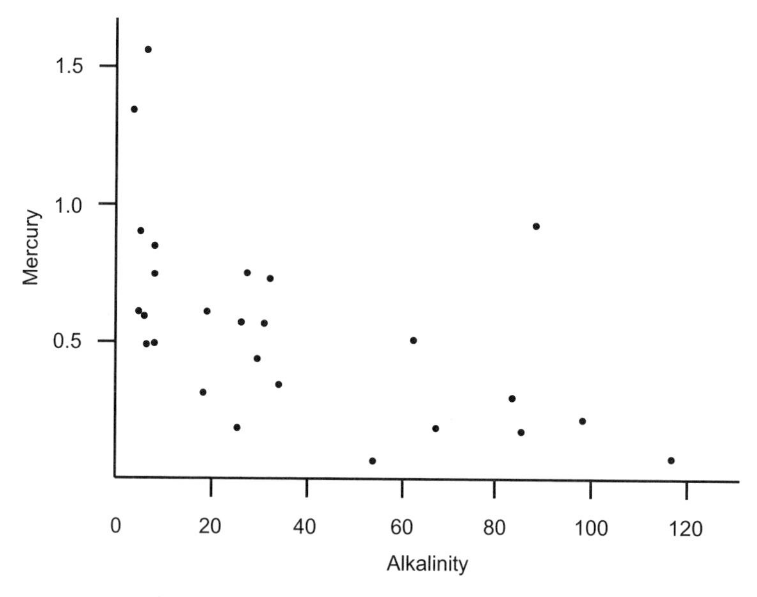

FIGURE 13-8. Mercury and alkalinity scatterplot.

Intrinsically linear models are of particular interest because the normal least squares estimating procedures can still be used to determine the parameters in the equation. How can we take scatterplots that are obviously nonlinear and convert them to scatterplots in which the plot is linear or near linear? This is done primarily by applying *transformations* to our two variables X and Y. By transformation we simply mean that we replace variable X for example, with another variable, say $\log_{10} X$ or $\sqrt{X}$ or X^2. Our scatter may be nonlinear when we graph X against Y but linear when we graph the points for $\log_{10} X$ against Y. Consider the following data and its scatterplot illustrated as Figure 13-9.

X	Y	$\sqrt{X}$
1.0	1.0	1.000
2.0	2.5	1.414
3.0	2.75	1.658
4.0	2.90	1.703

First, it is apparent from Figure 13-9a that the *rate* at which Y increases with X decreases with the value of X. Obviously, this cannot be a linear function since the slope or rate of change of Y with X, estimated by the parameter b in Equation 12-1 is not constant over the range 1–4. The last column of the table lists the values for the transformed variable $\sqrt{X}$. In the second row, for example, $1.414 = \sqrt{2}$. In Figure 13-9b, the values of the transformed variable $\sqrt{X}$ are graphed against Y. Note that the path of the

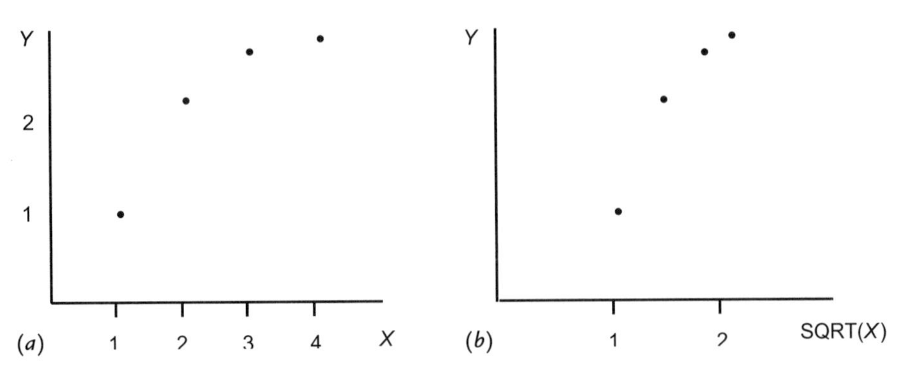

FIGURE 13-9. Transformation of independent variable to achieve linearity.

points is now nearly linear, and we would feel quite confident fitting a linear equation to this scatterplot.

Transformations of variables are also undertaken for reasons other than the apparent nonlinearity of the relationship. For example, we shall see that some transformations can be used to correct deficiencies in the residuals—to induce normality in their distribution or to reduce heteroscedasticity. Both of these issues are addressed below.

Transformations for Linearity

If the distribution of the error term is near normal and the error term is homoscedastic, then it is often possible to use a simple transformation of the independent variable X to correct for nonlinearity. Transformations of the Y variable are generally unnecessary and may create other problems. This is because any transformation of Y will invariably affect the distribution of the error term, and it may no longer be normally distributed or homoscedastic as required by the assumptions of least squares analysis (recall the required assumptions outlined in Section 12.2). The particular transformation of X to apply depends primarily on the shape of the function suggested by the scatterplot.

Figure 13-10 describes the three most common nonlinear patterns that can be corrected using simple transformations. Note that all patterns assume the error terms are both normal and homoscedastic, as reflected in the shape of the band around the relationship in each case. Notice also that in any one case at least two different transformations can be applied to the independent variable. Several different transformations can generate *virtually identically shaped* functional relations. In the absence of any theory suggesting the appropriate transformation to be taken, the choice of transformation rests with the analyst. This is sometimes difficult since the results of all the equations in terms of goodness of fit will also often be very similar.

Returning now to the mercury contamination of bass data depicted in Figure 13-8, we see that it follows the pattern of Figure 13-10c, and an appropriate transformation would be any one of $X' = \frac{1}{X}$, $X' = \exp(-X)$, or $X' = \log_{10}X$. Figure 13-11 illus-

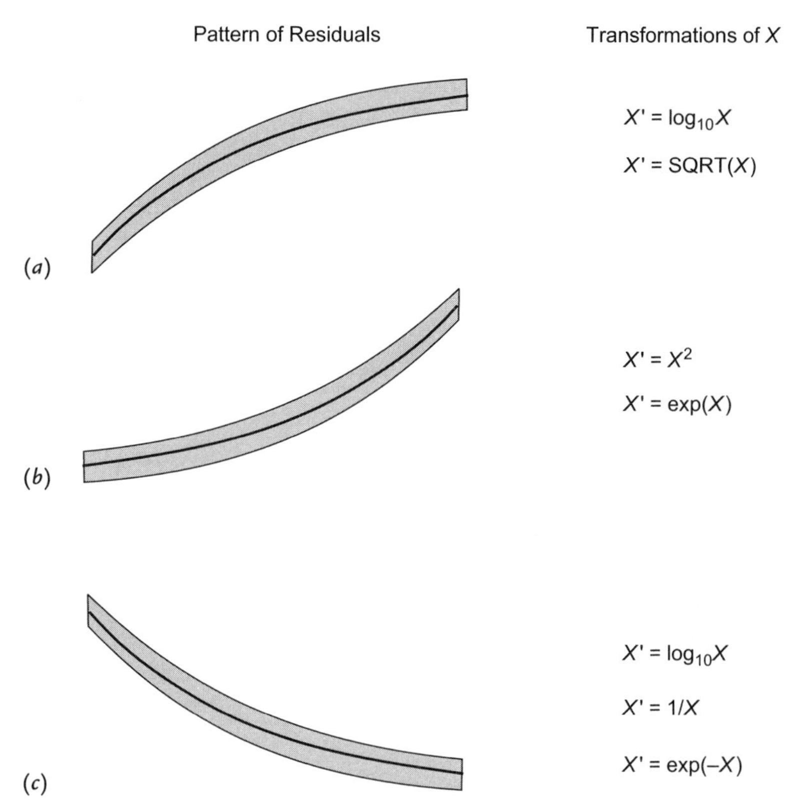

FIGURE 13-10. Simple transformations of X for nonlinear regression functions with constant error variance.

trates a graph of the relationship between mercury concentration and the logarithm to the base 10 of alkalinity. Though there is a wide scatter, it is clear that the relationship appears to be more linear than the one depicted in Figure 13-8, although in both cases the graphs are characterized by two or three prominent outliers.

Transformations to Correct Non-Normality and Unequal Error Variance

When we transform the independent variable X, we generally have little impact on the error term. But if the error term is non-normal or heteroscedastic, it often is useful to transform the dependent variable Y to remedy this departure from the simple regression model specified as Equation 12-1. Figure 13-12 illustrates several different heteroscedastic patterns of the residuals in which the variance of the error term varies with the value of the independent variable X. Any one of the three simple transformations $Y' = \frac{1}{Y}$, $Y' = \log_{10} Y$, or $Y' = \sqrt{Y}$ may be useful in this case. Plots of residuals such as Figure 12-15 may be used to determine the most effective transformation to undertake.

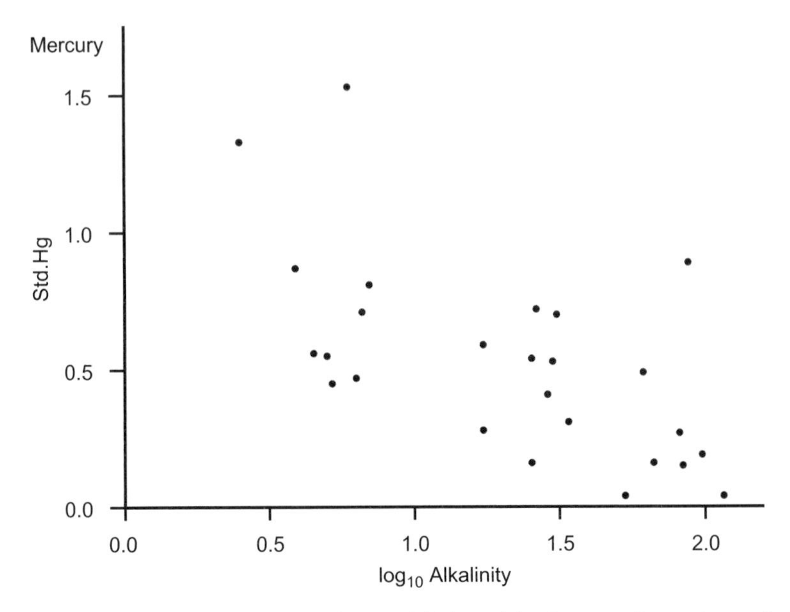

FIGURE 13-11. Mercury in bass with logarithmic transformation of alkalinity.

Functional Forms

Sometimes it is necessary to transform both the independent and dependent variables simultaneously in order to meet the assumptions inherent in the regression model. Consider the following multiplicative specification

$$Y_i = \alpha X_i^{\beta} \varepsilon_i \tag{13-9}$$

in which the dependent variable Y is equal to the product of a constant α, and the independent variable X to the power β and the error term ε. If we take the common logarithms of both sides of this equation and simplify, we obtain

$$\log_{10} Y_i = \log_{10} \alpha + \beta \log_{10} X_i + \log_{10} \varepsilon_i \tag{13-10}$$

This equation can be made consistent with the simple regression model of Equation 12-1 by substituting $Y' = \log_{10} Y_i$ and $X' = \log_{10} X_i$. There is one important difference between Equation 12-1 and Equation 13-10. For the assumptions of hypothesis-testing and confidence interval estimation, the *common logarithm of the error term must be normally distributed.*

One feature of the multiplicative power function specified by Equation 13-10 is that different values for the β parameter produce a wide variety of functional forms. As we see in Figure 13-13, when $\beta = 0$ the function is parallel to the X-axis and indicates no relation between X and Y. When $\beta = 1$, the function reduces to the ordinary simple linear regression model. Values of $\beta > 1$ result in power functions with in-

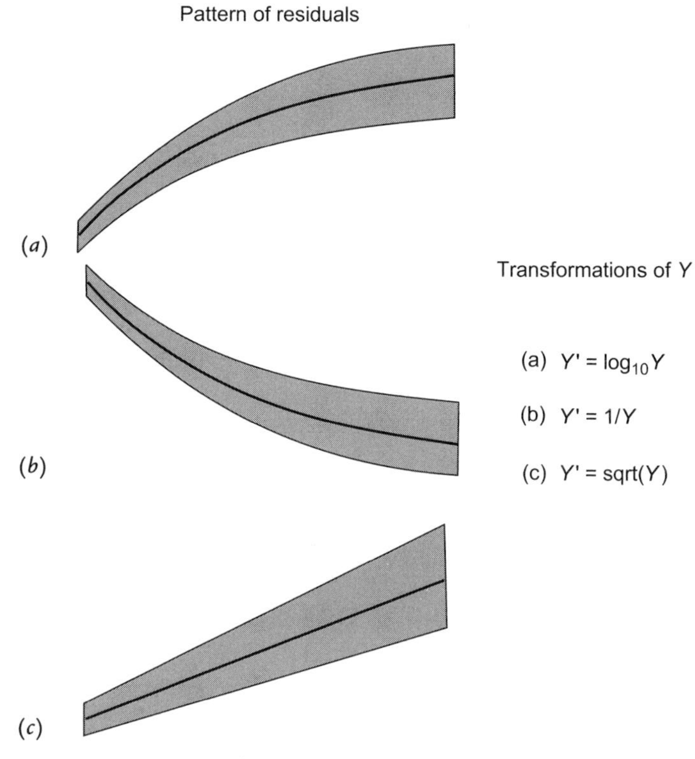

FIGURE 13-12. Simple transformations of *Y* for heteroscedastic regression plots.

creasing rates of change, and where $0 < \beta < 1$ the function is characterized by decreasing rates of change of Y with X. Finally, when $\beta = -1$, the function takes on the form of the reciprocal function. To find which functional form best describes the data, it is only necessary to fit the log-linear version of the regression equation (13-10). The value of the slope parameter β points to the appropriate form of the functional relation.

Transformations in Multiple Regression

Fitting a regression equation in which there are multiple independent variables does not preclude the use of variable transformations. For example, the multiplicative power function of Equation 13-10 can be extended to two variables:

$$Y_i = \alpha X_{1i}^{\beta_1} X_{2i}^{\beta_2} \varepsilon i \qquad (13\text{-}11)$$

In this case there is a regression coefficient for each variable X_1 and X_2, and the equation to be fit is

$$\log_{10} Y_i = \log_{10} \alpha + \beta_1 \log_{10} X_{1i} + \beta_2 \log_{10} X_{2i} + \log_{10} \varepsilon_i \qquad (13\text{-}12)$$

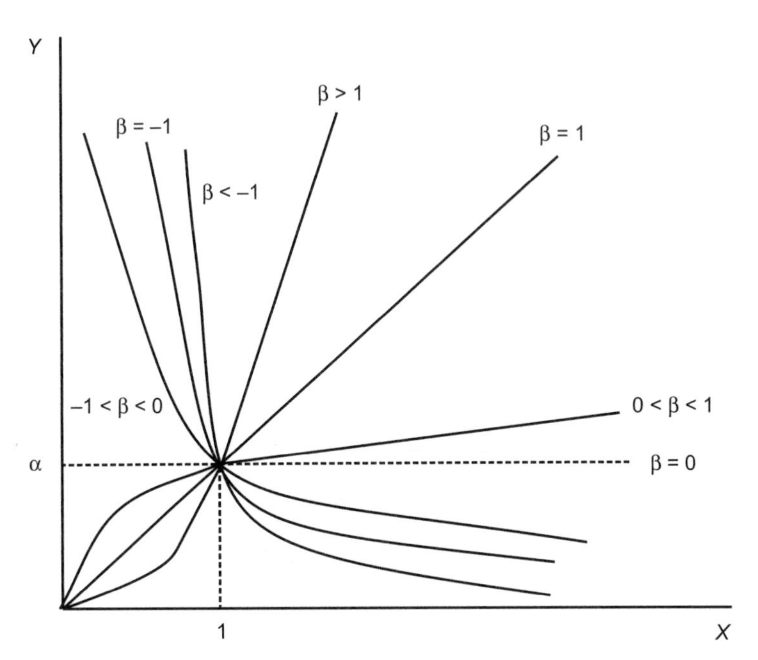

FIGURE 13-13. The family of curves for the power function $Y = \alpha X^{\beta}$.

so that the required transformations are $Y' = \log_{10}Y$, $X_1' = \log_{10}X_1$ and $X_2' = \log_{10}X_2$. This formulation can be extended to three or more variables.

Consider once again the data for mercury contamination in bass. Let us now extend our analysis by fitting a multiplicative power function for the data supplied in Problem 12-4. The results from a purely linear form of a predictive equation for mercury with independent variables alkalinity and chlorophyll is summarized in Table 13-4. In terms of goodness of fit, the two variables combine to explain almost 40% of the variation in mercury concentration levels, and the standard error is 0.2975 ppm of mercury. Note the significance of the overall equation of 0.0041, but the individual variables are not nearly so significant. This suggests some correlation exists between our two independent variables. Graphical analysis of the relationship between the independent variables and mercury concentration such as Figure 13-8 suggests some transformations may prove useful to linearize the relationships. It turns out that both individual variables have almost the identical form of nonlinear relationship, and so a logarithmic transformation seems appropriate.

Instead of simply transforming the two independent variables, a fully multiplicative model of the form of Equation 13-11 is now fit to the data. The results of the transformed model are summarized in Table 13-5. First, we note a distinct improvement in the goodness-of-fit statistics. The coefficient of multiple determination R^2 has risen by approximately 0.14, an improvement of over one-third. The ANOVA reveals an equation with an even higher level of significance (0.0002), and both of the individual variables are found to be significant with p-values less than 0.05. Of course, it

TABLE 13-4
Summary of Multiple Regression Output for Mercury in Bass Data

Multiple R	0.6168
Multiple R^2	0.3804
$S_{Y \cdot X}$	0.2975

ANOVA

	df	*SS*	*MS*	*F*	Sign.
Regression	2	1.2490	0.6245	7.0565	0.0041
Residual	23	2.0355	0.0885		
Total	25	3.2845			

Equation

	Coeff.	Std. error	*t*-stat.	*P*-value
Intercept	0.7503	0.0861	8.7099	0.0000
Alkalinity	–0.0037	0.0021	–1.7820	0.0880
Chlorophyll	–0.0033	0.0018	–1.8568	0.0762

is still necessary to examine the diagnostics on each of these equations, but the results appear promising.

Exploring the Shape of Regression Function

While scatterplots can often show the nature of the regression function to be fit, sometimes this is not the case. The scatterplot may be so complex that the actual nature of the regression function might not be readily apparent. In these cases we need some tools to assist us in uncovering the *shape* of the function that may link the two variables. Simple techniques that generate *smooth* curves are often used for this purpose since they might point us to a single model or perhaps a small number of different transformations that might be useful. Four commonly used methods are

1. The method of moving averages
2. The method of running medians
3. Band regression
4. The loess method

Considerable work has been done in developing smoothing techniques for time-series data, and these techniques are very useful when the X_i are equally spaced apart. The *method of moving averages* repeatedly calculates the average of successive, overlapping groups of observations.

TABLE 13-5
Regression Results for Multiplicative Power Function for Mercury in Bass Data

Multiple R	0.7218
Multiple R^2	0.5210
$S_{Y \cdot X}$	0.2853

ANOVA

	df	*SS*	*MS*	*F*	Sign.
Regression	2	2.0367	1.0184	12.51	0.0002
Residual	23	1.8724	0.0814		
Total	25	3.9091			

Equation

	Coeff.	Std. error	*t*-stat.	*P*-value
Intercept	.2525	0.1571	1.6073	0.1216
log(alkalinity)	–0.2925	0.1387	–2.1092	0.0460
log(chlorophyll)	–0.2667	.1108	–2.4065	0.0245

DEFINITION: MOVING AVERAGE

A moving average of Y for a sequence of equally spaced X values is found by replacing each sequence of T observations by the mean of the sequence. The first sequence contains the observations $y_1, y_2, \ldots, y_T$, the second sequence contains the observations $y_2, y_3, \ldots, y_{T+1}$, and so on. The value of T is the *term* of the moving average.

The greater the value of T, the smoother the curve will be. This method of smoothing has been well developed within *time-series analysis,* where the data exhibit the characteristic even spacing of values of X.

The *method of running medians* is quite similar to the method of moving averages, except that the smoothed value is calculated as the *median,* not the *mean,* of the series of consecutive values being smoothed. As we learned in Chapter 3, the median is much less sensitive to outlying values and thus is potentially more useful in this situation. Both of these first two smoothing techniques can be refined in a number of different ways. For example, we might weight observations closer to the value being smoothed more than observations further away. Or we might use a *compound* smoother that is applied successively to the series.

What do we do when our data for X is not evenly spaced and simple moving average or running median techniques cannot be used? One possibility is *band regression.* Consider the following set of 10 observations:

X	Y	Median X	Median Y
1.0	1.1		
1.5	2.0	1.5	2.0
2	3.7		
2.7	2.9		
3.1	5	3.1	4.8
3.5	4.8		
4	5.1		
4.2	4.9		
		4.45	5.3
4.7	5.5		
4.9	5.8		

As we see in Figure 13-14, the observations reveal a near linear pattern, but there is some question about whether there are any curvilinear portions of the relationship and whether some transformation might be useful. First, we divide our observations into groups or bands based on the value of X. But how many? Broken into classes at what values of X? There is considerable flexibility in our choice, but let us suppose we select the three groups indicated in the table based on the values $0 < X \leq 2.0$, $2.0 < X < 4.0$ and $X \geq 4.0$. Within each band or class, calculate the median value of X and the median value of Y. This identifies one point for each of the bands. Connect these by straight lines, and the result should help point to the shape of the regression curve. Figure 13-15 illustrates the form of the band regression for this small example. In this case,

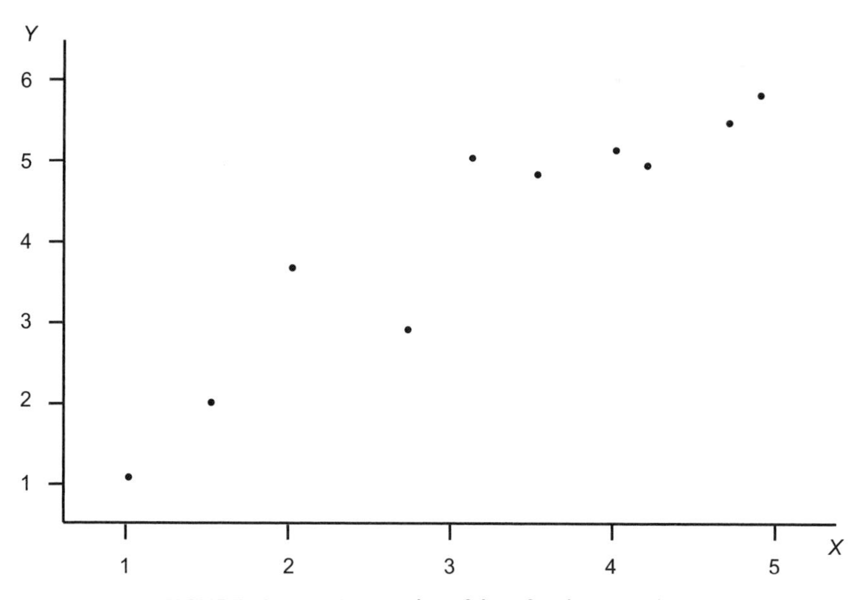

FIGURE 13-14. Scatterplot of data for shape analysis.

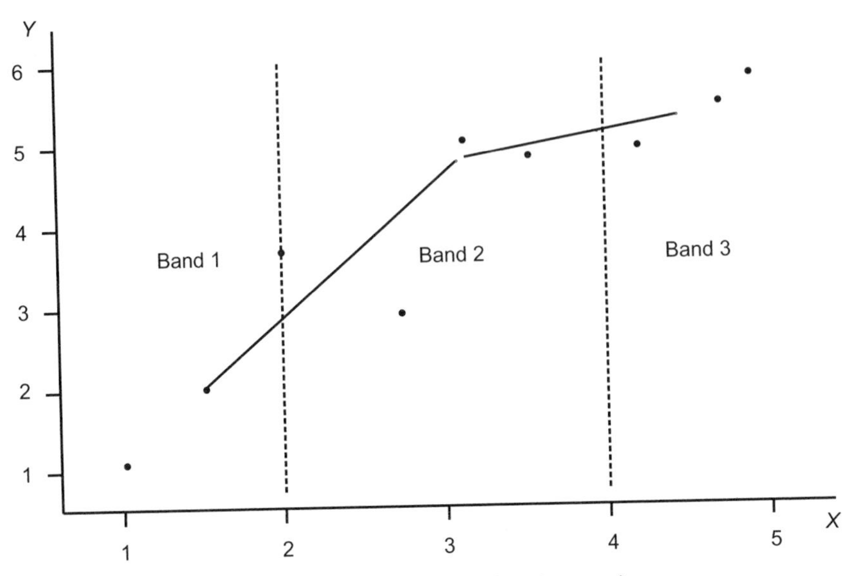

FIGURE 13-15. Example of band regression.

either of the transformations $X' = \log_{10}X$ or $X' = \sqrt{X}$ might represent an improvement over a simple linear model, since the shape of the band regression appears similar to these two models (see Figure 13-10).

The *loess* method is a slightly more complicated version of the band regression technique. This method obtains a smoothed curve by fitting successive linear regr*ess*ion functions in *lo*cal neighborhoods. Like the running medians and moving average models, a neighborhood or band is defined around each value of X. The smoothed value of Y for each value of X is then estimated as the predicted or fitted value of Y for that value of X from this local regression equation.

Consider the following simple case in which a neighborhood is defined by three points, the two nearest values of X to any observation. Suppose our values of X are ordered so that the value x_1 is the lowest valued X, x_2 is the second lowest value of X, and so on. The first local regression equation is fitted to the data

$$(x_1, y_1), (x_2, y_2), \text{ and } (x_3, y_3)$$

Using this equation, we then use $\hat{y}_2$, the value of Y predicted by the equation for the value $X = x_2$, as our smoothed value at this point. Proceed to point $X = x_3$ by fitting a regression to the points

$$(x_2, y_2), (x_3, y_3), \text{ and } (x_4, y_4)$$

and obtain a fitted value for $X = x_3$. The process is continued until a full set of values are obtained. They are then connected by a smooth curve. The loess curve for these

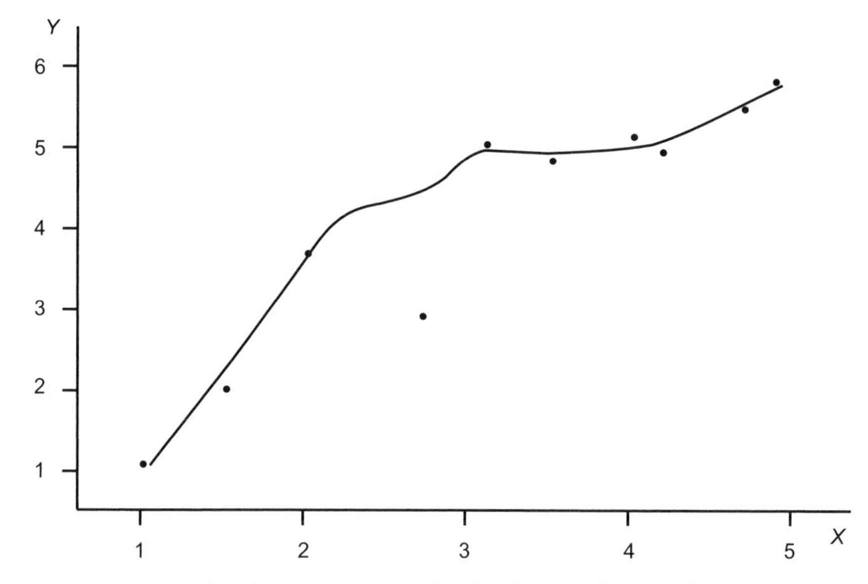

FIGURE 13-16. Loess function for shape exploration data.

data is illustrated in Figure 13-16. Note that it also suggests that the slope of the regression function appears to drop off above the value $X = 3$ just as we found for the band regression method.

The *loess* method can also be refined by giving observations further from the middle X level less weight, and also by giving observations with high residuals lower weights in a second fitting to the data. The method can also be adapted to multiple regression problems.

All of these methods can be used to *suggest* the functional form of the regression relationship. What they do not do is to provide a model or exact expression for the equation that should be fit to the data.

13.3. Validating a Regression Model

In Chapter 12, we noted the importance of *validating* our regression model by examining its predictive ability with new data. Now that we have seen that regression models can involve multiple independent variables and fairly complex functional forms, it is time to address the issue of model validation. Several different methods can be used to assist us in our model validation. Most are based on evaluating the model using either a holdout sample or entirely new data. Consider the logarithmic form of the mercury contamination of bass equation developed in Section 13.2. We found that the equation

$$\log_{10}\text{Merc} = 0.2525 - 0.2667 \log_{10}\text{Alk} - 0.2925 \log_{10}\text{Chloro} \qquad (13\text{-}13)$$

TABLE 13-6
Mercury Contamination of Bass: Holdout Sample

Name	Alkalinity	pH	Calcium	Chlorophyll	Mercury
Annie	3.5	5.1	1.9	3.2	1.33
Blue Cypress	39.4	6.9	16.4	3.5	0.44
Bryant	19.6	7.3	4.5	44.1	0.25
Crescent	71.4	8.1	55.2	33.7	0.16
Dias	4.8	6.4	4.6	22.5	0.81
Down	16.5	7.2	13.8	4.0	0.51
East Tohopelagia	7.1	5.8	5.2	5.8	1.00
Farm-13	128.0	7.6	86.5	71.1	0.05
Griffin	108.5	8.7	35.6	80.1	0.19
Hart	6.4	5.8	4.0	4.6	1.02
Iamonia	7.5	4.4	2.0	9.6	0.45
Jackson	12.6	6.1	3.7	21.0	0.41
Kingsley	10.5	5.5	6.3	1.6	0.42
Lochloosa	55.4	7.3	15.9	24.7	0.31
Miccasukee	5.5	4.8	1.7	14.8	0.50
Monroe	67.0	7.8	58.6	43.8	0.25
Ocean Pond	5.8	3.6	1.6	3.2	0.87
Okeechobee	119.1	7.9	38.4	16.1	0.16
Panasoffkee	106.5	6.8	90.7	16.5	0.23
Placid	8.5	7.0	2.5	12.8	0.56
Rodman	114.0	7.0	72.6	6.4	0.18
Sampson	11.8	5.9	24.2	1.6	0.44
Talquin	16.0	6.7	41.2	24.1	0.67
Tohopekaliga	25.6	6.2	12.6	27.7	0.58
Trout	1.2	4.3	2.1	6.4	0.98
Weir	15.5	6.9	5.2	16.5	0.43
Yale	71.8	7.9	20.5	8.8	0.25

had a reasonable value of $R^2 = 0.5221$, both independent variables were statistically significant at the $\alpha = .05$ level of significance, and the F-statistic for the ANOVA was also highly significant. But is this equation a useful valid model?

Table 13-6 contains a holdout sample of 27 additional lakes from the complete dataset reported by the authors. The holdout sample was created by simply assigning every other observation from the complete dataset to the holdout sample. If the model summarized in Equation 13-13 is indeed a useful explanatory model of bass contamination, then we would expect similar results from a multiple regression analysis of the holdout sample.

Examining the Regression Coefficients

First, we can examine the estimated regression coefficients and other characteristics of the fitted model in comparison to the those of the regression model based on the holdout data. Similar results, in both the magnitude of the coefficients and their signs, would tend to support the validity of the model; dissimilar values would not. The

TABLE 13-7
Summary of Regression Output from Holdout Sample

R	0.9153
R^2	0.7652
$S_{Y \cdot X}$	0.1605

ANOVA

	df	*SS*	*MS*	*F*	Sign.
Regression	2	2.0148	1.0074	39.1200	0.0000
Residual	24	0.6181	0.0258		
Total	26	2.6339			

Equation

	Coeff.	Std. error	*t*-stat.	*P*-value
Intercept	0.3009	0.0893	3.3694	0.0025
log alkalinity	–0.4415	0.0646	–6.8358	0.0000
log chlorophyll	–0.1203	0.0764	–1.5751	0.1283

results from the application of the model expressed in Equation 13-10 to the holdout sample are summarized in Table 13-7. First, we note that this model fits the holdout data even better than the original data. The R^2 value is over 90%, and the standard error is even smaller. Overall, the equation has a higher statistical significance, at least in terms of the ANOVA calculations. As far as the individual partial regression coefficients are concerned, the signs of both variables are negative and consistent, but the magnitude of the alkalinity coefficient has declined while the chlorophyll coefficient has increased. In fact, the chlorophyll variable is no longer as statistically significant in the second sample. However, the two samples are not identical, and significant differences between them may exist. It appears that chlorophyll levels in the original data are higher than those in the holdout sample. Nevertheless, the holdout data suggest that the multiplicative model may be reasonably applied to these data.

Calculating Prediction Errors

What about the predictive ability of our equation? If the equation developed from the original data is a valid model, then it should be able to offer reasonable predictions when used with other data. When we select a model for use, it is inextricably tied to the data on which it has been calibrated. It is successful as a model *because* it does fit the data with which it has been calibrated. It is therefore useful to predict the mercury levels for the lakes in the holdout sample using the equation from the original data. This is quite straightforward. To predict a value for any observation in the holdout sample, simply substitute the values for the independent variables into Equation 13-13 developed from the original data. These values can then be used to calculate the *mean squared prediction error* (MSPR):

$$\text{MSPR} = \frac{\sum_{i=1}^{n_h} (Y_i - \hat{Y}_i)^2}{n_h} \qquad (13\text{-}14)$$

where Y_i is the value of the dependent variable for the i-th observation in the holdout sample, $\hat{Y}_i$ is the predicted value for this observation using the equation developed from the original data, and n_h is the number of cases in the holdout sample.

Consider the first observation in the holdout sample for Annie Lake:

Alkalinity	3.5
Chlorophyll	3.2
Mercury	1.33
$\log_{10}$alkalinity	0.5441
$\log_{10}$chlorphyll	0.5051
$\log_{10}$mercury	0.1239

where the original data in the first three rows are transformed so that $.1239 = \log_{10}1.33$ for mercury and so forth. Therefore, our predicted value would be

$$\hat{Y}_i = 0.2525 - 0.2925(0.5441) - 0.2667(0.5051) = -0.04$$

This differs from the actual value by 0.1239 – (–0.04) = 0.1639. To generate our mean square prediction error, we simply square all these differences and divide by the number of cases in the holdout sample. The prediction is actually not as bad as it might seem. To convert our predicted value of –0.04 back to ppm, we simply calculate $10^{-.04} = 0.91$. This value can then be compared to the observed mercury concentration of 1.33. If our model is valid, our mean square prediction error (MSPR) should be close to the mean square error (MSE) of our model. In this case, MSPR = 0.0289 compares favorably to the MSE of 0.0257 for the holdout sample equation (see Table 13-7).

13.4. Summary

This chapter presents a simplified introduction to the fundamental concepts involved in multiple regression analysis. Multiple regression analysis can be used to describe and predict the relationships among two or more interval-scaled variables. Some of the complications arising from this extension, such as multicollinearity, reflect the consequences of moving from a simple univariate situation to a fully multivariate model. In addition, this chapter describes the transformations that can be applied to the multiple regression model so that it can accommodate nonlinear relationships. This greatly enhances the utility of multiple regression analysis. Diagnostic tools, particularly structured graphs of model components, are shown to be valuable in understanding the nature of the relationships apparent in a set of data. Finally, all models must be tested using other data to assess their *validity.*

FURTHER READING

W. S. Cleveland, "Robust Locally Weighted Regression and Smoothing Scatterplots," *Journal of the American Statistical Association* 74 (1979), 829–836.

J. F. Hair Jr., W. C. Black, Barry J. Babin, R. E. Anderson, and R. L. Tatham, *Multivariate Data Analysis, 6th ed.* (Upper Saddle River NJ: Prentice Hall, 2006).

M. H. Kutner, J. Neter, C. J. Nachtsheim, and W. Li, *Applied Linear Regression Models,* 4th edition (Chicago: Irwin, 2004).

PROBLEMS

1. Explain the meaning of the following terms:
   - Gross regression coefficients
   - Mulicollinearity
   - Intrinsically linear functions
   - Transformations
   - Coefficient of multiple determination
   - Dataspace
   - Running medians
   - Band regression
   - Mean square prediction error
   - Partial regression coefficients
   - Influential observation
   - Standardized regression coefficients
   - Functional forms
   - Multiple correlation coefficient
   - Outlier
   - Moving averages
   - Loess function

2. Develop simple linear regression equations between mercury concentration and each of the independent variables calcium, alkalinity, chlorophyll, and pH using any statistical software package. Use the combined data from Problem 12-4 and Table 13-6.
   a. Develop scatterplots for each of the relationships and graph the regression line within the plot.
   b. Try to improve each relationship by evaluating at least two appropriate transformations of each of the four variables (possibly transforming mercury as well if necessary). Before you select the most appropriate transformations, apply one of the four shape exploration tools described in Section 13.2.
   c. Select the most appropriate transformation in each case. Utilize graphs to assist you in making the case. In what ways is the equation improved by the application of the transformation? Which measures show improvement? How can you compare the different equations?

3. Create linear regression equations predicting mercury using the all possible regression techniques based on the four available independent variables using the data from Problem 12-4. Create a graph in the format of Figure 13-3 that summarizes your results.
   a. Repeat the analysis for the holdout data of Table 13-6.
   b. How do the results compare?

4. Consider again the data related to mercury contamination in bass presented in the data of Problem 12-4 and Table 13-6. Using any available statistical software package:
   a. Develop multiplicative models of the form of Equation 13-11 using the variables chlorophyll and alkalinity as predictor variables for each of the two datasets.
   b. Perform a validation of each separate model using the other dataset as the holdout sample.
   c. Create a multiple regression model using the combined dataset.
   d. Suppose you were given the job of coming up with the best predictive model for these data. Which equation would you suggest? Why?

5. You have been asked to examine the mercury data for the existence of potential multicollinearity problems. Using the results of your all possible regressions data, provide evidence showing the nature and significance of multicollinearity.

6. An environmental activist asks you to tell him which variables seem to contribute the most to high mercury levels in bass. What would you answer?

7. Consider the dataset from the Data and Story Library from Carnegie Mellon University available at *lib.stat.cmu.edu/DASL/Datafiles/homedat.html.* The data consist of several variables related to individual house sales in Albuquerque, New Mexico. In addition to the sale price for each of the homes, the data include several different independent variables that might be related to the sales price, including house square footage, number of features, and whether the house was on a corner lot or custom built.
   a. Examine scatterplots between each of the independent variables and the house sale price.
   b. Fit a multiple regression equation of house sale price as a function of the square footage the number of features. Interpret the meaning of the intercept and the partial regression coefficients. Which variable has the most impact on sales price?
   c. Change the dependent variable house sale price so that it is measured in thousands of dollars and not hundreds of dollars. For example, the first house would be coded as 205, not 2050. Estimate a new multiple regression equation. Explain what happens to the coefficients and why this is to be expected.
   d. Examine the residuals from your regression model. In particular, is there any relationship between the residuals and whether the house was on a corner lot, custom built, or in the northeast of the city.
   e. Specification error is said to occur if we are fitting the wrong equation to these data. Discuss several different ways in which the equation you have developed may be said to be improperly specified. Are there any variables omitted? Is this serious? Are the functional forms correct?

IV

PATTERNS IN SPACE AND TIME

14

Spatial Patterns and Relationships

Chapter 2 introduced different types of spatial data. In Chapter 3 we outlined a series of descriptive statistics used to summarize spatial data, and we highlighted a number of problems that complicate their use. As geographers we are particularly interested in the spatial distribution of various phenomena over the surface of the earth and of the processes that generate them. While many of the statistical concepts discussed so far can be applied to spatial data, more specialized techniques have been developed to explore spatial relationships. This chapter reviews some of these more specialized techniques, focusing on analysis of the spatial distribution of particular objects and spatial variation in the values assumed by a single variable. The spatial extent of our analysis is limited: the techniques we explore are typically deployed to investigate the properties of points lying on the plane, rather than points lying on the sphere or the ellipsoid.

In Section 14.1 we begin our analysis of spatial data by looking at point patterns and two techniques used to examine them, namely, quadrat analysis and nearest neighbor analysis. In Section 14.2 we discuss spatial autocorrelation, a method that describes the spatial distribution of values of a single variable. Section 14.3 extends this discussion, examining measures of local spatial association that are frequently used to detect geographical clusters or hot spots of particular activity. Section 14.4 identifies problems that spatial autocorrelation poses for inferential analysis using regression models, and it shows how these problems might be identified and solved. We provide a short introduction to geographically weighted regression in Section 14.5, and a brief conclusion in Section 14.6.

14.1. Point Pattern Analysis

Is there a spatial pattern or some form of geographical order in the location of cities, industrial factories, trees in a forest, earthquake epicenters, outbreaks of a disease, or the nests of a species of bird? Geographers and other scientists search for geographical patterns or spatial order across a broad range of phenomena, in the hope that this

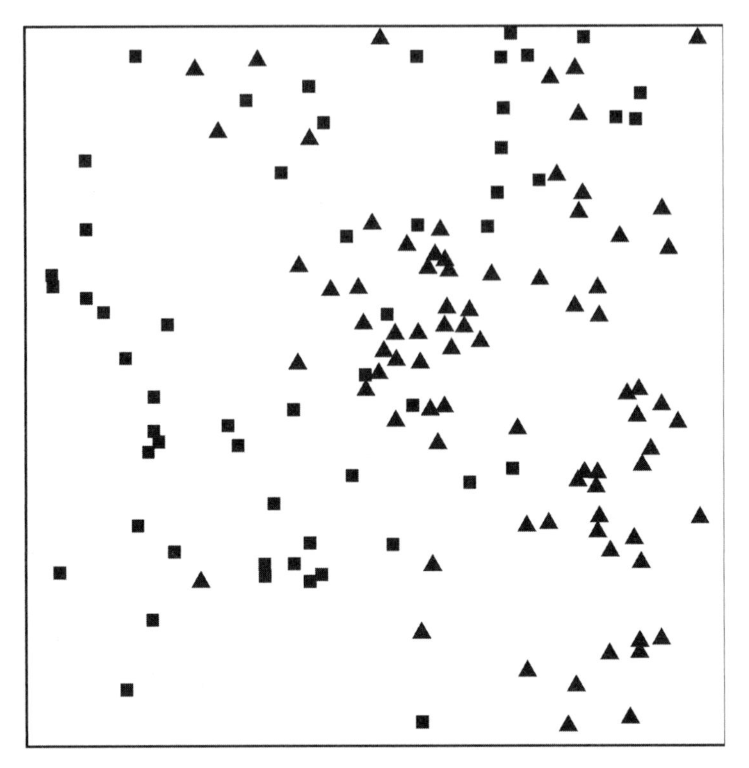

FIGURE 14-1. Spatial distribution of rainforest trees, Meliacae (squares) and Caesalpiniaceae (triangles) in Paracou, French Guiana. *Source:* Revised from Forget et al. (1999)

will lead to a better understanding of the processes that produced such patterns. We begin the search for pattern, or spatial relationship, by mapping the locations at which particular objects are located. The resulting maps, following the discussion of spatial data types in Chapter 2, are known as *point pattern maps.*

Figures 14-1 and 14-2 provide two illustrations of point patterns. As usual, our statistical analysis of point patterns begins with looking at the data. Figure 14-1 shows the geographical distribution of two rainforest trees in a sample plot from Paracau, French Guiana. The squares in the figure show the location of Meliacae trees, while the triangles show the location of Caesalpiniaceae trees. Biogeographers might examine point patterns such as these for evidence suggesting a particular mechanism of seed dispersal. In fact, both these species of tree have seeds dispersed by rodents. Figure 14-2 reveals the geographical distribution of lung cancer cases across a portion of Lancashire in northwest England. Epidemiologists and health geographers study point patterns of health events to gather clues on potential sources of a disease and/or mechanisms of transmission.

In general, when we examine a point map, we are looking to see whether the spatial distribution, or the geographical arrangement, of the variable of interest displays

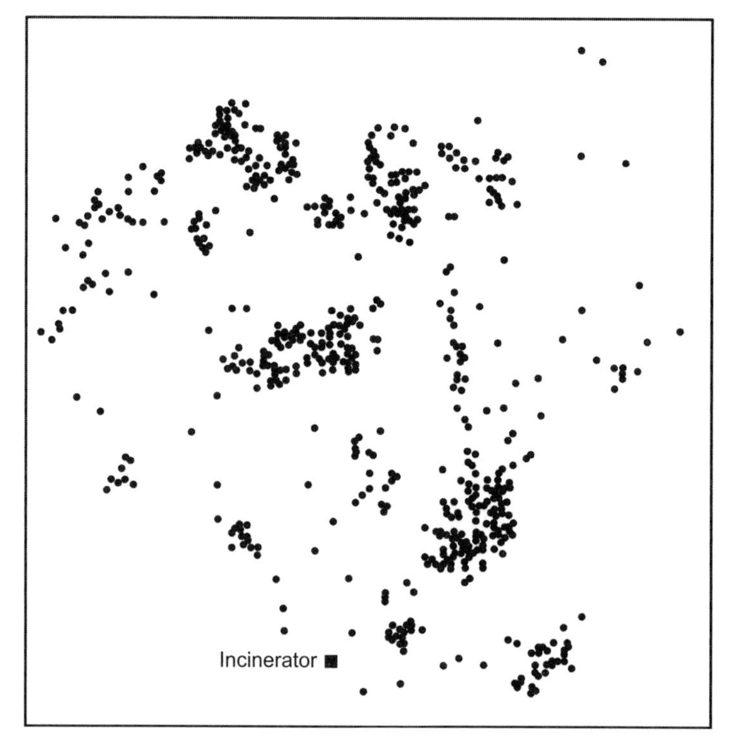

FIGURE 14-2. Spatial distribution of lung cancers in northwest Lancashire, 1974–1983. *Source:* Revised from Gatrell et al. (1995).

any sort of pattern. This is not always a straightforward exercise. In Chapter 3, we showed how to find measures of central tendency and dispersion for sets of points distributed across two dimensions, but how might we describe the absolute or relative locations of points within a study area? To make this task somewhat easier we typically look to see whether the observations in a point pattern, the locations of the points themselves, tend to cluster together, whether they are more uniformly distributed, or whether they appear to be arranged randomly. These different types of point patterns are illustrated in Figure 14-3. Figure 14-3a displays a *clustered* arrangement of points, where the objects of interest are found close to one another and where large areas of the study region contain no points. Clustered patterns tend to result from a contagion process where a particular location attracts a number of points. The *dispersed* point pattern in Figure 14-3c is commonly thought to result from some form of competition in space where points repel one another. The arrangement of points in the dispersed pattern is quite regular over the study area, and for this reason this arrangement is also referred to as a uniform spatial distribution.

Random point patterns, like those in Figure 14-3b, result from the operation of an *independent random process* (or a process consistent with *complete spatial randomness*). An independent random process is one in which every location (or small

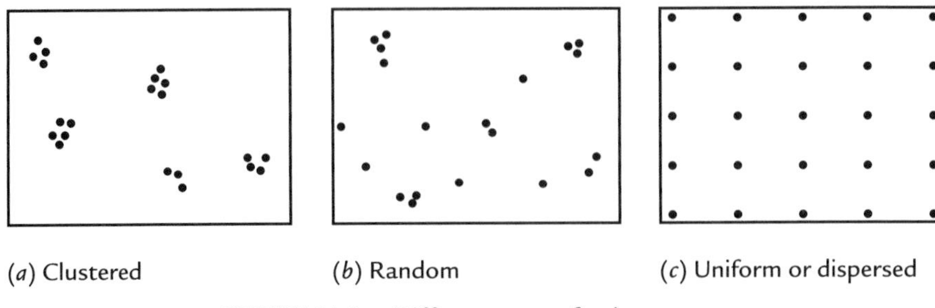

(*a*) Clustered (*b*) Random (*c*) Uniform or dispersed

FIGURE 14-3. Different types of point pattern.

area) of a study region has an equal probability of receiving an event or a point, and one for which the location of an event is independent of the location of all other events. Random point patterns are of little interest to the geographer because they are evidence that the underlying generating process has no spatial logic. For this reason, when we examine point patterns, we are often engaged in identification and calculation of departures from complete spatial randomness.

Visual comparison of a point pattern map to one of the fundamental point pattern arrangements of Figure 14-3 provides clues to the spatial distribution of a variable of interest and thus to the nature of the process that generated it. More precise investigation typically involves analysis of the frequency or the density of points across a study region (*quadrat analysis*) or of the distance between adjacent points (*nearest neighbor analysis*). We briefly review these methods next.

Quadrat Analysis

Quadrat analysis was initially developed by ecologists studying the spatial distribution of plants (see Greig-Smith, 1964). This technique focuses on changes in the density of points across a study region. The method is operationalized by overlaying a regular grid on the region of interest and then counting the number of points found in each *quadrat* (cell) of the grid. The observed frequency distribution of points per quadrat is then compared to a theoretical distribution with known properties. If the observed and theoretical distributions are similar, then we typically infer that the observed distribution could have been generated by a process consistent with the theoretical distribution. Let us now examine this procedure more carefully.

Given a point pattern to analyze, determining the size of the study area is a critical decision. In cases where the process involved is a social one, political boundaries of one form or another might make an appropriate frame for the study. In other cases, particularly those involving point patterns generated by physical processes, fixing the region of interest can be more difficult, often because the processes under consideration may be more or less continuous in space. Must the study region include all points representing a variable of interest, or should some stray or remote points be excluded if the analyst believes their inclusion alters the fundamental nature of the point distribution? These are difficult questions to answer. If study areas of different size can be utilized without excessive time and cost, the researcher can learn a great deal about the

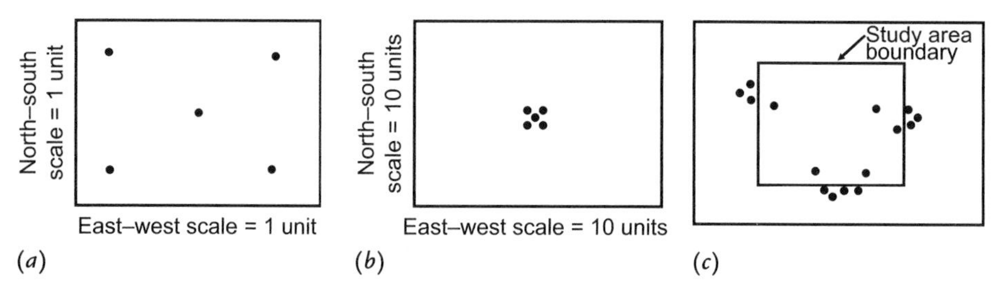

FIGURE 14-4. Study area boundaries.

influence of particular sets of points on the overall point pattern. Some of these issues are illustrated in Figure 14-4. On the one hand, enlarging the scale of the study area might make what appears to be a uniform distribution at one scale (Figure 14-4a) appear to be a clustered point pattern at another scale (Figure 14-4b). On the other hand, too small a study area boundary might truncate (or subset) what appears to be a clustered point distribution and turn it into a more regular spatial distribution (Figure 14-4c).

Once the study region is determined, the researcher has to decide the shape and size of quadrats. It has become common to use square quadrats (grid squares), though circular quadrats are also employed. The main advantage of square quadrats is that they pack together and completely cover an area, whereas circular quadrats leave some spaces uncovered unless they overlap. Such overlap leads to oversampling parts of the study area and introduces complex sampling issues. Other regularly shaped quadrats such as triangles and hexagons may also be used in place of squares. Figure 14-5 provides examples of a regular square quadrat census design and a circular quadrat sample applied to crime data covering a portion of the city of Portland, Oregon (*www.gis.ci.portland.or. us/maps/police*). By definition, the square quadrat *census* shown in Figure 14-5a covers the entire study region, whereas the circular quadrat *sample* of Figure 14-5b covers only a subset of the study area. The circular quadrats in the sample are positioned by drawing coordinates at random from the study region. As with all samples, the larger the number of observations, the more faithfully the sample data will represent the underlying population.

One additional factor complicates quadrat analysis: the appropriate size of the quadrats themselves. On the one hand, employing only a few, large quadrats means averaging counts of points that may be found in smaller quadrats. Thus, larger quadrats tend to smooth the heterogeneity that might exist over a study region. On the other hand, using relatively small quadrats tends to exacerbate differences in measures of point density across a spatial point pattern. Greig-Smith (1964) suggests that the optimal quadrat size is given as

$$\frac{2A}{n}$$

where A is the area of the study region and n is the number of points, representing the locations of the variable of interest.

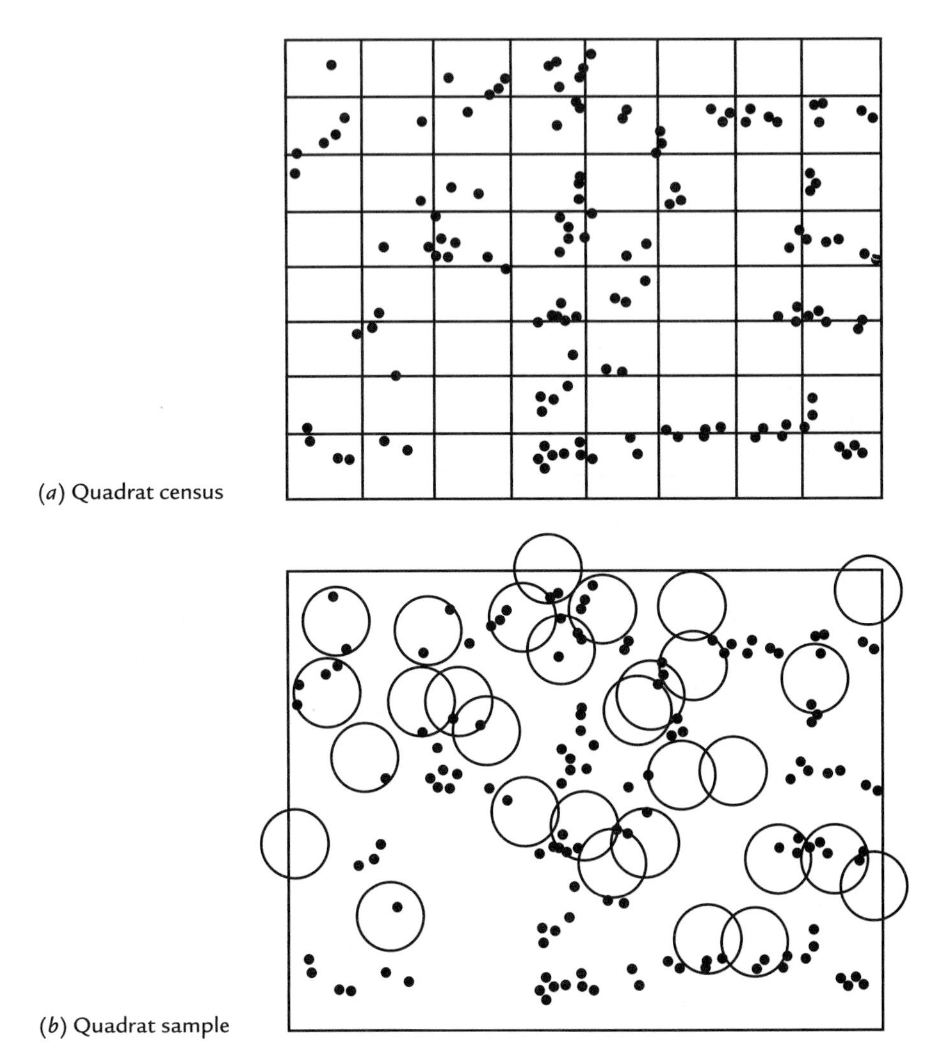

FIGURE 14-5. Robbery events in southeast Portland (July 2006–June 2007).

It should be clear from Figures 14-4 and 14-5 that a regular or uniform distribution would be characterized by a relatively similar number of points, or observations, within each quadrat of a study region. In turn, a clustered distribution would be characterized by relatively few quadrats that contain large numbers of points and many other quadrats that contain no points. Expressed somewhat differently, the variance in the number of points per quadrat across a study region will approach zero for a uniform distribution and will approach infinity for a clustered distribution. How much variation in the number of points per quadrat would we expect to see in a point pattern that was generated by an independent random process? We discussed this question in Chapter 5, where we showed that the Poisson probability distribution provides the probability that a single quadrat contains a specified number of points in an experiment where points are distributed over a surface in a process that is consistent with complete spatial randomness. The variance of a Poisson random variable, in this case the number of points per quadrat, is equal to the expected value of the Poisson random variable and that is given, in practice, by the average intensity of the point pattern:

$$\lambda = \frac{n}{q}$$

where n is the number of points in the study area and q is the number of quadrats.

This suggests that one way of classifying point patterns as clustered, random, or uniform is to divide the study area into quadrats and to examine the variance/mean ratio of the observed frequency distribution of points per quadrat. A random pattern will have a variance/mean ratio of one since it is described by the Poisson distribution for which the variance $[V(X)]$ and the mean $[E(X)]$ are equal. Dispersed patterns will have a variance/mean ratio that tends toward zero, and clustered patterns will have a variance/mean ratio tending toward infinity. To summarize:

$s^2/\bar{X} \to \infty$: indicates a clustered point pattern
$s^2/\bar{X} = 1$: indicates a random point pattern
$s^2/\bar{X} \to 0$: indicates a uniform point pattern

EXAMPLE 14-1. Quadrat Analysis. Let us examine how to perform quadrat analysis using the California earthquake data introduced in Chapter 2. Here we ignore the size of earthquakes and focus solely on their location. Figure 14-6 maps the epicenters of major earthquakes in California over the last 100 years or so. Quadrats of uniform size are superimposed over the study region. The number of earthquake epicenters within each quadrat represents the variable of interest. A quick glance at Figure 14-6 shows a cluster of earthquake epicenters extending from Mendocino on the northern coast of California into the Pacific Ocean. There is another cluster around Mammoth Lakes along the eastern Sierra Nevada. In Southern California, epicenters are more widely distributed. Overall, the pattern appears clustered, with many quadrats containing no earthquake activity.

Next, we calculate the sample mean and variance of the number of earthquake epicenters across the 35 quadrats that comprise the set of observations. Note that there are 111 earthquakes distributed across Figure 14-6, although some are difficult to identify because of the overplotting of some points. We find

$$\bar{X} = \sum_{i=1}^{q} X_i / q = 3.17 \quad \text{and} \quad s^2 = \frac{\sum_{i=1}^{q} (X_i - \bar{X})^2}{q - 1} = 41.73$$

where q is the number of quadrats, X_i is the number of points in quadrat i, and $\Sigma X_i = n$. Thus, the variance/mean ratio for the point pattern in Figure 14-6 is 41.73/3.17 = 13.16.

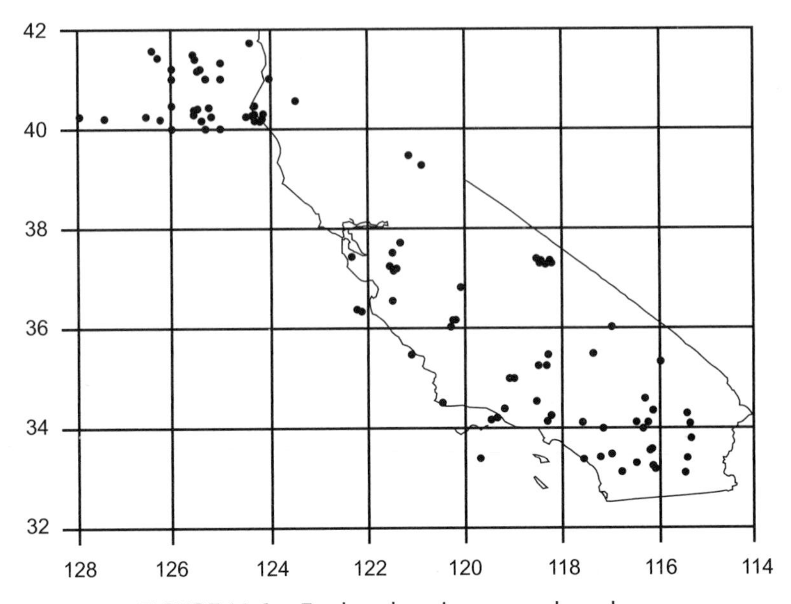

FIGURE 14-6. Earthquake epicenters and quadrats.

The variance/mean ratio is much larger than one indicating that the spatial distribution of earthquake epicenters in our sample is clustered.

A more formal goodness-of-fit test of the distribution of earthquake epicenters across the quadrats in Figure 14-6 can also be performed. If earthquake epicenters are randomly distributed in space they can be considered to have been generated by a Poisson process with a variance/mean ratio equal to one. We could then ask, what is the likelihood that a sample point pattern with a variance/mean ratio equal to 13.16 might have been selected by chance from a population point pattern that was truly random? Thomas and Huggett (1980) show that the sampling distribution of $s^2/\bar{X}$ about a Poisson prediction of one is approximated by the Student's t-distribution when the number of quadrats is reasonably large, say at least 30. Thus, to answer the question just posed, we calculate the following test statistic based on the Student's t-distribution

$$t = \frac{(s^2/\bar{X}) - 1}{\sqrt{2/(q-1)}} = \frac{13.16 - 1}{0.243} = 50.041$$

and where the denominator of this equation represents the standard error of the variance/mean ratio. From the tabulated values of the t-statistic (see Appendix Table A-6), with $q - 1$ *degrees of freedom,* the probability that a test statistic equal to 50.041 could occur by chance is essentially zero. We can therefore state with a considerable degree of confidence that the earthquake epicenters are not randomly distributed. Furthermore, because the test statistic is greater than one, we can state that the variable

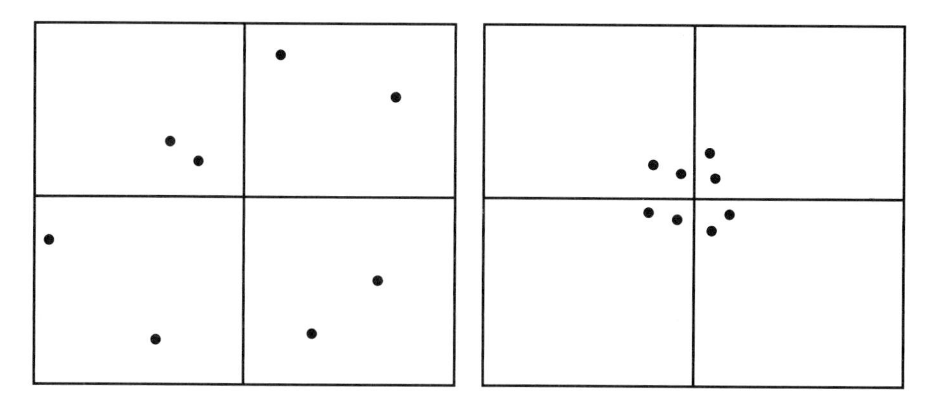

FIGURE 14-7. Uniform point patterns from quadrat analysis.

of interest is significantly clustered. A test statistic with a value significantly less than one would indicate a dispersed or uniform spatial distribution of points.

Nearest Neighbor Analysis

Quadrat analysis is insensitive to the spatial arrangement of points within quadrats. Thus, markedly different point patterns can give rise to identical frequency distributions of the number of points per quadrat. This issue is illustrated in Figure 14-7 where two point distributions are shown, each with 8 points in total evenly divided between the four quadrats. Quadrat analysis would reveal that both patterns are uniform, indicating the need for an alternative technique that takes into account the relationships between points themselves.

Nearest neighbor analysis provides this alternative technique to point pattern analysis. Developed by Clark and Evans (1954), nearest neighbor analysis focuses on the distances between points rather than on the density of points in a study region to determine whether the observed point pattern is clustered, random or dispersed. To begin, the distance d_{ij} between each pair of points i and j in a point pattern is calculated using Pythagoras's theorem. For each point $i = 1, 2, 3, \ldots, n$, the closest neighboring point is determined, that is, $\min_j d_{ij}$. The mean or average of these *observed* nearest neighbor points is denoted $\bar{d}_o$. Unfortunately, this statistic cannot be used to compare point pattern maps because it is measured in the same units as the map. What is needed is some standard against which the mean observed nearest neighbor distance can be compared. The obvious standard is the mean or expected distance between nearest neighbors in a *random* point pattern. The mean or expected nearest neighbor distance for a random point pattern is given by

$$\bar{d}_e = \frac{1}{2\sqrt{n/A}}$$

where n is the number of points in the pattern and A is the area of the study region. By convention, A is defined as the smallest rectangle that encloses all the points.

The ratio of observed to expected nearest neighbor distances, $R = \bar{d}_o / \bar{d}_e$, is known as the *nearest neighbor index.* The value of R can vary between 0 and 2.15. A value of R close to zero, that is, when the observed nearest neighbor distances are relatively small, indicates a clustered point pattern. A value of R close to 2.15 is found when the observed nearest neighbor distances are relatively large, indicating a dispersed pattern. A value $R = 1$ is consistent with a random pattern.

EXAMPLE 14-2. Nearest Neighbor Analysis. Figure 14-8 shows the location of donut stores in a neighborhood of Hamilton, Canada. Retail geographers might be interested in whether the donut stores are randomly located or whether the point pattern of donut stores is significantly different from random, perhaps indicating that some nonrandom process might explain their relative location.

Table 14-1 lists the 10 donut stores and the distance of each store to its nearest neighbor in km. The mean observed nearest neighbor distance between donut stores is

$$\bar{d}_o = \sum_{i=1}^{n} d_{ij} / n = \frac{12.134}{10} = 1.213 \text{ km}$$

The expected nearest neighbor distance, given a study area of $(4 \times 4) = 16$ km^2, based on the assumption that the point pattern exhibits complete spatial randomness is

$$\bar{d}_e = \frac{1}{2\sqrt{10/16}} = \frac{1}{1.581} = 0.633 \text{ km}$$

and thus the nearest neighbor index $R = 1.213/0.633 = 1.916$. This value of R is tending toward its upper limit, indicating that the point pattern is dispersed.

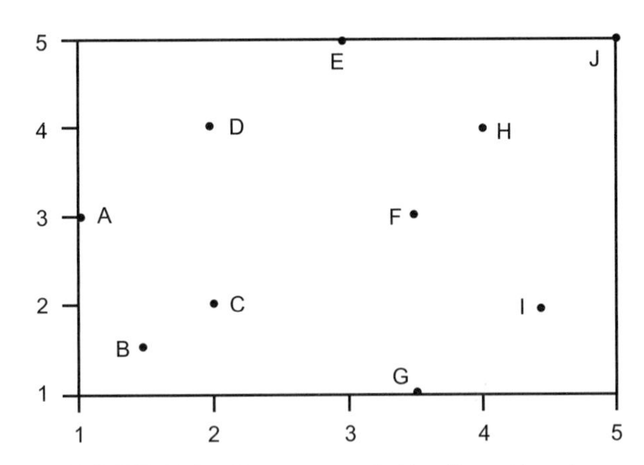

FIGURE 14-8. Donut stores in Hamilton, Canada.

TABLE 14-1
Nearest Neighbor Distance

Store	Nearest neighbor	Distance, (km)
A	D	1.414
B	C	0.707
C	B	0.707
D	A, E	1.414
E	D, H	1.414
F	H	1.118
G	I	1.414
H	F	1.118
I	F, G	1.414
J	H	1.414

A formal test of the hypothesis that donut stores are randomly distributed can be performed. This test involves calculating the following test statistic

$$z = \frac{\bar{d}_o - \bar{d}_e}{s_{\bar{d}_e}} = \frac{1.213 - 0.633}{0.1045} = 5.55$$

where $s_{\bar{d}_e} = \sqrt{0.0683A/n^2}$ represents the standard error of the expected mean nearest neighbor distance for a point pattern generated by a random process. The Z-scores of the test statistic are standard normal deviates and thus approximately normally distributed. Comparing the test statistic 5.55 to standard normal probabilities in Appendix Table A-3 reveals that there is less than a 1% chance that the test statistic could have been drawn from a normal distribution with a mean given by $\bar{d}_e$. Thus, in all likelihood, the nearest neighbor distances calculated for the point pattern of donut stores are not randomly distributed. The positive test statistic indicates a mean nearest neighbor distance larger than that expected for a random distribution. Thus, we may conclude that donut stores in this part of Hamilton are more uniformly distributed than random.

EXAMPLE 14-3. Nearest Neighbor Analysis of Earthquake Data. Using quadrat analysis, we discovered that the California earthquake point data are clustered. Let us now use nearest neighbor analysis to confirm this finding. All the earthquake data in Figure 14-6, including the geographical coordinates of all epicenters, can be found on the book's website. There are 111 earthquake epicenters in the point pattern, contained within an area equal to 107.16. The mean nearest neighbor distances are

$$\bar{d}_o = \sum_{i=1}^{n} d_{ij}/n = \frac{23.976}{111} = 0.216$$

$$\bar{d}_e = \frac{1}{2\sqrt{111/107.16}} = \frac{1}{2.036} = 0.491$$

Thus, the nearest neighbor index $R = 0.216/0.491 = 0.44$. The index value is less than one, indicating that the point pattern of earthquake epicenters is more clustered than random, confirming the results of the quadrat analysis.

Once more, a formal test of the hypothesis that earthquake epicenters are randomly distributed can be performed. This test involves calculating the following test-statistic

$$z = \frac{\bar{d}_o - \bar{d}_e}{s_{\bar{d}_e}} = \frac{0.216 - 0.491}{0.024} = -11.458$$

where $s_{\bar{d}_e} = \sqrt{0.0683A/n^2}$ represents the standard error of the expected mean nearest neighbor distance. Comparing the test statistic –11.458 to standard normal probabilities in Appendix Table A-3, we see that there is a very small probability that the test statistic could have been drawn from a normal distribution with a mean given by $\bar{d}_e$. Thus, in all likelihood, the nearest neighbor distances calculated for the earthquake data are not randomly distributed. A negative test statistic indicates a mean nearest neighbor distance smaller than that expected for a random distribution, and this is consistent with a clustered arrangement of points.

14.2. Spatial Autocorrelation

Autocorrelation was introduced in Chapter 4 as the correlation between the values of a variable and lagged values of that same variable. We can only understand the concept of autocorrelation when the observations on a variable of interest (X) can be ordered along one or more dimensions different from the measures of X itself. Such ordering is, of course, possible with time-series data, where observations on the variable X may be organized temporally. A similar ordering is possible with spatial data, where values of a variable X can be located using geographical coordinates.

DEFINITION: SPATIAL AUTOCORRELATION

Spatial autocorrelation is defined as the correlation of a variable with itself over space.

"Spatial autocorrelation exists whenever a variable exhibits a regular pattern over space in which its values at a set of locations depend on values of the same variable at other locations" (Odland, 1988, p. 7). When spatial autocorrelation is strong, nearby values of a variable are closely related to one another. When spatial autocorrelation is weak, or even nonexistent, the values of a variable are distributed randomly in space. Spatial autocorrelation can be negative or positive. If similar values of a variable tend to cluster in space, the geographic distribution of that variable is *positively spatially autocorrelated.* If very different values of a variable tend to cluster, that variable is *negatively spatially autocorrelated.* Positive spatial autocorrelation is quite common

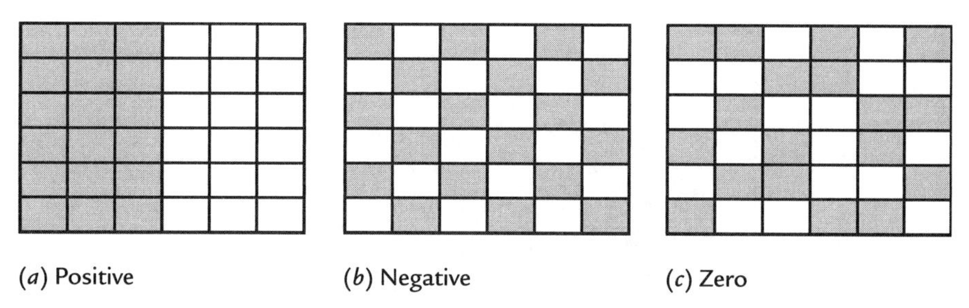

FIGURE 14-9. Positive, negative, and zero spatial autocorrelation.

in geography. The distribution of house prices or household income in a city is usually positively spatially autocorrelated. This is because wealthy households tend to live in exclusive neighborhoods, segregating themselves from lower-income households, relegating those households to other areas in the city. Many land-use activities are agglomerated in space and exhibit positive spatial autocorrelation. Variables collected by physical geographers can also display this pattern. Climatologists find similar temperatures and precipitation levels in adjacent areas. Biogeographers find that the plants of a particular species tend to cluster in some regions and not others. Negative spatial autocorrelation is not a common pattern in geography. Figure 14-9 shows examples of positive, negative and zero autocorrelation in binary (categorical) spatial data. The regions in these examples are the cells of a regular grid or *lattice.*

Autocorrelation is of concern to geographers for two reasons. First and foremost, the search for spatial pattern is one of the dominating themes of geographical research. Spatial distributions are rarely random. If there are any "laws" in geography, then surely the first should be, nearby locations are more alike than distant ones (Tobler, 1970). In statistical jargon this means that the geographical distributions of many phenomena exhibit positive spatial autocorrelation. Second, the inferential techniques presented in Chapters 7 and 8 are based on the assumption that the values of the observations in a sample are independent of one another. One of the ways in which spatial data might violate this assumption is if the data reveal a pattern of spatial autocorrelation. Our interest in spatial autocorrelation is thus twofold: first, to measure the strength of autocorrelation in geographical data, and second to test the assumption of independence or randomness. In this section we examine the *variogram,* a simple measure of how distance is related to differences in the value of a variable; *join count statistics,* that measure spatial autocorrelation for nominal variables measured across areas; and *Moran's index* of spatial autocorrelation for continuous variables that can be used with areal or point data.

Variograms

A simple way to examine *spatial continuity* in some data is to look at how values of a variable change over space. The contour map provides one way of visually exploring

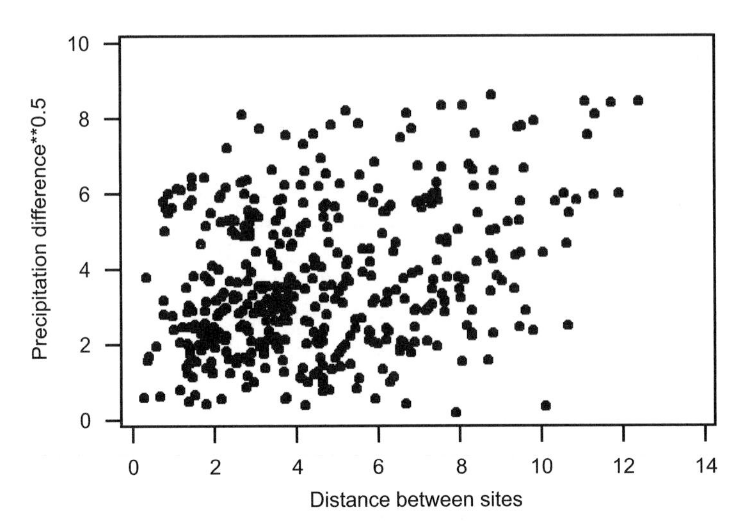

FIGURE 14-10. Variogram cloud for California precipitation data.

spatial continuity in a two-dimensional space, with lines connecting points that have equal values of a variable under study. How values of a variable change with distance can be demonstrated in a simple scatterplot known as a *variogram cloud.* The variogram cloud plots the difference in the value of a variable between two geographical locations against the distance between those locations.

Figure 14-10 shows a variogram cloud constructed from precipitation data measured at 30 sites across California (see Example 14-5). For each pair of sites the Euclidean distance was measured along with the square root of the absolute value of the precipitation difference between those sites. With 30 sample locations, the number of unique pairs of points is $n(n - 1)/2 = 30(29)/2 = 435$. The variogram cloud shows that as the distance between locations where precipitation was measured increases, so the difference in precipitation values also increases. This relationship is not very strong, however, with an R^2 value of only about 0.10. In general, precipitation increases in a northerly direction within California, and it decreases away from the coast in an easterly direction. Another significant influence on California precipitation is altitude, with more precipitation falling on locations at higher elevations. This relationship is complicated by a pronounced rain shadow effect that significantly reduces precipitation on the leeward (eastern) sides of most mountain ranges in the state.

Figure 14-11 shows another variogram cloud where observations between sample sites in the rain shadow and outside the rain shadow are distinguished. This scatterplot shows that regardless of the distance between points, locations within the rain shadow have precipitation values that are broadly similar ($r^2 = 0.09$). Outside the rain shadow, precipitation values vary more markedly, with larger differences in precipitation found between locations that are further apart ($r^2 = 0.39$).

One tool that summarizes the information in the scatterplots of Figures 14-10 and 14-11 is the variogram (more accurately called the semivariogram).

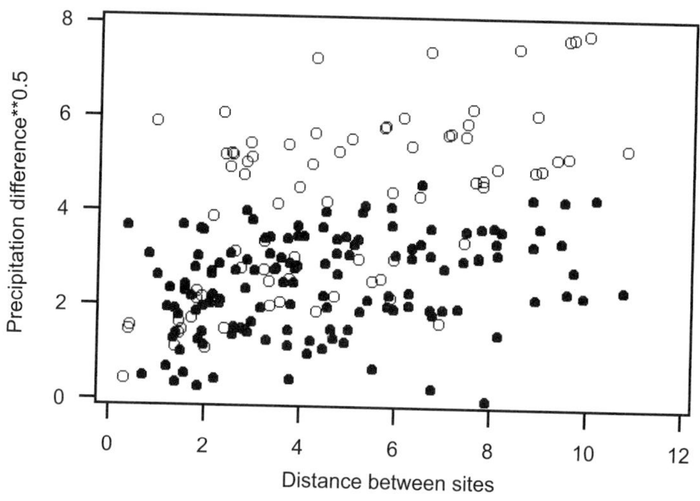

FIGURE 14-11. Variogram clouds for California precipitation data, in rain shadow (dark circles) and out of Rain Shadow (light circles).

DEFINITION: VARIOGRAM

The variogram is calculated as half the average squared difference between paired data values for a given distance (d)

$$\gamma(d) = \frac{1}{2n(d)} \sum_{d_{ij}} (x_i - x_j)^2 \tag{14-1}$$

where d_{ij} indicates summation across all pairs of points at distance d.

The variogram indicates how the average squared difference in values of a variable X measured at different points in space varies as the distance between those points changes. To understand what the variogram shows, imagine taking the point distribution presented in Figure 14-10 and splitting the 435 unique pairs of points illustrated into a series of groups each comprising observations that are a similar distance apart. It is unlikely that our sample data will contain many pairs of observations exactly the same distance from one another and so the pairs of observations are grouped in a series of distance bands (d) or lags. Table 14-2 provides an example in which six distance bands are used to classify the California precipitation data. The table shows the number of observations found in each band and the value of the variogram for the different distance bands. These values are plotted in Figure 14-12, which illustrates how the values of the variogram (the semivariance) increase as the distance separating the pairs of points increases.

The variogram in Figure 14-12 is constructed using the Euclidean distance between points and thus assumes that the underlying process that produced the values

TABLE 14-2
Variogram of California Precipitation Data

Distance band	d	n	$\gamma(d)$
0.0–1.99	1	73	138.1
2.0–3.99	3	140	192.1
4.0–5.99	5	98	262.6
6.0–7.99	7	69	305.1
8.0–9.99	9	40	535.4
10.0–11.99	11	15	1001.3

Note: d represents the distance band midpoint.

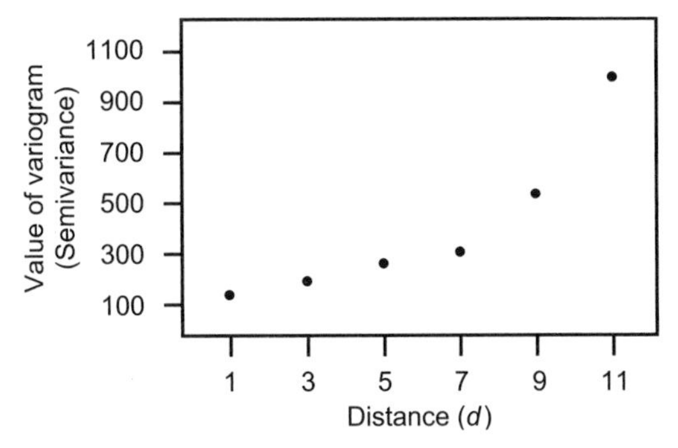

FIGURE 14-12. Variogram for California precipitation data.

of the variable of interest is *isotropic,* meaning that the process has no directional bias. Many processes that we examine in geography are anisotropic. For these processes directional variograms should be constructed where the distance between pairs of points is calculated along a one-dimensional ray, with sufficient directional tolerance to allow samples of reasonable size to be constructed. Variograms are also sensitive to the number of distance bands or lags used to partition the sample observations (see Isaaks and Srivastava, 1989).

Often, we might represent the scatterplot in a variogram with a continuous function. An example of such a function is indicated in Figure 14-13, along with its characteristic features, the nugget, sill, and range. The nugget indicates the value of the variogram for $d = 0$. We might expect this value to be zero, for paired observations at distance $d = 0$ apart from one another are in essence the same observations. However, the nugget effect typically expresses small-scale variability in values of the variable of interest. The sill is the largest value of the variogram that is reported. It is equal to the constant value of the semivariance that lies outside the range. The range itself indicates the distance (d) between observations at which the continuous variogram

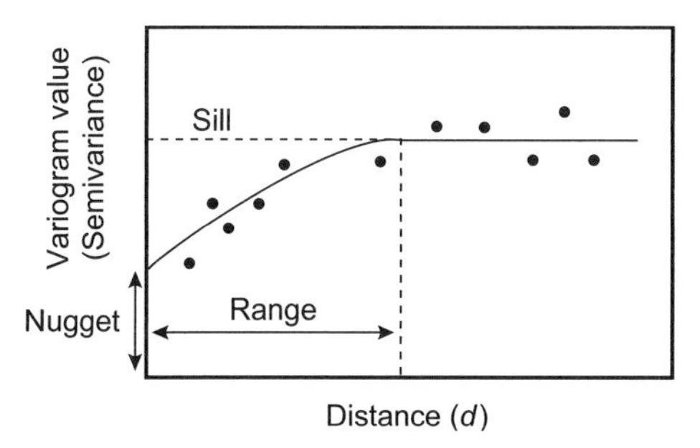

FIGURE 14-13. Continuous variogram function.

function plateaus. Although not detailed here, these terms can also be used to describe the correlogram, which shows how the value of the correlation coefficient changes across sets of observations located at varying distances apart. In general, the value of the correlation coefficient moves inversely to the semivariance across the range of distances (d) that separate observed values.

Join Count Statistics

The join count statistic is a simple measure of spatial autocorrelation for nominal-level, areal data. Here we consider the most straightforward case of a two-category or binary nominal variable. It has become convention to refer to the two binary classes as black and white. Figure 14-9 shows the spatial arrangement of 36 areas for which the variable of interest is coded either black or white. A simple index of spatial autocorrelation for this kind of binary areal data is to count the number of joins or borders between black and white areas or cells. In general, only joins of nonzero length are considered and thus diagonal cells are regarded as noncontiguous. This is sometimes called the Rook's case, after the possible row and column moves of a rook on a chessboard. In the Queen's case, diagonal cells are also regarded as contiguous. If areas with similar values tend to cluster, as in Figure 14-9a, there will be relatively few black–white joins. If areas with similar values are widely dispersed, as in Figure 14-9b, there will be a relatively large number of black–white joins. If the variable of interest is randomly distributed, as in Figure 14-9c, the result will be an intermediate number of black–white joins. The black–white join counts, under the Rook's case, for Figures 14-9a, b, and c are 6, 60, and 43, respectively.

EXAMPLE 14-4. Join Count Analysis. To illustrate the join count test for spatial autocorrelation, we analyze the results from the presidential election of 2000 shown in Figure 14-14. States are coded according to which party won the electoral college votes for that state, black for Republican and white for Democrat. Note that this does not always follow the popular vote, as the case of Florida amply demonstrates (compare Figure 14-14 with Figure 2-25). To begin, we must enumerate the number and type of joins between the contiguous U.S. states shown in Figure 14-14.

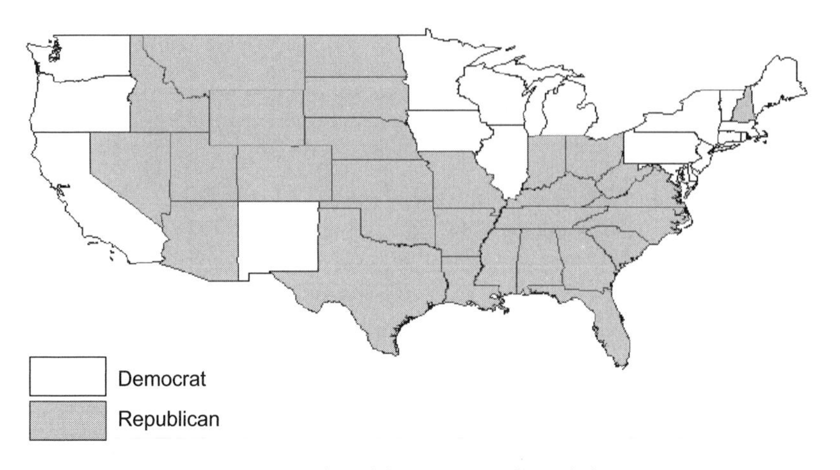

FIGURE 14-14. Results of the U.S. presidential election, 2000.

There are $n = 48$ states, of which 30 are black, $n_B = 30$, and 18 are white, $n_W = 18$. Table 14-3 collects the information necessary to compute the join count statistic. Each state is represented in the table by its two-digit postal prefix. Let L_i be the number of joins for state i. For example, the state of Washington has two joins, Oregon has four, Missouri eight, and so on. Note that Utah has six joins, and the corner boundary with New Mexico is regarded as a join of nonzero length. The total number of joins on the map is $L = \Sigma_{i=1}^{n} L_i/2 = 107$. The sum of L_i is divided by two to avoid the double-counting of joins. Table 14-3 reveals that the number of black–white joins for the electoral map is 26. The probability that a state is black or Republican is 0.625, and the probability that it is white or Democrat is 0.375.

In order to establish whether the black–white join count in Figure 14-14 indicates the presence or absence of spatial autocorrelation, we compare the observed count with one that represents an independent, random process. If the observed count is significantly different from that generated by a random process, then we could confidently predict that the election pattern was spatially autocorrelated.

If the color pattern on a binary map was random, as in Figure 14-9c, how many black–white joins would we expect? This depends on the way in which black and white cells or values are assigned across the study region. There are two possible methods of assignment for binary data. The first is called *sampling with replacement,* or *free sampling.* According to this method, the assignment of a black or white value to a cell or area within the overall study region corresponds to sampling from a binomial distribution (see Chapter 5), with a probability of π for assigning one of the values and a probability $(1 - \pi)$ of assigning the other. A sample is obtained once we have allocated a black or white value to each area. Under free sampling, repeated samples drawn from a binomial distribution might not produce the same number of black and white areas within the study region, simply by chance.

The second method of assignment is called *sampling without replacement* or *nonfree sampling.* According to this method, the number of black and white values to

TABLE 14-3
Join Count Statistics for Figure 14-14

		Joins				
State	Color	BB	WW	BW	L_i	$L_i(L_i - 1)$
AL	B	4			4	12
AR	B	6			6	30
AZ	B	3		2	5	20
CA	W		1	2	3	6
CO	B	6		1	7	42
CT	W		3		3	6
DE	W		3		3	6
FL	B	2			2	2
GA	B	5			5	20
ID	B	4		2	6	30
IL	W		2	3	5	20
IN	B	2		2	4	12
IA	W		3	3	6	30
KS	B	4			4	12
KY	B	6		1	7	42
LA	B	3			3	6
MA	W		4	1	5	20
MD	W		2	2	4	12
ME	W			1	1	0
MI	W		1	2	3	6
MN	W		2	2	4	12
MO	B	6	2		8	56
MS	B	4			4	12
MT	B	4			4	12
NE	B	5		1	6	30
NV	B	3		2	5	20
NH	B			3	3	6
NJ	W		3		3	6
NM	W			5	5	20
NY	W		5		5	20
NC	B	4			4	12
ND	B	2		1	3	6
OH	B	3		2	5	20
OK	B	5		1	6	30
OR	W		2	2	4	12
PA	W		4	2	6	30
RI	W		2		2	2
SC	B	2			2	2
SD	B	4		2	6	30
TN	B	8			8	56
TX	B	3		1	4	12
UT	B	5		1	6	30
VA	B	4		1	5	20
VT	W		2	1	3	6
WA	W		1	1	2	2
WI	W		4		4	12
WV	B	3		2	5	20
WY	B	6			6	30
Totals	$n_B = 30$ $n_W = 18$	116	46	52	214	860

$p = 30/48 = 0.625$
$q = 18/48 = 0.375$

$$L = (0.5)\sum_{i=1}^{n} L_i = (0.5)(214) = 107$$

$BB = (0.5)(116) = 58$ $WW = (0.5)(46) = 23$ $BW = (0.5)(52) = 26$

be assigned to the areas or cells of a study region is fixed. Repeated trials will generate different allocations of the black and white values to individual areas within the study region, but the number of black and white areas does not change.

Choosing between these sampling methods depends on available information. To free sample, we must have information independent of our study region that fixes the probabilities p and $q = (1 - p)$. When this information is unavailable, we must engage in nonfree sampling and base the probabilities p and q on the number of black and white values in the region under study. Choice of the appropriate sampling method is important because the mean and variance of the number of black–white joins differs between the two sampling strategies. Under the nonfree-sampling method, the variance of the number of black–white joins will always be smaller than under the free-sampling method, because fixing the numbers of black and white values reduces the number of different color patterns possible. When in doubt, it is safer to choose the nonfree-sampling method because it rests on less stringent assumptions.

Thus, there are two answers to the question of how many black–white joins we would expect in our study region if the distribution of black and white values was generated by an independent, random process. These expected values are:

$$\text{Under free sampling:} \quad E(BW) = 2pqL \tag{14-2}$$

$$\text{Under non-free sampling:} \quad E(BW) = \frac{2Ln_B n_W}{n(n-1)} \tag{14-3}$$

where p and q are the probabilities that an area or cell will be black and white, respectively, L is the total number of links or joins in the study region, n_B, and n_W denote the number of black areas and the number of white areas, respectively.

Of course, when we engage in free sampling or nonfree sampling, the number of black–white joins will not always be equal to the values given by Equation 14-2 or by Equation 14-3. As Chapter 5 demonstrated, when we draw a sample from a population, the characteristics of the sample are typically a little different from those of the underlying population. Thus, when we compare the observed and expected black–white join count, we must take this variability into consideration. For an independent random process the standard errors of the expected number of black–white joins are

Under free sampling

$$\sigma_{BW} = \sqrt{[2L + K]pq - 4[L + K]p^2q^2} \tag{14-4}$$

Under nonfree sampling

$$\sigma_{BW} = \sqrt{E(BW) + \frac{Kn_B n_W}{n(n-1)} + \frac{4[L(L-1) - K]n_B(n_B - 1)n_W(n_W - 1)}{n(n-1)(n-2)(n-3)} - E(BW)^2} \tag{14-5}$$

where $K = \Sigma_{i=1}^{n} L_i(L_i - 1)$.

It turns out that the sampling distribution of the number of black–white joins is *normal* under the following general conditions:

1. The number of cells or areas is relatively large, or greater than about 30.
2. Both $p, q > 0.2$.
3. The region is not elongated.
4. The join structure of the region is not dominated by a few cells.

Thus, we can *test* the significance of the difference between an observed black–white join count and the count generated by an independent random process using the following equation for a *standard normal deviate* or *Z-score*

$$Z(BW) = \frac{BW - E(BW)}{\sigma_{BW}} \tag{14-6}$$

where BW is the observed number of black-white joins.

In the case of our electoral map in Figure 14-14, we have no information about the distribution of black and white values other than those observed on the map. Thus we engage in nonfree sampling. Substituting the values from Table 14-3 into Equations 14-3 and 14-5 yields

$$E(BW) = \frac{2(107)(30)(18)}{48(47)} = 51.223$$

$$\sigma_{BW} = \sqrt{51.223 + 205.851 + 2390.191 - 2623.796} = 4.844$$

Now substituting these values into the formula for the Z-score

$$Z(BW) = \frac{26 - 51.223}{4.844} = -5.207$$

The $Z(BW)$ score $|-5.207|$ is too large to have occurred purely by chance in the sampling process. Thus, the election returns in Figure 14-14 show a significant departure from a random pattern. Because the observed number of black–white joins is much smaller than the expected number, we know that departure is in the direction of a clustered pattern. We can therefore conclude that the presidential election returns by state in 2000 are significantly and positively spatially autocorrelated.

This test can be used for nominal variables with more than two classes. The extension is quite straightforward. Details can be found in Cliff and Ord (1973, 1981).

Moran's Index of Spatial Autocorrelation

Examination of spatial autocorrelation in continuous data follows the same logic as the join count test for nominal data. First, a measure of spatial autocorrelation for an observed sample of data is constructed. Second, this measure is then compared in a test-statistic against an expected value that is consistent with a random spatial distribution of data values. Measures of spatial autocorrelation in continuous data can be generated for point data as well as areal data.

One of the most commonly used measures of spatial autocorrelation for continuous data is Moran's *I:*

$$I = \frac{n\sum_{i=1}^{n}\sum_{j=1}^{n} w_{ij}(X_i - \bar{X})(X_j - \bar{X})}{\sum_{i=1}^{n}\sum_{j=1}^{n} w_{ij}\sum_{i=1}^{n}(X_i - \bar{X})^2} \tag{14-7}$$

where n represents the number of points or areas, X_i is the value of the variable of interest for point (area) i, and w_{ij} is an element of a *spatial weights matrix* that represents the geographical relationship between all pairs of points (areas) i and j. Most of the terms in Equation 14-7 are familiar. Ignoring the spatial weights matrix, the numerator is a cross product similar to that used in the calculation of the correlation coefficient, and the denominator is based on the formula for the variance.

What about the spatial weights matrix? We have already encountered one simple spatial weighting system used to calculate the join count. That system for areal data uses a set of binary weights, the value one indicating the presence of a border between two areas or regions and the value zero denoting the absence of a common boundary. This binary weighting system is simple and based on the concept of contiguity. The underlying geographical logic is that contiguity is a measure of proximity, and we might expect areas that are contiguous to be more similar than areas that are noncontiguous and thus potentially distant from one another. Suggested improvements to this simple binary system might weight the length of borders. Under this scheme, areas with relatively long borders are assumed "closer" to one another than areas with relatively short borders. Similar spatial weighting schemes exist for point data. The most widely used system of spatial weights for point data is probably that based on the inverse of the distance between two points, as in $w_{ij} = 1/d_{ij}^b$, where larger values of b increase the *"friction" of distance.*

It is also useful to note that spatial autocorrelation between areas can use distance weights, for example, when the geography of a study region is represented by the distances between area centroids. When spatial weights between all pairs of points or areas are organized in the form of a matrix, with rows and columns representing different locations, the result is known as a *spatial weights matrix.*

Note that with areal data and binary spatial weights, Equation 14-7 simplifies to

$$I = \frac{n\sum_{i=1}^{n}{}_{(c)}(X_i - \bar{X})(X_j - \bar{X})}{L\sum_{i=1}^{n}(X_i - \bar{X})^2}$$

where L represents the number of joins between areas for the study region and the $\Sigma_{(c)}$ means sum the following values for all pairs of contiguous areas only.

A simplification of Equation 14-7 is also possible for point data if we row standardize the spatial weights matrix—that is, if we rescale the values along the rows in the weights matrix so that they each sum to one. In this case, $\Sigma\Sigma w_{ij} = n$ and Equation 14-7 can be rewritten as

$$I = \frac{\sum_{i=1}^{n}\sum_{j=1}^{n} w_{ij}(X_i - \bar{X})(X_j - \bar{X})}{\sum_{i=1}^{n}(X_i - \bar{X})^2}$$

Testing whether the observed value of Moran's I represents a significant departure from randomness demands a test statistic based on the expected or mean value of I and its standard deviation. As with the join count statistic, this test can take one of two forms depending on our assumption regarding the process by which values of the variable of interest are assigned over space. This is analogous to the free-sampling and nonfree-sampling assumptions of the join count test. Under the normality assumption, the observed values of the variable of interest are assumed to be the result of drawing a sample of size n from a normally distributed population of values. Under the randomization assumption, the observed values of the variable X are known, however, the spatial arrangement of those values is not. In this case, the distribution of the autocorrelation statistic is based on the number of different possible arrangements of the n values across a fixed set of points or areas. If in doubt, use the randomization assumption.

The mean or expected value of Moran's I under both normality and randomization assumptions is based solely on the number of observations (spatial units)

$$E(I) = \frac{-1}{n-1} \tag{14-8}$$

Clearly, as the number of observations gets large, the expected value of Moran's I approaches zero.

The variance of Moran's I under the assumption of normality is given by

$$\text{Var}(I) = \frac{n^2B - nC + 3A^2}{A^2(n^2 - 1)} \tag{14-9}$$

The variance of Moran's I under the randomization assumption is given by

$$\text{Var}(I) = \frac{n[(n^2 + 3 - 3n)B + 3A^2 - nC] - K[(n^2 - n)B + 6A^2 - 2nC]}{(n-1)(n-2)(n-3)A^2} \tag{14-10}$$

where

$$A = \sum_{i=1}^{n}\sum_{j=1}^{n} w_{ij}$$

$$B = \sum_{i=1}^{n}\sum_{j=1}^{n} w_{ij}^2$$

$$C = \sum_{i=1}^{n}\left(\sum_{j=1}^{n} w_{ij}\right)^2$$

$$K = \frac{n\sum_{i=1}^{n}(X_i - \bar{X})^4}{n\sigma^4}$$

Under both assumptions, the test statistic is based on the formula for a standard normal deviate or a Z-score:

$$Z = \frac{I - E(I)}{\sigma_I}$$

where, of course, $\sigma_I = \sqrt{Var(I)}$

The expected value of Moran's I is less than zero. Observed values of Moran's I that are greater than the expected value indicate a clustered distribution, consistent with positive spatial autocorrelation. Observed values of Moran's I that are less than the expected value indicate a dispersed pattern, consistent with negative spatial autocorrelation.

EXAMPLE 14-5. Precipitation in California. Precipitation was measured at 30 sites in California (see Figure 14-15). The sites are numbered, and those numbers correspond to the data values in Table 14-4. We wish to know whether the precipitation values are spatially autocorrelated or whether they are randomly distributed across the state. Because the data are measured at specific points, we will use distances between all pairs of points to represent the geography of the 30 observations. In Table 14-4,

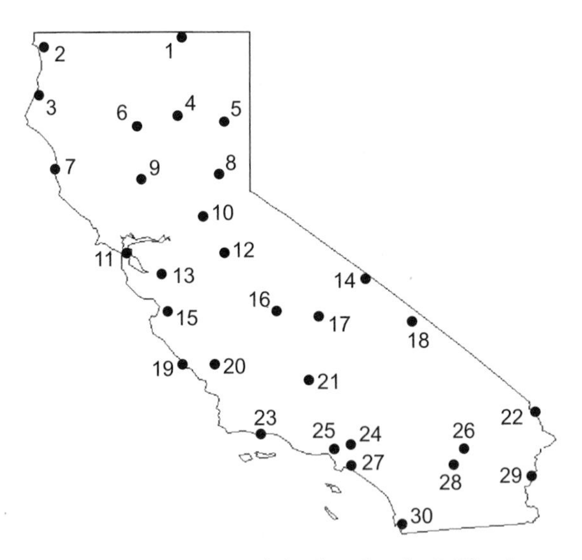

FIGURE 14-15. Precipitation sites in California.

TABLE 14-4
Precipitation Data for California

Site	Prcptn.	Long. W.	Lat. N.	Site	Prcptn.	Long. W.	Lat. N.
1	10.0	121.3	41.9	16	9.4	119.5	36.7
2	74.9	124.0	41.7	17	42.6	118.7	36.6
3	39.6	124.1	40.8	18	1.7	116.9	36.5
4	47.8	121.4	40.4	19	19.3	121.3	35.7
5	18.2	120.5	40.3	20	15.7	120.7	35.7
6	23.3	122.2	40.2	21	6.0	118.9	35.4
7	37.5	123.8	39.4	22	4.6	114.5	34.8
8	49.3	120.6	39.3	23	18.0	119.8	34.4
9	16.0	122.1	39.2	24	14.7	118.1	34.2
10	18.1	120.9	38.5	25	15.0	118.4	34.1
11	21.8	122.4	37.8	26	4.3	115.9	34.1
12	8.3	120.3	37.8	27	12.4	118.4	34.1
13	14.2	121.7	37.4	28	18.2	116.1	33.8
14	5.7	117.8	37.3	29	4.1	114.6	33.6
15	13.9	121.6	36.7	30	9.9	118.1	33.8

Note: Prcptn. is annual precipitation in inches; Long. W. is longitude west in decimal degrees; Lat. N. is latitude north in decimal degrees.
Source: Taylor (1980). The California precipitation values are reproduced in the dataset *caprecip.html.*

the precipitation values are shown along with the geographical coordinates of the sample locations. We calculate distances between points using Pythagoras's theorem and the geographical coordinates. Table 14-5a shows a subsample of the distances between points. The coordinates are used to construct a spatial weights matrix based on the inverse distance between points, $w_{ij} = 1/d_{ij}$, and where the distance between a point and itself is zero. Table 14-5b shows a portion of the weights matrix.

The observed value of Moran's I for the California precipitation data is

$$I = \frac{n\sum_{i=1}^{n}\sum_{j=1}^{n} w_{ij}(X_i - \bar{X})(X_j - \bar{X})}{\sum_{i=1}^{n} w_{ij} \sum_{i=1}^{n}(X_i - \bar{X})^2} = 0.0822$$

TABLE 14-5
A Sample of Distances and Spatial Weights

	a. d_{ij}					b. $w_{ij} = 1/d_{ij}$				
Sites	1	10	20	30	Sites	1	10	20	30	Row sum
1	0	3.423	6.229	10.113	1	0	0.292	0.161	0.099	6.759
10	3.423	0	2.807	6.934	10	0.292	0	0.356	0.144	11.408
20	6.229	2.807	0	4.686	20	0.161	0.356	0	0.213	10.739
30	10.113	6.934	4.686	0	30	0.099	0.144	0.213	0	7.346

Next, we must calculate the expected value E(I) and the variance Var(I) of Moran's I. There is good reason to believe that precipitation values gathered across climate stations are normally distributed, and so we use the assumption of normality.

$$E(I) = -0.034$$

$$Var(I) = 0.00199 \text{ and thus } \sigma_I = 0.0446$$

Finally, inserting these into the test-statistic yields

$$Z = \frac{0.0822 - (-0.034)}{0.446} = 2.605$$

Comparing the test statistic against the tabulated values of the normal distribution in Table A-3, it is apparent that there is less than a 1% chance of obtaining the observed value of Moran's I if the spatial distribution of precipitation in California is random. We thus conclude that the distribution of California precipitation is significantly spatially autocorrelated. Because the observed value of I is greater than that expected from a random process, the precipitation data are positively spatially autocorrelated.

EXAMPLE 14-6. The Geography of Patenting across U.S. States. Figure 14-16 maps the state level geography of patenting across the United States in 1975. The patent rate measures the number of patents granted within a state per 100,000 workers. The map appears to show some evidence of positive spatial autocorrelation with relatively high levels of patenting clustering around New Jersey and with relatively low

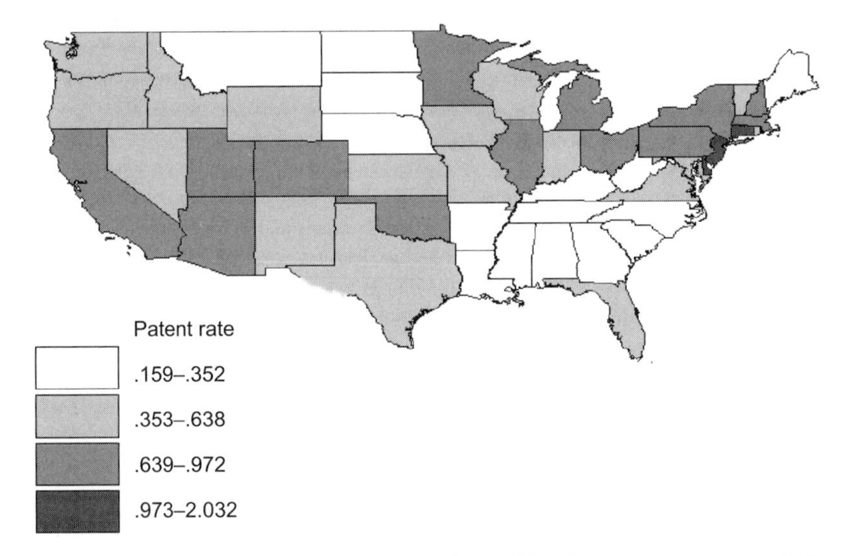

FIGURE 14-16. Patents per 100,000 workers, 1975. *Source: www.uspto.gov;* U.S. Bureau of the Census.

rates of patenting in the southeast of the country as well as in the northern Plains states and into the northern Mountain region. We can test the hypothesis that patenting exhibits some spatial autocorrelation at the state level by calculating Moran's I statistic and performing the usual test of significance. We do this below using two different spatial weighting schemes. In the first case, we use a binary contiguity matrix to represent the geographical arrangement of U.S. states. In the second case, we use inverse distance weights based on state centroids. The patent rate data and the different spatial weights matrices can be found on the website for the book, so the reader might replicate our results. In both cases we adopt the randomization assumption.

The null hypothesis that we are interested in testing states that the observed distribution of patent rates has a spatial autocorrelation value that is not significantly different from that we would expect if the rate of patenting across U.S. states was random. The alternative hypothesis states that the observed distribution of patent rates is different from random.

Case 1: Using a Binary Contiguity Matrix

$$I = 0.195$$

$$E(I) = -0.021 \qquad \sigma_I = 0.057$$

$$\text{Test statistic:} \quad Z = \frac{0.195 - (-0.021)}{0.057} = 3.793$$

Case 2: Using Inverse Distance Weights based on State Centroids

$$I = 0.024$$

$$E(I) = -0.021 \qquad \sigma_I = 0.018$$

$$\text{Test statistic:} \quad Z = \frac{0.024 - (-0.021)}{0.018} = 2.523$$

In both cases we would reject the null hypothesis at the 0.05 level of significance (Z-score = 1.96) and conclude that the spatial distribution of patenting rates across the United States exhibits positive spatial autocorrelation. It is interesting to note, however, the relatively large differences in the values of I and σ_I between the two cases. The patent rate data were the same in both cases, the only difference being in the representation of the geography of U.S. states. This example shows that considerable care must be taken before deciding on a spatial weighting scheme.

14.3. Local Indicators of Spatial Association

Section 14.2 examined what have become known as *global* indicators of spatial autocorrelation. These indicators are global in the sense that they provide one measure of spatial autocorrelation for all points or all areas of a study region. It is likely, however, that in different parts of a region under investigation, local measures of spatial association, measures generated by subsampling small parts of a broader region, will vary. Examination of local indicators of spatial autocorrelation underpins the search for clusters or hot spots—localized areas where values of a variable are significantly greater or significantly lower than average. In this section, we highlight two commonly used measures of local spatial autocorrelation, Anselin's (1995) local indicator of spatial association (LISA) and the G-statistic of Getis and Ord (1992). These indices are commonly employed in *exploratory spatial analysis* to identify locations where data values display particular patterns of interest.

LISA

LISA provides a measure of spatial association for each areal unit within a larger region of study. This measure is a local value of the global Moran's I statistic. The local Moran statistic for area i is

$$I_i = z_i \sum_j w_{ij} z_j \quad \text{for} \quad j \neq i \tag{14-11}$$

where the observations on the variable of interest are normalized ($z_i = [x_i - \bar{x}]/s$) and where the spatial weights matrix ($\boldsymbol{W}$) is row-standardized or scaled so that the sum of weights along any row of $\boldsymbol{W}$ equals 1. The local Moran statistic for area i is thus the product of the normalized value of the variable of interest in area i and the average normalized value of this variable in neighboring areas. This local statistic can be interpreted in the same way as the global Moran: high LISA values indicate spatial clustering of similar values of a variable of interest, while low LISA values indicate spatial clustering of dissimilar values of a variable of interest. As with the global Moran statistic, LISA values for an area vary with the designation of the spatial weights matrix.

EXAMPLE 14-7. Per Capita Income Variations across North Carolina Counties. Figure 14-17 illustrates variations in per capita income for 1999 across the 100 counties that comprise the state of North Carolina. Counties are assigned to one of four groups, with darker shading indicating higher per capita income. Cursory examination of Figure 14-17 reveals some clustering of relatively high income levels in the west of the state and in a crescent just west of the state's center. Pockets of relatively low per capita income are evident in the northeast of the state and in parts of the south. From Equation 14-7, using rook contiguity to capture the spatial association between North Carolina counties, Moran's global measure of spatial autocorrelation is I = 0.3936. With an expected value $E(I) = -0.01$, the observed value of Moran's I reveals that per capita income across North Carolina counties exhibits positive spatial autocorrelation: counties with similar values of per capita income tend to be located close to one another.

What about local indicators of spatial association for the county per capita income data in Figure 14-17? How do different counties in the state contribute to the

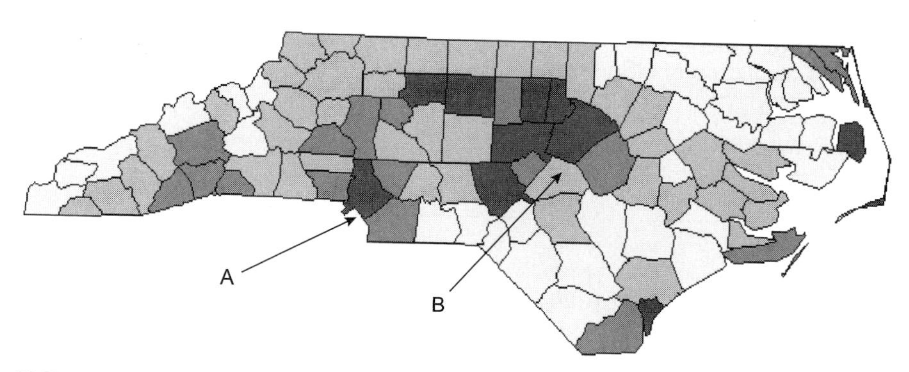

FIGURE 14-17. Per capita income by county in North Carolina, 1999. *Source:* American Fact Finder, U.S. Bureau of the Census.

global measure of Moran's *I*, and where do per capita incomes cluster the most? To help understand the calculation of LISA statistics, we focus attention on two counties labeled A and B in Figure 14-17. The county labeled A is Mecklenburg, which has a per capita income significantly above the county average. Mecklenburg has five contiguous neighbors, all with income levels above average for the 100 counties in the state and thus we anticipate a positive and relatively high LISA statistic for Mecklenburg. In contrast, the county labeled B, Harnett County, has a per capita income level that is below average. Harnett County is surrounded by two counties that also have below-average incomes and by five other counties that have above average levels of income. Because Harnett County has several neighbors with quite different income levels, it should have a LISA statistic that is negative.

We check these impressions in Table 14-6 which illustrates the calculation of local Moran indicators of spatial autocorrelation. Table 14-6a reveals that Mecklenburg County has a relatively high LISA value of 2.799, confirming our visual impression that this area is part of a cluster of counties all with similarly high incomes. Table 14-6b shows that Harnett County has two contiguous neighbors, Sampson and Cumberland counties, that have below-average levels of per capita income. Five other neighbors of Harnett County have above-average income levels, and thus Harnett County is not part of a group of neighboring counties with similar incomes. Indeed, it is this disparity in neighboring income levels that generates the negative LISA statistic of –0.331 for Harnett County.

Anselin (1995) shows that the local Moran statistics for different neighborhoods (local areas) of a study region sum to the value of the global Moran's *I* coefficient for that region (subject to a scaling coefficient). Anselin also shows that a regression of normalized values of a variable for all areas *i* of a study region against the average values of the same variable in neighbor areas has a slope coefficient equal to the global Moran coefficient (*I*). In other words,

$$\mathbf{z} = a + I\mathbf{W}\mathbf{z}$$

TABLE 14-6
Calculation of LISA for Mecklenburg and Harnett Counties

County	Per capita income ($000)	z_i	w_{ij} (row standardized)	$w_{ij}z_j$
(a) Mecklenburg County				
Mecklenburg	27,352	3.178		
Iredell	21,148	1.107	1/5 = 0.2	0.221
Cabarrus	21,121	1.098	0.2	0.220
Union	21,978	1.384	0.2	0.277
Gaston	19,225	0.464	0.2	0.093
Lincoln	18,877	0.348	0.2	0.070
				$\sum_{j\neq i} w_{ij}z_j = 0.881$
				$I_i = z_i\sum_{j\neq i} w_{ij}z_j = 2.799$
(b) Harnett County				
Harnett	16,755	–0.354		
Sampson	14,976	–0.955	1/7 = 0.143	–0.136
Cumberland	17,376	–0.153	0.143	–0.022
Moore	23,377	1.851	0.143	0.264
Lee	19,147	0.438	0.143	0.063
Chatham	23,355	1.844	0.143	0.263
Wake	27,004	3.062	0.143	0.437
Juohnston	19,225	0.464	0.143	0.066
				$\sum_{j\neq i} w_{ij}z_j = 0.935$
				$I_i = z_i\sum_{j\neq i} w_{ij}z_j = -0.331$

where z is a vector of normalized values of the variable of interest, $\mathbf{W}$ is the spatial weights matrix, and a and I represent the constant and slope coefficients to be estimated. This relationship can be illustrated in a Moran scatterplot.

The Moran scatterplot in Figure 14-18 shows the relationship between normalized values of per capita income in North Carolina counties and spatially lagged values of the same variable. The spatially lagged value for area i is the average of the per capita income values found in areas that are contiguous neighbors of i. This value is given as $\Sigma w_{ij}z_j$ in Table 14-6. To help understand the location of counties in Figure 14-18, the two counties from Table 14-6 are highlighted. Mecklenburg is again represented by the letter A and Harnett by B. Mecklenburg has a normalized value of per capita income that is relatively high, $z_i = 3.178$, and the spatially lagged value of per capita income for this county is also above average at 0.881. Harnett County has a below-average value of per capita income. It is left of zero on the horizontal axis, yet it has a spatially lagged income value that is above average because the majority of this county's neighbors have above-average incomes.

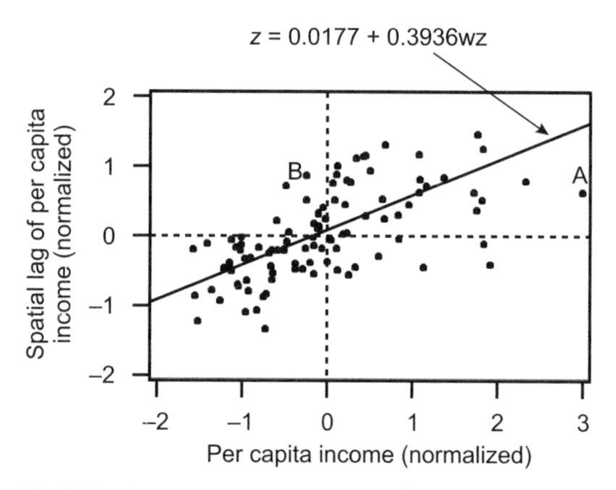

FIGURE 14-18. Moran scatterplot of per capita income in North Carolina counties.

The Moran scatterplot is useful in showing which subareas within a study region exert the most influence on the global Moran coefficient of spatial autocorrelation. These are the values that are located furthest away from the origin (where $z_i = w_{ij}z_j = 0$) in Figure 14-18. Thus, Mecklenburg County has a greater influence on the global Moran coefficient of spatial autocorrelation than does Harnett County. The orientation of the point distribution in the Moran scatterplot provides a general indication of whether the spatial distribution of the variable of interest is clustered or dispersed. When most points in the scatterplot are found in the northeast and southwest quadrants this is consistent with positive spatial autocorrelation, for both the normalized variable of interest and spatially lagged values of that variable covary about their respective means in the same direction. When most points in the scatterplot are located in the northwest and southeast quadrants, this is consistent with a pattern of negative spatial autocorrelation.

Finally, LISA statistics can help identify clusters of values of interest. Clusters might be thought of as groups of neighboring subareas in a broader study region where values of a variable show a discernible pattern. Most attention might normally be directed at identifying clusters where values of some variable of interest are significantly greater or smaller than average. Figure 14-19 provides an illustration, showing the geographical clustering of North Carolina counties with similar or dissimilar values of per capita income. These counties all have significant I_i scores, where significance is assessed by comparing the observed distribution of per capita income values with a conditional random assignment (see Anselin, 1995). The black shaded counties in Figure 14-19 have relatively high per capita income values and are surrounded by counties that also have, on average, high per capita incomes. The light gray counties have relatively low per capita incomes and are surrounded by counties with similarly low incomes. The two darker gray counties have low per capita incomes but are neighbored by counties that generally have higher than average incomes. The LISA cluster map of Figure 14-19 is a useful tool for identifying local instability within a broader region where there is evidence of significant global spatial autocorrelation.

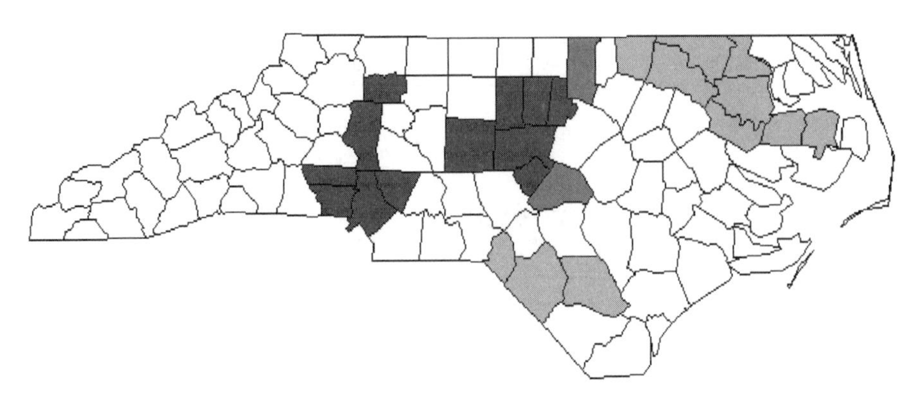

FIGURE 14-19. Spatial clusters of North Carolina counties in terms of per capita income. *Source:* GeoDa. *www.csiss.org/clearinghouse/GeoDa.*

The per capita income data for North Carolina counties can be found on the website for the book as *ncincome.html.*

G_i-Statistic

An alternative measure of local spatial association is provided by the G_i-statistic of Getis and Ord (1992). Similar to the LISA statistic discussed above, the G_i-statistic facilitates identification of local departures from the average value of a variable of interest defined over a broader region. The Getis–Ord measure of local spatial association is defined as

$$G_i(d) = \frac{\sum_{j=1}^{n} w_{ij}(d)x_j}{\sum_{j=1}^{n} x_j} \quad \text{for} \quad j \neq i \tag{14-12}$$

In Equation 14-12, w_{ij} is a spatial weights matrix comprising 0 or 1 values. The spatial weight is set at one if areal unit (or point) j is within a specific distance d of area (or point) i. All locations j that are further from i than distance d are given the spatial weight of zero. By convention, the distance of location i to itself is also set at zero. (Note that Getis and Ord [1992] identify the closely related G_i^*-statistic that differs only from G_i in that the subscript j may include point or area i.) Thus, the G_i-statistic measures the proportion of the sum of values of the variable X that are concentrated (no further than distance d) around location i. This proportion will be high when the x_j values around a location i tend to be large, and small when the x_j values around a location i tend to be small. Movements of this proportion over a study region yield information on the existence and location of clusters of the variable X with different characteristics.

To assess whether departures from the average concentration of values of the variable X within distance d of location i are statistically significant requires information about the distribution of the G_i-statistic. The G_i-statistic is approximately normal under the randomization assumption, and so we employ the following test statistic:

$$Z_i = \frac{G_i(d) - E[G_i(d)]}{\sqrt{\text{Var}[G_i(d)]}}$$

where

$$E[G_i(d)] = \frac{\sum_j w_{ij}(d)}{n-1}$$

$$\text{Var}[G_i(d)] = \left\{ \frac{\sum_j w_{ij}(d)[n-1-\sum_j w_{ij}(d)]}{(n-1)^2(n-2)} \right\} \left\{ \frac{\left(\frac{\sum_j x_j^2}{(n-1)}\right) - \left(\frac{\sum_j x_j}{n-1}\right)^2}{\left(\frac{\sum_j x_j}{n-1}\right)^2} \right\}$$

EXAMPLE 14-8. G_i-statistics for per capita income data for Mecklenburg and Harnett counties within North Carolina. Assuming that the five contiguous counties around Mecklenburg (see Table 14-6a) have centroids within distance d from the centroid of Mecklenburg, then for this area,

$$G_i(d) = 103{,}349/1{,}783{,}445) = 0.0574$$

and where these values can be checked from the ncincome data file on the website for the book. Then, the standardized test statistic for this area is

$$z_i = \frac{G_i(d) - E[G_i(d)]}{\sqrt{\text{Var}[G_i(d)]}} = \frac{0.0574 - 0.0505}{0.0047} = 1.894$$

which is larger than the standard normal score for a two-tailed test, with a significance level of 0.05. A positive and significant test statistic for the neighborhood d around Mecklenburg County indicates a pattern of positive spatial autocorrelation.

For Harnett county, assuming that the seven contiguous adjoining counties (see Table 14-6b) are within the distance d from the centroid of Harnett County, then the standardized $G_i(d)$ test statistic is

$$z_i = \frac{G_i(d) - E[G_i(d)]}{\sqrt{\text{Var}[G_i(d)]}} = \frac{0.0818 - 0.0707}{0.0043} = 2.581$$

For Mecklenburg, interpretation of the Z_i statistic is entirely consistent with the LISA statistic I_i: both values indicate positive spatial association. However, in the case of Harnett County, the local LISA and local G-statistics appear inconsistent. Thinking about the construction of both measures helps explain this divergence. The local LISA statistic compares the similarity of the value of a variable X at a given location, say A, to the value of the same variable in neighboring locations. If the neighboring locations have values of the variable of interest similar to the value found in A, then a positive local LISA statistic is generated. If the values of the variable X are quite different between A and its neighbors, a negative LISA statistic will result. (Think of a significant difference in this case as a departure from the mean in different directions.)

Turning to the G-statistic, there is no explicit comparison of the value of a variable X between one or more locations. What the G (or G^*)-statistic identifies is areas of a study region where a larger than average mass of the sum of all X values for that region are concentrated (positive spatial autocorrelation), or areas of a study region where a smaller than average mass of the sum of all X values are concentrated (negative spatial autocorrelation). Thus, for Harnett County, the LISA statistic (I_i) is negative because this county has a lower than average value of per capita income, and it is surrounded largely by neighbors with relatively high values of income. It is all these high-income values in the neighborhood of Harnett County that produce a local clustering of high income values and results in the positive G-statistic. The reader needs to understand this distinction between the different measures of local spatial association. Each is more or less useful in different circumstances (see Getis and Ord, 1992; Anselin 1995).

14.4. Regression Models with Spatially Autocorrelated Data

For geographers, regional scientists and others who work extensively with spatial data, spatial autocorrelation raises some analytical complications. Perhaps most important among these, when spatial autocorrelation is present in a dataset, the number of independent pieces of information is typically less than the number of observations. This, of course, reflects that fact that observed values of a variable in particular locations are dependent on values of the same variable in neighboring locations. Thus, the researcher does not typically have as many degrees of freedom as anticipated. Furthermore, for some statistical techniques such as regression, discussed in Chapters 12 and 13, spatial autocorrelation violates key assumptions and thus renders inferential analysis suspect. We discuss this issue below and present some alternative types of regression models that accommodate various forms of autocorrelation in geographical data.

From Chapters 12 and 13, the standard regression model is written as

$$Y_i = \alpha + \beta_1 X_{1i} + \beta_2 X_{2i} + \beta_3 X_{3i} + \varepsilon_i \tag{14-13}$$

Using β_1 to represent the intercept, α, of Equation 14-13, and providing that $X_{1i} = 1$, for all i, then Equation 14-13 can be rewritten

$$Y_i = \beta_1 X_{1i} + \beta_2 X_{2i} + \beta_3 X_{3i} + \beta_4 X_{4i} + \varepsilon_i \tag{14-14}$$

In turn, Equation 14-14 can be more conveniently expressed in matrix form as

$$\mathbf{Y} = \mathbf{X}\boldsymbol{\beta} + \boldsymbol{\epsilon} \tag{14-15}$$

where **Y** is an n-element column vector of observations on a dependent variable, **X** is a matrix of n observations across k independent variables, $\boldsymbol{\beta}$ is a k-element column vector of regression coefficients to be estimated, and $\boldsymbol{\epsilon}$ is an n-element column vector of errors, assumed to be independent of one another and normally distributed.

From Sections 14.2 and 14.3, we have seen that spatial data are often autocorrelated. What happens in our regression models when we examine data for geographical areas or points that exhibit significant spatial dependence? In part, the answer to this question depends on the nature of that dependence. In a first case, where the dependent variable in a regression model exhibits spatial autocorrelation, the assumption that observations are independent is violated, and consequently the errors in the regression are also correlated. In this situation, regression coefficients estimated using ordinary least squares (OLS) techniques will be biased and inefficient. In a second case, it is not the dependent variable in the regression that exhibits spatial autocorrelation, but rather the errors themselves. In this situation the regression coefficients estimated with OLS are inefficient.

Detection of spatial autocorrelation in spatial data makes use of many of the techniques discussed in Sections 14.2 and 14.3. When conducting analysis using regression techniques, perhaps the most straightforward way of identifying whether spatial autocorrelation might be influencing results is to estimate the regression model, capture the residuals from that regression, the difference between the observed values of the dependent variable and the values predicted from the regression equation, and plot those residuals on a map. Analysis of the resulting map, along with estimates of Moran's indicator of spatial autocorrelation, should confirm whether spatial autocorrelation is an issue that needs to be resolved.

EXAMPLE 14-9. Regression of the Republican Share of Registered Voters in North Carolina on Income and Race by County. We use county data in North Carolina once more to examine the characteristics of registered Republican voters. Past studies and theoretical insights suggest that Republican voters tend to have higher than average incomes and that they are more likely than not to be white. To examine this argument, we specify the following regression equation:

$$Y_i = \alpha + \beta_1 X_{1i} + \beta_2 X_{2i}$$

where Y, the dependent variable, measures the share of the registered voting population that is Republican, the first independent variable, X_1, is median household income, and the second independent variable, X_2, is the non-Hispanic white share of the population. The observations are the 100 counties of North Carolina, and the variables were measured in the year 2000. These data can be found in the file *ncvoters .html.*

Estimating this regression model using the ordinary least squares method yields the following results:

$$Y = -0.22 + 0.000005X_1 + 0.00518X_2 \qquad (14\text{-}16)$$

Standard errors	(0.0000014)(0.000458)
Standard error about the regression	$s = 0.0777$
Coefficient of determination	$R^2 = 0.642$

Student's t-tests of the hypotheses that the two partial regression coefficients are equal to zero produce t-scores of 3.63 for the first independent variable and 11.31 for the second independent variable, both statistically significant at the 0.01 level. However, can we be confident in these statistical tests?

Residuals from estimating regression Equation 14-16, were captured and mapped in Figure 14-20. In this figure, higher values of residuals are indicated by darker shading. The geography of counties was represented by rook contiguity. Analysis of these residual values generated a global Moran I-value of 0.3859 that is significantly greater than the expected value of Moran's I. Therefore, the residuals exhibit positive spatial autocorrelation with similar values clustering in specific parts of the state. When maps of residuals from a regression analysis reveal anything but a random pattern, it is likely that one or more of the assumptions underpinning statistical inference have been violated. A critical question is what to do in this situation.

If we run a regression model and diagnostic tests from that model reveal that the regression errors are spatially autocorrelated, the next step is to identify the nature of the spatial dependence in the data that we are analyzing. If we can identify the form that the spatial dependence takes, we can build that dependence into our regression model and thus remove its influence on the regression error term. As noted above, there are two general forms in which spatial autocorrelation enters the regression equation: the spatial lag form and the spatial moving average or spatial error form. Let us briefly examine each of these in turn.

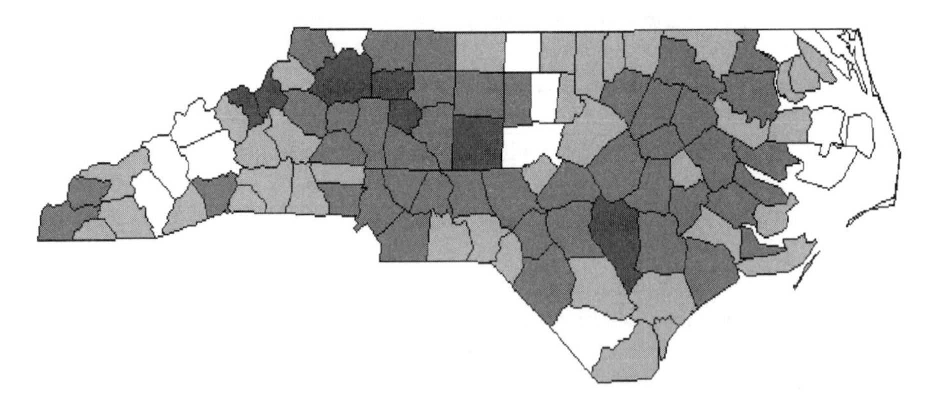

FIGURE 14-20. Map of residuals from regression of Republican share of voters on income and race, North Carolina. *Source: data.osbm.state.nc.us/pls/linc/dyn_linc_main.show.*

Spatial Lag Model

In North Carolina, the share of registered voters who are Republican in a specific county tends to be closely related to the share of Republican voters in neighboring counties. This is characteristic of a process that operates with varying intensity over space. Typically, that intensity changes relatively slowly across the geographical units for which we gather data, especially if those units are small in scale. Thus, in adjacent spatial units, the intensity of the process generating a particular pattern tends to be similar. We might capture this autocorrelation in a regression model by adding an independent variable that is specified as a spatial lag of the dependent variable. We encountered this spatial lag above in our discussion of the Moran scatterplot. The spatial lag of a variable X for location i is simply the average value of the variable X found in locations j that are neighbors of area i. Neighbors of area i might be contiguous areas with common boundaries of nonzero length (rook contiguity), they might be areas whose centroid is within a certain distance of the centroid of i, and so on. If we are dealing with point data, a neighborhood might be defined as all points within a certain distance of point i.

A spatial lag regression, or a spatial autoregressive, model takes the following form

$$\mathbf{y} = \mathbf{X}\boldsymbol{\beta} + \rho\mathbf{W}\mathbf{y} + \boldsymbol{\epsilon} \qquad (14\text{-}17)$$

where all variables are as defined above and $\boldsymbol{W}$ is a spatial weights matrix with elements $w_{ij} = 1$ indicating spatial units i and j are neighbors and $w_{ij} = 0$ otherwise. By convention $w_{ii} = 0$. ρ is the partial regression coefficient for the spatial lag variable. Unfortunately, OLS estimation techniques produce biased estimates of Equation 14-17, and so an alternative method, usually *maximum likelihood estimation,* is typically employed.

For the North Carolina data of Example 14-9, the spatial lag model takes the form

$$Y = -0.2073 + 0.0000039X_1 + 0.0035X_2 + 0.4502X_3 \qquad (14\text{-}18)$$

Standard errors (0.0000012) (0.0005) (0.0861)

where X_3 represents the spatial lag term $\mathbf{Wy}$. Note that the coefficient on the spatial lag, $\rho_3 = 0.4502$, is significantly different from zero at the 0.01 level. The partial regression coefficients on the first two independent variables in Equation 14-18 have not changed a great deal from those in Equation 14-16, but now we can be more confident in their estimation.

Spatial Moving Average or Error Model

The spatial error, or spatial moving average, model is used when the error terms in the regression model exhibit spatial autocorrelation. Spatial autocorrelation in errors rather than in the dependent variable usually result from measurement error or from the influence of spatially autocorrelated variables that are absent from the model and

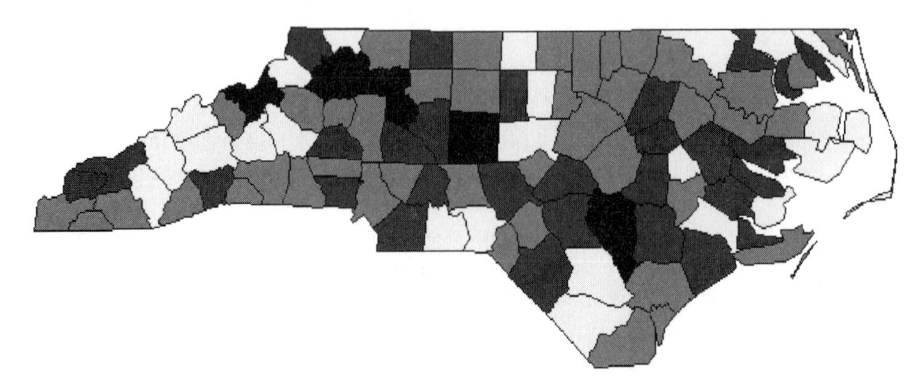

FIGURE 14-21. Residuals from spatial error model.

yet have some influence on other variables included in the regression. The spatial error model is specified as

$$\mathbf{y} = \mathbf{X}\boldsymbol{\beta} + \mathbf{U} \tag{14-19}$$

where $\boldsymbol{U}$ is a composite error term that includes the two components, $\rho\mathbf{Wu} + \boldsymbol{\epsilon}$—the first capturing spatially autocorrelated errors and the second a vector of errors assumed to possess the normal properties.

Estimating this spatial error model for the North Carolina voter data using maximum likelihood techniques yields

$$Y = -0.1868 + 0.0000045X_1 + 0.0049X_2 + 0.6014X_3 \tag{14-20}$$

Standard errors (0.0000016) (0.0006) (0.0922)

where X_3 represents the spatial error term.

In practice, it is difficult to determine whether the spatial lag or the spatial error regression model should be used. Anselin and Bera (1998) outline procedures that can be followed in making this decision. However, one further test can be performed to ensure that spatial autocorrelation has been accounted for within the regression model. After running the spatial lag or spatial error model, the residuals can be captured from the regression and then examined to see if they exhibit any remaining correlation over space. Figure 14-21 shows a map of the residuals from running the spatial error model in Equation 14-20. Moran's I coefficient of spatial autocorrelation for this residual map is 0.0156. This value is not significantly different from the expected value of Moran's I, consistent with a random spatial distribution.

14.5. Geographically Weighted Regression

Just as local indicators of spatial association calculate measures of spatial dependency in data for small areas of a broader study region, so the "global" regression model of

Chapters 12 and 13 and of Section 14.4 can be "localized" to provide estimates of regression parameters that vary over space. This is the focus of geographically weighted regression (GWR) developed by Fotheringham et al. (2002).

In the global regression model

$$\mathbf{Y} = \mathbf{X}\boldsymbol{\beta} + \boldsymbol{\epsilon}$$

and the k-element vector of partial regression coefficients, is estimated as

$$\hat{\boldsymbol{\beta}} = (\mathbf{X}^T\mathbf{X})^{-1}\mathbf{X}^T\mathbf{y}$$

In global models, this vector of regression coefficients is estimated across all observations in the study region and those observations are given equal weight. In the GWR model, estimates of partial regression coefficients are specific to subareas of the study region, and those coefficients are estimated using observations that are spatially weighted. Thus, in GWR models the regression parameters are estimated from the expression

$$\hat{\boldsymbol{\beta}} = (\mathbf{X}^T\mathbf{W}_i\mathbf{X})^{-1}\mathbf{X}^T\mathbf{W}_i\mathbf{y}$$

where **W** is the spatial weights matrix of dimension n × n, with elements on the principal diagonal denoting the spatial weights of all observations associated with location i and where the off-diagonal elements are zero, or

$$\mathbf{W}_i = \begin{pmatrix} w_{i1} & 0 & 0 & 0 & \cdots & 0 \\ 0 & w_{i2} & 0 & 0 & \cdots & 0 \\ 0 & 0 & w_{i3} & 0 & \cdots & 0 \\ \vdots & \vdots & \vdots & \vdots & \cdots & 0 \\ 0 & 0 & 0 & \vdots & \cdots & w_{in} \end{pmatrix} \qquad (14\text{-}21)$$

In the global regression model, the weights $\mathbf{W}_i$ in Equation 14-21 would all be equal to 1. In GWR models a variety of spatial weighting functions may be employed, those functions typically placing greater emphasis on values of variables that are located close to a chosen location. Thus, only observations within a certain distance of a particular location might be used to estimate the local regression model, or all observations in the study region might be used, but where a distance decay function reduces the weight of distant observations.

14.6. Summary

This chapter presented a series of tools for analysis of spatial data. The first of those tools show how to characterize spatial point patterns, focusing on the spatial arrangement of various phenomena. Typically, we are interested in whether spatial data are clustered, dispersed or randomly distributed. The geographer or regional scientist is interested in point patterns insofar as they may provide evidence as to the underlying process that might have generated a particular set of geographic data. Combining information on the geographical locations where data are gathered, together with information on the values of one or more variables of interest, allows us to examine spatial dependency in data using techniques of spatial autocorrelation. Spatial autocorrelation is a very real problem, and the lack of independence in much geographical data violates key assumptions of statistical inference and demands use of specialized techniques. Application of these techniques is explored within the context of regression. Localized measures of spatial autocorrelation are shown to provide important indicators of clustering and "hot spots" in data. Finally, geographically weighted regression is briefly discussed; it is a relatively new tool that, like local measures of spatial association, uses subsets of data from a broader study region to examine local departures from more aggregate spatial processes.

REFERENCES

L. Anselin, "Local Indicators of Spatial Autocorrelation—LISA," *Geographical Analysis* 27 (1995), 93–115.

L. Anselin, and A. Bera, "Spatial Dependence in Linear Regression Models with an Introduction to Spatial Econometrics." In A. Ullah and D. Giles (eds.), *Handbook of Applied Economic Statistics,* pp. 237–289 (New York: Marcel Dekker, 1998).

P. J. Clark and F. C. Evans, "Distance to Nearest Neighbor as a Measure of Spatial Relationship in Populations," *Ecology* 35 (1954), 445–453.

A. Cliff and A. K. Ord, *Spatial Autocorrelation* (London: Pion, 1973).

A. Cliff and A. K. Ord, *Spatial Processes: Models and Applications* (London: Pion, 1981).

P. F. Forget, F. Mercier, and F. Collinet, "Spatial Patterns of Two Rodent-Dispersed Rainforest Trees *Carapa procera* (Meliaceae) and *Vouacapova americana* (Caesalpiniaceae) at Paracou, French Guiana." *Journal of Tropical Ecology* 15 (1999), 301–313.

A. S. Fotheringham, C. Brunsdon, and M. E. Charlton, *Geographically Weighted Regression: The Analysis of Spatially Varying Relationships.* (Chichester: Wiley, 2002).

A. C. Gatrell, T. C. Bailey, P. J. Diggle, and B. S. Rowlingson, "Spatial Point Pattern Analysis and its Application in Geographical Epidemiology," *Transactions of the Institute of British Geographers* 21 (1996), 256–274.

A. Getis and J. K. Ord, "The Analysis of Spatial Association by Use of Distance Statistics," *Geographical Analysis* 24 (1992), 189–206.

P. Greig-Smith, *Quantitative Plant Ecology* (London: Butterworth, 1964).

E. H. Isaaks and R. Mohan Srivastava, *An Introduction to Applied Geostatistics* (Oxford: Oxford University Press, 1989).

J. Odland, *Spatial Autocorrelation* (Newbury Park, CA: Sage, 1988).

P. J. Taylor, "Pedagogic Application of Multiple Regression: Precipitation in California," *Geography* 65 (1980), 203–212.

R. W. Thomas and R. J. Huggett, *Modelling in Geography* (Totowa, NJ: Barnes and Noble Books, 1980).

W. R. Tobler, "A Computer Movie Simulating Urban Growth in the Detroit Region," *Economic Geography* 46 (1970), 234–240.

FURTHER READING

Very readable introductions to spatial data analysis are provided by Boots and Getis (1988) and Odland (1988). A more comprehensive, and still very readable, overview is offered by Bailey and Gatrell (1995). Haining (1990) provides an introduction that is a little more demanding. The classics for spatial data analysis remain Cliff and Ord (1973; 1981) and Upton and Fingleton (1985). A more recent overview is provided by Fotheringham et al. (2000).

T. C. Bailey and A. C. Gatrell, *Interactive Spatial Data Analysis* (Harlow, UK: Longman, 1995).

B. N. Boots and A. Getis, *Point Pattern Analysis* (Newbury Park, CA: Sage, 1988).

A. S. Fotheringham, C. Brundson, and M. E. Charlton, *Quantitative Geography: Perspectives on Spatial Data Analysis* (Newbury Park, CA: Sage, 2000).

R. Haining, *Spatial Data Analysis in the Social and Environmental Sciences* (Cambridge, UK: Cambridge University Press, 1990).

G. Upton and B. Fingleton, *Spatial Data Analysis by Rxample. Volume 1: Point Patterns and Quantitative Data* (Chichester: Wiley, 1985).

DATASETS USED IN THIS CHAPTER

Most of the datasets used in this chapter can be found at the website for the book. The data are stored in *html* format.

caprecip.html—California precipitation data
earthqk.html—California earthquake data
ncincome.html—North Carolina per capita income data by county
ncvoters.html—Republican registered voter data for North Carolina
patents.html—State patent data
statebincon.html—Binary contiguity matrix for (48) U.S. states
stateinverse.html—Inverse distance matrix for U.S. states

PROBLEMS

1. A grid of squares has been placed over a map and the number of points falling in each square is counted. The number in each quadrat of the 100-cell map illustrated here represents this frequency.

1	0	0	0	0	0	0	2	0	1
0	0	0	0	0	0	0	1	0	0
1	0	0	1	1	0	0	0	2	0
0	0	0	0	0	0	0	0	0	1
0	1	0	0	0	0	1	1	0	1
0	0	0	0	0	0	0	1	2	0
0	2	0	1	0	1	0	1	0	0
1	0	0	0	1	1	0	0	0	0
0	0	0	0	2	0	0	0	0	0
0	1	0	0	0	0	0	1	0	1

   a. Construct a frequency distribution of points per quadrat.
   b. Determine the mean and variance of the number of points per quadrat.
   c. Is the spatial distribution of points random?

2. One problem in quadrat analysis is that the results are highly dependent on the size of the quadrats chosen. Conclusions drawn from a study based on a certain quadrat size may be contradicted by conclusions of a second study of the same data based on a different quadrat size. This phenomenon might be called the scale problem, by analogy to the problem identified in descriptive spatial statistics (Chapter 3). To examine the impact of quadrat size, complete parts a–c of Problem 1, using the same base but with the quadrats grouped into fours, forming 25 larger square quadrats.

3. A second problem in quadrat analysis is that markedly different point patterns can give rise to identical frequency distributions of points by quadrats. This is because the frequency distribution cannot indicate whether quadrats with zero points are located close to or away from other quadrats with zero points. In Chapter 3 we called this the pattern problem. To see the significance of this problem, perform the following experiment. Construct two 5×5 grids of 25 quadrats each. Into each grid place a distribution of points with the following frequency distribution:

Number of points	Number of quadrats
0	12
1	9
2	3
3	1

   In the first map, place the points in the grid so that the distribution appears random. In the second map, place the points so that the map appears clustered. Think about what you learn from this exercise.

4. Coordinates for a set of $n = 20$ points of five maps A, B, C, D, and E, are given in the following table. The coordinates are based on a 10×10 km grid laid over an area A = 100 km^2.
   a. Draw maps of the five-point patterns using graph paper.
   b. For each map, calculate the distance between each point and its nearest neighbor. A spreadsheet program is a useful vehicle for this task, along with a function for finding the minimum.
   c. Determine $\bar{d}_o$ for each map, the observed mean nearest neighbor distance.
   d. Find $\bar{d}_e$ for a map with $n = 20$ points and $A = 100$ km^2.
   e. Calculate R for each map and classify each map as clustered, dispersed, or nearly random.
   f. Suppose the actual area on the map from which the coordinates are taken is 200 km^2. Does this alter any of your conclusions?

	Map coordinates				
Point	A	B	C	D	E
1	(2, 1)	(0.5, 0.5)	(2, 1)	(1, 3)	(2, 4)
2	(2, 3)	(4, 0.25)	(2, 3)	(1, 5)	(2, 4.5)
3	(2, 5)	(7, 0.75)	(1, 2)	(1, 9)	(2, 5)
4	(2, 7)	(9.5, 0.5)	(3, 2)	(2, 1)	(2, 5.5)
5	(2, 9)	(5, 2)	(5, 4)	(2, 4)	(2, 6)
6	(4, 1)	(2, 2.25)	(1, 5.5)	(3, 2)	(1.5, 5)
7	(4, 3)	(4, 4)	(4, 5.5)	(3, 6)	(1.5, 1.5)
8	(4, 5)	(5, 5)	(3, 6)	(3, 10)	(3, 5)
9	(4, 7)	(8, 4)	(3, 7)	(4, 8)	(3, 5.5)
10	(4, 9)	(8.5, 3.5)	(3.5, 6)	(5, 1)	(1.5, 6)
11	(6, 1)	(8.5, 4.5)	(6, 5)	(5, 4)	(7, 4)
12	(6, 3)	(2, 6.5)	(6, 5.5)	(6, 6)	(7, 4.5)
13	(6, 5)	(0.5, 9.5)	(6, 6)	(6, 9)	(7, 5)
14	(6, 7)	(3, 9)	(6, 6.5)	(7, 2)	(7, 6)
15	(6, 9)	(5, 9.5)	(9, 2)	(7, 4)	(6.5, 4)
16	(8, 1)	(9.5, 9.5)	(8, 9)	(8, 7)	(6.5, 6)
17	(8, 3)	(4, 7.5)	(8, 8)	(8, 9)	(8, 4)
18	(8, 5)	(4.5, 7)	(8, 7)	(9, 3)	(8, 5)
19	(8, 7)	(4.5, 8)	(9, 8.5)	(9, 8)	(8, 6)
20	(8, 9)	(5, 7.5)	(7, 8.5)	(10, 5)	(7, 5.5)

5. Figure 14-9b shows a pattern consistent with negative spatial autocorrelation. Why? Perform the join count test for randomness on Figure 14-9b when contiguity is measured using the "Rook's case." Repeat the calculations when contiguity is measured using the "Queen's case." Why do the results differ?

6. The following table shows the value of a variable of interest (X) at six locations. The coordinates of those locations are also provided.

Points	X_i	Coordinates
1	14	(1, 5)
2	12	(2, 4)
3	8	(5, 4)
4	8	(3, 2)
5	5	(4, 2)
6	4	(5, 1)

   a. Find the observed value of Moran's I for these data.
   b. Do the data points appear to be spatially autocorrelated?

7. Example 14-6 made use of a binary contiguity matrix for the contiguous 48 states of the United States, as well as a matrix of the inverse of the distances between state centroids. These matrices are included along with other datasets on the book's website. The binary contiguity matrix for U.S. states is *statebincon.html,* and the matrix of the inverse of distances between state centroids is *stateinverse.html.* Using this information and the teachers' wage data from Chapter 2, test the hypothesis that the state distribution of teachers' wages is random.

8. The software GeoDa is freely available from the website *www.csiss.org/clearinghouse/GeoDa.* You can download this software and use the sample datasets to construct Moran scatterplots, find LISA values for different variables and run spatial lag and spatial error regression models. You might link the data from the files *ncincome.html* and *ncvoters.html* to the North Carolina database files from GeoDa and examine results from Examples 14-7 and 14-9.

15

Time Series Analysis

Chapter 3 presented fundamental concepts of time series and described simple smoothing of observed series. In this chapter we extend the treatment of time series in two ways. First, we will take a more formal statistical approach, one based on modeling the time-varying behavior of an observed series. At the risk of stating the obvious, we emphasize that this modeling is empirical, meaning that the models are developed from observations rather than theory. We will not be assuming a priori knowledge of the mechanisms giving rise to the observations, but instead will attempt to discover statistical regularities consistent with whatever physical or social processes happen to be operating. Obviously, where theory is available, it should be used to restrict candidate models to those admitted by the theory. We also emphasize that, despite the central role played by the observations, researchers are by no means free from having to make assumptions about mechanisms underlying the observations. As a matter of fact, it will sometimes be necessary to make strong assumptions.

Second, this chapter will also return to the topic of smoothing, mainly with the idea of placing it in the larger context of time series filtering. As a first step, we will cover the simple running means described earlier in Chapter 3. As will be seen, the means are less than optimal in several regards. We will therefore discuss methods for designing filters better suited to smoothing, as well as filter design for other time series tasks.

Sections 15.1 and 15.2 introduce a fundamental concept called a *stochastic process* and its properties. Section 15.3 presents several important time series models. The following three sections discuss practical matters involving estimation and use of those models. Filtering is presented as a general process in Section 15.7. The next section shows how to analyze filters from a perspective known as the *frequency domain.* Section 15.9 provides methods for designing a filter to accomplish specific goals.

To keep the chapter concise, we will not provide a detailed treatment of these topics. The sections that follow should therefore be considered an introductory discussion, aimed primarily at presenting the rationale behind, and potential use of, some commonly used techniques. In addition, we want to provide enough coverage that should time series analysis be useful to students' own research, they will have at least a few tools at hand.

15.1. Time Series Processes

In Chapter 3, we drew a distinction between a time series *process* and an observed time series, called a *realization* of the process. Loosely speaking, we think of the process as giving rise to the observed series, in much the way we think of a random variable giving rise to a sample. Imagine, for example, a closed lake or reservoir subject to variable inflow gains and evaporative losses. In some years there is a net gain to the lake, whereas in other years evaporation exceeds inflow and lake level decreases. To simplify matters, we will collect gains and losses in a single term a_t, representing the addition for year t. Obviously, a_t will be positive or negative depending on whether or not gains are larger than losses for year t. Let us imagine that the climate is such that on average the gains equal losses. To be consistent with the climatic assumption, the average of a taken over all time must be zero. Let us further assume that the increments a_t are statistically independent of one another and that they come from the same probability distribution. (Random variables with these properties are said to be independently and identically distributed, or i.i.d.) Given some initial value for lake level, we can write the level for any year t as

$$Y_t = Y_{t-1} + a_t \tag{15-1}$$

The sequence above is called a *random walk* process, with a_t representing random steps, or shocks, acting to move Y up or down. This is a very simple example of a time series process, but it nevertheless illustrates a number of important concepts. First, we note that because the additions a_t are random, every sequence of Y will contain random variation. However, successive values of Y will not be uncorrelated, because of carryover from the previous year. Second, not only is there random variation from year to year, but the value in a particular year is a also a random variable. That is, because a_t is random, we must be willing to admit many values are possible for Y in a given year t. This is analogous to saying that many values are possible for the ith value in a sample, and that X_i is a random variable. The difference is that with regard to a sample, we take pains to be sure that the sample values are uncorrelated, whereas in time series analysis we are expressly interested in whatever temporal correlation might exist.

In time-series analysis, Y is called a *stochastic process*. It represents a family of random variables indexed by time. Each member of the family, a sequence of Y, is a separate realization of the process. When we observe a time series, we obtain one realization of the underlying stochastic process.

DEFINITION: STOCHASTIC PROCESS

A stochastic process Y_t is a family of random variables indexed by time. The process generates values of Y serially in time, with many sequences possible. Equivalently, many values of Y are possible for any given time t. An observed time series (or realization) is a particular sequence of Y, corresponding to one outcome of the stochastic process.

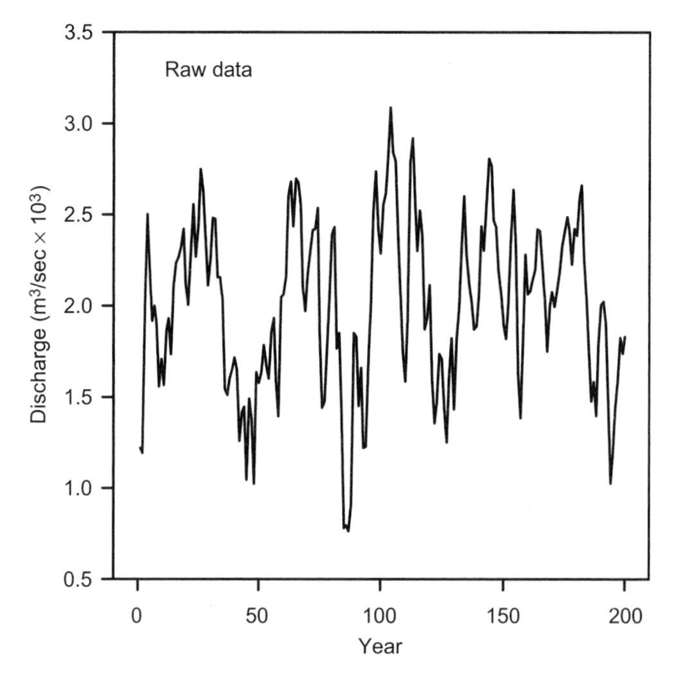

FIGURE 15-1. Time series of annual floods (or peak annual flows) in thousands of cubic meters per second.

As an example realization consider Figure 15-1, which shows a long sequence of maximum annual stream flows. Behind this particular series lies a stochastic process. Obviously, the realization inherits many of its statistical properties from the underlying stochastic process. Not surprisingly, time series have properties not found in nonsequential data. The next section describes some of these.

15.2. Properties of Stochastic Processes

Stationarity

Notice that a stochastic process generates multiple values of Y for any given t. It is quite natural, therefore, to ask questions about the probability distribution of Y at various t. For example, we might wonder if the mean and variance of Y are constant for all t. In the lake example, the "climate" (a_t) is clearly unchanging, but we might wonder if there is any upward or downward trend in the average lake level, or if there is a tendency for lake levels to become more variable as time advances. Though we can measure the mean and variance of a time series using the conventional formulas, we must also allow for the possibility that these values change over the period of our time series. With this small additional complication, we again rely on expected values for the definitions:

$$\mu_t = \mathrm{E}(Y_t)$$
$$\sigma_t^2 = E(Y_t - \mu_t)^2$$

To evaluate these functions, we need information about the probability distribution of Y for every t. Stochastic processes whose probability distributions are the same for all time are called *strictly stationary* processes. In this text we will work with a less stringent concept, called *weak stationarity.* A process is weakly stationary if its moments (e.g., mean, variance, etc.) are the same for all time. The distinction between strict and weak stationarity is subtle and of no real interest here; thus we will use stationarity to mean weak stationarity.

DEFINITION: STATIONARITY

A stochastic process is stationary if its statistical moments are invariant over time. If both the mean and variance are constant, the process is called second-order stationary.

Note that varying degrees, or order, of stationarity are possible. For example, a process might be stationary in the mean but not in variance. The random walk described earlier is one such process. Although the mean is constant for all time, the variance is proportional to the length of the series. Going back to the lake-level example, we would have unchanging *mean* lake levels, but with the variability of levels increasing through time. It is thus first-order stationary but not second-order stationary. It should be mentioned that stationarity at a given order requires stationarity at all lower orders. Unless we state otherwise, we will always assume second-order stationarity.

Autocorrelation

Another property of interest is temporal autocorrelation in Y, which was introduced in Chapter 4. Recalling that discussion, we are once again interested in the relationship between nearby values of Y. Here, of course, the observations are arranged in time; thus "nearby" means "near the same time." Considering adjacent observations first, we are led to correlate observations at time t with observations at time $t + 1$. This is the *lag-1* autocorrelation defined by

$$\rho(t, t+1) = \frac{E(Y_t - \mu_t)(Y_{t+1} - \mu_{t+1})}{\sigma_t \sigma_{t+1}}$$

As a covariance divided by the product of standard deviations, this conforms to the definition of a bivariate correlation coefficient. For a stationary process μ and σ are constants, and the definition simplifies to

$$\rho_1 = \frac{E(Y_t - \mu)(Y_{t+1} - \mu)}{\sigma^2}$$

Conceptually, it is as if we are correlating one column of Y with another identical column whose entries have been shifted upward one unit of time. Thus we put a subscript of unity on the correlation coefficient, writing ρ_1. For example, consider the first 10 years of the Georgia climate data shown in the following table. Observed annual mean temperatures for each year t are in the column labeled y_t. For the lag-1 autocorrelation, we pair every year's temperature with the following year's, as shown in column three (y_{t+1}). Assuming an infinite record (as needed for a population value), the lag-1 correlation could be obtained by correlating columns two and three.

Year	Annual temperature (°F)			
t	y_t	y_{t+1}	y_{t+2}	y_{t+3}
1895	62.2	64.6	64.4	63.8
1896	64.6	64.4	63.8	63.8
1897	64.4	63.8	63.8	63.9
1898	63.8	63.8	63.9	61.8
1899	63.8	63.9	61.8	63.8
1900	63.9	61.8	63.8	62.8
1901	61.8	63.8	62.8	62.9
1902	63.8	62.8	62.9	⋮
1903	62.8	62.9	⋮	⋮
1904	62.9	⋮	⋮	⋮
⋮	⋮	⋮	⋮	⋮

Like the ordinary correlation coefficient, ρ_1 ranges from –1 to 1 and measures linear dependency. A value near +1 indicates a strong positive association, whereas a large negative value means Y_t tends to oscillate, with values above average immediately followed by values below average. An autocorrelation of zero indicates no linear relationship between adjacent values.

Higher-order autocorrelations are defined in much the same way. In general, we will write ρ_k for the lag-k autocorrelation. It will also be useful to have a symbol for autocovariance, γ_k. Again assuming stationarity, these are formally defined by

DEFINITION: AUTOCORRELATION AND AUTOCOVARIANCE

The values for the kth lag are given by

$$\gamma_k = E(Y_t - \mu)(Y_{t+k} - \mu)$$

$$\rho_k = \frac{E(Y_t - \mu)(Y_{t+k} - \mu)}{\sigma^2} = \frac{\gamma_k}{\sigma^2}$$

Looking at the table of annual temperatures for Georgia, there is a column labeled y_{t+2} containing temperatures two years after t. Thus correlating that column with y_t would

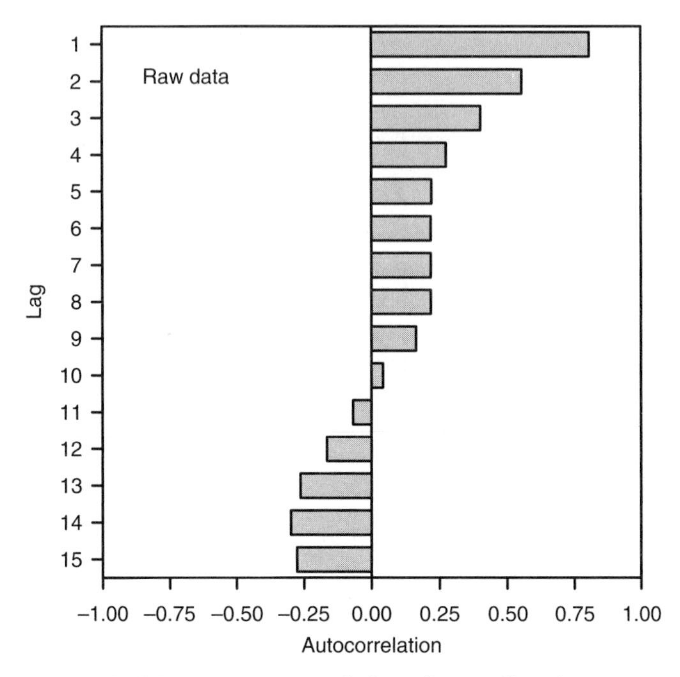

FIGURE 15-2. Autocorrelation of streamflow data.

give the lag-2 autocorrelation. Similarly, the lag-3 autocorrelation could be found by correlating y_t with the last column in the table.

We see that autocorrelation can be computed for any desired lag. The collection of all autocorrelations constitutes the so-called *autocorrelation function.* For example, Figure 15-2 depicts autocorrelations computed from the streamflow time series. (Because they were computed from a sample, the plotted values are actually estimates rather than population values. Estimation is discussed later in this chapter.) The graph shows that values near one another in time (small lag) are strongly correlated and that the strength of correlation decreases with increasing lag. Values about 10 years apart are virtually uncorrelated, and values at larger separations have weak negative correlation.

As an example of how autocorrelation might be used, consider a familiar problem of finding the variance of the sample mean. For independent *Y,* we have seen that $\sigma^2_{\bar{Y}} = \sigma^2/n$. In the case of a time series containing autocorrelation, the *Y*-values are clearly not independent of one another; thus the standard formula σ^2/n requires modification. In particular, we will need an additional term reflecting the aggregate effect of covariances at all lags. This additional term can be positive or negative, depending on the value of the individual covariances. The appropriate formula is

$$\sigma^2_{\bar{Y}} = \frac{\sigma^2}{n} + \frac{2}{n}\sum_{k=1}^{n-1}\left(1 - \frac{k}{n}\right)\gamma_k \tag{15-2}$$

Notice that if the autocovariances are positive, the sum will be positive, indicating that σ^2/n underestimates the standard error of $\bar{Y}$. In order to reliably estimate sampling error in $\bar{Y}$, we need a larger number of observations than would be needed in the absence of autocorrelation. This is not surprising. We expect a process with positive autocorrelation to have periods of unusually high values followed by periods of unusually low values. (After all, nearby values are positively correlated.) A sample occurring within one of these episodes will reveal within-episode variability but will not reflect the transition from one episode to another. Looking at this another way, the presence of autocorrelation means that we do not have n independent pieces of information. It is as if the effective sample size is smaller, and more observations are necessary to obtain an estimate of equal reliability.

If the autocovariances are negative, however, the sum in Equation 15-2 will be negative, indicating that we expect more variance in a short sample than would occur if the observations were independent. In either case, positive or negative γ, the implication is clear: when computing the standard error of $\bar{Y}$, we need to account for autocorrelation. The standard formula, σ^2/n, will not work in the presence of strong autocorrelation, whether negative or positive.

Partial Autocorrelation

The autocorrelation function ρ_k indexes the association between Y-values separated by k intervals of time. For example, ρ_3 measures the correlation at lag 3. It seems obvious that the correlation at lags 1 and 2 must influence the correlation at lag 3, and ρ_3 therefore includes the effects of intervening values. The *partial autocorrelation* function is useful for removing the influence of intermediate lags. It provides the correlation between values of Y at lag k, controlling for correlation at lag $k-1$, $k-2$, and so forth. The partial autocorrelation function for lag k is denoted ϕ_{kk} and can be understood by imagining a somewhat involved multiple regression procedure.

The procedure calls for values of Y_{t+k} to be regressed on preceding values of Y. For example, if we are interested in ϕ_{33}, we would regress Y_{t+3} on Y_{t+2} and Y_{t+1}. The residuals from this prediction equation correspond to variance in Y_{t+3} that cannot be explained by variance in the preceding two periods. We would also regress Y_t on Y_{t+2} and Y_{t+1}. Errors of this prediction represent variance in Y_t that cannot be explained by variance in the following two periods. The correlation of the two sets of residuals gives ϕ_{33} and represents correlation remaining at lag 3 after correlations at shorter lags have been removed.

The regression procedure above is useful for understanding how the partial correlation function comes about but is not likely to be used in actual practice, as much more convenient expressions exist. For example, an expression for ϕ_{22} is

$$\phi_{22} = \frac{\rho_2 - \rho_1^2}{1 - \rho_1^2} = \frac{\rho_2 - \phi_{11}\rho_1}{1 - \phi_{11}\rho_1}$$

By writing $\rho_1 = \phi_{11}$, we see that the partial correlation can be computed from the simple correlations up to order 2 and partial correlations of lower order. In other

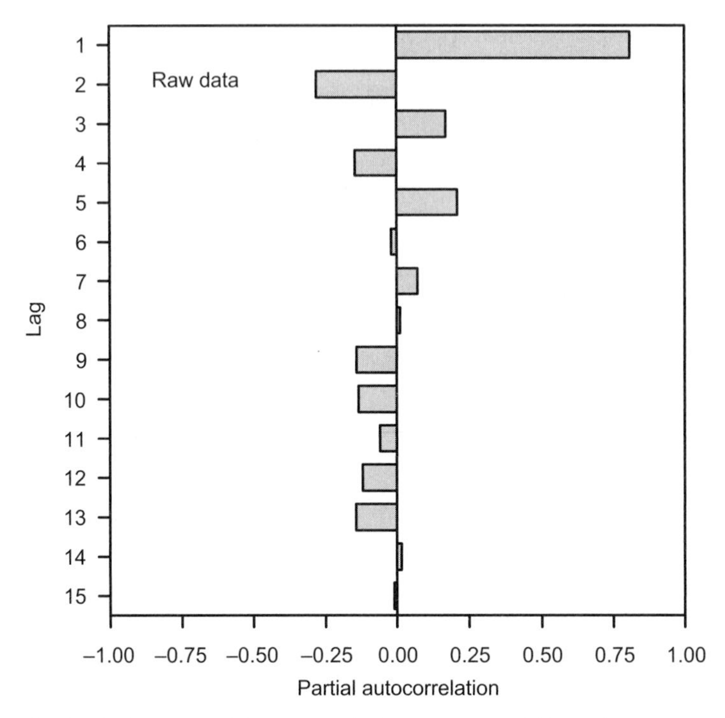

FIGURE 15-3. Partial autocorrelations computed from the streamflow data.

words, given the first-order partial coefficient, one can find the second-order partial coefficient without going through a regression procedure. For higher-order partial coefficients ϕ_{33}, ϕ_{44}, and so on, similar recursive equations can be used (see Cryer, 1986, p. 109).

Figure 15-3 shows partial autocorrelations for the streamflow series. Notice that the partial autocorrelation at lags 2 and higher is much smaller than the corresponding autocorrelation. This illustrates how the autocorrelation function allows associations at short lags to influence correlations at larger lags. The partial autocorrelation function, by contrast, is uncontaminated by correlation at shorter lags.

15.3. Types of Stochastic Processes

Having defined a stochastic process, we turn to several specific classes of process that form the basis of most time series modeling. These processes have the advantage of being easy to understand, yet provide for a very wide range of behavior, especially when combined. In order to simplify the mathematics, we assume that the mean has been subtracted, so that μ_Y is zero.

Autoregressive Processes

Autoregressive (AR) processes are those processes for which the current value Y_t can be found from past values plus a random shock a_t. The simplest is a first-order process, where only the value immediately preceding Y_t contributes:

$$Y_t = \phi Y_{t-1} + a_t$$

This is similar to the random walk model described earlier, except for the parameter ϕ. This coefficient determines how strongly the past (Y_{t-1}) influences the present (Y_t). Clearly, if ϕ is zero, Y is governed completely by the random component a_t. On the other hand, large values of ϕ imply that previous values carry over and are "remembered" by the system in subsequent years. The amount of memory is variable, but stationarity requires that $-1 < \phi < 1$. For ϕ outside this range, it can be shown that the variance of Y is a function of t, and it therefore does not meet the stationarity requirement. (The random walk process has ϕ exactly equal to 1 and thus is an example of a nonstationary process.) A first-order autoregressive process is abbreviated AR(1).

Figure 15-4 shows time series from two AR(1) processes. They are based on the same sequence of a_t but have different values for ϕ and μ. Both curves vary about their mean with no overall long-term trend, but the pattern of variance is very different. Notice how the large positive value for ϕ gives excursions from the mean that persist for considerable periods. It is easy to imagine that if we were to observe only a part of the record, say Y_{50} to Y_{75}, we might conclude that a deterministic trend is present. On the other hand, the process with negative ϕ shows a pronounced tendency to oscillate upward and downward from one period to the next.

The AR(1) process extends easily to higher orders.

DEFINITION: AR PROCESS

An autoregressive process models the present as a linear combination of p past values plus a random shock a_t. The general equation for an AR(p) process is

$$Y_t = \phi_1 Y_{t-1} + \phi_2 Y_{t-2} + \phi_3 Y_{t-3} + \ldots + \phi_p Y_{t-p} + a_t \qquad (15\text{-}3)$$

Although we are free to choose any order p, model orders higher than 3 are not frequently encountered. The stationarity requirements for higher order models are more complicated, and we will not state them here (see Wei, 1990, p. 46). Autoregressive models are a natural choice when a system contains components able to store quantities like energy or mass or information from one observing period to the next. The issues of how to choose the order and estimate coefficients will be covered later. For now we simply note that AR models treat the present as a linear combination of the past subject to random shocks. As might be expected, the size of the random shocks has a direct influence on the ability to predict Y from its past. If the random component is relatively small, Y will be highly predictable.

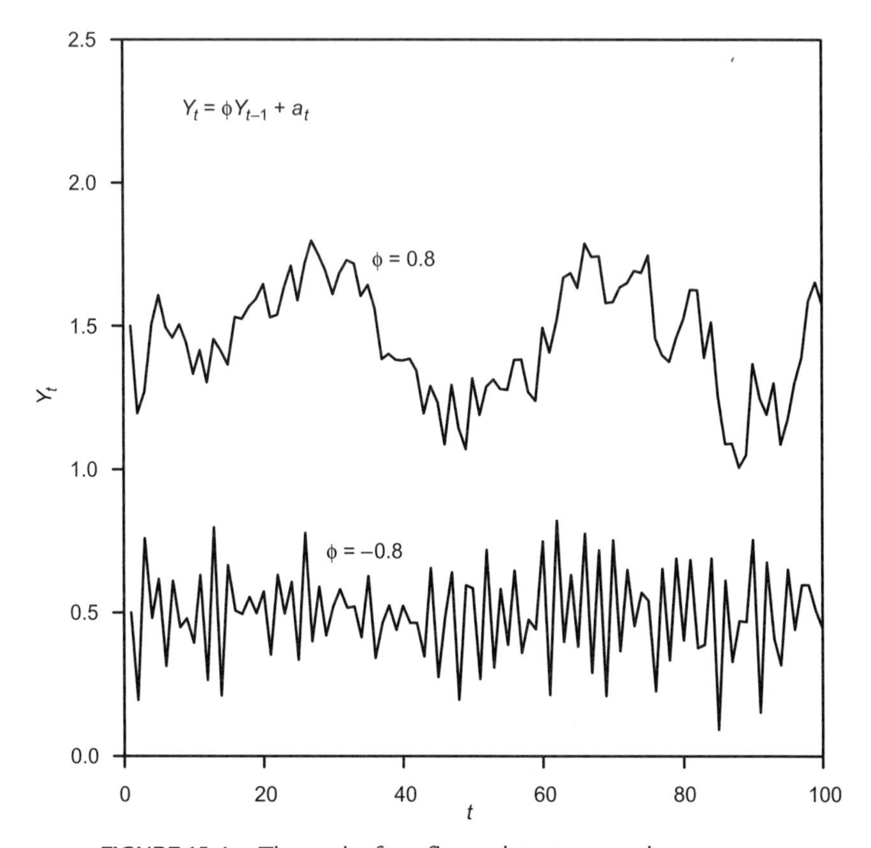

FIGURE 15-4. Time series from first-order autoregressive processes.

Moving Average Processes

A moving average (MA) model assumes Y_t is a linear combination of past shocks plus a new a_t. The first-order model is just

$$Y_t = a_t - \theta a_{t-1}$$

The name "moving average" seems ill-chosen, as it does not seem that we are averaging anything. To appreciate the moving average aspect, we must view the a_t as data, and notice that values of Y are weighted sums of nearby a. For the first-order moving average model, the weights in the running mean are 1 and $-\theta$. An example of an MA(1) process is shown in Figure 15-5 using θ of 0.5. The model produces a very noisy record, with large jumps possible in a single time step. We saw previously that the need for stationarity placed a restriction on ϕ in the AR(1) model. For MA models, the concern is to ensure *invertibility*. Looking ahead to the problem of identifying a model based on an observed series, we will want to choose models that can be uniquely identified from their autocorrelation function. Such models are called invert-

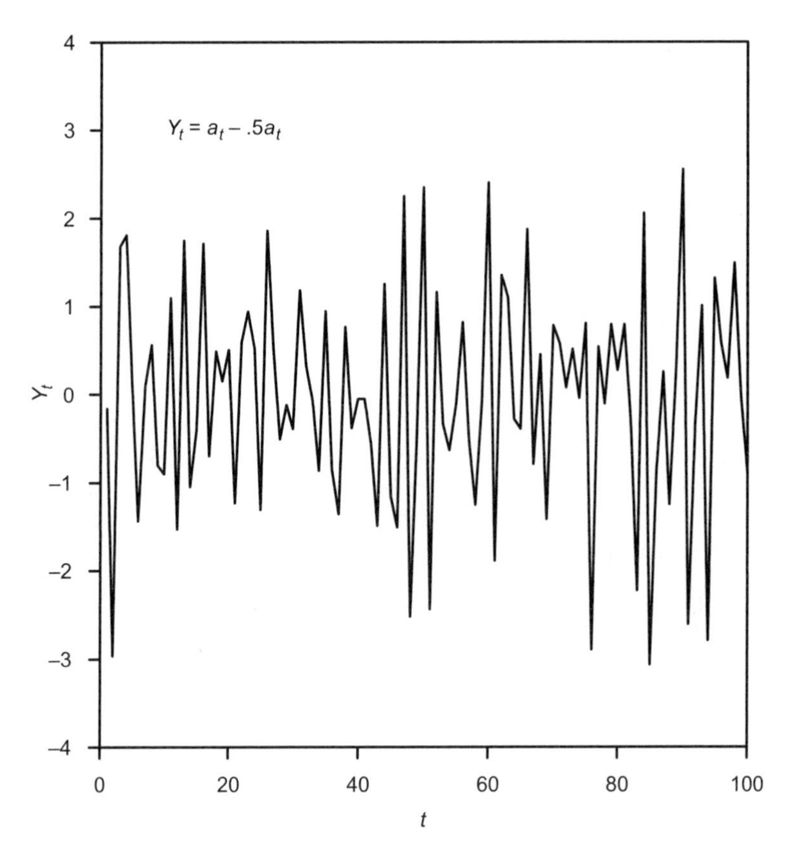

FIGURE 15-5. Time series from a first-order moving average process.

ible because their autocorrelation function can be "inverted" to find the underlying process. It so happens that if θ is outside the range $-1 < \theta < 1$, different processes (θ values) have the same autocorrelation structure and hence are not invertible. Knowing the autocorrelation would not allow one to identify the model. The need for invertibility therefore restricts the admissible values of θ, just as stationarity did for ϕ.

As is true for AR models, higher-order MA models are available. The order is symbolized by q, as given in the following definition.

DEFINITION: MA PROCESS

A moving average model treats the present as a combination of a random shock and q past random shocks. The general equation is

$$Y_t = a_t - \theta_1 a_{t-1} - \theta_2 a_{t-2} - \ldots - \theta_q a_{t-q} \tag{15-4}$$

In many applications, it will be difficult to interpret a moving average model in terms of physical or social processes. Some help is provided by thinking of a_t as a prediction error, the random shock at time t for which we have no predictive ability. With this view Equation 15-4 can be seen as

$$\text{Observed} = \text{Error} + \text{Predicted}$$

Notice that the predicted values are a linear combination of past values of a, each of which is itself a prediction error. In other words, moving average models make predictions for the future based on past errors of prediction. It is a bit like using an instrument known to give false readings. One could use past errors to guess about true values for the present. Obviously, this interpretation is somewhat tortured, and not easily related to commonly encountered processes. Given this seeming dearth of obvious applications, one might wonder why MA models are ever used. The answer lies in their ability to very succinctly encapsulate complicated temporal structure. In time series modeling, as in other modeling, one generally looks for models with as few parameters as possible. Faced with a choice between a low-order MA model or a high-order AR model with many coefficients, the MA model might be preferred despite the difficulty of interpretation.

Autoregressive–Moving Average Processes

To give still more flexibility, we can mix autoregressive and moving average terms in a single model. These are called autoregressive–moving average models, or ARMA (p, q) where p and q are the respective orders for the AR and MA parts.

DEFINITION: ARMA PROCESS

An autoregressive–moving average model combines p autoregressive terms and q moving average terms. The general equation is

$$\begin{aligned} Y_t = \phi_1 Y_{t-1} + \phi_2 Y_{t-2} + \phi_3 Y_{t-3} + \ldots + \phi_p Y_{t-p} \\ + a_t - \theta_1 a_{t-1} - \theta_2 a_{t-2} - \ldots - \theta_q a_{t-q} \end{aligned} \qquad (15\text{-}5)$$

For mixed models, it is necessary to ensure both stationarity and invertibility. As expected, this places restrictions on values for the parameters appearing in Equation 15-5.

15.4. Removing Trends: Transformations to Stationarity

Suppose one is working with a process known to be nonstationary. For example, perhaps siltation is causing reservoir levels to slowly increase, or inflation is affecting bank balances. These cause drift in the mean value, violating stationarity. Alternatively, perhaps the time series is subject to seasonal variation, so that the mean changes from one season to the next. Situations like this are obviously extremely common and call for special treatment before modeling. Fortunately, a number of techniques are available for transforming a nonstationary series into one that is suitable for modeling. Two basic methods are used: the *deterministic* approach and the *stochastic* approach.

The deterministic approach is appropriate when one believes that the disturbance from stationarity applies for all time without change. For example, we might believe that the seasonal temperature cycle operates throughout the record with unchanging amplitude. Or perhaps we believe that the siltation rate is constant. In these cases one is justified in fitting a trend equation to the data and using the equation to remove the trend. For example, one might fit a linear or quadratic function to a time series. By subtracting the fitted values from each observation, one obtains a "detrended" time series, presumably now stationary. Notice that the trend equation provides the mean for each observation time; thus we have estimated the so-called *mean value function* for the time series. Evaluating the function for a particular t gives the mean for that t.

In the case of a seasonal trend, we must believe that the mean for each season is the same for all years. Working with monthly data, we would calculate the mean January value for all years, the mean February value, and so on. By doing so we are effectively fitting another mean value function, this time one that varies seasonally. Once again the mean value function is subtracted from the original series, giving a detrended series. There are other ways of fitting a mean value function, but all are similar to the above in that they yield μ as a function of time. Moreover, one need not fit the mean value function prior to constructing the model. For example, we can include a constant δ in a general ARMA(p, q) model:

$$\begin{aligned} Y_t = \delta t + \phi_1 Y_{t-1} + \phi_2 Y_{t-2} + \phi_3 Y_{t-3} + \ldots + \phi_p Y_{t-p} \\ + a_t - \theta_1 a_{t-1} - \theta_2 a_{t-2} - \ldots - \theta_q a_{t-q} \end{aligned} \qquad (15\text{-}6)$$

At each time period δ increments Y; thus inclusion of the constant amounts to adding a linear trend.

Still other approaches can be used in special circumstances. Suppose, for example, that a measuring device is known to have changed at some point during the observing period. Perhaps the instrument was moved to a new location, or perhaps a different type of instrument was installed. Such changes might well lead to a discontinuity in the mean and provide justification for calculating the mean separately for the two recording periods. By expressing observations as deviations from their respective sub-period mean, the discontinuity is removed.

It is important to stress that one must have strong reasons for using a deterministic mean value function. It is not sufficient to look at the time series and conclude that the mean is changing and that detrending is required. As evidence, recall Figure 15-4, which showed several apparent "trends," all of them fictitious.

The alternative to a deterministic approach is a stochastic transformation, in which the source of nonstationarity is *not* assumed to be uniform over all time. It might be, for example, that the mean value changes slowly over time, with the changes determined by a random walk. Because changes in the mean are unpredictable, there is no deterministic function of time that can remove them. The solution is to subtract each observation from its neighbor, a process called differencing, denoted by the difference operator ∇. Instead of using the observations Y, one builds a time series for the differenced series.

DEFINITION: DIFFERENCING

Differencing is a stochastic method to achieve stationarity based on subtraction of observed values. For example, the first difference is just $Z_t = \nabla Y_t = Y_t - Y_{t-1}$ and removes a constant upward or downward drift over the series.

To see how differencing removes nonstationarity in the mean, imagine that μ has increased by a random amount b_t between times $t-1$ and time t. Differencing the Y's leaves behind the random increment, which is accommodated in a time series model for Z. In some cases, the time series will not be stationary after differencing, and it may be useful to take differences of the differenced series. Still higher differencing can be done as well, but in practical applications first or second differences will nearly always suffice. Models that are stationary after differencing are called *integrated* autoregressive–moving average models, or ARIMA models. They are denoted ARIMA(p, d, q), with d indicating the order of differencing.

In the ARIMA models, adjacent values of Y are differenced to produce a stationary series. In modeling nonstationary seasonal processes, one can use seasonal differences to produce a stationary series. Using monthly data for example, one can form the differences $Y_t - Y_{t-12}$. This is appropriate when the seasonal component is subject to random variation, as would be the case when the seasonal variation is governed by a random walk process. Once again, we are not restricted to first differencing, but can use second- or even higher-order differencing. Seasonal series sometimes present another challenge in that the value for a month might be related to the immediately preceding months and to the same month in preceding *years*. This leads to models that employ autoregressive and moving average terms from corresponding seasons in previous years. Such models are called *multiplicative seasonal* ARIMA models and are widely used in modeling economic time series. The order of the seasonal submodels can be different from the nonseasonal submodels; thus the number of possibilities is nearly endless. For details, see the Cryer, Graupe, and Wie references.

15.5. Model Identification

The foregoing clearly shows that a very broad range of time series models exist and that there will be few occasions where one has sufficient knowledge to choose an appropriate model on theoretical grounds. In time series analysis, one therefore almost always faces the problem of identifying a model from the data. By this we mean that one must somehow decide on the order of the moving average and autoregressive components. Note that choosing an order of zero amounts to excluding one or the other from the model. In addition, one must decide whether or not to detrend, and if

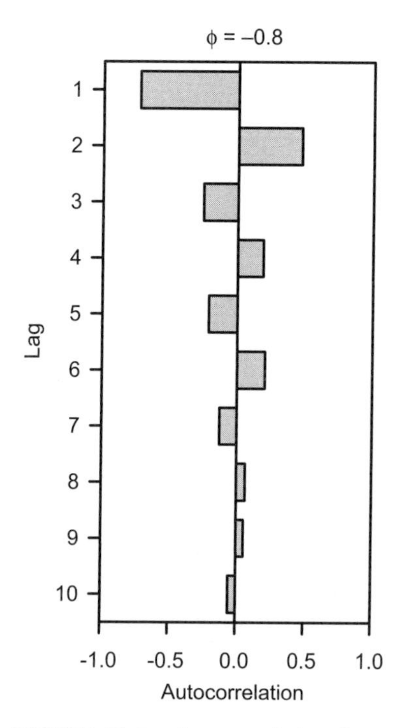

FIGURE 15-6. Autocorrelation functions for the AR(1) process $Y_t = -0.8Y_{t-1} + a_t$

so, how to do so (deterministic vs. stochastic, etc.). In this section we offer some general guidelines, along with the suggestion that one should always experiment with a number of candidate models.

The most powerful aids to model identification are the autocorrelation and partial autocorrelation functions. A quick look will often narrow the possibilities considerably. Consider, for example, Figure 15-6, which shows the autocorrelation function calculated from one of the AR(1) models plotted in Figure 15-4. We see the autocorrelation values alternating in sign from lag to lag and decreasing geometrically. This is diagnostic of all AR(1) processes having a negative coefficient. If we had only the graph to go on, we would certainly look closely at AR(1) models as a candidate for the time series.

As implied by the example, AR and MA models have their own characteristic pattern of autocorrelation and partial autocorrelation. The following relations are especially useful.

AR(1) Models

ρ_k The autocorrelation function is simply $\rho_k = \phi^k$. As a result, the autocorrelation decreases by a factor of ϕ for every increase in lag. Note that negative

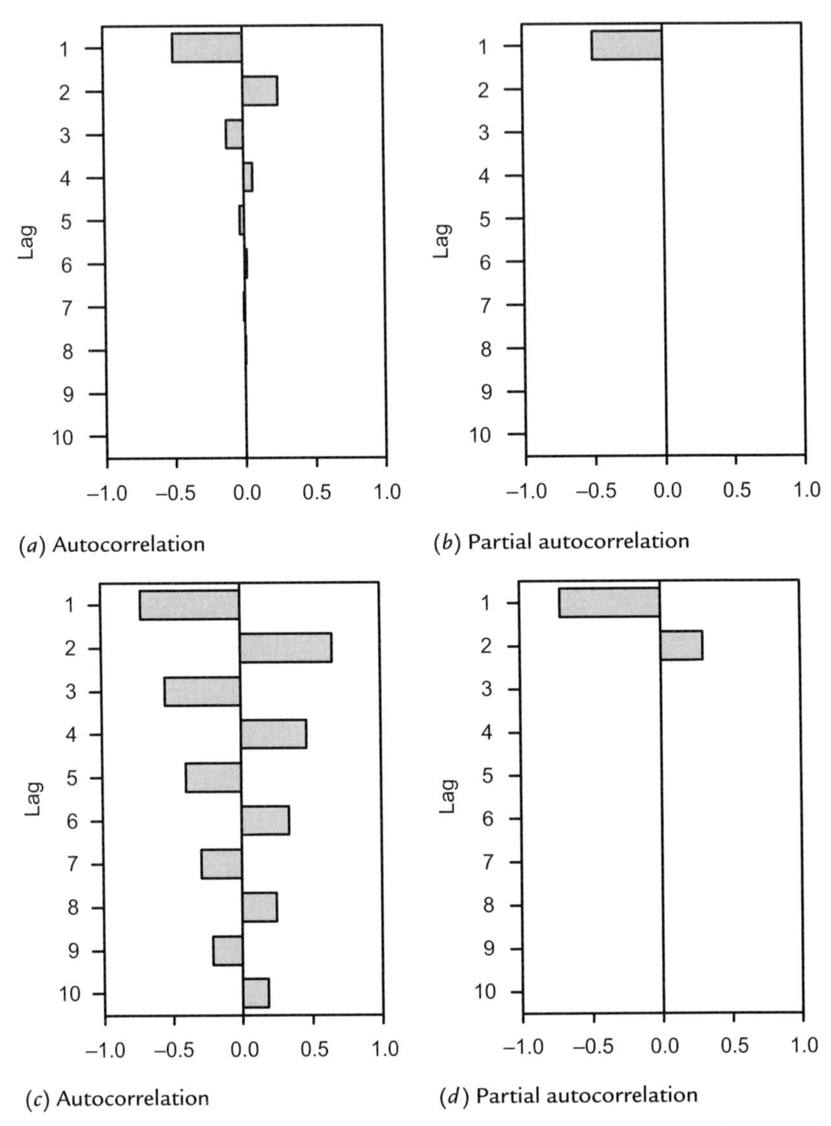

(*a*) Autocorrelation

(*b*) Partial autocorrelation

(*c*) Autocorrelation

(*d*) Partial autocorrelation

FIGURE 15-7. Autocorrelation and partial autocorrelation functions for two AR models and one MA model. The coefficients are $\phi_1 = -0.5$, $\phi_2 = 0.3$, $\theta_1 = 0.4$.

ϕ causes ρ_k to alternate in sign from lag to lag, with odd lags having negative autocorrelation. Figure 15-7a shows an example where $\phi_1 = -0.5$.

ϕ_{kk} The partial autocorrelation function is zero for all lags greater than 1. That is, $\phi_{kk} = 0$ for $k = 2, 3, \ldots$. This is expected, given that Y_t depends only on Y_{t-1}. Any correlation at higher lags arises from the first-order correlation, which is absent from ϕ_{22}, ϕ_{33}, etc. See Figure 15-7b.

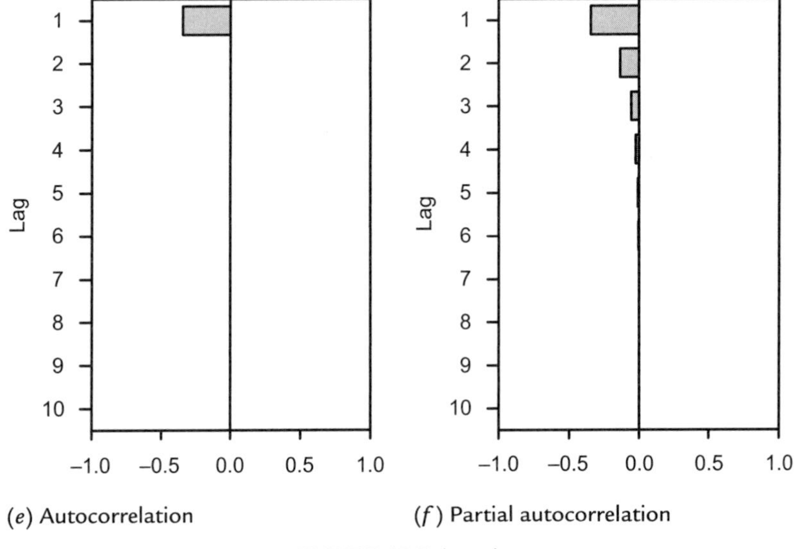

(*e*) Autocorrelation (*f*) Partial autocorrelation

FIGURE 15-7. (*cont.*)

AR(p) Models

ρ_k The autocorrelation function can be written $\rho_k = \phi_1\rho_{k-1} + \phi_2\rho_{k-2} + \ldots + \phi_p$. As for AR(1) models, ρ tails off geometrically. Figure 15-7c is an AR(2) example where $\phi_1 = -0.5$, $\phi_2 = 0.3$.

ϕ_{kk} The partial autocorrelation function is zero for all lags greater than p. See Figure 15-7d for the same AR(2) example.

MA(1) Models

ρ_k The autocorrelation function is zero for $k > 1$. Only Y_{t-1} is correlated with Y_t, the value of ρ_1 being $-\theta/(1 + \theta^2)$. See Figure 15-7e for an example where $\theta_1 = 0.4$.

ϕ_{kk} The partial autocorrelation function decreases slowly with increasing lag. See Figure 15-7f.

MA(q) Models

ρ_k The autocorrelation function is zero for all lags greater than q. For lower lags no general pattern exists, as ρ_k is a complicated function of the θ-values in the model. (ϕ_k is a rational polynomial in θ.) See Figure 15-8a for an example MA(2) model where $\theta_1 = 0.4$, $\theta_2 = -0.2$.

ϕ_{kk} The partial autocorrelation decreases slowly with increasing lag. See Figure 15-8b.

ARMA(p, q) Models

ρ_k After the first $q - p$ lags, the autocorrelation function decreases in a complicated fashion, with a mixture of geometric decrease superimposed on a sinusoid of decreasing amplitude. See Figure 15-8c for an ARMA(1,1) example where $\phi_1 = -0.5, \theta_1 = 0.4$.

ϕ_{kk} After the first $q - p$ lags, the partial autocorrelation shows the same sort of complex behavior as ρ_k. See Figure 15-8d.

In using the autocorrelation and partial autocorrelation functions, we will of course have to work with sample estimates of these functions. Because of sampling

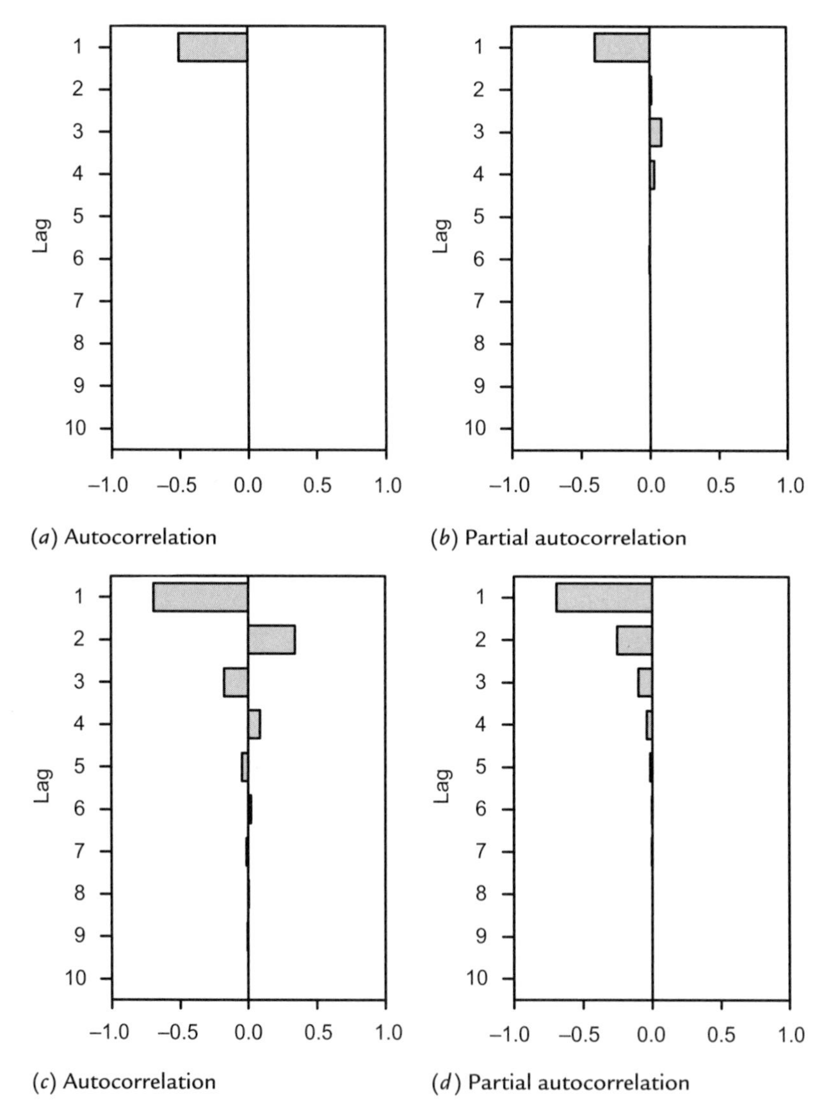

FIGURE 15-8. Autocorrelation and partial autocorrelation functions for an MA model and an ARMA model. The coefficients are $\phi_1 = -0.5, \phi_2 = 0.3, \theta_1 = 0.4, \theta_2 = -0.2$.

error, there will be departures from the patterns we have just described. For example, the sample partial correlations ϕ_{22}, ϕ_{33} . . . in an AR(1) model will be near zero but not exactly zero. To decide whether or not a function has decayed to zero, it will be necessary to consider the statistical significance of the sample values. Fortunately, the sampling distributions of ρ_k and ϕ_{kk} are known for large n. As a diagnostic aid, computer programs routinely provide confidence intervals along with estimates of ρ_k and ϕ_{kk}.

It should be pointed out that in arriving at the standard errors, it is necessary to assume that the a_t are independently and identically distributed. This will be the case for any stationary ARMA model, provided that the model has been correctly specified. After all, if the a_t are not i.i.d., some temporal structure remains in the residuals, which implies that we do not have the correct model form. This leads to another tool in model identification, residual analysis. Having fit a candidate model, one should examine the model residuals, looking for autocorrelation, changes in variance with t, and other departures from randomness. Many of the issues raised in the regression chapter are applicable here and can be used without modification if time t is substituted for the independent variable X. For example, the question of adding a higher-order lag is much like the question of adding another independent variable in regression analysis. Regression diagnostics—particularly residual analysis—are therefore an essential part of time series modeling, as will be illustrated later in the chapter.

The residual mean square is commonly used in comparing models, the idea being that a successful model will reproduce the observed series and thereby give a_t that are on average small. Thus, everything else being equal, one would choose a model giving a small mean square error over one with larger error. It must be remembered, however, that as the number of parameters increases, explained variance cannot decrease. As a result, adding terms to a model will necessarily improve the fit as measured by residual mean square. To avoid overfitting, it is necessary to "correct" residual error for the number of parameters present. For example, rather than minimize the raw mean square error, a number of authors prefer Akaike's index

$$\mathrm{AIC} = \hat{\sigma}_a^2\left(1 + \frac{p}{n}\right)$$

where p is the number of parameters, n is the number of observations, and $\hat{\sigma}_a^2$ is an estimate of the residual variance. Most computer programs report AIC or other similar indices, allowing the user to compare models with differing numbers of parameters.

15.6. Model Fitting

Assuming one has a candidate model, how does one go about estimating parameters? This is a question of model fitting, in which we must choose parameter estimates that are in some sense optimal. We faced this problem in regression analysis and relied on the principle of least squares. In time series several variants of least squares are used. To understand them, it is helpful to imagine that we wish to fit an AR(1) model. Let us suppose that the mean Y is zero or has been subtracted from the data prior to analysis.

Given a series of length n, we can predict Y for $t = 2, 3, \ldots n$. The least squares approach says we find $\hat{\phi}$, which minimizes

$$\text{RSS} = \sum_{t=2}^{n}(Y_t - \hat{\phi}Y_{t-1})^2$$

The solution is

$$\hat{\phi}_{\text{LS}} = \frac{\sum_{t=2}^{n} Y_{t-1}Y_t}{\sum_{t=2}^{n} Y_{t-1}^2}$$

The least squares estimate $\hat{\phi}_{LS}$ is unbiased but is not necessarily between –1 and 1. Consequently, we cannot be sure that the resulting model will be stationary. More commonly used is the Yule–Walker estimate

$$\hat{\phi}_{\text{YW}} = \frac{\sum_{t=2}^{n} Y_{t-1}Y_t}{\sum_{t=1}^{n} Y_t^2}$$

Notice that the only difference is in the denominator, where the sum runs from 1 to n. This gives a model that is always stationary, but at the expense of an estimate for ϕ that is biased low, toward zero. Another approach is the so-called maximum entropy (ME) algorithm, which gives unbiased estimates ϕ_{ME} such that $-1 \leq \phi_{\text{ME}} \leq 1$. The method is based on predicting both forward and backward in time rather than just forward as in the above methods. In other words, we compute an additional set of residuals, this time using Y_{t+1} to predict Y_t. The ME method in effect combines the forward and backward least squares estimates. For a first-order model, the formula for the estimate is

$$\hat{\phi}_{\text{ME}} = \frac{\sum_{t=2}^{n} Y_{t-1}Y_t}{Y_1^2/2 + \sum_{t=2}^{n-1} Y_{t-1}^2 + Y_n^2/2}$$

Notice that again only the denominator has changed from the least squares estimate. For a long time series, the differences will hardly affect the estimate at all; thus the methods are nearly equivalent for large samples. In addition to the least squares methods, many computer programs use something called the maximum likelihood method. For the AR(1) model, we find $\hat{\phi}$, which minimizes

$$l(\hat{\phi}_{\text{ML}}) = n \ln(\text{RSS}) - \ln(1 - \hat{\phi}_{\text{ML}})$$

Though not based on least squares directly, once again the residual sum of squares figures prominently. Not surprisingly, for reasonably large n, the maximum likelihood estimate will be very close to those mentioned earlier.

Fitting models with moving average terms is more complicated on two grounds. First, the moving average parameters θ_k enter nonlinearly into the residual sum of squares, so that MA models are intrinsically nonlinear. Iterative solution methods must be used, in which parameter values are repeatedly adjusted in a trial-and-error search procedure. A second complication is that values of a are needed for times prior to the observation period. To see this, recall that Y_1 is based partly on the prediction error a_0. There is no prediction error for time t_0, as we do not even have Y_0. One option is to simply take these to be zero, which is their expected (average) value. Another option is to generate forecasts backwards in time for the needed a_t. The method used to start the sequence of a is of little importance for long records but can make a significant difference when working with small samples.

In summary, time series modeling involves the following steps.

1. If necessary, transform the observed series to achieve stationarity, usually by differencing.
2. Examine the autocorrelation and partial autocorrelation functions to obtain candidate models.
3. Fit candidate models and compute diagnostics: find coefficients and their standard errors, MSE, R^2.
4. Examine residuals, including their autocorrelation.
5. Add or remove terms as appropriate, with the overall goal of achieving a good fit with the smallest number of terms.

The following example illustrates this general framework.

EXAMPLE 15-1. The streamflow plot (Figure 15-1) clearly reveals periods of consistently above- and below-average floods. From that we certainly get the impression of memory in the system, a fact that is confirmed by the autocorrelation plot (Figure 15-2). We will use the data to construct a time series model, hoping to learn more about temporal structure, especially how much predictability is present.

The data standard deviation is 431 cubic meters per second; this implies that if we were to use the mean as a forecast for every year, we would obtain a root-mean-square forecast error of 431 m^3/s. This corresponds to a mean square error of 185,761 (431^2) and provides a standard against which we will measure model performance. Any model that cannot improve on the mean is of little use in forecasting.

As a first step, we try an AR(1) model. The autocorrelation plot (Figure 15-2) does not suggest this will be completely successful, as ρ_k does not decay consistently toward zero with increasing lag. Nevertheless, for the sake of exposition, we persist. We obtain a model $Y_t = 0.919Y_{t-1} + a_t$ (Table 15.1). The residual mean square is 74,890, about 40% of the baseline figure. In other words, the AR(1) model explains about 60% of the variance in the original time series ($R^2 = 0.59$). The coefficient estimate $\hat{\phi}$ is about 35 times larger than its standard error, so it is highly significant. We have every confidence that ϕ is not zero, under the assumption that an AR(1) is appropriate. However, when the residuals from the model are examined (Figure 15-9), we see large partial autocorrelations at lags 2 and 4 that are significantly different from zero. We cannot conclude that the residuals are truly random.

TABLE 15-1
Statistics for Various Models Fit to Streamflow Data

Model	Parameter	Estimate	Standard error	Model MSE	Model R^2
AR(1)	($\hat{\phi}$)	0.919	0.027	74,890	0.59
MA(1)	($\hat{\theta}$)	–0.825	0.042	89,220	0.52
ARMA(1,1)	($\hat{\phi}$)	0.851	0.039	67,520	0.63
	($\hat{\theta}$)	–0.417	0.069		

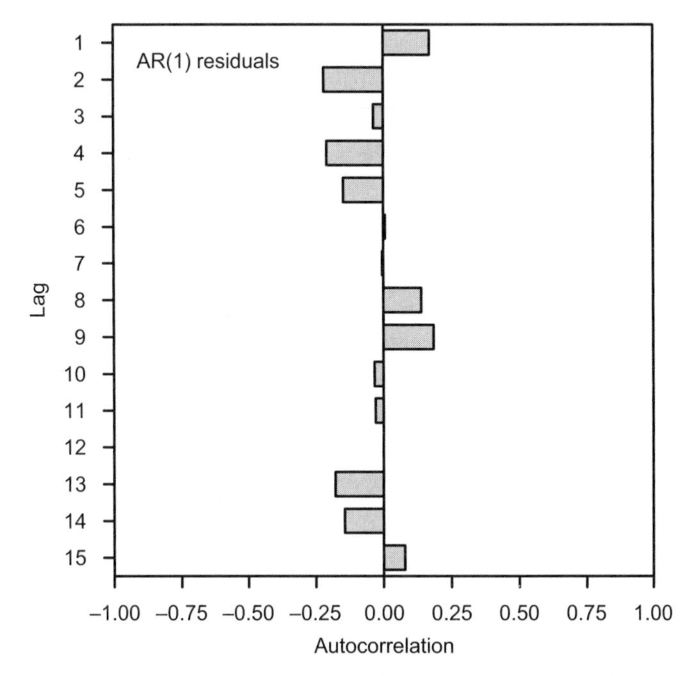

FIGURE 15-9. Partial autocorrelation function for residuals from AR(1) model fit to streamflow data.

Trying an MA(1) model again yields a significant parameter estimate, but the predictive power of the model is somewhat less than the AR(1) model. The residual MSE has increased, while R^2 falls slightly, to 0.52. The residual partial autocorrelations for lags 1 and 2 are both significant, implying that this model is not correct (Figure 15-10).

Finally, we turn to an ARMA(1,1) model. Both coefficients agree in sign with the previous models but are smaller in magnitude. In other words, like the others, this model finds a positive autoregressive and a negative moving average term. None of the residual partial autocorrelations is significantly different from zero (Figure 15-11). With this result, we would be hard-pressed to consider more complicated models.

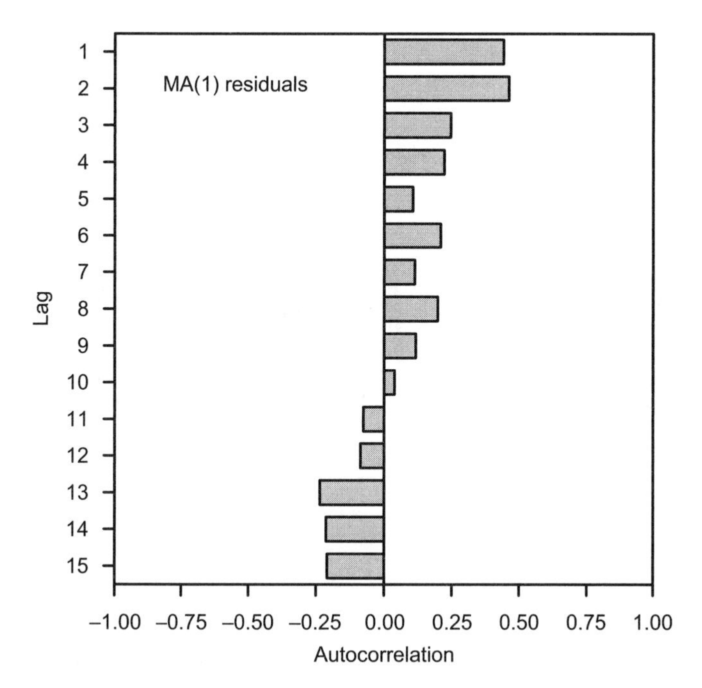

FIGURE 15-10. Partial autocorrelation function for residuals from MA(1) model fit to streamflow data.

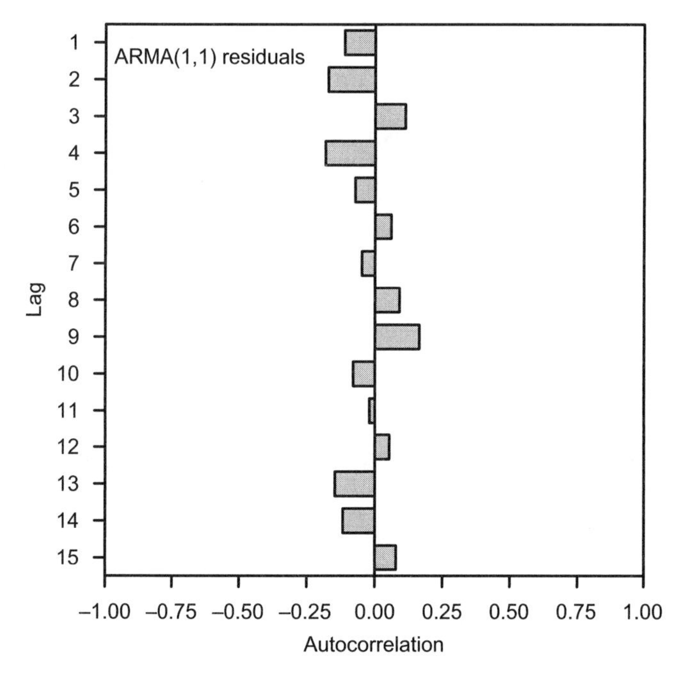

FIGURE 15-11. Partial autocorrelation function for residuals from ARMA(1,1) model fit to streamflow data.

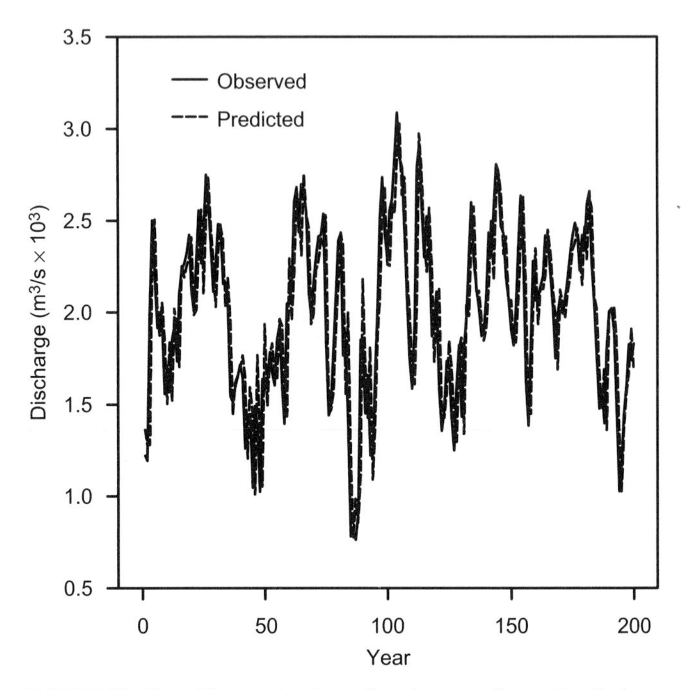

FIGURE 15-12. Observed and predicted stream flows. Predictions are based on an ARMA(1,1) model fit to observed series.

The residual MSE for the ARMA model is smaller than that for either of the other two models, and R^2 increases to 0.63. This figure sounds impressive and certainly represents real skill in forecasting floods. Comparing this to the AR(1) results leads to some interesting points. First, we see that, although the AR(1) model is wrong, replacing it with a (presumably) correct model does not lead to a dramatic improvement in forecast skill, as explained variance is only 4% larger. Looking at the time series of forecasts (Figure 15-12), we see large forecasts errors in many years.

Though not shown on the graph, the gap between the data and ARMA model is much larger than the difference between models. If forecast error alone is important, the models are nearly equivalent. On the other hand, if we are interested in temporal structure, the models are quite different, but MSE does not bring out the difference. In other words, on the basis of MSE alone, we have little reason to conclude there is a moving average component in the data. To detect the moving averaging component, we must examine the residuals.

A final point to make about prediction error is that we are computing "errors" using data from which the model was constructed. We must wonder how well the model will do when asked to forecast new values in the future. After all, our error statistics measure failure on the data used for fitting (calibrating) the model. We would have much more confidence in the performance measures if they were based on forecasts of *independent* data. This raises the issue of cross-validation, a topic covered in Chapter 13. Here we will simply point out that MSE, R^2, and other error measures based on the calibration data are likely to understate true forecast error.

15.7. Times Series Models, Running Means, and Filters

Though seemingly unrelated, there is a parallel between the time series models described above and the running means we used for smoothing in Chapter 3. In smoothing a time series, we apply an operation that is designed to remove the random component, leaving behind the systematic component. In time series modeling, on the other hand, we do just the reverse. From an observed time series we build a model designed to capture whatever systematic variation exists in the data. Given a sequence of data Y_t containing systematic and random variation, a time series model produces a new sequence a_t containing nothing but random variation. In addition to this conceptual similarity, there is a computational similarity as well. For example, recall the weighted moving average described in Chapter 3:

$$\hat{Y}_t = \frac{1}{8}Y_{t-2} + \frac{2}{8}Y_{t-1} + \frac{2}{8}Y_t + \frac{2}{8}Y_{t+1} + \frac{1}{8}Y_{t+2} \qquad (15\text{-}7)$$

where $\hat{Y}_t$ is the smoothed value at time t. Notice how observed values near t are multiplied by a coefficient and summed to give $\hat{Y}$, which is the systematic component Y. Now consider an AR(2) model

$$Y_t = \phi_1 Y_{t-1} + \phi_2 Y_{t-2} + a_t = \hat{Y}_t + a_t$$

where $\hat{Y}_t$ and a_t are the predicted and residual values at t. Considering just the predicted portion, which we hope is the systematic part of Y, we can write

$$\begin{aligned} \hat{Y}_t &= \phi_1 Y_{t-1} + \phi_2 Y_{t-2} \\ &= \phi_2 Y_{t-2} + \phi_1 Y_{t-1} + 0Y_t + 0Y_{t+1} + 0Y_{t+2} \end{aligned} \qquad (15\text{-}8)$$

Once again we are applying weights to values near t in hopes of uncovering the systematic component. In other words, the time series model can be seen as a type of running weighted mean. Of course, this particular running mean considers only values before t in obtaining $\hat{Y}_t$. We will return to the relationship between time series models and running means later. For the moment, we will focus on the smoothing operation. Equations 15-7 and 15-8 are known as *linear combinations* of Y. The observed values are combined with one another after multiplying by a coefficient. To write a more general expression for this operation, let c_k stand for a set of coefficients. The subscript k identifies each element in the array, just as i identified each sample element x_i. For our example (Equation 15-7) the coefficient array is $\boldsymbol{c}$ = [1/8, 2/8, 2/8, 2/8, 1/8]. Instead of going from 1 to 5, we will let the subscripts k run from –2 to 2. With this notation we can write the weighted mean as

$$\begin{aligned}\hat{Y}_t &= \tfrac{1}{8}Y_{t-2} + \tfrac{2}{8}Y_{t-1} + \tfrac{2}{8}Y_t + \tfrac{2}{8}Y_{t+1} + \tfrac{1}{8}Y_{t+2} \\ &= c_{-2}Y_{t-2} + c_{-1}Y_{t-1} + c_0Y_{t-0} + c_1Y_{t+1} + c_2Y_{t+2} \\ &= \sum_{k=-2}^{2} c_k Y_{t-k}\end{aligned}$$

To smooth the observed series, we move the coefficient array $\boldsymbol{c}$ through the data, forming the weighted sum above for every t. This operation is known as filtering the data, and the array $\boldsymbol{c}$ is called a filter. Technically speaking, the filter is *convolved* with the data. Notice that the AR(2) model is also a filter, one whose coefficients are [ϕ_2, ϕ_1, 0, 0, 0]. Many other operations can also be seen as filters, though we often do not think of them that way. For example, when we interpolate linearly between two points, we are using a filter with coefficients [1/2, 1/2]. Other examples of filtering include slope estimation, numerical integration, cartographic line generalization and edge detection, to name just a few.

DEFINITION: FILTERING

A filter is an array of constants applied to the values of a time series. Each element of the filter is paired with and multiplied by an element of a time series. The products are summed to give a filtered value.

We do not propose to discuss all types of filters here, but will consider only filters that can be expressed as a linear combination of the original data. Moreover, for reasons explained later, we will restrict our attention to filters that are symmetric about the central value, like the running mean above. This means that the coefficients will appear as pairs surrounding the central weight. If c_m is the element farthest away from the center, the array is

$$\boldsymbol{c} = [c_{-m}, c_{-m+1}, \ldots, c_{-1}, c_0, c_1, \ldots, c_m]$$

The filter contains a central value c_0, and m weights to either side. The length is therefore $2m + 1$. Because of symmetry, $c_1 = c_{-1}$, $c_2 = c_{-2}$ and so forth. This allows the filtering operation to be simplified:

$$\begin{aligned}Y_t &= \sum_{k=-m}^{m} c_k Y_{t-k} \\ &= c_0 Y_t + \sum_{k=1}^{m} c_k (Y_{t-k} + Y_{t+k}) \qquad (15\text{-}9)\end{aligned}$$

Because of symmetry in the filter, Y_{t-k} and Y_{t+k} are multiplied by the same value, c_k. This means they can be added together before multiplying, saving some work. If we were working by hand, or if computational efficiency were an issue, we would use the lower expression in Equation 15-9. The upper expression applies to both symmetric and asymmetric filters.

The way a filter affects a time series obviously depends on the coefficients that comprise the filter. But how can we know what a particular filter will do? For example, both the running mean and the AR(2) filters are supposed to isolate systematic variation in Y. Why would we prefer one to the other? How does a running mean with uniform weights compare to one like (15-7)? To answer these questions, we need some way of describing a filter in terms of the effects it produces. This requires that we adopt a different perspective, outlined in the next section.

15.8. The Frequency Approach

As mentioned in Chapter 3, a time series is smoothed with the goal of uncovering slowly varying trends. In smoothing the series, we are making the implicit assumption that the rapidly varying components are not of interest and so need to be removed. In other words, we view variations that last only a short period as *noise,* something that confounds our ability to detect the important, long-period changes. The running mean succeeds if it removes short-period variation while leaving long-period variation intact. This breakdown of variance into different lengths of persistence lies at the heart of the frequency approach to time series. It requires that we conceive of the time series as containing variance at different periods. To pick an obvious example, consider a time series of air temperature measured every hour. Most of the variance in temperature will be a result of the diurnal (daily) and annual cycles. To a good approximation, the temperature record could be reproduced by adding together two periodic functions, one representing the daily rise and fall of temperature, the other representing annual variations.

Alternatively, suppose we have a time series of monthly rainfall. We would not see any diurnal variation, but the annual cycle would be represented, as would variation at still longer periods. If we were interested in searching for solar cycles, we might want to remove the annual cycle so that any 11-year cycle could emerge. Figure 15-13 shows something similar for a hypothetical time series with variance at just two periods, 1 year and 10 years in duration. The 10-year oscillation, here labeled as signal, repeats itself every decade. We would therefore describe it as a *periodic* or *sinusoidal* function, with a fundamental period of 10 years. Equivalently, we would say its frequency is 1 cycle/decade, or 0.1 cycle per year. Note that the period of a function is measured in units of time (days, years, etc.), whereas frequency has dimensions of cycles per unit time (cycles/day, cycles/year, etc.). These are reciprocals of each other; thus low-frequency oscillations correspond to long-period oscillations, and vice versa.

Notice also that the amount of variation is different for signal and noise in Figure 15-13. The signal curve ranges from 5 to 25, a difference of 20. The variation in the noise curve is only half as large, running from –5 to 5, a difference of 10. The variation of a sinusoid is summarized by its amplitude, formally defined as the difference between the maximum and minimum divided by two. Our hypothetical signal and noise curves have amplitudes of 10 and 5, respectively.

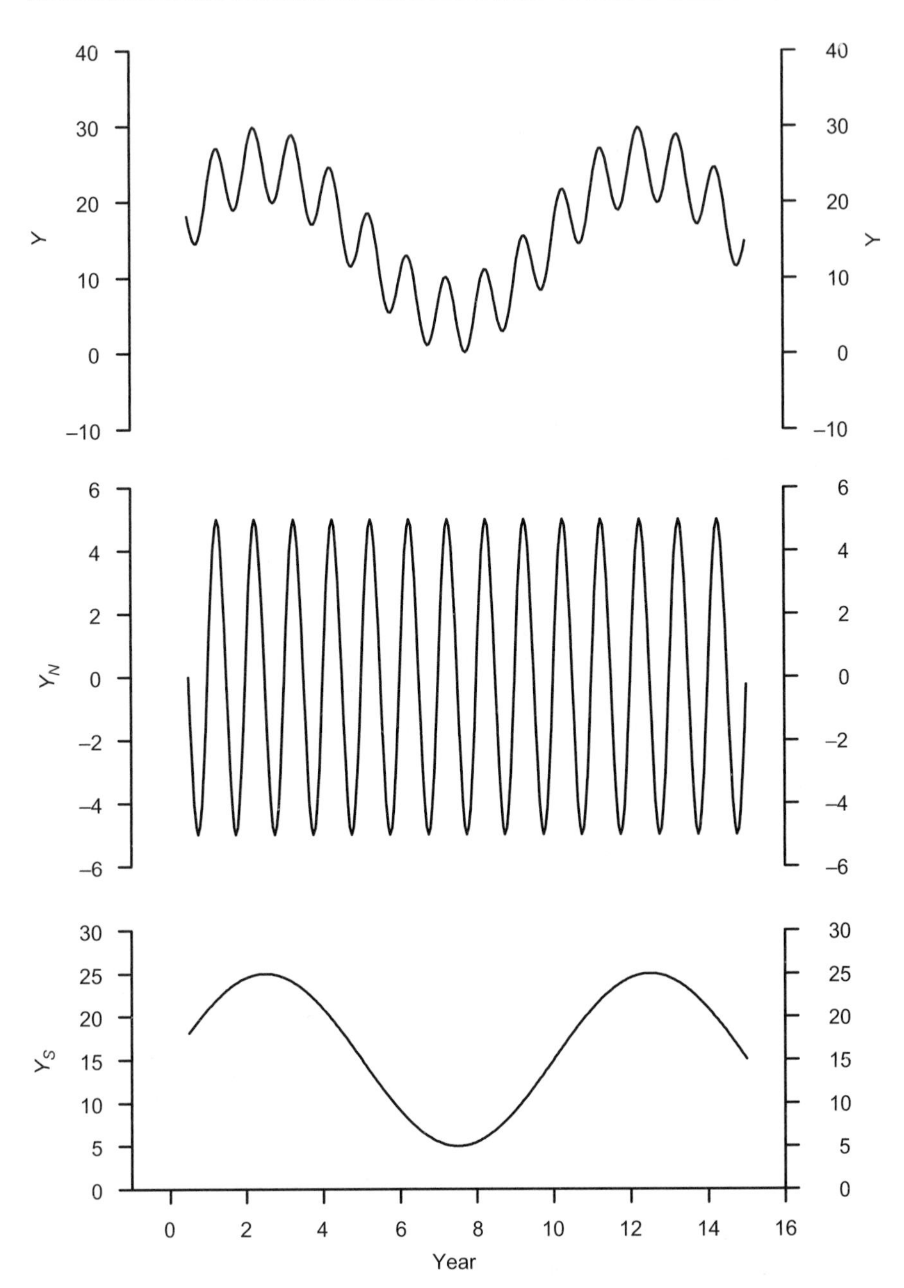

FIGURE 15-13. A hypothetical series Y, composed of rapidly varying noise Y_N superimposed on a slowly varying signal Y_S.

DEFINITION: PERIOD, FREQUENCY, AND AMPLITUDE
The period of a sinusoid is the length of time required for the function to complete one cycle. The frequency of a sinusoid is the number of cycles completed per unit time. Frequency and period are reciprocals: Frequency = 1/Period. The amplitude of a sinusoid is one-half the difference between maximum and minimum.

Suppose we filtered the upper curve of Figure 15-13 with a 5-year running mean. We would expect the annual cycle to be removed, leaving behind the 10-year cycle (bottom curve). A 20-year running mean would, of course, produce a different result. We see that different filters remove and pass different frequencies, and that knowing which frequencies are removed is helpful in knowing which filter to choose. Here, then, is the rationale for taking a frequency approach to filtering:

RULE: Filters selectively modify the frequencies present in a time series—only by understanding what a filter accomplishes from a frequency standpoint can we understand how the observed series is modified.

In adopting the frequency approach, we describe the action of a filter in terms of its effects on periodic functions. It must be emphasized that we need *not* believe that the time series itself is periodic, or that it arises from the action of periodic processes. The only requirement is that we desire to separate the time series into components with varying time scale. In most cases, the high-frequency components are considered noise, so that we seek smoothing filters that allow low-frequency variation to pass while removing high frequencies. There is, however, nothing about the filtering process that demands this posture. If appropriate, we can think of noise as appearing at low frequencies, and we can design filters that pass only high frequencies. Another alternative is a band-pass filter, which passes frequencies only within a specified range.

The Amplitude Response Function

As we have observed, a given filter will remove some frequencies and leave others untouched. When a frequency is removed, its amplitude is reduced to zero. If it is unaltered, the amplitude remains at 100% of the original amplitude. A filter's pattern of removal, reduction, and so on, is called the *amplitude response* of the filter. The response function gives the proportion of amplitude that remains after the filter is applied. Thus, if the response function is zero for some frequency, the filter will completely remove variation at that frequency. If the response is unity, variation is unchanged. If the response is 0.25, it means the amplitude would be reduced to only 25% of the original, a reduction of 75%.

DEFINITION: AMPLITUDE RESPONSE FUNCTION

The amplitude response for a filter provides information about which frequencies are passed by a filter and which frequencies are removed. If the value is unity for some frequency, the filter does not affect that frequency, and the amplitude at that frequency is unchanged by the filter. A response value of zero means that the amplitude is reduced to zero.

Before showing how the response function is calculated, we will describe the response for the 5-point weighted mean used before (Equation 15-7). The amplitude response for that filter is plotted in Figure 15-14.

Following common practice, frequency is plotted as increasing from left to right. The left-hand part of the diagram therefore refers to long-period, slowly varying components of variation, whereas the right-hand side refers to high-frequency variations. The frequency axis is plotted in units of cycles per observation, rather than cycles/year or cycles/day or some other time-based unit. This is because we are free to use the filter on yearly data, on daily data, or on data collected at any other interval. By plotting frequency as cycles/observation, we make the graph useful for all data intervals. Suppose we happen to have yearly data. In that case, a frequency value of 0.1 from the graph represents 0.1 cycle/year, or a period of 10 years. If we have daily data, 0.1 means 0.1 cycles/day, or a period of 10 days.

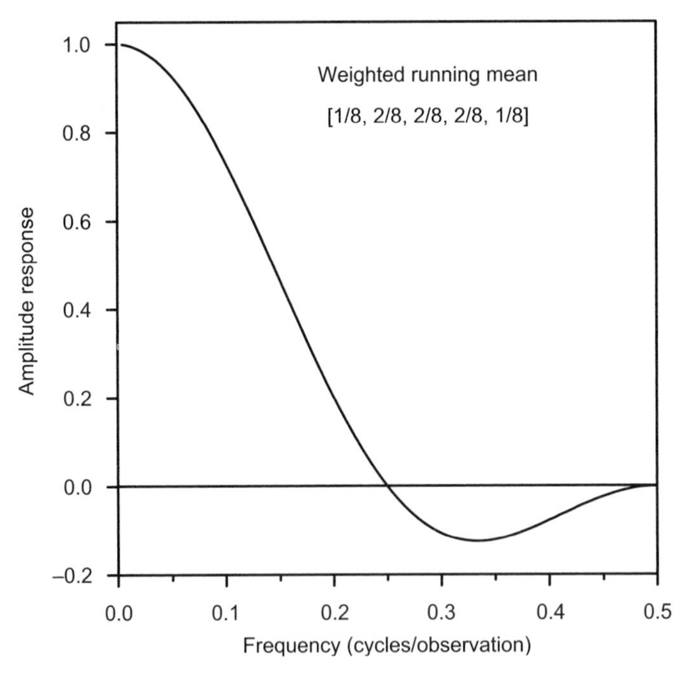

FIGURE 15-14. Amplitude response function for the 5-point weighted mean [1/8, 2/8, 2/8, 2/8, 1/8].

The highest frequency shown is 0.5 cycle/observation. Why is that? Does it mean the filter only affects lower frequencies? Or does it mean higher frequencies are completely removed? In point of fact, neither of these is true. The frequency range is restricted because only frequencies below 0.5 can be captured by sampling a process at discrete moments in time. There is a very important theorem, called the *sampling theorem,* which states that we must have at least two observations for every cycle present in the process. In other words, the shortest cycle we can detect has a period that is twice the sampling interval. A period of twice the sampling interval corresponds to a frequency of 1/2 cycle per observation.

DEFINITION: SAMPLING THEOREM FOR TEMPORAL DATA

Suppose a periodic function is sampled (measured) at uniform points in time. In order for the sample points to reproduce the function, at least two points are required for every cycle. The sampling frequency must be at least twice the frequency of the function.

The highest frequency detectable is known as the *Nyquist frequency* and is always 0.5 cycle/observation. It is interesting to think about what happens to frequencies above the Nyquist frequency. In other words, what happens if we violate the sampling theorem and do not sample often enough? Do we simply miss those undetectable frequencies? The answer, unfortunately, is no. Instead, oscillations at higher frequencies will appear in the sample as low-frequency oscillations. What we see as low-frequency variation will therefore be a combination of true variation and contamination from higher frequencies. This problem, called *aliasing,* is very important. It says that when we do not sample frequently enough, we don't just miss features, but rather we produce a sample series *distorted* by what is undetectable. There is no way to remove these distortions once they occur—no amount of processing can undo the damage done by aliasing.

Returning to Figure 15-14, we see the curve mostly decreases with increasing frequency, indicating that the filter is a smoothing filter. That is, as we hoped, the filter preferentially removes high frequencies but passes low frequencies.

Notice that the curve is unity only at frequency zero, so this is the only frequency that is not reduced in amplitude. A frequency of zero corresponds to a period of infinity; thus the figure is saying that only infinite-period sinusoids are unchanged. But what is an infinite-period sinusoid? It would seem to be a curve that changes infinitely slowly, so that it does not change at all, and is therefore a horizontal line. This is, in fact, the case. If we use the filter on a time series containing nothing but constant values, we will get the original time series back (see Problem 15.5).

For frequencies greater than zero, amplitude response falls from unity toward zero. The response is such that frequencies are damped out by various amounts, with the amount of damping increasing with increasing frequency. Finally, at a frequency of 0.25, the curve reaches zero. This is the first frequency completely removed by the filter—lower frequencies are only partly attenuated. Notice that amplitudes are negative at frequencies above 0.25. For example, the amplitude is about –0.1 at frequency

0.35. This means that oscillations are reduced to 10% of their original amplitude but are reversed! Where there is a peak in the original series, there will be a valley in the filtered series. For a smoothing filter, this is not at all what we are after. We would prefer a filter that does not pass *any* high-frequency information. If that isn't possible, we certainly do not want our filter to turn noise maxima into minima.

By examining the amplitude response of our filter, we see exactly what the filter is able to accomplish and what are its failings. The amplitude response is surprisingly easy to compute. Valid for any frequency f between 0 and 1/2, the equation is

$$R(f) = \sum_{k=-m}^{m} c_k \cos(2\pi k f), \quad 0 \le f \le 1/2 \tag{15-10}$$

When using this equation, it is very important that frequency f be expressed as cycles/observation. This will sometimes require conversion from a more natural unit of frequency. For example, perhaps we are interested in cyles/year but have data measured every month. To obtain the response at some frequency in cycles/year, we would divide f by 12 before applying (15-10).

Amplitude Response of Running Means

Equation 15-10 allows us to compute the amplitude response for any filter. Without question the most common type of filter is a running mean of length L, where L successive observations are averaged together (all $c_k = 1/L$). Given their widespread use, it is important to understand their behavior.

Figure 15-15 shows the amplitude response for 3-point through 11-point running means. All are low-pass smoothing filters, removing high frequencies much more than low frequencies. First, notice where each curve passes through zero. The 11-point mean reaches zero soonest, at $f = 1/11$, then the 9-point mean at $f = 1/9$, and so forth. The 3-point mean is zero only at $f = 1/3$. The implication is that a filter of length L *completely* removes variation at a frequency of $1/L$, equivalent to a period of L. This means, for example, that an 11-year average completely suppresses any 11-year cycle in the time series. In deciding on how long a filter to use, we would probably match our definition of "noise" to the frequencies removed. If an application called for "noise" to be anything at frequencies above 0.2 cycle/observation, we would want one of the longer filters shown. On the other hand, if a frequency of 0.2 were considered to be signal, we would have to choose the 3-point filter.

The graph shows that running means remove variation at a period of *exactly* L, but do not remove all variation at periods shorter than L. Frequencies higher than $1/L$ are damped but not removed. Sometimes the damping is such that maxima and minima are reversed in the filtered series (negative R). All of the running means show similar behavior, an amplitude response with wiggles for f above $1/L$. Depending on how large the noise is relative to the signal, the noise could dominate the filtered series. This is the primary objection to running means. Though simple to construct and use, they are far from optimum in regards to their amplitude response. In other words, they do not accomplish very well what we are after, namely, the removal of high frequencies.

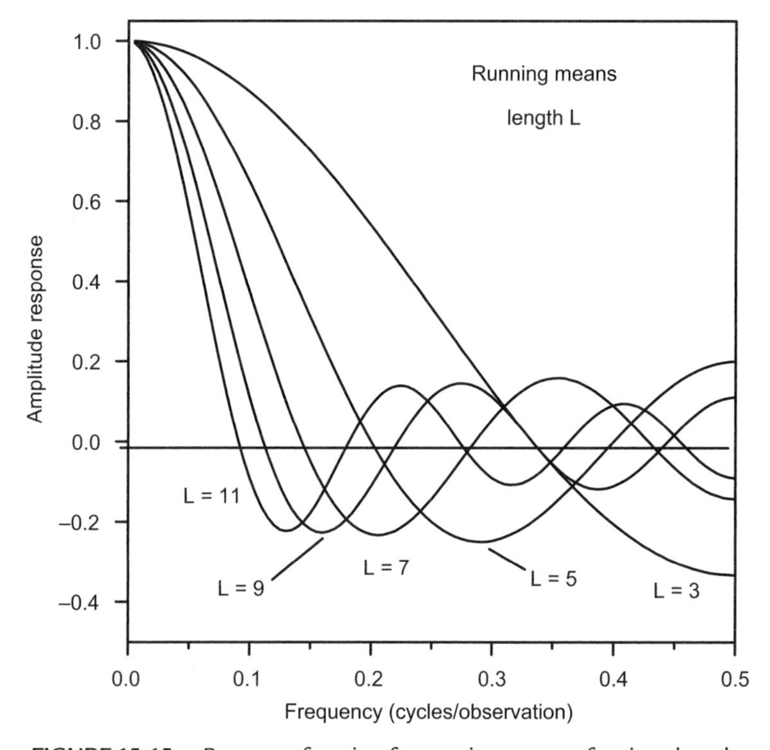

FIGURE 15-15. Response function for running means of various length.

The failure of running means suggests that we ought to turn our thinking around. Instead of looking at how close a filter comes to our goal, we ought to design a filter from the outset to meet a particular goal. Rather than compute the response function from the filter, we ought to begin with the response function and choose the filter coefficients accordingly.

15.9. Filter Design

In designing a filter, we are going to begin with a specification in terms of frequency. That is, we will specify what the filter is supposed to accomplish from the frequency standpoint. Having done so, we will then select filter weights that come as close as possible to our ideal. It will be seen that compromise is necessary in the design process, so that the filter obtained will not be perfect. Although we will not produce a perfect filter, we will have control over its failings, and we can make the compromises in a way that minimizes impact on the findings. Naturally, the compromises made will depend on the time series at hand: its length, the pattern of variation it contains, and what we hope to learn.

Several basic categories of filters exist for which we can state design goals quite easily:

1. **Low-pass.** A low-pass filter is a classic smoothing design that removes high-frequency variation. An ideal low-pass filter would have a response of zero for all frequencies *above* some cutoff frequency f_{cut}. The cutoff frequency is a design parameter, something we choose depending on our definition of noise. Frequencies below f_{cut} are called the *passband,* whereas those above f_{cut} are the *stopband.* Mathematically, the ideal low-pass filter is

$$R(f) = \begin{cases} 1 & \text{if } f \leq f_{\text{cut}} \\ 0 & \text{otherwise} \end{cases} \qquad (15.11)$$

2. **High-pass.** The ideal high-pass filter would have a response of zero for all frequencies *below* the cutoff frequency f_{cut}. The passband is $f > f_{\text{cut}}$, ideally having an amplitude response of unity. The corresponding equation is

$$R(f) = \begin{cases} 1 & \text{if } f > f_{\text{cut}} \\ 0 & \text{otherwise} \end{cases} \qquad (15\text{-}12)$$

3. **Band-pass.** We are interested only in frequencies within a band running from f_{low} to f_{high}. The ideal band-pass filter would have a response of zero for all frequencies below f_{low} *and* above f_{high}. Its response would be unity between these values:

$$R(f) = \begin{cases} 1 & \text{if } f_{\text{low}} \leq f \leq f_{\text{high}} \\ 0 & \text{otherwise} \end{cases}$$

Graphs of the three filter types are shown in Figure 15-16. As much as possible, we must choose filter coefficients that will produce those ideal response curves. Fortunately, we do not need a separate design strategy for each class of filter. Given a low-pass filter, one can easily construct a high-pass filter corresponding to its inverse. In addition, it will be seen that band-pass filtering can be performed using a low-pass and high-pass filter in tandem. The basic problem, therefore, is to come up with a good low-pass filter. At a minimum, we want to observe the following guidelines:

1. The coefficients should sum to unity. This will ensure that the response will be unity at $f = 0$ (the mean of Y will not be changed). A filter whose coefficients sum to unity is said to be *normalized.*

2. The coefficients should be symmetric, such that $c_k = c_{-k}$. This will ensure that the filter does not displace maxima and minima in the original series. Waveform amplitudes will be modified, but their position in time will be preserved. Filters with this property are called *zero phase* filters—they do not produce any phase shifts in Y.

3. The filter should be as long as is practical. Increasing L will improve the design but will also lead to greater data loss at the beginning and end of the series. For a short time series, where one can afford to sacrifice only a few points, it may be necessary to compromise on the design.

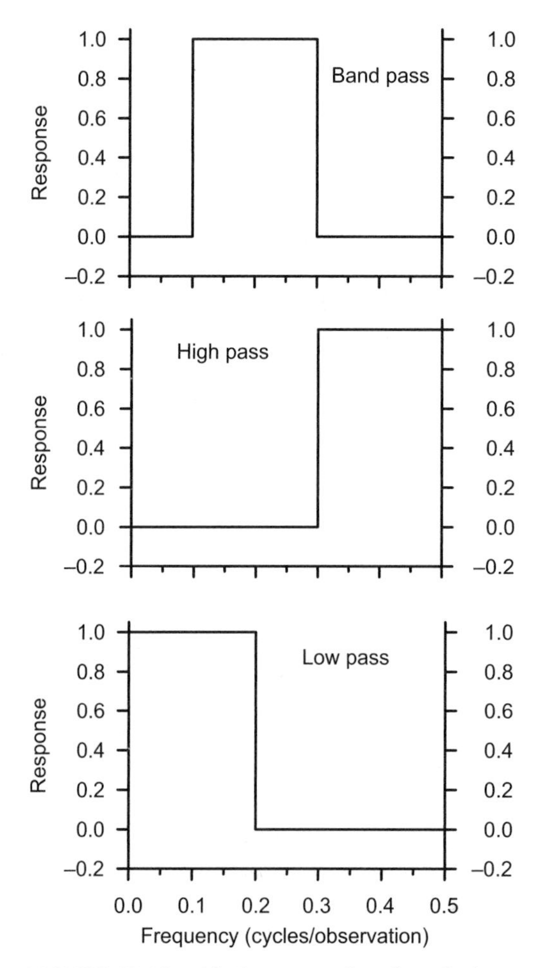

FIGURE 15-16. Ideal response functions for low-pass, high-pass, and band-pass filters.

A number of methods are available for constructing filters consistent with the above. We will describe one technique, the Lanczos method, which is both flexible and uncomplicated. Although its derivation requires advanced mathematics, the procedure used to construct a Lanczos filter is straightforward. For a low-pass design, the steps are:

1. Choose the cutoff frequency f_{cut}. This will mark the center of the transition between frequencies passed ($f < f_{cut}$) and frequencies removed ($f > f_{cut}$).
2. Choose the length of filter $L = 2m + 1$. There will be m pairs of coefficients, plus one central coefficient c_0. Filtered values will be unavailable for the beginning and ending m points.

3. Find the preliminary filter coefficients b_k from

$$b_k = \left[\frac{\sin(2\pi f_{\text{cut}} k)}{\pi k}\right]\left[\frac{\sin[\pi k/(m+1)]}{\pi k/(m+1)}\right] \tag{15-13}$$

for $k = 1, 2, \ldots m$. Use $b_0 = 2 f_{\text{cut}}$ for the central coefficient.
4. The preliminary filter coefficients will not sum to unity. To normalize the filter, we need to divide each preliminary coefficient by the filter sum. We must sum all the weights, remembering that the filter contains b_{-1}, b_{-2}, and so forth. Taking advantage of symmetry, the sum is

$$B = \sum_{k=-m}^{m} b_k = b_0 + 2\sum_{k=1}^{m} b_k$$

5. Compute the final coefficients

$$c_k = b_k / B \tag{15-14}$$

Lanczos filters have two primary advantages over running means. First, they let us adjust filter length independently of cutoff frequency. With running means, the only way to increase f_{cut} is to decrease L, giving a poorer amplitude response. Here we choose L on the basis of how much data we can afford to lose, without regard to f_{cut}. As L is increased, our filter comes closer and closer to the ideal. This effect is apparent in Figure 15-17, which shows three Lanczos filters all using a cutoff frequency of 0.25. Notice that for all three filters, the amplitude response is about 0.5 at $f = 0.25$. In the Lanczos design, the cutoff frequency becomes the midpoint of the transition from the passband ($R(f) = 1$) to the stopband ($R(f) = 0$). Increasing L causes the transition band to become narrower, so the actual response comes closer to the ideal response.

Another very important advantage of Lanczos filters is the nearly flat response in both the pass-and stopbands. Ripples at high frequencies, which are so prominent with running means, are nearly absent. This improvement is not happenstance but is explicitly part of the Lanczos design. Indeed, the second term in (15-13) appears solely to smooth out unwanted ripples in both the stopband and passband. Here also an increase in L improves the design, as the response function flattens with increasing filter length.

We see that the design strategy is fundamentally one of trading off loss of data against improvements in filter performance. Good practice calls for experimenting with various values of L and plotting the response function for each filter produced. The primary disadvantage of Lanczos filters is the rather slow transition from passband to stopband. This could be improved by increasing L, but for a given L Lanczos filters do not provide a very steep transition. If this is important in a particular application, some other design might be preferred. A number of other strategies exist, optimized in other ways. Many are described in Hamming (1982).

Earlier we suggested that a low-pass design could serve in high-pass and band-pass filtering. Suppose we are high-pass filtering a time series. We want a new series

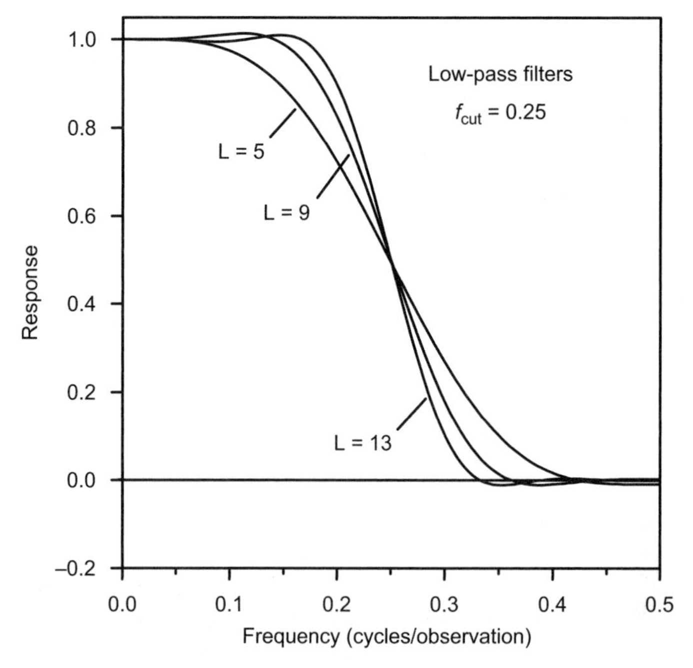

FIGURE 15-17. Response functions for three Lanczos filters of various length.

containing frequency components above f_{cut}, with low-frequency components removed. Looking at the equations for the ideal response functions, (15-11) and (15-12), we see that they add to unity at every frequency. Given the low-pass response $R(f)$, we could subtract from unity and obtain the high-pass response as $1 - R(f)$. In other words, the high-pass response is the inverse of the corresponding low-pass filter. There are two ways to arrive at the inversion.

One way is to invert the filter itself. The process is simple and calls for negating all but the central coefficient. To be specific, given a low-pass filter c_k, we obtain a high-pass filter d_k from

$$d_k = \begin{cases} 1 - c_0 & \text{for} \quad k = 0 \\ -c_k & \text{for} \quad k > 0 \end{cases}$$

If the response function for $\boldsymbol{c}$ is $R(f)$, the response function for the new filter $\boldsymbol{d}$ will be $1 - R(f)$, as it should be.

As an alternative, we can accomplish the same thing by inverting the time series. We design a low-pass filter and apply it to the data. We then subtract each filtered value from the corresponding original observation. This will produce a new series containing only high frequencies. The rationale is simple: if low frequencies are removed from the original series, one will be left with only high frequencies.

This same approach can be used to produce a band-pass filtered series. We filter the series with a low-pass filter *and* a high-pass filter. In both cases the original series are filtered. We subtract the two filtered series from the original, leaving behind band-pass filtered data. Writing this out in steps, the procedure is:

1. Filter the observed data with a low-pass filter using a cutoff frequency of f_{low}.
2. Filter the observed data with a high-pass filter using a cutoff frequency of f_{high}.
3. Subtract the series obtained in steps 1 and 2 from the original data. This will produce a filtered series containing frequencies between f_{low} and f_{high}.

Of course, the low-and high-pass filters should have the same length, so that the loss of data is the same in both steps and the filters are of comparable quality.

EXAMPLE 15-2. Let us suppose we want to filter the streamflow data shown in Figure 15-1. Recall that we used these data previously to construct several time series models. Because we can think of a time series model as a filter, it seems we have already filtered the streamflow data! The results using an ARMA(1,1) model were shown in Figure 15-12, labeled "Predicted." These are observed values, filtered by the model coefficients. Notice how many of the short-term excursions are absent in the forecast series. This shows that model is acting something like a low-pass filter, removing some of the fine-scale structure in the observations. Because it is one-sided, looking backward in time, the model obviously fails to meet our recommendation for symmetry in filter design. Lack of symmetry suggests that we should expect the filter to displace maxima and minima from their original locations. Looking carefully at the filtered series, we do see these phase shifts. The major excursions in the predicted values lag those in the original data. A one-sided filter like this is necessary for forecasting, but it is far from ideal for smoothing.

EXAMPLE 15-3. The time series model filtered the series but did not give us control over the amplitude response. We will therefore use the design procedure outlined above, constructing both a low-pass and high-pass filter for the streamflow data.

1. The time series shows a slowly varying component of variance having a period of about 25 years. We want a cutoff frequency that separates this from what appears to be largely random variation. Being conservative, we will choose $f_{\text{cut}} = 0.125$ cycle/year, corresponding to a period of 8 years.
2. We decide to sacrifice 10 observations at both ends, totaling about 10% of the sample values. This gives a filter length of $L = 2m + 1 = 21$.
3. We use Equation 15-13 to get the preliminary coefficients b_k (see Table 15.2).
4. Summing the preliminary filter gives the normalization value $B = 0.9915465$.
5. We dividing each b_k by B to obtain the final normalized low-pass filter c_k, $k = 0, 1, \ldots, m$.

TABLE 15-2
Coefficients of 21-Point Lanczos Filter

k	Preliminary b_k	Low-pass c_k	High-pass d_k
0	0.2500000	0.2521314	0.7478686
1	0.2220317	0.2239247	–0.2239247
2	0.1506405	0.1519248	–0.1519248
3	0.0661780	0.0667422	–0.0667422
4	0.0000000	0.0000000	0.0000000
5	–0.0312029	–0.0314689	0.0314689
6	–0.0306441	–0.0309054	0.0309054
7	–0.0146301	–0.0147548	0.0147548
8	0.0000000	0.0000000	0.0000000
9	0.0052602	0.0053050	–0.0053050
10	0.0031400	0.0031668	–0.0031668
Sum	0.9915465	1.0000000	0.0000000

Though we will not use it in this example, Table 15.2 also shows a high-pass filter constructed from the low-pass coefficients. Notice that for all but d_0, we have simply negated the low-pass coefficient. The response functions for both $\boldsymbol{c}$ and $\boldsymbol{d}$ are shown in Figure 15-18. Visual inspection confirms our design goal.

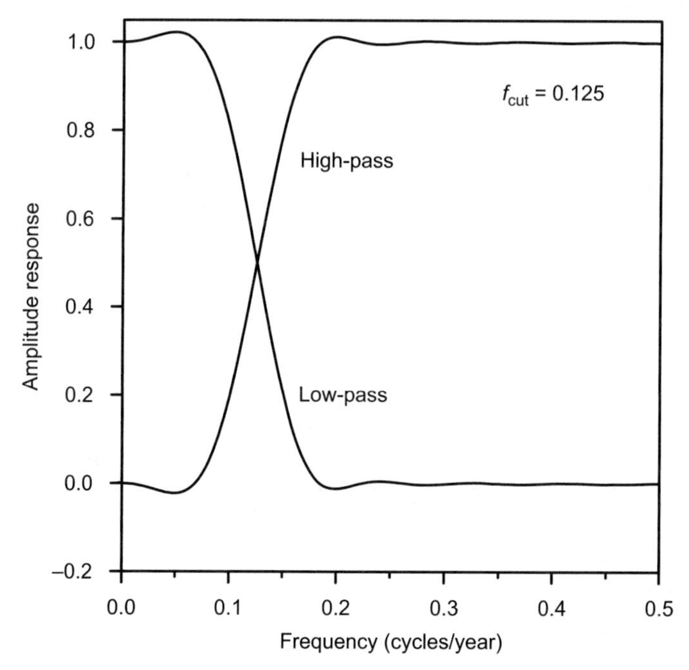

FIGURE 15-18. Amplitude response functions for the filters of Table 15-2.

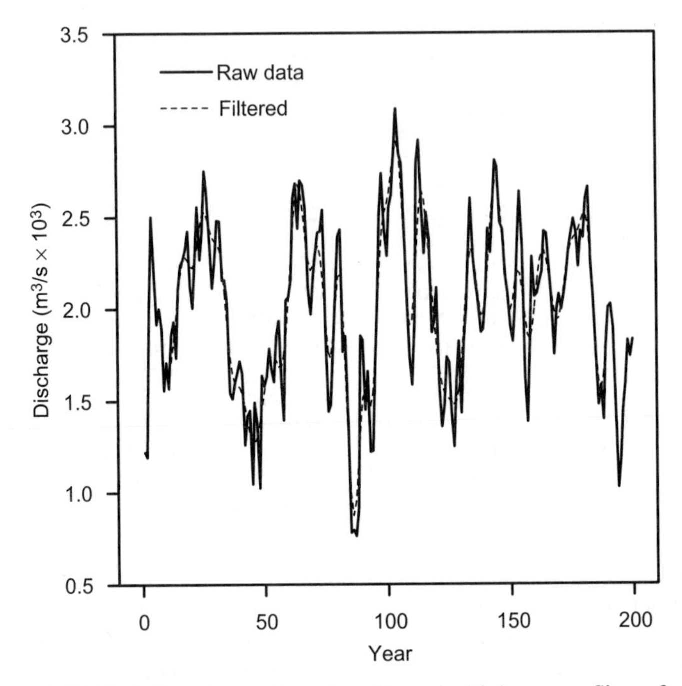

FIGURE 15-19. Streamflow data filtered with low-pass filter of Table 15-2.

Applying the low-pass filter to the streamflow data, we get the series shown in Figure 15-19. High-frequency components have been effectively removed, leaving behind the slowly varying components. Notice that the filter has modified amplitudes in the original series but has not introduced phase shifts. Peaks and valleys appear in the filtered series at the same times as in the raw data. In other words, low-flow and high-flow episodes maintain their position in the filtered series.

15.10. Summary

This chapter considered two related forms of time series analysis: time series modeling and time series filtering. Time series models attempt to capture the dynamic behavior of a series using a small number of parameters. The models are developed from the observed series and typically rely on some variant of least squares to estimate parameters. Two broad classes of models were considered, autoregressive and moving average models. Combining the two, possibly with differencing, provides a very general model framework, able to accommodate a wide range of observed series. These so-called ARIMA models are commonly used in human and physical geography, serving as forecast devices and as aids in understanding the structure of a time series.

Filtering is a process in which a set of coefficients is convolved with a time series, giving a new, filtered series. A special case of filtering is smoothing, where the goal is to remove noise associated with short-period departures from long-term trends. Many other operations, including time series forecasting, can also be written as a filtering operation. To understand how a given filter modifies a time series, we adopted the frequency approach. A key concept is the amplitude response of a filter, which shows how the filter modifies amplitudes at various frequencies. For example, by examining the amplitude response of simple running means, we saw that they are deficient in several regards.

The preferred approach in filtering is to begin with a design in the frequency domain—a specification of what the filter is supposed to accomplish from the frequency standpoint. Classic designs include low-pass, high-pass, and band-pass filters. Low-pass filters are most common, as they conform to the usual assignment of noise to high frequencies. A number of design strategies exist, one of which we covered, the so-called Lanczos method. It has the advantages of being easy to implement while still giving acceptable results. High-pass filtering is perhaps less common but is easily accomplished. One can either invert a low-pass filter or subtract a low-pass filtered series from the observed series. For band-pass filtering, two convolutions are performed, one using a low-pass filter and the other using a high-pass filter. By subtracting both filtered series from the original, one obtains a series that is effectively band-pass filtered.

This chapter has considered observations arranged serially in time. Many of the issues discussed apply equally well to spatial data. Rather than a time series, we might have data collected along a transect. Instead of using a times series model to study relations between future and past values, we would focus on relations between neighboring values. Similarly, rather than filter to selectively remove features according to their duration (period), we would filter in order to remove features according to their spatial scale, that is, their size.

Of course, true spatial data are typically two-dimensional, not one-dimensional like time series. It so happens that both the models and filters described above can be generalized to two dimensions, and thus can be used in common geographic contexts. In constructing a spatial time series, one might express observed values as a linear combination of surrounding values. The coefficients comprising the model encapsulate the spatial structure of the data, in the same way that temporal structure is encapsulated by the coefficients of a one-dimensional time series model. Similarly, when filtering geographic data, one will use a two-dimensional array of filter coefficients. The array is moved throughout the *area* of interest, much like one-dimensional filters are moved from end to end across the time series. Not surprisingly, two-dimensional models and filters are mathematically much more complicated than their one-dimensional analogs.

Interested readers are referred to the references for details.

FURTHER READING

R. J. Bennett, *Spatial Time Series* (London: Pion, 1979).

J. D. Cryer, *Time Series Analysis* (Boston: Duxbury Press, 1986).

D. Graupe, *Time Series Analysis, Identification and Adaptive Filtering* (Malabar, FL: Krieger, 1984).
R. W. Hamming, *Digital Filtering* (Englewood Cliffs, NJ: Prentice-Hall, 1983).
W. S. Wei, *Time Series Analysis: Univariate and Multivariate Methods* (Redwood City, CA: Addison-Wesley, 1990).

PROBLEMS

1. Explain the meaning of:
   - Random walk
   - Stationarity
   - Partial autocorrelation
   - Moving-average model
   - Symmetric filter
   - Low-pass filter
   - Stochastic process
   - Autocorrelation
   - Autoregressive model
   - Smoothing
   - Amplitude response function

2. Why are moving average models sometimes preferred to autoregressive models?

3. Explain why the familiar formula for the standard error of $\bar{X}$ ($\sigma_{\bar{X}} = \sigma/\sqrt{n}$) is inappropriate in the presence of strong autocorrelation.

4. In transforming for stationarity, what is the difference between the deterministic and stochastic approaches?

5. Suppose a stochastic process is AR(1) with a coefficient of –0.7. How would its autocorrelation function appear? How would its partial autocorrelation appear?

6. When identifying the appropriate time series model, does one always choose one with minimum residual error? Why not?

7. Using the Keokuk dataset, graph annual discharge as a time series. Do you think the sample lag-1 autocorrelation (r_1) is positive or negative? (*Hint:* what would a scatterplot of y_t vs. y_{t+1} show?) Compute r_1. Does its value confirm your suspicions?

8. Given the following time series of 100 values:

–0.7784	0.7500	–0.3678	–0.5595	–0.8068	0.6306	–0.2161	–0.5251
0.0090	0.3766	–0.3266	–0.2225	0.5015	0.1136	–0.3976	0.0820
0.2030	0.2598	–0.1462	0.3907	–0.0836	0.4817	–0.0666	0.4311
–0.0005	0.4766	–0.4059	–0.2351	–0.1060	0.1564	–0.6055	–0.1502
–0.4424	0.1580	0.0465	–0.6030	–0.2929	0.0327	0.0628	–1.2191
–0.4335	–0.4555	0.1592	–1.2033	–0.1421	–0.4881	0.6016	–1.2358
–0.0693	–0.3998	0.6816	–1.0855	–0.2658	–0.3522	0.4368	–0.1493
0.1064	–0.2848	0.6976	–0.1703	0.2378	–0.3518	0.6744	–0.5485
0.2665	–0.7421	0.5483	–0.3396	0.3229	–0.5831	0.1049	–0.7798
0.4224	–0.5528	–0.0297	–0.7729	0.1245	–0.9545	0.1793	–0.3034
0.0054	0.5092	0.2993	0.0193	0.2788	–0.6221	0.4148	0.5104
0.5560	–0.9757	0.4210	0.7366	0.2686	–0.3631	0.5365	0.4295
0.4316	–0.4213	–0.1576	0.2863				

a. Compute autocorrelations and partial autocorrelations out to lag 15.
b. Plot the sample autocorrelation and partial autocorrelation functions.
c. What type and order of time series models do the results above suggest?

9. Record maximum temperatures for the last 30 days. Compute autocorrelations and partial autocorrelations for lags 1 and 2. Explain the differences.

10. Using your temperature data from the previous question:
    a. Construct a first-order AR model.
    b. Construct a first-order MA model.
    c. How do the two models compare in terms of MSE?
    d. Is there reason to believe that an ARMA model is appropriate? Are higher order models called for?

11. What are the consequences of violating the sampling theorem?

12. Graph the amplitude response function for a 5-point running mean with coefficients [0.1, 0.2, 0.4, 0.2, 0.1].

13. Why are symmetric filters so common?

14. Construct a 21-point Lanczos filter with a cutoff frequency of 0.2.

15. Graph the amplitude response function of the above. What is the response at $f = 0.1$?

APPENDIX: STATISTICAL TABLES

TABLE A-1
Binomial Distribution

Entry is probability $P(X = x) = \binom{n}{x} p^x (1 - p)^{n-x}$

n	x	.01	.02	.03	.04	.05	.06	.07	.08	.09		
						p						
2	0	0.9801	0.9604	0.9409	0.9216	0.9025	0.8836	0.8649	0.8464	0.8281	2	
	1	0.0198	0.0392	0.0582	0.0768	0.0950	0.1128	0.1302	0.1472	0.1638	1	
	2	0.0001	0.0004	0.0009	0.0016	0.0025	0.0036	0.0049	0.0064	0.0081	0	2
3	0	0.9703	0.9412	0.9127	0.8847	0.8574	0.8306	0.8044	0.7787	0.7536	3	
	1	0.0294	0.0576	0.0847	0.1106	0.1354	0.1590	0.1816	0.2031	0.2236	2	
	2	0.0003	0.0012	0.0026	0.0046	0.0071	0.0102	0.0137	0.0177	0.0221	1	
	3	0.0000	0.0000	0.0000	0.0001	0.0001	0.0002	0.0003	0.0005	0.0007	0	3
4	0	0.9606	0.9224	0.8853	0.8493	0.8145	0.7807	0.7481	0.7164	0.6857	4	
	1	0.0388	0.0753	0.1095	0.1416	0.1715	0.1993	0.2252	0.2492	0.2713	3	
	2	0.0006	0.0023	0.0051	0.0088	0.0135	0.0191	0.0254	0.0325	0.0402	2	
	3	0.0000	0.0000	0.0001	0.0002	0.0005	0.0008	0.0013	0.0019	0.0027	1	
	4	0.0000	0.0000	0.0000	0.0000	0.0000	0.0000	0.0000	0.0000	0.0001	0	4
5	0	0.9510	0.9039	0.8587	0.8154	0.7738	0.7339	0.6957	0.6591	0.6240	5	
	1	0.0480	0.0922	0.1328	0.1699	0.2036	0.2342	0.2618	0.2866	0.3086	4	
	2	0.0010	0.0038	0.0082	0.0142	0.0214	0.0299	0.0394	0.0498	0.0610	3	
	3	0.0000	0.0001	0.0003	0.0006	0.0011	0.0019	0.0030	0.0043	0.0060	2	
	4	0.0000	0.0000	0.0000	0.0000	0.0000	0.0001	0.0001	0.0002	0.0003	1	
	5	0.0000	0.0000	0.0000	0.0000	0.0000	0.0000	0.0000	0.0000	0.0000	0	5
6	0	0.9415	0.8858	0.8330	0.7828	0.7351	0.6899	0.6470	0.6064	0.5679	6	
	1	0.0571	0.1085	0.1546	0.1957	0.2321	0.2642	0.2922	0.3164	0.3370	5	
	2	0.0014	0.0055	0.0120	0.0204	0.0305	0.0422	0.0550	0.0688	0.0833	4	
	3	0.0000	0.0002	0.0005	0.0011	0.0021	0.0036	0.0055	0.0080	0.0110	3	
	4	0.0000	0.0000	0.0000	0.0000	0.0001	0.0002	0.0003	0.0005	0.0008	2	
	5	0.0000	0.0000	0.0000	0.0000	0.0000	0.0000	0.0000	0.0000	0.0000	1	
	6	0.0000	0.0000	0.0000	0.0000	0.0000	0.0000	0.0000	0.0000	0.0000	0	6
7	0	0.9321	0.8681	0.8080	0.7514	0.6983	0.6485	0.6017	0.5578	0.5168	7	
	1	0.0659	0.1240	0.1749	0.2192	0.2573	0.2897	0.3170	0.3396	0.3578	6	
	2	0.0020	0.0076	0.0162	0.0274	0.0406	0.0555	0.0716	0.0886	0.1061	5	
	3	0.0000	0.0003	0.0008	0.0019	0.0036	0.0059	0.0090	0.0128	0.0175	4	
	4	0.0000	0.0000	0.0000	0.0001	0.0002	0.0004	0.0007	0.0011	0.0017	3	
	5	0.0000	0.0000	0.0000	0.0000	0.0000	0.0000	0.0000	0.0001	0.0001	2	
	6	0.0000	0.0000	0.0000	0.0000	0.0000	0.0000	0.0000	0.0000	0.0000	1	
	7	0.0000	0.0000	0.0000	0.0000	0.0000	0.0000	0.0000	0.0000	0.0000	0	7
8	0	0.9227	0.8508	0.7837	0.7214	0.6634	0.6096	0.5596	0.5132	0.4703	8	
	1	0.0746	0.1389	0.1939	0.2405	0.2793	0.3113	0.3370	0.3570	0.3721	7	
	2	0.0026	0.0099	0.0210	0.0351	0.0515	0.0695	0.0888	0.1087	0.1288	6	
	3	0.0001	0.0004	0.0013	0.0029	0.0054	0.0089	0.0134	0.0189	0.0255	5	
	4	0.0000	0.0000	0.0001	0.0002	0.0004	0.0007	0.0013	0.0021	0.0031	4	
	5	0.0000	0.0000	0.0000	0.0000	0.0000	0.0000	0.0001	0.0001	0.0002	3	
	6	0.0000	0.0000	0.0000	0.0000	0.0000	0.0000	0.0000	0.0000	0.0000	2	
	7	0.0000	0.0000	0.0000	0.0000	0.0000	0.0000	0.0000	0.0000	0.0000	1	
	8	0.0000	0.0000	0.0000	0.0000	0.0000	0.0000	0.0000	0.0000	0.0000	0	8
9	0	0.9135	0.8337	0.7602	0.6925	0.6302	0.5730	0.5204	0.4722	0.4279	9	
	1	0.0830	0.1531	0.2116	0.2597	0.2985	0.3292	0.3525	0.3695	0.3809	8	
	2	0.0034	0.0125	0.1262	0.0433	0.0629	0.0840	0.1061	0.1285	0.1507	7	
	3	0.0001	0.0006	0.0019	0.0042	0.0077	0.0125	0.0186	0.0261	0.0348	6	
	4	0.0000	0.0000	0.0001	0.0003	0.0006	0.0012	0.0021	0.0034	0.0052	5	
	5	0.0000	0.0000	0.0000	0.0000	0.0000	0.0001	0.0002	0.0003	0.0005	4	
	6	0.0000	0.0000	0.0000	0.0000	0.0000	0.0000	0.0000	0.0000	0.0000	3	
	7	0.0000	0.0000	0.0000	0.0000	0.0000	0.0000	0.0000	0.0000	0.0000	2	
	8	0.0000	0.0000	0.0000	0.0000	0.0000	0.0000	0.0000	0.0000	0.0000	1	
	9	0.0000	0.0000	0.0000	0.0000	0.0000	0.0000	0.0000	0.0000	0.0000	0	9
		.99	.98	.97	.96	.95	.94	.93	.92	.91	x	n
						p						

TABLE A-1 (continued)

n	x	.10	.15	.20	.25	.30	.35	.40	.45	.50		
						p						
2	0	0.8100	0.7225	0.6400	0.5625	0.4900	0.4225	0.3600	0.3025	0.2500	2	
	1	0.1800	0.2550	0.3200	0.3750	0.4200	0.4550	0.4800	0.4950	0.5000	1	
	2	0.0100	0.0225	0.0400	0.0625	0.0900	0.1225	0.1600	0.2025	0.2500	0	2
3	0	0.7290	0.6141	0.5120	0.4219	0.3430	0.2746	0.2160	0.1664	0.1250	3	
	1	0.2430	0.3251	0.3840	0.4219	0.4410	0.4436	0.4320	0.4084	0.3750	2	
	2	0.0270	0.0574	0.0960	0.1406	0.1890	0.2389	0.2880	0.3341	0.3750	1	
	3	0.0010	0.0034	0.0080	0.0156	0.0270	0.0429	0.0640	0.0911	0.1250	0	3
4	0	0.6561	0.5220	0.4096	0.3164	0.2401	0.1785	0.1296	0.0915	0.0625	4	
	1	0.2916	0.3685	0.4096	0.4219	0.4116	0.3845	0.3456	0.2995	0.2500	3	
	2	0.0486	0.0975	0.1536	0.2109	0.2646	0.3105	0.3456	0.3675	0.3750	2	
	3	0.0036	0.0115	0.0256	0.0469	0.0756	0.1115	0.1536	0.2005	0.2500	1	
	4	0.0001	0.0005	0.0016	0.0039	0.0081	0.0150	0:0256	0.0410	0.0625	0	4
5	0	0.5905	0.4437	0.3277	0.2373	0.1681	0.1160	0.0778	0.0503	0.0312	5	
	1	0.3280	0.3915	0.4096	0.3955	0.3601	0.3124	0.2592	0.2059	0.1562	4	
	2	0.0729	0.1382	0.2048	0.2637	0.3087	0.3364	0.3456	0.3369	0.3125	3	
	3	0.0081	0.0244	0.0512	0.0879	0.1323	0.1811	0.2304	0.2757	0.3125	2	
	4	0.0004	0.0022	0.0064	0.0146	0.0283	0.0488	0.0768	0.1128	0.1562	1	
	5	0.0000	0.0001	0.0003	0.0010	0.0024	0.0053	0.0102	0.0185	0.0312	0	5
6	0	0.5314	0.3771	0.2621	0.1780	0.1176	0.0754	0.0467	0.0277	0.0156	6	
	1	0.3543	0.3993	0.3932	0.3560	0.3025	0.2437	0.1866	0.1359	0.0938	5	
	2	0.0984	0.1762	0.2458	0.2966	0.3241	0.3280	0.3110	0.2780	0.2344	4	
	3	0.0146	0.0415	0.0819	0.1318	0.1852	0.2355	0.2765	0.3032	0.3125	3	
	4	0.0012	0.0055	0.0154	0.0330	0.0595	0.0951	0.1382	0.1861	0.2344	2	
	5	0.0001	0.0004	0.0015	0.0044	0.0102	0.0205	0.0369	0.0609	0.0938	1	
	6	0.0000	0.0000	0.0001	0.0002	0.0007	0.0018	0.0041	0.0083	0.0156	0	6
7	0	0.4783	0.3206	0.2097	0.1335	0.0824	0.0490	0.0280	0.0152	0.0078	7	
	1	0.3720	0.3960	0.3670	0.3115	0.2471	0.1848	0.1306	0.0872	0.0547	6	
	2	0.1240	0.2097	0.2753	0.3115	0.3177	0.2985	0.2613	0.2140	0.1641	5	
	3	0.0230	0.0617	0.1147	0.1730	0.2269	0.2679	0.2903	0.2918	0.2734	4	
	4	0.0026	0.0109	0.0287	0.0577	0.0972	0.1442	0.1935	0.2388	0.2734	3	
	5	0.0002	0.0012	0.0043	0.0115	0.0250	0.0466	0.0774	0.1172	0.1641	2	
	6	0.0000	0.0001	0.0004	0.0013	0.0036	0.0084	0.0172	0.0320	0.0547	1	
	7	0.0000	0.0000	0.0000	0.0001	0.0002	0.0006	0.0016	0.0037	0.0078	0	7
8	0	0.4305	0.2725	0.1678	0.1001	0.0576	0.0319	0.0168	0.0084	0.0039	8	
	1	0.3826	0.3847	0.3355	0.2670	0.1977	0.1373	0.0896	0.0548	0.0312	7	
	2	0.1488	0.2376	0.2936	0.3115	0.2965	0.2587	0.2090	0.1569	0.1094	6	
	3	0.0331	0.0839	0.1468	0.2076	0.2541	0.2786	0.2787	0.2568	0.2188	5	
	4	0.0046	0.0185	0.0459	0.0865	0.1361	0.1875	0.2322	0.2627	0.2734	4	
	5	0.0004	0.0026	0.0092	0.0231	0.0467	0.0808	0.1239	0.1719	0.2188	3	
	6	0.0000	0.0002	0.0011	0.0038	0.0100	0.0217	0.0413	0.0703	0.1094	2	
	7	0.0000	0.0000	0.0001	0.0004	0.0012	0.0033	0.0079	0.0164	0.0312	1	
	8	0.0000	0.0000	0.0000	0.0000	0.0001	0.0002	0.0007	0.0017	0.0039	0	8
9	0	0.3874	0.2316	0.1342	0.0751	0.0404	0.0207	0.0101	0.0046	0.0020	9	
	1	0.3874	0.3679	0.3020	0.2253	0.1556	0.1004	0.0605	0.0339	0.0176	8	
	2	0.1722	0.2597	0.3020	0.3003	0.2668	0.2162	0.1612	0.1110	0.0703	7	
	3	0.0446	0.1069	0.1762	0.2336	0.2668	0.2716	0.2508	0.2119	0.1641	6	
	4	0.0074	0.0283	0.0661	0.1168	0.1715	0.2194	0.2508	0.2600	0.2461	5	
	5	0.0008	0.0050	0.0165	0.0389	0.0735	0.1181	0.1672	0.2128	0.2461	4	
	6	0.0001	0.0006	0.0028	0.0087	0.0210	0.0424	0.0743	0.1160	0.1641	3	
	7	0.0000	0.0000	0.0003	0.0012	0.0039	0.0098	0.0212	0.0407	0.0703	2	
	8	0.0000	0.0000	0.0000	0.0001	0.0004	0.0013	0.0035	0.0083	0.0176	1	
	9	0.0000	0.0000	0.0000	0.0000	0.0000	0.0001	0.0003	0.0008	0.0020	0	9
		.90	.85	.80	.75	.70	.65	.60	.55	.50	x	n
						p						

(continued)

TABLE A-1 (continued)

		p										
n	*x*	.01	.02	.03	.04	.05	.06	.07	.08	.09		
10	0	0.9044	0.8171	0.7374	0.6648	0.5987	0.5386	0.4840	0.4344	0.3894	10	
	1	0.0914	0.1667	0.2281	0.2770	0.3151	0.3438	0.3643	0.3777	0.3851	9	
	2	0.0042	0.0153	0.0317	0.0519	0.0746	0.0988	0.1234	0.1478	0.1714	8	
	3	0.0001	0.0008	0.0026	0.0058	0.0105	0.0168	0.0248	0.0343	0.0452	7	
	4	0.0000	0.0000	0.0001	0.0004	0.0010	0.0019	0.0033	0.0052	0.0078	6	
	5	0.0000	0.0000	0.0000	0.0000	0.0001	0.0001	0.0003	0.0005	0.0009	5	
	6	0.0000	0.0000	0.0000	0.0000	0.0000	0.0000	0.0000	0.0000	0.0001	4	
	7	0.0000	0.0000	0.0000	0.0000	0.0000	0.0000	0.0000	0.0000	0.0000	3	
	8	0.0000	0.0000	0.0000	0.0000	0.0000	0.0000	0.0000	0.0000	0.0000	2	
	9	0.0000	0.0000	0.0000	0.0000	0.0000	0.0000	0.0000	0.0000	0.0000	1	
	10	0.0000	0.0000	0.0000	0.0000	0.0000	0.0000	0.0000	0.0000	0.0000	0	10
12	0	0.8864	0.7847	0.6938	0.6127	0.5404	0.4759	0.4186	0.3677	0.3225	12	
	1	0.1074	0.1922	0.2575	0.3064	0.3413	0.3645	0.3781	0.3837	0.3827	11	
	2	0.0060	0.0216	0.0438	0.0702	0.0988	0.1280	0.1565	0.1835	0.2082	10	
	3	0.0002	0.0015	0.0045	0.0098	0.0173	0.0272	0.0393	0.0532	0.0686	9	
	4	0.0000	0.0001	0.0003	0.0009	0.0021	0.0039	0.0067	0.0104	0.0153	8	
	5	0.0000	0.0000	0.0000	0.0001	0.0002	0.0004	0.0008	0.0014	0.0024	7	
	6	0.0000	0.0000	0.0000	0.0000	0.0000	0.0000	0.0001	0.0001	0.0003	6	
	7	0.0000	0.0000	0.0000	0.0000	0.0000	0.0000	0.0000	0.0000	0.0000	5	
	8	0.0000	0.0000	0.0000	0.0000	0.0000	0.0000	0.0000	0.0000	0.0000	4	
	9	0.0000	0.0000	0.0000	0.0000	0.0000	0.0000	0.0000	0.0000	0.0000	3	
	10	0.0000	0.0000	0.0000	0.0000	0.0000	0.0000	0.0000	0.0000	0.0000	2	
	11	0.0000	0.0000	0.0000	0.0000	0.0000	0.0000	0.0000	0.0000	0.0000	1	
	12	0.0000	0.0000	0.0000	0.0000	0.0000	0.0000	0.0000	0.0000	0.0000	0	12
15	0	0.8601	0.7386	0.6333	0.5421	0.4633	0.3953	0.3367	0.2863	0.2430	15	
	1	0.1303	0.2261	0.2938	0.3388	0.3658	0.3785	0.3801	0.3734	0.3605	14	
	2	0.0092	0.0323	0.0636	0.0988	0.1348	0.1691	0.2003	0.2273	0.2496	13	
	3	0.0004	0.0029	0.0085	0.0178	0.0307	0.0468	0.0653	0.0857	0.1070	12	
	4	0.0000	0.0002	0.0008	0.0022	0.0049	0.0090	0.0148	0.0223	0.0317	11	
	5	0.0000	0.0000	0.0001	0.0002	0.0006	0.0013	0.0024	0.0043	0.0069	10	
	6	0.0000	0.0000	0.0000	0.0000	0.0000	0.0001	0.0003	0.0006	0.0011	9	
	7	0.0000	0.0000	0.0000	0.0000	0.0000	0.0000	0.0000	0.0001	0.0001	8	
	8	0.0000	0.0000	0.0000	0.0000	0.0000	0.0000	0.0000	0.0000	0.0000	7	
	9	0.0000	0.0000	0.0000	0.0000	0.0000	0.0000	0.0000	0.0000	0.0000	6	
	10	0.0000	0.0000	0.0000	0.0000	0.0000	0.0000	0.0000	0.0000	0.0000	5	
	11	0.0000	0.0000	0.0000	0.0000	0.0000	0.0000	0.0000	0.0000	0.0000	4	
	12	0.0000	0.0000	0.0000	0.0000	0.0000	0.0000	0.0000	0.0000	0.0000	3	
	13	0.0000	0.0000	0.0000	0.0000	0.0000	0.0000	0.0000	0.0000	0.0000	2	
	14	0.0000	0.0000	0.0000	0.0000	0.0000	0.0000	0.0000	0.0000	0.0000	1	
	15	0.0000	0.0000	0.0000	0.0000	0.0000	0.0000	0.0000	0.0000	0.0000	0	15
20	0	0.8179	0.6676	0.5438	0.4420	0.3585	0.2901	0.2342	0.1887	0.1516	20	
	1	0.1652	0.2725	0.3364	0.3683	0.3774	0.3703	0.3526	0.3282	0.3000	19	
	2	0.0159	0.0528	0.0988	0.1458	0.1887	0.2246	0.2521	0.2711	0.2818	18	
	3	0.0010	0.0065	0.0183	0.0364	0.0596	0.0860	0.1139	0.1414	0.1672	17	
	4	0.0000	0.0006	0.0024	0.0065	0.0133	0.0233	0.0364	0.0523	0.0703	16	
	5	0.0000	0.0000	0.0002	0.0009	0.0022	0.0048	0.0088	0.0145	0.0222	15	
	6	0.0000	0.0000	0.0000	0.0001	0.0003	0.0008	0.0017	0.0032	0.0055	14	
	7	0.0000	0.0000	0.0000	0.0000	0.0000	0.0001	0.0002	0.0005	0.0011	13	
	8	0.0000	0.0000	0.0000	0.0000	0.0000	0.0000	0.0000	0.0001	0.0002	12	
	9	0.0000	0.0000	0.0000	0.0000	0.0000	0.0000	0.0000	0.0000	0.0000	11	
	10	0.0000	0.0000	0.0000	0.0000	0.0000	0.0000	0.0000	0.0000	0.0000	10	
	11	0.0000	0.0000	0.0000	0.0000	0.0000	0.0000	0.0000	0.0000	0.0000	9	
	12	0.0000	0.0000	0.0000	0.0000	0.0000	0.0000	0.0000	0.0000	0.0000	8	
	13	0.0000	0.0000	0.0000	0.0000	0.0000	0.0000	0.0000	0.0000	0.0000	7	
	14	0.0000	0.0000	0.0000	0.0000	0.0000	0.0000	0.0000	0.0000	0.0000	6	
	15	0.0000	0.0000	0.0000	0.0000	0.0000	0.0000	0.0000	0.0000	0.0000	5	
	16	0.0000	0.0000	0.0000	0.0000	0.0000	0.0000	0.0000	0.0000	0.0000	4	
	17	0.0000	0.0000	0.0000	0.0000	0.0000	0.0000	0.0000	0.0000	0.0000	3	
	18	0.0000	0.0000	0.0000	0.0000	0.0000	0.0000	0.0000	0.0000	0.0000	2	
	19	0.0000	0.0000	0.0000	0.0000	0.0000	0.0000	0.0000	0.0000	0.0000	1	
	20	0.0000	0.0000	0.0000	0.0000	0.0000	0.0000	0.0000	0.0000	0.0000	0	20
		.99	.98	.97	.96	.95	.94	.93	.92	.91	*x*	*n*
		p										

TABLE A-1 (continued)

		p										
n	*x*	.01	.02	.03	.04	.05	.06	.07	.08	.09		
10	0	0.3487	0.1969	0.1074	0.0563	0.0282	0.0135	0.0060	0.0025	0.0010	10	
	1	0.3874	0.3474	0.2684	0.1877	0.1211	0.0725	0.0403	0.0207	0.0098	9	
	2	0.1937	0.2759	0.3020	0.2816	0.2335	0.1757	0.1209	0.0763	0.0439	8	
	3	0.0574	0.1298	0.2013	0.2503	0.2668	0.2522	0.2150	0.1665	0.1172	7	
	4	0.0112	0.0401	0.0881	0.1460	0.2001	0.2377	0.2508	0.2384	0.2051	6	
	5	0.0015	0.0085	0.0264	0.0584	0.1029	0.1536	0.2007	0.2340	0.2461	5	
	6	0.0001	0.0012	0.0055	0.0162	0.0368	0.0689	0.1115	0.1596	0.2051	4	
	7	0.0000	0.0001	0.0008	0.0031	0.0090	0.0212	0.0425	0.0746	0.1172	3	
	8	0.0000	0.0000	0.0001	0.0004	0.0014	0.0043	0.0106	0.0229	0.0439	2	
	9	0.0000	0.0000	0.0000	0.0000	0.0001	0.0005	0.0016	0.0042	0.0098	1	
	10	0.0000	0.0000	0.0000	0.0000	0.0000	0.0000	0.0001	0.0003	0.0010	0	10
12	0	0.2824	0.1422	0.0687	0.0317	0.0138	0.0057	0.0022	0.0008	0.0002	12	
	1	0.3766	0.3012	0.2062	0.1267	0.0712	0.0368	0.0174	0.0075	0.0029	11	
	2	0.2301	0.2924	0.2835	0.2323	0.1678	0.1088	0.0639	0.0339	0.0161	10	
	3	0.0852	0.1720	0.2362	0.2581	0.2397	0.1954	0.1419	0.0923	0.0537	9	
	4	0.0213	0.0683	0.1329	0.1936	0.2311	0.2367	0.2128	0.1700	0.1208	8	
	5	0.0038	0.0193	0.0532	0.1032	0.1585	0.2039	0.2270	0.2225	0.1934	7	
	6	0.0005	0.0040	0.0155	0.0401	0.0792	0.1281	0.1766	0.2124	0.2256	6	
	7	0.0000	0.0006	0.0033	0.0115	0.0291	0.0591	0.1009	0.1489	0.1934	5	
	8	0.0000	0.0001	0.0005	0.0024	0.0078	0.0199	0.0420	0.0762	0.1208	4	
	9	0.0000	0.0000	0.0001	0.0004	0.0015	0.0048	0.0125	0.0277	0.0537	3	
	10	0.0000	0.0000	0.0000	0.0000	0.0002	0.0008	0.0025	0.0068	0.0161	2	
	11	0.0000	0.0000	0.0000	0.0000	0.0000	0.0001	0.0003	0.0010	0.0029	1	
	12	0.0000	0.0000	0.0000	0.0000	0.0000	0.0000	0.0000	0.0001	0.0002	0	12
15	0	0.2059	0.0874	0.0352	0.0134	0.0047	0.0016	0.0005	0.0001	0.0000	15	
	1	0.3432	0.2312	0.1319	0.0668	0.0305	0.0126	0.0047	0.0016	0.0005	14	
	2	0.2669	0.2856	0.2309	0.1559	0.0916	0.0476	0.0219	0.0090	0.0032	13	
	3	0.1285	0.2184	0.2501	0.2252	0.1700	0.1110	0.0634	0.0318	0.0139	12	
	4	0.0428	0.1156	0.1876	0.2252	0.2186	0.1792	0.1268	0.0780	0.0417	11	
	5	0.0105	0.0449	0.1032	0.1651	0.2061	0.2123	0.1859	0.1404	0.0916	10	
	6	0.0019	0.0132	0.0430	0.0917	0.1472	0.1906	0.2066	0.1914	0.1527	9	
	7	0.0003	0.0030	0.0138	0.0393	0.0811	0.1319	0.1771	0.2013	0.1964	8	
	8	0.0000	0.0005	0.0035	0.0131	0.0348	0.0710	0.1181	0.1647	0.1964	7	
	9	0.0000	0.0001	0.0007	0.0034	0.0116	0.0298	0.0612	0.1048	0.1527	6	
	10	0.0000	0.0000	0.0001	0.0007	0.0030	0.0096	0.0245	0.0515	0.0916	5	
	11	0.0000	0.0000	0.0000	0.0001	0.0006	0.0024	0.0074	0.0191	0.0417	4	
	12	0.0000	0.0000	0.0000	0.0000	0.0001	0.0004	0.0016	0.0052	0.0139	3	
	13	0.0000	0.0000	0.0000	0.0000	0.0000	0.0001	0.0003	0.0010	0.0032	2	
	14	0.0000	0.0000	0.0000	0.0000	0.0000	0.0000	0.0000	0.0001	0.0005	1	
	15	0.0000	0.0000	0.0000	0.0000	0.0000	0.0000	0.0000	0.0000	0.0000	0	15
20	0	0.1216	0.0388	0.0115	0.0032	0.0008	0.0002	0.0000	0.0000	0.0000	20	
	1	0.2702	0.1368	0.0576	0.0211	0.0068	0.0020	0.0005	0.0001	0.0000	19	
	2	0.2852	0.2293	0.1369	0.0669	0.0278	0.0100	0.0031	0.0008	0.0002	18	
	3	0.1901	0.2428	0.2054	0.1339	0.0716	0.0323	0.0123	0.0040	0.0011	17	
	4	0.0898	0.1821	0.2182	0.1897	0.1304	0.0738	0.0350	0.0139	0.0046	16	
	5	0.0319	0.1028	0.1746	0.2023	0.1789	0.1272	0.0746	0.0365	0.0148	15	
	6	0.0089	0.0454	0.1091	0.1686	0.1916	0.1712	0.1244	0.0746	0.0370	14	
	7	0.0020	0.0160	0.0545	0.1124	0.1643	0.1844	0.1659	0.1221	0.0739	13	
	8	0.0004	0.0046	0.0222	0.0609	0.1144	0.1614	0.1797	0.1623	0.1201	12	
	9	0.0001	0.0011	0.0074	0.0271	0.0654	0.1158	0.1597	0.1771	0.1602	11	
	10	0.0000	0.0002	0.0020	0.0099	0.0308	0.0686	0.1171	0.1593	0.1762	10	
	11	0.0000	0.0000	0.0005	0.0030	0.0120	0.0336	0.0710	0.1185	0.1602	9	
	12	0.0000	0.0000	0.0001	0.0008	0.0039	0.0136	0.0355	0.0727	0.1201	8	
	13	0.0000	0.0000	0.0000	0.0002	0.0010	0.0045	0.0146	0.0366	0.0739	7	
	14	0.0000	0.0000	0.0000	0.0000	0.0002	0.0012	0.0049	0.0150	0.0370	6	
	15	0.0000	0.0000	0.0000	0.0000	0.0000	0.0003	0.0013	0.0049	0.0148	5	
	16	0.0000	0.0000	0.0000	0.0000	0.0000	0.0000	0.0003	0.0013	0.0046	4	
	17	0.0000	0.0000	0.0000	0.0000	0.0000	0.0000	0.0000	0.0002	0.0011	3	
	18	0.0000	0.0000	0.0000	0.0000	0.0000	0.0000	0.0000	0.0000	0.0002	2	
	19	0.0000	0.0000	0.0000	0.0000	0.0000	0.0000	0.0000	0.0000	0.0000	1	
	20	0.0000	0.0000	0.0000	0.0000	0.0000	0.0000	0.0000	0.0000	0.0000	0	20
		.99	.98	.97	.96	.95	.94	.93	.92	.91	*x*	*n*
		p										

TABLE A-2
Poisson Distribution

Poisson probabilities

	λ									
X	.01	.02	.03	.04	.05	.06	.07	.08	.09	1.0
0	.9048	.8187	.7408	.6703	.6065	.5488	.4966	.4493	.4066	.3679
1	.0905	.1637	.2222	.2681	.3033	.3293	.3476	.3595	.3659	.3679
2	.0045	.0164	.0333	.0536	.0758	.0988	.1217	.1438	.1647	.1839
3	.0002	.0011	.0033	.0072	.0126	.0198	.0284	.0383	.0494	.0613
4		.0001	.0002	.0007	.0016	.0030	.0050	.0077	.0111	.0153
5				.0001	.0002	.0004	.0007	.0012	.0020	.0031
6							.0001	.0002	.0003	.0005
7										.0001

	λ									
X	1.5	2.0	2.5	3.0	3.5	4.0	4.5	5.0	6.0	7.0
0	.2231	.1353	.0821	.0498	.0302	.0183	.0111	.0067	.0025	.0009
1	.3347	.2707	.2052	.1494	.1057	.0733	.0500	.0337	.0149	.0064
2	.2510	.2707	.2565	.2240	.1850	.1465	.1125	.0842	.0446	.0223
3	.1255	.1804	.2138	.2240	.2158	.1954	.1687	.1404	.0892	.0521
4	.0471	.0902	.1336	.1680	.1888	.1954	.1898	.1755	.1339	.0912
5	.0141	.0361	.0668	.1008	.1322	.1563	.1708	.1755	.1606	.1277
6	.0035	.0120	.0278	.0504	.0771	.1042	.1281	.1462	.1606	.1490
7	.0008	.0034	.0099	.0216	.0385	.0595	.0824	.1044	.1377	.1490
8	.0001	.0009	.0031	.0081	.0169	.0298	.0463	.0653	.1033	.1304
9		.0002	.0009	.0027	.0066	.0132	.0232	.0363	.0688	.1014
10			.0002	.0008	.0023	.0053	.0104	.0181	.0413	.0710
11				.0002	.0007	.0019	.0043	.0082	.0225	.0452
12				.0001	.0002	.0006	.0016	.0034	.0113	.0264
13					.0001	.0002	.0006	.0013	.0052	.0142
14						.0001	.0002	.0005	.0022	.0071
15							.0001	.0002	.0009	.0033
16									.0003	.0014
17									.0001	.0006
18										.0002
19										.0001

Table entries are $P(X = x/\lambda)$

TABLE A-3
Standard Normal Probabilities

z	$P(-z < Z < z)$	$P(\|Z\| > z)$	$P(Z > z)$	$P(Z < -z)$	$P(Z < z)$	$P(Z > -z)$
0.50	0.383	0.617	0.309	0.309	0.691	0.691
0.60	0.451	0.549	0.274	0.274	0.726	0.726
0.70	0.516	0.484	0.242	0.242	0.760	0.758
0.80	0.576	0.424	0.212	0.212	0.788	0.788
0.90	0.632	0.368	0.184	0.184	0.816	0.816
1.00	0.683	0.317	0.159	0.159	0.841	0.841
1.28	**0.800**	**0.200**	**0.100**	**0.100**	**0.900**	**0.900**
1.50	0.866	0.134	0.067	0.067	0.933	0.933
1.60	0.890	0.110	0.055	0.055	0.945	0.945
1.65	**0.900**	**0.100**	**0.050**	**0.050**	**0.950**	**0.950**
1.70	0.911	0.089	0.045	0.045	0.955	0.955
1.80	0.928	0.072	0.036	0.036	0.964	0.964
1.90	0.943	0.057	0.029	0.029	0.971	0.971
1.96	**0.950**	**0.050**	**0.025**	**0.025**	**0.975**	**0.975**
2.00	0.954	0.046	0.023	0.023	0.977	0.977
2.10	0.964	0.036	0.018	0.018	0.982	0.982
2.20	0.972	0.028	0.014	0.014	0.986	0.986
2.30	0.979	0.021	0.011	0.011	0.989	0.989
2.40	0.984	0.016	0.008	0.008	0.992	0.992
2.50	0.988	0.012	0.006	0.006	0.994	0.994
2.58	**0.990**	**0.010**	**0.005**	**0.005**	**0.995**	**0.995**
2.60	0.991	0.009	0.005	0.005	0.995	0.995
2.70	0.993	0.007	0.003	0.003	0.997	0.997
2.80	0.995	0.005	0.003	0.003	0.997	0.997
2.90	0.996	0.004	0.002	0.002	0.998	0.998
3.00	0.997	0.003	0.001	0.001	0.999	0.999
3.10	0.998	0.002	0.001	0.001	0.999	0.999
3.20	0.999	0.001	0.001	0.001	0.999	0.999
3.30	0.999	0.001	0.000	0.000	1.000	1.000
3.40	0.999	0.001	0.000	0.000	1.000	1.000
3.50	1.000	0.000	0.000	0.000	1.000	1.000

TABLE A-4
t Distribution

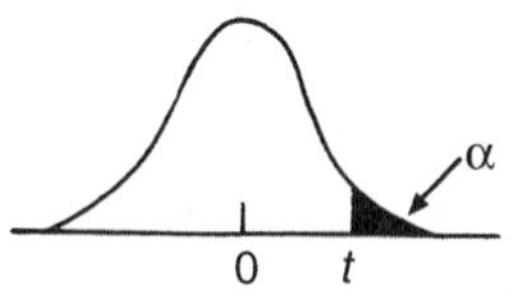

df \ α	.10	.05	.025	.01	.005
1	3.078	6.314	12.706	31.821	63.657
2	1.886	2.920	4.303	6.965	9.925
3	1.638	2.353	3.182	4.541	5.841
4	1.533	2.132	2.776	3.747	4.604
5	1.476	2.015	2.571	3.365	4.032
6	1.440	1.943	2.447	3.143	3.707
7	1.415	1.895	2.365	2.998	3.499
8	1.397	1.860	2.306	2.896	3.355
9	1.383	1.833	2.262	2.821	3.250
10	1.372	1.812	2.228	2.764	3.169
11	1.363	1.796	2.201	2.718	3.106
12	1.356	1.782	2.179	2.681	3.055
13	1.350	1.771	2.160	2.650	3.012
14	1.345	1.761	2.145	2.624	2.977
15	1.341	1.753	2.131	2.602	2.947
16	1.337	1.746	2.120	2.583	2.921
17	1.333	1.740	2.110	2.567	2.898
18	1.330	1.734	2.101	2.552	2.878
19	1.328	1.729	2.093	2.539	2.861
20	1.325	1.725	2.086	2.528	2.845
21	1.323	1.721	2.080	2.518	2.831
22	1.321	1.717	2.074	2.508	2.819
23	1.319	1.714	2.069	2.500	2.807
24	1.318	1.711	2.064	2.492	2.797
25	1.316	1.708	2.060	2.485	2.787
26	1.315	1.706	2.056	2.479	2.779
27	1.314	1.703	2.052	2.473	2.771
28	1.313	1.701	2.048	2.467	2.763
29	1.311	1.699	2.045	2.462	2.756
30	1.310	1.697	2.042	2.457	2.750
40	1.303	1.684	2.021	2.423	2.704
60	1.296	1.671	2.000	2.390	2.660
120	1.289	1.658	1.980	2.358	2.617
∞	1.282	1.645	1.960	2.326	2.576
two-tailed	.20	.10	.05	.02	.01

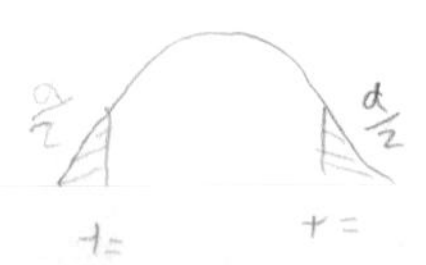

TABLE A-5
One-Tailed Probabilities of T

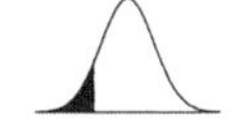

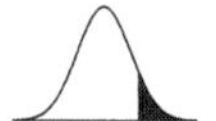

t-value	Degrees of freedom							
	1	2	3	4	5	6	7	8
0.5	0.352	0.333	0.326	0.322	0.319	0.317	0.316	0.315
0.6	0.328	0.305	0.295	0.290	0.287	0.285	0.284	0.283
0.7	0.306	0.278	0.267	0.261	0.258	0.255	0.253	0.252
0.8	0.285	0.254	0.241	0.234	0.230	0.227	0.225	0.223
0.9	0.267	0.232	0.217	0.210	0.205	0.201	0.199	0.197
1.0	0.250	0.211	0.196	0.187	0.182	0.178	0.175	0.173
1.1	0.235	0.193	0.176	0.167	0.161	0.157	0.154	0.152
1.2	0.221	0.177	0.158	0.148	0.142	0.138	0.135	0.132
1.3	0.209	0.162	0.142	0.132	0.125	0.121	0.117	0.115
1.4	0.197	0.148	0.128	0.117	0.110	0.106	0.102	0.100
1.5	0.187	0.136	0.115	0.104	0.097	0.092	0.089	0.086
1.6	0.178	0.125	0.104	0.092	0.085	0.080	0.077	0.074
1.7	0.169	0.116	0.094	0.082	0.075	0.070	0.066	0.064
1.8	0.161	0.107	0.085	0.073	0.066	0.061	0.057	0.055
1.9	0.154	0.099	0.077	0.065	0.058	0.053	0.050	0.047
2.0	0.148	0.092	0.070	0.058	0.051	0.046	0.043	0.040
2.1	0.141	0.085	0.063	0.052	0.045	0.040	0.037	0.034
2.2	0.136	0.079	0.058	0.046	0.040	0.035	0.032	0.029
2.3	0.131	0.074	0.052	0.041	0.035	0.031	0.027	0.025
2.4	0.126	0.069	0.048	0.037	0.031	0.027	0.024	0.022
2.5	0.121	0.065	0.044	0.033	0.027	0.023	0.020	0.018
2.6	0.117	0.061	0.040	0.030	0.024	0.020	0.018	0.016
2.7	0.113	0.057	0.037	0.027	0.021	0.018	0.015	0.014
2.8	0.109	0.054	0.034	0.024	0.019	0.016	0.013	0.012
2.9	0.106	0.051	0.031	0.022	0.017	0.014	0.011	0.010
3.0	0.102	0.048	0.029	0.020	0.015	0.012	0.010	0.009
3.1	0.099	0.045	0.027	0.018	0.013	0.011	0.009	0.007
3.2	0.096	0.043	0.025	0.016	0.012	0.009	0.008	0.006
3.3	0.094	0.040	0.023	0.015	0.011	0.008	0.007	0.005
3.4	0.091	0.038	0.021	0.014	0.010	0.007	0.006	0.005
3.5	0.089	0.036	0.020	0.012	0.009	0.006	0.005	0.004
4.0	0.078	0.029	0.014	0.008	0.005	0.004	0.003	0.002
4.5	0.070	0.023	0.010	0.005	0.003	0.002	0.001	0,001
5.0	0.063	0.019	0.008	0.004	0.002	0.001	0.001	0.001

(*continued*)

TABLE A-5 (continued)

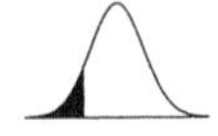

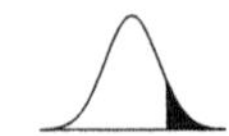

	Degrees of freedom							
t-value	9	10	15	20	30	50	100	Infinity
0.5	0.315	0.314	0.312	0.311	0.310	0.310	0.309	0.309
0.6	0.282	0.281	0.279	0.278	0.277	0.276	0.275	0.274
0.7	0.251	0.250	0.247	0.246	0.245	0.244	0.243	0.242
0.8	0.222	0.221	0.218	0.217	0.215	0.214	0.213	0.212
0.9	0.196	0.195	0.191	0.189	0.188	0.186	0.395	0.184
1.0	0.172	0.170	0.167	0.165	0.163	0.161	0.160	0.159
1.1	0.150	0.149	0.144	0.142	0.140	0.138	0.137	0.136
1.2	0.130	0.129	0.124	0.122	0.120	0.118	0.116	0.115
1.3	0.113	0.111	0.107	0.104	0.102	0.100	0.098	0.097
1.4	0.098	0.096	0.091	0.088	0.086	0.084	0.082	0.081
1.5	0.084	0.082	0.077	0.075	0.072	0.070	0.068	0.067
1.6	0.072	0.070	0.065	0.063	0.060	0.058	0.056	0.055
1.7	0.062	0.060	0.055	0.052	0.050	0.048	0.046	0.045
1.8	0.053	0.051	0.046	0.043	0.041	0.039	0.037	0.036
1.9	0.045	0.043	0.038	0.036	0.034	0.032	0.030	0.029
2.0	0.038	0.037	0.032	0.030	0.027	0.025	0.024	0.023
2.1	0.033	0.031	0.027	0.024	0.022	0.020	0.019	0.018
2.2	0.028	0.026	0.022	0.020	0.018	0.016	0.015	0.014
2.3	0.023	0.022	0.018	0.016	0.014	0.013	0.012	0.011
2.4	0.020	0.019	0.015	0.013	0.011	0.010	0.009	0.008
2.5	0.017	0.016	0.012	0.011	0.009	0.008	0.007	0.006
2.6	0.014	0.013	0.010	0.009	0.007	0.006	0.005	0.005
2.7	0.012	0.011	0.008	0.007	0.006	0.005	0.004	0.003
2.8	0.010	0.009	0.007	0.006	0.004	0.004	0.003	0.003
2.9	0.009	0.008	0.005	0.004	0.003	0.003	0.002	0.002
3.0	0.007	0.007	0.004	0.004	0.003	0.002	0.002	0.001
3.1	0.006	0.006	0.004	0.003	0.002	0.002	0.001	0.001
3.2	0.005	0.005	0.003	0.002	0.002	0.001	0.001	0.001
3.3	0.005	0.004	0.002	0.002	0.001	0.001	0.001	0.000
3.4	0.004	0.003	0.002	0.001	0.001	0.001	0.000	0.000
3.5	0.003	0.003	0.002	0.001	0.001	0.000	0.000	0.000
4.0	0.002	0.001	0.001	0.000	0.000	0.000	0.000	0.000
4.5	0.001	0.001	0.000	0.000	0.000	0.000	0.000	0.000
5.0	0.000	0.000	0.000	0.000	0.000	0.000	0.000	0.000

TABLE A-6
Two-Tailed Probabilities of T

	Degrees of freedom							
t-value	1	2	3	4	5	6	7	8
0.5	0.705	0.667	0.651	0.643	0.638	0.635	0.632	0.631
0.6	0.656	0.609	0.591	0.581	0.575	0.570	0.567	0.565
0.7	0.611	0.556	0.534	0.523	0.515	0.510	0.507	0.504
0.8	0.570	0.508	0.482	0.469	0.460	0.454	0.450	0.447
0.9	0.533	0.463	0.434	0.419	0.409	0.403	0.398	0.394
1.0	0.500	0.423	0.391	0.374	0.363	0.356	0.351	0.347
1.1	0.4'70	0.386	0.352	0.333	0.321	0.313	0.308	0.303
1.2	0.442	0.353	0.316	0.296	0.284	0.275	0.269	0.264
1.3	0.417	0.323	0.284	0.263	0.250	0.241	0.235	0.230
1.4	0.395	0.296	0.256	0.234	0.220	0.211	0.204	0.199
1.5	0.374	0.272	0.231	0.208	0.194	0.184	0.177	0.172
1.6	0.356	0.251	0.208	0.185	0.170	0.161	0.154	0.148
1.7	0.339	0.231	0.188	0.164	0.150	0.140	0.133	0.128
1.8	0.323	0.214	0.170	0.146	0.132	0.122	0.115	0.110
1.9	0.308	0.198	0.154	0.130	0.116	0.106	0.099	0.094
2.0	0.295	0.184	0.139	0.116	0.102	0.092	0.086	0.081
2.1	0.283	0.171	0.127	0.104	0.090	0.080	0.074	0.069
2.2	0.272	0.159	0.115	0.093	0.079	0.070	0.064	0.059
2.3	0.261	0.148	0.105	0.083	0.070	0.061	0.055	0.050
2.4	0.251	0.138	0.096	0.074	0.062	0.053	0.047	0.043
2.5	0.242	0.130	0.088	0.067	0.054	0.047	0.041	0.037
2.6	0.234	0.122	0.080	0.060	0.048	0.041	0.035	0.032
2.7	0.226	0.114	0.074	0.054	0.043	0.036	0.031	0.027
2.8	0.218	0.107	0.068	0.049	0.038	0.031	0.027	0.023
2.9	0.211	0.101	0.063	0.044	0.034	0.027	0.023	0.020
3.0	0.205	0.095	0.058	0.040	0.030	0.024	0.020	0.017
3.1	0.199	0.090	0.053	0.036	0.027	0.021	0.017	0.015
3.2	0.193	0.085	0.049	0.033	0.024	0.019	0.015	0.013
3.3	0.187	0.081	0.046	0.030	0.021	0,016	0.013	0.011
3.4	0.182	0.077	0.042	0.027	0.019	0.014	0.011	0.009
3.5	0.177	0.073	0.039	0.025	0.017	0.013	0.010	0.008
4.0	0.156	0.057	0.028	0.016	0.010	0.007	0.005	0.004
4.5	0.139	0.046	0.020	0.011	0.006	0.004	0.003	0.002
5.0	0.126	0.038	0.015	0.007	0.004	0.002	0.002	0.001

(*continued*)

TABLE A-6 (continued)

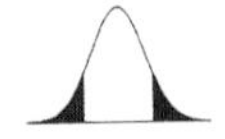

	Degrees of freedom							
t-value	9	10	15	20	30	50	100	Infinity
0.5	0.629	0.628	0.624	0.623	0.621	0.619	0.618	0.617
0.6	0.563	0.562	0.557	0.555	0.553	0.551	0.550	0.549
0.7	0.502	0.500	0.495	0.492	0.489	0.487	0.486	0.484
0.8	0.444	0.442	0.436	0.433	0.430	0.427	0.426	0.424
0.9	0.392	0.389	0.382	0.379	0.375	0.372	0.370	0.368
1.0	0.343	0.341	0.333	0.329	0.325	0.322	0.320	0.317
1.1	0.300	0.297	0.289	0.284	0.280	0.277	0.274	0.271
1.2	0.261	0.258	0.249	0.244	0.240	0.236	0.233	0.230
1.3	0.226	0.223	0.213	0.208	0.204	0.200	0.197	0.194
1.4	0.195	0.192	0.182	0.177	0.172	0.168	0.165	0.162
1.5	0.168	0.165	0.154	0.149	0.144	0.140	0.137	0.134
1.6	0.144	0.141	0.130	0.125	0.120	0.116	0.113	0.110
1.7	0.123	0.120	0.110	0.105	0.099	0.095	0.092	0.089
1.8	0.105	0.102	0.092	0.087	0.082	0.078	0.075	0.072
1.9	0.090	0.087	0.077	0.072	0.067	0.063	0.060	0.057
2.0	0.077	0.073	0.064	0.059	0.055	0.051	0.048	0.046
2.1	0.065	0.062	0.053	0.049	0.044	0.041	0.038	0.036
2.2	0.055	0.052	0.044	0.040	0.036	0.032	0.030	0.028
2.3	0.047	0.044	0.036	0.032	0.029	0.026	0.024	0.021
2.4	0.040	0.037	0.030	0.026	0.023	0.020	0.018	0.016
2.5	0.034	0.031	0.025	0.021	0.018	0.016	0.014	0.012
2.6	0.029	0.026	0.020	0.017	0.014	0.012	0.011	0.009
2.7	0.024	0.022	0.016	0.014	0.011	0.009	0.008	0.007
2.8	0.021	0.019	0.013	0.011	0.009	0.007	0.006	0.005
2.9	0.018	0.016	0.011	0.009	0.007	0.006	0.005	0.004
3.0	0.015	0.013	0.009	0.007	0.005	0.004	0.003	0.003
3.1	0.013	0.011	0.007	0.006	0.004	0.003	0.003	0.002
3.2	0.011	0.009	0.006	0.004	0.003	0.002	0.002	0.001
3.3	0.009	0.008	0.005	0.004	0.002	0.002	0.001	0.001
3.4	0.008	0.007	0.004	0.003	0.002	0.001	0.001	0.001
3.5	0.007	0.006	0.003	0.002	0.001	0.001	0.001	0.000
4.0	0.003	0.003	0.001	0.001	0.000	0.000	0.000	0.000
4.5	0.001	0.001	0.000	0.000	0.000	0.000	0.000	0.000
5.0	0.001	0.001	0.000	0.000	0.000	0.000	0.000	0.000

Two-Tailed *F*-Table

Table A-7 can be used in testing hypotheses about the variances of two populations. Most often, the null hypothesis will be that the population variances are equal, with the alternate hypothesis being they are unequal. The theory behind the test requires that the underlying variables are normally distributed. If so, and if the population variances are equal, the ratio of the sample variances, s_1^2/s_2^2, follows an *F* distribution. In a two-sided test, the null hypothesis $s_1^2 = s_2^2$ is rejected if the ratio of sample variances is far from unity in either direction. To reject H_0, at some level α, the probability of landing in *either* tail must be less than or equal to α. Alternatively, the *PROB-VALUE* of the same test is $P(F > s_1^2/s_2^2) + P(F < s_2^2/s_1^2)$. Table A-7 gives these two-tailed probabilities in the column labeled α. To use the table, *take the larger of the two sample variances as* s_1^2. (The observed ratio will never be less than unity.) Find the observed *F* ratio in the body of the table using the appropriate degrees of freedom:

$$r_1 = n_1 - 1 \quad \text{and} \quad r_2 = n_2 - 1$$

The probabilities are two-tailed probabilities, meaning that each tail contains α/2, as shown in the diagram below.

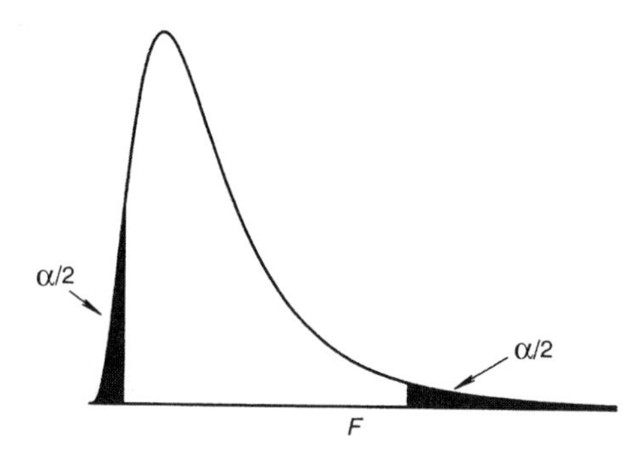

Example:

$H_0\colon s_1^2 = s_2^2$ $H_A\colon s_1^{2\ 1} s_2^2$

$n_1 = 25, s_1 = 75.5$ $n_2 = 13, s_2 = 25$

Observed *f*: $f = \dfrac{75.5}{25} = 3.02$ Degrees of freedom: $r_1 = 24, r_2 = 12$

PROB-VALUE: .05 (see page 000)

TABLE A-7
Two-Tailed F Table

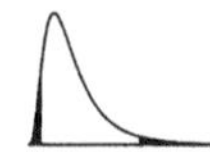

r_2	α	r_1 1	2	3	4	5	6	7	8	9
1	.2	39.9	49.5	53.59	55.83	57.24	58.2	58.91	59.44	59.86
	.1	161.4	199.5	215.71	224.58	230.16	233.99	236.77	238.88	240.54
	.05	647.8	799.5	864.16	899.58	921.85	937.11	948.22	956.66	963.28
	.02	4,052.2	4,999.5	5,403.35	5,624.58	5,763.65	5,858.99	5,928.36	5,981.07	6,022.47
	.01	16,210.7	19,999.5	21,614.74	22,499.58	23,055.8	23,437.11	23,714.57	23,925.41	24,091
	.001	1,621,138	2,000,000	2,161,518	2,250,000	2,305,620	2,343,750	2,371,494	2,392,578	2,409,138
2	.2	8.53	9.00	9.16	9.24	9.29	9.33	9.35	9.37	9.38
	.1	18.51	19.00	19.16	19.25	19.30	19.33	19.35	19.37	19.38
	.05	38.51	39.00	39.17	39.25	39.30	39.33	39.36	39.37	39.39
	.02	98.50	99.00	99.17	99.25	99.30	99.33	99.36	99.37	99.39
	.01	198.50	199.00	199.17	199.25	199.30	199.33	199.36	199.37	199.39
	.001	1,998.50	1,999.00	1,999.17	1,999.25	1,999.30	1,999.33	1,999.36	1,999.38	1,999.39
3	.2	5.54	5.46	5.39	5.34	5.31	5.28	5.27	5.25	5.24
	.1	10.13	9.55	9.28	9.12	9.01	8.94	8.89	8.85	8.81
	.05	17.44	16.04	15.44	15.10	14.88	14.73	14.62	14.54	14.47
	.02	34.12	30.82	29.46	28.71	28.24	27.91	27.67	27.49	27.35
	.01	55.55	49.80	47.47	46.19	45.39	44.84	44.43	44.13	43.88
	.001	266.55	236.61	224.70	218.25	214.20	211.41	209.38	207.83	206.61
4	.2	4.54	4.32	4.19	4.11	4.05	4.01	3.98	3.95	3.94
	.1	7.71	6.94	6.59	6.39	6.26	6.16	6.09	6.04	6.00
	.05	12.22	10.65	9.98	9.60	9.36	9.20	9.07	8.98	8.90
	.02	21.20	18.00	16.69	15.98	15.52	15.21	14.98	14.80	14.66
	.01	31.33	26.28	24.26	23.15	22.46	21.97	21.62	21.35	21.14
	.001	106.22	87.44	80.09	76.12	73.63	71.92	70.66	69.71	68.95
5	.2	4.06	3.78	3.62	3.52	3.45	3.40	3.37	3.34	3.32
	.1	6.61	5.79	5.41	5.19	5.05	4.95	4.88	4.82	4.77
	.05	10.01	8.43	7.76	7.39	7.15	6.98	6.85	6.76	6.68
	.02	16.26	13.27	12.06	11.39	10.97	10.67	10.46.	10.29	10.16
	.01	22.78	18.31	16.53	15.56	14.94	14.51	14.20	13.96	13.77
	.001	63.61	49.78	44.42	41.53	39.72	38.47	37.56	36.86	36.31
6	.2	3.78	3.46	3.29	3.18	3.11	3.05	3.01	2.98	2.96
	.1	5.99	5.14	4.76	4.53	4.39	4.28	4.21	4.15	4.10
	.05	8.81	7.26	6.60	6.23	5.99	5.82	5.70	5.60	5.52
	.02	13.75	10.92	9.78	9.15	8.75	8.47	8.26	8.10	7.98
	.01	18.63	14.54	12.92	12.03	11.46	11.07	10.79	10.57	10.39
	.001	46.08	34.80	30.45	28.12	26.65	25.63	24.89	24.33	23.88

TABLE A-7 (continued)

r_2	α	r_1 10	12	15	20	24	30	60	120	∞
1	.2	60.19	60.71	61.22	61.74	62	62.26	62.79	63.06	63.33
	.1	241.88	243.91	245.95	248.01	249.05	250.1	252.2	253.25	254.31
	.05	968.63	976.71	984.87	993.1	997.25	1,001.41	1,009.8	1,014.02	1,018.24
	.02	6,055.85	6,106.32	6,157.28	6,208.73	6,234.63	6,260.65	6,313.03	6,339.39	6,365.76
	.01	24,224.49	24,426.37	24,630.21	24,835.97	24,939.56	25,043.63	25,253.14	25,358.57	25,464.03
	.001	2,422,485	2,442,672	2,463,056	2,483,632	2,493,990	2,504,396	2,525,348	2,535,890	2,546,436
2	.2	9.39	9.41	9.42	9.44	9.45	9.46	9.47	9.48	9.49
	.1	19.4	19.41	19.43	19.45	19.45	19.46	19.48	19.49	19.5
	.05	39.4	39.41	39.43	39.45	39.46	39.46	39.48	39.49	39.5
	.02	99.4	99.42	99.43	99.45	99.46	99.47	99.48	99.49	99.5
	.01	199.4	199.42	199.43	199.45	199.46	199.47	199.48	199.49	199.5
	.001	1,999.4	1,999.42	1,999.43	1,999.45	1,999.46	1,999.47	1,999.48	1,999.49	1,999.5
3	.2	5.23	5.22	5.2	5.18	5.18	5.17	5.15	5.14	5.13
	.1	8.79	8.74	8.7	8.66	8.64	8.62	8.57	8.55	8.53
	.05	14.42	14.34	14.25	14.17	14.12	14.08	13.99	13.95	13.9
	.02	27.23	27.05	26.87	26.69	26.6	26.5	26.32	26.22	26.13
	.01	43.69	43.39	43.08	42.78	42.62	42.47	42.15	41.99	41.83
	.001	205.63	204.13	202.62	201.08	200.31	199.53	197.95	197.15	196.35
4	.2	3.92	3.9	3.87	3.84	3.83	3.82	3.79	3.78	3.76
	.1	5.96	5.91	5.86	5.8	5.77	5.75	5.69	5.66	5.63
	.05	8.84	8.75	8.66	8.56	8.51	8.46	8.36	8.31	8.26
	.02	14.55	14.37	14.2	14.02	13.93	13.84	13.65	13.56	13.46
	.01	20.97	20.7	20.44	20.17	20.03	19.89	19.61	19.47	19.33
	.001	68.35	67.42	66.48	65.53	65.05	64.56	63.58	63.08	62.58
5	.2	3.3	3.27	3.24	3.21	3.19	3.17	3.14	3.12	3.11
	.1	4.74	4.68	4.62	4.56	4.53	4.5	4.43	4.4	4.37
	.05	6.62	6.52	6.43	6.33	6.28	6.23	6.12	6.07	6.02
	.02	10.05	9.89	9.72	9.55	9.47	9.38	9.2	9.11	9.02
	.01	13.62	13.38	13.15	12.9	12.78	12.66	12.4	12.27	12.14
	.001	35.86	35.19	34.5	33.8	33.44	33.09	32.36	31.99	31.62
6	.2	2.94	2.9	2.87	2.84	2.82	2.8	2.76	2.74	2.72
	.1	4.06	4	3.94	3.87	3.84	3.81	3.74	3.7	3.67
	.05	5.46	5.37	5.27	5.17	5.12	5.07	4.96	4.9	4.85
	.02	7.87	7.72	7.56	7.4	7.31	7.23	7.06	6.97	6.88
	.01	10.25	10.03	9.81	9.59	9.47	9.36	9.12	9	8.88
	.001	23.52	22.96	22.4	21.83	21.54	21.25	20.65	20.35	20.04

(continued)

TABLE A-7 (continued)

r_2	α	r_1 = 1	2	3	4	5	6	7	8	9
7	.2	3.59	3.26	3.07	2.96	2.88	2.83	2.78	2.75	2.72
	.1	5.59	4.74	4.35	4.12	3.97	3.87	3.79	3.73	3.68
	.05	8.07	6.54	5.89	5.52	5.29	5.12	4.99	4.90	4.82
	.02	12.25	9.55	8.45	7.85	7.46	7.19	6.99	6.84	6.72
	.01	16.24	12.40	10.88	10.05	9.52	9.16	8.89	8.68	8.51
	.001	36.99	27.21	23.46	21.44	20.17	19.30	18.66	18.17	17.78
8	.2	3.46	3.11	2.92	2.81	2.73	2.67	2.62	2.59	2.56
	.1	5.32	4.46	4.07	3.84	3.69	3.58	3.5	3.44	3.39
	.05	7.57	6.06	5.42	5.05	4.82	4.65	4.53	4.43	4.36
	.02	11.26	8.65	7.59	7.01	6.63	6.37	6.18	6.03	5.91
	.01	14.69	11.04	9.6	8.81	8.3	7.95	7.69	7.5	7.34
	.001	31.56	22.75	19.39	17.58	16.44	15.66	15.08	14.64	14.29
9	.2	3.36	3.01	2.81	2.69	2.61	2.55	2.51	2.47	2.44
	.1	5.12	4.26	3.86	3.63	3.48	3.37	3.29	3.23	3.18
	.05	7.21	5.71	5.08	4.72	4.48	4.32	4.2	4.1	4.03
	.02	10.56	8.02	6.99	6.42	6.06	5.8	5.61	5.47	5.35
	.01	13.61	10.11	8.72	7.96	7.47	7.13	6.88	6.69	6.54
	.001	27.99	19.87	16.77	15.11	14.06	13.34	12.8	12.4	12.08
10	.2	3.29	2.92	2.73	2.61	2.52	2.46	2.41	2.38	2.35
	.1	4.96	4.1	3.71	3.48	3.33	3.22	3.14	3.07	3.02
	.05	6.94	5.46	4.83	4.47	4.24	4.07	3.95	3.85	3.78
	.02	10.04	7.56	6.55	5.99	5.64	5.39	5.2	5.06	4.94
	.01	12.83	9.43	8.08	7.34	6.87	6.54	6.3	6.12	5.97
	.001	25.49	17.87	14.97	13.41	12.43	11.75	11.25	10.87	10.56
12	.2	3.18	2.81	2.61	2.48	2.39	2.33	2.28	2.24	2.21
	.1	4.75	3.89	3.49	3.26	3.11	3	2.91	2.85	2.8
	.05	6.55	5.1	4.47	4.12	3.89	3.73	3.61	3.51	3.44
	.02	9.33	6.93	5.95	5.41	5.06	4.82	4.64	4.5	4.39
	.01	11.75	8.51	7.23	6.52	6.07	5.76	5.52	5.35	5.2
	.001	22.24	15.3	12.66	11.25	10.35	9.74.	·9.28	8.94	8.66
15	.2	3.07	2.7	2.49	2.36	2.27	2.21	2.16	2.12	2.09
	.1	4.54	3.68	3.29	3.06	2.9	2.79	2.71	2.64	2.59
	.05	6.2	4.77	4.15	3.8	3.58	3.41	3.29	3.2	3.12
	.02	8.68	6.36	5.42	4.89	4.56	4.32	4.14	4	3.89
	.01	10.8	7.7	6.48	5.8	5.37	5.07	4.85	4.67	4.54
	.001	19.51	13.16	10.76	9.48	8.66	8.1	7.68	7.37	7.11

TABLE A-7 (continued)

r_2	α	10	12	15	20	24	30	60	120	∞
		r_1								
7	.2	2.7	2.67	2.63	2.59	2.58	2.56	2.51	2.49	2.47
	.1	3.64	3.57	3.51	3.44	3.41	3.38	3.3	3.27	3.23
	.05	4.76	4.67	4.57	4.47	4.41	4.36	4.25	4.2	4.14
	.02	6.62	6.47	6.31	6.16	6.07	5.99	5.82	5.74	5.65
	.01	8.38	8.18	7.97	7.75	7.64	7.53	7.31	7.19	7.08
	.001	17.47	16.99	16.5	16	15.75	15.49	14.97	14.71	14.44
8	.2	2.54	2.5	2.46	2.42	2.4	2.38	2.34	2.32	2.29
	.1	3.35	3.28	3.22	3.15	3.12	3.08	3.01	2.97	2.93
	.05	4.3	4.2	4.1	4	3.95	3.89	3.78	3.73	3.67
	.02	5.81	5.67	5.52	5.36	5.28	5.2	5.03	4.95	4.86
	.01	7.21	7.01	6.81	6.61	6.5	6.4	6.18	6.06	5.95
	.001	14.01	13.58	13.14	12.69	12.46	12.22	11.75	11.51	11.26
9	.2	2.42	2.38	2.34	2.3	2.28	2.25	2.21	2.18	2.16
	.1	3.14	3.07	3.01	2.94	2.9	2.86	2.79	2.75	2.71
	.05	3.96	3.87	3.77	3.67	3.61	3.56	3.45	3.39	3.33
	.02	5.26	5.11	4.96	4.81	4.73	4.65	4.48	4.4	4.31
	.01	6.42	6.23	6.03	5.83	5.73	5.62	5.41	5.3	5.19
	.001	11.81	11.42	11.01	10.59	10.38	10.16	9.72	9.49	9.26
10	.2	2.32	2.28	2.24	2.2	2.18	2.16	2.11	2.08	2.06
	.1	2.98	2.91	2.85	2.77	2.74	2.7	2.62	2.58	2.54
	.05	3.72	3.62	3.52	3.42	3.37	3.31	3.2	3.14	3.08
	.02	4.85	4.71	4.56	4.41	4.33	4.25	4.08	4	3.91
	.01	5.85	5.66	5.47	5.27	5.17	5.07	4.86	4.75	4.64
	.001	10.32	9.94	9.56	9.17	8.96	8.76	8.34	8.12	7.91
12	.2	2.19	2.15	2.1	2.06	2.04	2.01	1.96	1.93	1.9
	.1	2.75	2.69	2.62	2.54	2.51	2.47	2.38	2.34	2.3
	.05	3.37	3.28	3.18	3.07	3.02	2.96	2.85	2.79	2.73
	.02	4.3	4.16	4.01	3.86	3.78	3.7	3.54	3.45	3.36
	.01	5.09	4.91	4.72	4.53	4.43	4.33	4.12	4.01	3.9
	.001	8.43	8.09	7.74	7.37	7.19	7	6.61	6.41	6.2
15	.2	2.06	2.02	1.97	1.92	1.9	1.87	1.82	1.79	1.76
	.1	2.54	2.48	2.4	2.33	2.29	2.25	2.16	2.11	2.07
	.05	3.06	2.96	2.86	2.76	2.7	2.64	2.52	2.46	2.4
	.02	3.8	3.67	3.52	3.37	3.29	3.21	3.05	2.96	2.87
	.01	4.42	4.25	4.07	3.88	3.79	3.69	3.48	3.37	3.26
	.001	6.91	6.59	6.26	5.93	5.75	5.58	5.21	5.02	4.83

(continued)

TABLE A-7 (continued)

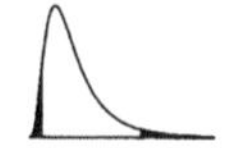

r_2	α	r_1 = 1	2	3	4	5	6	7	8	9
20	.2	2.97	2.59	2.38	2.25	2.16	2.09	2.04	2	1.96
	.1	4.35	3.49	3.1	2.87	2.71	2.6	2.51	2.45	2.39
	.05	5.87	4.46	3.86	3.51	3.29	3.13	3.01	2.91	2.84
	.02	8.1	5.85	4.94	4.43	4.1	3.87	3.7	3.56	3.46
	.01	9.94	6.99	5.82	5.17	4.76	4.47	4.26	4.09	3.96
	.001	17.19	11.38	9.2	8.02	7.27	6.76	6.38	6.09	5.85
24	.2	2.93	2.54	2.33	2.19	2.1	2.04	1.98	1.94	1.91
	.1	4.26	3.4	3.01	2.78	2.62	2.51	2.42	2.36	2.3
	.05	5.72	4.32	3.72	3.38	3.15	2.99	2.87	2.78	2.7
	.02	7.82	5.61	4.72	4.22	3.9	3.67	3.5	3.36	3.26
	.01	9.55	6.66	5.52	4.89	4.49	4.2	3.99	3.83	3.69
	.001	16.17	10.61	8.51	7.39	6.68	6.18	5.82	5.54	5.31
30	.2	2.88	2.49	2.28	2.14	2.05	1.98	1.93	1.88	1.85
	.1	4.17	3.32	2.92	2.69	2.53	2.42	2.33	2.27	2.21
	.05	5.57	4.18	3.59	3.25	3.03	2.87	2.75	2.65	2.57
	.02	7.56	5.39	4.51	4.02	3.7	3.47	3.3	3.17	3.07
	.01	9.18	6.35	5.24	4.62	4.23	3.95	3.74	3.58	3.45
	.001	15.22	9.9	7.89	6.82	6.13	5.66	5.31	5.04	4.82
60	.2	2.79	2.39	2.18	2.04	1.95	1.87	1.82	1.77	1.74
	.1	4	3.15	2.76	2.53	2.37	2.25	2.17	2.1	2.04
	.05	5.29	3.93	3.34	3.01	2.79	2.63	2.51	2.41	2.33
	.02	7.08	4.98	4.13	3.65	3.34	3.12	2.95	2.82	2.72
	.01	8.49	5.79	4.73	4.14	3.76	3.49	3.29	3.13	3.01
	.001	13.55	8.65	6.81	5.82	5.2	4.76	4.44	4.19	3.98
120	.2	2.75	2.35	2.13	1.99	1.9	1.82	1.77	1.72	1.68
	.1	3.92	3.07	2.68	2.45	2.29	2.18	2.09	2.02	1.96
	.05	5.15	3.8	3.23	2.89	2.67	2.52	2.39	2.3	2.22
	.02	6.85	4.79	3.95	3.48	3.17	2.96	2.79	2.66	2.56
	.01	8.18	5.54	4.5	3.92	3.55	3.28	3.09	2.93	2.81
	.001	12.8	8.1	6.34	5.39	4.79	4.37	4.06	3.82	3.62
∞	.2	2.71	2.3	2.08	1.95	1.85	1.77	1.72	1.67	1.63
	.1	3.84	3	2.61	2.37	2.21	2.1	2.01	1.94	1.88
	.05	5.02	3.69	3.12	2.79	2.57	2.41	2.29	2.19	2.11
	.02	6.64	4.61	3.78	3.32	3.02	2.8	2.64	2.51	2.41
	.01	7.88	5.3	4.28	3.72	3.35	3.09	2.9	2.75	2.62
	.001	12.12	7.6	5.91	5	4.42	4.02	3.72	3.48	3.3

TABLE A-7 (continued)

r_2	α	r_1 = 10	12	15	20	24	30	60	120	∞
20	.2	1.94	1.89	1.84	1.79	1.77	1.74	1.68	1.64	1.61
	.1	2.35	2.28	2.2	2.12	2.08	2.04	1.95	1.9	1.84
	.05	2.77	2.68	2.57	2.46	2.41	2.35	2.22	2.16	2.09
	.02	3.37	3.23	3.09	2.94	2.86	2.78	2.61	2.52	2.42
	.01	3.85	3.68	3.5	3.32	3.22	3.12	2.92	2.81	2.69
	.001	5.66	5.37	5.07	4.75	4.59	4.42	4.08	3.9	3.71
24	.2	1.88	1.83	1.78	1.73	1.7	1.67	1.61	1.57	1.53
	.1	2.25	2.18	2.11	2.03	1.98	1.94	1.84	1.79	1.73
	.05	2.64	2.54	2.44	2.33	2.27	2.21	2.08	2.01	1.94
	.02	3.17	3.03	2.89	2.74	2.66	2.58	2.4	2.31	2.21
	.01	3.59	3.42	3.25	3.06	2.97	2.87	2.66	2.55	2.43
	.001	5.13	4.85	4.56	4.25	4.09	3.93	3.59	3.41	3.22
30	.2	1.82	1.77	1.72	1.67	1.64	1.61	1.54	1.5	1.46
	.1	2.16	2.09	2.01	1.93	1.89	1.84	1.74	1.68	1.62
	.05	2.51	2.41	2.31	2.2	2.14	2.07	1.94	1.87	1.79
	.02	2.98	2.84	2.7	2.55	2.47	2.39	2.21	2.11	2.01
	.01	3.34	3.18	3.01	2.82	2.73	2.63	2.42	2.3	2.18
	.001	4.65	4.38	4.09	3.8	3.64	3.49	3.15	2.97	2.78
60	.2	1.71	1.66	1.6	1.54	1.51	1.48	1.4	1.35	1.29
	.1	1.99	1.92	1.84	1.75	1.7	1.65	1.53	1.47	1.39
	.05	2.27	2.17	2.06	1.94	1.88	1.82	1.67	1.58	1.48
	.02	2.63	2.5	2.35	2.2	2.12	2.03	1.84	1.73	1.6
	.01	2.9	2.74	2.57	2.39	2.29	2.19	1.96	1.83	1.69
	.001	3.82	3.57	3.3	3.02	2.87	2.71	2.38	2.19	1.98
120	.2	1.65	1.6	1.55	1.48	1.45	1.41	1.32	1.26	1.19
	.1	1.91	1.83	1.75	1.66	1.61	1.55	1.43	1.35	1.25
	.05	2.16	2.05	1.94	1.82	1.76	1.69	1.53	1.43	1.31
	.02	2.47	2.34	2.19	2.03	1.95	1.86	1.66	1.53	1.38
	.01	2.71	2.54	2.37	2.19	2.09	1.98	1.75	1.61	1.43
	.001	3.46	3.22	2.96	2.68	2.53	2.38	2.04	1.83	1.59
∞	.2	1.6	1.55	1.49	1.42	1.38	1.34	1.24	1.17	1.01
	.1	1.83	1.75	1.67	1.57	1.52	1.46	1.32	1.22	1.02
	.05	2.05	1.95	1.83	1.71	1.64	1.57	1.39	1.27	1.02
	.02	2.32	2.19	2.04	1.88	1.79	1.7	1.47	1.33	1.03
	.01	2.52	2.36	2.19	2	1.9	1.79	1.53	1.36	1.03
	.001	3.14	2.9	2.65	2.38	2.23	2.07	1.71	1.48	1.04

TABLE A-8
P-values for the Chi-square distribution. The tabled values are the probability of exceeding an observed value.

One-Tailed Chi-Square Distribution $P(\chi^2 > \chi^2)$

χ^2	*df* 1	2	3	4	5	6	7	8	9	10
2	0.157	0.368	0.572	0.736	0.849	0.920	0.960	0.981	0.991	0.996
3	0.083	0.223	0.392	0.558	0.700	0.809	0.885	0.934	0.964	0.981
4	0.046	0.135	0.261	0.406	0.549	0.677	0.780	0.857	0.911	0.947
5	0.025	0.082	0.172	0.287	0.416	0.544	0.660	0.758	0.834	0.891
6	0.014	0.050	0.112	0.199	0.306	0.423	0.540	0.647	0.740	0.815
7	0.008	0.030	0.072	0.136	0.221	0.321	0.429	0.537	0.637	0.725
8	0.005	0.018	0.046	0.092	0.156	0.238	0.333	0.433	0.534	0.629
9	0.003	0.011	0.029	0.061	0.109	0.174	0.253	0.342	0.437	0.532
10	0.002	0.007	0.019	0.040	0.075	0.125	0.189	0.265	0.350	0.440
12	0.001	0.002	0.007	0.017	0.035	0.062	0.101	0.151	0.213	0.285
15	0.000	0.001	0.002	0.005	0.010	0.020	0.036	0.059	0.091	0.132
20	0.000	0.000	0.000	0.000	0.001	0.003	0.006	0.010	0.018	0.029
25	0.000	0.000	0.000	0.000	0.000	0.000	0.001	0.002	0.003	0.005
30	0.000	0.000	0.000	0.000	0.000	0.000	0.000	0.000	0.000	0.001
35	0.000	0.000	0.000	0.000	0.000	0.000	0.000	0.000	0.000	0.000
40	0.000	0.000	0.000	0.000	0.000	0.000	0.000	0.000	0.000	0.000
45	0.000	0.000	0.000	0.000	0.000	0.000	0.000	0.000	0.000	0.000
50	0.000	0.000	0.000	0.000	0.000	0.000	0.000	0.000	0.000	0.000
60	0.000	0.000	0.000	0.000	0.000	0.000	0.000	0.000	0.000	0.000
70	0.000	0.000	0.000	0.000	0.000	0.000	0.000	0.000	0.000	0.000
80	0.000	0.000	0.000	0.000	0.000	0.000	0.000	0.000	0.000	0.000
90	0.000	0.000	0.000	0.000	0.000	0.000	0.000	0.000	0.000	0.000
100	0.000	0.000	0.000	0.000	0.000	0.000	0.000	0.000	0.000	0.000
110	0.000	0.000	0.000	0.000	0.000	0.000	0.000	0.000	0.000	0.000
125	0.000	0.000	0.000	0.000	0.000	0.000	0.000	0.000	0.000	0.000
150	0.000	0.000	0.000	0.000	0.000	0.000	0.000	0.000	0.000	0.000
175	0.000	0.000	0.000	0.000	0.000	0.000	0.000	0.000	0.000	0.000
200	0.000	0.000	0.000	0.000	0.000	0.000	0.000	0.000	0.000	0.000

TABLE A-8 (continued)

One-Tailed Chi-Square Distribution $P(\chi^2 > \chi^2)$

	df									
χ^2	10	20	30	40	50	60	80	100	150	200
2	0.996	1.000	1.000	1.000	1.000	1.000	1.000	1.000	1.000	1.000
3	0.981	1.000	1.000	1.000	1.000	1.000	1.000	1.000	1.000	1.000
4	0.947	1.000	1.000	1.000	1.000	1.000	1.000	1.000	1.000	1.000
5	0.891	1.000	1.000	1.000	1.000	1.000	1.000	1.000	1.000	1.000
6	0.815	0.999	1.000	1.000	1.000	1.000	1.000	1.000	1.000	1.000
7	0.725	0.997	1.000	1.000	1.000	1.000	1.000	1.000	1.000	1.000
8	0.629	0.992	1.000	1.000	1.000	1.000	1.000	1.000	1.000	1.000
9	0.532	0.983	1.000	1.000	1.000	1.000	1.000	1.000	1.000	1.000
10	0.440	0.968	1.000	1.000	1.000	1.000	1.000	1.000	1.000	1.000
12	0.285	0.916	0.999	1.000	1.000	1.000	1.000	1.000	1.000	1.000
15	0.132	0.776	0.990	1.000	1.000	1.000	1.000	1.000	1.000	1.000
20	0.029	0.458	0.917	0.997	1.000	1.000	1.000	1.000	1.000	1.000
25	0.005	0.201	0.725	0.969	0.999	1.000	1.000	1.000	1.000	1.000
30	0.001	0.070	0.466	0.875	0.989	1.000	1.000	1.000	1.000	1.000
35	0.000	0.020	0.243	0.695	0.947	0.996	1.000	1.000	1.000	1.000
40	0.000	0.005	0.105	0.470	0.843	0.978	1.000	1.000	1.000	1.000
45	0.000	0.001	0.039	0.271	0.674	0.925	0.999	1.000	1.000	1.000
50	0.000	0.000	0.012	0.134	0.473	0.818	0.997	1.000	1.000	1.000
60	0.000	0.000	0.001	0.022	0.157	0.476	0.954	0.999	1.000	1.000
70	0.000	0.000	0.000	0.002	0.032	0.177	0.780	0.990	1.000	1.000
80	0.000	0.000	0.000	0.000	0.004	0.043	0.479	0.930	1.000	1.000
100	0.000	0.000	0.000	0.000	0.000	0.001	0.065	0.481	0.999	1.000
120	0.000	0.000	0.000	0.000	0.000	0.000	0.003	0.084	0.966	1.000
130	0.000	0.000	0.000	0.000	0.000	0.000	0.000	0.024	0.879	1.000
150	0.000	0.000	0.000	0.000	0.000	0.000	0.000	0.001	0.485	0.997
170	0.000	0.000	0.000	0.000	0.000	0.000	0.000	0.000	0.126	0.939
200	0.000	0.000	0.000	0.000	0.000	0.000	0.000	0.000	0.004	0.487
250	0.000	0.000	0.000	0.000	0.000	0.000	0.000	0.000	0.000	0.009

TABLE A-9

P-Values for D, the Kolmogorov-Smirnov statistic*. The observed test statistic is $d = \max|S_i - F_i|$ where S_i and F_i are the sample and theoretical cumulative distributions. The tabled values are the probability of D exceeding an observed value d.

$P(D > d)$

	n										
d	6	8	10	12	15	20	25	30	35	40	50
0.10	1.000	1.000	1.000	0.999	0.995	0.976	0.943	0.896	0.841	0.782	0.662
0.12	1.000	0.999	0.995	0.987	0.964	0.903	0.823	0.736	0.651	0.571	0.434
0.14	0.998	0.990	0.974	0.947	0.892	0.778	0.660	0.552	0.458	0.378	0.256
0.16	0.991	0.967	0.926	0.872	0.782	0.629	0.494	0.385	0.299	0.231	0.138
0.18	0.969	0.919	0.847	0.770	0.652	0.481	0.350	0.253	0.183	0.132	0.069
0.20	0.932	0.848	0.749	0.653	0.522	0.353	0.236	0.158	0.106	0.070	0.031
0.22	0.878	0.759	0.643	0.536	0.404	0.249	0.153	0.094	0.057	0.035	0.013
0.24	0.809	0.662	0.535	0.427	0.302	0.169	0.094	0.053	0.029	0.016	0.005
0.26	0.728	0.566	0.435	0.332	0.221	0.111	0.056	0.028	0.014	0.007	0.002
0.28	0.642	0.475	0.346	0.252	0.156	0.070	0.032	0.014	0.006	0.003	0.001
0.30	0.555	0.390	0.271	0.187	0.108	0.043	0.017	0.007	0.003	0.001	0.000
0.32	0.473	0.315	0.207	0.136	0.073	0.025	0.009	0.003	0.001	0.000	0.000
0.34	0.400	0.250	0.156	0.097	0.047	0.014	0.004	0.001	0.000	0.000	0.000
0.36	0.335	0.196	0.115	0.067	0.030	0.008	0.002	0.001	0.000	0.000	0.000
0.38	0.276	0.151	0.083	0.046	0.019	0.004	0.001	0.000	0.000	0.000	0.000
0.40	0.224	0.115	0.059	0.030	0.011	0.002	0.000	0.000	0.000	0.000	0.000
0.42	0.179	0.086	0.041	0.020	0.007	0.001	0.000	0.000	0.000	0.000	0.000
0.44	0.142	0.063	0.028	0.012	0.004	0.000	0.000	0.000	0.000	0.000	0.000
0.46	0.111	0.045	0.019	0.008	0.002	0.000	0.000	0.000	0.000	0.000	0.000
0.48	0.086	0.032	0.012	0.005	0.001	0.000	0.000	0.000	0.000	0.000	0.000
0.50	0.066	0.023	0.008	0.003	0.001	0.000	0.000	0.000	0.000	0.000	0.000
0.52	0.049	0.015	0.005	0.002	0.000	0.000	0.000	0.000	0.000	0.000	0.000
0.54	0.037	0.010	0.003	0.001	0.000	0.000	0.000	0.000	0.000	0.000	0.000
0.56	0.027	0.007	0.002	0.000	0.000	0.000	0.000	0.000	0.000	0.000	0.000
0.58	0.019	0.004	0.001	0.000	0.000	0.000	0.000	0.000	0.000	0.000	0.000
0.60	0.014	0.003	0.001	0.000	0.000	0.000	0.000	0.000	0.000	0.000	0.000
0.62	0.009	0.002	0.000	0.000	0.000	0.000	0.000	0.000	0.000	0.000	0.000
0.64	0.006	0.001	0.000	0.000	0.000	0.000	0.000	0.000	0.000	0.000	0.000
0.66	0.004	0.001	0.000	0.000	0.000	0.000	0.000	0.000	0.000	0.000	0.000

*Computed by the method of Marsaglia et al., "Evaluating Kolmogorov's Distribution," *Journal of Statistical Software*, *8* (2003).

Index

I

J

K

L

M

O

P

About the Authors

James E. Burt is a professor in and former chair of the Department of Geography, University of Wisconsin–Madison. In addition to the previous editions of *Elementary Statistics for Geographers*, he is a coauthor of *Understanding Weather and Climate*, which won the 2001 Text and Academic Authors award for best physical science textbook. He teaches courses at all levels in physical geography, climatology, and spatial analysis, and he has a longstanding interest in applications of simulation models and digital technology in teaching. Currently his main research activities revolve around development of expert system and statistical approaches for quantitative prediction of soils information.

Gerald M. Barber is an associate professor of geography and teaches introductory and advanced courses in statistics at Queen's University, Kingston, Ontario, Canada. He is also the director of the program in Geographic Information Science and runs the GIS Laboratory. His principal interests are in the application of statistical and optimization models within GIS. He has previously taught at Northwestern University and the University of Victoria.

David L. Rigby is a professor in Geography and Statistics at the University of California, Los Angeles. His research interests focus on regional growth, technological change, evolutionary economic dynamics, and the impacts of globalization/trade on wage inequality. He is a coauthor of *The Golden Age Illusion: Rethinking Postwar Capitalism*.